Lecture Notes in Computer Science 4956

Commenced Publication in 1973
Founding and Former Series Editors:
Gerhard Goos, Juris Hartmanis, and Jan van Leeuwen

Editorial Board

David Hutchison
Lancaster University, UK

Takeo Kanade
Carnegie Mellon University, Pittsburgh, PA, USA

Josef Kittler
University of Surrey, Guildford, UK

Jon M. Kleinberg
Cornell University, Ithaca, NY, USA

Alfred Kobsa
University of California, Irvine, CA, USA

Friedemann Mattern
ETH Zurich, Switzerland

John C. Mitchell
Stanford University, CA, USA

Moni Naor
Weizmann Institute of Science, Rehovot, Israel

Oscar Nierstrasz
University of Bern, Switzerland

C. Pandu Rangan
Indian Institute of Technology, Madras, India

Bernhard Steffen
University of Dortmund, Germany

Madhu Sudan
Massachusetts Institute of Technology, MA, USA

Demetri Terzopoulos
University of California, Los Angeles, CA, USA

Doug Tygar
University of California, Berkeley, CA, USA

Gerhard Weikum
Max-Planck Institute of Computer Science, Saarbruecken, Germany

W0230088

Craig Macdonald   Iadh Ounis
Vassilis Plachouras   Ian Ruthven
Ryen W. White (Eds.)

Advances in Information Retrieval

30th European Conference on IR Research, ECIR 2008
Glasgow, UK, March 30-April 3, 2008
Proceedings

 Springer

Volume Editors

Craig Macdonald
Iadh Ounis
University of Glasgow
Department of Computing Science
Glasgow, UK, G12 8QQ, UK
E-mail: {craigm, ounis}@dcs.gla.ac.uk

Vassilis Plachouras
Yahoo! Research
Ocata 1, 1st floor, 08003 Barcelona, Spain
E-mail: vassilis@yahoo-inc.com

Ian Ruthven
University of Strathclyde
Department of Computing and Information Sciences
Glasgow, UK
E-mail: ir@cis.strath.ac.uk

Ryen W. White
Microsoft Research
One Microsoft Way, Redmond, WA 98052, USA
E-mail: ryenw@microsoft.com

Library of Congress Control Number: 2008922895

CR Subject Classification (1998): H.3, H.2, I.2.3, I.2.6-7, H.4, H.5.4, I.7

LNCS Sublibrary: SL 3 – Information Systems and Application, incl. Internet/Web
and HCI

ISSN   0302-9743
ISBN-10   3-540-78645-7 Springer Berlin Heidelberg New York
ISBN-13   978-3-540-78645-0 Springer Berlin Heidelberg New York

Springer is a part of Springer Science+Business Media

springer.com

© Springer-Verlag Berlin Heidelberg 2008
Printed in Germany

Typesetting: Camera-ready by author, data conversion by Scientific Publishing Services, Chennai, India
Printed on acid-free paper   SPIN: 12241804   06/3180   5 4 3 2 1 0

Preface

These proceedings contain the refereed technical papers and posters presented at the 30th Annual European Conference on Information Retrieval (ECIR 2008). ECIR is the annual conference of the British Computer Society's specialist group in Information Retrieval (BCS-IRSG). This year the conference was organised by the Department of Computing Science, University of Glasgow.

ECIR 2008 received 139 full paper submissions. Many of these submissions came from outside Europe, and 25 countries are represented in the final ECIR 2008 programme, reflecting the international popularity and reputation of the conference series. All submitted papers were reviewed by at least three members of the international programme committee. Thirty-three papers were selected as full research papers and a further 19 were accepted as short research papers. Of these 52 selected papers, 26 have a student as the primary author indicating that the traditional student focus of the conference is still relevant today. The accepted papers themselves come from a mixture of universities, research institutes, and commercial organisations.

The collection of papers presented in these proceedings demonstrates the commitment of Information Retrieval to sound theoretical research allied with strong empirical evaluation. The topics cover core IR problems, such as evaluation and retrieval, and emerging topics such as social media and expert search. We owe a great vote of thanks to our various committees – programme and poster committees, tutorial and workshop committee and award committees - for their hard work in ensuring the quality of the ECIR 2008 programme. We really appreciate the support, expertise and effort given so freely.

ECIR 2008 marked the 30th anniversary of the conference. The programme of the conference and its venue reflected the celebratory nature of ECIR 2008, with exceptionally rich and varied scientific and social events. The conference took place in the University of Glasgow's historic and most famous venue, the Bute Hall. For the first time in ECIR's history, we organised a programme of tutorials and workshops. We are very grateful to the workshop organisers and tutorial presenters for their contributions. Following the success of previous BCS-IRSG Industry Days, we organised an Industry Day as an intrinsic part of the ECIR 2008 programme, held in the same venue as the main conference.

We are grateful to our keynote speakers, Nicholas J. Belkin, Amit Singhal, and Bettina Berendt, for their stimulating contributions to the conference.

We are also grateful to our sponsoring institutions, Google, Matrixware Information Services, Microsoft Research, and Yahoo! Research, for their support of ECIR 2008.

Our final thanks go to Peter Dickman and Jon Ritchie, for dealing with all local arrangements, Claire Harper at the University of Glasgow's Conference & Visitor Services, who sorted out the registration process and the preparation of the conference venues with efficiency and tact, and the many local volunteers, for their huge contribution to the smooth running of ECIR 2008.

March 2008   Craig Macdonald
                                                              Iadh Ounis
                                                       Vassilis Plachouras
                                                              Ian Ruthven
                                                           Ryen W. White

Organisation

ECIR 2008 was organised by the Department of Computer Science, University of Glasgow, UK.

Organising Committee

Conference Chair	Iadh Ounis (University of Glasgow, UK)
Programme Chair	Ian Ruthven (University of Strathclyde, UK)
Poster Chair	Ryen W. White (Microsoft Research, USA)
Tutorial and Workshop Chair	Vassilis Plachouras (Yahoo! Research, Spain)
Industry Day Chair	Craig Macdonald (University of Glasgow, UK)
Local Arrangements Chair	Peter Dickman (University of Glasgow, UK)
Website and Publicity	Jie Peng (University of Glasgow, UK)
Local Organisation	Alasdair J. G. Gray (University of Glasgow, UK)
	Claire Harper (University of Glasgow, UK)
	Ben He (University of Glasgow, UK)
	Craig Macdonald (University of Glasgow, UK)
	Jie Peng (University of Glasgow, UK)
	Jon Ritchie (University of Glasgow, UK)

Programme Committee

Eugene Agichtein	Emory University, USA
Giambattista Amati	Fondazione Ugo Bordoni, Italy
Massih Amini	Université Pierre et Marie Curie, France
Einat Amitay	IBM Research, Haifa Lab, Israel
Javed Aslam	Northeastern University, USA
Anne Aula	Google, USA
Leif Azzopardi	University of Glasgow, UK
Richard Bache	University of Strathclyde, UK
Ricardo Baeza-Yates	Yahoo! Research, Spain
Alex Bailey	Google Switzerland GmbH, Switzerland
Peter Bailey	CSIRO, Australia
Mark Baillie	University of Strathclyde, UK
Alvaro Barreiro	University of A Coruña, Spain
Roberto Basili	University of Rome "Tor Vergata", Italy
Holger Bast	Max-Planck-Institute for Informatics, Germany
Micheline Beaulieu	University of Sheffield, UK

Preben Hansen	SICS, Sweden
Donna Harman	NIST, USA
David Hawking	CSIRO ICT Centre, Australia
Ben He	University of Glasgow, UK
Daqing He	University of Pittsburgh, USA
Djocrd Hiemstra	University of Twente, The Netherlands
Eduard Hoenkamp	Maastricht University, The Netherlands
Andreas Hotho	University of Kassel, Germany
Xiaohua Hu	Drexel University, USA
Jimmy Huang	York University, Canada
Juan Huete	University of Granada, Spain
Theo Huibers	Thaesis / University of Twente, The Netherlands
David Hull	Justsystems Evans Research, USA
Peter Ingwersen	Royal School of Library and Information Science, Denmark
Jim Jansen	Penn State University, USA
Frances Johnson	Manchester Metropolitan University, UK
Hideo Joho	University of Glasgow, UK
Gareth Jones	Dublin City University, Ireland
Joemon Jose	University of Glasgow, UK
Jaap Kamps	University of Amsterdam, The Netherlands
Gabriella Kazai	Microsoft Research, UK
Jaana Kekäläinen	University of Tampere, Finland
Diane Kelly	University of North Carolina, USA
Manolis Koubarakis	National and Kapodistrian University of Athens, Greece
Wessel Kraaij	TNO, The Netherlands
Udo Kruschwitz	University of Essex, UK
Ravi Kumar	Yahoo! Research, USA
Mounia Lalmas	Queen Mary, University of London, UK
Monica Landoni	University of Strathclyde, UK and University of Lugano, Switzerland
Birger Larsen	Royal School of Library and Information Science, Denmark
Ray Larson	University of California, Berkeley, USA
Hyowon Lee	Dublin City University, Ireland
Xuelong Li	Birkbeck College, University of London, UK
Christina Lioma	University of Glasgow, UK
David Losada	Universidad de Santiago de Compostela, Spain
Craig Macdonald	University of Glasgow, UK
Andrew MacFarlane	City University London, UK
Joao Magalhaes	Imperial College London, UK
Marco Maggini	University of Siena, Italy
Jean Martinet	National Institute of Informatics, Tokyo, Japan

Massimo Melucci	University of Padua, Italy
Donald Metzler	Yahoo! Research, USA
Alessandro Micarelli	Università Roma Tre, Italy
Gilad Mishne	Yahoo! Research, USA
Stefano Mizzaro	University of Udine, Italy
Marie-Francine Moens	KU Leuven, Belgium
Alistair Moffat	The University of Melbourne, Australia
Christof Monz	Queen Mary, University of London, UK
Javed Mostafa	University of North Carolina-Chapel Hill, USA
Josiane Mothe	IRIT, University of Toulouse, France
Vanessa Murdock	Yahoo! Research, Spain
Gheorghe Muresan	Microsoft Research, USA
G. Craig Murray	University of Maryland, USA
Jian-Yun Nie	University of Montreal, Canada
Michael Oakes	University of Sunderland, UK
Paul Ogilvie	mSpoke, USA
Arlindo Oliveira	INESC-ID/IST, Portugal
J. Scott Olsson	University of Maryland, USA
Gabriella Pasi	Università degli Studi di Milano Bicocca, Italy
Jan Pedersen	Yahoo! Research, USA
Nils Pharo	Oslo University College, Norway
Benjamin Piwowarski	Yahoo! Research, Chile
Andreas Rauber	Vienna University of Technology, Austria
Matthew Richardson	Microsoft Research, USA
Stephen Robertson	Microsoft Research, UK
Thomas Roelleke	Queen Mary, University of London, UK
Stefan Rueger	The Open University, UK
Tetsuya Sakai	NewsWatch, Inc., Japan
Mark Sanderson	University of Sheffield, UK
Ralf Schenkel	Max-Planck-Institut für Informatik, Germany
Hinrich Schütze	University of Stuttgart, Germany
Giovanni Semeraro	University of Bari, Italy
Milad Shokouhi	RMIT University, Australia
Stefan Siersdorfer	University of Sheffield, UK
Mário Silva	University of Lisbon, Portugal
Fabrizio Silvestri	CNR, Italy
Alan Smeaton	Dublin City University, Ireland
Mark Smucker	University of Massachusetts Amherst, USA
Vaclav Snasel	VSB-Technical University of Ostrava, Czech Republic
Ian Soboroff	NIST, USA
Dawei Song	The Open University, UK
Ruihua Song	Microsoft Research Asia, China
Eero Sormunen	University of Tampere, Finland
Amanda Spink	Queensland University of Technology, Australia

Nicola Stokes   NICTA, University of Melbourne, Australia
Jaime Teevan   Microsoft Research, USA
Martin Theobald   Stanford University, USA
Ulrich Thiel   Fraunhofer IPSI, Germany
Anastasios Tombros   Queen Mary, University of London, UK
Pertti Vakkari   University of Tampere, Finland
Olga Vechtomova   University of Waterloo, Canada
Robert Villa   University of Glasgow, UK
Dawid Weiss   Poznan University of Technology, Poland
Thijs Westerveld   Teezir Search Solutions, The Netherlands
Ross Wilkinson   CSIRO, Australia
Wensi Xi   Google, USA
Rong Yan   IBM Research, USA
Tao Yang   Ask.com and University of California, Santa
                           Barbara, USA
Elad Yom-Tov   IBM Research, Haifa Lab, Israel
Clement Yu   University of Illinois at Chicago, USA
Dell Zhang   Birkbeck College, University of London, UK
Justin Zobel   NICTA, Australia

Additional Reviewers

Jiang Bian   Georgia Institute of Technology, USA
Roi Blanco   University of A Coruña, Spain
Luigi Carnevale   Ask.com, USA
Wisam Dakka   Columbia University, USA
Erik Graf   University of Glasgow, UK
Martin Halvey   University of Glasgow, UK
Eric Heymann   Ask.com, USA
Evangelos Kanoulas   Northeastern University, USA
Pasquale Lops   University of Bari, Italy
Alessio Signorini   Ask.com, USA
Andrew Tjang   Rutgers University and Ask.Com, USA
Emine Yilmaz   Northeastern University, USA

Workshop and Tutorials Committee

Claudio Carpineto   FUB, Italy
Nick Craswell   Microsoft Research, UK
David Losada   Universidad de Santiago de Compostela, Spain
Josiane Mothe   IRIT, France
Jian-Yun Nie   University of Montreal, Canada
Andrew Trotman   University of Otago, New Zealand
Justin Zobel   NICTA, Australia

Best Paper Award Committee

Anne Aula   Google, USA
Udo Kruschwitz   University of Essex, UK
Thijs Westerveld   Teezir, The Netherlands

Best Student Paper Award Committee

Leif Azzopardi   University of Glasgow, UK
Massimo Melucci   University of Padova, Italy
Vanessa Murdock   Yahoo! Research, Spain

Best Poster Committee

Andrew MacFarlane   City University, UK
Nicola Stokes   University of Melbourne, Australia
Robert Villa   University of Glasgow, UK

Sponsoring Institutions

Organising Institutions

In Co-operation with

Table of Contents

Invited Presentations

Some(What) Grand Challenges for Information Retrieval 1
    Nicholas J. Belkin

Web Search: Challenges and Directions 2
    Amit Singhal

You Are a Document Too: Web Mining and IR for Next-Generation
Information Literacy 3
    Bettina Berendt

Evaluation

Discounted Cumulated Gain Based Evaluation of Multiple-Query IR
Sessions 4
    Kalervo Järvelin, Susan L. Price, Lois M.L. Delcambre, and
    Marianne Lykke Nielsen

Here or There: Preference Judgments for Relevance 16
    Ben Carterette, Paul N. Bennett, David Maxwell Chickering, and
    Susan T. Dumais

Using Clicks as Implicit Judgments: Expectations Versus
Observations 28
    Falk Scholer, Milad Shokouhi, Bodo Billerbeck, and Andrew Turpin

Web IR

Clustering Template Based Web Documents 40
    Thomas Gottron

Effective Pre-retrieval Query Performance Prediction Using Similarity
and Variability Evidence 52
    Ying Zhao, Falk Scholer, and Yohannes Tsegay

iCluster: A Self-organizing Overlay Network for P2P Information
Retrieval 65
    Paraskevi Raftopoulou and Euripides G.M. Petrakis

Social Media

Labeling Categories and Relationships in an Evolving Social Network . . . 77
    Ming-Shun Lin and Hsin-Hsi Chen

Automatic Construction of an Opinion-Term Vocabulary for Ad Hoc
Retrieval   89
    Giambattista Amati, Edgardo Ambrosi, Marco Bianchi,
    Carlo Gaibisso, and Giorgio Gambosi

A Comparison of Social Bookmarking with Traditional Search   101
    Beate Krause, Andreas Hotho, and Gerd Stumme

Cross-Lingual Information Retrieval

Effects of Aligned Corpus Quality and Size in Corpus-Based CLIR   114
    Tuomas Talvensaari

Exploring the Effects of Language Skills on Multilingual Web Search . . .   126
    Jennifer Marlow, Paul Clough, Juan Cigarrán Recuero, and
    Javier Artiles

A Novel Implementation of the FITE-TRT Translation Method   138
    Aki Loponen, Ari Pirkola, Kalervo Järvelin, and Heikki Keskustalo

Theory I

The BNB Distribution for Text Modeling   150
    Stéphane Clinchant and Eric Gaussier

Utilizing Passage-Based Language Models for Document Retrieval   162
    Michael Bendersky and Oren Kurland

A Statistical View of Binned Retrieval Models   175
    Donald Metzler, Trevor Strohman, and W. Bruce Croft

Video

Video Corpus Annotation Using Active Learning   187
    Stéphane Ayache and Georges Quénot

Use of Implicit Graph for Recommending Relevant Videos: A Simulated
Evaluation   199
    David Vallet, Frank Hopfgartner, and Joemon Jose

Representation I

Using Terms from Citations for IR: Some First Results   211
    Anna Ritchie, Simone Teufel, and Stephen Robertson

Automatic Extraction of Domain-Specific Stopwords from Labeled
Documents   222
    Masoud Makrehchi and Mohamed S. Kamel

Wikipedia and E-Books

Book Search Experiments: Investigating IR Methods for the Indexing
and Retrieval of Books   234
   Hengzhi Wu, Gabriella Kazai, and Michael Taylor

Using a Task-Based Approach in Evaluating the Usability of BoBIs in
an E-book Environment   246
   Noorhidawati Abdullah and Forbes Gibb

Exploiting Locality of Wikipedia Links in Entity Ranking   258
   Jovan Pehcevski, Anne-Marie Vercoustre, and James A. Thom

The Importance of Link Evidence in Wikipedia   270
   Jaap Kamps and Marijn Koolen

Expert Search

High Quality Expertise Evidence for Expert Search   283
   Craig Macdonald, David Hannah, and Iadh Ounis

Associating People and Documents   296
   Krisztian Balog and Maarten de Rijke

Modeling Documents as Mixtures of Persons for Expert Finding   309
   Pavel Serdyukov and Djoerd Hiemstra

Ranking Users for Intelligent Message Addressing   321
   Vitor R. Carvalho and William W. Cohen

Representation II

Facilitating Query Decomposition in Query Language Modeling by
Association Rule Mining Using Multiple Sliding Windows   334
   Dawei Song, Qiang Huang, Stefan Rüger, and Peter Bruza

Viewing Term Proximity from a Different Perspective   346
   Ruihua Song, Michael J. Taylor, Ji-Rong Wen,
   Hsiao-Wuen Hon, and Yong Yu

Extending Probabilistic Data Fusion Using Sliding Windows   358
   David Lillis, Fergus Toolan, Rem Collier, and John Dunnion

Theory II

Semi-supervised Document Classification with a Mislabeling Error
Model   370
   Anastasia Krithara, Massih R. Amini, Jean-Michel Renders, and
   Cyril Goutte

Improving Term Frequency Normalization for Multi-topical Documents
and Application to Language Modeling Approaches 382
    Seung-Hoon Na, In-Su Kang, and Jong-Hyeok Lee

Probabilistic Document Length Priors for Language Models 394
    Roi Blanco and Alvaro Barreiro

Short Papers

Applying Maximum Entropy to Known-Item Email Retrieval 406
    Sirvan Yahyaei and Christof Monz

Computing Information Retrieval Performance Measures Efficiently in
the Presence of Tied Scores 414
    Frank McSherry and Marc Najork

Towards Characterization of Actor Evolution and Interactions in News
Corpora 422
    *Rohan Choudhary, Sameep Mehta, Amitabha Bagchi, and
    Rahul Balakrishnan*

The Impact of Semantic Class Identification and Semantic Role
Labeling on Natural Language Answer Extraction.................... 430
    Bahadorreza Ofoghi, John Yearwood, and Liping Ma

Improving Complex Interactive Question Answering with Wikipedia
Anchor Text 438
    Ian MacKinnon and Olga Vechtomova

A Cluster-Sensitive Graph Model for Query-Oriented Multi-document
Summarization 446
    Furu Wei, Wenjie Li, Qin Lu, and Yanxiang He

Evaluating Text Representations for Retrieval of the Best Group of
Documents 454
    Xiaoyong Liu and W. Bruce Croft

Enhancing Relevance Models with Adaptive Passage Retrieval 463
    Xiaoyan Li and Zhigang Zhu

Ontology Matching Using Vector Space 472
    Zahra Eidoon, Nasser Yazdani, and Farhad Oroumchian

Accessibility in Information Retrieval 482
    Leif Azzopardi and Vishwa Vinay

Semantic Relationships in Multi-modal Graphs for Automatic Image
Annotation... 490
    Vassilios Stathopoulos, Jana Urban, and Joemon Jose

Conversation Detection in Email Systems 498
  Shai Erera and David Carmel

Efficient Multimedia Time Series Data Retrieval Under Uniform Scaling
and Normalisation 506
  Waiyawuth Euachongprasit and Chotirat Ann Ratanamahatana

Integrating Structure and Meaning: A New Method for Encoding
Structure for Text Classification 514
  Jonathan M. Fishbein and Chris Eliasmith

A Wikipedia-Based Multilingual Retrieval Model 522
  Martin Potthast, Benno Stein, and Maik Anderka

Filaments of Meaning in Word Space 531
  Jussi Karlgren, Anders Holst, and Magnus Sahlgren

Finding the Best Picture: Cross-Media Retrieval of Content 539
  Koen Deschacht and Marie-Francine Moens

Robust Query-Specific Pseudo Feedback Document Selection for Query
Expansion 547
  Qiang Huang, Dawei Song, and Stefan Rüger

Expert Search Evaluation by Supporting Documents 555
  Craig Macdonald and Iadh Ounis

Posters

Ranking Categories for Web Search 564
  *Gianluca Demartini, Paul-Alexandru Chirita, Ingo Brunkhorst, and
  Wolfgang Nejdl*

Key Design Issues with Visualising Images Using Google Earth 570
  Paul Clough and Simon Read

Methods for Augmenting Semantic Models with Structural Information
for Text Classification 575
  Jonathan M. Fishbein and Chris Eliasmith

Use of Temporal Expressions in Web Search 580
  Sérgio Nunes, Cristina Ribeiro, and Gabriel David

Towards an Automatically Generated Music Information System Via
Web Content Mining 585
  Markus Schedl, Peter Knees, Tim Pohle, and Gerhard Widmer

Investigating the Effectiveness of Clickthrough Data for Document
Reordering 591
  Milad Shokouhi, Falk Scholer, and Andrew Turpin

Analysis of Link Graph Compression Techniques 596
  David Hannah, Craig Macdonald, and Iadh Ounis

An Evaluation and Analysis of Incorporating Term Dependency for
Ad-Hoc Retrieval 602
  Hao Lang, Bin Wang, Gareth Jones, Jintao Li, and Yang Xu

An Evaluation Measure for Distributed Information Retrieval
Systems 607
  Hans Friedrich Witschel, Florian Holz, Gregor Heinrich, and
  Sven Teresniak

Optimizing Language Models for Polarity Classification 612
  Michael Wiegand and Dietrich Klakow

Improving Web Image Retrieval Using Image Annotations and Inference
Network 617
  Peng Huang, Jiajun Bu, Chun Chen, and Guang Qiu

Slide-Film Interface: Overcoming Small Screen Limitations in Mobile
Web Search 622
  Roman Y. Shtykh, Jian Chen, and Qun Jin

A Document-Centered Approach to a Natural Language Music Search
Engine 627
  Peter Knees, Tim Pohle, Markus Schedl, Dominik Schnitzer, and
  Klaus Seyerlehner

Collaborative Topic Tracking in an Enterprise Environment 632
  Conny Franke and Omar Alonso

Graph-Based Profile Similarity Calculation Method and Evaluation 637
  Hassan Naderi and Béatrice Rumpler

The Good, the Bad, the Difficult, and the Easy: Something Wrong with
Information Retrieval Evaluation? 642
  Stefano Mizzaro

Hybrid Method for Personalized Search in Digital Libraries............ 647
  Thanh-Trung Van and Michel Beigbeder

Exploiting Session Context for Information Retrieval - A Comparative
Study 652
  Gaurav Pandey and Julia Luxenburger

Structural Re-ranking with Cluster-Based Retrieval 658
  Seung-Hoon Na, In-Su Kang, and Jong-Hyeok Lee

Automatic Vandalism Detection in Wikipedia 663
  Martin Potthast, Benno Stein, and Robert Gerling

Evaluating Paragraph Retrieval for *why*-QA 669
   Suzan Verberne, Lou Boves, Nelleke Oostdijk, and
   Peter-Arno Coppen

Revisit of Nearest Neighbor Test for Direct Evaluation of
Inter-document Similarities 674
   Seung-Hoon Na, In-Su Kang, and Jong-Hyeok Lee

A Comparison of Named Entity Patterns from a User Analysis and a
System Analysis 679
   Masnizah Mohd, Fabio Crestani, and Ian Ruthven

Query-Based Inter-document Similarity Using Probabilistic
Co-relevance Model 684
   Seung-Hoon Na, In-Su Kang, and Jong-Hyeok Lee

Using Coherence-Based Measures to Predict Query Difficulty 689
   Jiyin He, Martha Larson, and Maarten de Rijke

Efficient Processing of Category-Restricted Queries for Web
Directories 695
   Ismail Sengor Altingovde, Fazli Can, and Özgür Ulusoy

Focused Browsing: Providing Topical Feedback for Link Selection in
Hypertext Browsing 700
   Gareth J.F. Jones and Quixiang Li

The Impact of Named Entity Normalization on Information Retrieval
for Question Answering 705
   Mahboob Alam Khalid, Valentin Jijkoun, and Maarten de Rijke

Workshop Summaries

Efficiency Issues in Information Retrieval Workshop 711
   Roi Blanco and Fabrizio Silvestri

Exploiting Semantic Annotations in Information Retrieval 712
   Omar Alonso and Hugo Zaragoza

Workshop on Novel Methodologies for Evaluation in Information
Retrieval 713
   Mark Sanderson, Martin Braschler, Nicola Ferro, and Julio Gonzalo

Tutorials

ECIR 2008 Tutorials 714

Author Index 717

Some(What) Grand Challenges for Information Retrieval

Nicholas J. Belkin

School of Communication, Information and Library Studies
Rutgers University
US

Although we see the positive results of information retrieval research embodied throughout the Internet, on our computer desktops, and in many other aspects of daily life, at the same time we notice that people still have a wide variety of difficulties in finding information that is useful in resolving their problematic situations. This suggests that there still remain substantial challenges for research in IR. Already in 1988, on the occasion of receiving the ACM SIGIR Gerard Salton Award, Karen Spärck Jones suggested that substantial progress in information retrieval was likely only to come through addressing issues associated with users (actual or potential) of IR systems, rather than continuing IR research's almost exclusive focus on document representation and matching and ranking techniques. In recent years it appears that her message has begun to be heard, yet we still have relatively few substantive results that respond to it. In this talk, I identify a few challenges for IR research which fall within the scope of association with users, and which I believe, if properly addressed, are likely to lead to substantial increases in the usefulness, usability and pleasurability of information retrieval.

C. Macdonald et al. (Eds.): ECIR 2008, LNCS 4956, p. 1, 2008.

Web Search: Challenges and Directions

Amit Singhal

Google Inc.

These are exciting times for the field of Web search. Search engines are used by millions of people every day, and the number is growing rapidly. This growth poses unique challenges for search engines: they need to operate at unprecedented scales while satisfying an incredible diversity of information needs. Furthermore, user expectations have expanded considerably, moving from "give me what I said" to "give me what I want". Finally, with the lure of billions of dollars of commerce guided by search engines, we have entered a new world of "Adversarial Information Retrieval". This talk will show that the world of algorithm and system design for commercial search engines can be described by two of Murphy's Laws: a) If anything can go wrong, it will; and b) Even if nothing can go wrong, it will anyway.

C. Macdonald et al. (Eds.): ECIR 2008, LNCS 4956, p. 2, 2008.
© Springer-Verlag Berlin Heidelberg 2008

You Are a Document Too: Web Mining and IR for Next-Generation Information Literacy

Bettina Berendt

Department of Computer Science, K.U. Leuven, B-3001 Heverlee, Belgium
http://www.cs.kuleuven.be/~berendt

Information retrieval and data mining often assume a simple world: There are people with information needs who search - and find - information in sources such as documents or databases. Hence, the user-oriented goals are (a) information literacy: the users' ability to locate, evaluate, and use effectively the needed information, and (b) tools that obviate the need for some of the technical parts of this information literacy. Examples of such tools are search-engine interfaces that direct each user's attention to only an individualised part of the "information overload" universe.

In this talk, I will argue that such simple-world assumptions are no longer justified, advocate a shift in focus, and outline concrete steps for using technology to further a more comprehensive form of information literacy. I will focus on data, documents, and information-related activities on the Web, which are analysed in Web mining and (Web) IR:

1. In today's (Web) information society,
   - the problem is not just information overload, but also information sparsity
   - information-related activities involve disclosing **and** withholding (the latter known under names such as "privacy" or "business secrets")
   - each information-related activity has (at least) one source, one manifestation as data/document, one user and one stakeholder; network effects abound.
   - most importantly, the dichotomy of information-seeking users and information-containing data/documents has vanished in a time when virtually every activity generates data/documents.
2. These considerations lead to a new operationalisation of information literacy, understood in its broader sense as a set of competencies that a citizen of an information society ought to possess to participate intelligently and actively in that society.
3. Based on a range of concrete examples, I will illustrate how tools can support this type of information literacy (and obviate the need to know some technical details).

C. Macdonald et al. (Eds.): ECIR 2008, LNCS 4956, p. 3, 2008.

Discounted Cumulated Gain Based Evaluation of Multiple-Query IR Sessions

Kalervo Järvelin[1], Susan L. Price[2], Lois M.L. Delcambre[2],
and Marianne Lykke Nielsen[3]

[1] University of Tampere, Finland
[2] Portland State University, USA
[3] Royal School of Library and Information Science, Denmark
kalervo.jarvelin@uta.fi, prices@cs.pdx.edu, lmd@cs.pdx.edu,
mln@db.dk

Abstract. IR research has a strong tradition of laboratory evaluation of systems. Such research is based on test collections, pre-defined test topics, and standard evaluation metrics. While recent research has emphasized the user viewpoint by proposing user-based metrics and non-binary relevance assessments, the methods are insufficient for truly user-based evaluation. The common assumption of a single query per topic and session poorly represents real life. On the other hand, one well-known metric for multiple queries per session, instance recall, does not capture early (within session) retrieval of (highly) relevant documents. We propose an extension to the Discounted Cumulated Gain (DCG) metric, the Session-based DCG (sDCG) metric for evaluation scenarios involving multiple query sessions, graded relevance assessments, and open-ended user effort including decisions to stop searching. The sDCG metric discounts relevant results from later queries within a session. We exemplify the sDCG metric with data from an interactive experiment, we discuss how the metric might be applied, and we present research questions for which the metric is helpful.

Keywords: Interactive IR, evaluation metrics, cumulated gain.

1 Introduction

IR research has a strong tradition of laboratory evaluation of IR systems. Such research is based on test collections, pre-defined test topics, and standard evaluation metrics. While recent research has emphasized the user viewpoint by proposing user-based metrics and non-binary relevance assessments, the methods are insufficient for truly user-based evaluation. Much of the evaluation literature assumes a single query per topic and session, which poorly represents real life.

User-based IR research seeks to attain more realism in IR evaluation [3]. For example, precision at recall = 10% or precision at various document cut-off values (DCV) both seek to account for searchers who choose to scan only a subset of the complete result list. The Discounted Cumulated Gain (DCG) [4] [5] takes a different approach by discounting the value of documents ranked further down in the result list. DCG also supports evaluation by graded relevance assessments. But these metrics as

C. Macdonald et al. (Eds.): ECIR 2008, LNCS 4956, pp. 4–15, 2008.
© Springer-Verlag Berlin Heidelberg 2008

well as traditional IR evaluation metrics assume one query per topic/session. In real life, interactive searchers often issue multiple queries using reformulation [1] and/or relevance feedback until they are satisfied or give up. Evaluation metrics that assume one query per topic are insufficient when the searcher's reformulation effort matters.

The TREC conferences introduced *instance recall* for evaluating interactive experiments [7]. This metric allows multiple queries per session as it rewards for the number of distinct relevant answers identified in a session of a given length. However, it does not reward a system (or searcher) for finding pertinent documents early in the session nor does it help to analyze which queries in a sequence are the most effective. The experiments based on instance recall set a fixed session time and a recall-oriented task. In real life, some tasks are precision-oriented due to time pressure. Stopping decisions often depend on the task, the context, personal factors, and the retrieval results [8]. In the present paper we address issues in session-based evaluation.

We approach session-based IR evaluation with the view that, for a given real search situation, (1) a searcher's information need may be muddled as there is no predefined topic to search on, (2) the initial query formulation may not be optimal, (3) his/her need may remain more or less stable, (4) he/she may switch focus, (5) he/she may learn as the session progresses, (6) highly relevant documents are desired, and (7) stopping decisions depend on search tasks and may vary among individuals [3]. Moreover, it is reasonable that (8) examining retrieval results involves a cost, (9) providing feedback or revising the query involves a cost, and (10) costs should be reflected as penalties in the evaluation. A metric allowing systematic testing under these conditions is needed.

We extend the Discounted Cumulated Gain (DCG) into a new, session-based metric for multiple interactive queries. DCG assigns a gain value to each retrieved document of a ranked result and then cumulates the gains from the first document position onwards to the rank of interest in each test design. DCG allows flexible evaluation under various evaluation scenarios through relevance weighting (see also [6] [9]) and document rank-based discounting. Unlike many traditional effectiveness measures, such as MAP, DCG can easily be extended to a session-based DCG (sDCG) metric, which incorporates query sequences as another dimension in evaluation scenarios and allows one to further discount relevant documents found only after additional searcher effort, i.e., feedback or reformulation. The contributions of this paper are to:

- Define the sDCG metric and describe a method for concatenating results from multiple queries into a single discounted cumulated gain for a session.
- Discuss the research questions that this metric can help answer for which there are no suitable existing metrics.
- Provide guidelines for when and how the sDCG metric can be applied.
- Exemplify the metric with data from a real interactive experiment.
- Discuss the contributions of the metric and the challenges of evaluating it.

Section 2 modifies the DCG metric slightly, making it more elegant and principled and then defines sDCG. Section 3 discusses the features, uses, and evaluation of sDCG and illustrates use of sDCG with data from an interactive IR experiment. Section 4 discusses our findings. The Appendix presents mathematical formulas used in defining sDCG. The focus of this paper is on methodological aspects, not on empirical findings per se, which instead serve as an illustration.

2 Cumulated Gain Based Metrics for Queries and Sessions

2.1 Discounted Cumulated Gain

Järvelin and Kekäläinen [4] [5] argue that highly relevant documents are more valuable than marginally relevant documents and that the searcher may reasonably be assumed to scan the result from its beginning up to some point before quitting. Accordingly, they define the cumulated gain (CG) metrics to produce a gain vector based on the ranked retrieved list, where each document is represented by its (possibly weighted) relevance score up to a ranked position n set for the experiment. The authors argue that the greater the ranked position of a relevant document, the less valuable it is for the searcher, because the searcher is less likely to examine the document due to time, effort, and cumulated information from documents already seen. This leads to "correcting" the readings provided by cumulated gain by a rank-based discount factor, the logarithm of the rank of each document. The normalized (discounted) cumulated gain is calculated as the share of ideal performance an IR technique achieves. The Appendix gives formulas for cumulated gain, and for discounting and normalizing it. The benefits of the CG, DCG, and nDCG metrics were discussed thoroughly in comparison to several earlier metrics in [5]. This discussion is not repeated here.

Compared with [5], the definition of the DCG presented here contains a notable modification making it more elegant and principled in discounting early relevant documents. The original formulation employed CG up to the rank of the base of the discounting logarithm and only thereafter discounted the value of relevant documents. The formulation presented here is simpler and systematic in discounting the value of all relevant documents including the early ones not discounted by the original DCG.

2.2 Discounting over a Query Sequence within a Session

A session consists of a sequence of queries, each producing a ranked result. Each query formulation requires some effort by the searcher and therefore the results gained by reformulated queries are progressively less valuable. A DCG vector representing the qth query in sequence is discounted by a factor, which is based on the position of the query. The base of the logarithm bq may be set to model varying searcher behavior: a small base, say $bq = 2$, for an impatient or busy searcher, who is unlikely to reformulate queries or issue novel ones, and a larger base, say $bq = 10$, for a patient searcher willing to probe the document space with several reformulations. sDCG uses the DCG metric to discount the gain within each query and further discounts its gain by a factor dependent on the sequence number of the query within the session. Let DCG be the ordinary DCG vector for the result of the qth query. The session-based discounted vector for the qth query is:

$$\text{sDCG}(q) = (1 + \log_{bq} q)^{-1} * \text{DCG}$$

where $bq \in \mathbf{R}$ is the logarithm base for the query discount; $1 < bq < 1000$ (1)
        q is the position of the query.

Each session-based discounted vector sDCG(q) is a vector representing query performance for one query in the session. Thus it may be normalized like any ordinary

DCG vector by the ideal vector and such normalized vectors can be concatenated to represent an entire session.

3 Application of sDCG: An Example

The sDCG metric evaluates entire interactive multiple query sessions. Because sessions are products of the search task, the searcher, the retrieval system, and the collection, experiments can be designed to evaluate any combination of these. For example, if search tasks and searchers are appropriately randomized and the collection is held constant, as we do below, one may evaluate the performance of search systems in interactive sessions. Current single query evaluation metrics require unnatural tricks (like freezing) in evaluation because there is no user in place to act.

We believe that something like the sDCG is needed because it has reasonable and intuitive behavior:

- documents at equivalent ranks are valued more highly if returned by an earlier query
- there is smooth discounting of both document rank and query iteration
- the parameters are understandable and reflect recognizable searcher and setting characteristics

Setting the parameters for each evaluation case must be based on either general findings on searcher behavior, specific findings on searcher behavior in the context of interest, or simulation scenarios where the sensitivity of retrieval performance to a range of searcher behaviors is of interest. The sDCG metric allows the experimenter to adjust the evaluation to reflect each setting evaluated.

The important contribution of sDCG is the new information and insight that other metrics do not deliver. We assess the validity of the metric by referring to its behavior in the light of available knowledge on real-life interactive searching. Note that there are no existing session-based metrics to compare to as a standard. For example, instance recall measures very different phenomena and requires tasks of a specific kind. There are no existing test collections for interactive searching with multiple queries per session. One needs a searcher to interact with the system and collection, and to produce the queries. Thus, one cannot test the metric on a known collection against a known metric to see if it produces the expected system ranking as might be done with metrics for traditional laboratory-based IR using the TREC collections.

We now illustrate use of the new metric by analyzing data from a user-based experiment. We introduce the test subjects, task and systems in Section 3.1. In Section 3.2, we use sDCG to analyze query effectiveness by query rank across the test searchers and systems. We also consider session effectiveness across the test searchers and systems, rewarding sessions for early finding of highly relevant documents in Section 3.3.

3.1 Sample Data

We show sample data from an empirical, interactive searching study that compared two search systems. Thirty domain experts (family practice physicians) each completed the same four realistic search scenarios that simulated a need for specific information required to make a decision in a short time frame of several minutes. Each scenario formed a separate session. The searchers had a mean of 21 years of experience in medicine and were also experienced searchers, with a mean of 7 years of Internet searching experience and over 2 year's experience with the test collection, sundhed.dk (Table 1). On a Likert scale (1-5), the average of their self-assessed searching skills was 2.4.

Table 1. Searcher features (N=30)

Feature	Average	Standard Deviation
Experience using Internet search engines (years)	7.2	± 2.8
Experience in using sundhed.dk (years)	2.4	± 1.4
Searching experience (low=1; high=5)	2.4	± 0.9
Professional experience in medicine (years)	21.4	± 7.6

We asked searchers to simulate a real-life situation by searching only as long as they would in a real setting. The searchers entered queries and examined results until either finding relevant documents that, in their opinion, satisfied the information need in the scenario or until they judged the search a failure. We also asked them to make graded relevance judgments when they viewed a document. All documents judged relevant by at least one user were judged by an independent domain expert to develop the reference standard we used for the calculations we show in this paper.

The two search systems, Systems 1 and 2, operated over the same collection of nearly 25,000 documents. System 1 used a combination of full text and keyword indexing. System 2 used the existing indexing plus a new form of supplemental document indexing, Semantic Components [8], that affected both the query language and document ranking. Each participant searched on two scenarios with each experimental search system, resulting in 15 sessions per scenario–system combination. The order of exposure to the search scenarios and the systems was randomized (a Latin Square design [2]). A more detailed description of the searching study, using traditional metrics, is available [8].

Table 2. The number of sessions (N=60 per system) issuing exactly 1 - 11 queries across four search tasks in Systems 1 and 2 and the average number of queries per session

Sessions in	Number of Queries											Avg per
(N=60)	1	2	3	4	5	6	7	8	9	10	11	Session
System 1	28	7	10	4	5	4	1	1	0	0	0	2.53
System 2	21	7	9	11	2	4	2	2	0	1	1	3.18

The test searchers constructed 1 to 11 queries for their search tasks for a total of 343 queries. The number varied by system and topic – most searchers quit as soon as they had a reasonably good result. Table 2 shows how many sessions constructed exactly 1 to 11 queries in each system and the average number of queries per session.

In the illustration below, we have several parameters at our disposal:

- Relevance assessments and weights: we use a four point scale (scores 0 to 3, from non-relevant to highly relevant) given by a domain expert; weighted as 0-1-10-100.
- The rank-based document discount – the log base b: a reasonable range is $1.5 \leq b \leq 10$ to reflect impatient to patient searcher; we use 2 to reflect impatient searchers or time-constrained task scenarios.
- The rank-based query discount – the log base bq: we believe that a reasonable range is $1.5 \leq b \leq 10$ to reflect impatient to patient searchers; we use 4 to reflect a impatient searchers or time-constrained task scenarios.
- Stopping – gain vector length: when looking at the effectiveness of queries by their rank, we examine the top-100. When we analyze the gain of an entire session, we concatenate the top-10 of each query in sequence, assuming that a searcher would rather switch to reformulation than continue scanning beyond 10 documents.

We test only some value combinations in the present study.

3.2 Effectiveness by Query Order

Most searchers quit after finding one reasonably good result; only a few continued beyond that point: sometimes they found more (or the same) relevant documents. This searcher behavior is shown clearly in Figure 1, which compares the discounted average gain of the last query in each session (LQ) to the average of the preceding ones (Non-last Q) in Systems 1 and 2. Until their last query the searchers gain little. The last query for searchers in System 2 tends to be somewhat better than in System 1. The last query performance levels off at rank 20 in both systems. The last query was nearly always the best – but not always; a corresponding analysis could be made on the best query of each session, but this does not change the basic result.

The initial query performance suggested by Figure 1 seems poor from the laboratory IR perspective. However, laboratory IR tests typically employ verbose, well-specified topic texts for automatic query construction whereas our sample data reflects real life: human professionals performing a typical search task as best they can. There are many possible explanations for the initial queries not delivering reasonable results. We observed the following problems in our study:

- Errors in using the syntax of the underlying search engine. Capitalizing a search key that is not a proper name when the engine is sensitive to capitalization.
- Using search keys that do not cover the topic appropriately or from the right angle.
- Applying an attribute/metadata based filter (e.g., location criterion) that was too restrictive when combined with content keywords.
- Incorrect controlled metadata value (e.g., wrong value for information type).

Other common reasons include typos, overly specific query formulations (which may return nothing), and far too broad formulations (which are too sparse for relevant documents) – all of which would require query reformulations. sDCG makes such performance variations visible. It shows the magnitude of the gain change due to reformulation and suggests which reformulation to focus on in the analysis. Thus sDCG helps us to analyze which initial formulations and reformulations work best. The query sequence discount penalizes systems that require more queries than others.

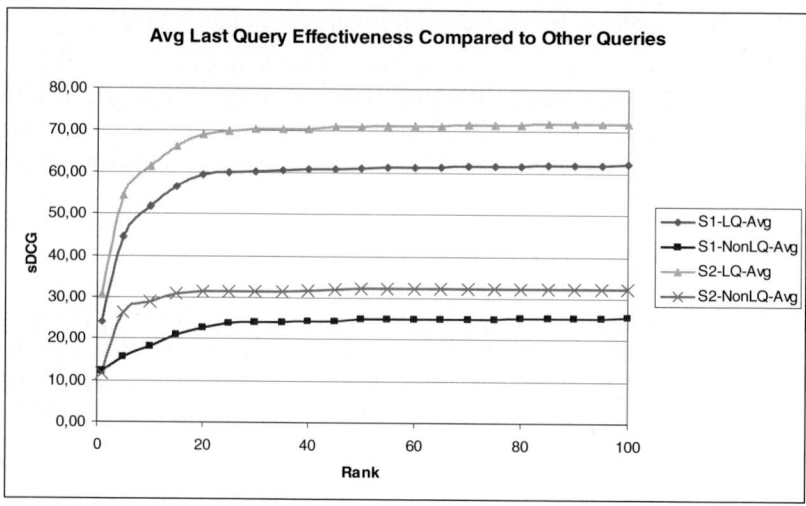

Fig. 1. Average session-based discounted query gain for last vs. non-last queries in Systems 1 and 2 across sessions ($b=2$; $bq=4$)

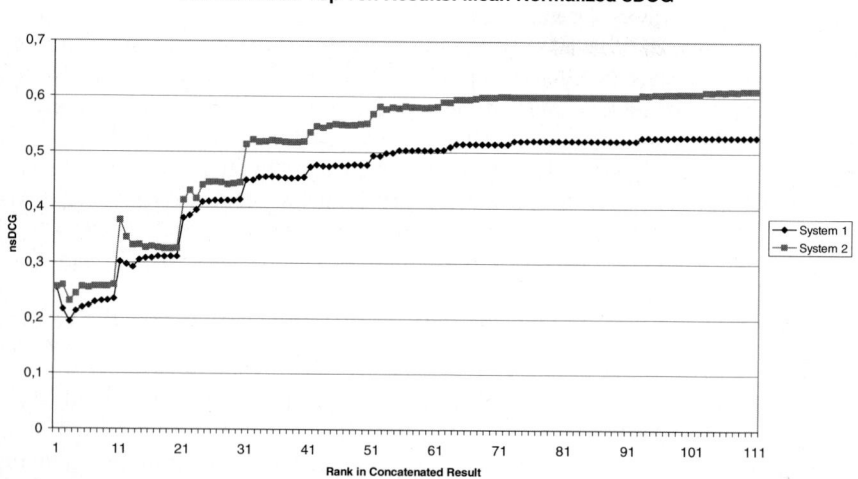

Fig. 2. Session performance by normalized sDCG based on concatenated Top-10 results averaged across all sessions in Systems 1 and 2 ($b=2$; $bq=4$)

3.3 Effectiveness by Sessions

In order to analyze the effectiveness of sessions in the two systems, we now represent each session by concatenating the discounted gains from each query result. This allows the analysis of individual sessions as well as of average sessions across searchers using a specific system. Having relevance assessments by an external judge allows us to analyze the gain vectors up to the last ranks of query results. Thus we may examine performance differences, assuming that the searchers would have scanned Top-X documents, regardless of whether they did in the actual experiment.

In order to *normalize* concatenated sDCG vectors, one needs the corresponding ideal vector and an approach to handle duplicate results. When the searchers are assumed to scan Top-X documents of each query in a sequence, we propose the ideal sDCG vector to be constructed as follows: First one constructs the ideal vector (see appendix) for a single query. Second, the Top-X components of this ideal result are concatenated n times to represent a query sequence of n queries of a session. Each repeated result is discounted using formula (1). This is justified because, in an ideal situation, the searcher issues only one optimal query, which retrieves sufficiently many relevant documents in the optimal order. Each new query in a real session is another attempt at the ideal result. Some documents are, however, returned multiple times by different queries in a session. One must therefore decide whether to cumulate value only the first time a document is returned or every time it is returned. In order to compare systems, and because of using the reference judgments, we chose the latter option because, in our study, some searchers overlooked relevant documents in early queries but recognized them in later ones and each appearance of such a document is a chance provided by the system for the user to recognize it.

Figure 2 reports normalized average performance analyses for concatenated Top-10 query results. In Figure 2, each lot of 10 ranks along the X-axis represents the discounted and normalized Top-10 sDCG of one query, from Q_1 to Q_{11}. The gains are summed progressively so that the gain for Q_n represents the total gain (in the Top-10 ranks of each query) from the beginning of the session. If a searcher stops at Q_n then the gain for that session is held constant up to Q_{11}, i.e., no more gain is amassed. We see that across all the sessions, System 2 has better average performance.

Figure 3 shows a clear trend in the data for one subset, Scenario D, with session length up to 6 queries. There are two pairs of graphs, the upper ones representing concatenated sDCG results averaged across all sessions for Scenario D and the lower ones representing the same for the subset of sessions issuing at least 4 queries. We did not normalize the data because all the queries are for the same scenario and therefore share the same ideal vector. It is clear that multi-query sessions are initially very ineffective. Graphs of the kind reported in Figure 3 may be created for any number of queries in a session and any query results length of interest. Additionally, one may experiment with the discounting parameters and observe their effects. Such graphs also show the performance up to n queries for any number of queries less than the final query.

Figure 4 displays some raw data, individual session performance by concatenated Top-10 results up to six queries in System 1 and across all 15 sessions of one search scenario. We have padded short sessions to the same length. When each graph turns horizontal, the searcher most often quit searching (and only sometimes found nothing more). We have not marked the actual quitting points on the graphs. This figure clearly demonstrates the great variability among sessions and that the last query was far more effective than earlier ones. Such a display is a very useful tool early in the evaluation at the individual session level.

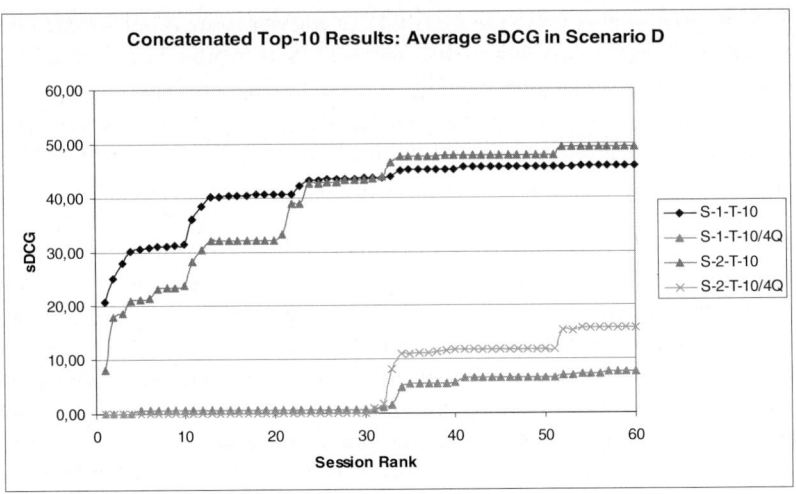

Fig. 3. Session performance by concatenated Top-10 results averaged across all sessions of Scenario D in Systems 1 (S-1-T-10) and 2 (S-2-T-10) and across the sessions that constructed at least four queries (S-1-T-10/4Q and S-2-T-10/4Q) (b=2; bq=4)

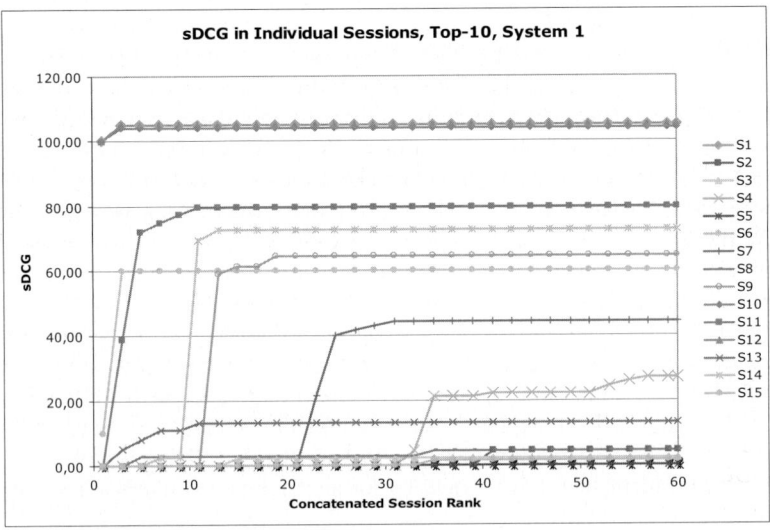

Fig. 4. Individual session performance by concatenated Top-10 results in System 1 across all 15 sessions, Scenario D (b=2; bq=4)

4 Discussion

Standard single query metrics, such as MAP, and interactive metrics, such as instance recall, are insufficient when IR systems and interaction are studied from a more realistic session-based perspective. We extended the Discounted Cumulated Gain metric

(DCG) [5] to a session-based sDCG metric by applying a query discount over a sequence of queries. The salient features of the new metric are:

- It uses graded relevance assessments and can reward highly relevant documents.
- It supports experimentation with relevance weighting from liberal, flat weights to sharp weighting of highly relevant documents.
- It supports experimentation using document and query discounts that can adjust the gain of documents retrieved late in a query or session. This supports modeling of various searcher/task scenarios regarding searcher impatience vs. persistence and regarding task/context dependent constraints and time limitations.
- For individual sessions, it supports identifying unsuccessful and successful reformulations, aiding searcher behavior analysis and system development.
- By selecting the assumed result scanning length (or recording observed searcher behavior) before stopping, it supports representing entire sessions by gain vectors that can be compared to each other or to ideal performance.
- sDCG can be normalized for comparisons across search scenarios. However, direct or averaged comparisons without normalization may be very informative as well and normalization is unnecessary when all queries are for the same scenario.

The sDCG metric and its application may appear complex compared to standard IR testing. This is unavoidable. Involving users and multiple query sessions introduces new variables into the test setting. The document and query discount parameters are important for realism because initial queries may not be successful and searching time may be limited. The new parameters allow assessing performance over a range of searcher/task scenarios. The complexity is a strength that allows bringing realism to evaluation and does not assume that all searchers or contexts are alike. Setting the parameters in each evaluation case depends on the evaluation purposes and should be based on relevant findings on searcher behavior or simulation scenarios exploring a range of searcher behaviors.

Initial queries may fail due to the widely known vocabulary problem. Our data shows that domain professionals have different interpretations and, consequently, construct differently behaving queries – even when facing the same scenario. This does not happen when the topics of test collections are used directly as queries. Using the sDCG we can evaluate systems and interfaces that may or may not be helpful in supporting good queries. For example, a plain engine with a keyword search box may be excellent in ranking documents for any given query. However, a domain specific interface may support the searcher in (re)formulating a much better query. New tools, such as sDCG, are essential for evaluating such interfaces.

sDCG can accept relevance scores derived from users, from independent relevance assessors, or from pre-existing test collections. For concatenating top-N results, there are three important issues to consider:

- Scanning length: In this study, we assumed most users consider the top ten results. Ideally we want to know how far each user scanned each result list, using eye-tracking data or having searchers mark the last document considered. Click data is a limited surrogate because it doesn't indicate which documents were rejected based on title or summary.
- Short results lists: Some queries return few or no hits. Concatenating short results lists "as is" ignores the time spent looking at a short or empty list and reformulating

the query. In this study we padded concatenated lists to a fixed length (10) but a hybrid approach might be used, assessing a minimum penalty per query.

- Duplicates: Documents are often returned multiple times by different queries in a session. In truly user-based evaluation, the searcher is free to score duplicates as relevant or non-relevant. For example, in our study some searchers overlooked relevant documents in early queries but recognized them in later ones. If judgments are supplied by external judges or a test collection, this option is not available. One must decide whether to cumulate value only the first time a document is returned or every time it is returned. We chose the latter option.

In the present paper we do not perform statistical testing as we are only using the data to illustrate the proposed metric. However, statistical testing may be applied on the sDCG findings. Appropriate statistical tests depend on the study design, and the sDCG metric may be used in many quite different designs. The final concatenated sDCG gain of a session, the sDCG gain of the best or last query, or the average gain within a session, are possible choices for evaluation. The *final concatenated gain* is insensitive to early stopping in some sessions, as the gain does not change after stopping. The *average gain,* i.e., the average position value of an sDCG vector (see [5] for the formula), represents the overall performance of all the queries in a session as a single number. When averaging gain across multiple search scenarios and using independent relevance judgments, normalizing (see Appendix) adjusts for the differing number of relevant documents across scenarios and represents performance in the range [0, 1] in relation to the ideal. Thence statistical testing is not affected by outliers, scenarios returning many highly relevant documents.

Acknowledgements

This work was supported in part by the Academy of Finland, under Project numbers 204978 and 1209960 and by the U.S. National Science Foundation, grant numbers 0514238, 0511050 and 0534762.

References

1. Bates, M.: The design of Browsing and Berrypicking Techniques for the Online Search Interface. Online Review 13(5), 407–424 (1989)
2. Beaulieu, M., Robertson, S., Rasmussen, E.: Evaluating Iinteractive Systems in TREC. Journal of the American Society for Information Science 47(1), 85–94 (1996)
3. Ingwersen, P., Järvelin, K.: The Turn: Integration of Information Seeking and Retrieval in Context. Springer, Dortrecht (2005)
4. Järvelin, K., Kekäläinen, J.: IR Evaluation Methods for Retrieving Highly Relevant Documents. In: 23rd Annual International ACM SIGIR Conference on Research and Development in Information Retrieval, pp. 41–48. ACM Press, New York (2000)
5. Järvelin, K., Kekäläinen, J.: Cumulated Gain-based Evaluation of IR Techniques. ACM Transactions on Information Systems 20(4), 422–446 (2002)
6. Kekäläinen, J.: Binary and Graded Relevance in IR Evaluations – Comparison of the Effects on Ranking of IR Systems. Inform. Processing & Management 41(5), 1019–1033 (2005)

7. Over, P.: TREC-7 interactive track report. In: NIST Special Publication 500-242: The Seventh Text Retrieval Conference, NIST, Gaithersburg (1999)
8. Price, S.L., Lykke Nielsen, M., Delcambre, L.M.L., Vedsted, P.: Semantic Components Enhance Retrieval of Domain-Specific Documents. In: 16th ACM conference on Conference on information and knowledge management, pp. 429–438. ACM Press, New York (2007)
9. Voorhees, E.: Evaluation by highly relevant documents. In: 24th Annual International ACM SIGIR Conference on Research and Development in Information Retrieval, pp. 74–82. ACM Press, New York (2001)

Appendix

The cumulated gain at ranked position i is computed by summing from position 1 to i, $i \leq 100$. By denoting the value in position i in the gain vector G by G[i], the cumulated gain vector CG is defined as the vector CG where:

$$CG[i] = \sum_{j=1}^{i} G[j] \tag{1}$$

For example, assuming G' = <3, 2, 3, 0, 0, 1, 2, 2, 3, 0, ... > we obtain CG' = <3, 5, 8, 8, 8, 9, 11, 13, 16, 16, ...>. The DCG metric also allows for weighting the relevance scores. For example, one can choose to replace the scores 0-3 by the weights of 0-1-10-100 to reward retrieval of highly relevant documents.

We define the vector DCG as follows:

$$DCG[i] = \sum_{j=1}^{i} G[j]/(1 + \log_b i) \tag{2}$$

For example, let b = 4. From G' given above we obtain DCG' = <3, 4, 5.67, 5.67, 5.67, 6.11, 6.94, 8.14, 9.30, 9.30, ...>. Note that the formulation is slightly different from the original [9] and more elegant.

The construction of average vectors requires vector sums and multiplication by constants. For this, let V = <v_1, v_2, ..., v_k> and W = < w_1, w_2, ..., w_k> be two vectors.

$$V + W = <v_1 + w_1, v_2 + w_2, ..., v_k + w_k>$$

$$\Sigma_{V \in \vartheta} \, V = V_1 + V_2 + ... + V_n \text{ when } \vartheta = \{V_1, V_2, ..., V_n\} \tag{3}$$

$$r * V = <r * v_1, \, r * v_2, \, ... , \, r * v_k> \text{ when } r \text{ is constant}$$

The average vector of vectors $\vartheta = \{V_1, V_2, ..., V_n\}$, is:

$$avg\text{-}vect(\vartheta) = |\vartheta|^{-1} * \Sigma_{V \in \vartheta} \, V \tag{4}$$

Given an (average) (D)CG vector V = <v_1, v_2, ..., v_k> for an IR technique, and the (average) ideal DCG vector I = <i_1, i_2, ..., i_k>, the normalized performance vector nDCG is obtained by the function [9]:

$$norm\text{-}vect(V, I) = <v_1/i_1, v_2/i_2, ..., v_k/i_k> \tag{5}$$

The ideal vector has the relevance scores of the recall base in descending order. The nDCG vector components have values in the range [0, 1], representing the share of the ideal discounted gain achieved by the DCG vector V.

Here or There
Preference Judgments for Relevance

Ben Carterette[1], Paul N. Bennett[2], David Maxwell Chickering[3],
and Susan T. Dumais[2]

[1] University of Massachusetts Amherst
[2] Microsoft Research
[3] Microsoft Live Labs

Abstract. Information retrieval systems have traditionally been evaluated over absolute judgments of relevance: each document is judged for relevance on its own, independent of other documents that may be on topic. We hypothesize that preference judgments of the form "document A is more relevant than document B" are easier for assessors to make than absolute judgments, and provide evidence for our hypothesis through a study with assessors. We then investigate methods to evaluate search engines using preference judgments. Furthermore, we show that by using inferences and clever selection of pairs to judge, we need not compare all pairs of documents in order to apply evaluation methods.

1 Introduction

Relevance judgments for information retrieval evaluation have traditionally been made on a binary scale: a document is either relevant to a query or it is not. This definition of relevance is largely motivated by the importance of *topicality* in tasks studied in IR research [1].

The notion of relevance can be generalized to a graded scale of absolute judgments. Järvelin and Kekäläinen [2] proposed doing so to identify very relevant documents in addition to relevant and non-relevant documents. They developed the *discounted cumulative gain* (DCG) measure to summarize performance taking into account both graded relevance and greater importance for items retrieved at the top ranks. DCG has been used to evaluate web search applications where the first few results are especially important. In web search applications, factors other than topical relevance, such as quality of information, quality of display, or important of the site, are often included in assessing relevance.

Although evaluations over graded relevance allow for finer distinctions among documents, adopting graded relevance has two significant drawbacks. First, the specifics of the gradations (i.e. how many levels to use and what those levels mean) must be defined, and it is not clear how these choices will affect relative performance measurements. Second, the burden on assessors increases with the complexity of the relevance gradations; when there are more factors or finer distinctions to consider, the choice of label is less clear. High measured levels of disagreement on binary judgments [3] suggests the difficulty of the problem.

C. Macdonald et al. (Eds.): ECIR 2008, LNCS 4956, pp. 16–27, 2008.
© Springer-Verlag Berlin Heidelberg 2008

When measurement is difficult in practice or not completely objective, judgments of *preference* may be a good alternative [4]. Instead of assigning a relevance label to a document, an assessor looks at two pages and expresses a preference for one over the other. This is a binary decision, so there is no need to determine a set of labels and no need to map judgments to a numeric scale.

Of course, using preference judgments poses a new set of questions: how do we use preference judgments to evaluate a search engine? The number of pairs of documents is polynomial in the number of documents; will it be feasible to ask for judgments on every pair? If not, which pairs do we choose? But these questions are more amenable to empirical investigation.

There is another advantage to direct preference judgments: algorithms such as RankNet [5] and ranking SVMs [6] are trained over preferences. Sometimes preferences are obtained by inference from absolute judgments [5]. By collecting preferences directly, some of the noise associated with difficulty in distinguishing between different levels of relevance may be reduced. Additionally, absolute judgments result in ties in inferred preferences; direct preferences may allow more data to be used for training.

In this work we follow three successive lines of investigation. First, we compare assessor agreement and time spent per judgment for preference and absolute judgments. Next, we consider the evaluation of search engines when judgments are preferences. Finally, we look at focusing assessor effort to collect sufficient preferences to be able to compare search engines accurately.

2   Previous Work

The idea of pairwise preference judgments has not been explored much in the IR literature. When the idea of preference judgments has arisen, the practice has typically been to infer preferences from existing absolute judgments (e.g. [7,8]), sidestepping questions about collecting preferences directly.

The most closely related previous work is that of Joachims, who first hypothesized that a click could be treated as a preference judgment (the document clicked being preferred to all ranked above it) [6], then used an eye-tracking study to verify that hypothesis [9]. Neither work touched on questions of evaluation.

Buckley and Voorhees's *bpref* evaluation measure [10] is calculated by summing the number of relevant documents ranked above nonrelevant documents. It suggests the idea of preferences, but it is defined over absolute judgments. The calculation of bpref entails inferring that each relevant document is preferred to every nonrelevant document, but all the relevant documents are "tied": none is preferred over any other.

Mizzaro has proposed measures of assessor agreement for both absolute and preference judgments [11], but we could find no work that empirically evaluated whether assessors tend to agree more on one or the other as we do here.

In a study that lends support to this work, Rorvig made the case for preference-based test collections using an idea from mathematical psychology known as "simple scalability" [12]. He argued that, despite their high cost, preference judgments are an imperative for tasks for which the goal is to find highly-relevant documents. Rorvig showed that necessary conditions for the application of simple scalability held in practice, but we were unable to find any follow-up studies on preferences versus absolute judgments.

Thus to the best of our knowledge this is the first comparison of absolute judgments versus preference judgments in terms of assessor performance. It is also the first investigation into making preference judgments cost-effective by reducing the total number needed for evaluation of search engines.

3   Assessor Study

Our study investigated whether preferences are "easier" to make than absolute judgments by measuring inter-assessor consistency and time spent on each judgment. All judgments will be made on web pages retrieved by the Yahoo!, Google, and Microsoft Live search engines.

We compared three types of judgments: (1) absolute judgments on a five-point scale (Bad, Fair, Good, Excellent, Perfect); (2) binary preference judgments as described above; and (3) a stronger version of preference judgment in which the assessor can additionally say that he or she *definitely* preferred one page over another. To mitigate against assessors abstaining from hard decisions, neither preference type allowed an "equal" or "same" judgment.[1]

3.1   Experimental Design

Measuring agreement requires that each query be seen by at least two different assessors for each of the three judgment types. Since an assessor cannot see the same query twice, we needed at least six assessors. Requiring that each assessor see every query imposed the constraint that assessors could not enter their own queries; the implications of this will be discussed in the next section.

Judging Interfaces. We designed interfaces for each of the three judgments types. Screenshots for two of them are shown in Figure 1; the binary preference interface is identical to Figure 1(b) but excludes the "Definitely Here" buttons.

The query was shown at the top of the screen. If the assessor did not understand the query, he or she could obtain context by clicking on the magnifying glass button to see snippets from the top 20 web pages retrieved by a search engine. The order of these pages was randomized so as not to influence the assessor's judgments.

We allocated the same area to each web page in all three interfaces, regardless of whether one or two pages were shown. We highlighted the query terms that we found in a simple parse of the web page to make it easier for the judge to find relevant content.

---

[1] A followup study that included a "duplicates" judgment on pairs showed results consistent with those described in the next section.

(a) Absolute judgments.

(b) Preference judgments.

Fig. 1. Screenshots of the judgment interfaces

Queries. We sampled 51 queries from Microsoft's Live Search query logs. We chose queries that had previously been judged for the purpose of assessing search engine quality; in particular, we selected a biased sample that had some diversity in existing judgments, but was in other respects random. Some of the queries had clear intent, but most were vague, underspecified, or had myriad possible intents. The queries can generally be considered "informational"; examples include "andie mcdowell", "binghamton", "soda pop and oral hygiene".

Assessors. The six assessors were Microsoft employees. All assessors had backgrounds in information retrieval or related fields and had experience in judging web pages for relevance.

Web Pages. For each query, we took the top five web pages retrieved by three large search engines. The number of unique pages for a query depended on the diversity of results retrieved by the engines: if all three retrieved the same documents, there would only be 5 pages to judge, but if they all retrieved different documents, there would be 15 pages. There were on average 11.9 unique pages per query, indicating a high degree of diversity among the top ranked results.

We did not remove web pages that had duplicate content but different URLs. We made this decision because there were some cases in which the URL provided some valuable additional information that helped judge relevance. For example, one query specified a product number. Two identical pages about the product were retrieved. Neither page contained the product number, but it was part of one of the URLs. It is important to note, however, that this will be the source of some disagreements in the preferences judgments.

In order to avoid time delays or temporary internet outages, we pre-captured web pages by saving images of them to disk. This also guaranteed that the pages were saved at a fixed point in time and the experiment could be reproduced.

Judgments. As shown in Figure 1, the three interfaces had buttons along the top for judging documents, as well as a "Bad" button at the top left of the displayed web pages (in the same location relative to the web page for consistency). In the preference interface, a "Bad" judgment could be used for pages that were clearly not relevant, had not been properly loaded, or were spam. A page labeled "Bad" would not be seen again for any subsequent preference judgment.

We gave assessors guidelines explaining differences between relevance labels. The guidelines included the topicality of the page as well as the ease of finding relevant content on the page, trust in the domain, the likelihood that the page reflects the intent of the query, and so on. Assessors used the same guidelines for both absolute and preference judgments.

We fixed pages in a random order prior to any judging. Assessors made absolute judgments in that order. For preference judgments, the first two pages in the fixed order were presented first. The next judgment retained the preferred page and asked for its preference over the next page in the fixed order. When all comparisons involving the preferred page were exhausted, judgments restarted with the next two pages in the fixed order.

3.2   Results and Analysis

Agreement. There are two types of agreement: agreement between two assessors over all judgments, or agreement about each judgment over all assessors. We chose to look at the latter in an attempt to average out differences in expertise, prior knowledge, or interpretation of the query.

Agreement for absolute judgments is shown in Table 1. Each cell (J_1, J_2) is the probability that one assessor would say J_2 (column) given that another said J_1 (row). They are normalized by row, which is why columns do not add to 1. The percentage of pages with each label is $20\%, 28\%, 25\%, 25\%, 2\%$, for Bad, Fair, Good, Excellent, and Perfect, respectively.

Agreement for preference judgments is shown in Table 2(a). For comparison, we inferred preferences from the absolute judgments: if the judgment on page A was greater than the judgment on page B, we inferred that A was preferred to B.

Table 1. Assessor agreement for absolute judgments

	Bad	Fair	Good	Excellent	Perfect	Total
Bad	0.579	0.290	0.118	0.014	0.000	221
Fair	0.208	0.332	0.309	0.147	0.003	307
Good	0.095	0.348	0.286	0.260	0.011	273
Excellent	0.011	0.167	0.264	0.535	0.022	269
Perfect	0.000	0.042	0.125	0.250	0.583	24

Table 2. Assessor agreement for actual (a) and inferred (b) preference judgments

	$A < B$	A, B bad	$A > B$	Total
$A < B$	0.752	0.033	0.215	2580
A, B bad	0.208	0.567	0.225	413
$A > B$	0.201	0.034	0.765	2757

(a) Preferences.

	$A < B$	A, B bad	$A > B$	Total
$A < B$	0.657	0.051	0.292	2530
A, B bad	0.297	0.380	0.323	437
$A > B$	0.278	0.053	0.669	2654

(b) Inferred preferences.

Table 3. Assessor agreement for definite preference judgments

	$A \ll B$	$A < B$	A, B bad	$A > B$	$A \gg B$	Total
$A \ll B$	0.247	0.621	0.000	0.132	0.000	219
$A < B$	0.059	0.661	0.043	0.221	0.015	2288
A, B bad	0.000	0.244	0.453	0.300	0.002	406
$A > B$	0.012	0.212	0.051	0.670	0.055	2389
$A \gg B$	0.000	0.180	0.005	0.680	0.134	194

To compare to true preferences, we had to assign some preference to pairs of pages that were given the same label ("ties"). Table 2(b) gives results when the assigned preference is random (i.e. the expected value of a coin flip), simulating an assessor that makes a random judgment about which of two similar pages is preferred.

Statistical significance between Tables 2(a) and 2(b) can be measured by a χ^2 test comparing the ratio of the number of pairs agreed on to the number disagreed on for both preferences and inferred preferences. The difference is significant ($\chi^2 = 143, df = 1, p \approx 0$).

We can also explore redistributing ties at different rates to model different levels of agreement. Up to about 70% agreement on ties, true preference agreement is still significantly greater than inferred preference agreement. Above 80%, inferred preference agreement is significantly greater.

Agreement for the two-level "definite" preferences is shown in Table 3. Assessors do not appear to have been very consistent in their use of the "definitely" judgment. When the definite judgments are pooled together with the preference judgments (i.e. $A < B$ and $A \ll B$ treated as identical), the agreement is slightly less than in Table 2(a), but more than Table 2(b).

Table 4. Median seconds per judgment by each assessor in each interface

	Preference	Definite	Absolute	Overall
Assessor 1	3.50	3.41	7.96	3.70
Assessor 2	3.24	3.67	6.12	3.55
Assessor 3	2.35	2.82	5.56	2.82
Assessor 4	4.13	4.30	8.78	4.71
Assessor 5	2.72	3.30	8.20	3.17
Assessor 6	2.09	2.40	3.21	2.31
Overall	2.87	3.15	6.33	3.23

Time. Table 4 shows the median number of seconds spent on each judgment by each assessor for each interface, along with overall medians for each assessor and for each interface.[2] Absolute judgments took about twice as long to make as preferences. As the table shows, there was little variance among assessors.

Two main variables affect the time it takes to judge a page or a pair of pages: time spent reading the page(s) and time spent deciding on the correct judgment. One reason that preferences could be faster is that the assessor "memorizes" the page, or at least forms an impression of it, so that he or she does not have to re-read it each time it appears. If this were the case, judgments would get faster as each document had been seen.

To investigate this, we looked at the time each assessor spent making a judgment the first time the page was shown. For the preference, definite, and absolute judgments, the median time spent on a judgment when seeing a document for the first time was 3.89, 5.40, and 6.33 seconds, respectively. Thus it seems that making a preference judgment is faster than making an absolute judgment even after taking reading time into account.

Additional Analysis. The "context search" button, which allowed assessors to see twenty search results, was used a total of 41 times, slightly under once per seven queries. There was no correlation between the judgment interface and use of context search.

After each query, assessors were presented with a feedback page to report their confidence in their judgments and their understanding of the query on a scale of 1 to 5, with 1 being least confident and 5 most. The median for both questions was 4, and despite not having "ownership" of queries there was no significant correlation between confidence and time spent judging.

4   Evaluating Engines

With preference judgments, standard evaluation measures like average precision and DCG can no longer be used. We must develop new evaluation measures.

---

[2] Median is reported instead of mean due to hours-long outlying inter-judgment times that skewed the means upward.

Table 5. Comparisons between evaluation measures defined over absolute judgments and measures defined over preferences

	NDCG	ppref	wpref
DCG	0.748	0.485	0.584
NDCG		0.662	0.738
ppref			0.950

(a) Correlation between evaluation measures.

	NDCG	ppref	wpref
DCG	1.000	0.873	0.866
NDCG		0.873	0.866
ppref			0.941

(b) Agreement on system differences.

A simple but intuitive measure is the proportion of pairs that are correctly ordered by the engine. We call this "precision of preferences" or *ppref* for short. More formally, over all pairs of pages i, j such that i is preferred to j by an assessor, ppref is the proportion for which the engine ranked i above j. If neither i nor j is ranked, ppref ignores the pair. If i is ranked but j is not, ppref considers i to have been ranked above j.

The pairs in ppref can be weighted for a measure we call *wpref*. We use a rank-based weighting scheme: for pages at ranks i and j such that $j > i$, let the weight $w_{ij} = \frac{1}{\log_2(j+1)}$. wpref is then the sum of weights w_{ij} over pairs i, j such that i is preferred to j and the rank of i is less than the rank of j. The normalizing constant is the sum of all weights w_{ij}.

4.1 Results

We compared evaluations between four different measures: DCG, normalized DCG (NDCG), ppref, and wpref. A common formulation of DCG is $DCG@k = \sum_{i=1}^{k} \left(2^{rel_i} - 1\right) / \log_2(i + 1)$ [5], where rel_i is the relevance of the document at rank i. NDCG@k is DCG@k divided by the DCG of the top k most relevant documents ranked in descending order of relevance.

DCG and NDCG were calculated over both sets of absolute judgments obtained for each query. Since assessor disagreement could be a source of variance in a comparison between absolute measures and preference measures, we calculated ppref and wpref over the preferences inferred from the absolute judgments.

Pearson correlations among the four measures calculated for each query are shown in Table 5(a). The absolute-based measures correlate well, and the preference-based measures correlate well. The correlation between wpref and NDCG is nearly as high as the correlation between DCG and NDCG.

We can also measure "agreement" among the measures in determining whether one system is better than another. We calculate each measure for each query and each system, then look at the sign of the difference between two measures on each query. If both measures say the difference is positive or negative, they agree; otherwise they disagree. As Table 5(b) shows, the measures agree at a fairly high rate, though preference measures agree more with each other than they do with absolute measures.

5   Efficient Judging

One of the biggest obstacles to the adaption of preference judgments, is that the number of document pairs increases polynomially with the number of documents. Although we had at most 15 documents for any query (105 preferences), in a large-scale evaluation there would likely be dozens or hundreds, as pages are drawn from different engines and different test algorithms. A polynomial increase in the number of judgments means much greater cost in assessor time, no matter how much faster assessors are at judging. In this section we look at ways to reduce the number of judgments required.

5.1   Transitivity

If assessors are consistently transitive, the full set of judgments is not necessary; this is the idea behind comparative sorting algorithms such as heapsort. The rate of growth in the number of comparisons needed by these algorithms is in $\mathcal{O}(n \lg n)$, much slower than the $\mathcal{O}(n^2)$ growth rate of all comparisons.

To evaluate transitivity, we iterated over all triplets of documents i, j, k in each set of preference judgments. We counted the number of times that, if i was preferred to j and j was preferred to k, the assessor also preferred i to k.

Transitivity holds for over 99% of triplets on average. Each individual assessor was consistently transitive at least 98% of the time. This suggests we can use a sorting algorithm with a minimum of information loss, and possibly improve assessor consistency at the same time. This agrees with Rorvig's finding that preference judgments of relevance are transitive [12]. Figure 2 shows the $\mathcal{O}(n \lg n)$ growth rate compared to the $\mathcal{O}(n^2)$ rate.

5.2   "Bad" Judgments

In Section 3 we discussed the use of "Bad" judgments in the preference interface. About 20% of absolute judgments were "Bad". Since we can reasonably assume that nothing will be preferred to these pages, we can additionally assume that every non-"Bad" page would be preferred to any "Bad" page. Therefore each "Bad" judgment gives us $\mathcal{O}(n)$ preferences.

The empirical reduction in judgments by inferring preferences in this way is shown in Figure 2. At $n = 15$, this has reduced the number of judgments to about 40 (averaged over all queries and all assessors that were assigned a preference interface for that query). The average decrease from $\mathcal{O}(n \lg n)$ over all values of n is 16 judgments.

The curve appears to be increasing at a rate of $n \lg n$, though it is not clear what it will do as n continues to increase beyond 15. Presumably increasing n results in a greater proportion of bad pages, so it may be that the curve asymptotically approaches a linear increase.

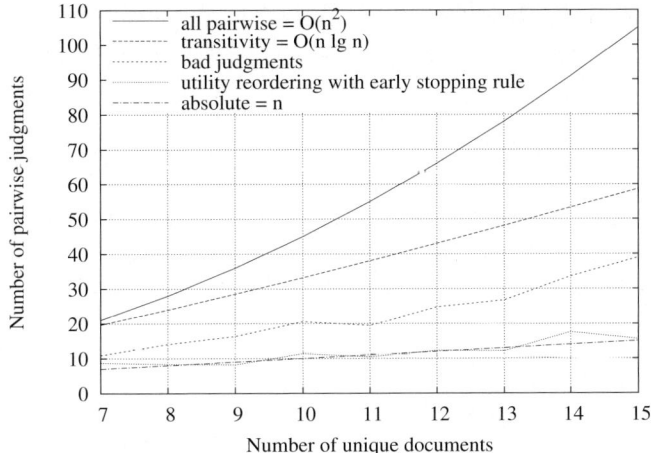

Fig. 2. Number of judgments made by assessors under different conditions

5.3   Cost-Effective Judgments

Applying transitivity and "Bad" judgments still gives us the full set of preference judgments, though some are inferred rather than asked of the assessor directly. There are additional steps we can take to increase the utility of the assessors' time, and, even with some judgments unmade, still prove that differences between systems exist on a particular set of queries. This is based on the work of Carterette et al. [13], who showed that by estimating the utility of a possible judgment and bounding performance differences, the relative performance of systems could be determined with very little effort.

Estimating Utility. Each judgment has a particular utility in helping us determine the sign of the difference in a measure over each engine. For example, if ppref is the measure of interest, and engines E_1 and E_2 both rank document A above document B, then whether A is preferred to B or not is of no consequence: the difference in ppref between the two engines will be the same regardless.

Furthermore, since transitivity holds, each judgment we make may bring additional transitive judgments along with it. For example, if we have already judged that A is preferred to B and we are debating whether to next judge (B, C) or (A, C), we should keep in mind that if we judge $B > C$, we can infer $A > C$ by transitivity; likewise, if we judge that $C > A$, we can infer $C > B$. As above, whether these transitive judgments are useful depends on how the documents are ranked by the systems.

The utility function for a preference measure is:

$$U(A, B) = p(A > B) \cdot gain(A > B) + p(B > A) \cdot gain(B > A), \text{ where}$$

$$gain(A > B) = |w_1(A, B)sgn\,(r_1(A) - r_1(B)) - w_2(A, B)sgn\,(r_2(A) - r_2(B))|$$
$$+ \sum_{i|B>i} |w_1(A, i)sgn\,(r_1(A) - r_1(i)) - w_2(A, i)sgn\,(r_2(A) - r_2(i))|$$
$$+ \sum_{i|i>A} |w_1(i, B)sgn\,(r_1(i) - r_1(B)) - w_2(i, B)sgn\,(r_2(i) - r_2(B))|$$

The sums are over pairs (i, B) such that we had previously judged that $B > i$ and (i, A) where we had judged $i > A$. These capture the transitive judgments discussed in the paragraph above. The weights $w_n(i, j)$ are set for an evaluation measure: $w_n(i, j) = 1$ gives the utility function for ppref, while $w_n(i, j) = 1/\log_2\,(\min\{r_n(i), r_n(j)\} + 1)$ (where $r_n(i)$ is the rank of document i by system n) produces the utility function for wpref.

Note that the expected utility relies on an estimate of the probability that A is preferred to B. We assume a priori that this probability is $\frac{1}{2}$. After we have made some judgments involving A and some judgments involving B, we may have more information. We can use a simple logistic regression model such as [14] to estimate these probabilities with no features; the model can easily be adapted to incorporate any feature.

By judging pairs in decreasing order of utility, we can ensure that after k judgments we have the most possible confidence in the difference between two systems. The next question is how big k has to be before we can stop judging.

Early Stopping Rule. Suppose after partially completing judgments, the ppref of E_1 is greater than that of E_2 (excluding unjudged pairs). If there is no possible set of judgments to the remaining pairs that would results in E_2 "catching up", we can safely stop judging.[3]

Although it is difficult to determine the exact point at which we are guaranteed that E_1 must be superior to E_2, we can easily compute bounds on $E_1 - E_2$ that allow us to stop judging before evaluating all pairs. A very simple bound iterates over all unjudged pairs and assigns them a judgment depending on how much they would "help" either engine. If we have a pair i, j such that E_1 ranked i above j but E_2 ranked j above i, then we want to know what happens if j is preferred to i, i.e. that pair helps E_2 and hurts E_1. We assign judgments in this way for all pairs, ignoring consistency of judgments. This gives us a loose bound.

The number of judgments that are required to differentiate between systems after applying dynamic reordering based on expected utility and the early stopping rule is shown in Figure 2. The number of judgments has effectively been reduced to n on average. Reordering and early stopping can be applied to absolute judgments as well, but the gain is not nearly as dramatic: on average it results in only 1–2 fewer judgments per query.

Although there is no guarantee our results would continue to hold as n increases, we can guarantee that using "Bad" judgments and transitivity will give us a slower rate of increase than making all preference judgments, and that using

---

[3] If we need all of the judgments in order to train a ranking algorithm, on the other hand, we may not want to stop.

dynamic reordering and the early stopping rule will give us an even slower rate of increase. Furthermore, utility-based reordering produces that a set of judgments is maximally useful no matter when the judging effort is stopped.

6   Conclusion

We have performed the first investigation into the direct acquisition of preference judgments for relevance and the first comparison of preference judgments to absolute judgments. We have also provided a suite of methods by which preference judgments can become practical to use for evaluation of search engines.

There are several clear directions for future work: choosing the correct evaluation measure for preferences, the robustness of these measures to missing preferences, and measuring the uncertainty in an evaluation when preferences are missing are three. Additionally, whether training ranking algorithms over preference judgments rather than inferred preferences results in more robust performance is an interesting open question.

References

1. Voorhees, E.M., Harman, D. (eds.): TREC. The MIT Press, Cambridge (2005)
2. Järvelin, K., Kekäläinen, J.: IR evaluation methods for retrieving highly relevant documents. In: Proceedings of SIGIR, pp. 41–48 (2000)
3. Voorhees, E.: Variations in relevance judgments and the measurement of retrieval effectiveness. In: Proceedings of SIGIR, pp. 315–323 (1998)
4. Kendall, M.: Rank Correlation Methods, 4th edn., Griffin, London, UK (1970)
5. Burges, C., Shaked, T., Renshaw, E., Lazier, A., Deeds, M., Hamilton, N., Hullender, G.: Learning to rank using gradient descent. In: Proceedings of ICML, pp. 89–96 (2005)
6. Joachims, T.: Optimizing search engines using clickthrough data. In: Proceedings of KDD, pp. 133–142 (2002)
7. Bartell, B., Cottrell, G., Belew, R.: Learning to retrieve information. In: Proceedings of the Swedish Conference on Connectionism (1995)
8. Frei, H.P., Schäuble, P.: Determining the effectiveness of retrieval algorithms. Information Processing and Management 27(2-3), 153–164 (1991)
9. Joachims, T., Granka, L., Pang, B., Hembrooke, H., Gay, G.: Accurately interpreting clickthrough data as implicit feedback. In: Proceedings of SIGIR, pp. 154–161 (2005)
10. Buckley, C., Voorhees, E.M.: Retrieval evaluation with incomplete information. In: Proceedings of SIGIR, pp. 25–32 (2004)
11. Mizzaro, S.: Measuring the agreement among relevance judges. In: Proceedings of MIRA (1999)
12. Rorvig, M.E.: The simple scalability of documents. JASIS 41(8), 590–598 (1990)
13. Carterette, B., Allan, J., Sitaraman, R.: Minimal test collections for retrieval evaluation. In: Proceedings of SIGIR, pp. 268–275 (2006)
14. Carterette, B., Petkova, D.: Learning a ranking from pairwise preferences. In: Proceedings of SIGIR, pp. 629–630 (2006)

Using Clicks as Implicit Judgments: Expectations Versus Observations

Falk Scholer[1], Milad Shokouhi[2,*], Bodo Billerbeck[2,**], and Andrew Turpin[1]

[1] School of Computer Science and IT, RMIT University, PO Box 2476v,
Melbourne 3001, Australia
[2] Microsoft Research, 7 J J Thompson Ave, Cambridge CB3 0FB, UK

Abstract. Clickthrough data has been the subject of increasing popularity as an implicit indicator of user feedback. Previous analysis has suggested that user click behaviour is subject to a quality bias—that is, users click at different rank positions when viewing effective search results than when viewing less effective search results. Based on this observation, it should be possible to use click data to infer the quality of the underlying search system. In this paper we carry out a user study to systematically investigate how click behaviour changes for different levels of search system effectiveness as measured by information retrieval performance metrics. Our results show that click behaviour does not vary systematically with the quality of search results. However, click behaviour does vary significantly between individual users, and between search topics. This suggests that using direct click behaviour—click rank and click frequency—to infer the quality of the underlying search system is problematic. Further analysis of our user click data indicates that the correspondence between clicks in a search result list and subsequent confirmation that the clicked resource is actually relevant is low. Using clicks as an implicit indication of relevance should therefore be done with caution.

1 Introduction

The behaviour of users as they interact with search systems has long been of interest to information retrieval practitioners. In particular, even subtle *implicit* indications of what users might like and dislike while engaged in a particular search activity could potentially be exploited to improve both the current and future search sessions.

With the popularity of web search, an implicit indicator that has received much attention is *clickthrough* data, which indicates which items in a search result list the user clicks. The underlying assumption is that users are able to infer with some degree of accuracy which items in a list are good candidates for relevance—those items that are clicked—and which are not. Clickthrough has been used among other things as a basis for re-ranking result lists [10] and document expansion [13], as well as a proxy for relevance judgements [2, 11].

---

* This research was carried out while the author worked at RMIT University.
** This research was carried out while the author worked at Sensis Pty Ltd.

C. Macdonald et al. (Eds.): ECIR 2008, LNCS 4956, pp. 28–39, 2008.

While the notion that click behaviour is indicative of user preferences is intuitively appealing, and has in previous work been shown to have some success in improving search results, many properties of clickthrough data have not been explored in detail. For example, if click data was an accurate proxy of relevance, then one might expect to observe significant differences in the click behaviour of users when engaged in a search task using a highly effective retrieval system with many relevant answers in the result list, compared with a poor retrieval system.

Previous work has suggested that quality of ranking list does influence user decisions. Joachims et al. observed two sources of bias in clickthrough data: a *trust* bias, and a *quality* bias [12]. Trust bias arises from the confidence that users have in a search system producing a sensible ordering of results. This behaviour is demonstrated through the order in which users view items on a search results page; in previous work this has been demonstrated to be an increasing function of rank. In other words, users generally read a result page from top to bottom [11, 19]. While trust bias might vary between different search systems (in which users could have different levels of confidence), this bias is unlikely to arise across a single perceived search system. Quality bias arises from the overall quality of search results; when the top 10 answers of a Google results list are reversed, there is a statistically significant change in the average rank of a clicked document [11].

Motivated by this result, we carried out a user study to investigate how click behaviour changes as the underlying quality of the search system is varied in a controlled way. Such a relationship could be used to try and infer the underlying quality of a search system directly from user behaviour. The research questions that we aim to address in this paper are:

1. how does the rank at which users click vary as the quality of the underlying search system changes (clicked ranks versus precision);
2. how does the frequency with which users click vary as the quality of the underlying search system changes (click frequency versus precision); and
3. how reliable are user judgements as a proxy for relevance?

2   Related Work

Clickthrough. Two early studies use clickthrough data to alter document rankings; one using weight functions and support vector machines [10], and the other modifying the vector space model [13]. The underlying assumption that clicks imply relevance, however, was not investigated in these papers.

Fox et al. [8] examined clickthrough and other user behaviour and found that a combination of implicit relevance judgements (clickthrough, time spent on search results page, and how a search session ended) correlated reasonably well with explicit judgements. Unfortunately, they found that usefulness of clickthrough as a relevance indicator was limited: only in 39% of instances where users clicked on a document were they satisfied with the document. In 29% of cases they were only partially satisfied, while 27% of the time they were dissatisfied.

As discussed in the introduction, Joachims et al. [11, 12] demonstrate trust and quality bias in clickthrough data. Accordingly, relevance is only unreliably deduced from clickthrough alone, but can be much more effective when used in combination with other user behaviour that may easily be collected along with clickthrough data, such as at what the position in the result set a click occurred, what other results the individual users clicked on, and in which order. So rather than considering clickthrough as *absolute* feedback, they use additional information to make *relative* feedback more robust. They reported around 80% agreement between the feedback obtained by the clickthrough and the human judges. This work has been extended by including query reformulation behavior in the form of *query chains* to add further insight on whether a click on a result is an indicator of relevance [16].

Craswell et al. apply random walks on the click graph to improve image search [6]. Their click graph contains nodes that are images and queries while the edges represent clicks. By using a Markov random walk, images that have not been clicked previously for a query can be effectively ranked. Instead of collecting clickthrough passively and then using this information to re-rank results in later search sessions for possibly different users, Radlinski et al. [17] gather implicit feedback in the form of clickthroughs about documents not necessarily ranked in the top spots. In order to maximise the learning opportunities provided by users in the form of clickthrough, they re-rank results while not impacting the quality of rankings significantly.

Agichtein et al. [1, 2] show that taking into account general user behaviour—in particular the tendency of users to click on the top ranked documents most heavily with a quickly decreasing frequency when lower ranks are inspected—as well as clickthrough information can lead to increased quality of rankings. They also show that by using a whole range of implicit user feedback such as page dwell time, the precision of rankings can be increased significantly.

Evaluation Metrics. Experimental work in information retrieval typically follows the "Cranfield" paradigm, where a search system is evaluated by running a set of queries over a fixed collection. A human judge assesses the relevance of each document for each query and, based on the position of relevant documents in the result list, a variety of performance metrics can be calculated. This experimental framework is also at the core of the ongoing annual TREC conferences [21]. Retrieval metrics are usually based on a calculation of the *precision* of the search system (the number of relevant document retrieved as a proportion of the total number of documents that have been retrieved) and the *recall* (the number of relevant documents retrieved as a proportion of the total number of relevant document for that query). Precision therefore reflects the accuracy of the search results, while recall reflects the completeness.

The *average precision* (AP) for a query is the mean of the precision for each relevant document that is returned in an answer list, where relevant documents that aren't found contribute a precision of zero. *Mean average precision* (MAP) is then the mean AP over a run of queries. MAP gives a single score for the overall performance of a retrieval system, and is one of the most widely-reported retrieval metrics [5]. In our experiments, we control the MAP of a search system to investigate how click behaviour varies with underlying system performance.

Precision at cutoff level N calculates the number of relevant documents that a retrieval system has found out of the top N items in the answer list. Commonly-reported values of N include 5 and 10; the latter corresponds to the default behaviour of popular web search engines that return 10 items per answer page.

User Behaviour and Search Performance. Several other studies have used the idea of controlling the precision of lists presented to users in order to observe behaviour in certain scenarios. Allan et al. [3] compared the performance of users of different retrieval systems at carrying out a passage-based question answering task. Their study investigated systems with different performance levels measured by the *bpref* metric (bpref has been proposed as an alternative to MAP for search scenarios where complete relevance judgements are not available). Their analysis showed that statistically significant differences in user performance do not occur over a broad range of system bpref levels; for easier topics, there were significant effects between bpref levels of 0.5 and 0.6, while for harder topics there were significant effects at bpref of 0.90 and 0.98.

Turpin and Scholer [18] investigated user performance on simple web search tasks, considering the time that a user takes to find a relevant document, and the number of relevant documents that a user can find within 5 minutes, across search systems operating at MAP in the range of 0.55 to 0.95. Results indicated that MAP level has no significant relationship with the time taken to find the first answer (a precision-oriented task), while there is a weak relationship with a recall-oriented task.

3   User Study

To investigate how the click behaviour of searchers varies as the quality of the underlying information retrieval system changes, we conducted a user study based on those used in previous papers [18]. The level of system performance is controlled, and user click behaviour is analysed based on click ranks (the positions in search results lists at which users view documents) and the click frequency (an aggregate of how many items users choose to view).

The different levels of system performance are expected to introduce a *quality* bias in clickthrough data; recall that quality bias refers to a change in click behaviour that arises as the result of differences in the quality of the ranking of search results [11]. Click behaviour may also be subject to *trust* bias; however, all of our users interact with what is (to their view) the same search system. Therefore, any trust bias would be the same across users and topics, and should not lead to differences in observed click behaviour.

Collections and Topics. Our study aims to investigate searcher behaviour in a web search setting. We therefore used the TREC WT10g collection, a 10 GB crawl of the Web in 1997 [4]. This collection was used for *ad hoc* retrieval in the TREC 9 and 10 Web tracks, and has 100 associated search topics and corresponding relevance judgements (topics 451–550). TREC topics have three components: a title field consisting of a few keywords that represent the information need; a description field giving a one-sentence explanation of the topic; and a narrative that further specifies what makes a document relevant for the topic. In our experiments, we investigate how click behaviour varies under different levels of search system effectiveness; we focus on controlling the level of MAP (defined in Section 2). To construct answer lists that include high levels of this metric, large numbers of relevant documents are required. We therefore use the 50 topics with the highest number of available relevance judgements for our search task.

Subjects. A total of 30 experimental subjects (the users of our search systems) were recruited from RMIT University, including a mixture of undergraduate and postgraduate students with a median age of 23. All subjects provided written informed consent, and the research was conducted under the guidelines of the RMIT Human Research Ethics Committee.

Participants were asked to complete a pre-experiment questionnaire, to establish their familiarity with online searching. Most users were very experienced with electronic search systems, including web search engines, with a mean rating of 4.7 on a scale of 1 (no experience) to 5 (a great deal of experience). The mean frequency of conducting a search was one or more times a day. Users also indicated they strongly enjoy carrying out information searches (a mean rating of 4.2 on a scale of 1–5).

Search Systems. To investigate the effect of system effectiveness on click behaviour, we control the level of system performance as measured by mean average precision (MAP). Based on the findings of Allan et al. [3], which suggested that the impact of varying system performance would only be likely to have an impact at high levels, we created search *systems* with MAP levels of 55%, 65%, 75%, 85% and 95%. To control the average precision (AP) of a system for a single query, we constructed answer lists, based on the available TREC relevance judgements. Relevant and irrelevant documents were chosen randomly from the judgements file so that the required level of AP was achieved. 200 such lists were created for each system and topic combination. Therefore, a user could enter more than one search query for a single topic, and be served with lists of a consistent AP level. The MAP of a system-topic combination is therefore the same as the AP of a single list. We note that, since we were investigating high levels of MAP, there are many relevant documents that occur towards the start of the search result lists.

Since we are interested in investigating click behaviour in a web search environment, the search interface presented to users was modelled closely on the interfaces of popular web search engines such as Yahoo! or Google. Queries are entered into a simple text-box input field in a web browser, and a list of 100 answer resources is returned. Answer list entries consist of the document title, which is a hyperlink leading to the underlying resource. Together with each title, a short query-biased summary is displayed. The summaries are pre-generated, using the title field of each search topic as the query terms. The summaries consist of up to three sentences containing query terms; more weight is given to sentences that contain a greater number of query terms, and where query terms occur in closer proximity to each other [20].

Experimental Setup. Users in our study were asked to carry out a traditional web search task: given an information need, find documents that contain relevant information. Users were given 5 minutes to find as many relevant documents as possible. To reduce ordering effects, any document that contained information contributing to the resolution of the information need was to be considered relevant, even if it contained information that had been seen before.

The user study followed a balanced design, to control for confounding factors such as user learning and order effects. Each subject conducted searches for 50 topics. However, due to unanticipated browser use, a small number of topics and sessions had be removed from the final analysis. This resulted in each system-topic pair being used an average of 4.9($\pm$0.1) times, with a mean number of users per topic of 24.3($\pm$3.4).

A search session proceeded as follows: first, a user was presented with an information need, consisting of the *description* and *narrative* field of a TREC topic. The user then had five minutes to interact with a search system (of a specific AP level, unknown to the user), identifying as many relevant documents as possible. In response to a user query, an answer list of the appropriate AP level was randomly selected from the pool of 200 lists created for each topic-system combination, and displayed. Users could browse the answer list, and view the underlying documents by clicking hyperlinks in the document title; this would open the document in a new window. After viewing a document, a user could choose to save it as being relevant to the information need using a save button, or simply close the document if it is not relevant. All interactions with the search system, in particular all clicks and views of documents, were stored in a log file and subsequently analysed.

4   Clicked Ranks Versus Precision

We first investigate whether the rank position at which users click is affected by the quality of the underlying search system that is being used. Figure 1 shows a boxplot of the system MAP level against the rank position at which users clicked on answer resources. For all boxplots used in this paper, the box shows the data falling between the 25th to 75th percentile, with the median indicated by a solid black line. Whiskers show the extent of the data range, and outlier values are shown as individual points. A multifactorial analysis of variance (ANOVA) indicates that there is no statistically significant relationship between the rank at which users click and the MAP level of the search system used ($p > 0.1$).

A similar lack of difference is observed when evaluating the clicked rank against system performance as measured by precision at 10 documents retrieved, as shown in Figure 2 ($p > 0.1$). We note that there are no observations for P@10 below the level of 0.4; this is due to the construction of our lists, which were designed to correspond to particular MAP levels. It is therefore possible that click behaviour might differ for very low levels of this metric. However, the results demonstrate that for a large range of system quality as measured by P@10, changing the proportion of relevant documents in the top ranks of a search

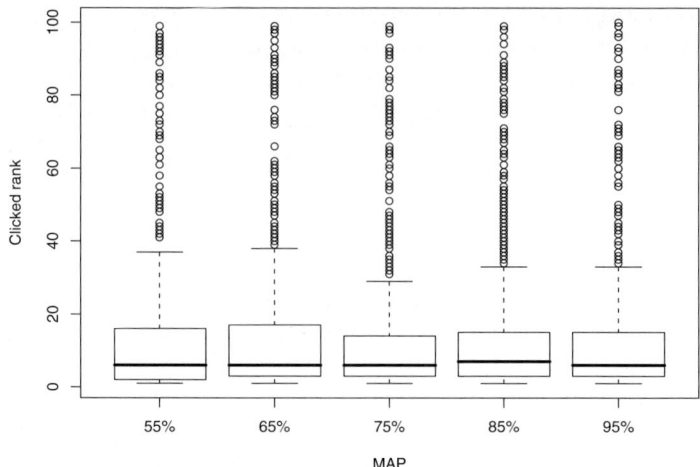

Fig. 1. Rank of the clicked documents for different systems. The x-axis shows the MAP value of systems according to TREC judgments.

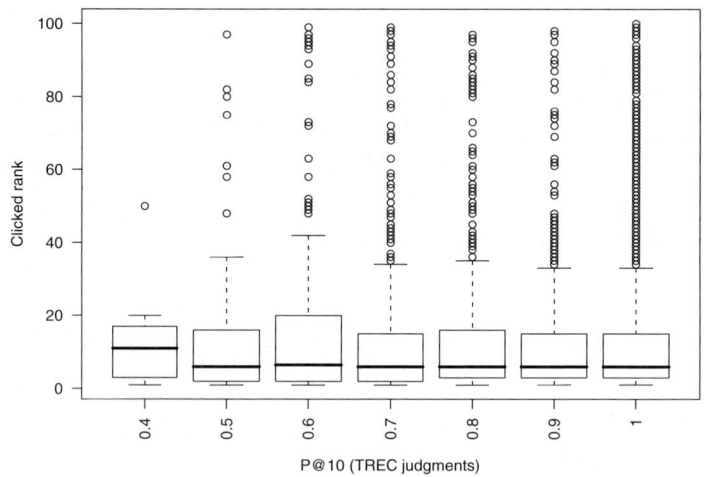

Fig. 2. Rank of the clicked documents for different system performance levels as measured by P@10. The x-axis shows the percentage of the top ten documents that are relevant according to TREC judgments.

results list has no impact on the overall rank position at which users click when viewing results.

As the users interacted with our search systems, for each document that a user viewed from a result list, they made a decision as to whether the document is relevant to the information need. An alternative way of viewing the level of system

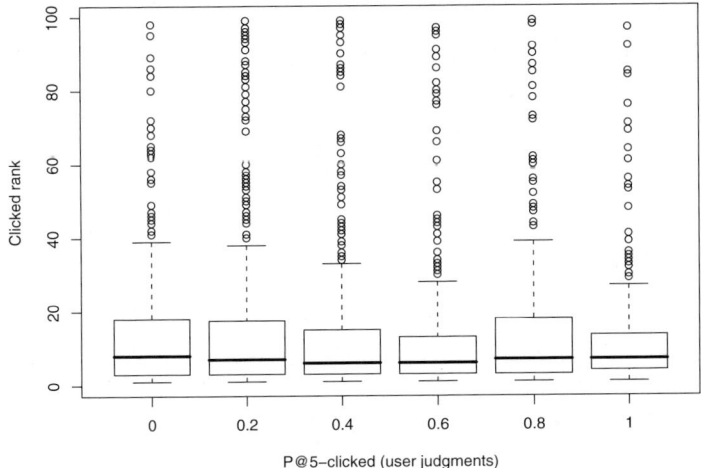

Fig. 3. Rank of the clicked documents for different system performance levels as measured by P@5-clicked. The x-axis shows the percentage of the top five *clicked* documents that users explicitly judged as relevant. In the rightmost bar, all the top five clicked documents are detected to be relevant by users.

effectiveness is therefore to calculate the precision based on explicit user relevance judgements only, rather than TREC relevance judgements. Figure 3 shows the rank position at which users clicked for different levels of "P@5-clicked"; that is, the precision is calculated from explicit user judgements for the top 5 clicked documents. The results again show no relationship between the level of system effectiveness and the rank position at which users click; there is no significant difference in the average click rank across all levels of user judgement-derived P@5-clicked ($p > 0.1$).

The lack of difference in the average clicked rank across starkly different levels of search system effectiveness is surprising. One possible explanation would be that click behaviour is subject to significant biases so that all systemic variation is hidden. If this was the case, we would also expect that there are no differences in click rank for other effects, such as between topics or between users. We therefore investigated how the clicked rank varies between topics. A boxplot of clicked rank for each search topic is shown in Figure 4, sorted by average clicked rank. ANOVA results indicate that there are highly significant user effects ($p < 0.0001$). A subsequent Tukey Honest Significant Differences test indicates 65 significant pairwise differences ($p < 0.05$). There are therefore strong and numerous differences in the average click rank between topics.

Possible user effects may also be present; the clicked ranks by individual users are shown in Figure 5. Again, ANOVA results indicate that there are significant differences in click ranks between users ($p < 0.0001$). A subsequent Tukey Honest Significant Differences test shows 110 significant differences ($p < 0.05$). Clicking behaviour varies strongly from user to user.

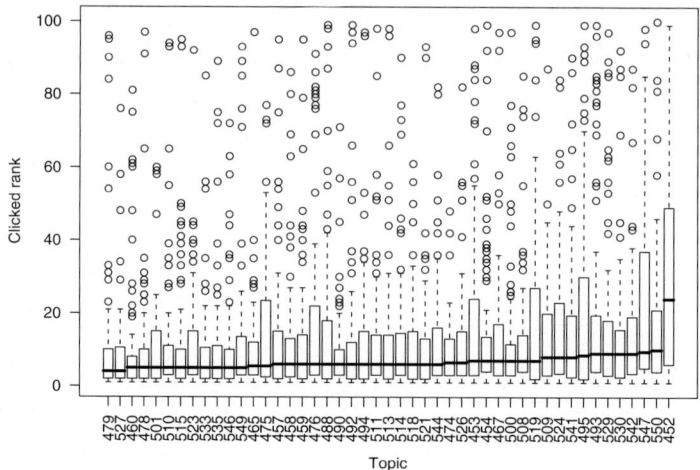

Fig. 4. The average rank of clicked documents across different topics

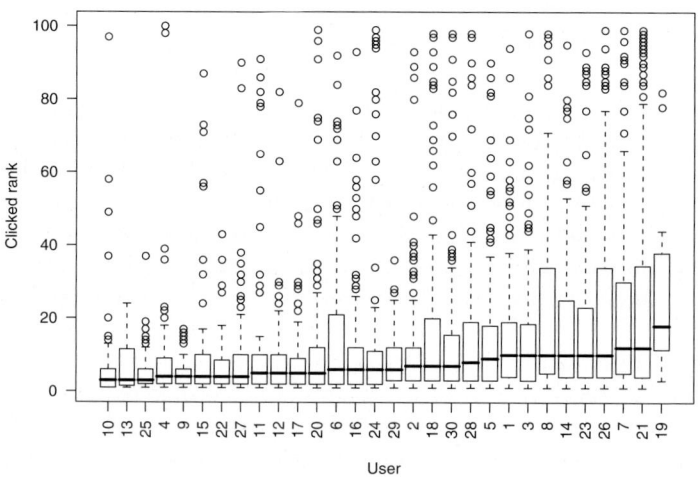

Fig. 5. The average rank of clicked documents across different users

5   Click Frequency Versus Precision

User click behaviour can be summarised in aggregate by two main measures:
the rank at which users click, and the overall frequency with which users click.
In general, we expect that a good search system (that is, one that returns a
better result list) would demonstrate a higher click frequency, as the result of
two complementary effects. First, a better result list will have more relevant
answer documents near the start of the list; therefore, users need to spend less

time reading down the list, and in a time-constrained search task, they would therefore have the opportunity to click on more items. A second, related, effect is that in a better result list, there are simply more "good" answer documents to look at. We compared the average click frequency of different search systems based on their MAP levels. According to our observations, the frequency remains largely invariant as the MAP level changes; an ANOVA detects no statistically significant differences ($p > 0.1$).

We also investigated the variation of the average click frequency for systems with varying levels of P@10. Again, the frequency is stable across different levels of search system effectiveness, and ANOVA detects no significant differences ($p > 0.1$).

6   User Judgments as a Proxy for Relevance

There is much interest in the IR community about using clickthrough data as an indication of relevance, either directly, or to generate preference relations. We therefore investigate the reliability of using clickthrough data as an indication of relevance, based on the data from our user-study.

Overall, our experimental subjects clicked 3,778 times in the answer lists. Each click would bring up a document, and lead to a subsequent decision about whether to save the document (because it contains relevant information), or to discard it. 1,980 clicked documents were saved; therefore, the proportion of clicked documents that are actually considered to be relevant is only 52%. This is surprisingly low, and indicates that it is not safe to infer relevance directly from recorded click information. The proportion of relevant to total clicks remains stable across different levels of system effectiveness, as shown in Table 1.

TREC relevance judgements are used as a ground truth for a large proportion of information retrieval experiments that are reported in the literature. As such, it is also interesting to compare how well user click behaviour corresponds with official relevance judgements. In total, user clicks agree with TREC judgements— including both agreement on relevance and non-relevance of documents—2,219 times. The rate of agreement is therefore 58%. Interestingly, the rate of agreement seems to decrease slightly as the effectiveness of the underlying system

Table 1. Agreement between user clicks and TREC relevance judgements. C: The total number of clicks, U: The total number of documents detected as relevant by users. A: The total number of agreements between TREC judgments and user judgments.

MAP	C	U	A	$\frac{U}{C}$	$\frac{A}{C}$
55%	692	348	315	0.50	0.61
65%	748	366	340	0.48	0.58
75%	795	432	417	0.54	0.58
85%	763	409	399	0.53	0.57
95%	780	425	415	0.54	0.56
Total	3778	1980	2219	0.52	0.58

gets higher, as indicated in Table 1. While not very high, the rate of agreement between overall clicks and the TREC judgements is greater than the underlying agreement between user and TREC judgements at the document level—when only unique document-topic combinations are counted—which is only 45% on the same data. These results support the view that TREC judgements are in general not directly transferable to other scenarios [9].

7   Conclusions

Clickthrough data is a popular implicit indicator of user preferences, and has been applied in a variety of situations to improve search results. Motivated by successes in previous work, we carried out a user study to investigate how click behaviour varies with changes in the quality of an underlying search system. To the best of our knowledge, this is the first study of how the click behaviour of users varies with controlled changes in system quality. Our results are surprising—we found no statistically significant relationship between clickthrough data and system performance, across different measures of click behaviour (click rank and click frequency) and across various measures of system performance, including metrics based on both TREC relevance judgements and user relevance judgements. This suggests that it is not safe to infer the quality of an underlying search system—at least as measured by currently popular IR system metrics—based on simple measures of click behaviour.

Analysis of our user click data further showed that the action of clicking is not strongly correlated with relevance—only 52% of clicks in a search result list led to a document that the user actually found to be relevant. Attempts to use clicks as an implicit indication of relevance should therefore be treated with caution. In future work, we plan to investigate how more complex interpretations of clickthrough behaviour may relate to system performance. For example, other studies have found a relationship between the experience of users and click behaviour. We therefore intend to conduct a larger-scale user study incorporating groups of user with markedly different levels of search ability, to investigate whether there are any interactions between this factor, click behaviour and system performance.

References

[1] Agichtein, E., Brill, E., Dumais, S.: Improving web search ranking by incorporating user behavior information. In: Efthimiadis, et al. (eds.) [7], pp. 19–26.

[2] Agichtein, E., Brill, E., Dumais, S., Ragno, R.: Learning user interaction models for predicting web search result preferences. In: Efthimiadis, et al. (eds.) [7], pp. 3–10.

[3] Allan, J., Carterette, B., Lewis, J.: When will information retrieval be "good enough"? In: Marchionini, et al. (eds.) [15], pp. 433–440.

[4] Bailey, P., Craswell, N., Hawking, D.: Engineering a multi-purpose test collection for web retrieval experiments. Information Processing and Management 39(6), 853–871 (2003)

[5] Buckley, C., Voorhees, E.M.: Retrieval system evaluation. In: TREC: experiment and evaluation in information retrieval [21]

[6] Craswell, N., Szummer, M.: Random walks on the click graph. In: Kraaij, et al. (eds.) [14], pp. 239–246

[7] Efthimiadis, E., Dumais, S., Hawking, D., Järvelin, K. (eds.): Proceedings of the 29th Annual International ACM SIGIR Conference on Research and Development in Information Retrieval, Seattle, WA (2006)

[8] Fox, S., Karnawat, K., Mydland, M., Dumais, S., White, T.: Evaluating implicit measures to improve web search. ACM Transactions on Information Systems 23(2), 147–168 (2005)

[9] Harman, D.K.: The TREC test collection. In: TREC: experiment and evaluation in information retrieval [21]

[10] Joachims, T.: Optimizing search engines using clickthrough data. In: Proceedings of the eighth ACM SIGKDD international conference on Knowledge discovery and data mining, Edmonton, Alberta, Canada, pp. 133–142. ACM Press, New York (2002)

[11] Joachims, T., Granka, L., Pan, B., Hembrooke, H., Gay, G.: Accurately interpreting clickthrough data as implicit feedback. In: Marchionini,, et al. (eds.) [15], pp. 154–161.

[12] Joachims, T., Granka, L., Pan, B., Hembrooke, H., Radlinski, F., Gay, G.: Evaluating the accuracy of implicit feedback from clicks and query reformulations in web search. ACM Transactions on Information Systems 25(2), 7 (2007)

[13] Kemp, C., Ramamohanarao, K.: Long-term learning for web search engines. In: Proceedings of the 6th European Conference on Principles of Data Mining and Knowledge Discovery, London, UK, pp. 263–274. Springer, Heidelberg (2002)

[14] Kraaij, W., de Vries, A., Clarke, C., Fuhr, N., Kando, N. (eds.): Proceedings of the 30th Annual International ACM SIGIR Conference on Research and Development in Information Retrieval, Amsterdam, The Netherlands (2007)

[15] Marchionini, G., Moffat, A., Tait, J., Baeza-Yates, R., Ziviani, N. (eds.): Proceedings of the 28th Annual International ACM SIGIR Conference on Research and Development in Information Retrieval, Salvador, Brazil (2005)

[16] Radlinski, F., Joachims, T.: Query chains: learning to rank from implicit feedback. In: Proceeding of the eleventh ACM SIGKDD international conference on Knowledge discovery in data mining, Chicago, Illinois, USA, pp. 239–248 (2005)

[17] Radlinski, F., Joachims, T.: Active exploration for learning rankings from clickthrough data. In: Proceedings of the 13th ACM SIGKDD international conference on Knowledge discovery and data mining, San Jose, California, pp. 570–579 (2007)

[18] Turpin, A., Scholer, F.: User performance versus precision measures for simple search tasks. In: Efthimiadis, et al. (eds.) [7], pp. 11–18.

[19] Turpin, A., Scholer, F., Billerbeck, B., Abel, L.: Examining the pseudo-standard web search engine results page. In: Proceedings of the 11th Australasian Document Computing Symposium, Brisbane, Australia, pp. 9–16 (2006)

[20] Turpin, A., Tsegay, Y., Hawking, D., Williams, H.E.: Fast generation of result snippets in web search. In: Kraaij, et al. (eds.) [14], pp. 127–134.

[21] Voorhees, E.M., Harman, D.K.: TREC: experiment and evaluation in information retrieval. MIT Press, Cambridge (2005)

Clustering Template Based Web Documents

Thomas Gottron

Institut für Informatik
Johannes Gutenberg-Universität Mainz
55099 Mainz, Germany
gottron@uni-mainz.de

Abstract. More and more documents on the World Wide Web are based on templates. On a technical level this causes those documents to have a quite similar source code and DOM tree structure. Grouping together documents which are based on the same template is an important task for applications that analyse the template structure and need clean training data. This paper develops and compares several distance measures for clustering web documents according to their underlying templates. Combining those distance measures with different approaches for clustering, we show which combination of methods leads to the desired result.

As more and more documents on the World Wide Web are generated automatically by Content Management Systems (CMS), more and more of them are based on templates. Templates can be seen as framework documents which are filled with different contents to compile the final documents. They are a standard (if not even essential) CMS technology. Templates provide the managed web sites with an easy to manage uniform look and feel. A technical side effect is that the source code of template generated documents is always very similar.

Several algorithms have been developed to automatically detect these template structures in order to identify and / or extract particular parts of a document such as the main content. These structure detection algorithms depend on training sets of documents which are all supposed to be based on the same template. Only few works though address the problem of actually creating these clean training sets or verifying that the documents in a given training set are all based on the same template. Approaches trying to handle this problem usually involve clustering the documents to group together those which have large structural similarities. However, to our knowledge this process has never been analysed or verified itself.

In this paper we take a closer look at web document distance measures which are supposed to reflect template related structural similarities and dissimilarities. We will evaluate the distance measures both under the aspect of computational costs and – given different clustering approaches – how suitable they are to cluster documents according to their underlying templates. The evaluation is based on a corpus of 500 web documents, taken from different sub-categories of five different web sites.

C. Macdonald et al. (Eds.): ECIR 2008, LNCS 4956, pp. 40–51, 2008.

We proceed as follows: In section 1 we give an overview over related works in this fields, focussing in particular on distance measures for web documents which take into account mainly structural information. In 2 and 3 we describe six different distance measures in more detail and some standard cluster analysis algorithms we used. The experiment setup and results are presented in section 4 before we conclude the paper in 5 with a discussion of the results.

1   Related Works

Several works address the challenge of recognising template structures in HTML documents. The problem was first discussed by Bar-Yossef and Rajagopalan in [1], proposing a template recognition algorithm based on DOM tree segmentation and segment selection. Yang et al. proved in [2] the general problem of finding an unambiguous schema representing a template to be NP complete.

Further practical solutions for template detection are discussed in various works. We mention only a few solutions: Lin and Ho developed InfoDiscoverer [3] which is based on the idea, that – opposite to the main content – template generated contents appear more frequently. To find these more frequent contents, they introduce the concept of block entropy to filter redundant DOM blocks. Debnath et al. used a similar assumption of redundant blocks in ContentExtractor [4] but take into account not only words and text but also other features like image or script elements. The Site Style Tree approach of Yi, Liu and Li [5] instead is concentrating more on the visual impression single DOM tree elements are supposed to achieve and declares identically formated DOM sub-trees to be template generated. A similar but slightly more flexible solution is presented by Reis et al. in [6]. They introduce the top down tree mapping algorithm RTDM to calculate a tree edit distance between two DOM trees. The RTDM tree edit distance is used as well to perform a cluster analysis in order to find clusters of different templates within the training set. Gibson et al. presented a site-level template detection algorithm in [7]. It is used as well by Chakrabarti et al. in [8] to automatise the building of training sets for a template detection algorithm.

Cruz et al. describe several distance measures for web documents in [9]. They distinguish between distance measures based on tag vectors, parametric functions or tree edit distances. In the more general context of comparing XML documents Buttler [10] stated tree edit distances to be probably the best but as well very expensive similarity measures. Therefore Buttler proposes the path shingling approach which makes use of the shingling technique suggested by Broder et al. in [11]. An approach to compare the structure of DOM trees by looking at the paths was already suggested earlier by Joshi et al. in [12]. Lindholm et al. [13] instead discuss possibilities to speed up calculating XML tree differences. Shi et al. [14] propose an alignment based on simplified DOM tree representation to find parallel versions of web documents in different languages.

The clustering techniques used in this paper are standard approaches. We found a good overview and discussion of the methods in [15]. Kruskal describes the problem of non-metric multidimensional scaling and an algorithm to solve it in [16].

2   Distance Measures for Template Structures

In this section we are going to describe some existing and some new, tag sequence based measures for calculating distances between template based web documents. We will focus on describing roughly the computation of the distance and the complexity of the approach. For the later purpose we assume D_1 and D_2 to be two HTML documents containing t_i tags and n_i DOM nodes of which l_i are leaf nodes, $i = 1, 2$. However, the number t_i of tags can roughly be estimated to be $2 \cdot n_i$, as in most cases a node in the DOM tree will correspond to two tags in the document's source code.

2.1   RTDM – Tree Edit Distance

Any tree edit distance measure is based on calculating the cost for transforming a source tree structure into a target tree structure. For this purpose elemental operations like inserting, deleting, replacing or moving nodes or entire sub-trees in the tree structure are associated with certain costs to perform these operations. When it comes to tree edit costs for HTML DOM trees the problem is usually simplified a bit, as the root node is known, the sibling nodes are ordered and as the sub-trees (especially as we are talking about documents based on the same template) are hardly ever changing their distance to the root node. Therefore the problem of tree matching is often simplified to top-down hierarchical tree matching. The RTDM algorithm [6] has proven to perform quite well in calculating a tree edit distance for HTML documents.

We used a slightly modified version of the original algorithm, which requires only linear space of degree $O(n_1)$, whereas the original algorithm needs quadratic space[1]. Even though RTDM is reported to usually behave better in practice, it still does have a worst case quadratic time complexity of order $O(n_1 \cdot n_2)$.

2.2   CP – Common Paths

Another way to compare the structure of web documents is to look at the paths leading from the root node to the leaf nodes in the DOM tree [12]. A path is denoted e.g. by concatenating the names of the elements it passes from root to leaf. For each document D it is then possible to represent it by the set $p(D)$ of paths it contains. A distance measure can be computed via the intersection of the two path sets of two documents:

$$d_{CP}(D_1, D_2) = 1 - \frac{|p(D_1) \cap p(D_2)|}{\max(|p(D_1)|, |p(D_2)|)} \tag{1}$$

Computing the paths for the documents can be done in linear time of degree $O(n_1 + n_2)$ with respect to the nodes. Using hashing, the intersection of two resulting sets can be computed in expected linear time as well, this time respect to the number of paths which corresponds to the number of leaf nodes.

---

[1] The improvements correspond to computing the Levenshtein distance with linear space, so we omit the details for the sake of brevity of the paper.

2.3   CPS – Common Path Shingles

A combination of the paths distance with a shingling technique was proposed by Buttler in [10]. The idea is not to compare complete paths but rather breaking them up in smaller pieces of equal length – the shingles. The advantage of this approach is that two paths which are differing only for a small part, but are quite similar for the rest, will have a large "agreement" on the shingles. The shingling can be realised in a way that it does not add any substantial cost to the computation compared to the CP distance.

So, if $ps(D)$ provides the path shingles for a document D, the path shingle distance can be computed similarly as above by:

$$d_{CPS}(D_1, D_2) = 1 - \frac{|ps(D_1) \cap ps(D_2)|}{\max(|ps(D_1)|, |ps(D_2)|)} \tag{2}$$

2.4   TV – Tag Vector

Thinking of the occurrence of tags as a typical feature of a document and in particular of a template based document leads to the tag vector distance measure [9]. Counting how many times each possible (i.e. complying with W3C's HTML recommendation) tag appears converts a document D in a vector $v(D)$ of fixed dimension N as the number of possible tags is limited. We used the Euclidean distance as it is a standard way to compute distances in vector spaces:

$$d_{TV}(D_1, D_2) = \sqrt{\sum_{i=1}^{N} (v_i(D_1) - v_i(D_2))^2} \tag{3}$$

A critic often mentioned when using the Euclidean distance for classification or clustering is that it is sensitive to vector length. When it comes to measuring templates of HTML documents this might instead be a desirable effect, as the vector length corresponds to the number of tags, which itself might be quite characteristic for a template. The computational costs correspond mainly to creating the tag vectors and are of order $O(t_1 + t_2)$.

2.5   LCTS – Longest Common Tag Subsequence

The tag vector approach neglects the order of the tags in the document. To overcome this drawback, the document's structure can be interpreted as a sequence of tags. The distance of two documents can then be expressed based on their longest common tag subsequence. The longest common tag subsequence $lcts(D_1, D_2)$ is the longest (but not necessarily continuous) sequence of tags which can be found in both of the documents. However, computation of this distance is expensive, as finding the longest common subsequence has quadratic complexity of $O(t_1 \cdot t_2)$. The longest common tag subsequence can be turned into a distance measure by:

$$d_{LCTS}(D_1, D_2) = 1 - \frac{|lcts(D_1, D_2)|}{\max(|D_1|, |D_2|)} \tag{4}$$

2.6   CTSS – Common Tag Sequence Shingles

To overcome the computational costs of the previous distance measure we utilise again the shingling technique. Breaking up the entire sequence in a set of shingles $ts(D)$ allows to maintain a certain context for each tag without having to look at the complete document. Thus, applying shingling reduces computational costs for this distance to $O(t_1 + t_2)$. The distance can then be computed similar to the path shingle distance:

$$d_{CTSS}(D_1, D_2) = 1 - \frac{|ts(D_1) \cap ts(D_2)|}{\max(|ts(D_1)|, |ts(D_2)|)} \tag{5}$$

3   Clustering Techniques

Cluster analysis is a vast field of ongoing research. We have applied three different techniques which we will describe briefly and straight away in the context of the given application.

3.1   Multidimensional Scaling

Multidimensional scaling (MDS) is a technique to find a configuration of data points in a (possibly low-dimensional) vector space which represent best a given distance matrix. MDS comes in two general flavours: metric for a distance matrix which is in fact based on a (usually Euclidean) metric and non-metric if the distances are computed in a different way or even estimated. It is commonly used to reduce dimensionality of data to the essential dimensions. The aim is often to achieve a 2D or 3D representation of the data allowing a visual analysis.

The latter was as well the intention of applying MDS on the distance matrices computed for the template based documents. As the distances are not all fulfilling the requirements of a metric, we had to apply non-metric MDS. Starting with the result of a metric Principal Component Analysis as starting configuration we used Kruskal's algorithm to obtain a stable configuration.

3.2   K-Median Clustering

k-means clustering is a classical approach for clustering data. The basic idea is to start with a configuration assigning randomly each of the documents to one of k cluster. For all the clusters a centroid is computed, i.e. a document in its centre. In an iterative process each document is now assigned to the cluster whose centroid is closest. This creates new clusters and thereby new centroids for the next step of the iteration. The iteration process is stopped if the configuration is not changing any more or the changes of the centroids become minimal.

In k-means clustering the mean of the data items in each cluster is computed and used as centroid. As in our case it is difficult to define a mean document, we used an adaptation of the method which uses the median document as centroid – hence the name k-median clustering. The median document is the document which has overall minimum distance to all other documents in the same cluster.

3.3   Single Linkage

Unlike the previous method, hierarchical clustering methods do not require to fix the number of clusters a-priori as a parameter. They start off with each document forming a cluster on its own. The clusters are iteratively merged until only one cluster remains. The information which clusters where merged at which step during the iteration is represented as a tree structure. This so-called dendrogram can be examined to determine different cluster configurations, e.g. due to cluster size, average distance or number of clusters.

There are several ways to decide which clusters are merged and thereby to compute the dendrogram. The single linkage approach is always merging those two clusters for which the distance between two of the documents from the two clusters is minimal over all inter-cluster distances.

4   Experiments

To evaluate the different distance measures we collected a corpus of 500 document from five different German news web sites. Each web site contributed 20 documents from five different topical categories: national and international politics, sports, business and IT related news. The idea for taking into the corpus not only documents from different web sites but to cover as well different sections within those is to see how well the measures can cope with the slight changes in the templates which usually occur within the categories of a web site.

While computing the distance matrices for the 500 documents we measured the time needed to compute the distance matrix for an increasing number of documents. The graph in figure 1 shows the time in seconds needed for computing (symmetric) distance matrices depending on the number of documents involved and using the different distance measures. While obviously the RTDM and tag sequence approach are very time consuming already for small document collections, the other measures can be computed reasonably fast. Tag sequence shingling is on average taking twice as long as the path shingle approach, which itself is slower than the paths distance measure by a factor of about 1.5. The tag vector distance can be computed fastest, probably because there is no need to handle sets and their intersections.

The resulting matrices for all 500 documents are interpreted in figure 2 in a graphical way. The documents were arranged in the same order from left to right for the columns and top down in the rows. The ordering grouped together documents from the same web site and within each site from the same topic category. Each pixel in the image represents the distance between two documents. The closer two documents are to each other, the brighter the pixels are coloured. Dark colours accordingly represent large distances between documents. In all cases the distances have been normalised to convert a distance of 0 to white pixels and the largest distance into black colouring.

The images confirm quite well that under all distance measures the documents based on the same template are having smaller distances than the documents

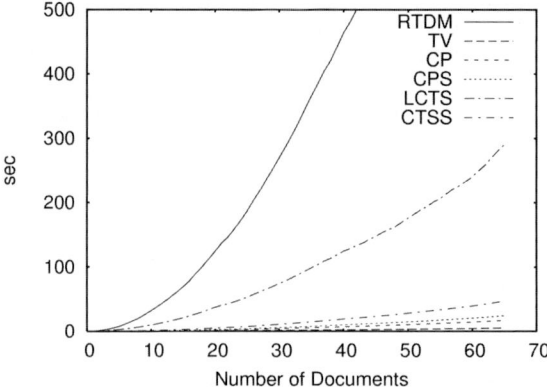

Fig. 1. Time needed to compute the distance matrix with the different distance measures, depending on the number of documents

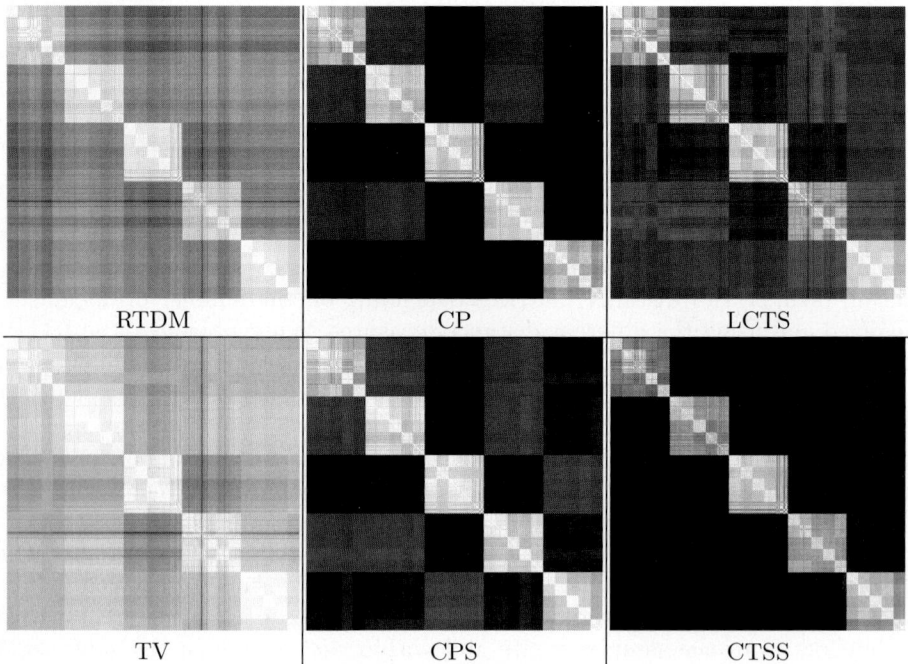

Fig. 2. Distance matrices for 500 template based documents from five web sites

from other web sites. As well the substructures of the different topic categories can be seen quite well for most distance measures and web sites.

To further analyse how well the different underlying templates are separated by the distance measures we computed the Dunn index. The Dunn index sets in comparison the maximum distance d_{max} which occurs between documents based on the same template with the minimum distance d_{min} which occurs between documents based on different templates and is defined as:

$$D = \frac{d_{min}}{d_{max}} \tag{6}$$

The higher the value of D, the better the distance measure separates the documents based on different templates. Table 1 shows the Dunn index for all distance measures. The best results are achieved by the two shingling measures and the paths method.

Table 1. Dunn index D for all distance measures

Distance Measure	RTDM	TV	CP	CPS	LCTS	CTSS
Dunn Index	0.4657	0.1031	1.1691	1.2272	0.6726	1.2901

Once the matrices had been computed the different cluster analysis methods were applied to each of them. To get a first idea on how the documents could be located relative to each other in a 2D vector space we used MDS as described above to generate the images in figure 3. Here as well the clusters of the different templates can be determined quite clear, but it becomes still more obvious that their separation is not always as distinctive as it could be expected. However, mapping the data into 2D space might reduce the dimensionality too much to allow more than a first visual analysis.

To further evaluate the clusters computed by the k-median and the single linkage algorithms we used different measures: the Rand index [17], cluster purity and mutual information (we found a good and brief introduction to those last two measures in [18]). We explain these measures shortly and once again straight away in terms of how they translate into the context of this paper of web documents and underlying templates.

Rand Index: Given a ground truth providing a "correct" clustering of the documents according to their underlying templates, the Rand index measures how often a computed cluster configuration "agrees" with the ground truth. In our case an agreement corresponds to the cluster analysis either claiming correctly two documents to be based on the same template (i.e. being grouped together in the same cluster) or to claiming correctly two of the documents having different underlying templates (i.e. putting them in different clusters). A disagreement accordingly corresponds to either putting documents together in a cluster which have different underlying templates or to separate them in different clusters though they are based on the same template. Therefore if A and D are the number of agreements and of disagreements, the Rand index is:

$$R = \frac{A}{A + D} \tag{7}$$

Purity: The purity of a single cluster compared to a ground truth is providing a measure of how many documents based on the same template are lying within

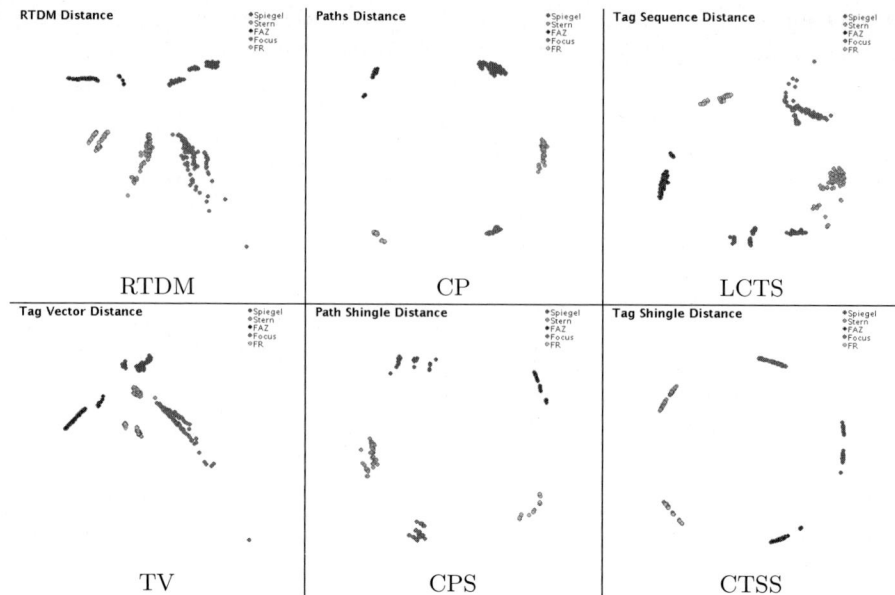

Fig. 3. Using MDS to map the documents onto data points in a 2D vector space

this cluster. Given a cluster c_l and $n_l^{(i)}$ the number of documents in cluster l which according to the ground truth actually belong to cluster i, the purity is:

$$P(c_l) = \frac{1}{\sum_i n_l^{(i)}} \cdot \max_i n_l^{(i)} \qquad (8)$$

Purity is a measure to evaluate one cluster only, so for an entire cluster analysis it is necessary to compute some kind of aggregation, e.g. the average purity.

Mutual Information: Mutual Information is another common measure to evaluate the consensus of a clustering with a ground truth. Given a collection of n documents based on g different templates and the cluster analysis grouped them in k different clusters, the mutual information is:

$$MI = \frac{1}{n} \sum_{l=1}^{k} \sum_{h=1}^{g} n_l^{(h)} \log_{g \cdot k} \frac{n_l^{(h)} \cdot n}{\sum_{i=1}^{k} n_i^{(h)} \cdot \sum_{j=1}^{g} n_l^{(j)}} \qquad (9)$$

As the results of the k-median algorithm depend on the random initial configuration we applied this algorithm 100 times and took the average performance for comparison with the single linkage algorithm. Table 2 shows the results for a clustering with k set to 5. RTDM is providing the best results, followed by the path measure. However, no distance measure allows the centroid based k-median approach to generate a perfect cluster configuration.

Single linkage clustering is performing far better. Extracting from the resulting dendrogram five clusters allows a perfect clustering under some measures as shown

Table 2. Evaluation of k-median clustering based on the different distance measures for $k = 5$ (Average of 100 repetitions)

Distance Measure	RTDM	TV	CP	CPS	LCTS	CTSS
Rand Index	0.9608	0.9399	0.9560	0.9140	0.9157	0.9293
Avg. Purity	0.9613	0.9235	0.9535	0.9057	0.8629	0.9218
Mutual Information	0.1444	0.1354	0.1432	0.1302	0.1250	0.1350

Table 3. Evaluation of single linkage clustering for five clusters

Distance Measure	RTDM	TV	CP	CPS	LCTS	CTSS
Rand Index	0.9200	0.9200	1.0000	1.0000	1.0000	1.0000
Avg. Purity	0.9005	0.9005	1.0000	1.0000	1.0000	1.0000
Mutual Information	0.1287	0.1287	0.1553	0.1553	0.1553	0.1553

in table 3. All measures except RTDM and tag vector group together exactly the documents based on the same templates. We can deduce that for those measures, single linkage is a better way to form clusters for template based documents.

As mentioned above, it is usually not known how many different templates occur within a set of documents. Therefore, we analysed the distributions of distances within each distance matrix. The graphs in figure 4 show the histograms of the distance distributions using a logarithmic scale. Some of the measures show gaps (distances which never occur) between higher and lower distances. In particular the paths shingle and the tag sequence shingle measure show clear gaps, the paths distance measure even two gaps. Tag sequence, RTDM and the tag vector distance do not have such a clear gap in their distance histogram, which corresponds to their lower scores for the Dunn index and the more problematic 2D configuration retrieved when using MDS.

Assuming these distributions to be typical we evaluated the single linkage method with a clustering threshold of 0.6 for the path shingle measure, a threshold

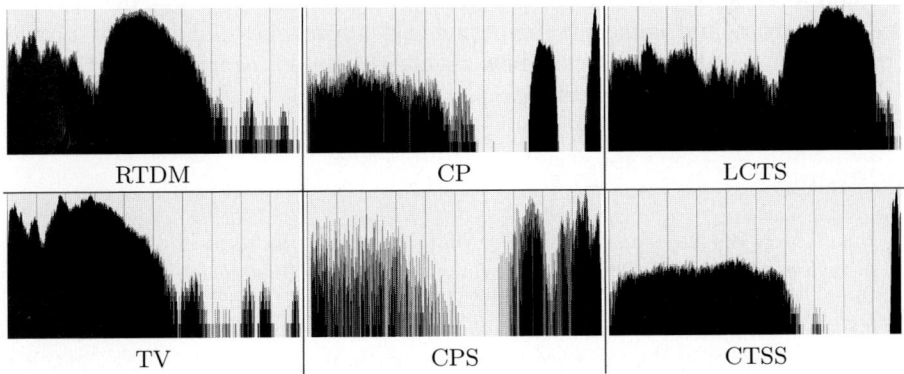

Fig. 4. Distribution of distances for all distance measures (logarithmic scale)

Table 4. Evaluation of single linkage with a distance threshold chosen according to the gaps in the distance histograms

Distance Measure	CP	CP	CTSS	CPS
Threshold	0.7	0.9	0.85	0.6
Clusters	5	3	5	5
Rand Index	1.0000	0.7600	1.0000	1.0000
Avg. Purity	1.0000	0.7778	1.0000	1.0000
Mutual Information	0.1553	0.1296	0.1553	0.1553

of 0.85 for the tag sequence shingle measure and with a thresholds of 0.7 and 0.9 for the paths measure. The latter threshold turned out to be a too low threshold and resulted in three clusters only. Table 4 shows that beside this exception the gaps do really correspond to a separation of the ground truth clusters. So choosing the distance threshold accordingly for a single linkage clustering results in perfect groups of documents which are based on the same template.

5   Conclusions and Future Works

We compared different distance measures and clustering methods for grouping together web documents which are based on the same template. Though tree edit distance measures are often referred to as the most suitable measures to compare HTML documents, it turned out that for the given purpose some simpler measures perform better. The paths, the path shingle and the tag sequence shingle measures in combination with a single linkage clustering deliver perfect results. While the computational cost for the tag sequence shingle approach is higher than for the other two, the gap in the distance histogram, the MDS analysis and the Dunn index hint to a better separation of the clusters.

We intend to confirm these results on a larger scale, incorporating more documents and more different underlying templates. Another issue for future research is to find a possibility for a more fine grained cluster analysis. The distance histogram analysis and some first experimental approaches did not yet provide a way of comparable quality on how to cluster the category caused template variations within the single web sites.

References

1. Bar-Yossef, Z., Rajagopalan, S.: Template detection via data mining and its applications. In: WWW 2002: Proceedings of the 11th International Conference on World Wide Web, pp. 580–591. ACM Press, New York (2002)
2. Yang, G., Ramakrishnan, I.V., Kifer, M.: On the complexity of schema inference from web pages in the presence of nullable data attributes. In: CIKM 2003: Proceedings of the twelfth International Conference on Information and Knowledge Management, pp. 224–231. ACM Press, New York (2003)

3. Lin, S.H., Ho, J.M.: Discovering informative content blocks from web documents. In: KDD 2002: Proceedings of the eighth ACM SIGKDD International Conference on Knowledge Discovery and Data Mining, pp. 588–593. ACM Press, New York (2002)

4. Debnath, S., Mitra, P., Giles, C.L.: Automatic extraction of informative blocks from webpages. In: SAC 2005, pp. 1722–1726. ACM Press, New York (2005)

5. Yi, L., Liu, B., Li, X.: Eliminating noisy information in web pages for data mining. In: KDD 2003: Proceedings of the ninth ACM SIGKDD International Conference on Knowledge Discovery and Data Mining, pp. 296–305. ACM Press, New York (2003)

6. Reis, D.C., Golgher, P.B., Silva, A.S., Laender, A.F.: Automatic web news extraction using tree edit distance. In: WWW 2004: Proceedings of the 13th International Conference on World Wide Web, pp. 502–511. ACM Press, New York (2004), doi:10.1145/988672.988740

7. Gibson, D., Punera, K., Tomkins, A.: The volume and evolution of web page templates. In: WWW 2005: Special Interest Tracks and Posters of the 14th International Conference on World Wide Web, pp. 830–839. ACM Press, New York (2005)

8. Chakrabarti, D., Kumar, R., Punera, K.: Page-level template detection via isotonic smoothing. In: WWW 2007: Proceedings of the 16th International Conference on World Wide Web, pp. 61–70. ACM Press, New York (2007)

9. Cruz, I.F., Borisov, S., Marks, M.A., Webbs, T.R.: Measuring structural similarity among web documents: preliminary results. In: Porto, V.W., Waagen, D. (eds.) EP 1998. LNCS, vol. 1447, pp. 513–524. Springer, Heidelberg (1998)

10. Buttler, D.: A short survey of document structure similarity algorithms. In: IC 2004: Proceedings of the International Conference on Internet Computing, pp. 3–9. CSREA Press (2004)

11. Broder, A.Z., Glassman, S.C., Manasse, M.S., Zweig, G.: Syntactic clustering of the web. Computer Networks 29(8-13), 1157–1166 (1997)

12. Joshi, S., Agrawal, N., Krishnapuram, R., Negi, S.: A bag of paths model for measuring structural similarity in web documents. In: KDD 2003: Proceedings of the ninth ACM SIGKDD International Conference on Knowledge Discovery and Data Mining, pp. 577–582. ACM Press, New York (2003)

13. Lindholm, T., Kangasharju, J., Tarkoma, S.: Fast and simple XML tree differencing by sequence alignment. In: DocEng 2006: Proceedings of the 2006 ACM Symposium on Document Engineering, pp. 75–84. ACM Press, New York (2006)

14. Shi, L., Niu, C., Zhou, M., Gao, J.: A DOM tree alignment model for mining parallel data from the web. In: ACL 2006: Proceedings of the 21st International Conference on Computational Linguistics and the 44th annual meeting of the ACL, Morristown, NJ, USA, Association for Computational Linguistics, pp. 489–496 (2006)

15. Liu, B.: Web Data Mining – Exploring Hyperlinks, Contents, and Usage Data. Springer, Heidelberg (2007)

16. Kruskal, J.B.: Nonmetric multidimensional scaling: A numerical method. Psychometrika 29(2), 115–129 (1964)

17. Rand, W.M.: Objective criteria for the evaluation of clustering methods. Journal of the American Statistical Association 66(336), 846–850 (1971)

18. Strehl, A., Ghosh, J., Mooney, R.: Impact of similarity measures on web-page clustering. In: AAAI 2000: Proceedings of the 17th National Conference on Artificial Intelligence: Workshop of Artificial Intelligence for Web Search, AAAI, pp. 58–64 (2000)

Effective Pre-retrieval Query Performance Prediction Using Similarity and Variability Evidence

Ying Zhao, Falk Scholer, and Yohannes Tsegay

School of Computer Science and IT, RMIT University, GPO Box 2476v, Melbourne, Australia
{ying.zhao,falk.scholer,yohannes.tsegay}@rmit.edu.au

Abstract. Query performance prediction aims to estimate the quality of answers that a search system will return in response to a particular query. In this paper we propose a new family of pre-retrieval predictors based on information at both the collection and document level. Pre-retrieval predictors are important because they can be calculated from information that is available at indexing time; they are therefore more efficient than predictors that incorporate information obtained from actual search results. Experimental evaluation of our approach shows that the new predictors give more consistent performance than previously proposed pre-retrieval methods across a variety of data types and search tasks.

1 Introduction

As the amount of electronic data continues to grow, the availability of effective information retrieval systems is essential. Despite a continuing increase in the average performance of information retrieval systems, the ability of search systems to find useful answers for individual queries still shows a great deal of variation [14].

An analysis of the chief causes of failure of current information retrieval (IR) systems concluded that, if a search system could identify in advance the problem associated with a particular search request, then the selective application of different retrieval technologies should be able to improve results for the majority of problem searches [8]. The ability to predict the performance of a query in advance would enable search systems to respond more intelligently to user requests. For example, if a user query is predicted to perform poorly, the user could be asked to supply additional information to improve the current search request. Alternatively, a search system could selectively apply different techniques in response to difficult and easy queries, for example the selective application of different retrieval models, or automatic relevance feedback.

Query performance prediction is the problem of trying to identify, without user intervention, whether a search request is likely to return a useful set of answers. The importance of the query difficulty prediction problem has been highlighted in the IR community in recent years; the Text REtrieval Conference (TREC) Robust tracks in 2004 and 2005 included an explicit query difficulty prediction task [14], and prediction has been the subject of specific workshops [4]. Despite this recent growth in attention, the prediction of query difficulty is an open research problem.

In this paper, we propose new pre-retrieval predictors based on two sources of information: the similarity between a query and the underlying collection; and the variability with which query terms occur in documents. Pre-retrieval predictors make use of

C. Macdonald et al. (Eds.): ECIR 2008, LNCS 4956, pp. 52–64, 2008.

information that is available at indexing-time, such as collection and term distribution statistics. They can be calculated without needing to first evaluate the query and obtain an answer set, and are therefore more efficient than post-retrieval predictors.

We evaluate the performance of our new predictors by considering the correlation between the predictors and the actual performance of the retrieval system on each query, as measured by mean average precision. We conduct experiments on both newswire and web data, and across informational and navigational search tasks. The results demonstrate that these new predictors show more consistent performance than previously published pre-retrieval predictor baselines across data collections and search tasks.

2   Background

Many different approaches for the prediction of query performance have been proposed. These can be divided into three broad categories: pre-retrieval predictors, post-retrieval predictors, and learning predictors. In this paper we focus on pre-retrieval predictors; the background section therefore concentrates on previous work in this area. We also provide brief descriptions of the other families of predictors for completeness.

Pre-retrieval predictors can be calculated from features of the query or collection, without requiring the search system to evaluate the query itself. The information that these predictors use is available at indexing-time; they are therefore efficient, and impose a minimal overhead on the retrieval system. Pre-retrieval predictors generally make use of evidence based on term distribution statistics such as the inverse document frequency, inverse collection term frequency, or the length of a query.

A range of pre-retrieval predictors were proposed and evaluated by He and Ounis [9]. Their experimental results showed the two best-performing predictors to be the average inverse collection term frequency (AvICTF), and the simplified clarity score (SCS). In their approach, the SCS is obtained by calculating the Kullback-Leibler divergence between a query model and a collection model. We use AvICTF and SCS as baselines in our experiments, and these approaches are explained in detail in Section 4.

Scholer et al. [11] describe results based on using the inverse document frequency (IDF) to predict query performance. They find that using the maximum IDF of any term in a query gives the best correlation on the TREC web data. We present results using the maximum IDF (MaxIDF) as a baseline in our experiments.

Post-retrieval predictors use evidence that is obtained from the actual evaluation of the underlying search query. These predictors can leverage information about the cohesiveness of search results, and can therefore show high levels of effectiveness. However, for the same reason they are less efficient: the search system must first process the query and generate an answer set, and the answer set itself is then usually the subject of further analysis, which may involve fetching and processing individual documents. This can impose a substantial overhead on a retrieval system.

Cronen-Townsend et al. [6] proposed a post-retrieval predictor based on language models: they calculate the divergence between a statistical model of the language used in the overall collection and a model of the language used in the query, to obtain an estimate of the ambiguity of the query. Unlike the simplified clarity score pre-retrieval predictor discussed previously, this approach estimates the query language model from the documents that are returned in the answer set of a retrieval system. The approach was demonstrated to be highly effective on newswire data. Post-retrieval predictors for web data were developed by Zhou and Croft [19], who use a weighted information gain approach that shows a high correlation with system performance for both navigational and informational web search tasks. Other post-retrieval predictors have considered factors such as the variability of similarity scores; for example, Kwok et al. divide a search results list into groups of adjacent documents and compare the similarity among these [10]. Zhou and Croft [18] introduced ranking robustness scores to predict query performance, by proposing noise channel from information theory. This approach has shown higher effectiveness than the clarity score [6].

Learning predictors incorporate a variety of statistical regression [10] or machine learning algorithms [16], such as neural networks or support vector machines, to train on labeled examples of easy and difficult queries. The learned estimator is then used to predict the difficulty of previously unseen queries. While some learning predictors have shown high levels of correlation with system performance, this family of predictors requires suitable training data to be available; a corresponding overhead is therefore incurred during the training stage.

3   Pre-retrieval Prediction of Query Performance

In this section, we present several predictors of query performance. The predictors are concerned with *pre-retrieval* prediction. The information required by such prediction is obtained from various collection, document and term occurrence statistics. These are all obtained at indexing time, and can be efficiently fetched from inverted index structures that are widely used in information retrieval [20]. The computation of these predictors can therefore be carried out prior to query evaluation. This has significant advantages in terms of simplicity and efficiency, factors whose importance increases as the size of collections continues to grow. We propose two broad classes of pre-retrieval predictors: first, predictors that are based on the similarity between queries and the collection; and second, predictors that are based on the variability of how query terms are distributed in the collection, by exploring the in-document statistics for the input queries.

3.1   The Similarity between a Query and a Collection

While many retrieval models have been proposed in the IR literature, at their core these rely to a greater or lesser extent on various collection, document and term distribution statistics, which are used as sources of evidence for relevance [20]. In particular, two of the most commonly-used sources of evidence are the term frequency (TF), and the inverse document frequency (IDF). The former represents the intuitive concept that the higher the frequency with which a query term occurs in a document, the more likely it is that the document is about that term. The latter is used to discriminate between query terms that are highly selective (they occur in fewer documents, and therefore have a high IDF), and those that are less selective (occurring in many documents, and therefore having a lower IDF).

An intuitive geometric interpretation of the similarity between documents and queries is provided by the vector space model [15]. Here, queries and documents are represented as vectors in n-dimensional space, where n is the number of unique terms that occur in a collection. The estimated similarity between a document and a given query is defined as the closeness of a document vector and a query vector, where the closeness is measured by degree of the angle between these two vectors. Documents whose vectors are closely aligned with a query vector are considered to have a high similarity, and are likely to be on similar subjects or topics to the query. In a similar manner, the collection itself can also be represented as an n-dimensional vector, and the similarity of a query to the collection as a whole can be calculated. Those query vectors which are more similar to the collection vector are considered to be easier to evaluate—the evidence suggests that the collection contains documents that are similar to the query. Such queries and therefore more likely to have higher performance.

Predictor 1 (SCQ): Given a query $Q(t_1, \ldots, t_n)$, the similarity score between the query and collection can be defined as:

$$SCQ = \sum_{t \in Q} (1 + \ln(f_{c,t})) \times \ln\left(1 + \frac{N}{f_t}\right) \tag{1}$$

where N is the total number of documents in the collection C, $f_{c,t}$ is the frequency of term t in the collection, and f_t is the number of documents that contain term t. In this version of the metric we simply sum up the contributions of the collection term frequencies and inverse document frequencies of all query terms. Such a process will be biased towards longer queries. We therefore calculate a normalised metric:

Predictor 2 (NSCQ): We define the NSCQ score as the SCQ score, divided by the query length, where only terms in the collection vocabulary are considered:

$$NSCQ = \frac{SCQ}{|Q|_{t \in \mathcal{V}}} \tag{2}$$

where $\mathcal{V}$ is the vocabulary (all unique terms in the collection).

Instead of normalising by query length, an alternative approach is to choose the maximum SCQ score of any query term. The intuition behind this approach is that, since web search queries tend to be short, if at least one of the terms has a high score then the query as a whole can be expected to perform well:

Predictor 3 (MaxSCQ): MaxSCQ considers that the performance of a query is determined by the "best" term in the query—the term that has the highest SCQ score:

$$MaxSCQ = max\left[\forall_{t \in Q} (1 + \ln(f_{c,t})) \times \ln\left(1 + \frac{N}{f_t}\right)\right] \tag{3}$$

We note that it is not rare to encounter a query term t that is missing in $\mathcal{V}$. For simplicity, we assign such terms with 0 scores in a query.

3.2   Variability Score

The previous group of predictors explored the surface features of a collection, such as the frequency with which terms occur in the collection as a whole. In this section, we propose alternative predictors which are concerned with the distribution of query terms over the collection, taking into account the variability of term-occurrences within individual documents. The standard deviation is a statistical measure of dispersion, reflecting how widely spread the values in a data set are around the mean: if the data points are close to the mean, then the standard deviation is small, while a wide dispersion of data points leads to a high standard deviation.

We incorporate such a mechanism in the prediction task, estimating the standard deviation of term occurrence weights across the collection. We hypothesise that if the standard deviation of term weights is high, then the term should be easier to evaluate. This is because the retrieval system will be able to differentiate with higher confidence between answer documents. Conversely, a low standard deviation would indicate that the system does not have much evidence on which to choose the best answer documents; such a query would therefore be predicted to perform less well.

In general, each query term t can be assigned with a weight value $w_{d,t}$ if it occurs in document d. From all the documents that contain term t in a collection, the distribution of t can then be estimated. We use a simple $TF.IDF$ approach to compute the term weight, $w_{d,t}$, within a document [20]:

$$w_{d,t} = 1 + \ln\left(f_{d,t}\right) \times \ln\left(1 + \frac{N}{f_t}\right)$$

Again, for query terms that are missing in $\mathcal{V}$, we assign $w_{d,t} = 0$.

Predictor 4 (σ_1): Given a query Q $(t_1, \ldots, t_n)$, the basic variability score is defined as the sum of the deviations:

$$\sigma_1 = \sum_{t \in Q} \sqrt{\frac{1}{f_t} \sum_{d \in \mathcal{D}_t} (w_{d,t} - \overline{w}_t)^2} \qquad \text{where} \qquad (4)$$

$$\overline{w}_t = \frac{\sum_{d \in \mathcal{D}_t} w_{d,t}}{|D_t|}$$

where $f_{d,t}$ is the frequency of term t in document d, and $\mathcal{D}_t$ is the set of documents that contain query term t. This predictor sums the deviations across query terms, and thus reflects the variability of the query as a whole. An alternative is to use a metric normalised for query length:

Predictor 5 (σ_2): Normalising the σ_1 score by the number of valid query terms, we obtain the σ_2 score for a given query Q:

$$\sigma_2 = \frac{\sigma_1}{|Q|_{t \in \mathcal{V}}} \qquad (5)$$

As for the SCQ score, a further intuitive alternative to simply normalising by query length is to select the largest variability score of any query term:

Predictor 6 (σ_3): σ_3 estimates the performance of a query based on the maximum deviation from the mean that is observed for any one query term:

$$\sigma_3 = max\left[\forall_{t \in Q} \sqrt{\frac{1}{f_t} \sum_{d \in \mathcal{D}_t} (w_{d,t} - \overline{w}_t)^2}\right] \qquad (6)$$

where $\overline{w}_{d,t}$ is defined as in **Predictor 4**.

The proposed SCQ score and variability score predictors are based on different sources of evidence—the former considers evidence at the high collection level, while the latter is based on the distribution of terms across documents. Combining both sources of information could therefore lead to additional prediction accuracy:

Predictor 7 (*joint*): For each query term t in query Q, this predictor combines both the MaxSCQ and σ_1 scores. We use a simple linear interpolation approach to combine the two scores; the computation is defined as:

$$joint = \alpha \cdot MaxSCQ + (1 - \alpha) \cdot \sigma_1 \tag{7}$$

where α is a parameter that determines the weight given to the SCQ and variability score components. The parameter is set using separate training data (for example, for the WT10g collection below, the parameter is trained using topics 501–550 for experiments on topics 451–500, and vice-versa). We find little variation in performance for a region of settings between 0.7 and 0.85.

4   Experimental Methodology

Query performance prediction aims to identify whether a set of search results is likely to contains useful answers. The established information retrieval methodology for this type of investigation involves testing the performance of a predictor across a set of queries that are run on a collection of documents. The performance of the predictor is measured by calculating the correlation between the predicted performance levels with actual system performance levels.

Correlation Coefficients.   In the query performance prediction literature, three different correlation coefficients are widely used (although individual papers often report only one or two of the three available variants): the Pearson product-moment correlation; Spearman's rank order correlation; and, Kendall's tau.

Although they make different assumptions about the data, each of the coefficients varies in the range $[+1, -1]$; the closer the absolute value of the coefficient is to 1, the stronger the correlation, with a value of zero indicating that there is no relationship between the variables. Moreover, each of the correlation coefficients can be used to conduct a hypothesis test to determine whether there is a significant relationship between the two variables, up to a specified level of confidence [12]. In this paper we report significance at the 0.05, 0.01 and 0.001 levels.

The *Pearson* correlation determines the degree to which two variables have a linear relationship, and takes the actual value of observations into account. *Spearman's* correlation coefficient is calculated based on the rank positions of observations; it therefore measures the degree to which a monotonic relationship exists between the variables (that is, a relationship where a change in the value of one variable is accompanied by a corresponding increase (decrease) in the value of the other variable). *Kendall's* tau is also calculated from rank information, but in contrast to Spearman's coefficient is based on the relative ordering of of all possible pairs of observations. For a comprehensive treatment of the properties of the different correlation coefficients the reader is referred to Sheskin[12]. We note that there is currently no consensus on which measure of correlation is the most appropriate for information retrieval experiments, with different measures being reported in different studies. However, the Pearson correlation assumes a linear relationship between variables [7]; there is no reason to assume that this assumption holds between various query performance predictors and retrieval system performance.

Retrieval Performance Metrics. Information retrieval experimentation has a strong underlying experimental methodology as used for example in the ongoing series of Text REtrieval Conferences (TREC): a set of queries is run on a static collection of documents, with each query returning a list of answer resources. Humans assess the relevance of each document-query combination, and from this a variety of system performance metrics can be calculated. *Precision* is defined as the proportion of relevant and retrieved documents out of the total number of documents that have been retrieved. The *average precision* (AP) of a single query is then the mean of the precision of each relevant item in a search results list. Mean average precision (MAP) is the mean of AP over a set of queries. MAP gives a single overall measure of system performance, and has been shown to be a stable metric [3]. For the purpose of evaluating predictors of query performance, we calculate the correlation between the predicted ordering of a run of queries, and the AP score for that run of queries.

For navigational search tasks (see below), where it is assumed that the user is looking for a single specific answer resource, the reciprocal rank (RR) at which the item is found in the result list is used to measure system performance [3].

Collections, Queries and Relevance Judgements. We evaluate our predictors using several different collections and query sets; the aim is to examine the performance of the predictors on different types of data and search tasks. For *web* data, we use two collections: the first is the TREC GOV2 collection, a 425Gb crawl of the `gov` domain in 2004, which contains HTML documents and text extracted from PDF, PS and Word files [5]. We also test our predictors on the WT10g collection, a 10Gb crawl of the web in 1998 [1]. For *newswire* data, we use the collection of the 2005 Robust track, consisting of around 528,000 news articles from sources such as the Financial Times and LA Times (TREC disks 4 and 5, minus the congressional record data).

Each of these collections has associated queries and relevance judgements; full details are provided in Table 1. In our experiments we use only the title fields of the TREC topics, which consists of a few key words that are representative of the information need. Being short, the title field are the most representative of typical queries that might be submitted as part of web search.

Users of web search engines engage in different types of search tasks depending on their information need. In an *informational* search task, the user is interested in learning about a particular topic, while in a *navigational* task the user is looking for a specific named resource [2]. A key difference between these two tasks is that for navigational tasks, it is generally assumed that the user is interested in a single named page. For an informational task, on the other hand it is assumed that there may be several resources that are relevant to the information need. We test our predictors on both types of task; in this paper, navigational queries are identified with the prefix NP.

Table 1. Experimental setup summary: collections and corresponding query sets

Task	Collection	Query Set
TB04	GOV2	701–750
TB05	GOV2	751–800
TB06	GOV2	801–850
TB05-NP	GOV2	NP601–NP872
Robust04	TREC 4+5 (minus CR)	301–450; 601–700
TREC-9	WT10G	451–500
TREC-2001	WT10G	501–550

Retrieval Models. We experiment with two retrieval models: a probabilistic model, and a language model. For the probabilistic model, we use the Okapi BM25 similarity function [13], with the recommended parameter settings of $k_1 = 1.2$, $k_3 = 7$ and $b = 0.75$. For language models, we use the Dirichlet smoothing approach which has been shown to be successful across a wide range of collections and queries [17], with the smoothing parameter set to a value of $\mu = 1000$ [6]. In our experiments, we use version 4.5 of the Lemur Toolkit, an information retrieval toolkit developed jointly by the University of Massachusetts and Carnegie Mellon University[1].

Baselines. We compare our proposed prediction approaches to the two best-performing pre-retrieval predictors evaluated by He and Ounis [9]: the average inverse collection term frequency (AvICTF), and the Simplified Clarity Score (SCS). In their approach, the SCS is obtained by calculating the Kullback-Leibler divergence between a query model (θ_q) and a collection model (θ_c): $SCS = \sum_{t \in Q} \theta_q \cdot \log_2 \frac{\theta_q}{\theta_c}$, where the query model is given by the number of occurrences of a query term in the query ($f_{q,t}$), normalised by query length ($|Q|$), $\theta_q = \frac{f_{q,t}}{|Q|}$. The collection model is given by the number of occurrences of a query term in the entire collection ($f_{c,t}$), normalised by the number of tokens in the collection ($|C|$): $\theta_c = \frac{f_{c,t}}{|C|}$. For a third baseline, we use the maximum inverse document frequency (MaxIDF), which was found to be an effective pre-retrieval predictor for web data by Scholer et al. [11].

5   Results and Discussion

In this section we present the results of our experiments, comparing the performance of our predictors across a range of data types and search tasks.

Web Data, Informational Topics. The correlations between our pre-retrieval predictors with topic-finding queries on the *GOV2* collection are shown in Table 2. We show results separately by TREC data sets (701–750[2], 751–800 and 801–850), as well as the performance over all 149 topics taken together.

Overall, it is apparent that the performance of all predictors varies depending on the query set. The data shows that the similarity between a query and the collection (SCQ) can provide useful information for the prediction of how well a query will perform; the most effective of the three proposed variants here is $MaxSCQ$, which considers

---

[1] http://www.lemurproject.org
[2] Topic 703 is excluded because there is no relevance judgement for this query.

Table 2. Pearson (Cor), Kendall (Tau), and Spearman (Rho) correlation test between average precision (AP) and predictors on the *GOV2* collection. Asterisk, dagger and double dagger indicate significance at the 0.05, 0.01 and 0.001 levels, respectively.

Query	Predictors	LM			Okapi		
		Cor	Tau	Rho	Cor	Tau	Rho
701–750	MaxIdf	0.343*	0.241*	0.328*	0.433†	0.311†	0.420†
	SCS	0.154	0.112	0.144	0.202	0.145	0.190
	AvICTF	0.345*	0.241*	0.331*	0.446†	0.321‡	0.439†
	SCQ	0.244	0.175	0.236	0.231	0.156	0.212
	NSCQ	0.388†	0.264†	0.352*	0.467‡	0.310†	0.431†
	MaxSCQ	0.412†	0.275†	0.399†	0.485‡	0.351‡	0.485‡
	σ_1	0.441†	0.310†	0.426†	0.477‡	0.294†	0.430†
	σ_2	0.401†	0.291†	0.442†	0.466‡	0.320†	0.476‡
	σ_3	0.418†	0.258†	0.394†	0.475‡	0.287†	0.448†
	joint	0.457‡	0.284†	0.399†	0.513‡	0.314†	0.447†
751–800	MaxIdf	0.267	0.238*	0.334*	0.308*	0.247*	0.354*
	SCS	0.082	0.068	0.131	0.094	0.094	0.155
	AvICTF	0.276	0.210*	0.302*	0.249	0.180	0.271
	SCQ	0.257	0.167	0.267	0.275	0.203*	0.313*
	NSCQ	0.395†	0.267†	0.399†	0.359	0.247	0.368†
	MaxSCQ	0.396†	0.251*	0.379†	0.448‡	0.287†	0.417†
	σ_1	0.379†	0.309†	0.449†	0.397‡	0.332‡	0.470‡
	σ_2	0.424†	0.321‡	0.470‡	0.420†	0.324‡	0.450†
	σ_3	0.373†	0.251*	0.391†	0.415†	0.290†	0.420†
	joint	0.423†	0.309†	0.461‡	0.466‡	0.336‡	0.486‡
801–850	MaxIdf	0.277	0.190	0.285*	0.293*	0.191	0.290*
	SCS	0.128	−0.048	−0.091	−0.111	−0.057	−0.094
	AvICTF	0.217	0.166	0.241	0.236	0.175	0.263
	SCQ	0.296*	0.241*	0.332*	0.280*	0.238*	0.327*
	NSCQ	0.094	0.113	0.167	0.090	0.108	0.161
	MaxSCQ	0.280*	0.172	0.257	0.278	0.180	0.265
	σ_1	0.367†	0.319†	0.414†	0.361†	0.317†	0.397†
	σ_2	0.230	0.234*	0.319*	0.227	0.219*	0.298*
	σ_3	0.304*	0.227*	0.316*	0.311*	0.221*	0.311*
	joint	0.369†	0.283†	0.383†	0.365†	0.270†	0.367†
701–850	MaxIdf	0.297‡	0.219‡	0.326‡	0.343‡	0.247‡	0.367‡
	SCS	0.041	0.053	0.076	0.067	0.064	0.094
	AvICTF	0.269‡	0.187‡	0.282‡	0.295‡	0.205‡	0.307‡
	SCQ	0.260†	0.191‡	0.277‡	0.254‡	0.189‡	0.273‡
	NSCQ	0.278‡	0.206‡	0.305‡	0.289‡	0.214‡	0.314‡
	MaxSCQ	0.357‡	0.231‡	0.347‡	0.395‡	0.266‡	0.388‡
	σ_1	0.392‡	0.285‡	0.407‡	0.401‡	0.290‡	0.411‡
	σ_2	0.384‡	0.291‡	0.411‡	0.396‡	0.293‡	0.412‡
	σ_3	0.359‡	0.247‡	0.369‡	0.392‡	0.266‡	0.390‡
	joint	0.410‡	0.287‡	0.415‡	0.436‡	0.297‡	0.430‡

the alignment between the most similar query term and the collection overall. The variability score predictors are extremely effective for the GOV2 data; in particular, the correlation of the σ_1 predictor is statistically significant ($p < 0.01$) for all topics sets. The *joint* predictor similarly shows consistent significant performance.

Considering performance over 149 topics, the *joint* predictor, which uses information from both the collection and the document level, consistently outperforms all baseline predictors, and shows highly significant correlation ($p < 0.01$) across correlation types and retrieval models.

Correlation results for the *WT10g* collection are shown in Table 3. Collection-level information is again useful for prediction; the most effective variant is $MaxSCQ$, which considers the alignment between the most similar query term and the collection overall; the $MaxSCQ$ predictor is statistically significant for all correlation coefficients,

Table 3. Pearson (Cor), Kendall (Tau), and Spearman (Rho) correlation test between average precision (AP) and predictors on the *WT10g* collection. Asterisk, dagger and double dagger indicate significance at the 0.05, 0.01 and 0.001 levels, respectively.

Query	Predictors	Okapi			LM		
		Cor	Tau	Rho	Cor	Tau	Rho
451-500	MaxIdf	0.086	0.221*	0.291*	0.090	0.227*	0.302*
	SCS	0.194	0.226*	0.333*	0.197	0.227*	0.332*
	AvICTF	−0.056	0.012	0.016	−0.057	0.020	0.028
	SCQ	0.124	0.148	0.204	0.130	0.151	0.211
	NSCQ	0.402†	0.347‡	0.516‡	0.403†	0.348‡	0.520‡
	MaxSCQ	0.443†	0.453‡	0.620‡	0.447†	0.456‡	0.624‡
	σ_1	0.252	0.272†	0.405†	0.259	0.275†	0.410†
	σ_2	0.253	0.281†	0.436†	0.257	0.286†	0.438†
	σ_3	0.347*	0.350‡	0.523‡	0.354*	0.358‡	0.529‡
	joint	0.337‡	0.352‡	0.510‡	0.344‡	0.358‡	0.514‡
501-550	MaxIdf	0.491‡	0.284†	0.396†	0.507‡	0.291†	0.408†
	SCS	0.155	0.084	0.128	0.162	0.089	0.132
	AvICTF	−0.007	0.008	0.002	0.007	0.007	0.010
	SCQ	0.260	0.136	0.224	0.235	0.135	0.215
	NSCQ	0.099	0.097	0.140	0.114	0.106	0.157
	MaxSCQ	0.399†	0.267†	0.364*	0.418†	0.274†	0.378†
	σ_1	0.654‡	0.425‡	0.612‡	0.652‡	0.435‡	0.619‡
	σ_2	0.282	0.269†	0.389†	0.304*	0.278†	0.407†
	σ_3	0.518‡	0.358‡	0.486‡	0.538‡	0.372‡	0.502‡
	joint	0.640‡	0.449‡	0.622‡	0.644‡	0.466‡	0.634‡

Table 4. Pearson (Cor), Kendall (Tau), and Spearman (Rho) correlation test between reciprocal rank (RR) and predictors on the *GOV2* collection, for navigational topics. Asterisk, dagger and double dagger indicate significance at the 0.05, 0.01 and 0.001 levels, respectively.

Query	Predictors	Okapi			LM		
		Cor	Tau	Rho	Cor	Tau	Rho
NP601-NP872	MaxIdf	0.531‡	0.466‡	0.625‡	0.562‡	0.510‡	0.678‡
	SCS	0.257‡	0.200‡	0.282‡	0.281‡	0.236‡	0.331‡
	AvICTF	0.411‡	0.355‡	0.488‡	0.446‡	0.380‡	0.523‡
	SCQ	0.361‡	0.279‡	0.389‡	0.383‡	0.313‡	0.436‡
	NSCQ	0.385‡	0.343‡	0.478‡	0.423‡	0.374‡	0.521‡
	MaxSCQ	0.426‡	0.415‡	0.571‡	0.440‡	0.453‡	0.623‡
	σ_1	0.457‡	0.423‡	0.582‡	0.516‡	0.470‡	0.643‡
	σ_2	0.318‡	0.405‡	0.555‡	0.380‡	0.448‡	0.608‡
	σ_3	0.409‡	0.430‡	0.591‡	0.453‡	0.479‡	0.649‡
	joint	0.478‡	0.445‡	0.608‡	0.522‡	0.490‡	0.666‡

across all queries ($p < 0.05$). The variability score predictors are extremely effective for the TREC-2001 topics ($p < 0.01$), but show less consistent performance for TREC-9, where the linear Spearman correlation is not significant. The *joint* predictor using both sources of information consistently performs well over both sets of topics, showing highly significant correlation ($p < 0.001$), and outperforming all three baselines.

We note that the *AvICTF* baseline performs particularly poorly for this collection; the reason appears to be that presence of queries that contain terms that do not occur in the collection. The *MaxIDF* predictor is highly effective on the TREC-2001 topics, but performs relatively poorly on the TREC-9 data. Basing prediction on just a single characteristic of queries therefore does not appear provide sufficient information to give consistent performance across informational searches on web data.

Table 5. Pearson (Cor), Kendall (Tau), and Spearman (Rho) correlation test between average precision (AP) and predictors on the *Robust* collection. Asterisk, dagger and double dagger indicate significance at the 0.05, 0.01 and 0.001 levels, respectively.

Query	Predictors	LM			Okapi		
		Cor	Tau	Rho	Cor	Tau	Rho
301-450;	MaxIdf	0.505‡	0.359‡	0.496‡	0.466‡	0.326‡	0.456‡
601-700	SCS	0.376‡	0.204‡	0.293‡	0.343‡	0.183‡	0.266‡
	AvICTF	0.386‡	0.234‡	0.336‡	0.355‡	0.203‡	0.294‡
	SCQ	0.062	0.102*	0.149*	0.058	0.090	0.132*
	NSCQ	0.338‡	0.258‡	0.375‡	0.304‡	0.227‡	0.333‡
	MaxSCQ	0.371‡	0.348‡	0.493‡	0.335‡	0.316‡	0.450‡
	σ_1	0.329‡	0.323‡	0.458‡	0.310‡	0.302‡	0.434‡
	σ_2	0.237†	0.368‡	0.514‡	0.223‡	0.353‡	0.495‡
	σ_3	0.478‡	0.382‡	0.528‡	0.444‡	0.356‡	0.496‡
	joint	0.379‡	0.370‡	0.514‡	0.284‡	0.363‡	0.505‡

Web Data, Navigational Topics. Table 4 shows the results for navigational topics 601–872 on the *GOV2* collection. The performance of the $MaxIDF$ baseline is very strong for named page finding topics; this predictor gives the highest performance for the task across correlation coefficients and retrieval models. While less strong than the $MaxIDF$ correlation, the performance of the *joint* predictor is consistently the second-highest, and is competitive for this task.

Newswire Data, Informational Topics. Experimental results for predictors on newswire data, using topics from the 2004 TREC *Robust* track, are shown in Table 5. The relative performance of the schemes shows more variation here than for other collections. In general, the two most effective predictors are $MaxIDF$, σ_3 and *joint*. The actual ordering varies depending on the correlation coefficient: $MaxIDF$ shows the highest correlation using the Pearson coefficient. Using the non-linear correlation coefficients leads to different conclusions, with σ_3 showing the highest correlation with the performance of the language model retrieval system, and *joint* giving the highest correlation with the Okapi model.

We note that our SCS and AvICTF baseline results are slightly lower than those reported by He and Ounis [9]. We believe that this is due to differences in retrieval systems (Terrier and Lemur) used to calculate the system MAP scores.

6   Conclusions

We have introduced two new families of pre-retrieval predictors, based on the similarity between a query and the overall document collection, and the variability in how query terms are distributed across documents. Our experimental results show that these sources of information are both important for different collections, and are significantly correlated with the mean average precision of two different retrieval models. The best performance is obtained when combining both sources of information in the *joint* predictor; this strongly outperforms three pre-retrieval baseline predictors for informational search tasks on web data, while giving comparable performance with the best baseline on newswire data, and for navigational search tasks. The new predictors offer a significant advantage over previously proposed pre-retrieval predictors, because the performance of the latter varies drastically between search tasks and data types.

In our results, it can be seen that different correlation coefficients may in some cases lead to different conclusions about the performance of predictors. In future work we intend to explore the methodology of query performance prediction, to investigate which are the more appropriate measures for this task. We also plan to consider a variety of post-retrieval predictors to complement the pre-retrieval approaches.

References

1. Bailey, P., Craswell, N., Hawking, D.: Engineering a multi-purpose test collection for web retrieval experiments. Information Processing and Management 39(6), 853–871 (2003)
2. Broder, A.: A taxonomy of web search. SIGIR Forum 36(2), 3–10 (2002)
3. Buckley, C., Voorhees, E.M.: Retrieval system evaluation. In: Voorhees, E.M., Harman, D.K. (eds.) TREC: experiment and evaluation in information retrieval, MIT Press, Cambridge (2005)
4. Carmel, D., Yom-Tov, E., Soboroff, I.: SIGIR workshop report: predicting query difficulty - methods and applications. SIGIR Forum 39(2), 25 28 (2005)
5. Clarke, C., Craswell, N., Soboroff, I.: Overview of the TREC, terabyte track. In: The Thirteenth Text REtrieval Conference (TREC 2004), Gaithersburg, MD, 2005. National Institute of Standards and Technology Special Publication 500-261 (2004)
6. Cronen-Townsend, S., Zhou, Y., Croft, W.B.: Predicting query performance. In: Proceedings of the ACM SIGIR International Conference on Research and Development in Information Retrieval, Tampere, Finland, pp. 299–306 (2005)
7. Freund, J.E.: Modern Elementary Statistics, 10th edn. (2001)
8. Harman, D., Buckley, C.: The NRRC reliable information access (RIA) workshop. In: Proceedings of the ACM SIGIR International Conference on Research and Development in Information Retrieval, Sheffield, United Kingdom, pp. 528–529 (2004)
9. He, B., Ounis, I.: Query performance prediction. Information System 31(7), 585–594 (2006)
10. Kwok, K.L.: An attempt to identify weakest and strongest queries. In: Predicting Query Difficulty, SIGIR 2005 Workshop (2005)
11. Scholer, F., Williams, H.E., Turpin, A.: Query association surrogates for web search. Journal of the American Society for Information Science and Technology 55(7), 637–650 (2004)
12. Sheskin, D.: Handbook of parametric and nonparametric statistical proceedures. CRC Press, Boca Raton (1997)
13. Sparck Jones, K., Walker, S., Robertson, S.E.: A probabilistic model of information retrieval: development and comparative experiments. Part 1. *Information Processing and Management* 36(6), 779–808 (2000)
14. Voorhees, E.M.: Overview of the TREC, robust retrieval track. In: The Fourteenth Text REtrieval Conference (TREC 2005), Gaithersburg, MD, 2006. National Institute of Standards and Technology Special Publication 500-266 (2005)
15. Witten, I., Moffat, A., Bell, T.: Managing Gigabytes: Compressing and Indexing Documents and Images, 2nd edn. Morgan Kaufmann, San Francisco (1999)
16. Yom-Tov, E., Fine, S., Carmel, D., Darlow, A.: Learning to estimate query difficulty: including applications to missing content detection and distributed information retrieval. In: Proceedings of the ACM SIGIR International Conference on Research and Development in Information Retrieval, Salvador, Brazil, pp. 512–519 (2005)

17. Zhai, C., Lafferty, J.: A study of smoothing methods for language models applied to information retrieval. ACM Transactions On Information Systems 22(2), 179–214 (2004)
18. Zhou, Y., Croft, W.B.: Ranking robustness: a novel framework to predict query performance. In: Proceedings of the ACM SIGIR International Conference on Research and Development in Information Retrieval, Arlington, Virginia, pp. 567–574 (2006)
19. Zhou, Y., Croft, W.B.: Query performance prediction in web search environments. In: Proceedings of the ACM SIGIR International Conference on Research and Development in Information Retrieval, Amsterdam, The Netherlands, pp. 543–550 (2007)
20. Zobel, J., Moffat, A.: Inverted files for text search engines. ACM Computing Surveys 38(2) (2006)

iCluster: A Self-organizing Overlay Network for P2P Information Retrieval

Paraskevi Raftopoulou and Euripides G.M. Petrakis

Department of Electronic and Computer Engineering,
Technical University of Crete (TUC),
Chania, Crete, GR-73100, Greece
{paraskevi,petrakis}@intelligence.tuc.gr

Abstract. We present iCLUSTER, a self-organizing peer-to-peer overlay network for supporting full-fledged information retrieval in a dynamic environment. iCLUSTER works by organizing peers sharing common interests into clusters and by exploiting clustering information at query time for achieving low network traffic and high recall. We define the criteria for peer similarity and peer selection, and we present the protocols for organizing the peers into clusters and for searching within the clustered organization of peers. iCLUSTER is evaluated on a realistic peer-to-peer environment using real-world data and queries. The results demonstrate significant performance improvements (in terms of clustering efficiency, communication load and retrieval accuracy) over a state-of-the-art peer-to-peer clustering method. Compared to exhaustive search by flooding, iCLUSTER exchanged a small loss in retrieval accuracy for much less message flow.

1   Introduction

Information sharing in a peer-to-peer (p2p) network requires searching in a distributed collection of peers [1]. Distributed Hash Tables (DHTs) [2,3] and Semantic Overlay Networks (SONs) [4,5] are common solutions to the problem of fast information search in p2p networks. DHTs provide fast lookup mechanisms facilitating information search over the network assuming that each peer is connected to other peers and is responsible for a part of the distributed index. SONs provide an alternative solution to the problem of decentralized indexing by relaxing the requirement of strict peer connectivity imposed by DHTs: peers are virtually linked together (forming clusters) based on the likelihood to contain similar content. The problem of finding the most relevant resources is then reduced to the one of locating clusters of peers similar to the query.

We present iCLUSTER, an approach towards efficient organization of p2p networks into SONs that supports Information Retrieval (IR) functionality: iCLUSTER is automatic (requires no intervention by the user), general (requires no previous knowledge of the peers' contents and works for any type of text contents), adaptive (adjusts to dynamic changes of the network contents), efficient (query processing is faster than existing solutions in the literature) and accurate (achieves high recall outperforming current approaches).

C. Macdonald et al. (Eds.): ECIR 2008, LNCS 4956, pp. 65–76, 2008.

Recent work on SONs by Loser [6] suggests combining information from all layers for scoring the peers. Spripanidkulchai [7] introduced the notion of peer clustering based on similar interests rather than similar documents. Edutella [8] uses metadata to arrange super-peers into the so called *HyperCup* topology [9]. Finally, Lu [10] suggests using content-based information to route query messages to a subset of neighboring peers. However, the work referred above assumes one interest per peer (peer specialization). Klampanos [11] proposed an approach for clustering peers holding information on more that one topics. Parreira [12] introduces the notion of "peer-to-peer dating" for allowing peers to decide which connections to create and which to avoid, based on various usefulness estimators. Additional work on peer organization using SONs is based on the idea of "small world networks" [13,14]. Schmitz [5] assumes that peers share concepts from a common ontology, and this information is used for organizing peers into communities (small worlds) with similar concepts.

iCLUSTER extends the idea of peer organization in small world networks by Schmitz [5] in the following ways: (a) Peers contribute documents in the network (rather than concepts from an ontology). To that end, peers are represented in the network by their interests (in fact document descriptions derived from their content by automatic text processing). Accordingly, query processing imposes document search operations over the network. (b) This organization allows for peers with multiple and dynamic interests (not known in advance). (c) iCLUSTER proposes new rewiring protocols for achieving dynamic (on the fly) organization of peers in clusters and also, effective information search in the derived clustered organization of peers.

The rest of this paper is organized as follows: iCLUSTER architecture and protocols are discussed in Sect. 2, experimental results are presented and discussed in Sect. 3, followed by conclusions and issues for future research in Sect. 4.

2   iCluster

Each peer is characterized by the content of the documents it contributes to the network. Peers with similar interests are grouped together into clusters. Peers may have more than one interests and belong to more than one clusters. Each peer maintains a *routing index* holding information for short- and long-range links to other peers:

short-range links correspond to *intra-cluster* information (i.e., links to peers with similar interests).

long-range links correspond to *inter-cluster* information (i.e., links to peers belonging to different clusters and thus, having different interests).

Entries in the routing index are of the form $(ip(p_j), c_{jk})$, where $ip(p_j)$ is the IP address of peer p_j the link points to and c_{jk} is the k-th interest of p_j. The number of routing indices maintained by a peer equals the number of peer's interests. Peers may merge or split their interests by merging or splitting their corresponding routing indices.

2.1 Peer Similarity

Initially, each peer organizes its documents into groups by applying a document clustering algorithm [15]. The documents of a peer may belong to more than one clusters (i.e., the peer may have more than one interests). Documents are represented by term vectors, and each cluster k, $k \in [1, L_i]$, is represented by its centroid c_{ik} (i.e., the mean vector of the vector representations of the documents it contains). Each peer p_i is represented by the list $\{c_{i1}, c_{i2}, \dots, c_{iL_i}\}$ with the cetroids of its clusters.

A peer p_i can be related to another peer p_j by virtue of more than one interests. The similarity between peers p_i and p_j with respect to interest k of p_i is defined as

$$\mathsf{S}_{ij}^k = S^k(p_i, p_j) = max_{\forall y}\{Sim(c_{ik}, c_{jy})\}, \tag{1}$$

where c_{ik}, c_{jy} are the interests of p_i and p_j, and $Sim(c_{ik}, c_{jy})$ is the similarity between their centroid document vectors. The overall similarity between two peers is defined as the maximum similarity over all pairs of cluster centroids:

$$S(p_i, p_j) = max_{\forall x, y}\{Sim(c_{ix}, c_{jy})\} \tag{2}$$

Finally, the similarity between a document (or query) d and a peer p_i is defined as the maximum similarity between the document (or query) and the peer's interests (centroids):

$$sim(d, p_i) = max_{\forall x}\{Sim(d, c_{ix})\}. \tag{3}$$

2.2 iCluster Protocols

The main idea behind iCLUSTER is to let peers self-organize into SONs, and then, search for similar answers (documents) by addressing the most similar clusters to a given query. The protocols regulating peer join, generation of peer clusters, and query processing in iCLUSTER are discussed next.

Peer Join: When a peer p_i connects to the network, it computes its description $\{c_{i1}, c_{i2}, \dots, c_{iL_i}\}$. For each interest c_{ik} in its description, p_i maintains a routing index RI_{ik}, which is constructed as follows: p_i issues a request to the network that, through a random walk, collects in RI_{ik} the IP addresses and descriptions from λ (randomly) visited peers, which form the initial neighborhood ν_{ik} of p_i. These (randomly selected) links will be refined according to p_i's k-th interest, using the peer organization protocol below.

Peer Organization: Peer organization proceeds by establishing new connections and by discarding old ones, producing this way groups of peers with similar interests. Each peer p_i periodically initiates a *rewiring procedure*. For each interest k, p_i computes the intra-cluster similarity NS_{ik} (as a measure of cluster cohesion) as

$$NS_{ik} = \frac{1}{|\nu_{ik}|} \cdot \sum_{\forall p_j \in \nu_{ik}} \mathsf{S}_{ij}^k, \tag{4}$$

Procedure Rewiring(p_i, k, t_F, θ, m)
A procedure initiated by a peer p_i whenever its neighborhood
similarity NS_{ik} drops below a predefined threshold θ.

input: peer p_i with interest c_{ik} and routing index R_{ik}
output: updated routing index R_{ik}

1: compute $NS_{ik} = \frac{1}{|\nu_{ik}|} \cdot \sum_{\forall p_j \in \nu_{ik}} \mathsf{S}_{ij}^k$
2: **if** $NS_{ik} < \theta$ update routing index R_{ik} as follows
3:    $P = \{ \}$
4:    initiate message FINDNODES= $(ip(p_i), c_{ik}, P, t_F)$
5:    issue FINDNODES to neighbors p_j, $j = 1, \ldots, m$, of p_i
      the issuing neighbors are with equal probability
      m random or the m most similar to p_i
6:    let $c_{j\kappa}$ the interest of p_j most similar to c_{ik}
7:    $P = P \cup \{(ip(p_j), c_{j\kappa})\}$
8: reduce message TTL t_F by 1
9: **do** the same for the neighbors of p_j
10: **repeat** until message TTL $t_F = 0$
11: return list P to p_i

Fig. 1. Peer organization procedure

where $|\nu_{ik}|$ is the number of peers in the neighborhood of p_i with respect to interest k. If NS_{ik} is greater than a threshold θ (θ is user defined), then p_i does not need to take any further action, since it is surrounded by peers with similar interests. Otherwise, p_i initiates a cluster refinement process by issuing FINDNODES= $(ip(p_i), c_{ik}, P, t_F)$ message, where $ip(p_i)$ is the IP address of p_i, c_{ik} is the centroid corresponding to k-th interest of p_i and t_F is the time-to-live (TTL) of the message (t_F is user defined). A peer p_j receiving the message computes the similarity between its interest c_{jy} with interest c_{ik} in FINDNODES message, appends to P the interest resulted in the maximum similarity value, reduces t_F by 1 and forwards FINDNODES message to its neighbors. When $t_F = 0$, FINDNODES message is sent back to the initial sender p_i. The message is forwarded with equal probability either to (i) a number m of randomly chosen peers contained in p_j's routing index, or (ii) to the m peers most similar to p_i (the sender of the message). The rationale of applying both forwarding solutions at the same time is not only to connect p_i directly to similar peers, but also indirectly, by enabling propagation of the forwarding message to other similar peers through non-similar peers in the neighborhood of p_i. Figure 1 summarizes the steps of the above rewiring process.

A peer p_j receiving FINDNODES message collects information about new peers with similar interests, and appends it in its routing index $RI_{j\kappa}$ by replacing old short-range links corresponding to less similar peers with new links corresponding to more similar peers. Additionally, p_j collects information about peers with non-similar interests in $RI_{j\kappa}$ updating its long-range links.

Query Processing: Queries are issued by free text or keywords and are formulated as term vectors. The peer issuing the query initiates a QUERY= (q, t_q) message, where q is the query vector and t_q is the TTL of the message (t_q is user defined). The initiator p_i of the message compares q against its interests and decides for the forwarding of the message to some or all of its neighbors according to the *query routing strategy* that follows. Similarly, peers receiving a QUERY message compare q against their interests and forward the message to neighboring peers.

A forwarding peer p_j compares q against its interests and forwards q to its short-range links (i.e., *broadcasts* the message to its neighborhood) if $sim(q, p_j) \geq \theta$. Otherwise, p_j forwards q to m peers, that are the most similar peers to q (*fixed forwarding*). At each step of the forwarding procedure, t_q is reduced by 1.

Apart from query forwarding, each peer p_j receiving q applies the following procedure for retrieving documents similar to q. The peer compares q against its interests and if $sim(q, p_j) \geq \theta$ the peer matches q against its locally stored content to retrieve similar documents. Pointers to these documents are sent to the initiator of the query p_i. When this process is completed, p_i produces a list R with results of the form $\langle d, Sim(q, d) \rangle$, where d is a pointer to a document and $Sim(q, d)$ is the similarity between q and d. The candidate answers are ordered by similarity to the query and returned to the user. Figure 2 summarizes the steps of the query processing algorithm.

2.3   Discussion

iCLUSTER is highly dynamic as it allows for random insertions or deletions of new documents in existing peers. Peers recompute their interests when their document collection has fairly changed. iCLUSTER is based solely on local interactions, requiring no previous knowledge of the network structure or of the overall content in the network. Each peer initiates a rewiring procedure every time the overall similarity of the peers in its neighborhood (intra-cluster similarity) drops below a predefined threshold. The cost of this organization results in extra message traffic, which increases with threshold θ. However, this extra message traffic is traded for faster and more efficient search at query time.

iCLUSTER maintains a fixed number of long-range links (i.e., links to other clusters) in the routing indices of the peers in addition to short-range links. This prevents clusters from becoming isolated and thus inaccessible by other clusters.

Methods such as [8,5] assuming one interest per peer (specialization assumption) might not perform well under this setting: the description of a peer would either reflect the contents of its strongest interest (e.g., the one with more documents) ignoring all other interests, or result in a single cluster corresponding to the averaging over the entire document collection of the peer. This in turn, would result in poor retrieval performance as queries (even very specific ones)

Algorithm Query_Routing(QUERY, p_i, t_q, θ, m)
A peer p_i compares the query q towards its interest, finds
similar documents and forwards the query to its neighbors.

input: query q issued by peer p_i and threshold θ,
output: document answer set R

1: search within the interests of p_i
2: **if** $sim(q, p_i) > \theta$ **then**
3:     $R_i = \{ \}$
4:     search for similar documents within p_i
5:     **if** $Sim(q, d) > \theta$ **then**
6:         include d in answer set $R_i = R_i \cup (d, Sim(q, d))$
7: **if** $sim(q, p_i) > \theta$ **then**
8:     forward q to all short-range links of p_i
        by issuing Query_Routing
9: **else**
10:     forward q to the m neighbors of p_i most similar to q
        by issuing Query Routing
11: reduce query TTL t_q by 1
12: search similarly within each visited peer p_j
13: **repeat** until query TTL $t_q = 0$
14: return answer sets R_j to p_i
15: rank results $R = \cup R_j$ on p_j by similarity with q

Fig. 2. Query routing algorithm

will be addressing highly incoherent clusters of peers. In iCLUSTER, each peer
identifies its interests by applying a local clustering process.

3   Evaluation

The experiments are designed to demonstrate the superiority of the proposed
iCLUSTER protocols over (a) a state-of-the-art approach for peer organization
and retrieval [5] and (b) exhaustive search by flooding [16].

3.1   Experimental Set-Up

iCLUSTER has been tested on a subset of the OHSUMED TREC[1] collection
with 30,000 medical articles and on the TREC6[2] data set with 556,078 documents. Each OHSUMED document belongs to one out of 10 classes while, each
document in the TREC6 collection belongs to one out of 100 classes.

---

[1] http://trec.nist.gov/data/t9_filtering.html
[2] http://boston.lti.cs.cmu.edu/callan/Data/

The network consists of 2,000 peers. Initially, each peer is assigned documents from one class (i.e., initially each peer has one interest). Each peer maintains one routing index with links to other peers (10% are long-range links).

Each peer periodically tries to find better neighbors by initiating the rewiring procedure. The base unit for time used is the period t. The start of the rewiring procedure for each peer is randomly chosen from the time interval $[0, 4K \cdot t]$ and its periodicity is randomly selected from a normal distribution of $2K \cdot t$ for each peer separately. We start checking the network at time $4K \cdot t$, when all peers have initiated at least once the rewiring procedure.

We experimented with different values of similarity threshold θ, message forwarding TTL t_F and query forwarding TTL t_q. In the following, we show how the critical values characterizing the network vary over time. We considered 5 different initial network topologies and for each topology the results were averaged over 5 runs.

3.2   Performance Measures

The performance of iCLUSTER is mainly evaluated in terms of peer organization, communication load and accuracy of retrieval (recall). The (*weighted*) *clustering coefficient* $\overline{w\gamma}$ is one of the common metrics [17,18] used to describe how well the peers are organized into groups with similar interests and is defined as:

$$\overline{w\gamma} = \frac{1}{N} \sum_{i=1}^{N} (\frac{1}{\lambda(\lambda - 1)} \sum_{\forall p_j, p_k \in \nu_i, p_k \in \nu_j} S(p_j, p_k)) \qquad (5)$$

The *network load* of a method is measured by the number of messages exchanged by the peers during rewiring or querying. In turn, the accuracy of retrieval is evaluated using *recall* (i.e., percentage of qualifying answers retrieved with respect to the total number of qualifying answers in the network). An organization (or search) strategy is better than another if it achieves better clustering coefficient (or retrieval accuracy) for less communication load.

3.3   Peer Organization

To evaluate the clustering effectiveness of iCLUSTER, we monitored how $\overline{w\gamma}$ varies over time for different values of θ and t_F. After a few iterations (after $9K \cdot t$), $\overline{w\gamma}$ stabilizes to 0.21 for the OHSUMED and 0.55 for the TREC6 data set. The variation in $\overline{w\gamma}$ is due to the variation in the number of document classes in the two data sets. Stability is achieved as peers are surrounded eventually by other peers with similar interests (i.e., $NS > \theta$). The experiments demonstrate that the values of $\overline{w\gamma}$ are slightly influenced by θ (less than 3%). Additionally, only a small number of organization messages are initially needed and are reduced to 0 after time $6K \cdot t$, when the network becomes coherent.

Additional experiments indicate that the network converges faster for higher values of t_F (i.e., high values of t_F address peers far apart from the peer originating the process). Although $\overline{w\gamma}$ do not vary significantly with t_F, the communication overhead increases by 86% (i.e., when t_F increases from 4 to 7), leading us to choose $t_F = 4$ for our setting.

The experiments above demonstrate that the rewiring protocol of iCLUSTER results in a effective peer organization at the expense of small communication load. The rewiring similarity threshold θ affects clustering cohesion, while the rewiring TTL parameter t_F has minimal effects on the convergence time of the network.

3.4 Performance of Retrieval

The purpose of this set of experiments is to evaluate the performance of the proposed query routing protocol as a function of (i) recall, (ii) communication overhead incurred by a query and (iii) recall per search message. We also examine the dependance of recall and communication overhead on t_q, θ and t_F. The plots below correspond to measurements on the OHSUMED data set (TREC6 produced similar results).

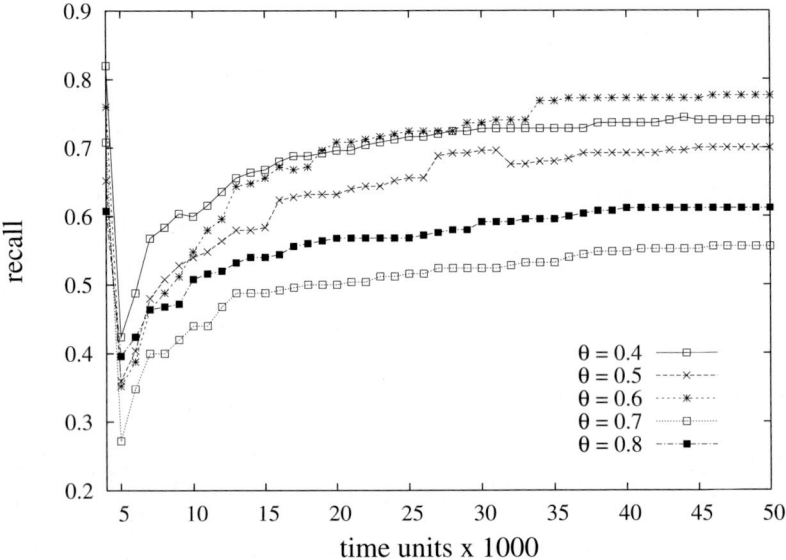

Fig. 3. Recall as a function of time for various values of θ

Figure 3 illustrates how recall varies with time for various values of θ. When no similarity structure is imposed in the network ($4K \cdot t$), the queries are flooded over the network reaching recall as high as 0.8. However, this value of recall is achieved for large communication overhead (1,200 messages per query). When the network becomes organized into cohesive clusters (after time $9K \cdot t$), iCLUSTER achieves the same high values of recall for much less communication overhead (500 messages per query).

As shown in Fig. 3, $\theta = 0.6$ achieves the highest recall on an organized network (after time $9K \cdot t$). For lower values of θ, there are many links from each peer towards non-similar peers, since the clusters are not coherent enough. For higher values of θ, the clusters are coherent but it becomes difficult for a query to be forwarded to similar clusters of peers through other non-similar peers. The optimal value of t_F achieving the best recall is 6.

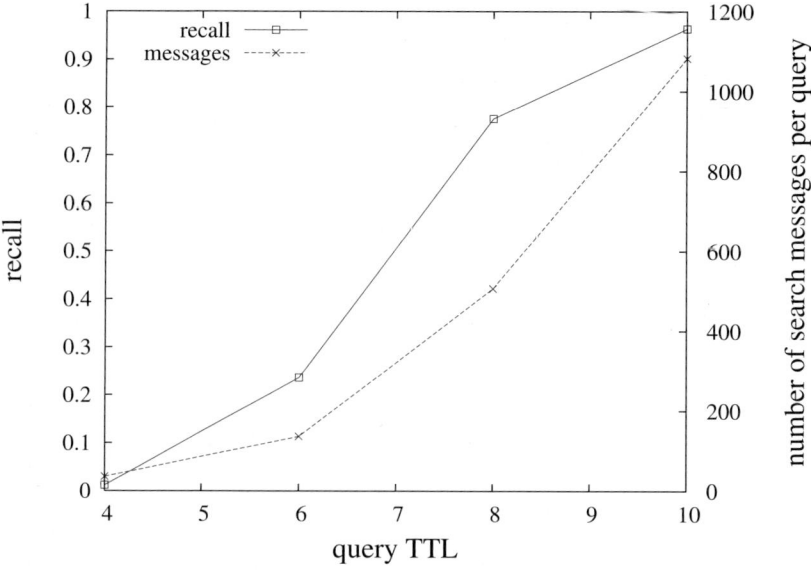

Fig. 4. Recall and query messages per query as a function of t_q

Figure 4 shows the dependence of recall and communication load incurred by the query (in number of messages sent) on t_q. Obviously, recall increases with t_q as more peers are receiving q, but communication load increases as well. For $t_q = 4$ or 6 the recall achieved is very low. Notice that $t_q = 10$ achieved almost 19% better recall (approaching recall 1) than $t_q = 8$ at the expense of 53% more communication load. Based on these observations, the suggested value of t_q is 8. Although $t_q = 8$ is relatively high, the communication overhead is low as iCLUSTER applies selective propagation of query messages to qualifying peers.

The experiments above showed that iCLUSTER achieves high values of recall for less communication load when the network becomes organized into cohesive clusters. We examined the dependance of the retrieval performance on the rewiring process and on t_q, and we suggested optimal values for the parameters (i.e., achieving high recall with small communication overhead).

3.5   Comparison to other Methods

The following methods are implemented and compared:

– iCLUSTER, the method proposed in this work for $\theta = 0.6$, $t_F = 6$ and $t_q = 8$.
– Query flooding, the method implemented by many p2p systems (e.g., Gnutella). It assumes no special network structure and the query is flooded over the network. For comparison with iCLUSTER we set $t_q = 8$.
– The peer organization approach proposed by Schmitz [5], using $\theta = 0.6$, $t_F = 6$ and $t_q = 8$ to have results that are comparable with iCLUSTER.

Notice that $\overline{w\gamma}$ is close to 0 for flooding, indicating no organization of peers into clusters. Compared to [5], iCLUSTER results in higher value of $\overline{w\gamma}$ (0.21 as opposed to 0.08) and therefore, better clustering quality.

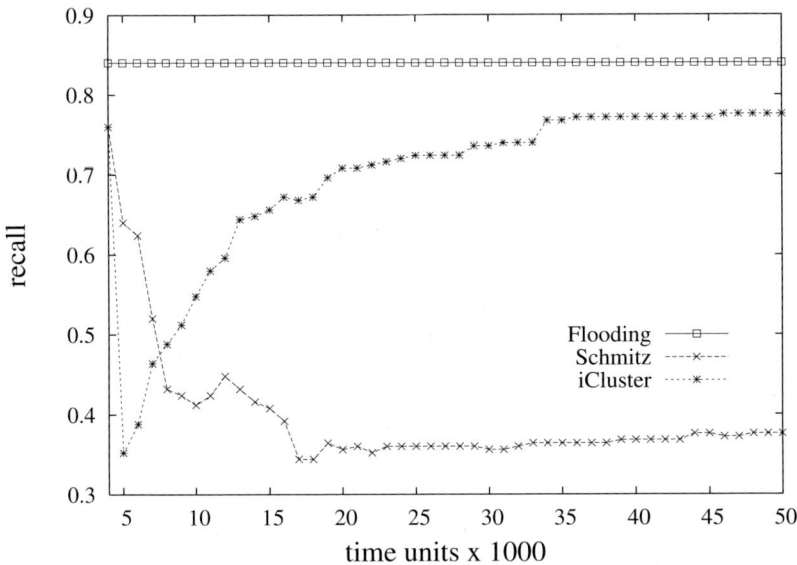

Fig. 5. Recall as a function of time for (a) iCluster (b) Flooding and (c) Schmitz [5]

Figure 5 shows how recall varies over time for all three approaches. The flooding approach achieves recall 0.85 as it searches the network almost exhaustively imposing high communication overhead. As Fig. 5 indicates, prior to imposing any similarity structure in the network (before time $6K \cdot t$), iCLUSTER and [5] achieves recall as high as 0.8. However, notice the high communication overhead incurred by both methods (i.e., almost 1,000 messages per query). When the network becomes organized, iCLUSTER (unlike [5]) achieves recall resembling that achieved by the flooding approach but for much less (up to 60%) communication overhead. Finally, Fig. 6 indicates that, in terms of message traffic per query, flooding is the worst method.

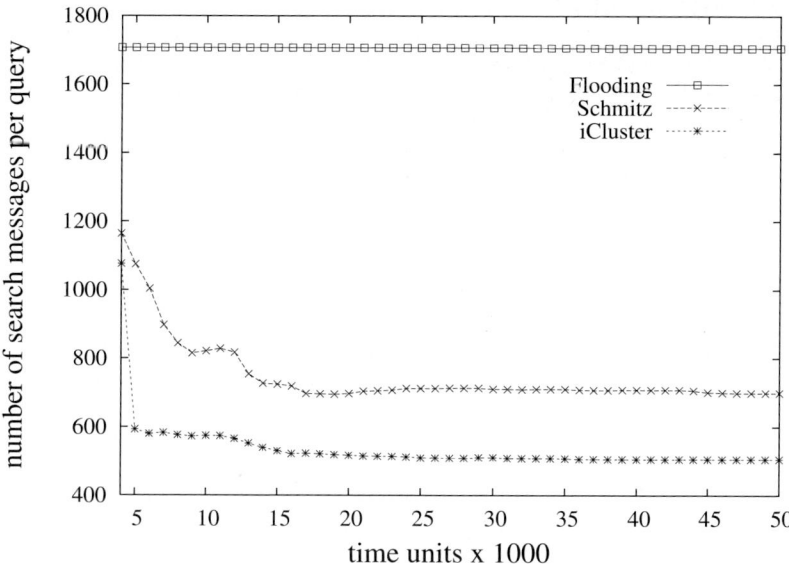

Fig. 6. Number of search messages as a function of time for (a) iCluster (b) Flooding and (c) Schmitz [5]

In this set of experiments, iCLUSTER is compared with the peer clustering approach by Schmitz [5] and the standard exhaustive search approach by flooding, in terms of both communication load and retrieval accuracy. The experiments showed that iCLUSTER benefited the most from creating coherent clusters of peers with similar interests, as it resulted in high recall for much less network load than all its competitor methods. In particular, iCLUSTER can be almost as effective as flooding for much less message flow (i.e., the communication load reduced by approximately 70%).

4   Conclusion

We present iCLUSTER, an approach for organizing peers into clusters and for supporting information retrieval functionality. iCLUSTER ensures clustering coherence while achieving high accuracy of retrievals by issuing a periodic rewiring procedure. The experimental results demonstrated that iCLUSTER outperforms other approaches for peer organization and retrieval, achieving higher clustering quality and higher recall for less communication overhead. Future work, includes experimentation with different query distributions, the study of the effect of churn (dynamic peer insertions/deletions) to network organization and data retrieval, and extension of the proposed protocols to support both information retrieval and filtering functionality (publish/subscribe).

Acknowledgements

This work was funded by project "Herakleitos" of the Greek Secretariat for Research and Technology. We are grateful to Dr. Gerhard Weikum and the members of the Databases and Information Systems group of Max-Planck-Institute for their support.

References

1. Milojicic, D.S., Kalogeraki, V., Lukose, R., Nagaraja, K., Pruyne, J., Richard, B., Rollins, S., Xu, Z.: Peer-to-Peer Computing. Technical report, HP Labs (2002)
2. Stoica, I., Morris, R., Liben-Nowell, D., Karger, D.R., Kaashoek, M.F., Dabek, F., Balakrishnan, H.: Chord: A Scalable Peer-to-Peer Lookup Protocol for Internet Applications. IEEE/ACM Trans. on Networking 11(1) (2003)
3. Ratnasamy, S., Francis, P., Handley, M., Karp, R., Shenker, S.: A Scalable Content-Addressable Network. In: SIGCOMM 2001 (2001)
4. Crespo, A., Garcia-Molina, H.: Semantic Overlay Networks for P2P Systems. Technical report, Stanford Univ. (2003)
5. Schmitz, C.: Self-Organization of a Small World by Topic. In: P2PKM 2004 (2004)
6. Loser, A., Tempich, C.: On Ranking Peers in Semantic Overlay Networks. In: Althoff, K.-D., Dengel, A., Bergmann, R., Nick, M., Roth-Berghofer, T.R. (eds.) WM 2005. LNCS (LNAI), vol. 3782, Springer, Heidelberg (2005)
7. Spripanidkulchai, K., Maggs, B., Zhang, H.: Efficient Content Location using Interest-Based Locality in Peer-to-Peer Systems. In: INFOCOM 2003 (2003)
8. Nejdl, W., Wolf, B., Qu, C., Decker, S., Sintek, M., Naeve, A., Nilsson, M., Palmer, M., Risch, T.: EDUTELLA: a P2P Networking Infrastructure based on RDF. In: WWW 2002 (2002)
9. Decker, S., Schlosser, M., Sintek, M., Nejdl, W.: HyperCuP - Hypercubes, Ontologies and Efficient Search on P2P Networks. In: Moro, G., Koubarakis, M. (eds.) AP2PC 2002. LNCS (LNAI), vol. 2530, Springer, Heidelberg (2003)
10. Lu, J., Callan, J.: Content-Based Retrieval in Hybrid P2P Networks. In: CIKM 2003 (2003)
11. Klampanos, I., Jose, J.: An Architecture for Information Retrieval over Semi-Collaborating Peer-to-Peer Networks. In: ACM SAC 2004 (2004)
12. Parreira, J.X., Michel, S., Weikum, G.: p2pDating: Real Life Inspired Semantic Overlay Networks for Web Search. Inf. Proc. and Manag. 43(1) (2007)
13. Crespo, A., Garcia-Molina, H.: Routing Indices for P2P Systems. In: ICDCS 2002 (2002)
14. Li, M., Lee, W.C., Sivasubramaniam, A.: Semantic Small World: An Overlay Network for Peer-to-Peer Search. In: ICNP 2004 (2004)
15. Steinbach, M., Karypis, G., Kumar, V.: A Comparison of Document Clustering Techniques. In: TextDM 2000 (2000)
16. Hughes, D., Coulson, G., Walkerdine, J.: Free Riding on Gnutella Revisited: The Bell Tolls? IEEE DS Online 6(6) (2005)
17. Schmitz, C., Staab, S., Tempich, C.: Socialisation in Peer-to-Peer Knowledge Management. In: I-KNOW 2004 (2004)
18. Hui, K., Lui, J., Yau, D.: Small-world Overlay P2P Networks: Construction, Mmanagement and Handling of Dynamic Flash Crowds. Computer Networks 50(15) (2006)

Labeling Categories and Relationships in an Evolving Social Network

Ming-Shun Lin and Hsin-Hsi Chen

Department of Computer Science and Information Engineering
National Taiwan University Taipei, Taiwan
mslin@nlg.csie.ntu.edu.tw, hhchen@ntu.edu.tw

Abstract. Modeling and naming general entity-entity relationships is challenging in construction of social networks. Given a seed denoting a person name, we utilize Google search engine, NER (Named Entity Recognizer) parser, and CODC (Co-Occurrence Double Check) formula to construct an evolving social network. For each entity pair in the network, we try to label their categories and relationships. Firstly, we utilize an open directory project (ODP) resource, which is the largest human-edited directory of the web, to build a directed graph, and then use three ranking algorithms, PageRank, HITS, and a Markov chain random process to extract potential categories defined in the ODP. These categories capture the major contexts of the designated named entities. Finally, we combine the ranks of these categories and tf*idf scores of noun phrases to extract relationships. In our experiments, total 6 evolving social networks with 618 pairs of named entities demonstrate that the Markov chain random process is better than the other two algorithms.

Keywords: Category Labeling, Relationships Labeling, and Evolving Social Network.

1   Introduction

Constructing social networks is important for many applications such as trust estimation, person name disambiguation, collaborative recommendation, prediction of dissemination behavior, etc. Aleman-Meza et al. [1] integrate DBLP bibliography (dblp.unitrier.de) and FOAF (Friend-of-a-Friend) documents which provide "knows" relationship of social network to detect conflict of interests. Rather than explicit relationships, many more social relationships are implicitly embedded in the cyberspace.

To tell if two given named entities have a relationship and further the relationship labeling is fundamental for social network construction. Two named entities mentioned in a same document may have a relationship due to that they work on a same group or same projects, have similar or opposing standpoints, etc. Of course, they may co-occur by chance. Matsuo et al. [8] use *Jaccard coefficient*

C. Macdonald et al. (Eds.): ECIR 2008, LNCS 4956, pp. 77–88, 2008.

to calculate the co-occurrence scores of two objects. This paper uses CODC (Co-Occurrence Double Check) [3] to determine the strengths of relationships.

Given two arbitrary named entities, their possible relationships are many. DBLP and FOAF are not enough to capture general relationships. Matsuo et al. [10] consult RELATIONSHIP[1], a vocabulary for describing relationships between people, and use decision tree to assign 4 types, including co-author, co-lab, co-project and co-conference, to members of two conferences. Mori et al. [11] analyze and cluster the snippets returned by a search engine to extract local contexts for representation relationships. The Open Directory Project (ODP) defines the most comprehensive ontology of the web including over 590,000 categories from 4,830,584 sites by 75,151 editors, and has bounteous taxonomy paths from which inferences of relationship can be obtained. Maguitman et al. [9] and Ferragina and Gulli [4] utilize those taxonomy nodes and paths to detect semantic relationships and cluster snippets. In this paper, we adopt Mori's postulation and ODP resource to extract potential categories by three ranking algorithms, PageRank [12], HITS [6], and a Markov chain random process [5]. After that, we combine the ranks of categories and tf*idf scores of noun phrases selected from snippets to identify their relationships.

This paper is organized as follows. Section 2 introduces how to construct an evolving social network from a search engine by given a seed. Section 3 describes how to build a directed graph by the ODP resource. Section 4 specifies three algorithms to extract potential categories, in particular, Markov chain random process. Section 5 shows and discusses the experimental results. Section 6 concludes the remarks.

2   Evolving Social Network

Given a seed, a person name, we introduce an algorithm to build a social network in this section. The algorithm produces relational named entities incrementally and it is an iterative procedure [7]. Our process will build a social network which is in terms of a tree structure. In the beginning, the tree only has one node, i.e., the seed. The process chooses a node P without any children from the tree, and submits it to a search engine. After parsing the top N returned snippets, the process extracts highly co-occurring named entities with the node P. The named entity of co-occurrence scores with P larger than a predefined threshold, θ, is added into the tree as a child of P. The process iterate the above procedure to evolve the social network. Hence, we call it "an evolving social network".

2.1   Co-occurrence Scores

We adopt our previous work "Co-Occurrence Double Check ($CODC$) [3]", a formula of measuring semantic similarity between two words by web pages, to verify the co-occurrence relationship of two named entities X and Y. If X and Y have a strong relationship, then we can find Y from X (a forward process denoted as $(Y@X)$) and find X from Y (a backward process denoted as $X@Y$).

---

[1] http://vocab.org/relationship

The $CODC(X, Y)$ is defined as follows:

$$CODC(X, Y) = \begin{cases} 0 & \text{if } f(Y@X) = 0 \text{ or } f(X@Y) = 0 \\ e^{\log(\frac{f(Y@X)}{f(X)} \times \frac{f(X@Y)}{f(Y)})^{\alpha}} & \text{otherwise} \end{cases}$$

(1)

where $f(Y@X)$ is total occurrences of Y in the top N snippets when query X is submitted to a search engine; similarly, $f(X@Y)$ is total occurrences of X in the top N snippets for query Y; $f(X)$ is total occurrences of X in the top N snippets of query X, and, similarly, $f(Y)$ is total occurrences of Y in the top N snippets of query Y.

To control the size of an evolving social network, we set some conditions. Each social network derives at most three levels. At level 0, i.e., root, it derives no more than N^* children. The maximal number of children of a node at level $(i+1)$ is bounded by the number at the level i multiplying by a decreasing rate, r.

2.2   An Example of Evolving Social Network

We utilize the Google search engine (http://www.google.com) and set $\theta=0.01$, $N^*=20$ and $r=0.5$ to construct and control our evolving social network. The process in each query crawls 200 snippets by the search engine, then we employ the "Stanford Named Entity Recognizer"[2] (*NER* parser) to parse the snippets and extract person name only.

Consider an example "Roger Federer" (a famous tennis player) to build his evolving social network. In beginning, the root node "Roger Federer" is regarded as a query term submitted to the search engine. Then, the *NER* parser and *CODC* checker extract 17 named entities from those returned snippets. Those named entities are shown in Figure 1(a). In the next iteration, we choose a node without any children, "Rafael Nadal", as a new query term by FIFO (First-In First-Out) criterion. The process repeats the above procedure to extract named entities. The result is shown in Figure 1(b). The final evolving social network is shown in Figure 1(c).

Some errors will appear in this network. (1) The node is not a named entity (NE). It is an NE recognizer error. (2) The node is an organization name. We only allow person names in the social network. Here we also regard it as an NE error. (3) The node has person name ambiguity problem. For example, "Lawrence" maybe is a researcher, an actor or a hockey player. This proposed approach does not deal with this problem. In the above example, we got 103 pairs. Among them, 90 pairs are related to tennis, 6 pairs are related to biographies, 1 pair is related to a soccer player and 1 pair is about entertainments. The remaining 5 pairs are NE errors. For each pair, their potential categories and possible relationships are many. In next two sections, we will utilize the ODP resource and three ranking algorithms to extract potential categories defined in the ODP and their possible relationships.

---

[2] http://www-nlp.stanford.edu/links/statnlp.html

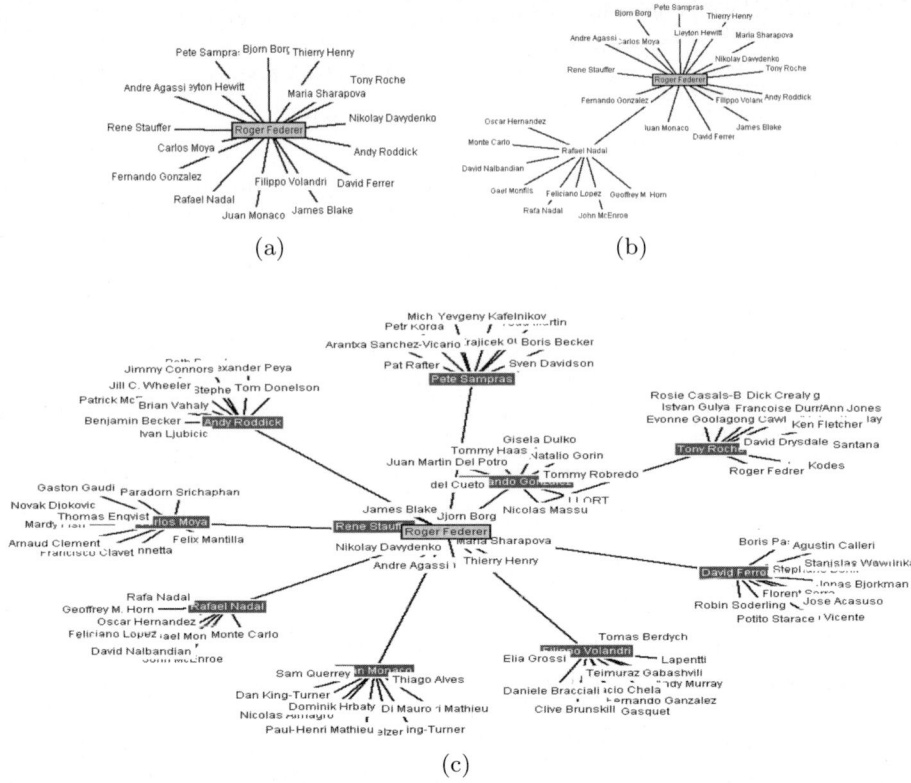

Fig. 1. The Process of Building Roger Federer's Evolving Social Network

3   Extracting Potential Categories and Relationships

It is a challenging to detect what the relationships are between named entities. Some existing methods employ the user profile, wikipedia or some websites which provide the friend of a friend (FOAF) information and interpersonal relations. However, those resources may not available in an open domain. Here, we adopt the postulation of Mori et al. [11]. That is, if an entity pair shares a similar relationship, then their local contexts are similar. In this paper, our local contexts include person names, organization names, location names, and some other noun phrases. These local contexts are called cue patterns hereafter. We further postulate that if the cue patterns identify the most representative categories, then they provide important information for relationship finding. Hence, labeling problem becomes how to determine potential categories by cue patterns. Our method is to build a directed graph from those cue patterns, and then to extract potential categories from the directed graph.

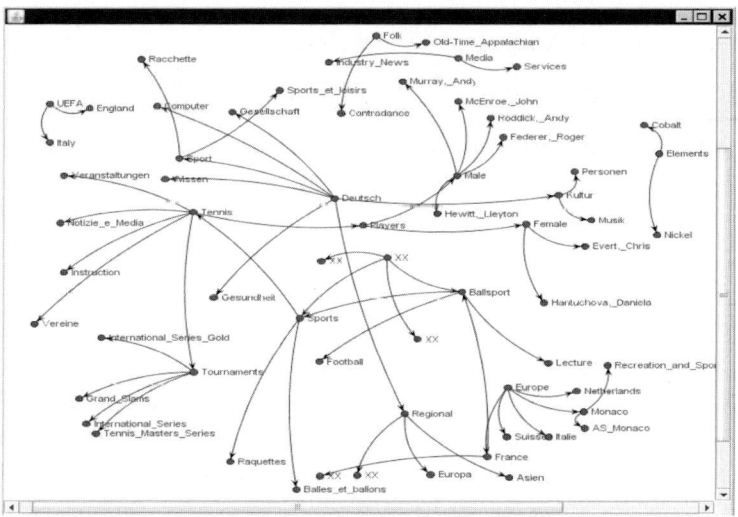

Fig. 2. A Directed Graph is Built by the ODP Resource

3.1   Building a Directed Graph by Cue Patterns

Given a named entity pair, identifying potential categories process firstly submit them to a search engine, and extract cue patterns from those returned snippets by an NER parse and a NP chunker. Maguitman et al. [9], and Ferragina and Gulli [4] utilize the ODP resource to detect semantic relationships and cluster snippets. Based on their ideas, we submits each cue pattern to an ODP search engine, and crawl top N taxonomy paths returned by the ODP search engine. For example, those taxonomy paths like "Sports > Tennis > Tournaments > Grand Slams (GS)", "Arts > Movies > Titles > Wimbledon", and so on, are retrieved for "Wimbledon". Next, the taxonomy nodes and edges in the paths are collected. In the above example, the set of the nodes is {Sports, Tennis, Tournaments, GS, Arts, Movies, Titles, Wimbledon} and the set of the edges is {(Sports, Tennis), (Tennis, Tournaments), (Tournaments, GS), (Arts, Movies), (Movies, Titles), (Titles, Wimbledon)}. Finally, a directed graph $G = \{V, E\}$ is built, where V is the set of the taxonomy nodes and E is the set of the edges between two nodes. Each edge, $u \to v$, has a weight in terms of its frequency.

Consider a named entity pair, "Roger Federer" and "Rafael Nadal". We treat "Roger Federer" and "Rafael Nadal" as a query term to Google search engine. After parsing the returned snippets, the process extracts a lot of cue patterns. Submitting those patterns to Google directory search engine[3] , an ODP search engine, the process can extract a set of taxonomy nodes, V, and a set of edges, E. If a node includes un-ASCII codes like, ä, ë, ÿ, and so on, we will filter it out and label it as "xx". This process could be termed text normalization. A directed graph $G = \{V, E\}$ is built. Figure 2 shows the directed graph. We can

---

[3] http://www.google.com/dirhp

find that "Tennis", "Male", "Ballsport", and "Players" have many in-degrees and our-degrees. In the graph, many nodes have zero in-degrees or out-degrees, or low frequencies. Most sub-graphs are small. The graph is similar to web page connection graphs. All ranking algorithms of web pages can be use to extract critical nodes, i.e., potential categories. In this approach, we adopt three ranking algorithms, PageRank (a surfer on a node with probability 0 to random jump to other nodes), HITS and a Markov chain random process, to extract those critical nodes from the directed graph. Section 4 specifies how the 3rd method works.

3.2   Relationships Labeling

After identifying some critical nodes by a ranking algorithm, we will select the most representative snippets and extract potential relationships according to those critical nodes. A function F with two parameters, a cue pattern and a critical node, is defined. If the cue pattern can identify the critical node, then F returns 1, else 0. Suppose that M critical nodes are extracted by the ranking algorithm, and a snippet has S cue patterns. Formula 2 computes a score for each snippet.

$$Snp_{score} = \sum_{i=1}^{S} \sum_{j=1}^{M} (tfidf(P_i) \times F(P_i, C_j)) \qquad (2)$$

where $tfidf(P_i)$ is $tf \times idf$ score of the cue pattern P_i and critical node C_j is extracted by the ranking algorithm. The tf is a pattern frequency and the idf is an invert document (snippet) frequency in those returned snippets.

4   Extracting Critical Nodes from a Directed Graph

A Markov chain random process is a probability based framework. At first, we transform a graph $G = \{V, E\}$ into an adjacency matrix M. Let V denote the set of nodes of matrix M. The value of an element m_{ij} is the weight of edge e_{ij}. Formula 3 transforms the matrix M into a stochastic matrix P with state space V. For each state i, $\sum_{j=1}^{|V|} p_{ij}$ equals to 1. If the probability of element m_{ii} equals to 1, the state i is an absorbing state. An absorbing state is a stop point in which out-degree is zero.

$$p_{ij} = \begin{cases} m_{ij}/\sum_{k=1}^{|V|} m_{ij} & \text{if } \sum_{k=1}^{|V|} m_{ij} \neq 0 \\ \\ 1 & \text{if } i = j \\ 0 & \text{if } i \neq j \end{cases} \quad \text{otherwise} \qquad (3)$$

In the Markov chain random process, P^n is an n-step stochastic matrix. p_{ij}^n is the probability of which the state i will pass to the state j in n steps. A state i could be a transient state, i.e., $p_{ii}^n = 0$, or an absorbing state, i.e., $p_{ii}^n = 1$, when n limits to infinite. Analyzing the limiting behavior of a stochastic matrix, each state can get a stationary distribution. The stationary distribution is similar to the PageRank for an indicator of a page weight [5]. For each transient state, the stationary distribution is an expected number in which the state was visited from other transient states and then end to an absorbing state. An important state will be visited many times by other states. If a stochastic matrix P is a nonnegative and irreducible matrix, then we can apply *Jordan form decomposition* to get the unique distribution for each i when n limits to infinite [2]. In general, our stochastic matrix is not an irreducible matrix. We need another method to calculate the stationary distribution.

Here, we partition those absorbing states and transient states into a set of mutually disjoint classes and check that each transient state will be absorbed on which absorbing states. We define a structure, $S =< R, T >$, where R is a set of absorbing states and T is a set of transient states which will be absorbed on R. The structure is an element for the partition. Suppose that the Markov chain random process has k absorbing states. For each absorbing state, x, we have a pair, $S_x =< R_x = \{x\}, T_x = \{$nodes: $x's$ ancestry in the directed graph $G\}>$. If state x is an isolate node, then T_x is assigned to be an empty set. Let the initial partition be $\{S_1, S_2, ..., S_x, ..., S_k\}$. The process which get mutually disjoint classes consists of two steps shown as follows:

Step1: For any two records, S_x and S_y, in the partition, if intersection of T_x and T_y is not empty, then we set a new record $S =< R = merge(R_x, R_y)$ and $T = merge(T_x, T_y) >$ to replace S_x and S_y in the partition.

Step2: We repeat the intersection and merge operations until no records in the partition can be merged.

Finally, the partition set is $\{..., S_z, ...\}$, where S_z is $< R_z, T_z >$. Now, we rearrange the stochastic matrix P into the following form:

$$
P = \begin{pmatrix} \begin{matrix} P_1 & & \\ & \ddots & \\ & & \boxed{P_z} \\ & & & \ddots \end{matrix} & \Large{0} \\ \hline \begin{matrix} & \ddots & \\ & & \boxed{Z} \\ & & & \ddots \end{matrix} & \begin{matrix} & \ddots & \\ & \boxed{Q} & \\ & & \ddots \end{matrix} \end{pmatrix},
$$

where Z is a sub-matrix with indexing $T_z \times R_z$, and P_z and Q are two sub-matrices with indexing $R_z \times R_z$ and $T_z \times T_z$, respectively.

Let i and j be two transient states in T_z and $E_i[N_j]$ be the expected number of state i visiting state j until the end comes to R_z. The $E_i[N_j] < \infty$, because i and j are transient states. The $P_{ii}^n \to 0$ implies $Q^n \to 0$, when n limits to infinite. That is, all eigenvalues of Q have absolute values less than 1. Hence, $(I - Q)$ is invertible, because the diagonal elements are not zero. Formula 4 specifies $E_i[N_j]$. The $1_j(X_n)$ is state i in time n visiting state j. The element matrix I

supports initial state as the first visit. Finally, the stationary distribution (SD) of the state i can be calculated by operation of an inverse matrix as Formula 5.

$$
\begin{aligned}
E_i[N_j] &= E_i[\textstyle\sum_{n=0}^{\infty} 1_j(X_n)] = \sum_{n=0}^{\infty} P_{ii}^n \\
&= (I + P + \ldots + P_n + \cdots) \\
&= (I + Q + \ldots + Q_n + \cdots) \\
&= (I - Q)_{ij}^{-1}
\end{aligned}
\tag{4}
$$

$$
SD(i) = P_{iR_z} = \sum_{j \in T_z} (I - Q)_{ij}^{-1}
\tag{5}
$$

5   Experiments

We prepare six seeds for generating their evolving social networks. They are two tennis players, "Rafael Nadal" and "Roger Federer", two baseball players, "Derek Jeter" and "Ichiro Suzuki", and two pioneers, "Bill Gates" and "Sergey Brin". They are shown in Table 1. In the table, the number of social chains for a seed denotes the number of pairs in their evolving social network. Totally, we have 618 named entity pairs to try to extract their potential categories and relationships. In extracting cue patterns, we employ the "Stanford Named Entity Recognizer" parser and a YamCha[4] chunker to extract them. When we submit each named entity pair to the Google web search engine, we can get a set of cue patterns for each pair. Each pattern in the cue pattern set will be submit to the Google directory search engine to crawl taxonomy nodes. We pick and choose top 10 nodes with high frequencies as our baseline model, and pick and choose top 120 nodes to ask 6 assessors to try to select about 5 critical nodes as answer keys. If NE error appears in the pair, assessors are asked to label it with 'NE-Error'. Besides, assessors are asked to label 'Non-Category' when no categories are applicable to the pair. Table 1 shows the numbers of NE-Errors, Non-Categories, and the numbers of categories by assessors for each evolving social network.

We adopt the reciprocal rank of the top relevant taxonomy nodes (abbreviated as recip rank) to evaluate the performance. There are four possible types of cue patterns, including noun phrase (NP), organization name (Org), location name (Loc), and person name (PN). We explore different combinations to determine which types are more important for extracting potential categories. A string with four bits is used to represent the 15 possible combinations. The 1st, 2nd, 3rd, and 4th bit from left to right of the string represent NP, Org, Loc, and PN, respectively. For example, a bit string "1111" (mixer type 15) denotes all the four types are used. Table 2 shows the results of the baseline model with various combinations. In the cases of the players, the most helpful type is PN, i.e., 0001. In pioneer cases, the most helpful type is Org, i.e., 0100. In summary, the order of helpful types is PN > Org > NP > Loc.

---

[4] http://chasen.org/ taku/software/yamcha/

Table 1. Statistics of 6 Evolving Social Networks

Seed	# of Social Chains	# of NE-Errors	# of non-Categories	# of Categories by Assessors
Roger Federer[*]	103	5	0	551 (5.62)
Rafael Nadal[*]	107	11	0	540 (5.63)
Derek Jeterx[x]	107	3	1	485 (4.70)
Ichiro Suzukix[×]	102	1	2	487 (4.92)
Bill Gates[+]	100	11	7	383 (4.67)
Sergey Brin[+]	99	20	7	295 (4.01)

Table 2. Recip Scores of the Baseline Model

N(Mixer)	Roger Federer[*]	Rafael Nadal[*]	Derek Jeterx[×]	Ichiro Suzukix[×]	Bill Gates[+]	Sergey Brin[+]
1 (0001)	**0.7882**	**0.7751**	0.4069	0.4096	0.2108	0.2005
2 (0010)	0.1772	0.2231	0.2503	0.2254	0.1666	0.148
3 (0011)	0.7506	0.7541	0.3949	0.4051	0.2314	0.2092
4 (0100)	0.6993	0.7051	0.3883	0.3655	0.2299	0.246
5 (0101)	0.7704	0.7714	0.4146	0.4209	0.2257	0.223
6 (0110)	0.4955	0.4742	0.3846	0.3167	0.2014	0.2446
7 (0111)	0.7739	0.7625	0.4126	0.3972	0.233	0.2273
8 (1000)	0.5718	0.505	0.2728	0.3013	0.2029	0.2221
9 (1001)	0.7597	0.7736	0.4123	0.4118	0.2179	0.218
10 (1010)	0.467	0.4342	0.3003	0.2777	0.2026	0.2382
11 (1011)	0.7508	0.7637	0.4087	0.396	0.2187	0.2433
12 (1100)	0.7145	0.6999	0.3802	0.3868	0.2232	0.2469
13 (1101)	0.7597	0.7637	**0.4225**	**0.4277**	**0.2333**	0.2242
14 (1110)	0.5831	0.5677	0.3883	0.3577	0.2111	**0.2543**
15 (1111)	0.766	0.762	0.4161	0.4145	0.2302	0.2374

5.1   Extracting Potential Categories

Different cue patterns crawl different taxonomy nodes, and different nodes can build different directed graphs. In a directed graph, each edge, $u \rightarrow v$, has a weight, which is a frequency of (u, v). Now, we analyze what weighting scheme in which combination can get a best score. If a link is weak, i.e., the weight of an edge is less than a predefined threshold, it will be removed from the directed graph. The three ranking algorithms, PageRank, HITS and the Markov chain random process (SD), are employed to extract 10 critical nodes from the directed graph.Figure 3 shows the recip rank scores of three evolving social networks with different thresholds. The numbers above the x-axis denote the combinations. Figures 3(a) and 3(c) show that the cue patterns, PN and Org, are most helpful, i.e., 5 (0101). We also find that the curve of the SD is similar to the PageRank. If a node does not have any in-degrees, the ranking score of its HITS is null. But those nodes are important. When we increase the thresholds, the curve of the

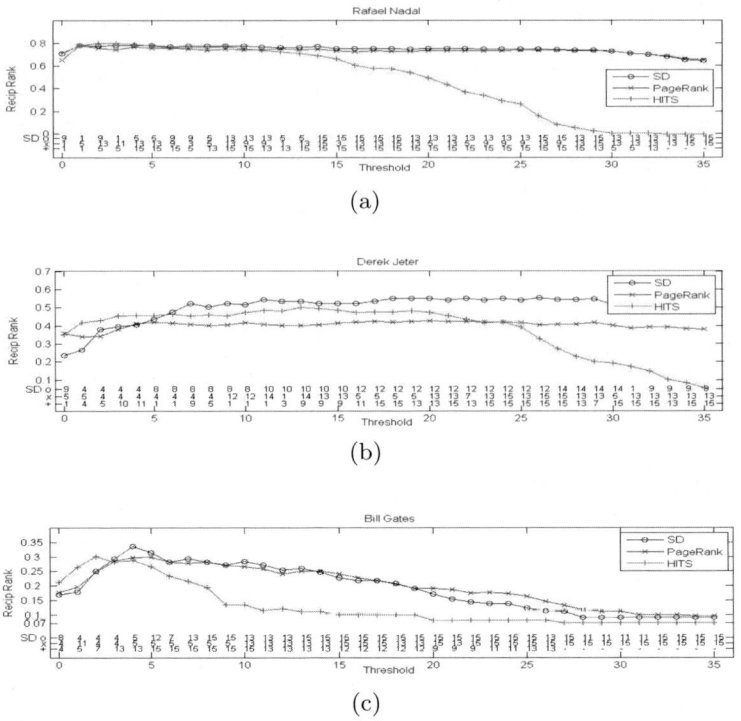

Fig. 3. Recip Ranks of HIT, PageRank, and Markov Chain Random Process on (a) Rafael Nadal, (b) Derek Jeter and (c) Bill Gates

Table 3. SD Compare with Baseline Model, HITS, and PageRank

Recip scores	Roger Federer*	Rafael Nadal*	Derek Jeter$^\times$	Ichiro Suzuki$^\times$	Bill Gates$^+$	Sergey Brin$^+$
Baseline	0.7882	0.7751	0.4225	0.4277	0.2333	**0.2543**
HITS	0.8107	**0.8002**	0.4953	0.4967	0.3025	0.2521
PageRank	0.7809	0.7792	0.4248	0.4673	0.2989	0.2445
SD	**0.8110**	0.7826	**0.5501**	**0.5621**	**0.3361**	0.2519

HITS will decrease more quickly than the PageRank and SD. It means that the PageRank and the SD are more robust than the HITS.

Most of cue patterns are NPs. Figure 3(b) shows the helpful type is NPs, i.e., 8 (1000). They can determine the correct critical nodes. Thus, when we increase the threshold, the reciprocal rank score decreases very slowly. Many absorbing nodes, i.e., those stop nodes without out-degree, are not important. The SD is different from the PageRank in that the SD filters out absorbing nodes (refer to Formula 4). Table 3 summarizes the experimental results. The SD is better than the PageRank in all cases, and outperform the HITS as well as baseline model in all cases except Nadal and Brin, respectively.

5.2   Labeling Relationships

Table 4 shows some results by using the Markov chain random process. The extracted categories are listed in the last column. The relationships are extracted by Formula 2 and are listed in the 4th column. In the 1st case, the named entity pair is two hottest tennis players "Roger Federer vs. Rafael Nadal." The relationship "*the hottest feud*" is quite good. In the 2nd case, Bill Gates and his wife Melinda Gates, their relationships are "*Gates Foundation*", "*The enormous endowment*", "*a charitable organization*" and "*Harvard University*". In the 3rd case, Micro is not a named entity. It is marked as an NE-Error by assessors. Our process cannot detect the NE error, and it still extracts their potential categories and relationships. The "Cruzer" is a storage facility. In the last case, "Sergey Brin vs. Larry Page", they all are the Google CEO. Using "*search engine*" or "*young men*" to describe them is reasonable. In summary, the cue patterns determine the correct categories, and provide important information for correct relationships finding.

Table 4. Four Examples for Potential Categories and Relationships Labeling

	Seed	Name Pair	Extracted Relationship	Potential Categories
1	Roger Federer	Roger Federer vs. Rafael Nadal	1. a worthy successor 2. **the tennis court.** 3. **the hottest feud**	Tennis Male Ballsport Players Sport
2	Bill Gates	Bill Gates vs. Melinda Gates	1. **Gates Foundation** 2. **The enormous endowment** 3. archival articles 4. **a charitable organization,** 5. **Harvard University**	Regional Massachusetts Pioneers
3	Bill Gates	Micro vs. Cruzer	1. prices, read reviews 2. Compare products, 3. comparison shop 4. the best place 5. **store ratings**	Hardware Business_and_Economy
4	Sergy Brin	Sergey Brin vs. Larry Page	1. **search engine,** 2. **young men,** 3. Steve Jobs, 4. Fujio Cho, 5. the first time	Reference Catholicism Denominations Ranking File_Sharing

6   Concluding Remarks

This paper introduces an evolving social network construction method. It not only sets up a network of named entities incrementally, but also labels the relationships between named entities. Two traditional link analysis algorithms, i.e., PageRank and HITS, and a probability-based framework, Markov chain random process, are employed to extract the potential categories. Markov chain random process performs the best using the reciprocal rank metric. Name disambiguation can reduce the errors in an evolving social network. In current experiments, we only consider the co-occurrence score of the parent-child pairs. Merging all the siblings will be investigated. In addition, we will study learning methods to decide important snippets and extract critical noun phrases.

Acknowledgments. Research of this paper was partially supported by Excellent Research Projects of National Taiwan University, under the contract 96R0062-AE00-02.

References

1. Aleman-Meza, B., Nagarajan, M., Ramakrishnan, C., Ding, L., Kolari, P., Sheth, A., Arpinar, I.B., Joshi, A., Finin, T.: Semantic Analytics on Social Networks: Experiences in Addressing the Problem of Conflict of Interest Detection. In: Proc. WWW Conference, pp. 407–416 (2006)
2. Cinlar, E.: Introduction to Stochastic Processes. Prentice Hall, Englewood Cliffs (1975)
3. Chen, H.H., Lin, M.S., Wei, Y.C.: Novel Association Measures Using Web Search with Double Checking. In: Proc. COLING-ACL Conference, pp. 1009–1016 (2006)
4. Ferragina, P., Gulli, A.: A Personalized Search Engine Based on Web Snippet Hierarchical Clustering. In: Proc. WWW Conference, pp. 801–810 (2005)
5. Golub, G.H., Greif, C.: Arnoldi-type Algorithms for Computing Stationary Distribution Vectors, with Application to PageRank. Technical Report, SCCM-04-15, Stanford University Technical Report (2004)
6. Kleinberg, J.M.: Authoritative Sources in a Hyperlinked Environment. Journal of the ACM 46(5), 604–632 (1999)
7. Lin, M.S., Chen, H.H.: Constructing a Named Entity Ontology from Web Corpora. In: Proc. LREC Conference, pp. 1450–1453 (2006)
8. Matsuo, Y., Tomobe, H., Hasida, K., Ishizuka, M.: Finding Social Network for Trust Calculation. In: Proc. ECAI Conference, pp. 510–514 (2004)
9. Maguitman, A.G., Menczer, F., Roinestad, H., Vespignani, A.: Algorithmic Detection of Semantic Similarity. In: Proc. WWW Conference, pp. 107–116 (2005)
10. Matsuo, Y., Junichiro, M., Masahiro, H.: Polyphonet: An Advanved Social Network Extraction System from the Web. In: Proc. WWW Conference, pp. 397–406 (2006)
11. Mori, J., Ishizuka, M., Matsuo, Y.: Extracting Keyphrases to Represent Relations in Social Networks from Web. In: Proc. IJCAI Conference, pp. 2820–2825 (2007)
12. Page, L., Brin, S., Motwani, R., Winograd, T.: The PageRank Citation Ranking: Bringing Order to the Web. Stanford Digital Libraries Working Paper (1998)

Automatic Construction of an Opinion-Term Vocabulary for Ad Hoc Retrieval

Giambattista Amati[1], Edgardo Ambrosi[2], Marco Bianchi[2],
Carlo Gaibisso[2], and Giorgio Gambosi[3]

[1] Fondazione Ugo Bordoni, Rome, Italy
gba@fub.it
[2] IASI "Antonio Ruberti" - CNR, Rome, Italy
{firstname.lastname}@iasi.cnr.it
[3] University of Tor Vergata, Rome, Italy
gambosi@mat.uniroma2.it

Abstract. We present a method to automatically generate a term-opinion lexicon. We also weight these lexicon terms and use them at real time to boost the ranking with opinionated-content documents. We define very simple models both for opinion term extraction and document ranking. Both the lexicon model and retrieval model are assessed. To evaluate the quality of the lexicon we compare performance with a well-established manually generated opinion-term dictionary. We evaluate the effectiveness of the term-opinion lexicon using the opinion task evaluation data of the TREC 2007 blog track.

1 Introduction

This work shows how to construct a subjective-word lexicon augmented by term-weights for real-time opinion retrieval. More generally, we address the problem of retrieving documents that contain opinions on a specific topic. Documents, like many posts of web logs (blogs), may contain authors' opinions on a specific subject, and user's information task may consist in retrieving different opinions, like reviews on films, products, books, or more simply people's opinions on public personalities.

The automatic construction of a sentiment and subjective lexicon, and how it can be used for topical opinion are very challenging problems. For example, several machine learning techniques (Naive Bayes, maximum entropy classification, and Support Vector Machines) have been shown to not perform as well on sentiment classification as on traditional topic-based categorization [15]. One of the difficulty of subjective analysis is that sentiment and subjective words distribute quite randomly or more uniformly in the set of relevant documents, while for retrieval or classification models the discriminating words instead occur non-randomly.

To assess the effectiveness of our automatic method we used a dictionary made up of 8221 words built by Riloff, Wiebe and Wilson [16,19]. The words of Riloff et al. dictionary are "clue" words for detecting topical opinions or subjective content, and were collected either manually from different resources or automatically using both annotated and unannotated data. Other opinion term lexicons were created by Mishne [12], and by Esuli and Sebastiani (SentiWordNet) [8]. In particular, Mishne extracted terms

C. Macdonald et al. (Eds.): ECIR 2008, LNCS 4956, pp. 89–100, 2008.
© Springer-Verlag Berlin Heidelberg 2008

from positive training data using information gain, removing terms appearing also in negative training data, and selecting manually a set of opinion terms.

In this work we show how to generate a sequence of dictionaries

$$OpinV = OpinV_1 \supset OpinV_2 \supset \ldots \supset OpinV_k \supset \ldots$$

that can be used to topical opinion retrieval. The surprising outcome of this work is that we are able to automatically select and weight the terms of a very small subset $OpinV_k$ (made up of about 50 words) of the entire dictionary $OpinV$ (made up of about 12250). This small list of subjective terms performs as good as the entire dictionary in terms of topical opinion retrieval. As a consequence, we are able to perform at real time topical-opinion retrieval with a negligible loss in performance. The reason why we obtain such a high performance is not only due to the technique that singles out the right set of subjective words, but is mainly due to the assigned weights of these subjective words.

This work is developed according to the three following steps:

1. We first define a learning algorithm based on a query expansion technique that selects a set of subjective-term candidates [3]. The selection is based on measuring the divergence of term frequencies in the set of opinionated and relevant documents and in the set of relevant-only documents. High divergence witnesses a potential subjective-term.
2. Then, we assume that the best subjective terms *minimize* the divergence of the within-document term-frequency with respect to the average term-frequency in the set of opinionated and relevant documents. In other words best subjective words spreads over the opinionated and relevant documents more uniformly than the informative words do.
3. We finally introduce a fusion methodology free from any parameter, that combines the content-only ranking with the opinion-only ranking.

Although we use a bag of words approach, we show that topic opinion retrieval performance is very high. Since retrieval, query expansion and ranking merging are obtained by parameter-free functions, our methodology is thus very effective, easy and efficient to implement.

2 Related Work

Topical opinion processing usually is conducted in three steps: extraction of opinion expressions from text (in general seen as a classification problem), document assessment by an opinionated score, and document ranking by opinionated content.

Hatzivassiloglou and McKeown propose data constraints on the semantic orientations of conjoined adjectives, to automatically construct a log-linear regression model predicting whether two conjoined adjectives are of same or different orientation. They further improve the classification of adjectives as positive or negative by defining a graph with orientation links [10]. Agreement on the orientation between adjectives is used as a link, and since positive adjectives tend to be used more frequently than negative ones, one of the two classes that has higher average frequency is classified as having positive semantic orientation.

Using Hatzivassiloglou and McKeown's semantic orientation of adjectives, Turney presents a simple semantic orientation method for phrases based on Mutual Information [6,9] of phrases with adjectives and verbs [18]. A document is classified as recommended if the average semantic orientation of its phrases is positive.

Classification on a whole collection is usually computationally expensive (e.g. Hatzivassiloglou and McKeown's method is NP-complete). A way to reduce the computational cost is to extract information by a topic-driven methodology similar to the query expansion process. For example, Skomowroski and Vechtomova [17] exploit the first-pass retrieval to extract a sample of topic relevant documents, from which co-occurence statistics about adjectives are more efficiently extracted. Nouns are counted when they are in the scope of an adjective, that is adjectives act like modalities or concepts. All subjective adjectives are ranked according to the standard normal score (Z-score) between expected and observed co-occurrences of adjectives with query-terms in the top R retrieved documents. Then, document ranking aggregates different scores, one of them being the opinionated probability of the query terms.

Skomowroski and Vechtomova's work has similar approach to the Local Context Analysis of Xu and Croft [20] who expand the original query taking into account text passages that contain both query-terms (concepts) and expanded-terms. Also Zhang and Yu [21] expand the original query with a list of concepts and a list of expanded words. A classifier for sentences based on Support Vector Machines is trained with some external resources, and then is applied to the set of returned documents. The final document ranking is obtained by removing the documents that do not contain opinions.

There is another approach based on language model that starts with a collection of ternary queries (sentiment, topical, polarity) and collects the statistics in the collection at the sentence level. Their estimate relies on a collection of paired observations, which represent statements for which they know which words are topic and sentiment words. To predict the sentiment value of a new sentence the two word frequencies (in sentence and in collection) are combined by cross-entropy [7].

A central problem for topical opinion document ranking is how to combine ad hoc retrieval scores with additional information on training data, in order to boost the ranks by opinion scores. A simple way to merge scores from different sources of evidence is the use of standard normal scores that has been shown to be very effective in some information tasks [4,17]. Our approach is parameter-free: first we obtain the document ranking by content, then we re-rank the documents taking into account the opinion score and the content rank as combining function.

3   Statistical Analysis of Subjective Terms

The logical representation of languages include three principal constituents: constants c, concepts C and relations R, that roughly, correspond to nouns, adjectives and verbs respectively. A context can be logically represented by $R(C_1(c_1), \ldots, C_k(c_k))$, that is a context is represented by relations among concepts expressed on constants.

However, Information Retrieval has a flat view of objects: the essence of words is their appearance and substance is quantified by probability of occurrence or by means of information theoretic notions like that of information content. It is a matter of fact that nouns provide the highest information content, while adjectives and verbs provide additional information to the context, but bringing less information content.

Our primary goal is to verify some hypotheses on subjective but non-informative terms only by means of information theoretic analysis of term-types and without a direct exploitation of term association and co-occurrence. This simplification will guarantee a faster implementation of opinion analysis. We process terms as for query expansion: we pool all relevant and opinionated documents with respect to all 50 topics of the blog track of TREC 2006, and use the set R of all relevant documents as population and the subset $O \subset R$ of opinionated documents as biased sample. Each term will have four frequencies:

- a relative frequency $\mathbf{p}_c$ in the set $\mathbf{D}$ of all documents;
- a relative frequency $\mathbf{p}_r$ in the set R of relevant documents;
- a relative frequency $\mathbf{p}_o$ in the set O of relevant and opinionated documents;
- a relative frequency $\mathbf{p_d}$ in the document $\mathbf{d}$.

A dictionary containing weighted terms is automatically generated on the basis of the following considerations:

- Since nouns describe better the content of documents, they possess the highest information content:

$$\text{Inf}(\mathbf{t}) = -\log_2 \text{Prob}(\mathbf{p_d}|\mathbf{p}_c)$$

The inverse of probability is used to provide the information content of a term in a document $\mathbf{d}$. The main property of Inf is that if $\mathbf{p_d} \sim \mathbf{p}_c$ then the document is a sample of the collection for the term $\mathbf{t}$ and thus it does not bring information, i.e. $\text{Inf}(t) \sim 0$. $\text{Inf}(\mathbf{t})$ will be used to provide the content score of a term in a document.
- Opinionated terms do not carry information content ($\text{Inf}(\mathbf{t})$ is low). However, they tend to appear more frequently in the opinionated set, O, rather than in the relevant one, R. Therefore, we maximize the opinionated entropy function, $OE(\mathbf{t})$:

$$OE(\mathbf{t}) = -\log_2 \text{Prob}(\mathbf{p}_o|\mathbf{p}_r)$$

to extract possible opinionated terms. On the other hand, information content terms tend to have a similar frequency in both relevant set R and opinionated set O, that is the function $OE(\mathbf{t})$ is minimized for information content terms.
- When nouns are in the scope of adjectives[1], adjectives possibly specify the polarity of opinions. Since verbs link nouns, verbs possibly testify presence of opinions. Concepts[2], adjectives, verbs and adverbs distribute more randomly in the set of opinionated documents. In other words, a high value $OE(\mathbf{t})$ can be due to peaks of frequencies in a restricted number of opinionated documents. The function $OE(\mathbf{t})$

---

[1] Skomowroski and Vechtomova [17] report that in English a noun follows an adjective the 57% of cases.

[2] According to Heiddeger (1957; Identity and Difference) things are either practical objects or abstracted from their context and reified as "objects" of our knowledge representation (concepts). Essence of objects, that is the permanent property of things, is the "substance" (understanding), that is the meaning. Nouns mainly represent such objects in our language.

is not robust since it does not consider if the maximization of $OE(\mathbf{t})$ is obtained with a more uniform distribution or not. To filter out noisy terms we use a second information theoretic function (average opinionated entropy, AOE(t)) which is the average divergence of document frequency from the expected frequency $\mathbf{p}_o$ in the set of opinionated documents:

$$AOE(\mathbf{t}) = -\frac{1}{|O|} \sum_{\mathbf{d} \in O} \log_2 \mathrm{Prob}(\mathbf{p_d}|\mathbf{p}_o)$$

We will use a very simple approximation of $AOE(\mathbf{t})$ that has not additional cost with respect to the computation of $OE(\mathbf{t})$. The approximation will act as a boolean condition for selecting terms with highest opinion entropy scores $OE(\mathbf{t})$.

The automatically generated dictionary will be further used at retrieval time to re-rank the set of retrieved documents by opinionated scores.

3.1 Distribution of Opinionated Terms in the Set of Opinionated Documents with Respect to Relevant Documents

We have assumed that those terms that occur more often in the set of opinionated documents rather than in the set of relevant documents are possible candidates to bring opinions. To give plausible scores to opinion-bearing terms, we compute an approximation of the opinion entropy $OE(\mathbf{t})$ by means of the asymmetric Kullback-Leibler (KL) divergence computed for all terms in the opinionated set O with respect to the set R of relevant documents, that is

$$OE(\mathbf{t}) = -\log_2 \mathrm{Prob}(\mathbf{p}_o|\mathbf{p}_r) \sim KL(\mathbf{p}_o||\mathbf{p}_r)$$

being $\mathbf{p}_o > \mathbf{p}_r$. We might have used the binomial distribution, or the geometric distribution instead of KL[3] to compute $\mathrm{Prob}(\mathbf{p}_o|\mathbf{p}_r)$, but for the sake of simplicity we prefer to support our arguments with the more intuitive KL measure.

We also anticipated that noise may be caused by some informative terms that appear more densely in a few set of opinionated documents, but the observation of a skewed frequency is mainly due to a more frequent occurrence in the set of documents that are relevant to a given topic. The asymmetric KL divergence is therefore a reliable measure when term-frequency is more randomly or uniformly distributed across all opinionated documents. The noise reduction problem is studied in the following section.

3.2 Distribution of Opinionated Terms in the Set of Opinionated Documents

We want to reduce the noise in opinion-term-selection, that is we want now to filter out those terms that show a distribution of their frequency that is skewed in a few number

---

[3] KL is an approximation of $\frac{-\log_2 \mathrm{Prob}(\mathbf{p}_o|\mathbf{p}_r)}{\mathrm{TotalFreq}(O)}$ where Prob is the binomial distribution. When weighting terms, the size $\frac{1}{\mathrm{TotalFreq}(O)}$ is a factor common to all words so we may assume that $-\log_2 \mathrm{Prob}(\mathbf{p}_o|\mathbf{p}_r) \sim KL(\mathbf{p}_o||\mathbf{p}_r)$ up to a proportional factor and a small error.

of opinionated documents. A skewed distribution is due to the type of our training data. The opinionated documents are also relevant with respect to a small set of topics (50 queries), and thus it may happen that informative terms might appear more frequently in opinionated documents because a topic may have all relevant documents that are also opinionated, that is when $O(\mathbf{q}) \sim R(\mathbf{q})$: such a situation is not an exception in the blogosphere. In such a case the $OE(\mathbf{t})$ of some non-opinionated terms may be large when compared with the set of all opinionated documents pooled from the set of all topics. We now show how to make a first noise reduction for such cases.

Let $\mathbf{p}_o = \frac{\mathbf{TF}_O}{\mathbf{TotalFreq(O)}}$ be the relative frequency of a term $\mathbf{t}$ in the set of opinionated documents, and $\mathbf{p_d} = \frac{\mathbf{tf}}{\mathbf{l(d)}}$ the relative frequency of the term in the document $\mathbf{d}$. Since the set of opinionated documents O is a large sample of the collection we may set $\mathbf{TotalFreq(O)} = |O| \cdot \bar{\mathbf{l}}$, where $\bar{\mathbf{l}}$ is the average document length and $|O|$ is the number of opinionated documents. The asymmetric KL divergence of the frequency of the term in the opinionated set of document with respect to the prior probability $\mathbf{p}_o = \frac{\mathbf{TF}_O}{|O| \cdot \bar{\mathbf{l}}}$ is:

$$AOE(\mathbf{t}) = \frac{1}{|O|} \sum_{\mathbf{d} \in O} KL(\mathbf{p_d} || \mathbf{p}_o) = \frac{1}{|O|} \sum_{\mathbf{d} \in O} \mathbf{p_d} \cdot \log \frac{\mathbf{p_d}}{\mathbf{p}_o}$$

We have assumed that opinionated terms do not carry information content, and this assumption translates into the assumption that opinion-bearing terms distribute more uniformly in the set of opinionated documents, that is when $\mathbf{p_d} \sim \mathbf{p}_o$, or more generally, when the KL divergence is minimized. If the term distributes uniformly $\mathbf{p_d}$ can be approximated by $\frac{\mathbf{TF}_O}{\mathbf{n_t} \cdot \bar{\mathbf{l}}}$, and we need to *minimize* the function:

$$AOE(\mathbf{t}) \propto - \sum_{\mathbf{d} \in O} \frac{\mathbf{TF}_O}{\mathbf{n_t}} \log \mathbf{n_t} = -\mathbf{n_t} \cdot \frac{\mathbf{TF}_O}{\mathbf{n_t}} \log \mathbf{n_t} = -\mathbf{TF}_O \cdot \log \mathbf{n_t}$$

Since we have to minimize $AOE(\mathbf{t})$ and the approximating expression is negative, and since we may suppose that all terms have a frequency $\mathbf{TF}_O$ of a similar order of magnitude in the set of opinionated documents, we may instead *maximize* the function

$$\log_2 \mathbf{n_t} \propto \mathbf{n_t}$$

where $\mathbf{n_t}$ is the set of opinionated documents containing the term $\mathbf{t}$. We define a term of level k if it appears in at least k relevant and opinionated documents[2]. Therefore the higher the number of documents containing a term, the higher is the probability that the term is opinionated. The larger k is chosen, the less the number of terms that are selected. Therefore, we need to find an optimal level k that generates a vocabulary as small as possible to reduce the computational cost, and in the meantime to be as effective as possible in terms of retrieval performance. The efficiency/effectiveness problems of the automatic generation of an opinionated vocabulary is studied in the following sections.

3.3 Opinion-Term Vocabulary

In summary the information theoretic methodology consists of three steps:

1. Terms with the highest divergence $OE(\mathbf{t})$ between the frequency in the set of opinionated-relevant documents and the frequency in the set of all relevant-only

Table 1. The number of words of the dictionary SCD after the application of the weak Porter stemming is 6,352. The precision and the recall of the automatically generated dictionary $OpinV_k$ are measured with respect to a semi-manually generated dictionary SCD.

Level	$OpinV_k \cap SCD$	$OpinV_k$	Prec.	Rec.	F-Measure
1	2,325	12,263	0.1896	0.3660	0.2498
100	1,528	4,022	0.3800	0.2406	0.2946
250	994	2,504	0.3970	0.1565	0.2245
500	642	1,625	0.3951	0.1011	0.1610
750	466	1,209	0.3854	0.0734	0.1233
1,000	349	927	0.3765	0.0734	0.1228
3,000	77	219	0.3516	0.0121	0.0234
4,000	47	128	0.3672	0.0074	0.0145
6,000	16	42	0.3809	0.0025	0.0050
8,000	5	12	0.4167	0.0008	0.0016

documents are selected, and then weighted by the same opinion-entropy score $OE(t)$. This step generates a list $CandV$ of weighted opinion-term candidates.

2. Terms of $CandV$ are then filtered. All terms of $CandV$ with the lowest average divergence $AOE(t)$ (average divergence between term-frequency in O and the term-frequency within each opinionated-relevant document $\mathbf{d} \in O$), are selected from the list of all terms with positive $OE(t)$ scores. We simply use the minimal number k of opinionated-relevant documents containing the term as fast and effective implementation of the $AOE(t)$ scores.
3. A sequence of weighted dictionaries $OpinV$ is obtained at different level of k. The optimal level is obtained when the performance is maintained stable while the dictionary size is kept small enough to be used at real-time retrieval.

The $OpinV_k$ vocabulary is submitted to the system as a standard query and each document obtains an opinionated score. At this aim, in our experiments, we assess the precision of the obtained lexicon and its performance in opinion task retrieval, using a parameter free model of IR (DPH is a variant of the model by Amati [2]) for first pass retrieval and a parameter-free model for query expansion [3]. Using this parameter-free setting for the experiments, we can only concentrate on the methodology to assess the potentiality of the proposed approach. However, other models can be used to enhance initial ranking, because better initial rankings generates better topical opinion rankings.

4   A Computationally Lightweight Algorithm for Topical Opinion Retrieval

The opinionated and relevant document ranking is obtained in three steps:

1. We use the parameter free model DPH as retrieval function to provide the content score of the documents $content\_score(\mathbf{d}||\mathbf{q}) = score_{DPH}(\mathbf{d}||\mathbf{q})$. We obtain a content rank for all documents: $content\_rank(\mathbf{d}||\mathbf{q})$.

Table 2. The list of terms of $OpinV_{6000}$. The table also presents terms of $OpinV_{6000} \cap SCD$ (italicized terms), terms of $OpinV_{8000}$ (underlined terms) and $OpinV_{8000} \cap SCD$ (italicized and underlined terms). A weak Porter stemmer is applied to terms.

<u>am</u>	0,0893	*just*	0,0672	people	0,1094	*view*	0,0139
<u>archive</u>	0,0368	*know*	0,0514	pm	0,1185	wai	0,0303
back	0,0113	last	0,0161	post	0,0326	*want*	0,0395
call	0,0253	left	0,0104	read	0,0293	*well*	0,0187
<u>can</u>	0,0353	*like*	0,0782	*right*	0,0530	<u>who</u>	0,1261
come	0,0193	<u>link</u>	0,0341	sai	0,1124	*will*	0,0070
comment	0,0056	*look*	0,0157	see	0,0350	work	0,0031
dai	0,0247	mai	0,0023	*show*	0,0229	world	0,0286
don	0,0640	*mean*	0,0110	state	0,0049		
first	0,0057	*need*	0,0101	*think*	0,0748		
help	0,0013	<u>now</u>	0,0289	time	0,0407		

2. We submit the entire dictionary $OpinV$ as a query and weight the set of retrieved documents: $opinion\_score(\mathbf{d}||OpinV) = score_{DPH}(\mathbf{d}||OpinV)$. The opinion-ated score with respect to a topic $\mathbf{q}$ is defined as follows[4]:

$$opinion\_score(\mathbf{d}||\mathbf{q}) = \frac{opinion\_score(\mathbf{d}||OpinV)}{content\_rank(\mathbf{d}||\mathbf{q})}$$

We thus obtain an *opinion rank* for all documents: $opinion\_rank(\mathbf{d}||\mathbf{q})$.

3. We further boost document ranking with the dual function of $opinion\_score(\mathbf{d}||\mathbf{q})$:

$$content\_score^+(\mathbf{d}||\mathbf{q}) = \frac{content\_score(\mathbf{d}||\mathbf{q})}{opinion\_rank(\mathbf{d}||\mathbf{q})}$$

The final topical opinion ranking is obtained re-ranking the documents by $content\_score^+(\mathbf{d}||\mathbf{q})$.

5   Experiments and Results

Our experimentation is based on TREC BLOG track dataset [14]. The blog collection was crawled over a period of 11 weeks (December 2005 - February 2006). The total size of the collection amounts to 148 GB with three main different components: feeds (38.6 GB), permalinks (88.8GB), and homepages (20.8 GB). The collection contains

---

[4] Ranking is a mixture of a normal distribution for relevant documents and an exponential distribution for non-relevant documents [11]. Since the non relevant documents are almost all the documents of the collection for a given query, ranking roughly follows the power law, that is the probability of relevance of a document is inversely proportional to its document rank. Therefore:

$$opinion\_score(\mathbf{d}||\mathbf{q}) \propto opinion\_score(\mathbf{d}||OpinV) \cdot p(\mathbf{d}||\mathbf{q})$$

spam as well as possibly non-blogs and non-English pages. For our experimentation we considered only the permalinks component, consisting of 3.2 millions of Web pages, each one containing a post and the related comments.

We preprocess data with the aim to remove not English documents from the collection. This goal is achieved by a text classifier, implemented using Lingpipe [1], a suite of Java libraries for the linguistic analysis of human language, and trained using the Leipzig Corpora Collection [5].

We obtain our baseline performing a topic relevance retrieval. For the indexing and retrieval tasks we adopt Terrier [13]. As already stated in Section 3.3, in our experimentation we use DPH, a parameter free retrieval model. This choice has two main consequences: at first we can ignore the tuning issue and focus our efforts on the proposed methodology, evaluating the gain obtained with respect to the baseline; on other hand, all results presented in this Section could be boosted adopting (and properly tuning) a parameter dependent retrieval model.

We use the semi-manual subjectivity clues dictionary [16,19], that we denote here by SCD, to study and enhance the performance of the automatically generated dictionaries, $OpinV$ in what follows.

Results are shown in Table 3. The first outcome of our work is very surprising: using a set of only 5 subjective and weighted words, that are filtered at the level with $k = 8,000$, we improve both the MAP with respect to relevance (all relevant documents), from 0.3480 of the baseline to 0.3833 (+10%), and the opinionated relevance MAP (only opinionated and relevant documents) from 0.2740 to 0.3128 (+ 14%). Similarly, relevance precision at 10 retrieved documents improves from 0.6480 to 0.7140, while opinionated relevance improves from 0.4480 (0.3031 is the median run of blog track) to 0.5180. It is quite a surprising that a small number of query independent words can improve so largely the quality of ad hoc retrieval. Thus, we may boost both relevance and topical opinion retrieval at real-time with a negligible computational cost.

The best performance values of relevance MAP (0.3938) is obtained with 16 unweighted subjective words (+18% over the median run of TREC 2007 blog track), relevance Precision at 10 (0.7240, +12%) with 349 weighted words, opinionated relevance MAP (0.3213, +33%) with 77 unweighted subjective words, opinionated relevance Precision at 10 (0.5420 , +81%) with 1,528 weighted words. The whole semi manual dictionary SCD containing more than 6,000 of subjective words does not perform as good as its smaller subset $SCD \cap OpinV_k$ for any level k. This support the idea that it is not the exhaustivity of the dictionary but the subjectivity strength of the words that improves both relevance and topical opinion. More specifically, modalities, conditional sentences or verbs that express possibilities (as the words *can*, *may*) or that relates directly the content to its author (as the words *(I) am*, *like*, *think*, *want*, *agree*) are better predictors of opinions than subjective adjectives. Modal words tend to appear very often in the blogosphere and they alone are almost sufficient to achieve best performance in topical opinion retrieval.

It is worth to note that the $OpinV_k$ dictionary still contains noisy words due to the fact that we have not used linguistic or lexical tools. As a consequence we did not remove geographical adjectives (e.g. "American") and other words produced by spam or by blog dependent text in the permalinks (e.g. "post" or "comment"). On the

Table 3. Performance of relevance and topical opinion retrieval by using the semi-manual dictionary SCD and the fully automatic $Opin\mathbf{V}_k$. Test data are from the set 50 queries of the new blog track of TREC 2007. Training data are from the blog track 2006.

	Relevance		Opinion	
Level k	MAP	P@10	MAP	P@10
Baseline	0.3480	0.6480	0.2704	0.4440
Median run of the Blog track 2007	0.3340	0.6480	0.2416	0.3031
SCD	0.3789	0.7000	0.3046	0.5280

$Opin\mathbf{V}_k \cap SCD$, $Opin\mathbf{V}_k$ weighted					$Opin\mathbf{V}_k \cap SCD$, $Opin\mathbf{V}_k$ not weighted				
	Relevance		Opinion			Relevance		Opinion	
Level k	MAP	P@10	MAP	P@10	Level k	MAP	P@10	MAP	P@10
1	0.3862	0.7160	0.3173	0.5420	1	0.3801	0.7040	0.3113	0.5340
100	0.3862	0.7160	0.3172	**0.5420**	100	0.3807	0.7100	0.3118	0.5380
250	0.3864	0.7160	0.3171	0.5380	250	0.3817	0.7100	0.3126	0.5380
500	0.3867	0.7160	0.3172	0.5380	500	0.3825	0.7100	0.3125	0.5340
750	0.3865	0.7160	0.3168	0.5320	750	0.3821	0.7000	0.3110	0.5340
1000	0.3871	**0.7240**	0.3167	0.5380	1000	0.3836	0.7040	0.3107	0.5340
3000	0.3910	0.7140	**0.3213**	0.5320	3000	0.3889	0.7120	0.3135	0.5280
4000	0.3909	0.7180	0.3193	0.5300	4000	0.3913	0.7120	0.3144	0.5180
6000	0.3911	0.7140	0.3204	0.5160	6000	**0.3938**	0.7200	0.3123	0.5160
8000	0.3833	0.7140	0.3128	0.5180	8000	0.3874	0.7120	0.3060	0.4960

Full weighted $Opin\mathbf{V}_k$					$Opin\mathbf{V}_k \cup SCD$, $Opin\mathbf{V}_k$ weighted				
	Relevance		Opinion			Relevance		Opinion	
Level k	MAP	P@10	MAP	P@10	Level k	MAP	P@10	MAP	P@10
1	0.3846	0.7000	0.3080	0.5260	1	0.3856	0.7100	0.3168	0.5400
100	0.3848	0.7000	0.3082	0.5260	100	0.3856	0.7100	0.3168	0.5400
250	0.3851	0.7000	0.3084	0.5260	250	0.3856	0.7100	0.3168	0.5400
500	0.3853	0.7020	0.3083	0.5260	500	0.3857	0.7100	0.3170	0.5380
750	0.3856	0.6980	0.3086	0.5220	750	0.3860	0.7080	0.3172	0.5360
1000	0.3862	0.7020	0.3103	0.5220	1000	0.3857	0.7100	0.3165	0.5360
3000	0.3885	0.7040	0.3109	0.5120	3000	0.3902	0.7140	0.3202	0.5300
4000	0.3879	0.7060	0.3090	0.5080	4000	0.3903	0.7180	0.3211	0.5380
6000	0.3869	0.7120	0.3103	0.5100	6000	0.3899	0.7140	0.3205	0.5360
8000	0.3863	0.7060	0.3087	0.5140	8000	0.3871	0.7160	0.3166	0.5380

other hand, removing words is a challenging task, since $Opin\mathbf{V}_k$ contains words that are exclamations, slang or vulgar words that express emotions or opinions but that do not belong to a clean dictionary like SCD. Furthermore some words are missing (e.g. "good" or "better") because the collection has been indexed using the default stopword list provided by the Terrier framework.

6   Conclusions

We have automatically generated a dictionary of subjective words and we have introduced a method to weight the words of the dictionary through information theoretic measures for topical opinion retrieval. In contrast to term-association or co-occurrence techniques, we have used the training collection as a bag of words. We have first learned all possible subjective words candidates by measuring the divergence of opinionated term-frequencies from only-relevant term-frequencies. Then, we have made the assumption that the best (most discriminating) subjective words are the most frequent ones, and that they distribute non-randomly in the set of opinionated documents. Following this hypothesis, we built a sequence of refined dictionaries, each of them shows to keep almost unaltered the performance for both retrieval tasks (relevance and opinionated relevance), up to the limit point of using a very small number of words of the dictionary. Our opinionated relevance ranking formula is also very robust and does not need any parameter tuning or learning from relevance data. Because of the small size of these dictionaries, we may boost opinionated and relevant documents at real-time with a negligible computational cost. Further refinements of the dictionary are possible, for example using lexical or other external resources. Also minimization of the average divergence $AOE(t)$, that filters out good subjective words, can be computed more accurately than the first approximation we have used for these experiments.

References

1. Alias-i. Lingpipe named entity tagger, http://www.alias-i.com/lingpipe/
2. Amati, G.: Frequentist and Bayesian approach to Information Retrieval. In: Lalmas, M., MacFarlane, A., Rüger, S.M., Tombros, A., Tsikrika, T., Yavlinsky, A. (eds.) ECIR 2006. LNCS, vol. 3936, pp. 13–24. Springer, Heidelberg (2006)
3. Amati, G., Carpineto, C., Romano, G.: Query difficulty, robustness, and selective application of query expansion. In: McDonald, S., Tait, J.I. (eds.) ECIR 2004. LNCS, vol. 2997, pp. 127–137. Springer, Heidelberg (2004)
4. Amati, G., Carpineto, C., Romano, G.: Merging xml indices. In: Fuhr, N., Lalmas, M., Malik, S., Szlávik, Z. (eds.) INEX 2004. LNCS, vol. 3493, pp. 253–260. Springer, Heidelberg (2005)
5. Biemann, C., Heyer, G., Quasthoff, U., Richter, M.: The leipzig corpora collection - monolingual corpora of standard size. In: Proceedings of Corpus Linguistic 2007, Birmingham, UK (2007)
6. Church, K.W., Hanks, P.: Word association norms, mutual information, and lexicography. In: Proceedings of the 27th. Annual Meeting of the Association for Computational Linguistics, pp. 76–83. Association for Computational Linguistics, Vancouver, B.C (1989)
7. Eguchi, K., Lavrenko, V.: Sentiment retrieval using generative models. In: Proceedings of the 2006 Conference on Empirical Methods in Natural Language Processing, July 2006, pp. 345–354. Association for Computational Linguistics, Sydney, Australia (2006)
8. Esuli, A., Sebastiani, F.: SentiWordNet: A publicly available lexical resource for opinion mining. In: Proceedings of LREC-06, the 5th Conference on Language Resources and Evaluation (2006)
9. Fano, R.M.: Transmission of Information: A Statistical Theory of Communications. MIT Press, Cambridge, Wiley, New York (1961)
10. Hatzivassiloglou, V., McKeown, K.: Predicting the semantic orientation of adjectives. In: acl97, pp. 174–181 (1997)
11. Manmatha, R., Rath, T., Feng, F.: Modeling score distributions for combining the outputs of search engines. In: SIGIR 2001: Proceedings of the 24th annual international ACM SIGIR conference on Research and development in information retrieval, pp. 267–275. ACM, New York (2001)

12. Mishne, G.: Multiple ranking strategies for opinion retrieval in blogs. In: The Fifteenth Text REtrieval Conference (TREC 2006) Proceedings (2006)
13. Ounis, I., Amati, G., Plachouras, V., He, B., Macdonald, C., Lioma, C.: Terrier: A High Performance and Scalable Information Retrieval Platform. In: Proceedings of ACM SIGIR'06 Workshop on Open Source Information Retrieval (OSIR 2006) (2006)
14. Ounis, I., de Rijke, M., Macdonald, C., Mishne, G., Soboroff, I.: Overview of the trec-2006 blog track. In: Proceedings of the Text REtrieval Conference (TREC 2006), National Institute of Standards and Technology (2006)
15. Pang, B., Lee, L., Vaithyanathan, S.: Thumbs up?: sentiment classification using machine learning techniques. In: EMNLP 2002: Proceedings of the ACL-02 conference on Empirical methods in natural language processing, pp. 79–86. Association for Computational Linguistics, Morristown, NJ, USA (2002)
16. Riloff, E., Wiebe, J.: Learning extraction patterns for subjective expressions. In: Proceedings of the 2003 conference on Empirical methods in natural language processing, pp. 105–112. Association for Computational Linguistics, Morristown, NJ, USA (2003)
17. Skomorowski, J., Vechtomova, O.: Ad hoc retrieval of documents with topical opinion. In: Amati, G., Carpineto, C., Romano, G. (eds.) ECIR 2007. LNCS, vol. 4425, pp. 405–417. Springer, Heidelberg (2007)
18. Turney, P.: Thumbs up or thumbs down? Semantic orientation applied to unsupervised classification of reviews. In: acl2002, pp. 417–424 (2002)
19. Wilson, T., Wiebe, J., Hoffmann, P.: Recognizing contextual polarity in phrase-level sentiment analysis. In: Proceedings of HLT-EMNLP (2005)
20. Xu, J., Croft, W.B.: Query expansion using local and global document analysis. In: Proceedings of ACM SIGIR, Zurich, Switzerland, August 1996, pp. 4–11 (1996)
21. Zhang, W., Yu, C.: Uic at trec 2006 blog track. In: The Fifteenth Text REtrieval Conference (TREC 2006) Proceedings (2006)

A Comparison of Social Bookmarking
with Traditional Search

Beate Krause[1,2], Andreas Hotho[1], and Gerd Stumme[1,2]

[1] Knowledge & Data Engineering Group, University of Kassel
Wilhelmshöher Allee 73, D-34121 Kassel, Germany
[2] Research Center L3S, Appelstr. 9a, D-30167 Hannover, Germany

Abstract. Social bookmarking systems allow users to store links to internet resources on a web page. As social bookmarking systems are growing in popularity, search algorithms have been developed that transfer the idea of link-based rankings in the Web to a social bookmarking system's data structure. These rankings differ from traditional search engine rankings in that they incorporate the rating of users.

In this study, we compare search in social bookmarking systems with traditional Web search. In the first part, we compare the user activity and behaviour in both kinds of systems, as well as the overlap of the underlying sets of URLs. In the second part, we compare graph-based and vector space rankings for social bookmarking systems with commercial search engine rankings.

Our experiments are performed on data of the social bookmarking system Del.icio.us and on rankings and log data from Google, MSN, and AOL. We will show that part of the difference between the systems is due to different behaviour (e. g., the concatenation of multi-word lexems to single terms in Del.icio.us), and that real-world events may trigger similar behaviour in both kinds of systems. We will also show that a graph-based ranking approach on folksonomies yields results that are closer to the rankings of the commercial search engines than vector space retrieval, and that the correlation is high in particular for the domains that are well covered by the social bookmarking system.

Keywords: social search, folksonomies, search engines, ranking.

1 Introduction

Collaborative tagging systems such as Del.icio.us[1], BibSonomy[2], or Flickr[3] have become popular among internet users in the last years. Taggers actively index and describe Web resources by adding keywords to interesting content and storing them in a so-called *folksonomy* on a shared platform. Over the last years, a significant number of resources has been collected, offering a personalized, community driven way to search and explore the Web.

---

[1] http://del.icio.us/
[2] http://www.bibsonomy.org/
[3] http://flickr.com/

C. Macdonald et al. (Eds.): ECIR 2008, LNCS 4956, pp. 101–113, 2008.

As these systems are growing, the currently implemented navigation by brows-ing tag clouds with subsequent lists of bookmarks that are represented in chrono-logical order may not be the best arrangement for concise information retrieval. Therefore, a first ranking approach based on the graph structure of the under-lying system was proposed in [8].

In this paper, we will compare search in social bookmarking systems with traditional Web search. After a brief presentation of related work (Section 2) and of the used datasets (Google, MSN, AOL and Del.icio.us; Section 3) we will concentrate on an analysis of tagging and traditional search behaviour consider-ing tagging and search interest: Are query terms and tags used similarily (Sec-tion 4.1)? Is tagging and search behaviour correlated over time (Section 4.2)? How strong is the overlap of the content in social bookmarking systems and search engines (Section 4.3)?

In Section 5, we turn to the comparison of the different ranking paradigms. We compare graph-based and vectors space rankings for social bookmarking systems with the rankings of commercial search engines.

2   Related Work

Search engine rankings and folksonomies have been analyzed separately in several studies. Different aspects of search were classified by [4]. In [6], temporal correla-tion based on the Pearson correlation coefficient is used to find similar queries. [1] calculated cross-correlation and dynamic time warping to visualize rises and falls of different terms in blogs, search engine click data and news. In [13], time series data from query logs of the MSN search engine is analyzed. A comparison of traditional search engine rankings using correlation coefficients was carried out by [3].

The vision of folksonomy-based systems and a first analysis of Del.icio.us is presented in [7]. Several studies consider social annotations as a means of im-proving web search. [10] conducted a user study to compare the content of social networks with search engines. [2,14] propose to use data from social bookmark-ing systems to enhance Web search: [2] introduces two algorithms to incorporate social bookmarking information into Web rankings. [14] considers popularity, temporal and sentiment aspects. In [8], two of the authors presented a ranking algorithm for folksonomies, the FolkRank. It adopts the idea of PageRank [11] to the structure of folksonomies. To the best of our knowledge no work examines differences and similarity of user interactions with folksonomy and search engine systems, coverage and rankings as done in this work.

3   Experimental Setup

3.1   Basic Notions

Tags in Folksonomies. The central data structure of a social bookmarking system like Del.icio.us is called *folksonomy*. It can be seen as a lightweight clas-sification structure which is built from *tag* annotations (i. e., freely chosen key-words) added by different users to their resources. A folksonomy consists thus of a set of users, a set of tags, and a set of resources, together with a ternary relation between them.

Query Terms in Search Engines. For comparing tagging and search behaviour, we need similar structures on both sides. In the search engine log data, we will therefore split up each query into the terms that constitute it. *Query terms* are thus all substrings of a query that are separated by blanks.

Items. We will use the term *item* to subsume tags and query terms.

3.2   Data Collection

We consider a MSN log data set and data of the social bookmarking system Del.icio.us to compare the search behaviour with tagging. To compare folksonomy rankings to search engine rankings, we use crawls of commercial search engines (MSN, Google). A log dataset from AOL [12] is further used to find out about overlaps between different systems. Table 1 presents an overview of the datasets' dates, numbers of queries and numbers of different URLs.

Social bookmarking data. In summer 2005 and November 2006 we crawled Del.icio.us to obtain a comprehensive social bookmarking set with tag assignments from the beginning of the system to October 2006. Based on the time stamps of the tag assignments, we are able to produce snapshots. In this paper, we use a snapshot of May 2006 for Section 4 and the entire dataset to compute rankings in Section 5. The first 40,000 tags of the Summer 2005 dataset served as queries in our search engine crawls.

Click data. We obtained a click data set from Microsoft for the period of May 2006. To make it comparable to tags, we decomposed a query into single query terms, removed stop words and normalized them. Sessions which contained more than ten queries with the same query terms in a row were not included into our calculations. A second click data set was obtained from AOL.

Search engine data. Two crawls from MSN and Google are used. While we retrieved 1000 URLs for each query in the MSN dataset, we have 100 URLs for each query in Google.

All query terms and all tags were turned to lowercase.

4   Tagging and Searching

Both search engines and bookmarking systems allow users to interact with the Web. In both systems, the fundamental resources are URLs. In search engines, a user's information need is encoded in a query being composed of one or more terms. In social bookmarking systems, the users themselves assign in a proactive fashion the tags – which later will be used in searches — to the resources.

In this section, we will compare the search and tagging behaviour. Search behaviour is described by the query terms submitted to a search engine. We use the number of occurrences of a term in the queries of a certain period of time as

Table 1. Overview of datasets

Dataset Name	Date	Terms/Tags	Nb. of different URLs
Del.icio.us 2005	until July 2005	456,697	3,158,435
Del.icio.us May only	May. 06	375,041	1,612,405
Del.icio.us complete	until Oct. 06	2,741,198	17,796,405
MSN click data	May 06	2,040,207	14,923,285
MSN crawl	Oct. 06	29,777	19,215,855
AOL click data	March – May 06	1,483,186	19,215,858
Google crawl	Jan. 07	34,220	2,783,734

an indicator for the users' interests. The interests of taggers are described by the tags they assigned to resources during a certain period. We start our exploration with a comparison of the overlap of the set of all query terms in the MSN log data with the set of all tags in Del.icio.us in May 2006. This comparison is followed by an analysis of the correlation of search and tagging behaviour in both systems over time. Query log files were not available for bookmarking systems, hence we study the tagging (and not the search) behaviour only.

The section ends with an analysis of the coverage of URLs considering again the bookmarking system Del.icio.us, and the search engines Google, MSN and AOL. As we do not have access to the indexes of the search engines, we approximate their content by the results of the most prominent queries.

4.1 Query Term and Tag Usage Analysis

By comparing the distribution of tags and query terms we will get some first insights into the usage of both systems. The overlap of the set of query terms with the set of tags is an indicator of the similarity of the usage of both systems. We use the Del.icio.us data from May 2006 to represent social bookmarking systems and the MSN 2006 click data to represent search engines.

Table 2 shows statistics about the usage of query terms in MSN and tags in Del.icio.us. The first row reflects the total number of queried terms, and the total number of used tags in Del.icio.us. The following row shows the number of distinct items in all systems. As can be seen, both the total number of terms and the number of distinct terms is significantly larger in MSN compared to the total number of tags and the number of distinct tags in Del.icio.us. Interestingly, the average frequency of an item is quite similar in all systems (see third row). These numbers indicate that Del.icio.us users focus on fewer topics than search engine users, but that each topic is, in average, equally often addressed.

Figure 1 shows the distribution of items in both systems on a log-log scale. The x-axis denotes the count of items in the data set, the y-axis describes the number of tags that correspond to the term/tag occurrence number. We observe a power law in both distributions.

Power law means in particular that the vast majority of terms appears once or very few times only, while few terms are used frequently. This effect also explains

Table 2. Statistics of Del.icio.us and MSN in May, 2006

	MSN	Del.icio.us	MSN - Del.
items	31,535,050	9,076,899	—
distinct items	2,040,207	375,041	96,988
average	15.46	24.20	—
frequent items	115,966	39,281	18,541
frequent items containing "."	90	1,840	1
frequent items containing "-"	1,643	1,603	145
frequent items cont. "www.", ".com", ".net" or ".org"	17,695	136	30

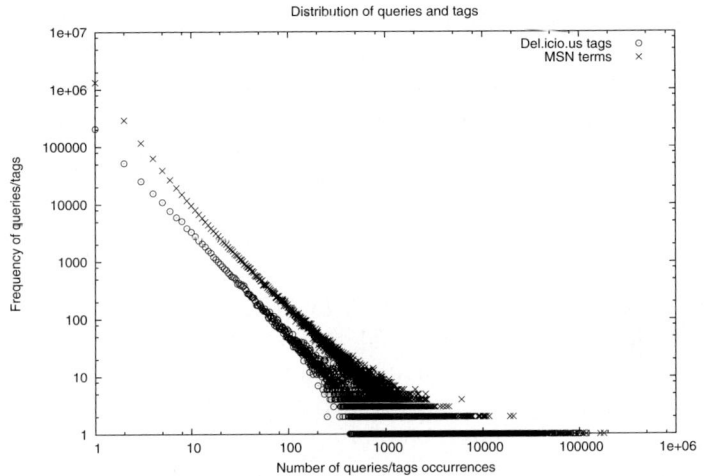

Fig. 1. Item distribution

the relatively small overlap of the MSN query terms with the Del.icio.us terms, which is given in the 2nd row/3rd column of Table 2. In order to analyse the overlap for the more central terms, we restricted both sets to query terms/tags that showed up in the respective system at least ten times.[4] The resulting frequencies are given in the first line of the second part of Table2. It shows that the sizes of the reduced MSN and Del.icio.us datasets become more equal, and that the relative overlap increases.

When browsing both reduced data sets, we observed that the non-overlapping parts result very much from the different usages of both systems. In social bookmarking systems, for instance, people frequently encode multi-word lexems by connecting the words with either underscores, hyphens, dots, or no symbol at all. (For instance, all of the terms 'artificial_intelligence', 'artificial-intelligence', 'artificial.intelligence' and 'artificialintelligence' show up at least ten times in Del.icio.us). This behaviour is reflected by the second and third last rows in

---

[4] The restriction to a minimum of 5 or 20 occurrences provided similar results.

Table 3. Top items in May 2006

Tags Del	Frequency	Query terms MSN	Frequency
design	119,580	yahoo	181,137
blog	102,728	google	166,110
software	100,873	free	118,628
web	97,495	county	118,002
reference	92078	myspace	107,316

Table 2. Underscores are basically used for such multi-word lexemes only, whereas hyphens occur also in expressions like 'e-learning' or 't-shirt'. Only in the latter form they show up in the MSN data.

A large part of the query terms in MSN that are not Del.icio.us tags are URLs or part of URLs, see the last row of Table 2. This indicates that users of social bookmarking systems prefer tags that are closer to natural language, and thus easier to remember, while users of search engines (have to) anticipate the syntactic appearance of what they are looking for.

The top five tags of Del.icio.us and the top five terms of MSN in May 2006 can be seen in Table 3 with their frequencies. One can see that Del.icio.us has a strong bias towards IT related terms. Eleven of the 20 top tags are computer terms (such as web, programming, ajax or linux). The top terms of MSN are more difficult to interpret. "yahoo" and "google" may be used when people have the MSN search interface as a starting point in their internet explorer, or when they leave Microsoft related programs such as hotmail, and want to use another search engine. "county" is often part of a composed query such as "Ashtabula county school employees credit union" or "county state bank". We lack a good explanation for the high frequency of this term. This might result from the way Microsoft extracted the sample (which is unknown to us).

4.2 Correlation of Search and Tagging Behaviour over Time

Up to now, we have considered both data collections as static. Next we analyze if and how search and tagging behaviour are correlated over time. Again we use the MSN query data and the Del.icio.us data of May 2006. Each data set has been separated into 24-hour bins, one for each day of May 2006. As the unit of analysis, we selected those tags from Del.icio.us that also appeared as a query term in the MSN click data. In order to reduce sparse time series, we excluded time series which had fewer than five daily query or tagging events. In total, 1003 items remained.

For each item i, we define two time series. The Del.icio.us time series is given by $X_i^d = (x_{i,1}^d, ..., x_{i,31}^d)$, where $x_{i,t}^d$ is the number of assignments of tag i to some bookmark during day $t \in \{1, ..., 31\}$. For MSN, we define $X_{i,t}^m = (x_{i,1}^m, ..., x_{i,31}^m)$, where $x_{i,t}^m$ is the number of times this term was part of a query on day t according to the MSN data.

To reduce seasonal effects, we normalized the data. We chose an additive model for removal of seasonal variation, i.e., we estimated the seasonal effect for a particular weekday by finding the average of each weekday observation minus the corresponding weekly average and substracted this seasonal component from the original data [5]. The model underlies the assumption that no substantial (i. e., long-term) trend exists which otherwhise would lead to increasing or decreasing averages over time. As our time period is short, we assume that long term trends do not influence averages. We also smoothed the data using simple average sine smoothing [9] with a smoothing window of three days to reduce random variation. Other smoothing techniques delivered similar results.

In order to find out about the similarity of the two time series of an item i, we used the correlation coefficient between the two random variables $x_{i,t}^d$ and $x_{i,t}^m$ which is defined as $r = \frac{\sum_t (X_{i,t}^d - \mu(X_i^d))(X_{i,t}^m - \mu(X_i^m))}{\sigma(X_i^d)\sigma(X_i^m)}$ where $\mu(X_i^d)$ and $\mu(X_i^m)$ are the expected values and $\sigma(X_i^d)$ and are $\sigma(X_i^m)$ the standard deviations.

We applied the t-test for testing significance using the conventional probability criterion of .05. For 307 out of 1003 items, we observed a significant correlation. We take this as indication that tagging and searching behaviour are indeed triggered by similar motivations.

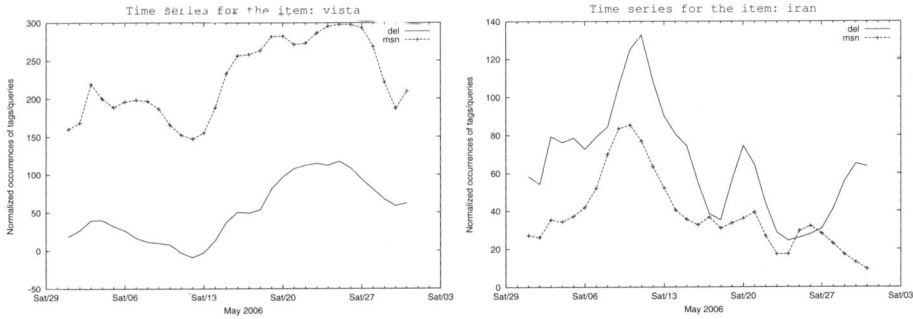

Fig. 2. Time series of highly correlated items

The highest correlation has the item 'schedule' ($r = 0.93$), followed by 'vista' ($r = 0.91$), 'driver', 'player' and 'films'. While both 'schedule' time series are almost constant, the following item 'vista' has a higher variance, since a beta 2 version of Microsoft's Vista operating system was released in May 2006 and drew the attention of searchers and taggers. The 'vista' time series are given in the left of Figure 2. Another example where the peaks in the time series were triggered from an information need after a certain event is "iran" ($r = 0.80$), which has the 19th highest correlation of all tags. The peaks show up shortly after the confirmation of the United States White House that Iran's president sent a letter to the president of the US on May 08, 2006; and are strongly correlated. A similar peak for 'iran' can be observed in Google Trends[5] showing Google's search patterns in May 2006. These examples support the hypothesis that popular events trigger both search and tagging close to the event.

---

[5] http://www.google.com/trends?q=Iran&geo=all&date=2006-5

Table 4. Averages of all Del.icio.us URLs (full / normalised) with the search datasets

Dataset	top 25	top 50	top 75	top 100
Google	19.91 / 24.17	37.61 / 47.83	54.00 / 71.15	69.21 / 85.23
MSN	12.86 / 20.20	22.38 / 38.62	30.93 / 56.47	39.09 / 74.14
AOL	— / 19.61	— / 35.57	— / 48.00	— / 57.48

4.3   Coverage of Del.icio.us with MSN, Google and AOL

In this section we shift our focus from query terms and tags to the underlying resources, i. e., the URLs. Considering today's size of the Web, both search engines
(in particular the part we can crawl) and folksonomies constitute only a small
fraction of the Web. An interesting question is thus if there is any significant
overlap between the URLs provided by both systems.

To compare the coverage of the different data sets, we compute the overlaps
between MSN crawl, Google crawl, AOL click data and the Del.icio.us dataset
of October 2006. As we had no access to the indices of the search engines, we
crawled all search engines with 1,776 queries to obtain comparable datasets.
These queries were determined by taking the 2000 most popular tags of the
Del.icio.us 2005 dataset and intersecting them with the set of all AOL items.
The resulting datasets are described in more detail in Section 3.2.

In order to see whether Del.icio.us contains those URLs that were judged relevant by the traditional search engines, we computed a kind of "recall" for
folksonomy-URLs on the other data sets as follows: First we cut each of the 1,776
rankings of each search data set after the first 25, 50, 75 and 100 URLs. For each
ranking size, we computed the intersection with all Del.icio.us URLs. As the AOL
log data consist of domain names only (and not of full URLs), we also pruned the
URLs of the other systems in a second step to the domain names.

Table 4 shows the results. The first number in each cell is the average number
of overlaps for the original URLs, the second for the pruned URLs. Google
shows the highest overlap with Del.icio.us, followed by MSN and then AOL. For
all systems, the overlap is rather high. This indicates that, for each query, both
traditional search engines and folksonomies focus on basically the same subset
of the Web. The values in Table 4 will serve as upper bounds for the comparison
in Section 5.

Furthermore, the top rankings show more coverage: While in average 24.17
URLs in the top Google 25 ranking are represented in Del.icio.us, only 85.23 are
represented in the top 100 URLs in average. This indicates that the top entries
of search engine rankings are – in comparison with the medium ranked entries –
also those which are judged more relevant by the Del.icio.us users.

4.4   Conclusions of Section 4

The overlap of the whole set of the MSN query terms with the set of all Del.icio.us
tags is only about a quarter of the size of the latter, due to a very high number
of very infrequent items in both systems (Section 4.1, Table 2). Once the sets

are reduced to the frequent items, the relative overlap is higher. The remaining differences are due to different usage, e. g., to the composition of multi-word lexems to single terms in Del.icio.us, and the use of (parts of) URLs as query terms in MSN.

In Section 4.2, we have seen that for a relatively high number of items the search and tagging time series were significantly correlated. We have also observed that important events trigger both search and tagging without significant time delay, and that this behaviour is correlated over time.

Considering the fact that both the available search engine data and the folksonomy data cover only a minor part of the WWW, the overlaps of the sets of URLs of the different systems (as discussed in Section 4.3) are rather high, indicating that users of social bookmarking systems are likely to tag web pages that are also ranked highly by traditional search engines. The URLs of the social bookmarking system cover over-proportionally the top results of the search engine rankings. A likely explanation is that taggers use search engines to find interesting bookmarks.

5   Comparison of Social and Traditional Rankings

In the previous section we compared the user interaction in social bookmarking systems and search engines and the coverage of URLs of folksonomies in search engines. In this section we focus on ranking algorithms. Are overlapping results different when we introduce a ranking to the folksonomy structure? Are important URLs in search engines similar to important URLs in social bookmarking systems? Is the ranking order within the overlap the same? These questions will be answered below.

For the commercial search engines, we rely on our crawls and the data they provided, as the details of their ranking algorithms are not published (beside early papers like [11]). To rank URLs in social bookmarking systems, we used two well-known ranking approaches: the traditional vector space approach with TF-IDF weighting and cosine similarity, and FolkRank [8], a link-based ranking algorithm similar to PageRank [11], which ranks users, resources or tags based on the tripartite hypergraph of the folksonomy.

5.1   Overlap of Ranking Results

To compare the overlap of rankings, we start with an overview of the average intersection of the top 50 URLs calculated for all of our datasets. In this case we based the analysis on the normalized URLs of the same datasets as used in Section 4.3. Table 5 contains the average overlap calculated over the sets of normalized URLs and the TF, TF-IDF and FolkRank rankings of the Del.icio.us data. We see that the overlap of Del.icio.us Oct. 2006 with the result sets of the three commercial search engines is low. The average overlap of the MSN and Google crawl rankings is considerably bigger (11.79) – also compared to the AOL results which are in a similar range with the Del.icio.us data. The two major

Table 5. Average overlap of top 50 normalized URLs

	Google	MSN	Del FolkRank	Del TF-IDF	Del TF
AOL	2.39	1.61	2.30	0.30	0.21
Google		11.79	6.65	1.60	1.37
MSN			3.78	1.20	1.02
Del FolkRank				1.46	1.79
Del TF-IDF					49.53

Table 6. Average overlap with top 100/1,000 normalized Del.icio.us URLs

	Google top 50	MSN top 50	AOL top 50
Del 100	9.59	5.00	1.65
Del 1000	22.72	13.43	5.16

search engines therefore seem to have more in common than folksonomies with search engines.

The TF and TF-IDF based rankings show a surprisingly low overlap with Google, MSN and AOL, but also with the FolkRank rankings for Del.icio.us. This indicates that – as for web search – graph-based rankings provide a view on social bookmarking systems that is fundamentally different to pure frequency-based rankings.

Although the graph-based ranking on Del.icio.us has a higher overlap with the search engine rankings than TF-IDF, it is still very low, compared to the potential values one could reach with a 'perfect' folksonomy ranking, e. g., an average overlap of 47.83 with the Google ranking as Table 4 shows. The remaining items are contained in the Del.icio.us data, but FolkRank ranked them beyond the top 50.

To investigate this overlap further, we have extended the Del.icio.us result sets to the top 100 and top 1,000, resp.

Table 6 shows the average overlap of the top 100 and the top 1,000 normalized URLs of the FolkRank computations in Del.ico.us data of Oct. 2006 to the top 50 normalized URLs in the Google crawl, MSN crawl and AOL log data. It extends thus the middle column of Table 5. For Google, for instance, this means that the relative average overlap is $\frac{6.65}{50} \approx 0.13$ for the top 50, $\frac{9.59}{100} \approx 0.10$ for the top 100, and only $\frac{22.7}{1000} \approx 0.02$ for the top 1000. This supports our finding of Section 4.3, that the similarity between the FolkRank ranking on Del.icio.us and the Google ranking on the Web is higher for the top than for the lower parts of the ranking.

5.2   Correlation of Rankings

After determining the coverage of folksonomy rankings in search engines, one question remains: Are the rankings obtained by link analysis (FolkRank) and term frequencies / document frequencies (TF-IDF) correlated to the search engine

Table 7. Correlation values and number of significant correlations

Datasets	# overlap > 20)	Avg. corre-lation	Avg.of signif-icant correla-tions	# correlated rankings	# significant correlated rankings
			pos/neg	pos/neg	pos/neg
Google/FolkRank	361	0.26	0.4/-0.17	326/37	176/3
Google/TF-IDF	17	0.17	0.34/0	15/2	5/0
MSN/FolkRank	112	0.25	0.42/-0.01	99/13	47/1
MSN/TF-IDF	6	-0.21	-/-	2/4	0/0
AOL/FolkRank	1	0.25	-/-	1/0	0/0
AOL/TF-IDF	1	0.38	0.38/-	1/0	1/0

Table 8. Top Correlations of Delicious FolkRank with Google (left) and MSN (right), based on top 100 of Del.icio.us.

Item	Intersection	Correlation	Item	Intersection	Correlation
technorati	34	0.80	validator	21	0.64
greasemonkey	34	0.73	subversion	22	0.60
validator	34	0.71	furl	23	0.59
tweaks	22	0.68	parser	27	0.58
metafilter	24	0.67	favicon	28	0.57
torrent	29	0.65	google	25	0.57
blender	22	0.62	blogosphere	21	0.56
torrents	30	0.62	jazz	26	0.56
dictionaries	21	0.62	svg	23	0.55
timeline	21	0.62	lyrics	25	0.54

rankings? Again, we use the rankings of the 1,776 common items from Section 4.3. As we do not have interval scaled data, we select the Spearman correlation coefficient $r_s = 1 - \frac{6 \sum d^2}{n(n^2-1)}$, where d denotes the difference of ranking positions of a specific URL and n the size of the overlap.[6]

In Section 5.1 we showed that the overlap of the rankings is generally low. We therefore only compared those rankings having at least 20 URLs in common. For each such item, the Spearman coefficient is computed for the overlap of the rankings. Table 7 shows the results. The AOL comparisons to Del.icio.us (using the link-based method as well as TF-IDF) do not show sufficient overlap for further consideration. The Google and MSN comparisons with the link-based FolkRank ranking in Del.icio.us yield the highest number of ranking intersections containing more than 20 URLs (Google 361, MSN 112). Both Google and MSN show a large number of positive correlations. For instance, in Google, we have

---

[6] In [3], enhancements to Kendall's tau and Spearman are discussed to compare rankings with different URLs. These metrics are heavily influenced if the intersection between the rankings is small. Because of this we stick to the Spearman correlation coefficient.

326 positive correlations, whereby 176 are significant. This confirms our findings from Section 5.1.

From the results above we derive, that if overlap exists, a large number of rankings computed with FolkRank are positively correlated with the corresponding search engine rankings. In order to find out the topics on which the correlation is high, we extracted the top ten correlations of the Del.icio.us FolkRank with Google and MSN, resp., see Table 8. We found that most items in this set are IT related. As a major part of Del.icio.us consists of IT related contents, we conclude that link-based rankings for topics that are specific and sufficiently represented in a folksonomy yield results similar to search engine rankings.

5.3 Conclusions of Section 5

In Section 5.1, we have seen that a comparison of rankings is difficult due to sparse overlaps of the data sets. It turned out that the top hits of the rankings produced by FolkRank are closer to the top hits of the search engines than the top hits of the vector based methods. Furthermore we could observe that the overlap between Del.icio.us and the search engine results is larger in the top parts of the search engine rankings.

In Section 5.2 we observed that the folksonomy rankings are stronger correlated to the Google rankings than to MSN and to AOL, whereby the graph-based FolkRank is closer to the Google rankings than TF and TF-IDF. Again, we assume that taggers preferably use search engines (and most of all Google) to find information. A qualitative analysis showed that the correlations were higher for specific IT topics, where Del.ico.us has a good coverage.

6 Discussion and Outlook

In this paper, we conducted an exploratory study to compare social bookmarking systems with search engines. We concentrated on information retrieval aspects by analyzing search and tagging behaviour as well as ranking structures. We were able to discover both similar and diverging behaviour in both kinds of systems, as summarized in Sections 4.4 and 5.3. An open question is whether, with more data available, the correlation and overlap analyses could be set on a broader basis. A key question to be answered first though is what is to be considered a success? Is it desirable that social search tries to approximate traditional web search? Is Google the measure of all things? Computing overlap and comparing correlations helped us finding out about the similarities between systems. However, we have no information which approach offers more relevant results from a user's perspective. A user study in which users create a benchmark ranking and performance measures might be of benefit. Further investigation also has to include a deeper analysis of where URLs show up earlier and the characteristics of both system's URLs not being part of the overlap.

Acknowledgment. This work has partially been supported by the Microsoft Live Labs Award "Accelerating Search in Academic Research".

References

1. Adar, E., Weld, D., Bershad, B., Gribble, S.: Why we search: Visualizing and predicting user behavior. In: Proc. WWW 2007, Banff, Canada, pp. 161–170 (2007)
2. Bao, S., Xue, G., Wu, X., Yu, Y., Fei, B., Su, Z.: Optimizing web search using social annotations. In: Proc. WWW 2007, Banff, Canada, pp. 501–510 (2007)
3. Bar-Ilan, J., Mat-Hassan, M., Levene, M.: Methods for comparing rankings of search engine results. Comput. Networks 50(10), 1448–1463 (2006)
4. Broder, A.Z.: A taxonomy of web search. SIGIR Forum 36(2), 3–10 (2002)
5. Chatfield, C.: The analysis of time series: an introduction, 6th edn. CRC Press, Florida (2004)
6. Chien, S., Immorlica, N.: Semantic similarity between search engine queries using temporal correlation. In: Proc. WWW 2005, pp. 2–11. ACM Press, New York (2005)
7. Golder, S., Huberman, B.A.: The structure of collaborative tagging systems. Journal of Information Science 32(2), 198–208 (2006)
8. Hotho, A., Jäschke, R., Schmitz, C., Stumme, G.: Information retrieval in folksonomies: Search and ranking. In: Sure, Y., Domingue, J. (eds.) ESWC 2006. LNCS, vol. 4011, pp. 411–426. Springer, Heidelberg (2006)
9. Ivorix. Sine-weighted moving average (2007),
   `http://www.ivorix.com/en/products/tech/smooth/smooth.html`
10. Krishna, A.M., Gummadi, P., Druschel, P.: Exploiting social networks for internet search. In: Proc. HotNets-V, pp. 79–84 (2006)
11. Page, L., Brin, S., Motwani, R., Winograd, T.: The PageRank citation ranking: Bringing order to the web. In: Proc. WWW 1998, Brisbane, Australia, pp. 161–172 (1998)
12. Pass, G., Chowdhury, A., Torgeson, C.: A picture of search. In: Proc. 1st Intl. Conf. on Scalable Information Systems, ACM Press, New York (2006)
13. Vlachos, M., Meek, C., Vagena, Z., Gunopulos, D.: Identifying similarities, periodicities and bursts for online search queries. In: Proc. SIGMOD 2004, ACM Press, New York (2004)
14. Yanbe, Y., Jatowt, A., Nakamura, S., Tanaka, K.: Can social bookmarking enhance search in the web? In: Proc. JCDL 2007, pp. 107–116. ACM Press, New York (2007)

Effects of Aligned Corpus Quality and Size in Corpus-Based CLIR

Tuomas Talvensaari

Department of Computer Sciences, University of Tampere
Kanslerinrinne 1, FIN-33014, University of Tampere, Finland
tuomas.talvensaari@uta.fi

Abstract. Aligned corpora are often-used resources in CLIR systems. The three qualities of translation corpora that most dramatically affect the performance of a corpus-based CLIR system are: (1) topical nearness to the translated queries, (2) the quality of the alignments, and (3) the size of the corpus. In this paper, the effects of these factors are studied and evaluated. Topics of two different domains (news and genomics) are translated with corpora of varying alignment quality, ranging from a clean parallel corpus to noisier comparable corpora. Also, the sizes of the corpora are varied. The results show that of the three qualities, topical nearness is the most crucial factor, outweighing both other factors. This indicates that noisy comparable corpora should be used as complimentary resources, when parallel corpora are not available for the domain in question.

1 Introduction

In Cross-Language Information Retrieval (CLIR), the aim is to find documents that are written in a different language from the query. Consequently, besides the usual information retrieval (IR) issues, in CLIR one has to address the problem of crossing the language barrier. Usually, the query is translated from the *source language* into the *target language*, i.e., the language of the documents, after which a normal monolingual retrieval process can take place. The query translation approaches can be categorized according to the linguistic resources employed. The main approaches use either machine-readable dictionaries, machine translation (MT) systems, fuzzy cognate matching, multilingual corpora, or a combination of these resources [1,2].

In approaches based on multilingual corpora, the translation knowledge is extracted statistically from the corpora used. These methods can further be categorized based on the relatedness of the corpora. A *parallel corpus* consists of document pairs that are more or less exact translations of each other. In a *comparable corpus*, the document pairs are not exact translations but have similar vocabulary [3]. The aligned documents can be, e.g., accounts of the same news event written independently in different countries. Therefore, the *alignment quality* of the corpus at hand can vary significantly – a parallel corpus and a noisy comparable corpus represent the extremes of this characteristic.

C. Macdonald et al. (Eds.): ECIR 2008, LNCS 4956, pp. 114–125, 2008.

Besides alignment quality, the *topical nearness* between the corpus and the translated queries is also a significant factor. For example, a parallel corpus consisting of sports news is not likely to provide dependable translation knowledge if the queries concern quantum physics. Many of the query words would be out-of-vocabulary (OOV) for the system.

Naturally, the *size* of the corpus is also an important factor. The more aligned documents we have, the more reliable the translation knowledge is. Therefore, a large parallel corpus with good coverage of domain vocabulary would be ideal for CLIR. The availability of such corpora is often a problem, however. This is especially true with rarer languages and special domains. Consequently, noisier but more easily available comparable corpora may have to be used.

In this paper, the aim is to examine the effects of the above three factors to the performance of a CLIR system. This is done by applying translation corpora of varying alignment quality and size to retrieval topics of different domains. To our knowledge, studies where all the three factors are simultaneously experimented with, do not exist previously. The results show that although the alignment quality and size are important factors, it is essential that the corpus covers the vocabulary of the domain in question.

The rest of this paper is organized as follows. In Sec. 2, we take a look at related work done previously. Section 3 introduces the corpora used in the experiments, and also the methods that are used to utilize them. Section 4 describes the test runs and results, and Sec. 5 provides a brief conclusion.

2   Previous Work

Franz et al. [4] varied the size of various parallel corpora and found that the performance of a CLIR system based on parallel corpora is inversely proportional to the query OOV rate of the system. This would mean that the OOV rate would be a handy and easily calculated measure of system performance. Of the three qualities discussed in Sec. 1, topical nearness is the one closest related to the OOV rate. However, when comparable corpora are concerned, things are not that straightforward. It is not enough for the words just to appear in the corpus – they also have to appear frequently and the document alignments have to be of good quality.

Zhu and Wang [5] degraded a rule-based MT system by decreasing the size of the rule base and the dictionary. They found that removing dictionary entries (i.e., increasing the OOV rate) impaired the performance of the system more dramatically than removing rules. This seems to be in line with [4].

Xu and Weischedel [6] studied CLIR performance as a function of parallel corpus and dictionary size. They found that a large dictionary can compensate for a small parallel corpus and vice versa, and that the combination of these resources always performs better than either of them alone. As the size of the corpus was increased, the performance also improved, up to a point. Again, though, it remains unclear whether the same would be true when a comparable corpus is used.

3   Data and Methods

Two topic sets of different domains and languages were used in the experiments. The Swedish topics were provided by the CLEF consortium [7] and cover news events from the mid-90's. The German topics were the German translations of a TREC genomics track. The topics themselves are more closely introduced in Sec. 4. For both of the topic sets, two translation corpora were applied. Figure 1 presents the corpora in relation to their alignment quality and topical nearness to the topics.

All of the corpora and the queries were preprocessed by removing stopwords and using a lemmatizer program to normalize the words to their base form. Next, the translation corpora are introduced in more detail.

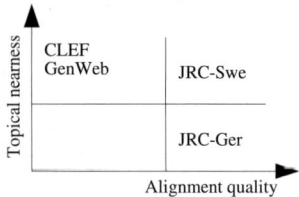

Fig. 1. Alignment quality and topical nearness of the used translation corpora

3.1   The JRC-Acquis Parallel Corpus

For both of the language pairs and topic sets, the JRC-Acquis parallel corpus [8] was applied. The corpus consists of official EU documents in all official EU languages. The alignments in the corpus were made on paragraph or sentence level, and the aligned documents were exact translations of each other – as one would expect from a parallel corpus. With respect to the German genomics topics, JRC-Acquis is topically distant (see Fig. 1), while the distance to the Swedish news topics is smaller. The ultimate measure for the topical distance is the corpus' OOV rate with respect to the queries, which is discussed in Sec. 4.

3.2   The CLEF Swedish-English Comparable Corpus

The CLEF comparable corpus consists of Swedish news-wire reports by the news agency TT from 1994-95, aligned with news articles by the Los Angeles Times from 1994. Both of the collections are provided by CLEF [7]. The CLEF corpus is topically very near the topics (Fig. 1), since it covers news events from the same period as the topics do. The alignments were created by Talvensaari et al. [9] in the following way.

Let $d_S \in C_S$ and $d_T \in C_T$ be documents in the source and target collections, respectively. The aim was to produce a set of alignments $A = \{\langle d_S, D \rangle \mid D \neq \emptyset\}$, where $D = \{d_T | sim(d_S, d_T) > \theta\}$, in other words, to map each source document to a set of target documents whose similarity with the source document exceeds some threshold θ. Each set D is called a *hyper document*. It is not realistic to expect that a satisfying counterpart for every source language document could be found. Thus, $|A| < |C_S|$.

To find the similar counterparts in the target collection, queries were first formed from each source document. Second, the queries were translated into the target language (English) with a smallish dictionary. Third, the translated queries were run against the L.A. Times corpus with the InQuery retrieval system [10]. The InQuery score was used as an indication of the similarity between the source document and the target documents. For each source document, at most 20 target documents whose similarity exceeded a score threshold were chosen into the set D.

3.3 The Genomics Web Corpus

The genomics Web corpus (GenWeb) consists of paragraphs extracted from German and English genomics-related Web pages. They were acquired by Talvensaari et al. [11] by means of focused Web crawling, which refers to the acquisition of Web pages that cover a specific topic [12]. The alignments between the German and English paragraphs were made in the same manner as for the CLEF corpus. The Gen-Web corpus is topically quite near the topics that is is used to translate (see Fig. 1).

Table 1 presents the translation corpora in more detail. The alignments in the JRC corpora are all 1-to-1, that is, each source document is aligned to exactly one target document.

Table 1. The translation corpora in detail

Corpus	Source language	$|A|$	Avg. $|D|$	Unique target documents	Avg. source document length (words)	Avg. target document length (words)
GenWeb	German	39,143	6.5	39,190	114	139
CLEF	Swedish	12,579	4.3	7,732	183	421
JRC-Ger	German	282,417	1	282,417	21	23
JRC-Swe	Swedish	277,735	1	277,735	21	24

3.4 COCOT: Employing the Corpora

Cocot, a Comparable Corpus Translation program [9], uses the aligned corpus as a *similarity thesaurus*, which implies calculating similarity scores between a source language word and the words in the target documents. The similarity thesaurus' similarity score can be calculated by using traditional IR weighting approaches, reversing the roles of documents and words. A source language word is thought of as the query, and target language words are retrieved as the answer.

For a document d_j, in which a word t_i appears, the Cocot system calculates the weight w_{ij} as follows:

$$w_{ij} = \begin{cases} 0 & \text{if } tf_{ij} = 0 \\ \left(0.5 + 0.5 \cdot \frac{tf_{ij}}{Maxtf_j}\right) \cdot \ln\left(\frac{NT}{dl_j}\right) & \text{otherwise} \end{cases}, \qquad (1)$$

where tf_{ij} is the frequency of word t_i in document d_j, $Maxtf_j$ the largest term frequency in document d_j, dl_j the number of unique words in the document. NT can be the number of unique words in the collection, or its approximation.

For a hyper document D_k in which a word t_i appears, the weight is

$$W_{ik} = \sum_{d_j \in D_k} \frac{w_{ij}}{\ln(rank_{jk} + 1)} \quad , \tag{2}$$

where $rank_{jk}$ is the rank of the document d_j in the hyper document D_k, i.e. the rank calculated by InQuery in the alignment phase. The lower the rank, the less similar the target document is to the source document, according to InQuery. Thus, the lower rank documents can be trusted less as a source of translation knowledge. This is echoed in the equation above.

Finally we can calculate Cocot's similarity score between a word s_i appearing in the source documents, and a word t_j appearing in the target hyper documents:

$$sim(s_i, t_j) = \frac{\sum_{\langle d_k, D_k \rangle \in A} w_{ik} \cdot W_{jk}}{\|\mathbf{s_i}\| \cdot \left((1 - slope) + slope \cdot \frac{\|\mathbf{t_j}\|}{avg\_trg\_vlength} \right)} \quad , \tag{3}$$

where A is the set of alignments, $\mathbf{s_i}$ and $\mathbf{t_j}$ are the feature vectors representing s_i and t_j, and $avg\_trg\_vlength$ the average length of the target word vectors. The formula employs the pivoted vector length normalization scheme, introduced by Singhal et al. [13]. The *slope* value is a parameter of this scheme, and its range is $[0, 1]$. The scheme was applied because standard cosine normalization favors words with short feature vectors, i.e. rare words.

When the above score is calculated between a source language word and every word appearing in the target documents, we get a rank of the target words. Table 2 shows Cocot ranks for three genomics-related German words. Score thresholding and word cut-off values (WCV) can be used as translation parameters to define Cocot's query translation behavior. For example, the parameters $WCV = 4, \theta = 6.0$ mean that for the word *dna*, the four highest ranking words would be returned, whereas, for *reparatur*, only the first word would be used as the translation. For the word *oxidativer*, no words would be returned, and the word would be effectively out-of-vocabulary.

Table 2. Example Cocot translations

Rank	dna		reparatur		oxidativer	
1	dna	13.8	repair	8.8	oxidative	2.3
2	sequence	8.1	damage	5.3	ros	2.3
3	base	7.4	dsb	5.2	superoxide	2.1
4	strand	7.2	excision	4.9	dismutase	1.9
5	rna	6.7	dsbs	4.4	peroxide	1.7

4   Experiments

Two different topic sets and target collections were used in the experiments. Table 3 provides an overview of them. The TREC genomics collection [14] is a subset of the MEDLINE database of about 4.6 million documents. This collection was used in the German-English experiments. The 50 topics were translated into German by a molecular biologist, who is also a native speaker of German. The CLEF collection used in this study consists of newspaper articles by the Los Angeles Times, published in 1994. This collection was used in the Swedish-English experiments. The collection is the same as the target collection in the Swedish-English CLEF comparable corpus. The 7,732 documents that were part of the comparable corpus (see Tab. 1) were removed from the test database. The 70 topics were provided by CLEF in Swedish and English. The topic sets were further divided into two subsets, one for training the parameters of COCOT (i.e., θ, WCV and the slope value of Eq. 3), and one for the actual tests. InQuery [10] retrieval system was used in the experiments.

Table 3. Test collections

Collection	Source	Documents	Training topics	Test topics
TREC Genomics	MEDLINE	4,591,008	20	30
CLEF English	L.A. Times	113,005	30	40

4.1   Tests on Corpus Quality and Topical Nearness

For both of the languages pairs, COCOT was used to translate the queries with two different translation corpora, the JRC parallel corpus and a comparable corpus (GenWeb for the German queries, CLEF for the Swedish ones). Besides COCOT, we also applied Utaclir [15], a dictionary-based query translation program, in the experiments. In the German-English experiments, Utaclir used a German-English dictionary of 29,000 entries, while the Swedish-English dictionary had 36,000 entries.

Utaclir was applied to provide a baseline CLIR performance. Utaclir was also used in collaboration with COCOT. In these experiments, the concatenated output of the two programs was used as the target language query. This represents a more realistic approach – in a working CLIR system different translation approaches are likely to be used in collaboration (see, e.g., [16]). It should be noted, however, that the aim in these experiments was not to build a fully functional CLIR system, but to experiment with the qualities of aligned corpora. Consequently, performance-enhancing techniques such as pre-translation query expansion were not used.

The *title* and *need* parts of the TREC topics were used in the experiments – the longer *context* field was ignored. Table 4 presents an example topic in English; the same topic in German after stopword removal and word form normalization; and the translations provided by Utaclir and COCOT that uses the GenWeb comparable corpus. Both programs enclose multiple translation alternatives inside InQuery's syn-operator. This causes InQuery to treat the enclosed words as synonyms. This kind of query structuring reduces the translation ambiguity brought by multiple translation alternatives [17]. Initial experiments showed that this approach performed better than approaches based on weighting the query words according to their COCOT scores. OOV words were left unchanged in all of the approaches.

Table 4. From English topic to German query to English query

English topic	DNA repair and oxidative stress. Find correlation between DNA repair pathways and oxidative stress.
German query	dna reparatur oxidativer stress hang zusammen zusammenhang dna weg reparatur reparaturweg oxidativem stress
Utaclir	#sum(dna repair oxidativer stress hang #syn(context coherence) dna #syn(lane path road course way track channel walk) repair oxidativem stress)
COCOT (*slope* = 0.6,θ = 2.0,WCV= 5)	#sum(#syn(dna sequence base strand rna) #syn(repair damage dsb excision dsbs) #syn(oxidative ros superoxide) stress hang zusammen zusammenhang #syn(dna sequence base strand rna) #syn(cell protein gene pathway stem) #syn(repair damage dsb excision dsbs) #syn(ner nhej dsb repair dsbs) #syn(antioxidant oxidative p4502e1 roi peroxide) stress)

Tables 5 and 6 present the results of the Swedish-English and German-English runs respectively. In addition to mean average precision and precision after 10 documents, the OOV rate and the type of translation resource is provided for each CLIR approach. Figures 2 and 3 depict the recall-precision curves of the runs.

Table 5. Results for the German-English runs, '> X' means the translation method is significantly better ($p < 0.05$) than method X according to the Friedman test (UC=Utaclir)

CLIR approach	MAP	P@10 docs	OOV %	Translation quality
JRC-Ger	0.087	0.210	35.0	Parallel
GenWeb	0.137	0.297	29.1	Comparable
UC	0.170	0.270	56.6	Dictionary
JRC-UC	0.136	0.303	29.3	Combined
GenWeb-UC	0.225 (> JRC-Ger,JRC-UC)	0.407	17.3	Combined

In the German-English runs, the combined GenWeb-Utaclir approach performs significantly better than approaches based on the JRC parallel corpus. This happens because the GenWeb corpus is topically closer to the genomics topics than JRC. The OOV rate of GenWeb is surprisingly high, though, considering that GenWeb consists of genomics-related text. However, the GenWeb OOV words are mostly general, non-topical words. This is proven by the low OOV rate

Table 6. Results for the Swedish-English runs, '> X' means the translation method is significantly better ($p < 0.05$) than method X according to the Friedman test (UC=Utaclir)

CLIR approach	MAP	P@10 docs	OOV %	Translation quality
JRC-Swe	0.247	0.245 (> UC)	27.6	Parallel
CLEF	0.204	0.190	15.2	Comparable
UC	0.186	0.168	24.2	Dictionary
JRC-UC	0.294 (> CLEF, UC)	0.288 (> CLEF, UC)	4.6	Combined
CLEF-UC	0.282 (> CLEF, JRC-Swe, UC)	0.293 (> CLEF,UC)	3.2	Combined

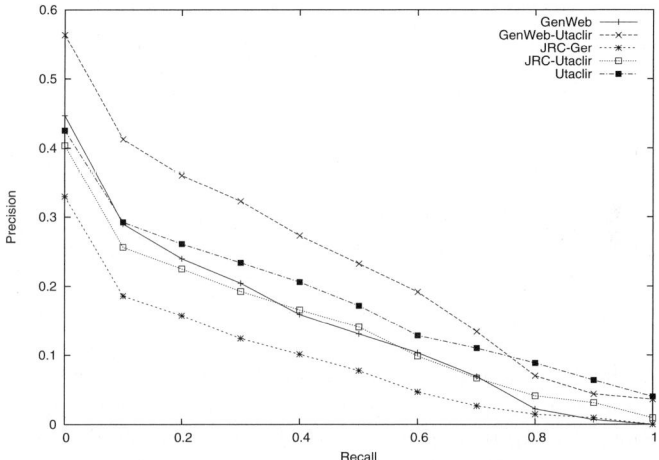

Fig. 2. Precision at 11 recall points for the German-English runs

of GenWeb-Utaclir. In the combined approach, Utaclir, with its general-purpose dictionary, can cover much of GenWeb's OOV words. Therefore, GenWeb seems to fulfill its purpose to a certain extent. The role of a noisy comparable corpus is usually to complement other, more general, resources. Therefore, the relatively low performance of GenWeb alone is not alarming.

In the Swedish-English runs, the CLEF comparable corpus is arguably topically even closer to the topics than GenWeb to the genomics topics. The topics are news events from 1994-95, and the CLEF corpus consists of news articles from the same period. The vocabulary of the news domain, however, is much more general than genomics vocabulary. Consequently, the JRC corpus fares much better than in the German runs. When JRC is used as the sole translation corpus, it quite clearly outperforms CLEF. The combined approaches perform evenly, and they both achieve very low OOV rates.

Table 7 presents a closer look at the performance of individual queries. For each CLIR approach, the average precision of each query was compared to the

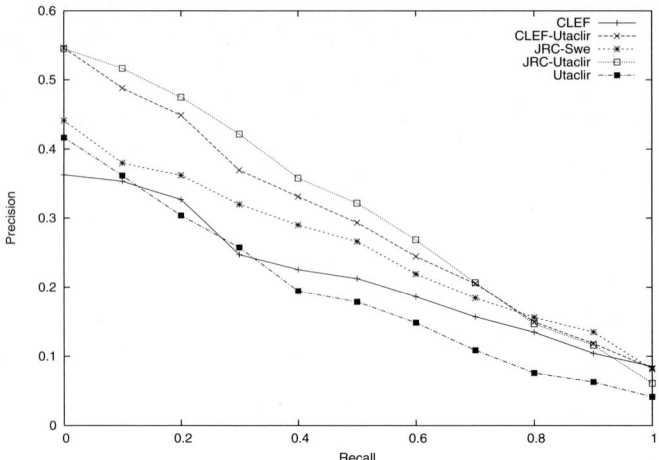

Fig. 3. Precision at 11 recall points for the Swedish-English runs

median performance of the query. The table depicts the number of queries for whom average precision was significantly greater or smaller (absolute difference of 5 %) than the median, for each approach. This analysis was adopted instead of a more complete query-by-query analysis because of its compactness.

The analysis echoes the results presented earlier. It is notable that in the Swedish runs, the approach JRC-Utaclir has more queries that perform significantly below median than CLEF-Utaclir (5 against 1), although JRC-Utaclir had higher MAP. This indicates that JRC-Utaclir performed very well on some individual queries, while CLEF-Utaclir was better overall. Also the stability of GenWeb-Utaclir is notable – not one its queries perfomed significantly worse than the median.

Table 7. The number of queries that perform significantly better or worse (absolute difference of 5% in average precision) than the (language-specific) median of each query

	German(30 queries)				
	JRC-Ger	GenWeb	Utaclir	JRC-Utaclir	GenWeb-Utaclir
Better	2	7	6	4	11
Worse	7	6	3	2	0
	Swedish(40 queries)				
	JRC-Swe	CLEF	Utaclir	JRC-Utaclir	CLEF-Utaclir
Better	7	5	4	13	13
Worse	9	12	13	5	1

4.2   Tests on Corpus Size

The effect of corpus size on the performance level of COCOT was tested by increasing the sizes of the translation corpora step-by-step from 500,000 source document words onwards, until they reached their full sizes (see Tab. 1) . The shrinked corpora were created by removing alignments from the sets A (see Sec. 3.2) randomly. On each size level, COCOT was first trained with the same set of training topics as in Sec. 4. Then, the actual test queries were run with the learned parameters. Figure 4 presents the mean average precision of the runs plotted against the increasing corpus size.

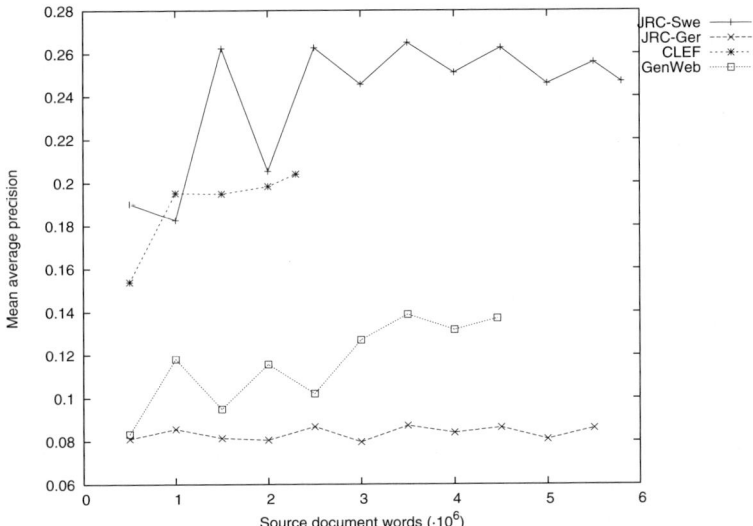

Fig. 4. MAP plotted against corpus size for four translation corpora

For each corpus, save for JRC-Ger, there seems to be a significant difference in performance between the smallest level and the full size. However, from about 1M words onwards, the corpora seem to reach a performance level comparable to that of the full corpus quite quickly. This is somewhat surprising and puzzling. The JRC-Ger corpus performs badly on all size levels. This is an example of topical distance weighing more than alignment quality or corpus size. The fluctuations in performance (especially for JRC-Swe) are perhaps due to unoptimal COCOT parameters on some of the levels.

5   Conclusion

The performance of CLIR systems based on aligned corpora are affected by three qualities of the corpora: 1. The topical nearness of the corpus to the translated queries. 2. The alignment quality of the corpus – a parallel corpus is better than a noisy comparable corpus. 3. The size of the corpus. Based on the experiments discussed in this paper, the topical nearness seems to be the most crucial factor. The JRC-Ger corpus was of high alignment quality and sufficiently large – yet it performed badly, when it was used to translate queries from the genomics domain. This indicates that noisier, but easily available comparable corpora should be used for special domain vocabulary, if parallel corpora are not available. Comparable corpora are most effective as a complimentary resource.

The effect of alignment quality was also shown quite clearly. The JRC-Swe corpus was topically not as close to the topics as the CLEF corpus. However, JRC-Swe outperformed CLEF because it is a high-quality parallel corpus. The effect of corpus size was not as clear, though. The results seem to indicate that after a certain threshold in corpus size, the performance does not increase significantly.

Acknowledgements. This research was partly funded by the Tampere Graduate School in Information Science and Engineering (TISE).

References

1. Oard, D.W., Diekema, A.R.: Cross-Language Information Retrieval. Annual Review of Information Science and Technology (ARIST) 33, 223–256 (1998)
2. Pirkola, A., Hedlund, T., Keskustalo, H., Järvelin, K.: Dictionary-Based Cross-Language Information Retrieval: Problems, Methods, and Research Findings. Inf. Retr. 4, 209–230 (2001)
3. Sheridan, P., Ballerini, J.P.: Experiments in Multilingual Information Retrieval Using the SPIDER System. In: SIGIR 1996: Proceedings of the 19th Annual International ACM SIGIR Conference on Research and Development in Information Retrieval, pp. 58–65. ACM Press, New York (1996)
4. Franz, M., McCarley, J.S., Ward, T., Zhu, W.J.: Quantifying the Utility of Parallel Corpora. In: SIGIR 2001: Proceedings of the 24th Annual International ACM SIGIR Conference on Research and Development in Information Retrieval, pp. 398–399. ACM Press, New York (2001)
5. Zhu, J., Wang, H.: The Effect of Translation Quality in MT-Based Cross-Language Information Retrieval. In: ACL 2006: Proceedings of the 21st International Conference on Computational Linguistics and the 44th Annual Meeting of the ACL, pp. 593–600. Association for Computational Linguistics, Morristown, NJ (2006)
6. Xu, J., Weischedel, R.: Empirical Studies on the Impact of Lexical Resources on CLIR Performance. Inf. Process. Manage. 41, 475–487 (2005)
7. Peters, C.: What Happened in CLEF 2006. In: Peters, C., Clough, P., Gey, F.C., Karlgren, J., Magnini, B., Oard, D.W., de Rijke, M., Stempfhuber, M. (eds.) CLEF 2006. LNCS, vol. 4730, pp. 1–10. Springer, Heidelberg (2007)
8. Steinberger, R., Pouliquen, B., Widiger, A., Ignat, C., Erjavec, T., Tufiş, D., Varga, D.: The JRC-Acquis: A Multilingual Aligned Parallel Corpus with 20+ Languages. In: LREC 2006:Proceedings of the 5th International Conference on Language Resources and Evaluation, pp. 2142–2147. European Language Resources Association, Paris (2006)
9. Talvensaari, T., Laurikkala, J., Järvelin, K., Juhola, M., Keskustalo, H.: Creating and Exploiting a Comparable Corpus in Cross-language Information Retrieval. ACM Trans. Inf. Syst. 25, 4 (2007)

10. Allan, J., Callan, J.P., Croft, W.B., Ballesteros, L., Broglio, J., Xu, J., Shu, H.: Inquery at TREC-5. In: TREC-5: The Fifth Text Retrieval Conference, pp. 119–132. National Institute of Standards and Technology (1996)
11. Talvensaari, T., Järvelin, K., Pirkola, A., Juhola, M., Laurikkala, J.: Focused Web Crawling in Acquisition of Comparable Corpora. Information Retrieval (submitted, 2007)
12. Chakrabarti, S., van den Berg, M., Dom, B.: Focused Crawling: a New Approach to Topic-specific Web Resource Discovery. In: WWW 1999: Proceeding of the Eighth International Conference on World Wide Web, pp. 1623–1640. Elsevier North-Holland, Inc (1999)
13. Singhal, A., Buckley, C., Mitra, M.: Pivoted Document Length Normalization. In: SIGIR 1996: Proceedings of the 19th Annual International ACM SIGIR Conference on Research and Development in Information Retrieval, pp. 21–29. ACM Press, New York (1996)
14. Hersh, W.R.: Report on the TREC 2004 Genomics Track. SIGIR Forum 39, 21–24 (2005)
15. Keskustalo, H., Hedlund, T., Airio, E.: Utaclir: General Query Translation Framework for Several Language Pairs. In: SIGIR 2002: Proceedings of the 25th Annual International ACM SIGIR Conference on Research and Development in Information Retrieval, pp. 448–448. ACM Press, New York (2002)
16. McNamee, P., Mayfield, J.: Comparing Cross-language Query Expansion Techniques by Degrading Translation Resources. In: SIGIR 2002: Proceedings of the 25th Annual International ACM SIGIR Conference on Research and Development in Information Retrieval, pp. 159–166. ACM Press, New York (2002)
17. Pirkola, A.: The Effects of Query Structure and Dictionary Setups in Dictionary-based Cross-language Information Retrieval. In: SIGIR 1998: Proceedings of the 21st Annual International ACM SIGIR Conference on Research and Development in Information Retrieval, pp. 55–63. ACM Press, New York (1998)

Exploring the Effects of Language Skills on Multilingual Web Search

Jennifer Marlow[1], Paul Clough[1], Juan Cigarrán Recuero[2], and Javier Artiles[2]

[1] Department of Information Studies, University of Sheffield, UK
{j.marlow,p.d.clough}@shef.ac.uk
[2] Departamento de Lenguajes y Sistemas Informáticos,
Universidad Nacional de Educación a Distancia, Spain
{juanci,javart}@lsi.uned.es

Abstract. Multilingual access is an important area of research, especially given the growth in multilingual users of online resources. A large body of research exists for Cross-Language Information Retrieval (CLIR); however, little of this work has considered the language skills of the end user, a critical factor in providing effective multilingual search functionality. In this paper we describe an experiment carried out to further understand the effects of language skills on multilingual search. Using the Google Translate service, we show that users have varied language skills that are non-trivial to assess and can impact their multilingual searching experience and search effectiveness.

Keywords: Cross-language web search, user study, design, language skills.

1   Introduction

As globalisation and the Internet have facilitated the exchange and accessibility of ideas and information in a variety of languages, the field of Cross-Language Information Retrieval (CLIR) has emerged as an area of focus in the IR community. Many experiments have been conducted under the Cross-Language Evaluation Forum (CLEF[1]), mostly focusing on evaluating the retrieval of news articles from an unknown language collection based on a query submitted in a user's native language.

However, in reality, individuals' needs are not always so simplistic or limited only to this type of situation. There are other scenarios in which cross-language information requirements can vary. For example, users may wish to access multilingual material that is not plain text (e.g. web pages or images). Furthermore, with regards to language skills, individuals can have a range of both passive (e.g. comprehension) and active (production) abilities based on their mother tongue and other languages they may have studied for any length of time.

The present experiment was designed to expand upon previous CLIR research by focusing on the role language skills play in a multilingual web searching context, whilst also considering the importance of other factors inherent to the interactive search process (such as user satisfaction). Participants were asked to find a variety of

---

[1] http://www.clef-campaign.org/ (site visited: 06/01/2008)

C. Macdonald et al. (Eds.): ECIR 2008, LNCS 4956, pp. 126–137, 2008.

web pages in three different languages: their native language, one that could be passively understood, and one that was completely foreign. The Google web search engine and associated Google Translate[2] service for search results were chosen as representative systems for testing. Search behaviours, functionalities used, and overall performance was compared in each of the three language conditions. As expected, many of these varied depending on the target language and the type of query submitted. However, the findings provide useful input into future design of cross-language support in information retrieval systems.

2 Background

Any study examining cross-language search must consider its users' language skills. Unknown and native languages are the two endpoints of a spectrum of language knowledge; foreign language ability can vary greatly within these two extremes. While Ringbom [1] points out the distinction between passive and active ability, Laufer & Goldstein [2] suggest that this dichotomy is too simplistic, and propose a continuum of knowledge strength that also includes recall and recognition. According to Gibson & Hufeisen [3], prior knowledge of a language has been shown to assist understanding of an unfamiliar but related one (e.g. German and Swedish).

As argued by Gonzalo [4], there are two different situations relating to a user's language skills that carry different design implications for cross-language systems. If a user is monolingual, full translation assistance is needed in a CLIR context (e.g. back translation of query terms and document translation). If the user has some passive language skills, then document translation is less likely to be used or desired. Language ability, therefore, is an important variable to consider when designing a system that will cater to a range of users with different needs.

Other studies have focused on user behaviour when performing cross-language search for text or images. Zazo Rodriguez et al. [5] examined the effect of users' language abilities on the types of functionalities they used for a question-answering exercise. Compared to individuals with "good" foreign language skills, users with poor skills were found to be more likely to enter queries in their native language and then have them automatically translated to the document language. These "poor" users were also more likely to use and appreciate a functionality which translated the document summaries into their native language.

Petrelli et al. [6] also acknowledged that users are not always monolingual and looked beyond this typical view by investigating how polyglots interacted with a cross-language text retrieval system. However, completely bilingual users with excellent language skills were studied, and thus little insight was given into how the system could have served users with moderate or passive language abilities.

Artiles et al. [7] studied which CLIR functionalities were employed when searching for images with a system that offered three query translation options: no translation, automatic translation, and assisted translation (where the machine translated result could be viewed and edited). Translation was typically selected in cases where the search was precision-oriented and geared towards finding something specific. Overall, the assisted translation mode was the most popular, although the

---

[2] http://www.google.com/translate_s?hl=en (site visited: 06/01/2008)

possibility of changing the default translations of the system was largely unexploited (perhaps partly due to the tasks assigned).

Research by Kralisch & Berendt [8] found that the linguistically-determined cognitive effort involved in processing information in a foreign language can be mediated or lessened in cases where domain knowledge is high. Similarly, Gaspari [9] asserted that some users may understand specialized terms relevant to their field of interest, even if their general foreign language ability is limited.

Other studies have looked at how users interact with cross language functionality, even if language skills are not explicitly considered as a variable. Dorr et al. [10] noted that letting users edit machine translated output led to a more satisfying overall experience (although this control was still not as effective as query reformulation). What this study did not examine, however, was the role that knowledge of the target language played and how this could have affected users' behaviour.

To examine the best way of displaying machine translations to a monolingual user, He et al. [11] tested two different approaches: pure back translations and more contextual translations (showing the keyword in the context of a sample sentence). Overall, the potential utility of each approach was deemed to depend on factors such as the characteristics of the topic, the collection, and the translation resources. Even if query translation is offered, it may not necessarily be used if it is not perceived as providing some benefit. For example, research conducted as part of the European TRANSLIB project revealed "people made little use of the title translation capabilities in TRANSLIB because they tended to use the system only to find documents in languages that they could read." (cited in [12]).

Many of the aforementioned studies focused on individuals searching for text-only articles. Web pages are different from texts because they often contain images or other cues to help provide additional (non-verbal) information about the content. Little is known about how people may conduct cross-lingual search using mainstream Web-based systems, especially in a variety of languages; hence, these areas will be the focus of the present investigation.

3   Methodology

Due to time and resource limitations, 12 participants were involved in this study. They were predominantly computer science postgraduate students or researchers with a mean age of 30 years and a median age of 27.5. To investigate the influence of cross-linguistic similarity, individuals with Romance language skills were specifically recruited. Because these languages share a common origin; it was assumed that each participant would have some latent, interchangeable passive knowledge of the others.

Five of the participants were native (or near-native) speakers of Spanish, four of Portuguese, two of French, and one of Italian. Before starting, participants completed a questionnaire relating to search engine use and reading/writing ability levels in all languages with which they had experience active (L1), passive (L2), and unknown (L3) languages. L1 was counted as the native language or a language spoken at near-native fluency. L2 was defined as a Romance language similar to the individual's L1, but for which their self-rated reading/writing abilities were "beginner" or below. L3 was a language the participant could not understand, selected at random from the possibilities of German, Japanese, and Russian. Three options were necessary because some people were familiar with at least two of the languages.

The Google Translate "search results" translation service was used for these experiments. It was chosen over other similar systems because Google's search engine draws upon a large index, and its widespread use means it is familiar to most individuals. This system provides a wide range of functionalities, including automatic query translation, snippet translation, web page translation, and possibility of viewing and editing the query's translation. As such, it provides the set of "ideal" cross-language search functionalities advocated by Zhang & Lin [12].

Participants were asked to imagine they were high school teachers looking for web pages to show non-English speaking students. They were given a list of topics and asked to find and bookmark 3 relevant pages for each one (within a set time of 5 minutes for each topic). To find this information, they could use the Google search engine (including any localised versions, e.g. google.es) or the Google Translate site. None of the participants had used Google Translate before; therefore, the basic functionalities and features of the site were demonstrated to them beforehand.

There were 12 topics in total (4 for each language). This study was conducted in the context of a project focusing on cultural heritage and designed to focus on the common behaviour of focused web search. Search topics were chosen from a list of popular queries submitted to cultural sites, ranging from proper names and titles (e.g. Hamlet, The Last Supper) to more general subjects (modern art, still life) and fairly specific terms (Gothic cathedrals, Etruscan tombs). Half of the topics were considered "hard" for translation (that is, they were incorrectly translated by Google Translate), and half considered "easy" for translation. Hard topics were not always identical across languages because the automatic translation system did not make the same types of mistakes in all languages. Nonetheless, types of errors leading to hard queries had characteristics corresponding to one of three main categories of "performance issues" in CLIR (cf. [13]): lack of coverage (out of vocabulary terms - e.g. Etruscan,) translation ambiguity (Hamlet being translated as "small village" instead of the title of a play) and incorrect translation of phrases ("still life" translated word-for-word).

The language orders and the task-language combinations were assigned based on a Latin-square arrangement, with 2 hard and 2 easy topics for each language. After each set of 4 questions (one language set), the participants filled out a brief questionnaire to assess the difficulty of the task and their confidence with finding relevant sources for each topic. At the end of the experiment, participants filled out a language test for their passive language to assess the correspondence with their self-reported levels. They were also asked to rate the usefulness of functionalities of Google Translate and comment on potential future improvements to the system.

4   Results

4.1   Languages Used for Web Search

Except for one individual, none of the participants were native speakers of English. However, they reported using English to search on the Internet between 48 and 95% of the time (mean 75.5%). This may be because all participants were currently studying or working in the UK and therefore may have needed to, grown accustomed to, felt more confident, or had more success using English to search on a regular basis. Responses indicated that users predominantly search in English or their native language, using other foreign languages relatively infrequently.

Foreign language abilities in reading and writing were self-reported on a scale from 1 (beginner) to 5 (advanced). Across all responses in all languages, the mean value of reading skills (3.59) was slightly higher than that of writing skills (2.96), suggesting that people judged themselves to be better at reading than at writing (this difference was not statistically significant).

Table 1. Frequency of use of Google Translate functionalities for each topic, by language

	Query translation	Translated query editing	Original links viewed	Translated links viewed	Both links viewed
L1	13 (27.1%)	0	4 (8.3%)	1 (2.1%)	1 (2.1%)
L2	37 (77.1%)	3 (6.3%)	26 (54.2%)	2 (4.2%)	4 (4.2%)
L3	46 (96.0%)	0	14 (29.2%)	19 (39.6%)	9 (18.8%)

As shown in Table 1, reliance upon query translation functionalities increased with language unfamiliarity: users were more likely to look at the translated versions of pages for L3, and the original versions for L2. Query editing occurred only 3 times out of all 144 topics, and these were exclusively in the L2 condition. Based on the tools available (which offered limited editing assistance for translated queries), users were much more likely to reformulate or edit the query in the source language than to deal with the machine translation, behaviour also noted by Dorr et al. [10].

Table 2. Number of topics searched with each site, by language

	Google only	Google Translate only	Google and Google Translate
L1	35 (72.9%)	4 (8.3%)	9 (18.8%)
L2	9 (18.8%)	15 (31.3%)	22 (45.8%)
L3	2 (4.1%)	26 (54.2%)	20 (41.7%)

4.2   Sites and Functionalities Used

In general, as language unfamiliarity increased, the use of Google Translate also increased. Many searches were conducted with a combination of Google and Google Translate (Table 2). Often, participants switched from one to the other after a few unsatisfactory query modifications, thinking that the second system would yield different results (in reality there was no difference; Google Translate results were the same as those obtained from using Google). Because the search topics were given in English, 27.1% of participants utilized Google Translate in the L1 condition to find (or verify) the corresponding term in their native language.

4.3   Performance Measures

The following quantitative measures were used to assess user performance on the tasks: **Relevant Items**: the number of pages bookmarked (0-3); **Time**: the length of time taken to do so; **Modifications**: the number of times the query was modified (something else was typed into the search box) per task; **Links viewed**: the number of page links selected (in original language and in target language); **Percent Chosen**: the number of links bookmarked as a proportion of total links clicked on; **Success**: a relative indication of how easy the task was, determined by dividing the number of bookmarks by time (a higher number means the person was more "successful" at completing the task); **Difficulty**: a rating of task difficulty supplied by the user (this referred to all four searches for a given language) (1=very difficult, 7=very easy); and **Confidence**: a rating of user confidence that sources found were relevant (1=not at all confident, 7=very confident).

Perhaps not surprisingly, more people successfully completed the task of book-marking three pages in the L1 condition (67%) as opposed to the L2 and L3 conditions (33% and 19%, respectively). Within each language, more bookmarks were made for the easy topics than the hard topics. However, nearly 30% of participants found three bookmarks they felt were relevant using easy queries in the L3 condition.

Language Effects

A one-way ANOVA was carried out to determine the effect of search language on the quantitative measures mentioned above. The tasks in L1 were self-rated as significantly easier than those of L2, which were in turn rated as significantly easier than those of L3. The significant differences between the language groups with

Table 3. Effects of search language on various measures

Measure	Language	Mean	Sig.
Relevant Items**	L1	2.458[ab]	[a] =.001
	L2	1.646[a]	[b] =.000
	L3	1.437[b]	
Time**	L1	3.985[ab]	[a] =.002
	L2	4.584[a]	[b] =.000
	L3	4.766[b]	
Modifications	L1	1.655	
	L2	2.000	
	L3	2.313	
Links Viewed	L1	4.479	
	L2	4.333	
	L3	3.812	
Percent Chosen**	L1	0.588[ab]	[a] =.006
	L2	0.395[a]	[b] =.010
	L3	0.408[b]	
Success**	L1	0.7193[ab]	[a] =.000
	L2	0.4077[a]	[b] =.000
	L3	0.3433[b]	
Difficulty**	L1	5.667[ab]	[a] =.000
	L2	4.000[ac]	[b] =.000
	L3	3.167[bc]	[c] =.003
Confidence**	L1	5.958[ab]	[a] =.000
	L2	4.213[a]	[b] =.000
	L3	4.106[b]	

** differences significant at p<.01.

respect to mean values for relevant items found, time, percent chosen, success, and confidence were between L1-L2 and L1-L3 (as highlighted by the letter superscripts in Table 3). The differences in time and success seem to be in accordance with findings by Kralisch & Berendt [8] that non-L1 information processing requires more cognitive effort than L1 information processing.

Subsequent ANOVA analysis comparing these measures across the L3 conditions yielded no significant differences across performance measures, although German search was rated as significantly easier than either Russian or Japanese search (due presumably to the more familiar alphabet). German searchers were also significantly more confident in their findings than Russian searchers.

Effects of Topic Difficulty

An independent samples t-test was performed to compare mean results between easy and hard topics (see Table 4). Significant differences were found between these two groups with respect to the number of pages bookmarked, number of query modifications made, success, and confidence. The significantly reduced number of modifications made for easy queries corresponds with an assertion by Och et al. [14] that better quality machine translations result in reduced post-editing effort.

The effect of topic difficulty on confidence was also significant on the results for L1, L2, and L3 when analysed separately using independent samples t-tests (see Table 5). Within each language, users were significantly more confident with the results they found for the easy queries as opposed to the hard queries. This easy-hard distinction also emerged, surprisingly, in the L1 condition (in which occurrences of query translation were much lower). Since Google Translate exploits the web as a parallel corpus, perhaps what helps to make a query easily translatable or not is influenced by the number of pages available on that topic. If the hard topics were less well represented even in English, then the likelihood or speed of finding relevant results could be reduced compared to more popular, "easy" topics. There was no significant interaction between language and difficulty.

Table 4. Effects of topic difficulty on various measures

Measure	Topic type	N	Mean	Sig. (2-tailed)
Relevant Items*	E	76	2.105	.003
	H	68	1.559	
Time	E	76	4.310	.061
	H	68	4.596	
Modifications*	E	76	1.526	.000
	H	68	2.500	
Links Viewed	E	76	4.461	.065
	H	68	3.927	
Percent Chosen	E	76	0.506	.078
	H	68	0.414	
Success*	E	76	.5678	.016
	H	68	.4033	
Difficulty	E	76	4.408	.303
	H	68	4.132	
Confidence*	E	76	5.461	.000
	H	68	3.970	

* differences significant at $p < .05$.

Table 5. Mean confidence rating for easy vs. hard topics, by language

Language	Topic type	N	Mean	Sig. (2-tailed)
L1*	E	28	6.393	.007
	H	20	5.350	
L2*	E	24	5.083	.002
	H	23	3.304	
L3*	E	24	4.750	.024
	H	23	3.435	

* differences significant at p<.05.

Quantitative Measures from Final Questionnaire

The average mean ratings of the usefulness of the three translation aids offered by Google Translate ranged from 3.90 for query editing to 5.08 for query translation to 5.50 for translated snippets (with 7 being most useful). The rated usefulness of the various features corresponded with their frequency of use (as shown in Table 2). That is, since both query and snippet translations were actually used more often than query editing, it is not surprising that they were also rated as more useful.

The usefulness ratings of the various functionalities (query translation, snippet translation, and query editing) varied based on the language being considered. For non-native languages L2 and L3, opinion on the most useful feature was split equally between query translation and translated snippets. The mean usefulness ratings of proposed additional functionalities (dictionary support and greater control over the query) were 6.25 and 5.41, respectively. However, dictionary support with back translations or pictures was viewed as more helpful than just showing the alternative translations in the target language with no further explanation or assistance.

Passive Language Reading Test

The final element of the questionnaire was a short analytic test of basic passive L2 reading ability. Only reading comprehension abilities were tested since this was deemed to be the main skill being tested in the experiment (the ability to skim and understand the content of the results summaries and the web pages). This was adapted from the BBC Languages site[3] and consisted of 12 questions of increasing difficulty. Scores on these tests (taken in French, Italian, or Spanish), ranged between 6 and 12, with a mean value of 8.4. It should be noted that none of the participants self-reported any knowledge of the L2 they were assigned.

Comments and Observations

Responses to the post-experiment question: "Would you use Google Translate again? Why or why not?" revealed three common attitudes:

1. Not a useful tool (5): "Interesting tool but not sure I need to use it", "Just when no other means to get the information are available."
2. Helpful in some situations (4): "Useful to translate words into different languages" "The searching environment is very useful"

---

[3] http://www.bbc.co.uk/languages/ (site visited: 06/01/2008)

3. Translation quality could be improved (3): "I'm not quite sure about how accurate the translations were."   "It's not very reliable when doing translations"

What emerges here is a feeling that many people could not normally envision a reason to use a system like Google Translate. However, none of them were aware of the system's existence prior to using it in the study. They may have not realized its potential use in some situations (i.e. when planning foreign travel, to broaden the scope of a search, etc). The experiment by its very nature created a somewhat artificial, restricted situation in which users were only allowed to use two specific sites. It is unclear to what extent they would voluntarily use (or need to use) Google Translate in their everyday search behaviour. Further research could examine this question in a more open-ended and naturalistic context.

Other observations made of search behaviour indicate that people using a machine translation system expect it to operate in the same way as an ordinary search engine (with regards to query syntax and formulation). In the case of Google Translate, this was not so. Adding quotation marks to mark phrases and refining queries with supplementary terms, while conventions for Google search, did not have an effect on the machine translation system. Individuals employed creative strategies to find information when the Google Translate results were unsatisfactory. For example, some users were seen to exploit the multilingual structure of Wikipedia pages. This was carried out to bypass the potentially inaccurate query translations of Google.

This suggests that the automatic translation, while beneficial, still produces many errors (and this was recognised by the users). The means of dealing with these errors was not sufficiently developed in Google Translate for the "hard" queries, leading to a lower level of search precision. It should be noted that shortly after the present experiment was conducted, a dictionary service was added to the Google Translate pages to allow the lookup of words or phrases in a limited set of language pairs. This no doubt can help the user to identify the correct translation for their query. However, the dictionary service is located in a separate tab and thus is still not as user-friendly as it could be if it were integrated into the main "translated search" interface or integrated into the search service to seamlessly display alternative translations.

5 Discussion and Design Implications

Overall, it was encouraging to see that given the appropriate tools for assistance, people can still find basic relevant information in a partially or completely unknown foreign language. Despite this, however, there are clear differences in the level of functionality required from a CLIR system based on the users' language skills. This is in line with the findings of Gonzalo [4]. The main observations (and implications for improved system design) included the following:

- A query-editing feature does not appear to be helpful for L2 and L3 conditions if editing assistance is not provided. Users with passive reading skills (L2) still struggle to write queries themselves and therefore may not be able to correct a translation they identify as erroneous.

- Suggestions of alternative terms are needed when a query is ambiguous or in-correctly translated (e.g. dictionary support is needed to supplement "pure" machine translation). Particularly for L1 and L2 conditions, providing functionality to assist the user in selecting between alternative translations would be helpful. This could be achieved by ranking the term translations in some way (i.e. the most likely translation is ranked highest or highlighted in some way). Depicting terms pictorially (language-independently) is an alternative and novel approach that warrants further investigation.
- Both query and document translation is essential when a user searches in an unknown language. The former of these is still important when an individual has passive knowledge, although the need for the latter may be less.
- Users employ the same strategies for search with or without query translation and expect that adding extra query terms or using common web search query syntax (e.g. using quotes to mark phrases) will be effective in a query translation situation as well. Such syntax should therefore be supported by cross-language search interfaces.
- Searches are more successful when the query terms are correctly translated; therefore, the continued improvement of machine translation (and appropriate lexical resources) is important.
- Users are not always accurate reporters of their own language abilities and tended to under-estimate their passive skills in this experiment. Creating personalised CLIR interfaces (based either around results of an objective test or on a self report) could help to appropriately target support to users based on their spectrum of knowledge.

6 Conclusions

This study expanded upon previous work in cross-language information retrieval by examining the effect of language skills on web search behaviour using Google Translate. Whereas the majority of CLIR-based research has focused only on how people retrieve material in unknown languages, the present study indicates that many individuals also have passive language skills. They behaved closer to native language ability when using a passive language as opposed to one that was unknown, although these differences were not statistically significant. Overall, as might be expected, the perceived and actual difficulty of the task increased as language unfamiliarity increased. However, the accuracy of query translation also seemed to have an effect across all the language conditions, so that it was harder to find relevant information (in any language) for queries that were incorrectly translated. This problem was further compounded when queries were modified by adding extra terms.

One limitation of the study may have been the five-minute time limit for each task. Whilst this was put into place to reduce fatigue effects and keep the experiment down to a reasonable length of 1.5 hours, some users felt it was "artificial" and it may have led them to bookmark some less appropriate sites just to feel that they were able to complete the task in time. Google Translate was clearly able to provide enough support to help participants locate at least some relevant material in both passive and unknown languages, however there are ways in which it (or any similar cross-language searching system) could be improved. Aside from creating translation systems that produce fewer mistakes, it would be beneficial to offer: (1) phrase recognition and translation (either automatically detected or manually indicated) and (2) integrated dictionary support to identify alternative translations for ambiguous terms, with some means of displaying these in an understandable way.

As the associated pictures and visual cues of the web pages helped the participants to make relevance judgments, future work could focus on cross-language functionalities that would assist users searching for other types of media (e.g. images or video), as these may differ from those used in a purely text-based situation. Overall, the present experiment provided insight into the behaviours and strategies of individuals searching for material in a variety of languages. Findings can help to influence the design of personalized cross-language searching support based on an individual's varying abilities and language needs.

Acknowledgments

Work partially supported by European Community under the Information Society Technologies (IST) programme of the 6th FP for RTD - project MultiMATCH contract IST-033104. The authors are solely responsible for the content of this paper.

References

[1] Ringbom, H.: Cross-linguistic similarity in foreign language learning. Multilingual Matters, Clevedon (2007)
[2] Laufer, B., Goldstein, Z.: Testing vocabulary knowledge: Size, strength and computer adaptiveness. Language Learning 54(3), 399–436 (2004)
[3] Gibson, M., Hufeisen, B.: Investigating the role of prior foreign language knowledge: Translating from an unknown into a known foreign language. In: Cenoz, J., Hufeisen, B., Jessner, U. (eds.) The Multilingual Lexicon, Kluwer Academic Publishers, Dordrecht (2003)
[4] Gonzalo, J.: Scenarios for interactive cross-language information retrieval systems. In: Proceedings of SIGIR 2002 Workshop on Cross-Language IR (2002)
[5] Zazo Rodriguez, A.F., Figuerola, C., Alonso Berrocal, J.L., Fernandez Marcial, V.: iCLEF 2005 at REINA-USAL: Use of free on-line machine translation programs for interactive cross-language question answering (2005)
[6] Petrelli, D., Levin, S., Beaulieu, M., Sanderson, M.: Which user interaction for cross-language information retrieval? Design issues and reflections. JASIST 57, 709–722 (2006)
[7] Artiles, J., Gonzalo, J., Lopez-Ostenero, F., Peinado, V.: Are users willing to search cross-language? An experiment with the Flickr image sharing repository. In: Peters, C., Clough, P., Gey, F.C., Karlgren, J., Magnini, B., Oard, D.W., de Rijke, M., Stempfhuber, M. (eds.) CLEF 2006. LNCS, vol. 4730, pp. 195–204. Springer, Heidelberg (2007)
[8] Kralisch, A., Berendt, B.: Language-sensitive search behaviour and the role of domain knowledge. New Review of Hypermedia and Multimedia 11, 221–246 (2005)
[9] Gaspari, F.: Online machine translation services and real users' needs: An empirical usability evaluation. In: Frederking, R.E., Taylor, K.B. (eds.) AMTA 2004. LNCS (LNAI), vol. 3265, pp. 74–85. Springer, Heidelberg (2004)

[10] Dorr, B., He, D., Luo, J., Oard, D.: iCLEF at Maryland: Translation selection and document selection. In: Peters, C., Gonzalo, J., Braschler, M., Kluck, M. (eds.) CLEF 2003. LNCS, vol. 3237, Springer, Heidelberg (2004)

[11] He, D., Wang, J., Oard, D., Nossal, M.: Comparing User-Assisted and Automatic Query Translation. In: Peters, C., Braschler, M., Gonzalo, J. (eds.) CLEF 2002. LNCS, vol. 2785, pp. 400–415. Springer, Heidelberg (2003)

[12] Zhang, J., Lin, S.: Multiple language supports in search engines. Online Information Review 31(4), 516–532 (2007)

[13] Hull, D., Grefenstette, G.: Experiments in multilingual information retrieval. In: Proceedings of 19th ACM International Conference on Research and Development in Information Retrieval (SIGIR 1996), pp. 49–57 (1996)

[14] Och, F., Zens, R., Ney, H.: Efficient search for interactive statistical machine translation. In: Proceedings of 38th Annual Meeting of Association for Computational Linguistics (ACL), pp. 440–447 (2003)

A Novel Implementation of the FITE-TRT Translation Method

Aki Loponen, Ari Pirkola, Kalervo Järvelin, and Heikki Keskustalo

Department of Information Studies, University of Tampere, Finland
{Aki.Loponen,Kalervo.Jarvelin,Heikki.Keskustalo}@uta.fi,
pirkola@cc.jyu.fi

Abstract. Cross-language Information Retrieval requires good methods for translating cross-lingual spelling variants which are not covered by the available dictionary resources. FITE-TRT is an established method employing frequency-based identification of translation equivalents received from transformation rule based translation. This study further develops and evaluates the FITE-TRT method. The paper contributes on four areas. First, an efficient implementation for the FITE-TRT method is discussed. Secondly, a novel iterative FITE-TRT translation approach is developed in order to further improve the effectiveness of the method. Thirdly, the effectiveness of FITE-TRT is assessed in three classes of source-target word similarity. FITE-TRT was found to be very strong in the class of the most similar source and target words and only becomes unsuccessful when the words were dissimilar. Fourthly, in comparison to n-gram and s-gram matching methods, FITE-TRT is shown consistently stronger. All in all, FITE-TRT clearly outperforms the fuzzy string matching methods under comparable conditions. Therefore it is the method of choice for the identification of translation equivalents of cross-lingual spelling variants when the requirements for the result quality are high.

Keywords: Approximate string matching, cross-language information retrieval, cross-lingual spelling variants, fuzzy matching, out-of-vocabulary words, transformation rules.

1 Introduction

Frequency-based identification of translation equivalents received from transformation rule based translation (FITE-TRT) is a method which addresses the out-of-vocabulary (OOV) problem in cross-language information retrieval (CLIR) by providing an effective method for translating spelling variants [5]. FITE-TRT consists of two consecutive methods: the TRT and the FITE. TRT receives a *keyword* in source language and outputs a list of *translation candidates* in a target language. FITE processes these candidates and outputs a result - the proposed translation or a flag indicating that a translation cannot be identified.

The TRT phase, originally presented in [6], generates a number of translation candidates by applying suitable *transformation rules* to a given keyword. A transformation rule contains source language characters that are transformed into the

C. Macdonald et al. (Eds.): ECIR 2008, LNCS 4956, pp. 138–149, 2008.

target language characters given in the rule, and their context characters. A rule also has two numerical factors: *confidence factor* and *frequency*, which determine the importance of a rule. Frequency refers to the number of occurrences of the rule in the dictionary data that was used in rule generation and confidence factor is defined as the frequency of a correct rule application divided by the number of source words where the source substring of the rule occurs.

The FITE phase scans through a list of candidates generated by TRT and gives either exactly one translation or an empty output (no translation cases) in contrast to approximate string matching [4][7] where the source word always matches some target words (thus there are no 'no translation' cases). The FITE phase has three conditions and if those are fulfilled then a translation can be given. The first one is the *beta condition* which checks that the frequency of the candidate with the highest frequency value in a target language is more than a predefined *beta*-value (β) times the frequency of the second best candidate. If the first and second best candidates do not fulfil the beta-condition requirements, the second best candidate is compared with the third best. If the comparison meets the beta condition, then the first candidate is selected, because the most common among similar candidates is the most probable candidate for translation. An example of beta condition is presented below with following words and their frequency values:

- lucille   20,000
- lucile   5000
- lusille   200

If β=2, then the first comparison between the words 'lucille' and 'lucile' satisfies the condition (20,000 > 2*5000) and thus the 'lucille' is qualified. If β=10, then the comparison between first two words fails (20,000 ≤ 10*5000). Next the frequency values of words 'lucile' and 'lusille' are compared and now the condition is satisfied (5000 > 10*200) and again the 'lucille' can be qualified. Both stages fail if β=25.

The second condition checks that the relation between the frequency of a candidate in a target language and the frequency of a source word in a source language is valid. The frequencies are normalized using predefined parameter *alpha*, thus the condition is called *alpha-condition*. FITE takes the frequency information from word frequency lists specifically constructed for this purpose. The third condition (the length factor) checks that the length difference between the key and candidate is reasonable.

The present paper focuses on four issues. First, an efficient implementation for the FITE-TRT method is developed for the first time. Secondly, novel iterative FITE-TRT translation strategies are studied in order to further improve the effectiveness and efficiency of the method. The idea is to translate the source words stepwise, gradually relaxing the FITE-TRT parameters. By first applying stringent criteria, the number of target word candidates remains small. If a translation is not identified for some word, then more relaxed criteria are employed – and more candidates generated. Thirdly, the effectiveness of FITE-TRT is assessed in three classes of source-target word similarity. In this paper it is shown that FITE-TRT handles well translations of source words that are at least moderately similar to their translations – better than known alternatives. Fourthly, FITE-TRT is compared to n-gram and s-gram matching [1] methods in a large-scale test which demonstrates that FITE-TRT is highly competitive. Finally, an analysis on how many fuzzy translations are required to achieve the recall of FITE-TRT is performed. All in all, the tests are to show that FITE-TRT clearly outperforms the fuzzy matching methods under comparable conditions and can be implemented efficiently. Therefore it is the method of choice for the identification of translation equivalents of cross-lingual spelling variants when the requirements for the result quality are high.

The paper is organized as follows: First, an efficient implementation for the FITE-TRT method is considered in Section 2. The novel iterative FITE-TRT translation strategies are presented in Section 3. The effectiveness of FITE-TRT method is evaluated and also compared to *n-gram* and *s-gram* matching in Section 4, followed by conclusions in Section 5.

2   New Features of FITE-TRT

In this section an example keyword along with few suitable TRT-rules are used to illustrate the FITE-TRT process. The keyword is Spanish word "aditivo" which has an English translation "additive". The TRT-rules utilized are {adi addi b 6 42.86}, {ti tai c 2 0.08}, {tivo t e 1 0.69} and {vo ve e 123 62.44}, where the first character string is the source language substring (which is replaced in the key), the second character string is the substring for target language (which replaces the source language substring in the key) and the third separate character indicates the position of the rule: *b* means that the rule targets the beginning of the key, *c* means center and *e* means that the end of the key is targeted. The integer represents the frequency value of the key and the decimal number is rule's confidence factor.

2.1   New Implementation

The basic FITE-TRT implementation [5] was effective but not very efficient as such. It introduced the *windowing of rules* by their confidence factor and frequency as a means to reduce the number of generated translation candidates, but it can still create vast numbers of candidates. Resources can be saved when the consecutive TRT and FITE processes are merged into a joint process (Figure 1).

Retrieving frequency information from external data storage for each generated candidate is very ineffective. In the basic implementation each candidate requires one query operation in target language frequency data. For example, using all Spanish-English TRT-rules defined in [6] for the Spanish the keyword "aditivo" yields 16,400 translation candidates and as many frequency data queries.

Optimization is achieved by trimming the number of frequency data queries in two ways. First the number of generated candidates is reduced. Direct limit to the number of generated candidates will not work, because the correct candidate can be any of the created. While windowing the ruleset is quite an effective pruning method, efficiency is a bit arbitrary, because still unknown numbers of rules become selected: long keys can have plenty of suitable rules while short keys can only have a few if any.

A method for *weighting of the rules* is adopted to ensure a fixed upper limit for the number of applied rules. Suitable rules are ordered into descending order by their *weight* which for single rule r is calculated using formula (1), where f is the frequency of the rule, cf its confidence factor, af is the sum frequency of all the rules, and pf is

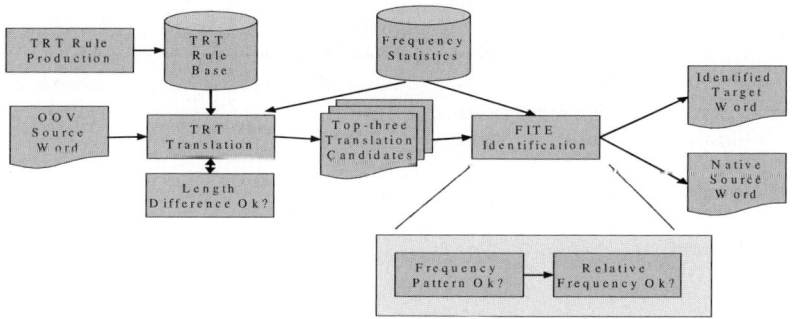

Fig. 1. The merged FITE-TRT process

the sum frequency of all the rules affecting the same source language characters as *r*. Now the number of generated candidates is more controllable than with windowing and presumably the quality of utilized rules is better. Weighting of the rules adds the *rule number* parameter, which defines the number of best rules to be selected from all rules compatible with a given key. However, a key which has several rules fitting the same part of the key and has several of such parts will still produce lots of candidates.

$$weight(rule) = \frac{f * cf}{af * pf} \tag{1}$$

Merging some of FITE's functionality into candidate generation also reduces the queries into the data storage structures. As described above, the FITE method has three conditions that a candidate must fulfil to be accepted as a translation. *Length factor* is the third condition, but there is no valid reason for not to utilize it while generating candidates. If a generated candidate does not meet the length criteria, then target language frequency for it is not retrieved and processing continues to the next candidate. In the optimized implementation the candidates are generated recursively in preorder. The main root of the recursive key generation tree is the source key itself and it has as many child nodes as there are possible single rule adaptations. Adaptations for the key's rule slots are done from left to right. Further adaptations for a generated candidate are done to the part right from previous adaptations window. The maximum depth of the recursion tree is the highest number of rules that can be reconciled with the keyword. An example of the recursive candidate tree for the key and rules given in the beginning of this section is presented in Figure 2. The bold section of each candidate string represents the part of the candidate which has been influenced by a rule. When traversed in pre-order, the candidates generated are in following order: `aditivo, additivo, additaivo, additaive, addit, additive, aditaivo, aditaive, adit, aditive.` (In this example the rules have not been windowed or weighted.)

Only three candidates bearing the highest frequency values from target language frequency data are kept in memory instead of storing all the generated candidates. When all the candidates have been created, the two remaining FITE conditions are applied into the top-three list and the topmost candidate in the list is accepted or rejected.

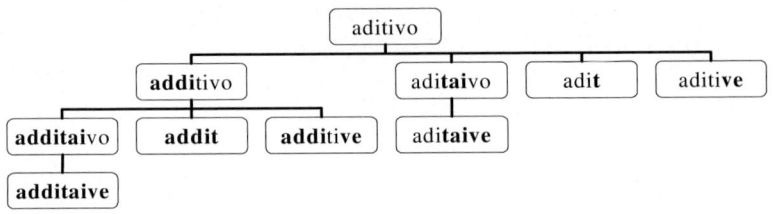

Fig. 2. An example of the recursive candidate generation tree

2.2 Division between Short and Long Words

When experiments with FITE-TRT were made, it was noticed that the same combination of parameters was not necessarily optimal for words with significant difference in their lengths. Better results were achieved when shorter words were processed with a different parameter combination than the longer ones. For all the languages used, the length of 7 letters was found to be as the best dividing value. Every keyword shorter than, or exactly, seven letters in length was handled as a short word and each longer keyword as a long one. This word length factor is genuine because longer words have more possible positions for rule transformations than shorter ones.

2.3 Frequency Data

FITE needs an associated program that collects Web data and constructs *word frequency lists* for words of a given language. The list can become very large. In particular the English list covers extensively the English lexicon but it is also littered with misspelled words and some foreign language words (because language filtering is not 100% accurate). Sometimes a candidate obtained in TRT matches with a litter word contained in the frequency list. By further filtering the lists the effectiveness of FITE-TRT is expected to improve. The number of unique words contained in the frequency lists gathered for the experiments is as follows:

- English: 1,643,000 words
- French: 771,000 words
- German: 1,422,000 words
- Spanish: 957,000 words

2.4 Performance Gains

The novel implementation was programmed in Java which enabled portability to several platforms as well as good modularity for different kinds of applications. When the frequency data and the TRT-rules are loaded into memory prior to translations, the typical resolution time for a key was around 25-35 milliseconds (using Windows XP in a PC with Intel DualCore2 6300 processor with 2 GB's of RAM and 1600 MB's allocated to Java Runtime Environment).

While this resolution time is lengthy for real time use, the keys to be translated are essential in (short) queries and therefore it makes sense to pay the effort. Also when compared to previous way of implementing the FITE-TRT, this novel implementation is justified. Implementing the FITE-TRT manually could take days, even weeks for a single key. A brute-force method where TRT- and FITE-phases are applied successively, can easily take hours for a single long key.

3   Iterative FITE-TRT

The original FITE-TRT processing utilizes a single combination of parameters, which are alpha- and beta-values, length criteria for a key and candidates, and the number of best rules employed. This language dependent combination is trained to produce the highest recall and precision values and therefore the best possible overall result for FITE-TRT. Some number of words will be left untranslated, because there is no single parameter combination that could reliably translate every valid word. Therefore an iterative method is presented here as an extension to the original method.

The iterative FITE-TRT will apply different parameter combinations consecutively until the candidate is accepted or until it becomes certain enough, that the key cannot be translated. Parameter combinations are evaluated with the numbers of correct and incorrect translations, and the number of untranslated words. During iteration incorrect translations are avoided by translating carefully in each level of iteration. Each parameter combination perfectly translates some words while it can also translate other words incorrectly. Therefore effective iteration requires such a sequence of parameter combinations, that most of the words are translated as correctly as possible.

3.1   Strategies for the Identification of Iteration Parameters

In order to identify the best sequence of parameter combinations three strategies are presented, because it was not obvious, at the outset, how a good iteration could be formed. All strategies have the risk of overfitting to the training data and all strategies are also *static* in a way that they have to be tuned each time the TRT-rules or language frequency data changes. In the next section the training of these strategies is evaluated.

The original FITE-TRT used simply the best single parameter combination that resulted in as high recall and precision values as possible. This strategy is used as a *baseline* in evaluating the iterative strategies.

The first strategy is a *manual* search for a suitable sequence of parameter combinations. This translates words in consecutive steps, or in iteration. The words that do not translate in the first step are handled in the second step, and so on. In each step the aim is to achieve as high a recall as possible while at the same time maintaining 100% precision. If 100% precision is not achieved the parameters of the steps are changed, or the process is restarted from a previous step whose parameters are changed in such way that 100%, or as high as possible, precision is still achieved ("back-tracking"). The process continues until words cannot be translated any more. The difficult words accumulate to the end of the process, and these words remain untranslatable. This strategy's main problem is the huge manual effort required because of the back-tracking.

The second strategy is called the *best&iterative* strategy. It tries to improve the baseline strategy by applying it in subsequential steps until all translatable words are translated. Here the combination which yields the highest recall value with high precision is selected. The untranslated training words are the new keywords for the next step, where all reasonable parameter combinations are again used to translate the new keywords and - just like in the baseline strategy - the best combination is selected. This is continued until no more new translations with reasonable results can be made. The choice between high valued combinations is done manually, since the cases can be fairly different. This strategy takes less time to prepare than the first one.

The third strategy - the *min-error* strategy - is to select the combination that yields the minimal number of incorrect translations. In case of a tie, the combination with highest recall is selected. This strategy results in a long chain of combinations and it takes a lot of time to be found, but is fully automatic. Untranslated keywords are mined again as in the second strategy. The "back-tracking" step of the first strategy is not made, since it can result in explosive growth of recursion.

3.2   Iteration Tests and Results

The iterative translation strategies were developed using three *training word sets*, one set for each language pair. These word lists were received from a researcher who has investigated cross-lingual spelling variant matching. The training lists contained 463 (French-English), 468 (German-English), and 546 (Spanish-English) word pairs.

In the final iterative experiments, the effectiveness of iterative translation was evaluated using three sets of test words, again one set for each language pair. The test words were extracted from the Multilingual Glossary of Medical Terms by Heymans Institute of Pharmacology, University of Gent [3]. All entries where the entry words in the considered languages were single words were extracted from the Glossary. Because FITE-TRT is intended to translate similar words, for the final test word lists only word pairs whose words were sufficiently similar with one another were selected. As a threshold the similarity value of LCS/LW=0.60 was used (LCS/LW is a longest common subsequence based similarity measure, see Section 4.1.). In addition, the words that were used in the training experiments were removed from the final lists. The final lists contained 1013 (French-English), 1014 (German-English), and 1009 (Spanish-English) unique word pairs.

Obtaining training and test word sets from dictionaries doesn't create a conflict with the objective to create a reliable method for spelling variant OOV words. For technical terms, the transformation rules of the spelling variant word translations between two languages are similar whether the words are in or out of vocabulary, i.e. orthographical conventions are constructed with quite homogenous linguistic rules.

Because the manual strategy takes huge amounts of time to prepare, it was only done for the Spanish-English language pair. However the manual iteration sequence was applied to other language pairs to make a point that universal iteration is not likely to exist, or at least it is not easy to find.

The results for all strategies and language pairs are presented in Table 1. Column "best&iter" shows the results for the second strategy. The columns "manual" and "min error" stand for the other strategies, where the former is the manually searched

Table 1. Recall and precision values for all strategies and baseline translation in all language pairs

		baseline	best&iter	min error	manual
Spa-Eng	Recall (%)	80.2	82.2	82.0	86.4
	Precision (%)	86.6	86.0	83.1	88.1
Ger-Eng	Recall (%)	73.3	73.8	74.3	73.5
	Precision (%)	85.3	85.3	85.3	81.7
Fre-Eng	Recall (%)	75.5	77.2	78.3	77.1
	Precision (%)	82.9	82.0	82.8	81.2

iteration sequence for Spanish to English translation employed on all language pairs and the latter is the sequence generated with fully automatic parameter mining, which is trained to maximise precision at the expense of recall.

The baseline values were as expected. Spanish yielded the highest recall and German the lowest. This was the case for all other strategies as well. The German rules and frequency data contained more characters and created more diversity to rules and source words. Precision values were quite close to each other cross-lingually.

The thoroughness of the baseline was proved by the best&iterative strategy, since the improvements in recall were marginal (Spanish gained the most by two percent units while German only gained half a unit) and strategy's precision either remained or got slightly worse. The baseline strategy already translated most words and left only difficult words without translation. Translating a wide variety of problem words without overfitting parameters to those specific words is a difficult task.

The minimal error -strategy tried to beat the baseline strategy by translating words in smaller groups while avoiding incorrect translations. Training this strategy resulted in long iterations: from 25 to around 30 levels. Recall improved slightly from baseline for all three languages, but precision either remained or got slightly worse. The manually constructed iteration for Spanish to English translation outperformed the baseline. Recall was 86.4% and precision 88.1% against the baseline's 80.2% and 86.6%. The manual strategy also beat other strategies in Spanish to English -pair mainly because it utilized backtracking steps when combining the iteration parameter sequence.

4 FITE-TRT vs. Fuzzy Matching

The effectiveness of FITE-TRT is compared to that of fuzzy matching (n-gram and skip-gram matching). N-grams have been found effective for fuzzy matching in IR [4], [7] and its generalization, the skip-grams (s-gram for short), consistently outperformed n-grams in the identification of translation equivalents of cross-lingual spelling variants [1]. Here the term n-gram refers to di-grams formed of consecutive characters of words. The s-gram fuzzy matching technique constructs di-grams both of consecutive and non-consecutive characters of words [1]. The generated di-grams are put into comparable categories based on the number of skipped characters as di-grams are constructed. The character combination index (CCI) indicates the number of skipped characters as well as the comparable categories. Here the CCI= {{0}, {1, 2}} was used. This means that di-grams formed of consecutive characters form one comparable category and di-grams with one and two skipped characters the other. S-grams formed in this way consistently outperformed conventional di-grams in [1].

For example, with string S = abcd and CCI={{0},{1,2}}, two digrams sets are formed, namely DS{0}(S) = {ab, bc, cd} (by zero skipping) and DS{1,2}(S) = {ac, ad, bd} (by skipping both one and two characters).

The strength of n-/s-grams is that they do not need frequency lists and may be operated in the target database index, thus they are more efficient than FITE-TRT.

4.1 The Experiments

The dependence of FITE-TRT, n-gram and s-gram effectiveness on the similarity degree of the source and target words was considered in these experiments. The test words were extracted from the Multilingual Glossary for Art Librarians [2] which is well suited for this test because it contains both similar and dissimilar words. FITE-TRT translates individual words and only such entries from the Glossary were extracted where English, French, German, and Spanish entry words appear as individual words. This is no loss of generality as OOV phrases can be TRT'ed word by word.

Three test word sets were constructed, one for each language pair. All the word sets contained the same word pairs (n=123) albeit in different languages. All the 3 x 123 word pairs were looked at, and if the original target word was not the most similar one to a source word, it was removed and a new target word, the most similar one, was taken from the dictionary as the equivalent of a source word.

A simple measure of longest common subsequence divided by the length of the longer word of the word pair (LCS/LW) was used as a similarity measure. The closer to 1.00 LCS/LW is, the more similar the words are: LCS/LW = 1.00 refers to identical words. As an example, for the Spanish word *suplemento* LCS/LW = 9/10 = 0.90 w.r.t. the English equivalent *supplement*. For the native Spanish word *vinculante* LCS/LW = 0.30 w.r.t. its equivalent *binding*. The test words were classified into three categories based on their LCS/LW values: (1) 0.80-100, (2) 0.60-0.79, and (3) 0.00-0.59. The FITE-TRT effectiveness w.r.t. n-gram and s-gram matching effectiveness is analyzed within these categories.

The test words in French, German, and Spanish were translated into English by FITE-TRT and were matched against the English word list using n-grams and s-grams. The English word list used in fuzzy matching was the same list as used by FITE-TRT, except that it only contained words without frequency figures. In the first experiment the effectiveness of FITE-TRT vs. fuzzy matching was considered from the viewpoint of FITE-TRT in that only the highest ranked word in the ranked result list of fuzzy matching was considered. The highest ranked word yielded by fuzzy matching was either a correct or incorrect translation for a source word. The matching results were used to compute recall / precision for n-grams and s-grams.

FITE-TRT was run without iteration using two parameter combinations (PC): PC1: 10 best rules (by weight), $\alpha = 2$, and $\beta = 2$; and PC2: 3 best rules, $\alpha = 10$, and

$\beta = 10$. In both cases the length difference between the source word and target word was set at 0-3 characters for long words and 0-2 characters for short words.

The recall of fuzzy matching can be improved by increasing the number of target words that are considered in the result list. The second experiment determined what is the minimum number of target words (MNW) that need to be collected to obtain a recall higher than the one obtained by FITE-TRT. For example, MNW=7 means that on average fuzzy matching achieves higher recall than FITE-TRT when the best seven words are considered, and with six words recall in fuzzy matching remains lower than in FITE-TRT. In this experiment the performance of s-grams (which consistently outperformed n-grams) was compared against the performance of FITE-TRT with parameter combination 2 (the worse of two parameter combinations regarding recall). In this test, the recall and precision of fuzzy matching are not equal since more than one target words per source word are considered. In this experiment s-gram matching is compared with FITE-TRT on the similarity level of LCS/LW=0.80-100.

4.2 Findings

Table 2 reports the results of the experiment, described in Section 4.1, where FITE-TRT's performance was compared against the performance of fuzzy matching. On each similarity level the number of test words varies by the language pair. Note that for each language pair the total number of test words was n=123. The figures suggest that French and English words are more similar to each other than German-English words and Spanish-English words since on the level of LCS/LW=0.80-100 the number of test words is higher for French-English than for the other pairs. The recall

Table 2. FITE-TRT, s-gram and n-gram effectiveness

Translation / matching method	LCS/WL > 0.79			0.6<LCS/WL<0.8			0.0<LCS/WL<0.6			Overall
	N	R%	P%	N	R%	P%	N	R%	P%	F-measure%
FITE-TRT/PC1										
Fre-Eng	84	78.6	88.0	23	34.8	44.4	16	0	0	64.0
Ger-Eng	52	71.2	78.7	26	26.9	38.9	45	2.2	10.0	40.4
Spa-Eng	53	79.2	87.5	45	44.4	58.8	25	8.0	14.3	56.6
FITE-TRT/PC2										
Fre-Eng	84	66.7	96.6	23	30.4	58.3	16	0	0	61.5
Ger-Eng	52	69.2	90.0	26	23.1	60.0	45	2.2	20.0	43.6
Spa-Eng	53	62.3	91.7	45	22.2	43.5	25	4.0	16.7	44.5
S-gram										
Fre-Eng	84	65.5	65.5	23	4.3	4.3	16	0	0	45.5
Ger-Eng	52	44.2	44.2	26	0	0	45	0	0	18.7
Spa-Eng	53	32.1	32.1	45	0	0	25	0	0	13.8
N-gram										
Fre-Eng	84	60.7	60.7	23	0	0	16	0	0	41.5
Ger-Eng	52	42.3	42.3	26	0	0	45	0	0	17.9
Spa-Eng	53	32.1	32.1	45	0	0	25	0	0	13.8

and precision values are equal for s- and n-grams because, in this test, the grams always produce one translation candidate which is either correct or incorrect.

It can be seen in Table 2 that FITE-TRT performs much better than fuzzy matching on similarity level of LCS/LW > 0.79. FITE-TRT/PC1 achieves 71.2%-78.6% recall (R) and 78.7%-88.0% precision (P). For FITE-TRT/PC2 recall is lower but precision is higher than for FITE-TRT/PC1. S-grams outperform n-grams. Regarding recall, only French-English/s-grams is competitive with FITE-TRT achieving a 65.5% recall. However, FITE-TRT/PC1's precision value (88.0%) is clearly higher than s-grams' value (65.5%). Other language pairs cannot really meet up with French-English.

Importantly, on the similarity level of LCS/LW=0.60-0.79 fuzzy matching loses its ability to identify target language equivalents whereas FITE-TRT performs fairly well. FITE-TRT loses its ability to translate words only on the similarity level of LCS/LW=0.00-0.59.

The overall column reports the F-measure for the entire word sets and for each method. In calculating the F-measure, recall and precision were held equally important. The overall tendencies already discussed in the three word similarity classes are confirmed.

Table 3 shows the results of the experiment where MNW was determined for s-grams. As shown, for French MNW is 2, for German 6, and for Spanish 3. In other words, to obtain the recall level of the worst case FITE-TRT two (Fre-Eng), six (Ger-Eng), and three (Spa-Eng) words have to be scanned in the result list of skip-gram matching. In each s-gram case precision is much lower than FITE-TRT's precision. French-English fuzzy matching comes closest to FITE-TRT in MNW and precision but remains still clearly inferior.

Table 3. MNW for s-grams (LCS/WL = 0.80-1.00)

Translation / matching method	MNW	Recall %	Precision %
FITE-TRT/PC2			
Fre-Eng (n=84)	-	66.7	96.6
Ger-Eng (n=52)	-	69.2	90.0
Spa-Eng (n=53)	-	62.3	91.7
S-gram			
Fre-Eng (n=84)	2	69.0	50.0
Ger-Eng (n=52)	6	71.2	16.7
Spa-Eng (n=53)	3	69.8	33.3

5   Conclusions

In this study, an effective FITE-TRT translation system was implemented and evaluated. The merge of FITE and TRT made the FITE-TRT process straightforward and more efficient as an actual computer implementation. The system was used in iterative translation experiments and in FITE-TRT vs. fuzzy matching translation / matching experiments. The results of iterative translation experiments serve as a basis to further develop iterative FITE-TRT. The manually constructed iteration for Spanish to English translation proved that an iterative method can be better than the baseline method. Still, the costs of building a reliable iteration approach seem prohibitive compared to its benefits. It was also shown that FITE-TRT performs considerably better than n- or s-gram matching which further lose on effectiveness and efficiency if translation recall is emphasized (i.e. a low precision tolerated). Based on these results the FITE-TRT technique is recommended for handling untranslatable technical terms in cross-language retrieval and other information systems where automatic translation is part of the system and when the requirements for the result quality are high.

References

1. Keskustalo, H., Pirkola, A., Visala, K., Leppänen, E., Järvelin, K.: Non-adjacent digrams improve matching of cross-lingual spelling variants. In: SPIRE 2003 Conf. Manaus, pp. 252–265 (2003)
2. Multilingual Glossary for Art Librarians, http://www.ifla.org/VII/s30/pub/mg1.htm
3. Multilingual Glossary of Medical Terms by Heymans Institute of Pharmacology, University of Gent, http://users.ugent.be/~rvdstich/eugloss/welcome.html
4. Navarro, G.: A guided tour to approximate string matching. ACM Computing Surveys 33(1), 31–88 (2001)
5. Pirkola, A., Toivonen, J., Keskustalo, H., Järvelin, K.: FITE-TRT: A high quality translation technique for OOV words. In: Proceedings of the 21st Annual ACM Symposium on Applied Computing, Dijon France, April 23-27, 2006, pp. 1043–1049 (2006)
6. Pirkola, A., Toivonen, J., Keskustalo, H., Visala, K., Järvelin, K.: Fuzzy translation of cross-lingual spelling variants. In: Proc. 26th ACM SIGIR Conf. Toronto, pp. 345–352 (2003)
7. Zobel, J., Dart, P.: Phonetic string matching: lessons from information retrieval. In: Proc. 19th ACM SIGIR Conf. Zurich, pp. 166–172 (1996)

The BNB Distribution for Text Modeling

Stéphane Clinchant[1] and Eric Gaussier[2]

[1] Xerox Research Centre Europe, 6 chemin de Maupertuis, F-38240, Meylan, France
Stephane.Clinchant@xrce.xerox.com
[2] University Joseph Fourier (LIG). BP 53 - 38041 Grenoble cedex 9, France
Eric.Gaussier@imag.fr

Abstract. We first review in this paper the burstiness and aftereffect of future sampling phenomena, and propose a formal, operational criterion to characterize distributions according to these phenomena. We then introduce the Beta negative binomial distribution for text modeling, and show its relations to several models (in particular to the Laplace law of succession and to the *tf-itf* model used in the Divergence from Randomness framework of [2]). We finally illustrate the behavior of this distribution on text categorization and information retrieval experiments.

1 Introduction

The goal of this paper is to introduce the Beta negative binomial distribution as a new tool for modeling texts in information access applications (as text categorization or information retrieval). This distribution was derived from two considerations: one based on word burstiness, the other on the good empirical fit to data provided by the negative binomial distribution.

Several recent works have proposed different distributions that can model word burstiness, or the aftereffect of future smapling ([2,8,4]). However, no formal characterization for these phenomena has been proposed. Introducing a clear definition for these phenomena and proposing a formal, operational criterion for characterizing word distributions wrt them is the first problem we address in this paper, in section 2. On a different aspect, [3], and more recently [1,14], have emphasized the adequacy of the negative binomial distribution for text modeling. Despite its nice properties, few attempts have been made to make use of it in text categorization or information retrieval. In the second part of this paper, we come back to this distribution and derive from it the Beta negative binomial distribution, which can be used to re-interpret several aspects of the Divergence from Randomness framework presented in [2]. This is the subject of section 3. Lastly, in section 4, we present several categorization and information retrieval experiments that illustrate the behavior of the Beta negative binomial Distribution.

2 Burstiness and the Aftereffect of Future Sampling

An important phenomenon reported in [3,7] is the one of *burstiness*. The term "burstiness" describes the behavior of words which tend to appear in bursts, ie,

C. Macdonald et al. (Eds.): ECIR 2008, LNCS 4956, pp. 150–161, 2008.
© Springer-Verlag Berlin Heidelberg 2008

once they appear in a document, they are much more likely to appear again. The notion of burstiness is similar to the one of *aftereffect* of future sampling ([5]), which describes the fact that the more we find a word in a document, the higher the expectation to find new occurrences. [2] makes use of the Laplace law of succession (which is called *normalization L*) as well as a ratio of Bernoulli processes (*normalization B*) to model the aftereffect of sampling.

Let x_i be the number of occurrences of word i in a given document. For a word probability distribution $P(x_i)$, [3] measures the burstiness through the quantity:

$$B_P = \frac{E_P[x_i]}{P(x_i \geq 1)}$$

where E_P denotes the expectation with respect to P. This measure provides a way to compare two different word distributions with respect to burstiness, but does not give a clear measure on whether a given word distribution accounts or not for bursty and non-bursty words. In order to do so, we introduce the following definition:

Definition 1. *We say that a word i is bursty at level n_0 under a distribution P iff there exists an integer $n_0, 1 \leq n_0$, such that for all integers $(n', n), n' \geq n \geq n_0$:*

$$P(x_i \geq n' + 1 | x_i \geq n') \geq P(x_i \geq n + 1 | x_i \geq n)$$

This definition directly translates the fact that a word is bursty if it is easier to generate it again once it has been generated a certain number of times (passed a certain level). The introduction of a burstiness level (n_0) in this definition allows one to capture finer-grain behaviors. In pratice, however, it is not always easy to compute $P(x_i \geq n + 1 | x_i \geq n)$ and determine whether a particular word distribution can account for burstiness. The following property (the proof of which is given in appendix A) can be used to do so:

Property 1. **Characterization of Burstiness**

Let $P(x_i)$ be a frequency distribution for word i and let $a_n = \frac{P(x_i = n+1)}{P(x_i = n)}$.

(i) If there exists n_o such that a_n is increasing from n_0 on, then i is bursty (at level n_0) under P

(ii) If there exists n_o such that a_n is decreasing from n_0 on, then i is not bursty (at any level) under P

Using this property, it is easy to see that the binomial, Poisson and geometric distributions cannot account for burstiness (we skip the derivation which is mainly technical). Other distributions can however model burstiness and the aftereffect of future sampling:

- **Normalization L (Laplace Law of succession)** In this case, a_n is equivalent to the quantity Prob$_2$ of [2]:

$$a_n = \frac{n}{(n + 1)}$$

which is increasing. The Laplace law of succession thus models burstiness and the aftereffect of future sampling.

- **Normalization B** We have:

$$a_n = 1 - \frac{F + 1}{D(n+1)}$$

where F and D are constant here (respectively defined, in [2], as the total number of occurrences of the word and the number of documents in which the word occurs). Here again a_n is increasing, so that *normalization B* fully models the aftereffect of future sampling.

In addition to normalizations L and B, other distributions have been considered to model burstiness. In particular, the fact that the multinomial distribution does not model burstiness (as its marginals are binomial distributions) has led several researchers to develop a new model, namely the DCM (*Dirichlet Compound Multinomial*) model, first introduced in [10], then studied in [8] and extended in [4] under the name EDCM. Both the DCM and EDCM models can account for word burstiness. We show here that it is the case for EDCM (the proof for DCM is similar). Marginalizing all the words in the document but word i leads, for the EDCM model, to:

$$P(x_i^d | \beta) \propto \frac{\Gamma(s - \beta_i + l_d - x_i^d)}{\Gamma(s - \beta_i)} \frac{\beta_i}{x_i^d}$$

where $\beta_i(1 \leq i \leq M)$ are the parameters of the model, $x_i^d(1 \leq i \leq M)$ the number of occurrences of word i in document d, l_d the length of document d and $s = \sum_{i:x_i^d \geq 1} \beta_i$. From this, we have:

$$\frac{a_{n+1}}{a_n} = \left(\frac{n^2 + 2n + 1}{n^2 + 2n} \right) \times \left(\frac{s - \beta_i + l_d - n - 1}{s - \beta_i + l_d - n - 2} \right) > 1$$

which shows that EDCM accounts for word burstiness.

Of course, both normalizations L and B were chosen in [2] because of their capacity to model the aftereffect of future sampling. Similarly, the DCM and EDCM models were partly chosen for their ability to model burstiness. Our development here simply confirms this on the basis of property 1. However, property 1 can be used as a formal, operational criterion to select new distributions for modeling burstiness and the aftereffect of future sampling. We will see below that indeed both the negative binomial and beta negative binomial distributions can be used to this effect, distributions that we now introduce.

3   The Negative Binomial and Beta Negative Binomial Distributions

Church and Gale ([3]) were the first ones, to our knowledge, to provide a complete study of the number of documents in which a given word i occurs exactly x_i times (a quantity we denote by $\#(d, x_i)$). Their work led to several findings on how words behave in document collections. In particular, they compared the binomial and Poisson distributions with mixtures of Poissons to model $P(\#(d, x_i))$. Their results indicate that the negative binomial distribution, which is an infinite mixture of Poisson distributions, fits the data better than the other distributions, which are however used in many document probabilistic models. The family of negative binomial distributions is a two parameter family, and supports several equivalent parametrizations. A commonly used one employs two real valued parameters, β and r, with $0 < \beta < 1$ and $0 < r$, and leads to the following probability mass function:

$$\text{NegBin}(\text{x;r},\beta) = \frac{\Gamma(r + x)}{x!\Gamma(r)}(1 - \beta)^r \beta^x$$

for $x = 0, 1, 2, \cdots$ (Γ is the gamma function).

The good behavior of the negative binomial distribution for text processing has also been observed in different, recent works. [1] uses respectively a binomial, a Poisson and a negative binomial distribution to model the probability of words given classes in a Naïve Bayes classifier. [14] reproduces the experiments reported in [3] on different collections. Again, the negative binomial is shown to provide a better fit to the data. One disadvantage however of the negative binomial distribution lies in the fact that maximum likelihood estimators for its parameters do not exist when only one observation is available, which makes this distribution not suited to model the probability of occurrences in a single document, a quantity used in eg the language model for information retrieval ([12]), or the divergence from randomness model ([2]).

An interesting extension to the negative binomial distribution consists in considering that the parameter β arises from a prior Beta distribution. In this case, the resulting distribution has the form:

$$BNBGen(x; r, a, b) = \frac{\Gamma(r + x)\Gamma(a + x)}{x!\Gamma(r)\Gamma(a)\Gamma(b)} \times \frac{\Gamma(a + b)\Gamma(r + b)}{\Gamma(a + b + r + x)} \tag{1}$$

where $x = 0, 1, 2, \cdots$, and a and b represent the two parameters of the prior Beta distribution. Assuming that this prior is uniform (ie $a = b = 1$), one obtains the following, simple, one-parameter distribution, which we will refer to as the **Beta Negative binomial distribution**, or **BNB** in short[1]:

$$BNB(x; r) = \frac{r}{(r + x + 1)(r + x)} \tag{2}$$

Again, this distribution is defined for $x = 0, 1, 2, \cdots$. Figure 1 shows probability distributions of the BNB for several values of the parameter. Furthermore, as we are going to see, maximum likelihood estimates exist for the BNB distribution even in the case where only one observation is available.

---

[1] This distribution is sometimes referred to as the Johnson distribution, inasmuch as it was studied by N. Johnson in [6].

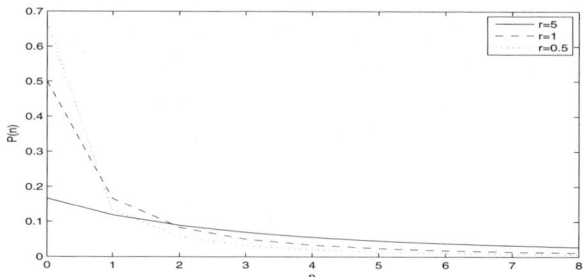

Fig. 1. BNB probability distribuions for several values of r

3.1 Burstiness

In the case of the negative binomial distribution, we have (for a_n as defined in property 1):

$$\forall n, a_n = \frac{\beta(r+n)}{n+1}$$

a_n is strictly increasing *iff* $r < 1$, strictly decreasing *iff* $r > 1$ and constant otherwise. This shows that the negative binomial can account for both bursty and non-bursty words, depending on the value of the parameter r.

For the BNB distribution, $a_n = \frac{r+n}{r+n+2}$ is strictly increasing. The BNB distribution can thus model burstiness and the aftereffect offuture sampling. More generally, the family of distributions given by equation 1 can be used to model word burstiness and the aftereffect of future sampling if the parameter r is such that $0 < r \leq 1$.

3.2 Parameter Estimation

We assume here that each word $i (1 \leq i \leq M)$ in a collection of N documents is modeled, independently of the other words, by a BNB distribution with parameter r_i. As before, the number of occurrences of word i in document d will be noted x_i^d. The maximum likelihood estimated for each r_i is defined as:

$$\hat{r}_i = \text{argmax}_{r_i} L(\mathcal{D}, r_i) = \text{argmax}_{r_i} \prod_d \frac{r_i}{(r_i + x_i^d)(r_i + x_i^d + 1)}$$

The derivative of L wrt r_i is:

$$\frac{\partial \log L}{\partial r_i} = \sum_d \frac{1}{r_i} - \frac{1}{r_i + x_i^d} - \frac{1}{r_i + x_i^d + 1}$$

Setting this derivative to 0 leads to: $\hat{r}_i = \frac{1}{N} \frac{1}{\sum_d \frac{1}{\hat{r}_i + x_i^d} - \frac{1}{\hat{r}_i + x_i^d + 1}}$ which defines a fixed-point equation for $\hat{r}_i$. Furhtermore, in the case where we have a single document, or where $\forall d, x_i^d = x_i$, the above equation leads to:

$$\hat{r}_i = \sqrt{x_i(x_i + 1)} \tag{3}$$

3.3   Relations to other Distributions

The probability of presence of a word in a document under the BNB model is provided by:

$$P_{\text{BNB}}(x_i \geq 1 | r_i) = \frac{r_i}{r_i + 1}$$

which, under the setting $r_i = x_i$, amounts to the Laplace law of succession. Hence, one can re-interpret the quantity Prob_2 in the normalization L of [2] as the probability of presence of a word in a document, under a BNB model the parameter of which is set to x_i. This last setting almost corresponds to the one obtained by maximum likelihood. Indeed, from a Taylor expansion of equation 3, one gets:

$$r_i \sim x_i(1 + \frac{1}{2x_i}) = x_i + 0.5$$

this approximation being already valid for small x_is, as the approximation error for $x_i = 2$ is around 2.5%. Thus, when x_i is sufficiently large, $r_i \sim x_i$ (the approximation error being around 5% for $x_i = 10$).

Let F_i be the total number of occurrences of word i in the collection: $F_i = \sum_d x_i^d$. Setting r_i to $\frac{F_i}{N}$ (ie the number of occurrences of i in any document on the basis of a random distribution) yields:

$$P_{\text{BNB}}(x_i \geq 1 | r_i) = \frac{F}{F + N} \approx \frac{F}{N}, \text{ for } \frac{F}{N} \text{ small or moderate}$$

which is the quantity Prob_1 used in the *tf-itf I(F)* model of [2]. Hence, the *tf-itf I(F)* model of [2] can be directly derived from a BNB distribution, the parameter of which is set to $\frac{F_i}{N}$, ie the number of occurrences of word i under a random distribution.

Lastly, based on the representation: $P(d) \propto \prod_{1 \leq i \leq M} P(x_i \geq 1)^{x_i^d}$, in the context of categorization, the decision for probabilistic classifiers is to assign d to the category c which maximizes:

$$F(c) = \log(P(c)) + \sum_{1 \leq i \leq M} x_i^d \log(P(x_i \geq 1 | c))$$

where $P(x_i \geq 1 | c)$ denotes the probability of presence of word i in category c. Assuming that this probability is given by a BNB model, the parameter of which (r_i^c) is set to $\frac{x_i^c}{l_c}$ (that is the number of occurrences of i in c normalized by the length of c), we have:

$$F(c) = \log(P(c)) + \sum_{1 \leq i \leq M} x_i^d (\log(\frac{x_i^c}{l_c}) - \log(\frac{x_i^c}{l_c} + 1))$$

In general, $x_i^c << l_c$, so that: $F(c) \sim \log(P(c)) + \sum_{1 \leq i \leq M} x_i^d \log(\frac{x_i^c}{l_c})$, which is exactly the function used by a multinomial Naive Bayes classifier, the parameters of which are set through maximum likelihood (see for example [9]). Hence, the multinomial Naive Bayes classifier can be approximated by a model based on BNB distributions for words, the parameters of which are set to the number of occurrences of the words in the category, normalized by the length of the category. The validity of this approximation will be confirmed in the experimental section (4).

4   Experiments

We want to illustrate the behavior of the BNB distribution in the context of text categorization and information retrieval. Regarding categorization, we have compared BNB models to state-of-the-art probabilistic models which use the same type of information, with different distributions. The models we have retained are the multinomial Naive Bayes classifier and the DCM ([8,4]) model. Several works have compared these classifiers with discriminative approaches such as Support Vector Machines (see for example [11,13]). On several collections, probabilistic and discriminative classifiers are on par. We will not replicate such a comparison here. Regarding information retrieval, we have used the general Divergence from Randomness, which we will refer to as *DFR*, framework defined in [2] to build information retrieval models. Our results suggest that a model based on the BNB distribution yields state-of-the-art results, while being simpler to implement as it dispenses from one normalization.

Categorization. We used the 20NewsGroup and Industry corpora for our categorization experiments. These corpora have been processed following [8] (both corpora are part of the DCM toolbox, available at http://www.imm.dtu.dk/rem/). However, we did not filter out the 500 most common words as done in [8].

For the multinomial model, we used a standard Laplace smoothing. For the DCM model, we directly used the code available in the DCM toolbox. For the BNB, we used the decision function mentioned above:

$$F(c) = \log(P(c)) + \sum_{1 \leq i \leq M} x_i^d \log(P(x_i \geq 1|c))$$

and considered here two BNB models for $P(x_i \geq 1|c)$. In the first one, the parameter r_i^c is estimated through maximum likelihood ($r_i^c = \sqrt{x_i^c(x_i^c + 1)}$). We refer to this model as *BNB-mle*. In the second one, we used the approximation leading to the Laplace law of succession ($r_i^c = x_i^c$). We refer to this model as *BNB-laplace*. In both cases, x_i^c represents the number of occurrences of word i in category c. To ensure that the quantities $P(d|c)$ are comparable across categories, we re-normalized the parameters r_i^c by constraining their sum s to be equal to a fixed value.

In a first experiment, we used 20 splits (80% training, 20% test for 20NewsGroup, 50% training 50% test for Industry) and averaged the precision obtained with the different models, with s set to 1. The results obtained are displayed in table 1. These results confirm that the approximation of the maximum likelihood estimate of the parameter of the BNB distribution by the number of occurrences is valid, as the two BNB models perform similarly. They also confirm the similarity between the BNB and multinomial distributions, established for document

Table 1. Average precision, on 20 splits, of the different models (best results in bold)

	20NewsGroup	Industry
Multinomial	0.875	**0.805**
BNB-laplace $(s = 1)$	**0.878**	0.804
BNB-mle $(s = 1)$	0.876	**0.805**
DCM $(s = 1)$	**0.878**	0.785

categorization in 3.3: the two BNB models and the multinomial one yield here the same level of performance. Comparing the different models, one can see that the DCM model does not behave here as well as the other models on Industry. This is in accordance with the results presented in [11]. Setting s to different values in the DCM model will yield different results. However, we were not able, on this corpus, to get results significantly higher than the ones reported here. In addition, the DCM is a rather complex model, which is very slow to train.

The quantity s used as a normalization in the BNB models defines an additional parameter that can be tuned to the collection considered. In order to test its influence, we let it vary from 0.5 to 10,000. The results obtained for different values of s on the BNB-laplace model are reported in table 2. As one can note, larger values of s yield a higher precision on the Industry corpus, whereas the best results for the 20NewsGroup are obtained with small values of s. Furthermore, the gain, over the previous models, is significant on the Industry corpus. We then estimated the best value for s on a development set consisting of 10% of each collection, and used the value obtained on the test set. The best values for s were 1 for 20NewsGroup, and $1,000$ for Industry, respectively yielding an average precision of 0.878 for 20NEwsGroup and 0.8225 on Industry. Again, the gain over the other models on Industry is significant.

Table 2. Influence of s on the precision for BNB-laplace (results are averaged over 5 splits; best results are in bold)

	0.5	1	10	50	100	500	1000	2000	5000	10000
20NewsGroup	**0.8789**	**0.8789**	**0.8789**	0.8784	0.8784	0.8775	0.8768	0.8754	0.8732	0.8701
Industry	0.8003	0.8006	0.8051	0.8119	0.8157	0.8214	**0.8226**	0.8219	0.8208	0.8194

Information Retrieval. Our experiments were carried out on the English part of the CLEF-2003 corpus[2], which contains around 160,000 documents. The indexing was done using the Lemur toolkit[3]. The resulting vocabulary contains ca. 80,000 distinct terms (types). We used the 60 queries of CLEF-2003 and their associated relevance judgements for evaluation. We tested the different models on short and long queries, respectively containing the title, and the title and the description.

---

[2] http://www.clef-campaign.org/
[3] http://www.lemurproject.org

In order to illustrate the behavior of the BNB distribution, we tested different versions of the two quantities used in the DFR framework of [2]. For $Prob_1$, we used (the notations are the same as before, with r_i^c the parameter of the BNB estimated on the collection):

- The model $I(F)$ and the geometric approximation of the Bose-Einstein distribution (denoted G), both used in [2];
- $Prob_1(x_i^d) = P_{\text{BNB}}(x_i \geq 1 | r_i^c)^{x_i^d}$, with r_i^c either set to $\frac{F_i}{N}$ or to its MLE value: $\frac{\sqrt{F_i^2 + N \times F_i}}{N}$. The first setting corresponds to the BNB approximation of $I(F)$ mentioned in 3.3. We will denote it $I(F)_{BNB}$. The second setting directly corresponds to a BNB distribution estimated through maximum likelihood. We will denote it $I(F)_{MLE}$;
- $Prob_1(x_i^d) = P_{\text{BNB}}(x_i^d | r_i^c)$, with r_i^c either set to its MLE value, or to: $r_i^c = \frac{F_i}{\sum_j F_j} \frac{\sum_d l_d}{N}$. The second setting allows one to take into account the effect of the document length on the number of occurrences of a word and corresponds to the proportion of word i in the collection times the average document length. However, as these two settings yielded very comparable results in our experiments, we retained only the second one, which will be denoted IM.

For $Prob_2$, we used another BNB model, $P_{\text{BNB}}(x_i \geq 1 | r_i^d)$, with r_i^d either set to x_i^d or to its MLE value: $\sqrt{x_i^d(x_i^d + 1)}$. As shown in 3.3, the first setting corresponds to the Laplace normalization, and will be denoted $L2$. We will denote the second setting L_{MLE}.

We then considered the following models for combining $Prob_1$ and $Prob_2$: $GL2$, $I(F)L2$, $I(F)_{BNB}L2$, $I(F)_{MLE}L_{MLE}$, $IML2$, IM, G. The first two combinations are used in [2] and serve here as gold standards. The third approximation aims at assessing the validity of the approximation of the $I(F)$ distribution by a BNB distribution, whereas the fourth one investigates the behavior of the MLE of the BNB, for both $Prob_1$ and $Prob_2$. The fifth model ($IML2$) aims at assessing a different use of the BNB distribution, namely the one directly based on the the probability of occurrence of a word. The last two models (IM and G) do not make use of $Prob_2$ so as to assess the impact and necessity of the $Prob_2$ normalization. For all models, we resized the term frequency using equation (42) of [2].

The results we obtained are presented in table 3. As one can see, the approximatin of $I(F)$ by a BNB distribution is valid: the three models $I(F)L2$, $I(F)_{BNB}L2$ and $I(F)_{MLE}L_{MLE}$ provide comparable results on both short and long queries. Furthermore, model $I(F)_{MLE}L_{MLE}$ is based on two BNB distributions directly estimated from the data, so that the sole assumption made is the one of a BNB modeling of the data. As the $L2$ normalization can be seen as a particular BNB distribution (as shown in 3.3), we obtain here two models, $I(F)_{BNB}L2$ and $I(F)_{MLE}L_{MLE}$, which are directly based on the BNB distribution for their two components ($Prob_1$ and $Prob_2$), and which yield state-of-the-art performance. This suggests that the DFR framework could be simplified by relying on a single distribution, the parameters of which may be estimated through maximum likelihood.

Table 3. Mean average precision (MAP), R-precision (R-PREC), precision at 5 documents (P5) and precision at 10 documents (P10) for the different models on the CLEF2003 English corpus (best results in bold)

		MAP	R-PREC	P5	P10
query-title	GL2	0.3610	0.3419	0.3367	0.2833
	I(F)L2	0.3609	**0.3422**	0.3367	**0.2883**
	I(F)$_{BNB}$L2	**0.3618**	0.3407	0.3367	0.2850
	I(F)$_{MLE}$L$_{MLE}$	0.3590	0.3398	0.3400	0.2850
	IML2	0.0865	0.0870	0.0833	0.0633
	IM	0.3617	0.3385	**0.3433**	0.2767
	G	0.1661	0.1589	0.1600	0.1433
query-desc	GL2	0.4905	0.4524	0.4233	**0.3433**
	I(F)L2	0.4925	0.4555	0.4267	0.3400
	I(F)$_{BNB}$L2	**0.4950**	0.4575	**0.4367**	0.3400
	I(F)$_{MLE}$L$_{MLE}$	0.4937	**0.4652**	0.4267	0.3350
	IML2	0.1328	0.1280	0.1317	0.1183
	IM	0.4682	0.4144	0.4300	0.3327
	G	0.2479	0.2260	0.2433	0.2000

Comparing the different complete models with *IM*, one can note that the performance of *IM* is comparable to the others on short queries, and slightly below on long queries (for the MAP and R-precision, the precision at 5 and 10 documents being comparable). However, model *IM* does not make use of any renormalization based on Prob$_2$, so that it achieves a good performance with only part of the information used by the other complete models. More surprising here is the fact that model *IM* behaves very poorly when coupled with the Laplace normalization, whereas this normalization seems necessary for other models: model *G*, which is based on *GL2* but does not make use of the Laplace normalization, yields very bad results. The relationship between the distribution used and the *first normalization of the informative content* of [2] remains to be better understood.

5   Conclusion

The goal of this work was to introduce the Beta negative binomial distribution as a possible alternative to other distributions used for probabilistic text modeling. As we mentioned, we derived this distribution from two considerations: one based on word burstiness, the other on the good empirical fit to data provided by the negative binomial distribution. Regarding burstiness and the aftereffect of future sampling, we have presented in this paper a formal, operational characterization criterion. This criterion helped us validate previous distributions retained to model these phenomena. In addition, it allowed us to characterize new distributions. We have then introduced the Beta negative Binomial (BNB) distribution, and shown how it was related to other models: the *I(F)* model and the Laplace normalization of [2], and the multinomial Naive Bayes classifier.

Lastly, we have illustrated the behavior of this distribution on text catego-rization and information retrieval experiments. Regarding text categorization, a first version of the BNB distribution behaved similarly to the multinomial one, as predicted theoretically. However, as the BNB model we have considered for categorization has an additional parameter, we were able, by tuning this param-eter, to outperform both the multinomial and the DCM models on the Industry corpus. Of course, other models can be used for categorization. The recently pro-posed Smoothed Dirichlet (SD) model ([11]) respectively yielded 0.89 and 0.85 on the two collections we retained. However, this model uses a different kind of information than the one used by the BNB, multinomial and DCM models. Extending these latter models to take into account other information is a work that remains to be done.

For information retrieval, we have shown that a model based on the BNB distribution for both $Prob_1$ and $Prob_2$) was equivalent (on both theoretical and experimental grounds) to the *I(F)L2* model of [2], a state-of-art model in prob-abilistic information retrieval. Furthermore, a version of this model estimated through ML yielded similar results. We have also introduced a simple BNB model (model *IM*) which yielded results comparable to the other models on short queries. However, it does so without the *first normalization of the informative content*, a normalization necessary for other models, eg *GL2*. The importance of this normalization in the DFR framework, its impact on various models as well as possible normalizations to be used with the BNB distribution remain to be studied thoroughly.

Acknowledgements

We wish to thank the IR Glasgow group (in particular Keith van Rijsbergen) for useful discussions on these topics. We also thank the three reviewers for useful comments.

References

1. Airoldi, E., Cohen, W., Fienberg, S.: Statistical models for frequent terms in text. CMU-CLAD Technical Report (2004), `http://reports-archive.adm.cs.cmu.edu/cald2005.html`
2. Amati, G., van Rijsbergen, C.: Probabilistic models of information retrieval based on measuring the divergence from randomness. ACM Transactions on Information Systems 20(4) (2002)
3. Church, K., Gale, W.: Poisson mixtures. Natural Language Engineering 1(2) (1995)
4. Elkan, C.: Clustering documents with an exponential-family approximation of the dirichlet compound multinomial distribution. In: ICML 2006: Proceedings of the 23rd international conference on Machine learning, ACM Press, New York (2006)
5. Feller, W.: An Introduction to Probability Theory and Its Applications, vol. I. Wiley, New York (1968)
6. Johnson, N., Kemp, A., Kotz, S.: Univariate Discrete Distributions. John Wiley, Chichester (1993)

7. Katz, S.: Distribution of content words and phrases in text and language modeling. Natural Language Engineering 2(1) (1996)
8. Madsen, R.E., Kauchak, D., Elkan, C.: Modeling word burstiness using the dirichlet distribution. In: ICML 2005: Proceedings of the 22nd international conference on Machine learning, ACM Press, New York (2005)
9. McCallum, A., Nigam, K.: A comparison of event models for naive bayes text classification. In: Proceedings of AAAI-98 Workshop on Learning for Text Categorization (1998)
10. Minka,T.: Estimating a Dirichlet Distribution. PhD thesis (2003) Unpublished paper available at: www.research.microsoft.com/~minka
11. Nallapati, R., Minka, T., Robertson, S.: The smoothed-dirichlet distribution: a new building block for generative models. CIIR Technical Report (2006), http://www.cs.cmu.edu/~nmramesh/sd_tc.pdf
12. Ponte, J.M., Croft, W.B.: A language modeling approach to information retrieval. In: Research and Development in Information Retrieval, SIGIR 1998 (1998)
13. Rennie, J., Shih, L., Teevan, J., Karger, D.: Tackling the poor assuptions of naive bayes classifiers. In: ICML 2003 (2003)
14. Rigouste, L.: Modéthodes probabilistes pour l'analyse exploratoire de données textuelles. PhD thesis, Thèse de l'ENST, Télécom Paris (2006)

A   Proof of Property 1

Let us recall what property 1 states:

Let $P(x_i)$ be a frequency distribution for word i and let $a_n = \frac{P(x_i=n+1)}{P(x_i=n)}$.

(i) If there exists n_o such that a_n is increasing from n_0 on, then i is bursty (at level n_0) under P

(ii) If there exists n_o such that a_n is decreasing from n_0 on, then i is not bursty (at any level) under P

Proof We have: $P(x_i \geq n+1 | x_i \geq n) = \frac{P(x_i \geq n+1)}{P(x_i \geq n)} = \frac{1}{\frac{P(x_i=n)}{P(x_i \geq n+1)}+1}$

But:

$$\frac{P(x_i \geq n+1)}{P(x_i = n)} = a_n + a_n a_{n+1} + \cdots \; ; \; \frac{P(x_i \geq n+2)}{P(x_i = n+1)} = a_{n+1} + a_{n+1} a_{n+2} + \cdots$$

Comparing the right members of the above equations term by term leads for (i) to: $\forall n \in N, n \geq n_0$

$$\frac{P(x_i \geq n+2)}{P(x_i = n+1)} \geq \frac{P(x_i \geq n+1)}{P(x_i = n)}$$

and hence: $\forall n \in N, n \geq n_0, P(x_i \geq n+2 | x_i \geq n+1) \geq P(x_i \geq n+1 | x_i \geq n)$ which establishes (i).

Similarly, for (ii) we obtain: $\forall n \in N, n \geq n_0, P(x_i \geq n+2 | x_i \geq n+1) \leq P(x_i \geq n+1 | x_i \geq n)$ which proves (ii).

Utilizing Passage-Based Language Models for Document Retrieval

Michael Bendersky[1] and Oren Kurland[2]

[1] Center for Intelligent Information Retrieval, Department of Computer Science,
University of Massachusetts, Amherst, MA 01003
bemike@cs.umass.edu
[2] Faculty of Industrial Eng. & Mgmt., Technion, Israel
kurland@ie.technion.ac.il

Abstract. We show that several previously proposed *passage-based* document ranking principles, along with some new ones, can be derived from the same probabilistic model. We use language models to instantiate specific algorithms, and propose a *passage language model* that integrates information from the ambient document to an extent controlled by the estimated *document homogeneity*. Several document-homogeneity measures that we propose yield passage language models that are more effective than the standard passage model for basic document retrieval and for constructing and utilizing *passage-based relevance models*; the latter outperform a document-based relevance model. We also show that the homogeneity measures are effective means for integrating document-query and passage-query similarity information for document retrieval.

Keywords: passage-based document retrieval, document homogeneity, passage language model, passage-based relevance model.

1 Introduction

The ad hoc retrieval task is to rank documents in response to a query by their assumed relevance to the information need it represents. While a document can be compared as a whole to the query, it could be the case (e.g., for long and/or heterogeneous documents) that only (very few, potentially small) parts of it, i.e., *passages*, contain information pertaining to the query. Thus, researchers have proposed different approaches for utilizing passage-based information for document retrieval [1,2,3,4,5,6,7,8].

We show that some of these previously proposed passage-based document-ranking approaches can in fact be derived from the same probabilistic model. Among the methods we derive are ranking a document by the highest query-similarity score that any of its passages is assigned [2,4,8], and by interpolating this score with the document-query similarity score [2,4].

We instantiate specific retrieval algorithms by using *statistical language models* [9]. In doing so, we propose a *passage language model* that incorporates information from the ambient document to an extent controlled by the estimated

C. Macdonald et al. (Eds.): ECIR 2008, LNCS 4956, pp. 162–174, 2008.

document homogeneity. Our hypothesis is that (language) models of passages in highly homogeneous documents should pull a substantial amount of information from the ambient document; for passages in highly heterogeneous documents, minimal such information should be used.

Several document-homogeneity measures that we propose yield passage language models that are more effective than the standard passage model [8] — as experiments over TREC data attest — for basic passage-based document ranking and for constructing and utilizing *passage-based relevance models* [8]; the latter also outperform a document-based relevance model [10].

We also derive, and demonstrate the effectiveness of, a novel language-model-based algorithm that integrates, using document-homogeneity measures, the query-similarity of a document and of its passages for document ranking.

2   Retrieval Framework

In what follows we show that some previously-proposed passage-based document-ranking approaches, and some new ones, can be derived from the same model.

Notation and conventions. Throughout this section we assume that the following have been fixed: a query q, a document d, and a corpus of documents $\mathcal{C}$ ($d \in \mathcal{C}$). We use g to denote a passage, and write $g \in d$ if g is one of d's m passages. (Our algorithms are not dependent on the type of passages.) We write $p_x(\cdot)$ to denote a (smoothed) unigram language model induced from x (a document or a passage); our language model induction methods are described in Sec. 2.2.

2.1   Passage-Based Document Ranking

We rank document d in response to query q by estimating the probability $p(q|d)$ that q can be generated[1] from a model induced from d, as is common in the language modeling approach to retrieval [12,9]. We hasten to point out, however, that our framework is not committed to any specific estimates for probabilities of the form $p(q|x)$, which we often refer to as the "query-similarity" of x.

Since passages are smaller — and hence potentially more focused — units than documents, they can potentially "help" in generating queries. Thus, assuming that *all* passages in the corpus can serve as proxies (representatives) of d for generating *any* query, and using $p(g_i|d)$ to denote the probability that passage g_i (of some document in the corpus) is chosen as a proxy of d, we can write

$$p(q|d) = \sum_{g_i} p(q|d, g_i)p(g_i|d) \ . \tag{1}$$

---

[1] While it is convenient to use the term "generate" in reference to work on language models for IR [9], we do not think of text items as literally generating the query. Furthermore, we do not we assume an underlying generative theory in contrast to Lavrenko and Croft [10], and Lavrenko [11], *inter alia.*

If we assume that d's passages are much better proxies for d than passages not in d, then we can define $\hat{p}(g_i|d) \stackrel{def}{=} \frac{p(g_i|d)}{\sum_{g_j \in d} p(g_j|d)}$ if $g_i \in d$, 0 otherwise, and use it in Eq. 1 to rank d as follows:

$$Score(d) \stackrel{def}{=} \sum_{g_i \in d} p(q|d, g_i)\hat{p}(g_i|d) \ . \tag{2}$$

To estimate $p(q|d, g_i)$, we integrate $p(q|d)$ and $p(q|g_i)$ based on the assumed *homogeneity* of d: the more homogeneous d is assumed to be, the higher the impact it has as a "whole" on generating q. Specifically, we use the estimate[2] $h^{[\mathcal{M}]}(d)p(q|d) + (1 - h^{[\mathcal{M}]}(d))p(q|g_i)$, where $h^{[\mathcal{M}]}(d)$ assigns a value in $[0, 1]$ to d by homogeneity model $\mathcal{M}$. (Higher values correspond to higher estimates of homogeneity; we present document-homogeneity measures in Sec. 2.3.) Using some probability algebra (and the fact that $\sum_{g_i \in d} \hat{p}(g_i|d) = 1$), Eq. 2 then becomes

$$Score(d) \stackrel{def}{=} h^{[\mathcal{M}]}(d)p(q|d) + (1 - h^{[\mathcal{M}]}(d)) \sum_{g_i \in d} p(q|g_i)\hat{p}(g_i|d) \ , \tag{3}$$

with more weight put on the "match" of d as a whole to the query as d is considered more homogeneous.

If we consider d to be highly heterogeneous and consequently set $h^{[\mathcal{M}]}(d)$ to 0, and in addition use the relative importance (manually) attributed to g_i as a surrogate for $\hat{p}(g_i|d)$, Eq. 3 is then a previously proposed ranking approach for (semi-)structured documents [4]; if a uniform distribution is used for $\hat{p}(g_i|d)$, instead, we score d by the mean "query-similarity" of its constituent passages, which yields poor retrieval performance that supports our premise from Sec. 1 about long (and heterogeneous) documents.

Alternatively, we can bound Eq. 3 by

$$Score_{inter-max}(d) \stackrel{def}{=} h^{[\mathcal{M}]}(d)p(q|d) + (1 - h^{[\mathcal{M}]}(d)) \max_{g_i \in d} p(q|g_i) \ . \tag{4}$$

This scoring function is a generalized form of approaches that interpolate the document-query similarity score and the maximum query-similarity score assigned to any of its passages using fixed weights [14, 2, 15, 4]; hence, such methods (implicitly) assume that all documents are homogeneous to the same extent. Furthermore, note that assuming that d is highly homogeneous and setting $h^{[\mathcal{M}]}(d) = 1$ results in a standard document-based ranking approach; on the other hand, assuming d is highly heterogeneous and setting $h^{[\mathcal{M}]}(d) = 0$ yields a commonly-used approach that scores d by the maximum query-similarity measured for any of its passages [2, 7, 4, 8]:

$$Score_{max}(d) \stackrel{def}{=} \max_{g_i \in d} p(q|g_i) \ . \tag{5}$$

---

[2] This is reminiscent of some work on cluster-based retrieval [13].

2.2 Language-Model-Based Algorithms

Following standard practice in work on language models for IR [9], we estimate $p(q|d)$ and $p(q|g_i)$ using the unigram language models induced from d and g_i, i.e., $p_d(q)$ and $p_{g_i}(q)$, respectively. Then, Eq. 4 yields the novel **Interpolated Max-Scoring Passage** algorithm, which scores d by $h^{[\mathcal{M}]}(d)p_d(q) + (1 - h^{[\mathcal{M}]}(d)) \max_{g_i \in d} p_{g_i}(q)$. Using language models in Eq. 5 yields the **Max-Scoring Passage** algorithm, which scores d by $\max_{g_i \in d} p_{g_i}(q)$ as was proposed by Liu and Croft [8].

Language Model Induction. We use $\widetilde{p}_x^{MLE}(w)$ to denote the maximum likelihood estimate (MLE) of term w with respect to text (or text collection) x, and smooth it using corpus statistics to get the standard (basic) language model [16]:

$$\widetilde{p}_x^{[basic]}(w) = (1 - \lambda_{\mathcal{C}})\widetilde{p}_x^{MLE}(w) + \lambda_{\mathcal{C}}\widetilde{p}_{\mathcal{C}}^{MLE}(w) \; ; \tag{6}$$

$\lambda_{\mathcal{C}}$ is a free parameter.

We extend the estimate just described to a sequence of terms $w_1 w_2 \cdots w_n$ by using the unigram-language-model term-independence assumption

$$p_x^{[basic]}(w_1 w_2 \cdots w_n) \stackrel{def}{=} \prod_{j=1}^{n} \widetilde{p}_x^{[basic]}(w_j) \; . \tag{7}$$

Passage Language Model. Using $p_{g_i}^{[basic]}(q)$ in the above-described algorithms implies that document d is so heterogeneous that in estimating the "match" of each of its passages with the query we do not consider any information from d, except for that in the passage itself.

Some past work on question answering, and passage and XML retrieval [17, 18, 19, 20, 21, 22] uses a passage language model that exploits information from the ambient document to the same fixed extent for all passages and documents. In contrast, here we suggest to use the document estimated homogeneity to control the amount of reliance on document information. (Recall that homogeneity measures are used in the Interpolated Max-Scoring Passage algorithm for fusion of similarity scores.) Hence, for $g \in d$ we define the passage language model

$$\widetilde{p}_g^{[\mathcal{M}]}(w) \stackrel{def}{=} \lambda_{psg}(g)\widetilde{p}_g^{MLE}(w) + \lambda_{doc}(d)\widetilde{p}_d^{MLE}(w) + \lambda_{\mathcal{C}}\widetilde{p}_{\mathcal{C}}^{MLE}(w) \; ; \tag{8}$$

we fix $\lambda_{\mathcal{C}}$ to some value, and set $\lambda_{doc}(d) = (1 - \lambda_{\mathcal{C}})h^{[\mathcal{M}]}(d)$ and $\lambda_{psg}(g) = 1 - \lambda_{\mathcal{C}} - \lambda_{doc}(d)$ to have a valid probability distribution. We then extend this estimate to sequences as we did at the above

$$p_g^{[\mathcal{M}]}(w_1 w_2 \cdots w_n) \stackrel{def}{=} \prod_{j=1}^{n} \widetilde{p}_g^{[\mathcal{M}]}(w_j) \; . \tag{9}$$

Setting $h^{[\mathcal{M}]}(d) = 0$ — considering d to be highly heterogeneous — we get the standard passage language model from Eq. 7. On the other hand, assuming d is highly homogeneous and setting $h^{[\mathcal{M}]}(d) = 1$ results in representing each of d's passages with d's standard language model from Eq. 7; note that in this case the Max-Scoring Passage algorithm amounts to a standard document-based language model retrieval approach.

2.3   Document Homogeneity

We now consider a few simple models $\mathcal{M}$ for estimating document homogeneity. We define functions $h^{[\mathcal{M}]} : \mathcal{C} \to [0,1]$ with higher values corresponding to (assumed) higher levels of homogeneity.

Long documents are often considered as more heterogeneous than shorter ones. We thus define the normalized length-based measure

$$h^{[length]}(d) \stackrel{def}{=} 1 - \frac{\log|d| - \min_{d_i \in \mathcal{C}} \log|d_i|}{\max_{d_i \in \mathcal{C}} \log|d_i| - \min_{d_i \in \mathcal{C}} \log|d_i|} \ ,$$

where $|d_j|$ is the number of terms in d_j.[3]

The length-based measure does not handle the case of short heterogeneous documents. We can alternatively say that d is homogeneous if its term distribution is concentrated around a small number of terms [23]. To model this idea, we use the *entropy* of $d's$ unsmoothed language model and normalize it with respect to the maximum possible entropy of *any* document with the same length as that of d (i.e., $\log|d|$):[4]

$$h^{[ent]}(d) \stackrel{def}{=} 1 + \frac{\sum_{w' \in d} \widetilde{p}_d^{MLE}(w') \log(\widetilde{p}_d^{MLE}(w'))}{\log|d|} \ .$$

Both homogeneity measures just described are based on the document as a whole and do not explicitly estimate the variety among its passages. We can assume, for example, that the more similar the passages of a document are to each other, the more homogeneous the document is. Alternatively, a document with passages highly similar to the document as a whole might be considered homogeneous. Assigning d's passages with unique IDs, and denoting the tf.idf[5]

---

[3] Normalizing the length with respect to documents in several corpora (including the ambient corpus) yields very similar retrieval performance to that resulting from normalization with respect to documents in the ambient corpus alone.

[4] $Entropy(d) \stackrel{def}{=} - \sum_{w' \in d} \widetilde{p}_d^{MLE}(w') \log(\widetilde{p}_d^{MLE}(w'))$; higher values correspond to (assumed) lower levels of homogeneity. A document d with all terms different from each other has the maximum entropy ($\log|d|$) with respect to documents of length $|d|$. If $|d| = 1$, we set $h^{[ent]}(d)$ to 1.

[5] Modeling these two homogeneity notions using the KL divergence between language models yields substantially-inferior retrieval performance to that of using the proposed vector space representation with the cosine measure.

vector-space representation of text x as $\boldsymbol{x}$, we can define these homogeneity notions using the functions $h^{[interPsg]}(d)$ and $h^{[docPsg]}(d)$, respectively, where

$$h^{[interPsg]}(d) \stackrel{def}{=} \begin{cases} \frac{2}{m(m-1)} \sum_{i<j;g_i,g_j \in d} \cos(\boldsymbol{g_i}, \boldsymbol{g_j}) & \text{if } m > 1 \text{ ,} \\ 1 & \text{otherwise ;} \end{cases}$$

$$h^{[docPsg]}(d) \stackrel{def}{=} \frac{1}{m} \sum_{g_i \in d} \cos(\boldsymbol{d}, \boldsymbol{g_i}) \text{ .}$$

3 Related Work

There is a large body of work on utilizing (different types of) passages for document retrieval [1, 2, 3, 4, 5, 6, 7, 8]. We showed in Sec. 2 that several of these methods can be derived and generalized from the same model.

Utilizing passage language models is a recurring theme in question answering [24, 25, 18], sentence and passage retrieval [26, 20, 22], document retrieval [3, 6, 8] and XML retrieval [19, 21]. As mentioned in Sec. 2.2, some prior work [17, 18, 19, 20, 21, 22] smooth the passage (sentence) model with its ambient document's statistics, by using interpolation with fixed weights. We present in Section 4.1 the relative merits of our approach of using document homogeneity measures for controlling the reliance on document statistics.

Liu and Croft's work [8] most resembles ours in that they use the Max-Scoring Passage algorithm with the basic passage model from Eq. 7; they also use a *passage-based relevance model* [10] to rank documents. We demonstrate the merits in using their methods with our passage language model in Sec. 4.

4 Evaluation

We conducted our experiments on the following four TREC corpora:

corpus	# of docs	avg. length	queries	disk(s)
FR12	45,820	935	51-100	1,2
LA+FR45	186,501	317	401-450	4,5
WSJ	173,252	263	151-200	1-2
AP89	84,678	264	1-50	1

FR12, which was used in work on passage-based document retrieval [2, 8], and LA+FR45, which is a challenging benchmark [27], contain documents that are longer on average (and often considered more heterogeneous) than those in WSJ and AP89.

We used the Lemur toolkit (www.lemurproject.org) to run our experiments. We applied basic tokenization and Porter stemming, and removed INQUERY stopwords. We used titles of TREC topics as queries.

To evaluate retrieval performance, we use the mean average (non-interpolated) precision (MAP) at 1000, and the precision of the top 10 documents (p@10). We determine statistically significant differences in performance using the two-tailed Wilcoxon test at the 95% confidence level.

Passages. While there are several passage types we can use [7], our focus is on the general validity of our retrieval algorithms and language-model induction techniques. Therefore, we use *half overlapping fixed-length windows* (of 150 and 50 terms[6]) as passages and mark them *prior* to retrieval time. Such passages are computationally convenient to use and were shown to be effective for document retrieval [2], specifically, in the language model framework [8].

Table 1. Performance numbers of the Max-Scoring Passage algorithm (MSP) with either the basic passage language model (MSPbase) or our passage language model (MSP[$\mathcal{M}$]) that utilizes homogeneity model $\mathcal{M}$. Document-based language-model (DOCbase) retrieval performance is presented for reference. Boldface: best result per colomn; underline: best performance for a corpus per evaluation measure. d and p mark statistically significant differences with DOCbase and MSPbase, respectively.

| | FR12 | | | | LA+FR45 | | | |
| | PsgSize 150 | | PsgSize 50 | | PsgSize 150 | | PsgSize 50 | |
	MAP	p@10	MAP	p@10	MAP	p@10	MAP	p@10
DOCbase	22.0	13.3	22.0	13.3	22.7	26.4	22.7	**26.4**
MSPbase	28.4	14.8	30.1^d	14.8	21.9	25.5	21.7	25.7
MSP[length]	**29.6**^{d}	15.7	**31.8**^d_p	15.7	23.1$_p$	27.5	**23.6**$_p$	26.0
MSP[ent]	29.3^d	**16.2**	30.1^d	**16.2**	22.2	26.2	21.8	26.0
MSP[interPsg]	29.1^d	15.7	30.7^d	**16.2**	22.8$_p$	26.6	21.9	25.3
MSP[docPsg]	29.3^d	**16.2**	31.0^d	15.7	**23.2**^{d}	**27.9**	23.0	25.5

| | WSJ | | | | AP89 | | | |
| | PsgSize 50 | | PsgSize 150 | | PsgSize 50 | | PsgSize 150 | |
	MAP	p@10	MAP	p@10	MAP	p@10	MAP	p@10
DOCbase	28.4	39.6	28.4	39.6	**20.0**	**24.1**	**20.0**	24.1
MSPbase	28.8	41.8	26.1^d	40.4	18.8^d	23.0	17.7^d	22.4
MSP[length]	**29.3**^{d}	**43.0**^{d}	29.0$_p$	**44.8**^d_p	19.3$_p$	23.7	18.7$_p$	**24.6**
MSP[ent]	**29.3**$_p$	41.6	27.9$_p$	41.8	19.1$_p$	22.8	18.2^d_p	22.6
MSP[interPsg]	29.2^d	42.4^d	28.2$_p$	43.2$_p$	19.5$_p$	23.7	18.4^d_p	23.9
MSP[docPsg]	29.1^d	42.6^d	**29.2**$_p$	**44.8**^d_p	19.8$_p$	23.3	19.1$_p$	**24.6**

4.1   Experimental Results

Passage Language Model. To study the performance of our passage language model independently of score-integration (as performed by Interpolated Max-Scoring Passage), we use it in the Max-Scoring Passage algorithm, which was previously studied with the basic passage model [8].

Specifically, let $MSP[\mathcal{M}]$ denote the implementation of Max-Scoring Passage with our passage model $p_g^{[\mathcal{M}]}(\cdot)$, and $MSPbase$ denote its implementation with the basic passage model $p_g^{[basic]}(\cdot)$ [8]. Since our passage model leverages information from the ambient document, we also use as a reference comparison a standard document-based language-model retrieval arpproach ($DOCbase$) that scores document d by $p_d^{[basic]}(q)$.

All tested algorithms incorporate a single free parameter λ_C, which controls the extent of corpus-based smoothing. We fix λ_C to 0.5, because this

---

[6] Passages of 25 terms yield degraded performance as in some previous reports [2,8].

results in (near) optimal (MAP) performance for *both* our reference comparisons (*MSPbase* and *DOCbase*) with respect to values in $\{0.1, 0.2, \ldots, 0.9\}$.[7]

We present the performance numbers in Table 1. Our first observation is that the Max-Scoring Passage algorithm is consistently more effective (many times to a statistically significant degree) when utilizing our new passage language model ($MSP[\mathcal{M}]$) than when using the basic passage language model (*MSPbase*).

We can also see in Table 1 that the most effective homogeneity measures for inducing our passage model are *length* — demonstrating its correlation with heterogeneity — and *docPsg*; the latter measures the similarity between a document and its passages, and is thus directly related to the balance we want to control of using document-based vs. passage-based information. Furthermore, $MSP[length]$ and $MSP[docPsg]$ yield performance that is superior to that of document-based retrieval (*DOCbase*) in many of the relevant comparisons, espeically for FR12 and WSJ. For AP89, however, document-based retrieval is superior (in terms of MAP) to using (any) passage-based information, possibly due to the high homogeneity of the documents.

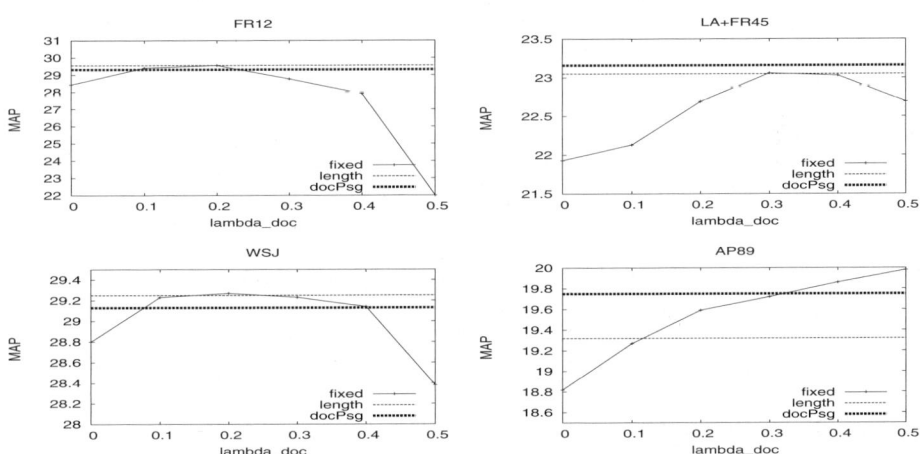

Fig. 1. The MAP performance curve of Max-Scoring Passage (PsgSize=150) when setting $\lambda_{doc}(d)$ (see Eq. 8) to the same *fixed* value in $\{0, 0.1, \ldots, 0.5\}$ for *all* documents. (0 and 0.5 correspond to *MSPbase* and *DOCbase*, respectively.) The performance of using the homogeneity measures *length* and *docPsg* is plotted for comparison with thin and thick horizontal lines, respectively. Note: figures are not to the same scale.

Further Analysis. Our passage model incorporates information from the ambient document to an extent controlled by the estimated document homogeneity. We now study the alternative of fixing the reliance on document information to the same extent for all documents and passages, as proposed in some past work [17, 18, 20, 22]. We do so by fixing $\lambda_{doc}(d)$ in Eq. 8 to a value in $\{0, 0.1, \ldots, 0.5\}$. (Recall that $\lambda_{doc}(d) = (1 - \lambda_C) h^{[\mathcal{M}]}(d)$ and $\lambda_C = 0.5$; also, setting $\lambda_{doc}(d)$ to 0

---

[7] Similar relative-performance patterns are observed for $\lambda_C = 0.3$.

and 0.5 corresponds to *MSPbase* and *DOCbase*, respectively.) We depict the resultant MAP performance curve (for passages of 150 terms) of the Max-Scoring Passage algorithm in Fig. 1. We plot for comparison the performance of using our best-performing homogeneity measures *length* and *docPsg*.

We can see in Fig. 1 that using homogeneity measures improves performance over a poor choice of a fixed $\lambda_{doc}(d)$; furthermore, for FR12, LA+FR45 and WSJ, the measures yield performance that is sometimes better than the best performance obtained by using some fixed $\lambda_{doc}(d)$, and always better than that of using either passage-only information or document-only information (see the end points of the curves). Many of the performance improvements posted by the homogeneity measures over a fixed $\lambda_{doc}(d)$ are also statistically significant, e.g., $MSP[length]$ and $MSP[docPsg]$'s performance is better to a statistically significant degree than setting $\lambda_{doc}(d)$ to (i) 0 for LA+FR45 and AP89, (ii) 0.5 for FR12 and WSJ, and (iii) $\{0.1, 0.2\}$ for AP89.

Table 2. Performance numbers of a *passage-based relevance model* [8]. We use either the originally suggested basic passage language model (relPsgBase) or our passage language model (relPsg[$\mathcal{M}$]). Document-based relevance-model performance is presented for reference (relDoc). Best result in a colomn is boldfaced, and best result for a corpus (per evaluation measure) is underlined; statistically significant differences with relDoc and relPsgBase are marked with d and p, respectively.

	FR12				LA+FR45			
	PsgSize 150		PsgSize 50		PsgSize 150		PsgSize 50	
	MAP	p@10	MAP	p@10	MAP	p@10	MAP	p@10
relDoc	10.7	9.1	10.7	9.1	20.7	23.8	20.7	23.8
relPsgBase	**31.7**^{d}	**14.3**^{d}	31.1^d	16.2^d	**22.4**	26.0	21.9	24.7
relPsg[length]	28.0^d	14.8^d	30.7^d	**18.1**^{d}	21.8$_p$	**26.6**	**23.3**^d_p	25.3
relPsg[docPsg]	26.9^d	**15.7**^{d}	**34.2**^{d}	**18.1**^{d}	20.4$_p$	25.1	22.8^d_p	**25.7**

	WSJ				AP89			
	PsgSize 150		PsgSize 50		PsgSize 150		PsgSize 50	
	MAP	p@10	MAP	p@10	MAP	p@10	MAP	p@10
relDoc	33.9	48.4	33.9	48.4	25.6	28.5	**25.6**	28.5
relPsgBase	34.5	47.2	34.0	45.0	24.1	29.8	22.2	25.9
relPsg[length]	35.4^d	50.0	37.5^d_p	49.0$_p$	25.1	**30.4**	24.3$_p$	30.0$_p$
relPsg[docPsg]	**35.9**^{d}	**50.2**	**37.6**^d_p	**50.2**$_p$	**25.7**	29.8	25.1$_p$	**31.3**$_p$

Relevance Model. The most effective *passage-based relevance model* approach for ranking documents that was suggested by Liu and Croft [8] is to construct a relevance model [10] only from passages and use it to rank documents. We compare their original implementation *relPsgBase*, which utilizes the basic passage model, to an implementation *relPsg[$\mathcal{M}$]*, which utilizes our passage language model $p_g^{[\mathcal{M}]}(\cdot)$. We also use a document-based relevance model (*relDoc*) [10] as a reference comparison.

We optimize the performance of *each* of our reference comparisons (*relPsgBase* and *relDoc*) with respect to the number of top-retrieved elements (i.e., passages or documents) and the number of terms used for constructing the relevance models;

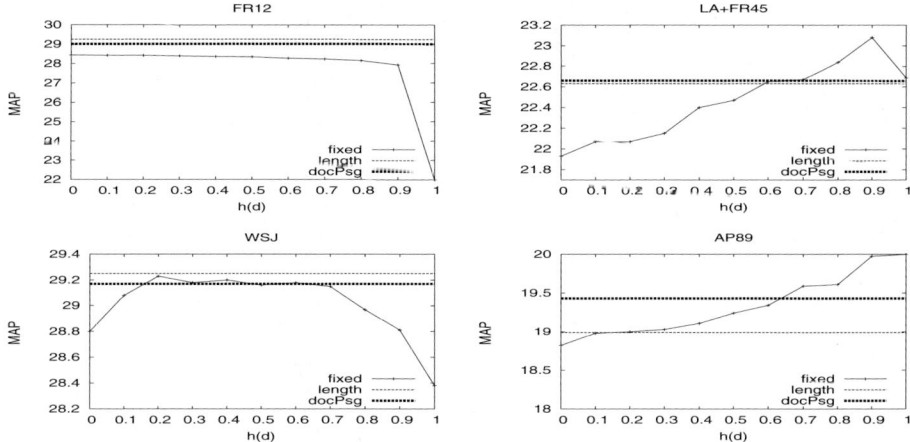

Fig. 2. The MAP performance curve of the Interpolated Max-Scoring Passage algorithm (PsgSize=150) when setting $h^{[\mathcal{M}]}(d)$ (see Eq. 4) to the same *fixed* value in $\{0, 0.1, \ldots, 1\}$ for *all* documents. (0 and 1 correspond to $MSPbase$ and $DOCbase$, respectively.) We also plot the performance of setting $\mathcal{M}$ to *length* and *docPsg* with thin and thick horizontal lines, respectively. Note: figures are not to the same scale.

specifically, we select these parameters' values from $\{25, 50, 75, 100, 250, 500\}$ — i.e., 36 parameter settings — so as to optimize MAP performance. We set $\lambda_{\mathcal{C}} = 0.5$ (as at the above) except for estimating top-retrieved elements' language models for constructing relevance models, wherein we set $\lambda_{\mathcal{C}} = 0.2$ following past recommendations [10]. Our $relPsg[\mathcal{M}]$ ($\mathcal{M} \in \{length, docPsg\}$) algorithms use the parameter values selected for the $relPsgBase$ reference comparison; therefore, their performance is not necessarily the optimal one they can achieve.

Table 2 shows that in most of the relevant comparisons using our passage language model yields passage-based relevance models ($relPsg[\mathcal{M}]$) that outperform both the original implementation ($relPsgBase$) [8] — which utilizes the basic passage model — and the document-based relevance model ($relDoc$). (Note, for example, that underlined numbers that constitute the best performance for a corpus per evaluation metric appear only in $relPsg[\mathcal{M}]$ rows.) In many cases, the performance differences are also statistically significant.

Interpolated Max-Scoring Passage. The algorithm scores document d by interpolation (governed by the homogeneity-based interpolation weight $h^{[\mathcal{M}]}(d)$) of the document-based language model score ($DOCbase$) with the score assigned by Max-Scoring Passage. (See Sec. 2.2.) To focus on this score integration, rather than combine it with information integration at the language model level[8], which we explored at the above, we use the basic passage language model $p_g^{[basic]}(\cdot)$; the Max-Scoring Passage implementation is then the $MSPbase$ defined above.

---

[8] Experiments show that such combination yields additional performance gains.

In Fig. 2 we present the MAP performance of Interpolated Max-Scoring Passage (with passages of 150 terms). We either use $h^{[\mathcal{M}]}(d)$ with the *length* and *docPsg* homogeneity measures[9], or set $h^{[\mathcal{M}]}(d)$ to a fixed value in $\{0, 0.1, \dots, 1\}$ for *all* documents (0 and 1 correspond to $MSPbase$ and $DOCbase$, respectively), which echoes some past work [2, 4].

We see in Fig. 2 that homogeneity measures yield performance that is (i) better than that of several fixed values of $h^{[\mathcal{M}]}(d)$, (ii) always better than the worse performing among $MSPbase$ and $DOCbase$ (see the end points of the curves), and (iii) sometimes (e.g., for FR12 and WSJ) better than the best performance attained by using some fixed $h^{[\mathcal{M}]}(d)$ for all documents[10]. Many of the improvements obtained by our homogeneity measures over a fixed $h^{[\mathcal{M}]}(d)$ are also statistically significant, e.g., *length* is significantly better than setting $h^{[\mathcal{M}]}(d)$ to (i) 0 for LA+FR45, WSJ and AP89, (ii) 0.9 for FR12, (iii) $\{0.1, \dots, 0.4\}$ for LA+FR45, and (iv) $\{0.1, 0.3\}$ for WSJ.

5   Conclusions

We derived some previously-proposed and new passage-based document-ranking approaches from the same model. We proposed an effective *passage language model* that incorporates information from the ambient document to an extent controlled by the estimated *document homogeneity*. Our homogeneity measures are also effective for integrating document and passage query-similarity information for document retrieval.

Acknowledgments. We thank the anonymous reviewers for their helpful comments. This paper is based upon work done in part while the first author was at the Technion and the second author was at Cornell University, and upon work supported in part by the Center for Intelligent Information Retrieval and by the National Science Foundation under grant no. IIS-0329064. Any opinions, findings and conclusions or recommendations expressed in this material are the authors' and do not necessarily reflect those of the sponsoring institutions.

References

1. Salton, G., Allan, J., Buckley, C.: Approaches to passage retrieval in full text information systems. In: Proceedings of SIGIR, pp. 49–58 (1993)
2. Callan, J.P.: Passage-level evidence in document retrieval. In: Proceedings of SIGIR, pp. 302–310 (1994)
3. Mittendorf, E., Schäuble, P.: Document and passage retrieval based on hidden Markov models. In: Proceedings of SIGIR, pp. 318–327 (1994)

---

[9] *ent* and *interPsg* yield inferior performance to that of *length* and *docPsg*, and are omitted to avoid cluttering of the figure.

[10] These observations also hold if Dirichlet-smoothed [16] language models are used for both passages and documents.

4. Wilkinson, R.: Effective retrieval of structured documents. In: Proceedings of SI-GIR, pp. 311–317 (1994)
5. Kaszkiel, M., Zobel, J.: Passage retrieval revisited. In: Proceedings of SIGIR, pp. 178–185 (1997)
6. Denoyer, L., Zaragoza, H., Gallinari, P.: HMM-based passage models for document classification and ranking. In: Proceedings of ECIR, pp. 126–135 (2001)
7. Kaszkiel, M., Zobel, J.: Effective ranking with arbitrary passages. Journal of the American Society for Information Science 52(4), 344–364 (2001)
8. Liu, X., Croft, W.B.: Passage retrieval based on language models. In: Proceedings of the 11th International Conference on Information and Knowledge Managment (CIKM), pp. 375–382 (2002)
9. Croft, W.B., Lafferty, J. (eds.): Language Modeling for Information Retrieval. Information Retrieval Book Series, vol. 13. Kluwer, Dordrecht (2003)
10. Lavrenko, V., Croft, W.B.: Relevance-based language models. In: Proceedings of SIGIR, pp. 120–127 (2001)
11. Lavrenko, V.: A Generative Theory of Relevance. PhD thesis, University of Massachusetts Amherst (2004)
12. Ponte, J.M., Croft, W.B.: A language modeling approach to information retrieval. In: Proceedings of SIGIR, pp. 275–281 (1998)
13. Kurland, O., Lee, L.: Corpus structure, language models, and ad hoc information retrieval. In: Proceedings of SIGIR, pp. 194–201 (2004)
14. Buckley, C., Salton, G., Allan, J., Singhal, A.: Automatic query expansion using SMART: TREC3. In: Proceedings of of the Third Text Retrieval Conference (TREC-3), pp. 69–80 (1994)
15. Cai, D., Yu, S., Wen, J.R., Ma, W.Y.: Block-based web search. In: Proceedings of SIGIR, pp. 456–463 (2004)
16. Zhai, C., Lafferty, J.D.: A study of smoothing methods for language models applied to ad hoc information retrieval. In: Proceedings of SIGIR, pp. 334–342 (2001)
17. Abdul-Jaleel, N., Allan, J., Croft, W.B., Diaz, F., Larkey, L., Li, X., Smucker, M.D., Wade, C.: UMASS at TREC 2004 — novelty and hard. In: Proceedings of the Thirteenth Text Retrieval Conference (TREC-13) (2004)
18. Hussain, M.: Language modeling based passage retrieval for question answering systems. Master's thesis, Saarland University (2004)
19. Ogilvie, P., Callan, J.: Hierarchical language models for XML component retrieval. In: Proceedings of INEX (2004)
20. Murdock, V., Croft, W.B.: A translation model for sentence retrieval. In: Proceedings of HLT/EMNLP, pp. 684–695 (2005)
21. Sigurbjörnsson, B., Kamps, J.: The effect of structured queries and selective indexing on XML retrieval. In: Proceedings of INEX, pp. 104–118 (2005)
22. Wade, C., Allan, J.: Passage retrieval and evaluation. Technical Report IR-396, Center for Intelligent Information Retrieval (CIIR), University of Massachusetts (2005)
23. Kurland, O., Lee, L.: PageRank without hyperlinks: Structural re-ranking using links induced by language models. In: Proceedings of SIGIR, pp. 306–313 (2005)
24. Corrada-Emmanuel, A., Croft, W.B., Murdock, V.: Answer passage retrieval for question answering. Technical Report IR-283, Center for Intelligent Information Retrieval, University of Massachusetts (2003)
25. Zhang, D., Lee, W.S.: A language modeling approach to passage question answering. In: Proceedings of the Twelfth Text Retrieval Conference (TREC-12), pp. 489–495 (2004)

26. Jiang, J., Zhai, C.: UIUC in HARD 2004 — passage retrieval using HMMs. In: Proceedings of the Thirteenth Text Retrieval Conference (TREC-13) (2004)
27. Kurland, O., Lee, L., Domshlak, C.: Better than the real thing? Iterative pseudo-query processing using cluster-based language models. In: Proceedings of SIGIR, pp. 19–26 (2005)

A Statistical View of Binned Retrieval Models

Donald Metzler[1], Trevor Strohman[2], and W. Bruce Croft[2]

[1] Yahoo! Research, Santa Clara, CA
[2] University of Massachusetts, Amherst, MA

Abstract. Many traditional information retrieval models, such as BM25 and language modeling, give good retrieval effectiveness, but can be difficult to implement efficiently. Recently, document-centric impact models were developed in order to overcome some of these efficiency issues. However, such models have a number of problems, including poor effectiveness, and heuristic term weighting schemes. In this work, we present a statistical view of document-centric impact models. We describe how such models can be treated statistically and propose a supervised parameter estimation technique. We analyze various theoretical and practical aspects of the model and show that weights estimated using our new estimation technique are significantly better than the integer-based weights used in previous studies.

1 Introduction

Most of the information retrieval models developed recently fall into a class of models known as parameterized retrieval models. Examples of these models are BM25 [1], language modeling [2], the axiomatic model [3], the divergence from randomness model [4], and linear discriminative models [5,6]. At the very core of these models is some term weighting function that is composed of one or more free parameters and standard information retrieval features, such as term frequency, inverse document frequency, and document length. These term weighting functions are responsible for *quantitatively* assigning importance values to document and query terms. The standard procedure for training or tuning a parameterized model of this form is to learn a set of parameters using either supervised or unsupervised methods that maximizes some information retrieval metric.

It is important to note that the importance values (weights) are quantitative variables, and therefore, their absolute and relative values are indeed important. If term A is given double the weight of term B then we must conclude that A is two times as important as term B. This is very different than concluding that term A is *more* important than term B. Such a conclusion would require us to assume that term importance is an *ordinal* variable, rather than a quantitative one. While term weighting functions impose an implicit ordering on terms according to importance, they do not explicitly model the ordinal nature of term importances.

Recently, Anh and Moffat introduced the document-centric impact model, which represents a paradigm shift in the design of retrieval models [7]. The

C. Macdonald et al. (Eds.): ECIR 2008, LNCS 4956, pp. 175–186, 2008.

model, which was experimentally shown to be both effective and highly efficient, moves away from complex parameterized term weighting functions. Instead, a method is proposed by which document terms are partitioned into a small number (e.g., fewer than 16) of bins. Each bin contains a set of terms of equal importance. For example, there may be a bin that contains all of the most important terms, another that contains less important terms, and a third that contains the least important terms. This imposes an explicit *ordering* of sets of terms (bins), instead of the implicit ordering of terms imposed by classic term weighting functions.

In this work, we present a model that can be considered to be a statistical interpretation of the document-centric impact model. Like the document-centric impact model, our model also requires binning of document terms in order to estimate term weights. However, in our model, we take a probabilistic approach that allows many of the techniques available in the language modeling literature to be used. Without a statistical interpretation, such techniques would not be as easily applied. Furthermore, such an interpretation allows us to use non-integral impacts, and estimate parameters more formally in a supervised fashion, thereby eliminating one of the more heuristic pieces of the document-centric impact model. As we will show, this statistical interpretation, along with the newly devised estimation technique consistently and significantly improves retrieval effectiveness relative to existing impact-based retrieval models.

The remainder of this paper is laid out as follows. In Section 2 we describe related models. Section 3 lays out the theoretical foundations of our model. In Section 4 the results from our experimental evaluation are presented. Finally, Section 5 concludes the paper and presents potential areas of future work.

2   Related Models

In this section we review the language modeling framework for information retrieval and the document-centric impact model, both of which are closely related to our proposed model.

2.1   Language Modeling for IR

The language modeling framework for information retrieval was first proposed by Ponte and Croft [2]. Language models attempt to model *language* or *topicality*. Although there are many different variants of language modeling, we will only describe one of the most robust and commonly used formulations [8]. In this formulation, we are tasked with estimating document and query models. Models are defined as multinomial distributions over some fixed vocabulary $\mathcal{V}$. Due to their very nature, document and query models are often estimated differently. Documents are typically estimated using some smoothed maximum likelihood estimate [9]. Query models are either estimated according to their maximum likelihood estimate or using a more complex pseudo-relevance feedback-based formulation, such as model-based feedback [10,11] or relevance-based models [12,13].

Given a query, documents are ranked according the negative KL divergence between the query and document model, which is computed as:

$$- KL(\theta_Q \| \theta_D) \; = \; H(\theta_Q) - CE(\theta_Q, \theta_D)$$
$$\stackrel{rank}{=} \sum_{w \in \mathcal{V}} \theta_{w,Q} \log \theta_{w,D} \tag{1}$$

where H is the entropy, CE is cross-entropy, θ_Q is the query model, and θ_D is the document model. Here, $\theta_{w,Q}$ and $\theta_{w,D}$ are shorthand for $P(w|\theta_Q)$ and $P(w|\theta_D)$, respectively. Although this sum is shown to go over the entire vocabulary $\mathcal{V}$, it is very often the case that terms that do not occur in the query are assigned a zero probability in the query model, thus significantly reducing the number of terms in the sum.

Pros. Language modeling has several appealing characteristics. First, the model is formally motivated and based on a strong statistical foundation. This allows estimation and learning techniques from statistics and machine learning to be easily applied. Examples of such techniques are Bayesian smoothing [9], translation models [14], mixture models [15], cluster-based models [16,17], and topic models [18]. Second, the model has proven to be highly effective over a wide range of retrieval tasks. Finally, the model is relatively easy to understand and implement.

Cons. As with all models, language modeling also has several unappealing characteristics. One of the most fundamental theoretical issues with the framework concerns how query and document models are estimated and compared. Document models are estimated using techniques such as maximum likelihood estimation and Bayesian smoothing. These models, at their core, are modeling term *occurrences*. When sampling terms from a model estimated in this way, we expect our sample to exhibit term occurrence statistics similar in nature to those observed in the document they were estimated from. However, queries and documents exhibit very different term occurrence statistics, as was pointed out in the past [19]. For example, documents contain many function words, while queries rarely do. Therefore, it is theoretically unsettling to compare a query model with a document model, given that the statistical properties of term occurrences in queries and documents are fundamentally different.

Smoothing and high model complexity are also concerns. It is well known that smoothing does more than overcome the zero frequency problem. It also results in an implicit IDF factor in the query likelihood retrieval model, which ultimately results in a very *tf.idf*-like ranking function [9]. In terms of model complexity, a general multinomial model over a vocabulary of size $|\mathcal{V}|$ requires $|\mathcal{V}|-1$ parameters to be estimated. This is a very large number of free parameters to estimate for a model. Fortunately, most types of smoothing wash away these many degrees of freedom (although they are still implicitly there) down to one or two parameters.

2.2   Document-Centric Impact Model

Anh and Moffat's document-centric impact retrieval model has been shown to be relatively effective and highly efficient [20,7]. The model moves away from using quantitative parameterized term weighting functions. Instead, the model ranks terms according to their importance and imposes a very simple, pre-defined term weighting function to the sorted terms. As we discussed earlier, this type of model captures the notion of ordinal importance between terms, rather than trying to explicitly quantify importance, as is done in most other retrieval models.

Term weighting within the model is accomplished in two stages. First, terms are sorted according to some importance criteria. After sorting, the terms are then partitioned and assigned to bins. Each bin is assigned an integral impact ranging from 1 to k, where k is the total possible number of bins. The result of this process is that every term in every document is assigned a term weight in the set $\{1, \ldots, k\}$. Typical values of k include 4, 8 and 16 [7]. Each document is binned in the same way. We describe the details of Anh and Moffat's sorting and binning technique in Section 3.2.

Query terms are weighted differently, for several reasons. Anh and Moffat suggest that applying the strategy just described to queries will fail, due to the small number of query terms [7]. In addition, properly setting the query term weights is critical in order to achieve reasonable effectiveness. Our preliminary experiments showed that using uniform term weights results in poor effectiveness. The details of Anh and Moffat's query binning technique are given later.

Ranking within the model is done via a simple dot product between the document and query impacts (weights). This is computed according to:

$$RSV(D;Q) = \sum_{w \in \mathcal{V}} I_{w,Q} I_{w,D} \qquad (2)$$

where $I_{w,Q}$ is the impact value assigned to query term w and $I_{w,D}$ is the impact value assigned to document term w. Terms not occurring in the query are often assigned an impact value of zero, although this is not required.

Pros. Previous studies have shown that document-centric impact models are highly efficient, especially on large collections [21]. Impact-ordered indexes can reduce the amount of disk storage necessary compared to standard inverted list indexes. Furthermore, the model is amenable to efficient query processing [22]. This makes the model more attractive, from an efficiency standpoint, than language modeling and BM25.

Cons. Despite the efficiency of the model, the effectiveness is often not as strong as language modeling or BM25 baselines. This trade-off between efficiency and effectiveness can be controlled by choosing an appropriate number of bins. As expected, as fewer bins are used, efficiency increases, but effectiveness decreases.

Another issue with the model is the fact that there is no formal justification or motivation for the various binning strategies previously proposed in the literature. These strategies are typically heuristic and built from intuition.

Furthermore, using integral impact values is a matter of convenience and efficiency. However, there again is no formal motivation for choosing impacts in such a way. In Section 3.3 we describe a less heuristic estimation technique for choosing our model's equivalent of impact values.

3   Model

Our model is designed to combine the best aspects of language modeling and the document-centric impact model. The model is probabilistic like language modeling, thus allowing it to be incorporated into more complex statistical techniques, such as those described in Section 2.1. However, unlike language modeling, we do not model the generation of text. Instead, we model the importance of bins of terms (or arbitrary features).

The first step of our model, much like the document-centric impact model, requires us to bin the terms according to their importance. We assume that there is some fixed set of bins defined by $\mathcal{B}$, where each $B \in \mathcal{B}$ is an ordinal variable indicating relative importance. For example, B_1 may denote "most important", B_2 may denote "medium importance", and B_3 may denote "least important". Binning is performed on each document. There are many different ways to bin terms. We describe several approaches in the next section. The final result of this process is, for each document, a partitioning of the vocabulary into $|\mathcal{B}|$ bins.

After binning, we must estimate a model for each document. Rather than estimating text generation models, as is done in language modeling, we define importance models. These models attempt to capture the likelihood that the terms in a certain bin are important. We define a document importance model as a multinomial distribution over bins. As a matter of shorthand, we write $P(B|\theta_D)$ as $\theta_{B,D}$ and interpret $\theta_{B,D}$ as the probability that the terms in bin B (for document D) are important. This is fundamentally different than the language modeling interpretation. Indeed, we believe this interpretation is philosophically more appealing, as it does not assume that queries and documents are generated from the same underlying model. Instead, we model a fundamental, yet difficult to define, notion of term importance which is consistent across models.

Now that we have all of the pieces of our model, the final step is to describe how documents are ranked in response to a query. We rank documents using a generalized likelihood ranking function that allows query term weighting. We call this the *weighted likelihood* ranking function. It is defined as:

$$
\begin{aligned}
P(Q|D) &= \prod_{w \in \mathcal{V}} \theta_{b_D(w)}^{wt_{w,Q}} \\
&\overset{rank}{=} \sum_{w \in \mathcal{V}} wt_{w,Q} \log \theta_{b_D(w),D}
\end{aligned} \tag{3}
$$

where $wt_{w,Q}$ defines a weight for query term w and $b_D(w)$ is the bin that term w is assigned to in document D. A more formal definition of $b_D(w)$ is provided in the next section. This ranking function assigns high weights to documents that contain query terms that are both highly weighted and highly likely to be important. We note that this ranking function is reduced to the standard likelihood function when query term weights are set proportionally to the number of times they occur in the query. In the remainder of this section we describe various binning and weighting strategies for both queries and documents.

3.1   Query Binning and Weighting

IDF-Weighted. Anh and Moffat propose a query binning strategy based on query term $idfs$. Their strategy has two steps. First, each query term is assigned a weight according to:

$$wt_w = (1 + \log tf_{w,Q}) \log \left(1 + \frac{maxtf_w}{cf_w}\right) \tag{4}$$

where $tf_{w,Q}$ is the number of times w occurs in the query, $maxtf_w$ is the maximum number of times w occurs in any document in the collection and cf_w is the total number of times w occurs in the collection.

The final step bins terms linearly according to their weight. This results in query terms with very large idf values being assigned to the "important" bins and those with low idf being assigned to bins of lower importance. Query term weights are then assigned according to:

$$wt_w = I_{b_Q(w),Q} \tag{5}$$

where it is assumed that some *a priori* set of impacts have been assigned to each query bin and $I_{b_Q(w),Q}$ is the impact assigned to query bin $b_Q(w)$. In this work, we follow Anh and Moffat and assume integral impacts. That is, query terms in the least important bin are assigned an impact value of 1, those in the next least important bin are assigned impact value 2, and so on.

Other Methods. Other methods for computing query weights are possible, although not explored in detail here. For example, relevance-based language models estimate query weights by mixing together the language models of a set of relevant or pseudo-relevant documents [13]. An analogous technique could be used within our model to estimate better query weights.

One criticism of relevance-based language models is that they assign large probabilities to function words due to their prevalence in the top ranked documents. Such models do not try to separate out the meaningful terms from the background, as is done with parsimonious language models [23]. Indeed, we suspect that relevance-based query models estimated using our model will behave as the parsimonious language models do. This is due to the fact that function terms will be given very low probability, as they are assigned to "unimportant" bins, and give topical terms larger probability, as they are assigned to "important" bins. Investigating this phenomenon further is part of future work.

3.2   Document Binning

For each document, we define a binning function $b_D : \mathcal{V} \rightarrow \mathcal{B}$ that maps the original vocabulary ($\mathcal{V}$) onto a set of bins $\mathcal{B}$. For a document D, the binned document representation is generated by applying b_D to each term. This results in a new document that only consists of bins from $\mathcal{B}$.

The bin vocabulary $\mathcal{B}$ can be thought of as a surrogate vocabulary that captures some latent aspect of the original vocabulary. The purpose of binning is to cluster or combine terms that are similar under some criteria. In this work we aim at binning terms according to their importance. However, it is possible that other binning criteria may be more appropriate for other applications.

Another important consequence of binning is the significant reduction of the effective vocabulary size when we choose $|\mathcal{B}| \ll |\mathcal{V}|$. The binning process reduces the dimensionality of our document representation. This results in significantly reduced model complexity which can minimize the effects of overfitting and significantly improve query processing efficiency.

(TF,IDF) Binning. Anh and Moffat propose a number of document-centric binning strategies [7]. Each of their proposed strategies have a *sorting* and *assignment* stage. In the sorting stage, document terms are sorted according to some criteria. In the assignment stage, the sorted terms are assigned to bins.

Anh and Moffat report that the *(TF, IDF)* sorting method results in the best effectiveness [7]. This method sorts terms in descending order using term frequency as the primary key and inverse document frequency as a tie breaking secondary key.

The sorted terms are assigned to bins according to a geometric sequence. That is, a small number of terms (i.e. those at the beginning of the sorted list) are assigned to the "most important" bin, a larger number of terms are assigned to the next most important bin, with the least important bin containing the largest number of terms (i.e. those terms at the end of the sorted list). More formally, the number of terms in bin b_i is given by:

$$x_i = (|D| + 1)^{1/k} x_{i+1} \tag{6}$$

$$x_{|\mathcal{B}|} = (|D| + 1)^{1/k} - 1 \tag{7}$$

where the least important terms are assigned to bin b_1 and the most important to bin $b_{|\mathcal{B}|}$.

Other Binning Strategies. Although not explored in this work, we note that there are a number of reasonable strategies for binning terms. In particular, index pruning strategies [24,25] and probabilistic indexing techniques [26] may be useful. These methods share the same goal as binning by term importance. However, instead of explicitly creating a binning, these methods only choose to index those terms that are likely to be important within a given document.

3.3   Document Model Estimation

Document-Centric Impact Estimate. If some pre-defined impact (integral or real-valued) value is assigned to each term bin, as in the document-centric impact model [7], then we can convert the impacts to probabilities as follows:

$$\hat{\theta}_{B,D} = \frac{\exp\left[I_{B,D}\right]}{\sum_{B' \in \mathcal{B}} \exp\left[I_{B',D}\right]} \tag{8}$$

Unfortunately, it is unclear how to optimally set the impact weights given some binning. While the integral assignment proposed by Anh and Moffat is simple, it is likely not to be optimal. Therefore, a more informed, well-founded method for estimating the document model probabilities is required.

It is straightforward to show that when document models are estimated in this way and query term weights are computed using the IDF-Weighted method described in Section 3.1 our ranking function (Equation 3) is equivalent to the impact ranking function (Equation 2). This provides a probabilistic interpretation of the document-centric impact model.

Discriminative Estimation. As described previously, one paradox of the language modeling approach to information retrieval is that document models are estimated so as to maximize the likelihood (or the *a posteriori*) of generating the terms in the document, while the overarching goal is to maximize some evaluation metric, such as mean average precision. Therefore, we propose to choose document model probabilities in such a way that they maximize some retrieval metric, instead of properly modeling term occurrence statistics. We acknowledge that it is common practice in language modeling to train a model by tuning the smoothing parameter in order to maximize some metric. However, this is typically a single, coarse grained parameter that has very specific interactions with the model. We are proposing to tune the model in a radically different way that allows finer control and results in parameter settings that can be interpreted more intuitively than Dirichlet or Jelinek-Mercer smoothing parameters.

Given a set of bins $\mathcal{B}$, our goal is to estimate $\theta_{B,D}$ by maximizing some retrieval metric. This optimization problem involves setting $|\mathcal{B}| - 1$ parameters for each document in the collection. Even when a small number of bins is chosen, this problem is infeasible. However, if we make the simplifying assumption that $\theta_{B,D} = \theta_{B,D'}$ for all D and D' in the collection, then the problem becomes more reasonable. This assumption ties all of the bin importance probabilities together. That is, it assumes that the likelihood a term in some bin j is important is the same across all documents. While this assumption may be overly simplistic, it significantly reduces the number of free parameters in our optimization problem to $|\mathcal{B}| - 1$, which is easily solved for most reasonable bin settings. Another side effect of our assumption is that it allows for very efficient query processing.

Formally, our discriminative estimation technique requires the following optimization problem to be solved:

$$\left[\hat{\theta}_1 \ldots \hat{\theta}_{|\mathcal{B}|} \right] = \arg\max E(\mathcal{R}; \theta_1 \ldots \theta_{|\mathcal{B}|}) \tag{9}$$

where $\mathcal{R}$ is the set of relevance judgments and E is some evaluation metric.

Since most information retrieval metrics are not amenable to standard optimization techniques, we choose to solve this optimization problem using greedy hill climbing, which is a local search technique. This hill climbing approach is reasonable for small numbers of bin, even on very large collections, because of the low cost of evaluating large numbers of queries.

4   Evaluation

In this section we evaluate our proposed binning and estimations techniques in terms of effectiveness. Although efficiency is important, the evaluation of the efficiency of integral vs. non-integral impacts is beyond the scope of this work.

Table 1. Overview of collections and topics used

Collection	# Docs	Train Topics	Test Topics
TREC Disks 1,2	741,856	51-150	151-200
TREC Disks 4,5	528,155	301-450	601-700
WT10g	1,692,096	451-500	501-550

All binning-related experiments are carried out using Galago[1], a new indexing and retrieval system developed to test our new probabilistic model. In addition, the Indri search system is used for the query likelihood runs [27]. We evaluate our methods on three TREC data sets with varying characteristics. Table 1 provides an overview of each data set. The TREC Disks 1 and 2 (TREC12) and TREC Disks 4 and 5 (TREC45) data sets consist of newswire articles from several sources. The WT10G data set is much larger and is made up of web documents. The queries associated with each data set are split into a training and test set. The training set is used to tune parameters (smoothing parameters and document importance model probabilities). The test set is used solely for evaluation purposes.

Documents are stemmed using the Porter stemmer and stopped using the same list of stopwords used by Anh and Moffat [7]. Queries are constructed using only the title portion of the TREC topic. Finally, we use 8 bins when IDF-weighted query term binning is employed.

4.1   Integral vs. Discriminative Weights

We now scrutinize the optimality of choosing document model probabilities based on integral impacts. Therefore, we wish to compare the results of (TF,IDF) binning and integral document estimates with (TF,IDF) binning and discriminative document model estimates. Recall that the integral weights are set in a completely unsupervised fashion, whereas the discriminative weights are learned from training data.

As we see from Table 2, the discriminatively trained weights are consistently better than the integral weights across various document bin sizes. These improvements are statistically significant for over half of our test cases.

While this result is not necessarily surprising, it does allow us to quantitatively evaluate the optimality of the naïvely chosen integral weights. Indeed, the results of our experiments show that integral weights, while not being optimal, achieve results that are often close to optimal. This is a more interesting and surprising

---
[1] http://www.galagosearch.org

Table 2. Mean average precision for various combinations of document model estimation techniques, query weight estimation strategies, and number of document bins. A query likelihood language modeling run using Dirichlet smoothing is also included for comparison. A † superscript indicates statistically significant improvements in effectiveness over the cell immediately above it using a one-tailed t-test with $p < 0.05$.

Data	θ_D Estimation	$wt_{w,Q}$ Estimation	2 bins	4 bins	8 bins	16 bins
TREC12	Integral	IDF	0.2067	0.2241	0.2273	0.2273
	Discriminative	IDF	0.2105	0.2269	0.2315	0.2336†
	Language Modeling		0.2633			
TREC45	Integral	IDF	0.2325	0.2417	0.2427	0.2459
	Discriminative	IDF	0.2430†	0.2494†	0.2577†	0.2567†
	Language Modeling		0.2920			
WT10g	Integral	IDF	0.1522	0.1598	0.1863	0.1886
	Discriminative	IDF	0.1570	0.1692†	0.1879†	0.1887
	Language Modeling		0.1861			

result, as it was expected that such weights would be far from optimal. The reason why such weights may be so close to optimal may be the result of the particular binning strategy used, and therefore our analysis does not extend beyond (TF,IDF) binning. It is unclear whether these results will hold for more complex binning strategies. It is likely that in more complex cases the divide between the discriminatively trained model and the integral weight model will increase.

4.2   Language Modeling vs. Impact-Based Models

We now briefly investigate how well the impact-based models perform when compared to a strong language modeling baseline. The language modeling baseline significantly outperforms the best impact-based formulation for the TREC12 and TREC45 data sets. Interestingly, the two models demonstrate comparable effectiveness on the WT10G collection.

Our results seem to contradict those described by Anh and Moffat [7], which showed that the impact-based model significantly outperformed language modeling and BM25. However, most of the language modeling results outlined in their work were quoted from previous work that had very weak language modeling baseline numbers. Indeed, our rigorously tuned language modeling approach shows significantly stronger performance on the newswire data sets compared to the impact-based model.

Our experience with impact-based models suggest that they strongly prefer documents that contain all of the query terms. We believe that this is an asset to the model in large collections where there are likely to be many documents that contain all the query terms [28]. Furthermore, some recent work suggests that relevance judgments in large TREC collections are biased toward those documents that contain all of the query terms. We believe that this may explain why impact methods perform strongly compared to language modeling on larger collections while there is a large effectiveness gulf on smaller collections, which presumably have a higher percentage of relevant documents that do not contain all of the query terms.

5   Conclusions

In this paper, we presented a probabilistic retrieval model that can be considered a statistical view of the document-based impact model. Our model achieves good effectiveness and efficiency by combining the strengths of the language modeling and document-centric impact models.

In addition, we described a supervised method for discriminatively learning document importance model weights. Rather than using integral weights, as was done in previous work, we find the set of weights that maximize some underlying retrieval metric. Our results showed consistent and significant improvements in effectiveness when weights were learned in this way.

Acknowledgments

This work was supported in part by the Center for Intelligent Information Retrieval, in part by NSF grant #CNS-0454018, and in part by Advanced Research and Development Activity and NSF grant #CCF-0205575. Any opinions, findings and conclusions or recommendations expressed in this material are the authors' and do not necessarily reflect those of the sponsor.

References

1. Robertson, S.E., Walker, S.: Some simple effective approximations to the 2-poisson model for probabilistic weighted retrieval. In: Proc. 17th SIGIR, pp. 232–241. Springer, New York (1994)
2. Ponte, J.M., Croft, W.B.: A language modeling approach to information retrieval. In: Proc. 21st SIGIR, pp. 275–281 (1998)
3. Fang, H., Zhai, C.: An exploration of axiomatic approaches to information retrieval. In: Proc. 28th SIGIR, pp. 480–487 (2005)
4. Amati, G., Rijsbergen, C.J.V.: Probabilistic models of information retrieval based on measuring the divergence from randomness. ACM Transactions on Information Syststems 20(4), 357–389 (2002)
5. Nallapati, R.: Discriminative models for information retrieval. In: Proc. 27th SIGIR, pp. 64–71 (2004)
6. Gao, J., Qi, H., Xia, X., Nie, J.Y.: Linear discriminant model for information retrieval. In: Proc. 28th SIGIR, pp. 290–297 (2005)
7. Anh, V.N., Moffat, A.: Simplified similarity scoring using term ranks. In: Proc. 28th SIGIR, pp. 226–233 (2005)
8. Song, F., Croft, W.B.: A general language model for information retrieval. In: Proc. 8th CIKM, pp. 316–321 (1999)
9. Zhai, C., Lafferty, J.: A study of smoothing methods for language models applied to ad hoc information retrieval. In: Proc. 24th SIGIR, pp. 334–342 (2001)

10. Tao, T., Zhai, C.: Regularized estimation of mixture models for robust pseudo-relevance feedback. In: Proc. 29th SIGIR, pp. 162–169 (2006)
11. Zhai, C., Lafferty, J.: Model-based feedback in the language modeling approach to information retrieval. In: Proc. 10th CIKM, pp. 403–410 (2001)
12. Diaz, F., Metzler, D.: Improving the estimation of relevance models using large external corpora. In: Proc. 29th SIGIR, pp. 154–161 (2006)
13. Lavrenko, V., Croft, W.B.: Relevance based language models. In: Proc. 24th SIGIR, pp. 120–127 (2001)
14. Berger, A., Lafferty, J.: Information retrieval as statistical translation. In: Proc. 22nd SIGIR, pp. 222–229 (1999)
15. Ogilvie, P., Callan, J.: Combining document representations for known-item search. In: Proc. 26th SIGIR, pp. 143–150 (2003)
16. Liu, X., Croft, W.B.: Cluster-based retrieval using language models. In: Proc. 27th SIGIR, pp. 186–193 (2004)
17. Kurland, O., Lee, L.: Corpus structure, language models, and ad hoc information retrieval. In: Proc. 27th SIGIR, pp. 194–201 (2004)
18. Wei, X., Croft, W.B.: Lda-based document models for ad-hoc retrieval. In: Proc. 29th SIGIR, pp. 178–185 (2006)
19. Jones, K.S.: Language modelling's generative model: Is it rational? Technical report, University of Cambridge (2004)
20. Anh, V.N., Moffat, A.: Collection-independent document-centric impacts. In: Proc. Australian Document Computing Symposium, pp. 25–32 (2004)
21. Anh, V.N., Moffat, A.: Melbourne university 2004: Terabyte and web tracks. In: Proceedings of TREC 2004 (2004)
22. Anh, V.N., Moffat, A.: Pruned query evaluation using pre-computed impacts. In: Proc. 29th SIGIR, pp. 372–379 (2006)
23. Hiemstra, D., Robertson, S., Zaragoza, H.: Parsimonious language models for information retrieval. In: Proc. 27th SIGIR, pp. 178–185 (2004)
24. Büttcher, S., Clarke, C.L.A.: A document-centric approach to static index pruning in text retrieval systems. In: Proc. 15th CIKM, pp. 182–189 (2006)
25. Carmel, D., Cohen, D., Fagin, R., Farchi, E., Herscovici, M., Maarek, Y.S., Soffer, A.: Static index pruning for information retrieval systems. In: Proc. 24th SIGIR, pp. 43–50 (2001)
26. Fuhr, N.: Two models of retrieval with probabilistic indexing. In: Proc. 9th SIGIR, pp. 249–257 (1986)
27. Strohman, T., Metzler, D., Turtle, H., Croft, W.B.: Indri: A language model-based serach engine for complex queries. In: Proceedings of the International Conference on Intelligence Analysis (2004)
28. Buckley, C., Dimmick, D., Soboroff, I., Voorhees, E.: Bias and the limits of pooling. In: Proc. 29th SIGIR, pp. 619–620 (2006)

Video Corpus Annotation Using Active Learning

Stéphane Ayache and Georges Quénot

Laboratoire d'Informatique de Grenoble (LIG)
385 rue de la Bibliothèque - BP 53
38041 Grenoble - Cedex 9 - France

Abstract. Concept indexing in multimedia libraries is very useful for users searching and browsing but it is a very challenging research problem as well. Beyond the systems' implementations issues, semantic indexing is strongly dependent upon the size and quality of the training examples. In this paper, we describe the collaborative annotation system used to annotate the High Level Features (HLF) in the development set of TRECVID 2007. This system is web-based and takes advantage of *Active Learning* approach. We show that *Active Learning* allows simultaneously getting the most useful information from the partial annotation and significantly reducing the annotation effort per participant relatively to previous collaborative annotations.

1   Introduction

Semantic content-based access to image and video documents is a strong need for many industrial applications. Indexing concepts in images and in video segments is the main key to enable it and it is still a research challenge. Due to the so called *semantic gap* between the raw image or video contents and the elements that makes sense to human beings, indexing concepts in image or video documents is a very hard task. It is most often carried out using classifiers or networks of classifiers [10, 14, 3] trained using supervised learning. Systems' performance depends a lot upon the implementation choices and details but it also strongly depends upon the size and quality of the training examples. While it is quite easy and cheap to get large amounts of raw data, it is usually very costly to have them annotated because it involves human intervention for the judging of the "ground truth".

Many research works on content-based image and video indexing are conducted in the context of the TRECVID campaigns [13]. These campaigns provide to the participants a complete framework with data collections, well defined tasks, ground truth and metrics for the evaluation of indexing and/or retrieval systems. Additionally, annotated data are provided for some tasks like the "High Level Feature (HLF) extraction task" which is actually a concept indexing task. Large annotation efforts were organized in 2003 [8] and 2005 [15, 9] in order to produce a complete annotation of the development set for a series of target concepts. These initiatives produced very valuable resources but at a very high cost.

While the volume of data that can be manually annotated is limited due to the cost of manual intervention, there remains the possibility to select the data

C. Macdonald et al. (Eds.): ECIR 2008, LNCS 4956, pp. 187–198, 2008.

samples that will be annotated so that their annotation is "as useful as possible" [1]. Deciding which samples will be the most useful is not trivial. *Active learning* is an approach in which an existing system is used to predict the usefulness of new samples. This approach is a particular case of *incremental learning* in which a system is trained several times with a growing set of samples. The objective is to select as few samples as possible to be manually indexed and to get from then the best possible classification performance.

In this paper, we describe the use of active learning technique for annotation of unlabeled video corpus. In order to provide manually annotation on the TRECVID 2007 development set at cheapest cost, we organized a web-based collaborative annotation tool in the spirit of what was done in the 2003 and 2005 [15]. Active learning has been used in order to simultaneously get the most useful information from the partial annotation and significantly reduce the annotation effort per participant relatively to previous collaborative annotations. In the following of this paper, we first describe previous active learning experiments and then present the principles and the organization of this project and the lessons learnt from it.

2   Simulated Active Learning

In a previous work, [2] simulated an active learning process using the TRECVID 2005 fully annotated development set and the TRECVID 2006 test set and metrics. By progressively including annotations in the training set, various active learning strategies have been evaluated in a variety of conditions. Results have been obtained using a particular corpus (TRECVID 2005/2006), a particular type of concepts (LSCOM-lite) and using a particular learning system (network of SVM classifiers). They might not transpose directly to other types of contents, target concepts or learning system though we expect the observed general trends to still be valid.

Three strategies were compared: "relevance sampling", "uncertainty sampling", and "random sampling". The two first strategies respectively select the most probable and the most uncertain samples [7]. The third one is a random choice. Here are the main conclusions:

- For easy concepts, the "relevance sampling" strategy is the best one when less than 15% of the dataset is annotated and the "uncertainty sampling" strategy is the best one when 15% or more of the dataset is annotated.
- The "relevance sampling" and "uncertainty sampling" strategies are roughly equivalent for moderately difficult and difficult concepts. In all cases, the maximum performance is reached when 12 to 15% of the whole dataset is annotated.
- The previous results depend upon the step size and the training set size. $1/40^{\text{th}}$ of the training set size is a good value for the step size.
- The size of the subset of the training set that has to be annotated in order to reach the maximum achievable performance varies with the square root of the training set size.

- The "relevance sampling" strategy is more "recall oriented" while the "uncertainty sampling"' strategy is more "precision oriented".

Figure 1 shows the evolution of the system Mean Average Precision (MAP, actually inferred average precision as it was introduced in TRECVID 2006) with the number of annotated samples for the three strategies and with an active learning step size of $1/40^{th}$ of the training set size. The active learning process was initialized with a set of 10 positive samples and 20 negative samples randomly chosen (the assumption is that the user has at least a few positive examples of what he is looking for and that negative examples are easy to find). What is remarkable is that the maximum system performance is reached when only a small fraction of the development set is annotated if this fraction is carefully chosen. Here the fraction is of about 12-15% for a development set size of 36014 samples. Other experiments (not shown here) indicate that this is also the case for different development set sizes and that the optimal fraction varies with the square root of the development set (it is of about 25-30% of the development set if its size is reduced to 9003 samples).

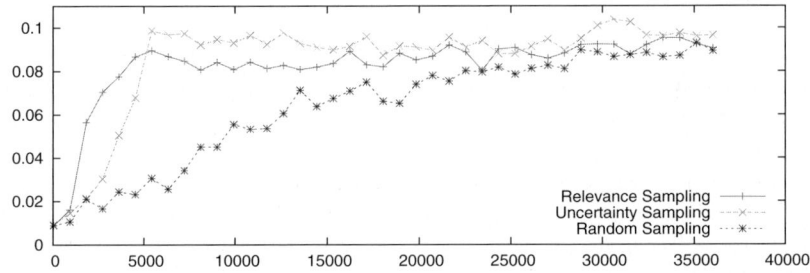

Fig. 1. Evolution of system MAP with the number of annotated samples for the three strategies, all concepts

Figure 2 shows the evolution of the number of positive samples found (average on all concepts) as a function of the number of annotated samples for the three strategies. The rate of finding of positive samples near the beginning are of about 2.4:1 and 4.5:1 for "uncertainty sampling" and "relevance sampling"' strategies respectively relatively to the "random" choice.

3   Collaborative Annotation System

For the TRECVID 2003 annotation effort, [8] provided a tool to facilitate multimedia annotation tasks for general users. This tool generated MPEG-7 compatible outputs and included various features from video shot segmentation to ontology editing and region based annotation. However, Videoannex was a standalone system, thus each user needs to get possession of the entire collection and the annotation data must be collected afterwards. Moreover, this tool was

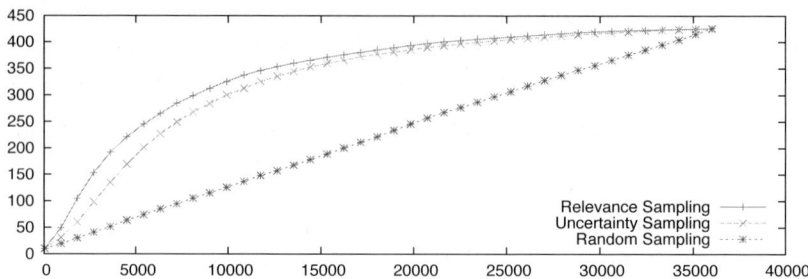

Fig. 2. Evolution of the number of positive samples found with the number of annotated samples for the three strategies, all concepts

not user centered as it forced to annotate all available concepts from the ontology simultaneously. The TRECVID 2005 collaborative annotation system was a
web-based application that allowed users to annotate using a web browser [15].
Thanks to the centralized architecture, the system was able to display a set of
overall statistics during an annotation session.

Our system is web-based and relies on an active learning approach. Similar approaches have already been considered for image and video indexing or retrieval
[5, 11] but not yet in the context of a web based collaborative annotation. As this
was done in the previous collaborative annotation, we produced samples at the
subshot level since these are much more likely to have a homogeneous and non
ambiguous content. In order to ease the annotation process, annotation is consists to judge one key frame per subshot. We finally extracted 21532 key frames
using the video segmentation tool described in [12]. The following subsections
describe the interface and organization of the collaborative annotation system.

3.1   Web Interface

The TRECVID 2007 Collaborative Annotation system has been designed to be
efficient and easy to use. Like the TRECVID 2005 collaborative annotation system [15], it operates through the Web and requires no local software installation.
Participation is restricted to groups that are registered TRECVID participants
and that have signed a license agreement to access the video data.

The system has two modes of operation: a sequential mode in which the images
to annotate are displayed one by one and a parallel mode (Figure 3) in which the
images are displayed by groups in a two-dimensional array. In the parallel mode,
users can define the dimensions of the array in order and adapt visualization to
his screen size.

Users were required to annotate only one concept at a time. The system
gave priority to the concept which had the less annotated samples. For the
current concept to annotate, images are displayed, either one by one or by group
depending upon the mode chosen, and for each image the user has three choices

Fig. 3. Parallel interface of the annotation system

for its annotation: POSITIVE (the concept is clearly there), NEGATIVE (the concept is clearly not there), SKIPPED (any other case, whatever the cause of uncertainty).

In the parallel mode, users see by default an image at a smaller resolution than the video one (160×120 instead of 352×288). By passing the mouse over one of the small images they can get an enlarged view of it in a corner of the screen. In both modes, users also have the possibility to play the whole video shot if they feel that this can help them to make a better decision. This is often the case for "dynamic" concepts like "Walking_Running".

3.2   Organization

TRECVID participants register as teams and each team may have several users doing the annotation. In order to encourage participation to the collaborative annotation, the resulting annotation is available only to the teams that have completed a minimum amount of annotations, as this was also the case in previous TRECVID collaborative annotations. The minimum annotation effort was set to 3% of the total number of annotations that should be done in order to annotate each key frame/subshot for each concept once. This amounts to 23255 annotations per team and can be completed in about 13 hours considering an average annotation time of 2 seconds per key frame × concept.

4   Active Learning System

We implemented the same system described in [2]: an iterative process which use samples score from previous iteration to sort samples depending of the strategies. The active learning process was running permanently during the whole annotation period (over two months). It has been optimize in order to run with a parallel implementation on 10 bi-processor (3 GHz P4) servers. The process continuously computed (training/prediction) one concept at once. Hence, in order to have similar annotation progress for the concepts, the system continuously chooses the concept which received the largest number of annotations since its last training. Consequently, there is not any step size as iterations occur when a concept has been selected by the system.

The collaborative annotation system also runs permanently and independently of the active learning process. The Collaborative annotation process uses the last version of the classification system produced by the active learning system in order to select the samples for annotation. Similarly, the active learning system uses the last available set of annotations to re-train the classification systems.

4.1   Classification System

The classification system used for the active learning process is derived from the one used for our participation the TRECVID 2006 high level feature extraction task. Since the language used in both collections is different and since the English machine translation was not available yet, we used two variants, one using the text input and the other not using it.

The system is detailed in [3]. It uses visual and text features when available. Visual features include local and global features and both include color, texture and motion low-level features. The system uses network of SVM classifiers [4] and implements a mix of early and late fusion schemes. Its performance on the TRECVID 2006 HLF extraction task was slightly above the median with an Inferred Average Precision of 0.088.

4.2   Cold Start and Strategies

Since the concepts to annotate are the same in 2005/2006 and 2007, we can use a system trained only on 2005 data for starting the selection of the samples on the 2007 data. This is a challenge since the 2005 and 2007 corpora are quite different on visual, sound and text modalities. The "cold start" strategy was finally to begin the training with only 2005 samples and then to progressively replace as many as possible of them by 2007 samples. This was done until enough 2007 positive and negative samples were found. This was quite hard to judge but we finally decided to remove the last 2005 samples and therefore switch to "2007 only" training when 25% of the development set was annotated.

During the mixed training phase, using both 2005 and 2007 samples, it was not possible to use the text features in the classification system since no common representation was possible (English vs. Dutch language). This phase was therefore completed using only the visual content. The text was finally added as an additional feature for classification after the switch to 2007 only. It was actually introduced when about 40% of the development set was annotated both because we wanted to observe and distinguish both effects.

We started with the "relevance sampling" strategy as it was identified as the most efficient for the beginning of the process. Switching to the "most uncertain" strategy was considered at a time but we finally did not activate it as the expected gain was low and because we still wanted to observe other effects that might have interacted with it.

We implemented an additional strategy in order to boost annotation of positive samples, we call "neighborhood sampling". It consists in looking for new positive samples in the temporal neighborhood of already found positive samples. Each time a positive sample has been found, the preceding and following samples (previous and next subshots in the same video file) are selected with the highest priority for annotation. This additional strategy was used jointly with the "relevance sampling" strategy and it was activated early, when about 1.5% of the development set was annotated.

5   Quality

From the TRECVID 2005 collaborative annotation study [15], it was observed that disagreement among annotators occurred for about 3% of the annotated key frame × concepts. These are due sometimes to obvious mistakes, to misunderstanding of the concept or to subjective interpretation of the key frame/subshot contents. We had an additional source of inconsistency that is that some users apparently failed sometimes to notice the change of the concept to annotate despite the displayed warning. Such changes occur quite frequently since they are required by the active learning framework. Those various wrong annotations introduced some false positive and negatives which could affect the active learning process.

Since we wanted to keep the annotation effort reasonable, we did not want to have most of the concept being annotated several times. We decided to have a multiple check of only the most suspect annotations. We used for that the active learning approach by re-proposing the samples that have been predicted as most misclassified (i.e. positive annotated samples that were most probably predicted as negative and vice versa). All samples marked as skipped were also proposed for a second annotation. In case of disagreement between the first and second annotation of a key frame × concept, this one was proposed for a third judgment and a majority voting was used for making the final decision. As indicated in the following section, only a small fraction of the samples have been annotated twice, an even smaller fraction was annotated three times and so on while these were done as cleverly as possible to clean up as much as possible the collaborative annotation.

6   Analysis

32 teams participated to the 2007 TRECVID collaborative annotation effort and produced a total of 711566 annotations. Table 1 gives some statistics on these annotations. "Pass 1", "Pass 2", "Pass 3" and "Pass 4" corresponds to the

Table 1. Annotation statistics by pass, average on all concepts

	Annotated	% Annotated	Negative	Skipped	Positive	% Positive
Pass 1	641223	82.7	578299	13163	49761	7.76
Pass 2	46864	6.05	11904	7478	27482	58.6
Pass 3	21987	2.84	9383	4040	8564	39.0
Pass 4	1492	0.19	324	940	228	15.3
Synthesis	641223	82.7	578683	15348	47192	7.36

number of annotations that were done respectively at least once, at least twice, at least three times and at least four times for a given key frame × concept. The "Synthesis" correspond to the global annotation when a "majority" rule is applied if there is more than one annotation for a key frame × concept.

Table 2 indicates the frequency of the concepts in the collection. These figures come from incomplete data and this may cause a bias. Thanks to the active learning approach and to the fact that 75-90% of the corpus has been annotated, the bias is expected to be negligible except for the most frequent concepts like "Face" or "Person".

Table 2. Frequency of concepts (in percent)

Flag-US	0.06	Maps	0.60	Military	2.31	Crowd	8.56
Prisoner	0.15	Mountain	0.65	TV-screen	2.99	Walking_Run.	9.69
Weather	0.18	Truck	0.67	Car	3.68	Urban	9.70
Explosion_Fire	0.24	Court	0.73	Studio	4.22	Building	12.1
Natural-Disaster	0.25	Snow	0.75	Meeting	4.42	Vegetation	14.3
Airplane	0.30	Police_Security	1.40	Animal	4.63	Sky	17.4
Bus	0.30	People-Marching	1.43	Waterscape	5.07	Outdoor	39.3
Desert	0.35	Sports	1.50	Road	5.92	Face	56.3
Charts	0.60	Boat_Ship	1.58	Office	7.25	Person	72.4
		Median	1.95	Average	7.36		

The annotation finally reached a level of about 82% in average varying from about 75 to 90% depending upon the concepts, some having been more often multiply annotated than others. Figure 4 shows how the collaborative effort was spread over time. Horizontal units correspond to the days of May 2007 between 1 and 31 included and extrapolated outside. The effort started slowly with only the organizers (LIG) participating in order to control the size of the first active learning steps and to keep them small for an efficient start. Other users were asked to participate after a few days and to do their main effort during the following 15 days. Additional teams joined from time to time afterwards and contributed with a small but sustained effort which was mainly used for cleaning up the collaborative annotation with double and triple checks of suspect or inconsistent annotations.

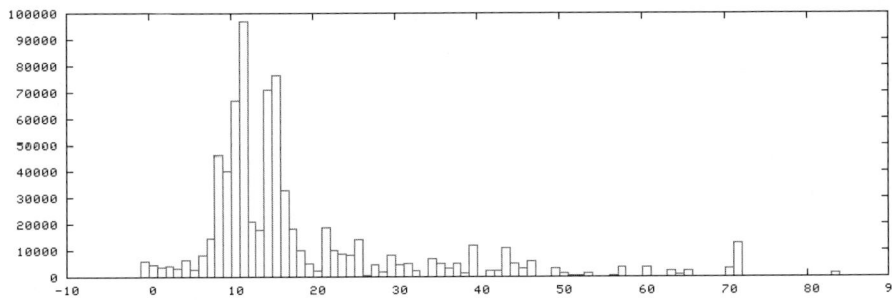

Fig. 4. Daily annotations in the collaborative annotation project (GMT time, May 2007 days)

Evolution of the number of positive samples found with the fraction of annotated samples gives idea of the reduction of effort provided by active learning method. Figure 5 shows this evolution (average for all concepts) for the TRECVID 2007 collaborative annotation. The prediction of what would have been the case for a random or sequential scan is shown as the diagonal. The shape is similar and the scale of the active learning effect is comparable. Three particular behaviors can be observed though the effects are small:

- Near the origin, at about 0.015, an increase in the finding rate is probably due to the activation of the "neighborhood sampling" strategy.
- After 0.25, an increase in the finding rate is probably due to the closing of the "cold start". Before this point, active learning uses a mix of 2005 and 2007 data; after this point, it uses only 2007 data.
- After 0.40 an increase in the finding rate is probably due to the inclusion of text feature in the classification system.

Though all these events have small effects of the overall finding rate, they may have larger effects for individual concepts. This is the case for example for the "Prisoner" concept when text features are included.

Figure 5 only shows the general trend of the evolution of the positive annotations with the total of annotations but this evolution is highly variable according to the considered concept. Figure 6 shows a superposition of the same curve for each of the 36 target concepts. The active learning effect is visible everywhere but it is more important for some concepts and sometimes more important in different regions.

Some effect linked to the fact that the cold start was done using a different collection can be observed. For instance, in the "Court", "Charts" and "Studio" concepts, the visual aspect is quite different in both collections and the active learning has first a negative effect (less positive samples are found than what a random choice would provide) and then, when a few are finally encountered (possibly by chance) the effect becomes positive and quite strong. In figure 6, the first and second curves close to the upper left corner have these behavior

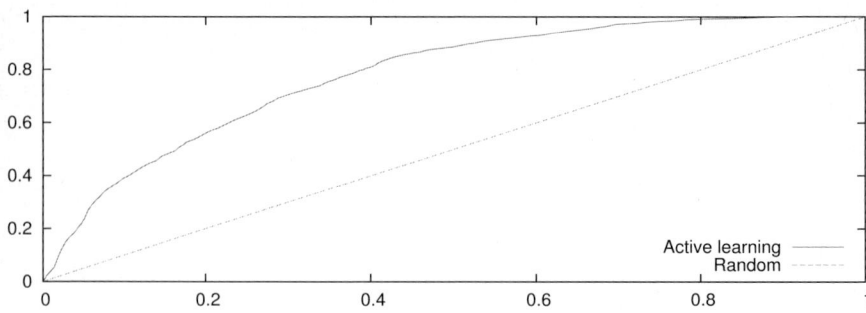

Fig. 5. Evolution of the fraction of positive samples found with the fraction of annotated samples; comparison between active learning and random annotation, all concepts

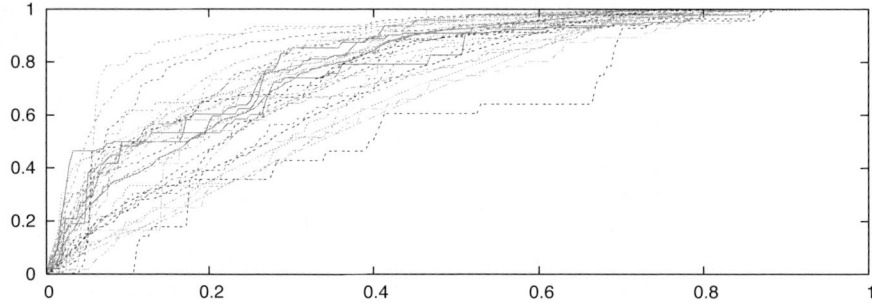

Fig. 6. Evolution of the fraction of positive samples found with the fraction of annotated samples for the 36 concepts individually

and correspond respectively to "Court" and "Studio" concepts. Furthermore, we observe some "step" shapes for some concepts, this effect typically happens for some visually heterogeneous concepts. When a positive sample is found, the system possibly finds many others positives in his temporal neighborhood. In figure 6, the lower curve corresponds to the "Prisoner" concept.

In order to study the benefit provided the quality and diversity of the samples selected by the active learning process, we computed classification of the test set with several fraction of the learning set from 5% to 90%. Figure 7 shows the evolution of the Inferred Average Precision (IAP) (average for the 20 concepts selected by TRECVID2007 for evaluation) with the number of annotated samples. The experiment has been conducted with two different systems: one from LIG wich is close to the one used for active learning during the annotation process and another from Helsinki University [6]. For the LIG system, it appears that the most useful samples are quickly selected: classification based on the 15% first annotated samples gives satisfying performance, while the

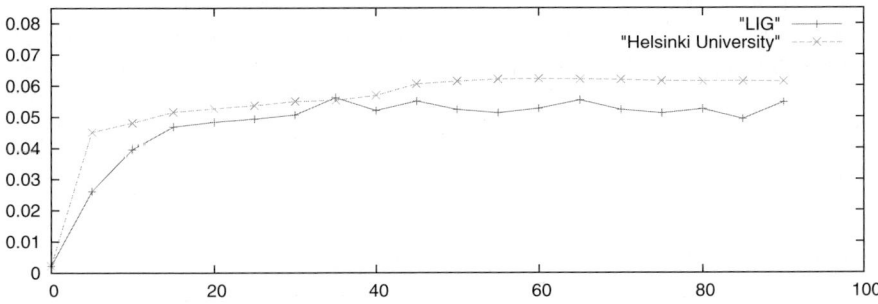

Fig. 7. Evolution of the mean of IAP of the 20 evaluated concepts with the fraction of annotated samples

classification based on the 35% first annotated samples gives the best performance. For the Helsinki University system, the best performance is reached slightly afterwards when about 50% of the samples have been annotated.

7   Conclusion

We organized the collaborative annotation of the High Level Features (HLF) in the development set of TRECVID 2007. These annotations have been used by the TRECVID 2007 participants to train their systems for the HLF extraction task. The annotation system is web-based and takes benefits of the *Active Learning* approach. This system allows participants to simultaneously get the most useful information from the partial annotation and significantly reduce the annotation effort relatively to previous collaborative annotations. We described the principles and the organization of this project and the lessons learnt from it. Previous experiments indicated that annotating only 20 to 30% of the development set would not hurt the systems' performances if these are carefully chosen. A similar behavior in the finding rate of positive samples was observed in the TRECVID 2007 collaborative annotation. While the development collection of TRECVID 2007 was quite small compared to the TRECVID 2003 and 2005 development collections, the benefits of the active learning approach for corpus annotation would be even more visible on a larger corpus to be annotated. Such an annotation system would be valuable in other machine-learning based areas, but not necessarily take benefit of the neighborhood sampling.

References

[1] Angluin, D.: Queries and concept learning. Machine Learning 2, 319–342 (1988)
[2] Ayache, S., Quénot, G.: Evaluation of active learning strategies for video indexing. Signal Processing: Image Communication (2007)
[3] Ayache, S., Quénot, G., Gensel, J.: CLIPS-LSR Experiments at TRECVID 2006. In: Proceedings of the TRECVID 2006 (2006)

[4] Chang, C.-C., Lin, C.-J.: LIBSVM: A Library for Support Vector Machines, Software (2001), available at http://www.csie.ntu.edu.tw/~cjlin/libsvm

[5] Gosselin, P.H., Cord, M.: A comparison of active classification methods for content-based image retrieval. In: CVDB 2004: Proceedings of the 1st international workshop on Computer vision meets databases, ACM Press, New York (2004)

[6] Koskela, M., Sjberg, M., Viitaniemi, V., Laaksonen, J., Prentis, P.: PicSOM Experiments in TRECVID 2007. In: Proceedings of the TRECVID 2007, November 5-6 (2007)

[7] Lewis, D.D., Gale, W.A.: A sequential algorithm for training text classifiers. In: Croft, W.B., van Rijsbergen, C.J. (eds.) Proceedings of SIGIR 1994, 17th ACM International Conference on Research and Development in Information Retrieval, Dublin, IE, pp. 3–12. Springer, Heidelberg (1994)

[8] Lin, C.-Y., Tseng, B.L., Smith, J.R.: Video collaborative annotation forum: Establishing ground-truth labels on large multimedia datasets. In: Proceedings of the TRECVID 2003, November 17–18 (2003)

[9] Naphade, M., Smith, J.R., Tesic, J., Chang, S.-F., Hsu, W., Kennedy, L., Hauptmann, A., Curtis, J.: Large-scale concept ontology for multimedia. IEEE Multi-Media 13(3), 86–91 (2006)

[10] Naphade, M.R., Smith, J.R.: On the detection of semantic concepts at trecvid. In: MULTIMEDIA 2004: Proceedings of the 12th annual ACM international conference on Multimedia, pp. 660–667. ACM Press, New York (2004)

[11] Qi, G.-J., Song, Y., Hua, X.-S., Zhang, H.-J., Dai, L.-R.: Video annotation by active learning and cluster tuning. In: CVPRW 2006: Proceedings of the 2006 Conference on Computer Vision and Pattern Recognition Workshop, Washington, DC, USA, p. 114. IEEE Computer Society, Los Alamitos (2006)

[12] Quénot, G.: Computation of Optical Flow Using Dynamic Programming. In: IAPR Workshop on Machine Vision Applications, November 12-14, 1996, pp. 249–252 (1996)

[13] Smeaton, A.F., Over, P., Kraaij, W.: Evaluation campaigns and trecvid. In: MIR 2006: Proceedings of the 8th ACM international workshop on Multimedia information retrieval, pp. 321–330. ACM Press, New York (2006)

[14] Snoek, C.G.M., Worring, M., Hauptmann, A.G.: Learning rich semantics from news video archives by style analysis. ACM Trans. Multimedia Comput. Commun. Appl. 2(2), 91–108 (2006)

[15] Volkmer, T., Smith, J.R., Natsev, A.P.: A web-based system for collaborative annotation of large image and video collections: an evaluation and user study. In: MULTIMEDIA 2005: Proceedings of the 13th annual ACM international conference on Multimedia, pp. 892–901. ACM Press, New York (2005)

Use of Implicit Graph for Recommending Relevant Videos: A Simulated Evaluation

David Vallet[1,2], Frank Hopfgartner[2], and Joemon Jose[2]

[1] Universidad Autónoma de Madrid, Madrid, Spain
[2] University of Glasgow, Glasgow, UK
david.vallet@uam.es, {hopfgarf,jj}@dcs.gla.ac.uk

Abstract. In this paper, we propose a model for exploiting community based usage information for video retrieval. Implicit usage information from a pool of past users could be a valuable source to address the difficulties caused due to the semantic gap problem. We propose a graph-based implicit feedback model in which all the usage information can be represented. A number of recommendation algorithms were suggested and experimented. A simulated user evaluation is conducted on the TREC VID collection and the results are presented. Analyzing the results we found some common characteristics on the best performing algorithms, which could indicate the best way of exploiting this type of usage information.

1 Introduction

In recent years, the rapid development of tools and systems to create and store private video enabled people to build their very own video collections. Besides, the easy to use Web applications such as YouTube and Google Video, accompanied by the hype produced around social services, motivated many to share video, leading to a rather uncoordinated publishing of video data. Despite the ease with which data can be created and published the tools that exist to organise and retrieve are insufficient in all terms (effectiveness, efficiency and usefulness). Hence, there is a growing need to develop new retrieval methods that support the users in searching and finding videos they are interested in. However, video retrieval is affected by the semantic gap [5] problem, which is the lack of association between the data representation based on the low-level features and the high-level concepts users associate with video.

One promising approach taken from the textual domain is the integration of relevance feedback to improve retrieval results. However, as in text retrieval, giving explicit relevance feedback is a cognitively demanding task and can affect the search process. A solution is to take implicit relevance feedback into account. However, which of these feedback possibilities in video retrieval are positive indicators about the relevance of a result has rarely been analysed.

In this paper, we are interested in using implicit relevance feedback from previous users of a digital video library to form a collaborative model of user behaviour, helping users find results which match their information need. We believe that the combined implicit relevance feedback of a larger group can be used to provide users

C. Macdonald et al. (Eds.): ECIR 2008, LNCS 4956, pp. 199–210, 2008.
© Springer-Verlag Berlin Heidelberg 2008

with positive recommendations. Although part of the data used in our evaluation comes from a user study, our main interest was evaluating a relatively high number of recommendation algorithms, which made the possibility to extend the user study to all the algorithms highly costly. Our evaluation required the possibility of being repeatable, allowing the study of different variables within a reasonable amount of time. Therefore, we introduce an approach of analysing implicit relevance feedback mechanisms based on a simulation-based evaluation.

The remainder of the paper is structured as follows. A brief summary of related work on implicit feedback applied to Multimedia Information Retrieval (MIR) and simulation-based evaluation is presented in section 2. Section 3 introduces a graph-based implicit pool representation, along with different recommendation strategies and subsequently in section 4, we describe the simulation based evaluation methodology. Section 5 will discuss the simulation results and will conclude in section 6 with some final thoughts.

2 Background

2.1 Implicit Feedback in Multimedia Information Retrieval

Deviating from the method of explicitly asking the user to rate the relevance of retrieval results, the use of implicit feedback techniques helps learning user interests unobtrusively. The main advantage is that users are relieved from providing feedback. While the techniques have been studied intensively in the textual domain [7], rarely anything is known in the multimedia domain. Hopfgartner and Jose [4] identified various implicit indicators of relevance in video retrieval when comparing the interfaces of state-of-the-art video retrieval tools. They introduced a simulation framework to analyse the effect of implicit relevance feedback in video retrieval, concluding that the usage of implicit indicators can influence retrieval performances. However, which of these implicit measures are useful to infer relevance has rarely been analysed in detail. Kelly and Belkin [6] criticise the use of display time as relevance indicator, as they assume that information-seeking behaviour is not influenced by contextual factors such as topic, task and collection. Therefore, they performed a study to investigate the relationship between the information-seeking task and the display time. Their results cast doubt on the straightforward interpretation of dwell time as an indicator of interest or relevance.

Usage information from a community of previous users can aid multimedia information retrieval. Usage information in the form of click-through data has been exploited [1]. When a user enters a query, the system can exploit the behaviour of previous users that issued a similar query. In this work, we are interested in approaches regarding MIR and graph-based representations of usage information. White et al. [10] introduced the concept of query and search session trails, where the interaction between the user and the retrieval system is seen as a path that leads from the first query to the last document of the query session or the search session (i.e. multiple queries). They argue that the last document of these trails is more likely to be relevant for the user. In our approach, we adopt this introduced concept of search trails. Furthermore, we are interested in representing and exploiting the whole interaction process. In video retrieval, the interaction sequence is a reasonable way to track the user's information need. Craswell and Szummer [1] represent the clickthrough data of an image retrieval system as a graph, where queries and documents are the nodes and links are the clickthrough data. We adopt also a graph-based approach, as it facilitates the representation of interaction sequences. While the authors limit the graph to clickthrough data, we propose to integrate other sources of implicit relevancy into the representation, as following [4].

2.2 Simulation Frameworks

In the de facto standard evaluation methodology known as Cranfield evaluation, users interact with a system searching for given search topics in a limited dataset. An analysis of recorded transaction log files and the retrieval results is then used to evaluate the research hypothesis. An alternative way of evaluating such user feedback is the use of simulated interactions. In such an approach, a set of possible steps are assumed when a user is performing a given task with the evaluated system [3,4 ,11].

Finin [2] introduced one of the first user simulation modelling approaches. This *"General User Modelling System"* (GUMS) allowed software developers to test their systems in feeding them with simple stereotype user behaviour. White et al. [11] proposed a simulation-based approach to evaluate the performance of implicit indicators in textual retrieval. They simulated user actions as viewing relevant documents, which were expected to improve the retrieval effectiveness. In the simulation-based evaluation methodology, actions that a real user may take are assumed and used to influence further retrieval results. Hopfgartner et al. [3] introduce a simulation framework to evaluate adaptive multimedia retrieval systems. In order to develop a retrieval method, they employed a simulated evaluation methodology which simulated users giving implicit relevance feedback. Hopfgartner and Jose [4] extended this simulation framework and simulated users interacting with state-of-the-art video retrieval systems. They argue that a simulation can be seen as a pre-implementation method which will give further opportunity to develop appropriate systems and subsequent user-centred evaluations. In this work, we will use the concept of simulated actions, although we will simulate user actions based on the past history and behaviour of users, trying to mimic the interaction of past users with an interactive video retrieval system.

3 Implicit Graph Recommendation Approaches

In this section, we present a set of recommendation algorithms on the graph representation. The approaches have been adapted to exploit the implicit graph, introduced in this section. The implicit graph models the historical data of interaction across all users and sessions. The main two characteristics of this graph model are 1) the representation of all the user interactions with the system, including the interaction sequence and 2) a scalable aggregation of the implicit information into a single representation. The implicit graph facilitates the analysis and exploitation of past implicit information, resulting in a model that is easy to build on top of different recommendation algorithms.

3.1 Implicit Graph Representation

The representation of the implicit graph can be seen in two different layers: the first one, a Labelled Directed Multigraph (LDM), gives a full detailed representation of the implicit information, and the second, a Weighted Directed Graph (WDG), is inferred from the previous, simplifying the interpretation of the LDM. It is on top of the WDG where the different recommendation rankings will be defined. Note that the WDG is not dependent on the LDM, and can be computed directly.

A user session s can be represented as a set of queries Q_s, which were input by the user, and the set of multimedia documents D_s the user accessed during the session. Queries and documents are therefore the nodes $N_s = \{Q_s \cup D_s\}$ of our graph representation $G_s = (N_s, A_s)$, in which the arcs are the set of actions $A_s(G) = \{n_i, n_j, a, u, t\}$ indicating that, at a time t, the user u performed an action of type a that lead the user from node n_i to node n_j, and $n_i, n_j \in N_s$. Note that n_j is the object of the action and that actions can be reflexive, for instance when a user clicked to view a video and then navigate through it. Actions types depend on the kind of actions recorded by the video retrieval system, like clicking, playing for an interval, navigating through the video or browsing to the next keyframe etc... Links can contain extra associated metadata, as type specific attributes, e.g. length of play in a play type action. The graph is multilinked, as different actions can have same source and destination nodes. All the session-based graphs are aggregated into a single graph $G = G(N, A)$, $N = \cup_s N_s$, $A = \cup_s A_s$ which can be seen as an overall pool of implicit information.

In order to enable the exploitation of the previous representation by the recommendation algorithms, we simplify the LDM by using no-labelled weighted links and collapsing all links interconnecting two nodes into one. This process is done in two steps: the first step computes a weighted graph $G_s = (N_s, W_s)$ that represents the user interactions during a single session. Links $W_s = \{n_i, n_j, w_s\}$ indicate that at least one action lead the user from node n_i to n_j. The weight value w_s represents the final relevance value calculated for node $n_{j,}$, its *local relevance* $lr(n_{j,})$. This value is obtained from the accumulation of implicit relevance evidences, given by the function $lr(n) = 1 - \frac{1}{x(n)}$, where $x(n)$ is the total of added weights associated to each type of action in which node n is object of. This subset of actions is defined as $A_s(G_s, n) = \{n_i, n_j, a, u, t | n_j = n\}, n \in N_s$. The x(n) weights are natural positive values returned by a function $f(a): A \to \mathbb{N}$, which returns higher values as the action are understood to give more evidence of implicit relevance. For instance, a user navigating through a video is a somehow good indication of implicit relevance. On the other hand, playing duration has proved to be a not as good indication [6], thus having a lower weight.

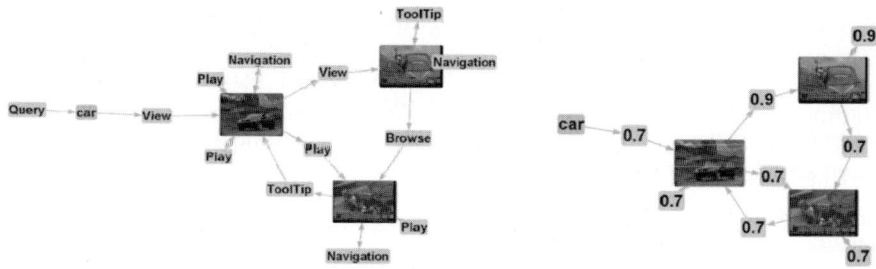

Fig. 1. Correspondence between the LDM (left) and WDG (right) representations

This analysis on the impact of implicit feedback importance weights is based on a previous work by Hopfgartner et al. [4]. The accumulation of implicit relevance weights can thus be calculated as $x(n) = \sum_{a \in A_s(G_s,n)} f(a)$. Figure 1 depicts an example of *LDM* and its correspondent *WDG* for a given session.

In the second step all the session-based *WDG*s are aggregated into a single overall graph $G = (N, W)$, which represents the implicit relevance pool, as collects all the implicit relevance evidence of all users across all sessions. The nodes of the implicit pool are all the nodes involved in any past interactions, $N = \bigcup_s N_s$, whereas the weighted links are a simple aggregation of the session-based values $W = \{n_i, n_j, w\}$, $w = \sum_s w_s$. These links represent the overall implicit relevance that users, which actions lead from node n_i to n_j, gave to node n_j. Figure 2 shows an example of implicit relevance pool.

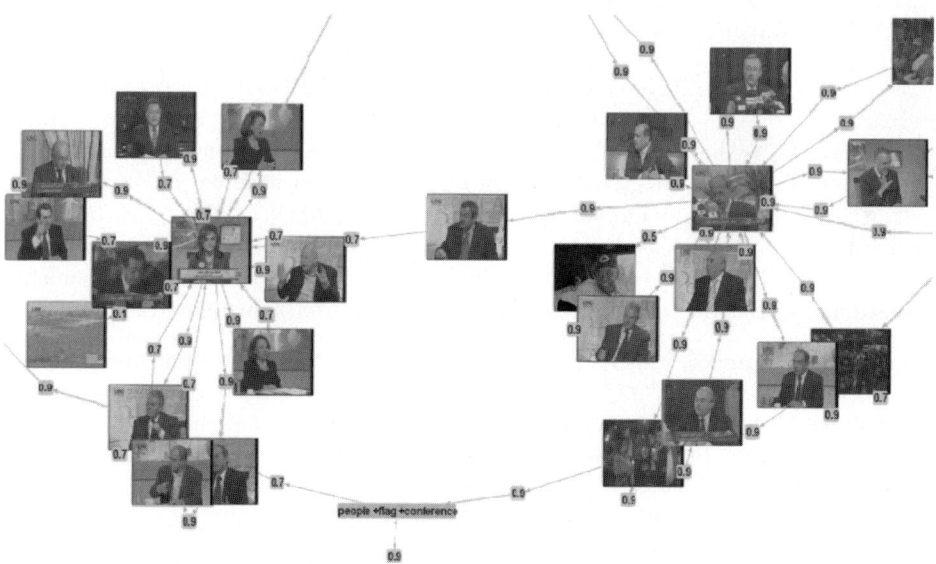

Fig. 2. Typical graph structure, where some relevant nodes receive a large number of links

3.2 WDG Based Recommendation Algorithms

As the user interacts with the system, a session-based WDG is constructed. The current user's session is thus represented by $G_{s'} = (N_{s'}, W_{s'})$. This graph is the starting point of the recommendation algorithms presented next, and can be seen as a form of actual context for the user.

Neighbourhood. As a way of obtaining related nodes, we define the node neighbourhood of a given node n as:

$$NH(n) = \{n_1, \dots, n_M | distance(n, n_m) < D_{MAX}, n_m \in N\}$$

which are the nodes that are within a distance D_{MAX} of n, without taking into consideration the link directionality. These nodes are somehow related to n by the actions of the users, either because the users interacted with n after interacting with the neighbour nodes, or because there are the nodes the user interacted with after interacting with n.

Using the properties derived from the implicit graph, we can calculate the overall relevance value for a given node, this value indicates the aggregation of implicit relevance that users gave historically to n, when was involved in the users' interactions. Given all the incident weighted links of n, defined by the subset $W_s(G_s, n) = \{n_i, n_j, w | n_j = n\}, n \in N_s$, the overall relevance value for n is calculated as follows:

$$or(n) = \sum_{w \in W_s(G_s, n)} w$$

Given the current session of a user and the implicit relevance pool we can then define the node recommendation value as:

$$nr(n, N_{s'}) = \sum_{n_i \in N_{s'}} lr'(n_i) \cdot or(n) | n \in NH(n_i)$$

where $lr'(n_i)$ is the local relevance value for the current session of the user, using the subset of actions $A_s(G_{s'}, n)$. We can then define two different recommendation values: the query neighbourhood $nh_q(n, N_{s'}) = nr(n, Q_{s'}) | Q_{s'} \in N_{s'}$, which recommends nodes related to the actual queries of the user and, similarly, the document neighbourhood $nh_d(n, N_{s'}) = nr(n, D_{s'}) | D_{s'} \in N_{s'}$, which recommends instead nodes related to the documents involved in the user's interactions.

Interaction Sequence. This recommendation approach tries to take into consideration the interaction process of the user, with the scope of recommending those nodes that are following this sequence of interactions. For instance, if a user has opened a video of news highlights, the recommendation could contain the more in-depth stories that previous users found interesting to view next. It is defined as follows:

$$is(n, N_{s'}) = \sum_{n_i \in N_{s'}} \left((lr'(n_i) \cdot \xi^{l-1} \cdot w) \middle| \begin{array}{l} \exists\, p = n_i \rightsquigarrow n_j \rightarrow n \\ w \in \{n_j, n, w\} \\ l = length(p) \\ l < L_{MAX} \end{array} \right)$$

where p is the path between any node n_i and node n, taking into consideration the link directionality. l is the length of the path (counted as the number of links), having a distance lower than a maximum length L_{MAX}. Finally, ξ is a length reduction factor, set to 0.8 in our experiments.

Query Destination. This algorithm is adapted from the work of White et al. [11] on query and search trails. White suggests that the last documents that a user visits within a search or query session has a high relevancy. We choose the query destination measure, which they proved that was best for explorative tasks (used in the evaluation process). The query destination value ranks by popularity the query trails' destinations. In our own representation is defined as:

$$ qd(q,d) = S(d,q) \cdot \sum_p w \left| \begin{array}{l} \exists\, p = q \rightsquigarrow d_j \rightarrow d \rightarrow n_q \\ d_j, d \in D_s\,, n_q \in Q_s \\ w \in \{d_j, d, w\} \end{array} \right. $$

where $S(d,q)$ is the *tf.idf* similarity measure between document d and the last query $q \subset last(Q_{st})$ input by the user. Note that the links between documents in the WDG are essentially trail links, but we don't limit these trails to clicks, but extended them with more types of actions. The popularity value is defined by the weight aggregation of all incident links within the paths of the different historical query trails defined between q and d.

Random Walk. Craswell and Szummer [1] exploit the clickthrough data with a random walk algorithm. The random walk computation will end, in theory, with a higher probability on those nodes that previous users found (implicitly) relevant after issuing the query (forward walk approach) or on those documents that represent the information need of the query (backward walk approach). For this computation, a probability of going from node n_k to n_j is needed:

$$ t_{+1|t}\left(n_k|n_j\right) = \begin{cases} (1-s)\,C_{jk} / \sum_i C_{ji} \;\; \forall k \neq j \\ s \text{ when } k = j \end{cases} $$

where s is the probability of staying in the same node (set to 0.9) and the click count is $C_{ij} = w \in \{n_i, n_j, w\}$, thus taking into considerations the aggregation of implicit evidences. Using these probabilities, we compute a backwards random walk $rw_B(q)$ and a forward random walk $rw_F(q)$, $q \in last(Q_{st})$. Both random walks were computed using 11 steps.

4 Simulated User Behaviour for Interactive Retrieval Evaluation

To analyse the performance of each recommendation methodology we had to construct a graph pool with implicit data from previous users and evaluate the performance of each recommendation algorithm. The graph pool was constructed by monitoring the interaction of 24 users, mostly postgraduate students and research assistants, with a video retrieval system introduced by Urban et al. [9]. The participants' group consisted of 18 males and 6 females with an average age of 25.2 years and an advanced proficiency with English. Each of the users performed the same selection of four explorative tasks from TRECVID 2006 [8], spending 15 minutes for each task. We decided to use those tasks that performed the worst in TRECVID, mostly due to their multifaceted and ambiguous nature, while still being quite specific, therefore being the most challenging for current multimedia retrieval systems. The four tasks were:

- Find shots with a view of one or more tall buildings (more than four stories) and the top story visible (Task 1)
- Find shots with one or more soldiers, police, or guards escorting a prisoner (Task 2)
- Find shots of a group including at least four people dressed in suits, seated, and with at least one flag (Task 3)
- Find shots of a greeting by at least one kiss on the cheek (Task 4)

Our intention was to analyse if the recommendation algorithms are able to improve the performance of these difficult tasks. As advanced retrieval techniques such as search-by-concept of search-by-example did not perform well on these tasks within TRECVID, here implicit feedback could be a promising approach to aid users with their search. A post search questionnaire confirmed that the tasks were, in general, indeed perceived as difficult for the users, with a special mention for Task 4.

We therefore constructed the implicit pool WDG, which contained the interaction information of each user, including also noisy data, obtained from two training tasks which users performed for ten minutes each. Once we filled the implicit pool with the user data, a natural next step would be to use users to evaluate each system. However, having six different recommendation strategies makes this evaluation step too costly in both time and human resources. Instead of this, we opted to create a simulation framework that used the statistical data mined from the original 24 users. Using this data, we simulated users that interact with a hypothetical extension of the original retrieval system, with the addition of both query and video recommendations.

The evaluation system thus simulates a user interacting with this extension of the original video retrieval system and receiving recommendation from the evaluated algorithm. We used the statistical information from the 24 training users in order to simulate probabilities of the user performing certain types of actions. A new interaction was added: selecting a recommended query. In order to evaluate the recommendation algorithms, we made the following assumption: after a query is launched, users first review the five top recommended results before they continue to look into the query result set. Therefore, the five recommended results are added on top of the result set. Note that there are various recommendation approaches that can be updated as soon as new implicit information is obtained. However, in order to evaluate the algorithms evenly we choose to update the recommenda-tion by issued query. Table 1 shows the probability values obtained from the user study.

Table 1. Probability and normal distribution measures for observed action types

Action type	Probability	Action type	μ	σ
Click relevant result	0.8	Navigation	0.5	2
Click irrelevant result	0.2	Play duration (3 sec interval)	2	3
Tooltip results[1]	0.8	Browsing near keyframes	0.25	1

The simulation system, based on a system introduced by Hopfgartner et al. [3], simulates a user performing one of the four tasks, using ten interactions (i.e. queries) for each task, and interacting with ten documents per query, which were the averages observed during the user experiments. Given the generic recommendation algorithm ra, the steps of each interaction for task t are as follows:

1) With probability p_q (fixed to 0.6 in our experiments) execute first recommended query $q \in ra(WDG_{s_t})$, otherwise execute a random query $q \in Q$ from task t.
2) Collect $\{top5(ra(WDG_{s_t})), top20(query\ results)\}$ as the result set of the interaction, and until the user has clicked ten results:
   - With probability p(tooltip result) tooltip result
   - With probability p(click|relevant) click result
   - If result clicked
     - Simulate browsing steps: $N(\mu(browsing), \sigma(browsing))$
     - Simulate navigation actions: $N(\mu(navigation), \sigma(navigation))$
     - Simulate playing duration: $N(\mu(play), \sigma(play))$

The recommendation algorithm has access to the current session information, i.e. WDG_{s_t}. Therefore, the recommendation algorithms has access to the interaction sequence, the last input query and the last accessed documents. There is one exception with the query destination algorithm, which does not recommend queries, in this case the queries are always chosen at random.

5   Experiments

The simulations results are discussed in this section. Each recommendation strategy was simulated through 50 runs, which proved to be statistically relevant. Figure 3 depicts the overall performance of each system, including the baseline system, which is a simulation with no recommendation whatsoever. The evaluation measure is the average of the P@N points for every run. Following an interactive evaluation methodology, we take as final result set the rank-based merge of the results sets for each of the 10 interactions, which include on top the first five recommended results.

The recommendation strategy that overall appears to perform best is the query destination recommendation, followed by the interaction sequence and the forward random walk. One singular characteristic of the query destination approach is that the

---

[1] In the retrieval system, when the user leaves the mouse on top of a result for one second, a tooltip appears showing the nearby keyframes for the video.

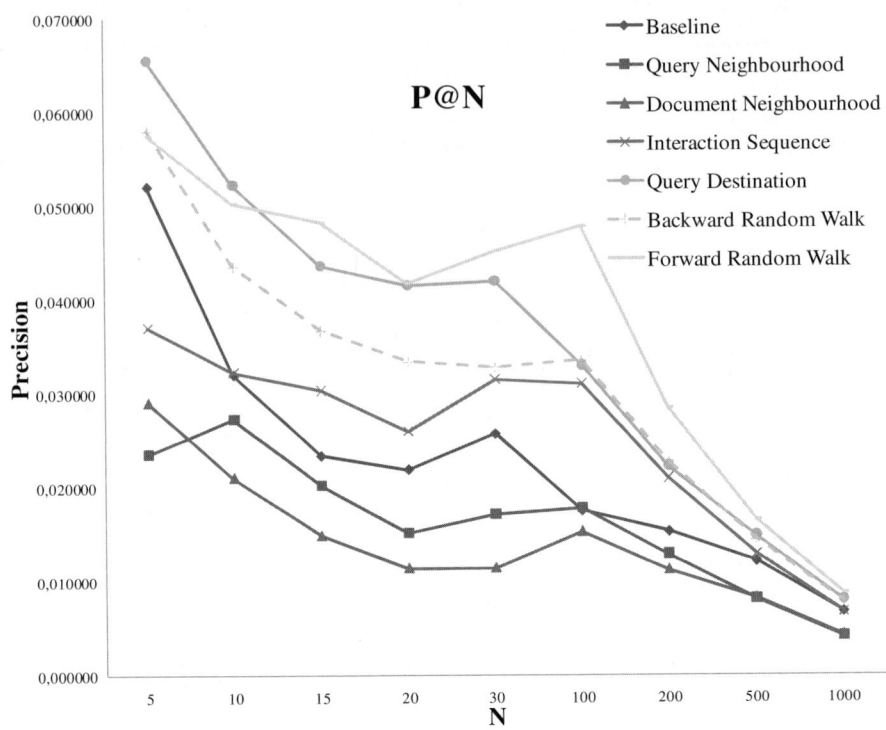

Fig. 3. Precision cut-off points for each recommendation strategy

similarity between the last query and the recommended documents is taken into consideration, apart from the popularity measure. The interaction sequence algorithm performance does highlight the importance of exploiting the search and query trails similarities. The random walk approach also exploits these trails. This could be the reason why the forward random walk performance is close to the interaction sequence. Surprisingly, the backward random walk has a sensible loss of performance against the forward approach, although Craswell and Szummer report the contrary. The poor performance of the neighbourhood based strategies suggests that the link directionality has indeed to be taken into consideration, as well as the density of the paths that point from the node to its neighbours.

Although the query destination performs the best on average, the results per topic show that the performance of each algorithm varies meaningfully for each task. Figure 4 shows the performance of this four recommendation strategies for each topic.

Note that there is a different algorithm that performs better in the tree first tasks: query destination in Task 1, interaction sequence in Task 2 and forward random walk in Task 3. Finally, no recommendation approach was able to outperform the baseline in Task 4. The reason was probably that users showed an erratic behavior in this task, as they confessed a great difficulty on meeting the semantic of the task at hand with the videos' textual metadata.

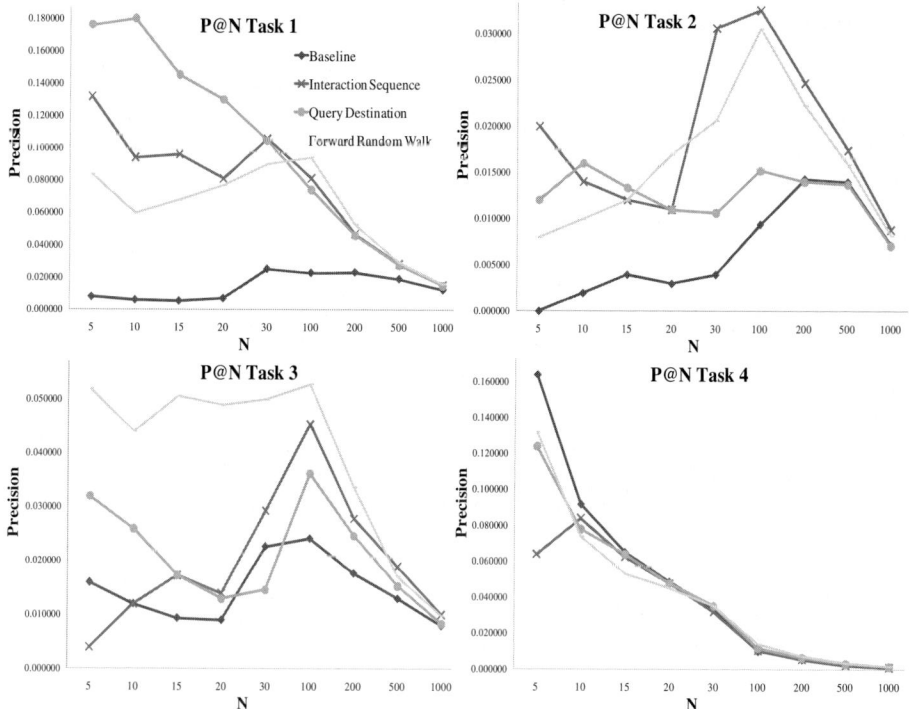

Fig. 4. Precision cut-off point for the best four strategies and the four evaluated tasks

6   Conclusion

In this work, we have explored the exploitation of community usage feedback information to aid users in difficult video retrieval tasks. The presented integrated model includes an efficient and scalable way of representing this past information and, even more important, eases the use of any desired recommendation strategy. The implicit graph representation has proven to facilitate the analysis of the diverse types of implicit actions that a video retrieval system can provide, thus allowing an easy extension. In addition, an evaluation framework is introduced, of which the main goal is to facilitate evaluation of new recommendation strategies.

Using the presented evaluation framework, we have reported a set of experiments on different recommendation approaches, either created by us, or adapted from related work. We have observed that the performance of each evaluated strategy varied significantly with each specific task, indicating that there could be different complementary approaches for video retrieval recommendation. The use of the overall popularity of the document, the exploitation of interaction trails and taking into consideration the last submitted query were some of the characteristics of the evaluated recommendation strategies that performed the best.

Acknowledgements

This research work was partially supported by the European Commission under contracts: K-Space (FP6-027026), SALERO (FP6- 027122) and SEMEDIA (FP6 - 045032). Authors acknowledge Martin Halvey for help with the experiments and for fruitful discussions.

References

1. Craswell, N., Szummer, M.: Random walks on the click graph. In: IGIR 2007: Proceedings of the 30th annual international ACM SIGIR 2007, pp. 239–246. ACM, New York (2007)
2. Finin, T.W.: GUMS: A General User Modeling Shell. In: User Models in Dialog Systems, pp. 411–430. Springer, Heidelberg (1989)
3. Hopfgartner, F., Urban, J., Villa, R., Jose, J.: Simulated Testing of an Adaptive Multimedia Information Retrieval System. Content-Based Multimedia Indexing, 2007. In: CBMI 2007. International Workshop, Bourdeaux, France, pp. 328–335 (2007)
4. Hopfgartner, F., Jose, J.: Evaluating the Implicit Feedback Models for Adaptive Video Retrieval. In: ACMMIR 2007, pp. 323–331 (2007)
5. Jaimes, A., Christel, M., Gilles, S., Ramesh, S., Ma, W.: Multimedia Information Retrieval: What is it, and why isn't anyone using it? In: ACM MIR 2005, pp. 3–8. ACM Press, New York (2005)
6. Kelly, D., Belkin, N.J.: Display time as implicit feedback: understanding task effects. In: ACM SIGIR 2004., pp. 377–384. ACM, New York (2004)
7. Kelly, D., Teevan, J.: Implicit Feedback for Inferring User Preference: A Bibliography. SIGIR Forum 37(2), 18–28 (2003)
8. Over, P., Ianeva, T.: TRECVID 2006 Overview. In: TRECVid 2006 – Text Retrieval Conference TRECVID, Gaithersburg, MD (2006)
9. Urban, J., Hilaire, X., Hopfgartner, F., Villa, R., Jose, J., Chantamunee, S., Goto, Y.: TRECVID 2006. TRECVID, Glasgow University, Gaithersburg, MD (2006)
10. White, R., Bilenko, M., Cucerzan, S.: Studying the use of popular destinations to enhance web search interaction. In: ACM SIGIR 2007, pp. 159–166. ACM Press, New York (2007)
11. White, R., Jose, J.M., van Rijsbergen, C.J., Ruthven, I.: A Simulated Study of Implicit Feedback Models. In: McDonald, S., Tait, J.I. (eds.) ECIR 2004. LNCS, vol. 2997, pp. 311–326. Springer, Heidelberg (2004)

Using Terms from Citations for IR: Some First Results

Anna Ritchie[1], Simone Teufel[1], and Stephen Robertson[2]

[1] University of Cambridge, Computer Laboratory, 15 J J Thompson Avenue, Cambridge, CB3 0FD, U.K.
`{ar283,sht25}@cl.cam.ac.uk`
[2] Microsoft Research Ltd, Roger Needham House, 7 J J Thomson Avenue, Cambridge, CB3 0FB, U.K.
`ser@microsoft.com`

Abstract. We present the results of experiments using terms from citations for scientific literature search. To index a given document, we use terms used by citing documents to describe that document, in combination with terms from the document itself. We find that the combination of terms gives better retrieval performance than standard indexing of the document terms alone and present a brief analysis of our results. This paper marks the first experimental results from a new test collection of scientific papers, created by us in order to study citation-based methods for IR.

1 Introduction

There has been a recent resurgence of interest in using citations between documents. However, while the potential usefulness of the text used in association with citations has been noted in relation to, e.g., text summarization [1,2], thesaurus construction [3] and other tasks, recent work in IR has focused on statistical citation data, like citation counts and PageRank-style methods, e.g., [4,5,6]. We test whether term-based IR on scientific papers can be improved with citation information, by using terms from the citing document to additionally describe (i.e., index) the cited document. This idea of using terms external to a document for indexing, coming from a 'citing' document, is also used in Web IR. Citations are quite like hyperlinks and link structure, including anchor text, has been used to advantage in retrieval tasks [7,8]. In this comparable situation, Web pages are often poorly self-descriptive [9], while anchor text is often a higher-level description of the pointed-to page [10].

We explore whether using terms from citations to a paper in combination with terms from the paper itself can improve the retrieval performance achieved when only the paper terms are indexed. Some work has been done in this area but no previous experiments have used both citing and cited papers. Previous experiments have indexed cited papers using terms from citing papers but no terms from the cited papers themselves: Bradshaw used terms from a fixed window around citations [11], while Dunlop and van Rijsbergen used the abstracts of citing papers [12].

In this paper, we first motivate our use of citations for term-based IR. Then, Section 3 describes our experimental setup; in Section 4, we present and analyse our

C. Macdonald et al. (Eds.): ECIR 2008, LNCS 4956, pp. 211–221, 2008.

hyperlink: The <a href="*http://www.google.com*">Google</a> search engine...

citation: "Dictionaries can be constructed in various ways - see *Watson (1993a, 1995)* for a taxonomy of (general) finite-state automata construction algorithms."

Fig. 1. Similarity between Hyperlinks and Citations

results, which show that using citation terms can indeed improve retrieval; Section 5 concludes and outlines future work.

2   Motivation and Related Work

There are definite parallels between the Web and scientific literature: "hyperlinks... provide semantic linkages between objects, much in the same manner that citations link documents to other related documents" [13]. However, there are also fundamental differences. An important and widespread factor in the use of hyperlinks is the additional use of their anchor text (i.e., the text enclosed in the ⟨a⟩ tags of the HTML document (see Fig. 1)). It is a well-documented problem that Web pages are often poorly self-descriptive [9]. Anchor text, on the other hand, is often a higher-level description of the pointed-to page. Davison discusses just how well anchor text does this and provides experimental results to back this claim [10]. Thus, beginning with McBryan [7], there is a trend of propagating anchor text along its hyperlink to associate it with the linked page, as well as that in which it is found (as in Fig. 2). Google, for example, includes anchor text as index terms for the linked page [9].

Returning to the analogy between the Web and scientific literature, the anchor text phenomenon is also observed with citations: citations are usually introduced purposefully alongside some descriptive reference to the cited document (see Fig. 1). However, no anchor text exists in scientific papers, unlike in Web pages, where there are HTML tags to delimit the text associated with a link. The question is raised, therefore, of what is the anchor text equivalent for formal citations. Bradshaw calls the concept *referential text*, using it as the basis of his *Reference-Directed Indexing* (RDI), whereby a scientific document is indexed by the text that refers to it in documents that cite it [11], instead of by the text in the document itself, as is typical in IR. The theory behind RDI is that, when citing, authors describe a document in terms similar to a searcher's query for the information it contains. Thus, this referential text should contain good index terms for the document and Bradshaw shows an increase in retrieval precision over a standard vector-space model implementation; 1.66 more relevant documents are retrieved in the top ten in a small evaluation on 32 queries.

However, a number of issues may be raised with RDI. Firstly, it only indexes referential text so a document must be cited at least once (by a document available to the indexer) in order to be indexed. Bradshaw's evaluation excluded any documents that were not cited and does not disclose how many of these there were. Secondly, referential text is extracted using CiteSeer's *citation context* (a window of around one hundred words around the citation). This method is simplistic: the terms that are definitely associated with a citation are variable in number and in distance from the citation, so a fixed

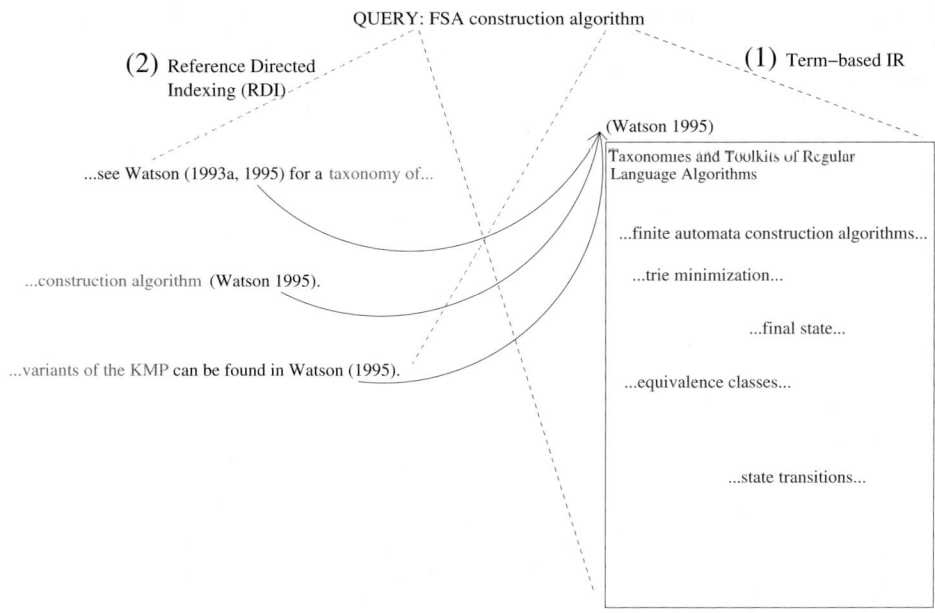

Fig. 2. Use of citation terms in IR: how does (1)+(2) compare to just (1) or (2) alone?

window will not accurately capture the citation terms for all citations. In a much earlier study, O'Connor noted the inherent difficulty in identifying which terms belong with a citation [14] and Bradshaw too states the difficulty in extracting good index terms automatically from a citation.

Dunlop investigated a similar technique with a different application in mind (i.e., retrieval of non-textual documents, such as image, sound and video files [12]). Dunlop's retrieval model uses clustering techniques to create a description of a non-textual document from terms in textual documents with links to that document. In order to establish how well descriptions made using the model represent documents, the method was applied to textual documents, indeed, to the CACM test collection, where the links between documents were citations. The experiment compared retrieval performance using the cluster-based descriptions against using the documents themselves; the cluster-based descriptions achieved roughly 70% of the performance from using the document content. Again, however, Dunlop did not measure the performance using the cluster-based descriptions in combination with the document content.

Thus, there is a gap in the research: retrieval performance from using the combination of terms from citing and cited documents has not been measured. How will this compare to using the document terms alone? How will it compare to using terms from the citing documents alone? We could make these comparisons on the CACM collection of abstracts. However, a test collection with the full text of a substantial number of citing and cited papers will allow broader experimentation, e.g., comparisons between using the full cited paper versus the abstract alone. We previously introduced such a test collection [15,16], since no existing test collection satisfies our requirements. The newswire articles in traditional collections do not contain citations. CACM contains abstracts and the GIRT collection [17], likewise, consists of content-bearing fields, not full documents. The earlier TREC Genomics collections consist of MEDLINE records, containing abstracts but not full papers [18,19]. In the 2006 track, a new collection of full-text documents was introduced but this was designed for passage retrieval for a QA task, not document retrieval [20]. Our test collection allows us to conduct not only the experiments we report here, but a wider range of experiments using different combinations of information from citing and/or cited documents.

3   Experimental Setup

3.1   Data and Tools

Our test collection is centred around the ACL Anthology[1] digital archive of Computational Linguistics (CL) research papers. The document collection is a ~9800 document subset of the archive; roughly, all documents published in 2005 or earlier, with non-papers (e.g., letters to the editor, tables of contents) removed. Our query set consists of 82 research questions from CL papers, with an average of 11.4 judged relevant documents per query [16], such as:

- *Does anaphora resolution improve summarization (based on latent semantic analysis) performance?*
- *Can knowledge-lean methods be used to discourse chunk a sentence?*

The test collection was built using a methodology based on the Cranfield 2 design [21]. The principle behind the method is that research papers are written in response to research questions, i.e, information needs, and that the references in a paper are a list of documents that are relevant to that need. Thus, papers are a source of queries (the research questions) and relevant documents (the references). For our queries, the authors of accepted papers for two upcoming CL conferences were asked, by email, for the research question or questions underlying their papers. By asking for queries from recent conference authors, we aimed for a query set that is a realistic model of searches that representative users of the document collection would make; genuine information needs from many different people with many different research interests from across the CL domain. They were also asked to make relevance judgements for their references. Due to the relative self-containedness of the CL domain[2], we expected a significant proportion of the relevance judgements gathered in this way to be for documents in the ACL Anthology and, thus, useful as test collection data.

Based on some analytical experiments [15], there were too few relevance judgements at this stage; we executed a second stage to obtain more judgements for our queries. The Anthology is too large to make complete judgements feasible. Therefore, we used the pooling method to identify potentially relevant documents in the Anthology for each of our queries, for the query authors to judge. First, for each query, we manually searched

---

[1] http://www.aclweb.org/anthology/
[2] We empirically measured the proportion of collection-internal references in Anthology papers to be 0.33.

the Anthology using its Google search facility. We then ran the queries through three standard retrieval models, as implemented in the Lemur Toolkit[3]: Okapi BM25, KL-divergence (both with relevance feedback using the existing relevant documents) and Cosine similarity. We pooled the results from the manual and automatic searches, including all manual search results and adding non-duplicates from each of the automatic rankings in turn until fifteen documents were in the list. Our pool was very shallow compared to TREC-style pools; our method relies on volunteer judges and therefore we needed to keep the effort asked of each judge to a minimum. The list of potentially relevant documents was sent to the query author, with an invitation to judge them and materials to aid the relevance decision.

For the experiments in this paper, we index our documents using Lemur, specifically Indri [22], its integrated language-model based component, using stopping and stemming. Our queries are likewise stopped and stemmed. We use the Cosine, Okapi, KL-divergence and Indri retrieval models with standard parameters to test our method. We also performed retrieval runs using KL[4] with relevance feedback (KL FB). In each run, 100 documents were retrieved per query; this is already far greater than the number of relevant documents for any query. For evaluation, we use the TREC evaluation software, trec_eval[5].

3.2   Citation Method

We firstly carry out some pre-processing of the documents: we use methods based on regular expressions to annotate the reference list, to identify citations in the running text and to associate these with items in the reference list. This is a non-trivial task, for which high precision methods have been developed independently [23]. Our approach is more simplistic but nevertheless performs well: from a small study of ten journal papers, we found and correctly matched 388 out of 461 citations with their corresponding reference (84.2%). Errors mostly occur due to noise from the PDF to XML conversion.

Next, we use the reference information to identify which references are to documents in the ACL Anthology; we extract terms from the citations associated with these references to a database. Specifically, we use the tokeniser from a statistical natural language parser [24] to segment the text into sentences, then extract all terms from the sentence that contains the citation. Identifying which terms are associated with a citation is an interesting problem, which we have discussed elsewhere [25]; our method is only one of many possibilities and we do not claim that it is optimal. Our database contains terms from over 23,000 citations to over 3300 papers. Finally, we add these terms to an XML representation of the document before indexing. We build one index from the XML documents alone and another from the documents plus citation terms. In order to investigate the effect of weighting citation terms differently relative to document terms, we build separate indexes where the citation terms are added in duplicate to the XML document, to achieve the desired weight. The method is resource-hungry, however, and we investigate only a small range of weights in this way.

---

[3] http://www.lemurproject.org/

[4] We do not report results using Okapi with relevance feedback: the Lemur documentation notes a suspected bug in the Okapi feedback implementation.

[5] http://trec.nist.gov/trec_eval/trec_eval.8.1.tar.gz

Table 1. Retrieval Performance With versus Without Citations (W = weight of citation terms. Differences in **bold** are significant for p≤0.05 and those **underlined** for p≤0.01)

Retrieval Model (W)	MAP W/out	With	p	P(5) W/out	With	p	R-precision W/out	With	p	GMAP W/out	With	bpref W/out	With	p
Okapi (1)	.083	.084	.582	.110	.120	.251	.094	.098	.313	.004	.004	.218	.227	.118
(2)		.084	.786		.127	.070		.103	.133		.005		**.234**	**.018**
(3)		.085	.636		.127	.070		.108	.053		.005		**.234**	**.016**
(4)		.084	.794		**.129**	**.045**		.104	.228		.004		.230	.130
Cosine (1)	.140	.143	.454	.185	.188	.567	.141	.146	.223	.041	.046	**.313**	**.328**	**.001**
(2)		.146	.148		.190	.418		**.156**	**.002**		.048		**.333**	**.001**
(3)		.143	.528		.185	1.000		**.156**	.028		.048		**.335**	**.001**
(4)		.146	.326		.190	.596		**.155**	**.044**		.049		**.338**	**.001**
Indri (1)	.158	**.172**	**.000**	.254	**.285**	**.001**	.188	**.199**	**.025**	.056	.072	.366	**.379**	**.014**
(2)		**.176**	**.000**		**.298**	**.000**		**.210**	**.000**		.077		**.383**	**.005**
(3)		**.180**	**.000**		**.305**	**.000**		**.213**	**.000**		.080		**.387**	**.006**
(4)		**.182**	**.000**		**.302**	**.000**		**.220**	**.000**		.082		**.385**	**.019**
KL (1)	.166	**.174**	**.026**	.256	**.273**	**.019**	.192	**.206**	**.028**	.065	.074	.373	.379	.420
(2)		**.180**	**.003**		.271	.159		**.213**	**.003**		.077		.387	.095
(3)		**.183**	**.001**		.278	.072		**.215**	**.006**		.080		.389	.059
(4)		**.184**	**.004**		.283	.070		**.216**	**.006**		.082		**.393**	**.028**
KL FB (1)	.238	**.250**	**.004**	.332	.349	.090	.251	**.264**	**.015**	.157	.177	.483	.493	.199
(2)		**.259**	**.000**		.346	.259		**.268**	**.009**		.189		**.504**	**.020**
(3)		**.263**	**.000**		.349	.195		**.279**	**.000**		.195		**.511**	**.008**
(4)		**.267**	**.000**		.354	.095		**.282**	**.000**		.199		**.515**	**.006**

There are alternatives to this weighting method: the Indri query language allows terms to be weighted according to which part of the document they occur in. However, this method can only be used with the Indri retrieval model; we cannot use weighted queries to investigate the effects of citation term weighting on all models' performance. Furthermore, the two methods are not equivalent, in terms of document scoring and ranking, for multiple reasons. Firstly, the weighted query method calculates scores using term statistics across individual fields, rather than across whole documents, as in the case of unweighted queries. Thus, the ranking produced by a weighted query where the fields are weighted equally and the ranking produced by its unweighted counterpart on the same index will not necessarily be the same. Secondly, in the term duplication method, the statistics for a given term will be different in each index, as it is altered by the citation 'weight': there will be an additional occurrence of that term in the index for every duplicate citation term that is added. This is not the case in the weighted query method, where each citation term is added exactly once to the index. The differences between these weighting methods opens the door for comparative experimentation between them; we intend to investigate this in the future.

4   Results and Analysis

Table 1 summarizes the results. In each row, we compare the performance of a given retrieval model on the index without citation terms to its performance on one index with citation terms (i.e., with a particular citation term weight). We consider the values of a range of standard performance measures and t-test for statistical significance of with-versus without-citation performance; differences highlighted in bold are significant for $p \leq 0.05$.

Performance is uniformly higher with citations than without, for all models, for all measures, with the exception of two Okapi runs where GMAP is unchanged. The general trend is for performance to increase as citation terms are weighted more highly. Notably, the performance increases on all Indri runs for all measures are statistically significant. All MAP and R-precision increases are significant for Indri, KL and KL FB. Cosine and Okapi show the smallest and least significant performance increases. The results for Okapi, in particular, do not appear to follow the trend of increasing performance with increasing citation term weight. One possible explanation for this is that the weights investigated here are too small to have any significant effect on Okapi's performance; in the comparable situation of Web IR, optimal Okapi performance has been achieved by weighting anchor text as much as 35 times higher than Web page body text [26]. This may also be the case for the Cosine model. Similarly, the narrow range of citation weights does not show a plateau in the performance of any of the other models. Further investigation is required to discover the optimal weighting of citation terms to document terms.

To try to better understand the overall results, we studied in detail the retrieval rankings for a few queries and observed the effects of adding citation terms at the individual document level. We selected queries with the most marked and/or anomalous performance changes. Judged relevant and irrelevant documents are denoted by R and I, respectively; ? denotes an unjudged document.

Query #34. *Given perspective, opinion, and private state words, can a computer infer the hierarchy among the different perspectives?*
```
{perspective opinion private state word compute infer
hierarchy perspective}
```
This query exhibits a drop in Okapi performance while the majority of the models' performance increases or stays the same. Okapi's MAP drops from 0.7037 to 0.5913, while bpref, R-precision and P(5) are all unchanged. The pertinent ranking changes are summarised as follows:

Doc ID	Rel	Rank	Cits	Query Terms in Doc+Cits
W03-04_LONG	R	1→1	4	opinion 18+1, private 5+0, perspective 17+0, compute 1+0
C04-1018	R	2→3	4	opinion 23+0, private 86+0, perspective 4+0, state 93+0, word 5+0, compute 1+0
P99-1017	I	3→4	4	private 8+0, perspective 2+0, infer 1+0, state 1+0, compute 1+0, hierarchy 1+0
W05-0308	?	4→2	6	opinion 15+2, private 72+0, perspective 1+1, infer 1+0, state 79+0, word 10+0, compute 2+0
W03-0404	?	5→5	8	opinion 8+3, private 10+0, perspective 2+0, state 12+0, word 48+0, compute 3+0, hierarchy 1+0
C90-2069	R	27→28	0	private 50+0, perspective 1+0, state 51+0, compute 4+0

The query has only three judged relevant documents in total (versus 16 judged irrelevant), all of which are retrieved both with and without citation terms. The relevant documents are retrieved at ranks 1, 2 and 27 without citations and at ranks 1,3 and 28 with citations, respectively. This accounts for the drop in MAP. The new document at rank 2 is an unjudged document with six citations added to it, resulting in two additional occurrences of the query term *opinion* and one of *perspective* in the with-citations index. This overtakes the relevant document previously at rank 2, which gains only one *opinion* from its four citations. Because the document is unjudged, bpref is not affected. Its title is *'Annotating Attributions and Private States'*, suggesting it might indeed be relevant to the query, in which case MAP would increase (from 0.725 to 0.786) not decrease. The relevant document at rank 1 has no citations in the database and, thus, no new terms in the with-citations index. However, it has a high occurrence of the query terms *opinion*, *private* and *state*, as well as some occurrences of other query terms, and retains a high enough score to remain at rank 1. This document would not be retrieved if only citation terms (and not document terms) were used for indexing. The one judged irrelevant document in the top 5 is moved from rank 3 to 4 since its citations add no query terms.

Therefore, it appears that the citations are, in fact, contributing useful terms and helping relevant documents be retrieved. In the case of Okapi, however, the positive effects are not shown by the evaluation measures, for several reasons: a) the movements in the rankings are small because there are few citations for the judged documents, b) there are very few judged relevant documents for the query and c) some of the (potentially positive) effect of the citation terms affects unjudged documents, which affects MAP negatively. Given the incompleteness of the judgements, it is likely that measures which do not assume completeness, such as bpref, will be more reliable indicators of retrieval performance.

Query #15. *How can we handle the problem of automatic identification of sources of opinions?*
`{handle problem automatic identify source opinion}`

This query shows one of the largest increases in performance: the values of all measures increase (or stay the same) for all models. Considering Indri in detail, MAP increases from 0.4492 to 0.5361, R-precision from 0.5263 to 0.5789 and bpref from 0.6645 to 0.8224. The number of relevant documents retrieved increases from thirteen to sixteen, out of a possible nineteen in total. These newly retrieved documents largely account for the overall performance increases, summarised as follows:

Doc ID	Rel	Rank	Cits	Query Terms in Doc+Cits
H05-1044	R	_→74	7	opinion 12+7, identify 12+4, source 0+3, automatic 4+2
P02-1053	R	_→81	40	opinion 6+5, handle 1+0, identify 1+6, source 1+0, automatic 2+7, problem 2+1
W02-1011	R	_→87	31	opinion 7+1, identify 1+43, source 3+0, automatic 7+3, problem 14+2

Now considering Okapi, MAP increases from 0.0999 to 0.1028, bpref from 0.4934 to 0.5263 and eleven relevant documents are retrieved with citations, versus ten without. The remaining measures remain unchanged. Again, the overall performance increase is mainly due to the newly retrieved document:

Doc ID	Rel	Rank	Cits	Query Terms in Doc+Cits
W03-0404	R	_→93	8	opinion 8+3, identify 15+2, source 2+0, automatic 12+2, problem 2+0

Thus, Okapi's performance on this query does improve, following the trend of the other models. However, its increase is somewhat smaller than that of the other models. If this is generally the case, for queries with less marked performance increases than this example query, in addition to queries such as #34 where Okapi's performance drops slightly, then it is unsurprising that the overall measured differences in performance with and without citations are statistically insignificant.

Query #46. *What is the state-of-the-art on semantic role labeling using real syntax?*
{state art semantic role label real syntax}

This query shows a general increase in performance across measures and models, with the exception that bpref drops for KL, KL FB and Cosine and stays the same for Okapi and Indri. We consider the KL rankings in detail, where the decrease is the most marked (0.6207 to 0.1034). This is accounted for by the fact that the one judged irrelevant document for this query is retrieved at rank 10 with citations, whereas it was previously unranked without citations, overtaking judged relevant documents:

Doc ID	Rel	Rank	Cits	Query Terms in Doc+Cits
C04-1100	I	_ →10	9	state 0+1, semantic 0+7, role 0+3, label 0+1

Similarly, this document was retrieved at rank 64 by Cosine, where it was previously unranked. Neither Okapi nor Indri retrieved the document, with or without citations, so their bpref values do not change.

This is an example where the citation terms, again, have a definite effect on the retrieved documents but this time result in an irrelevant document, as well as relevant documents, being ranked higher. The citation terms that cause this do, indeed, match the query terms. However, closer inspection of the citing sentences that the terms were taken from reveals that they do not match the particular sense of the terms in the query, e.g., one of the occurrences of the term *role* comes from the phrase *to explore the role of semantic structures in question answering*. Indeed, the document is titled *'Question Answering Based on Semantic Structures'* and is not about semantic role labeling, the topic of the query. This is an inherent danger in term-based IR and not a product of our citation method: such semantic mismatches can occur with document terms as well as citation terms.

5   Conclusions and Future Work

We have conducted some first experiments using the combination of terms from citations to a document and terms from the document itself to index a given document. Performance, as gauged by a range of standard evaluation measures, generally increases when citation terms are used in addition to document terms. Furthermore, performance generally increases as citations are weighted higher relative to document terms. The Okapi and Cosine retrieval models, as implemented in the Lemur toolkit, do not appear to follow this trend. It may be a characteristic of these models, however, that the citation terms need to be weighted much higher relative to document terms than we have investigated here: for Okapi, weights of up to 35 have been found to be optimal for weighting anchor text relative to Web page body text, in the comparable situation of Web IR. Likewise, our results do not allow us to surmise an optimal citation term weight for the Indri and KL retrieval models. We intend to investigate a wider range of weights in future work. We also intend to investigate alternative methods of weighting the citation terms.

These are the first reported experimental results from a new test collection of scientific papers, created by us in order to more fully investigate citation-based methods for IR. The relevance judgements are known to be incomplete; we have noted the definite effects of this incompleteness on perceived retrieval performance, according to measures such as MAP. It is likely that measures which do not assume complete judgements, such as bpref, will be more reliable indicators of retrieval performance. Each of the retrieval models investigated here shows a statistically significant increase in bpref, when citation terms are added, for some citation term weight.

In conclusion, our experiments indicate that indexing citation terms in addition to document terms improves retrieval performance on scientific papers, compared to indexing the document terms alone. It remains to be seen what the optimal weighting of citation terms relative to document terms is and what the best way of implementing this weighting might be. It will also be interesting to investigate alternative methods of extracting citation terms, e.g., how using citing sentences compares to using fixed windows or more linguistically motivated techniques. Finally, we intend to compare against the performance when only citation terms are indexed. This method, however, restricts the documents that can be retrieved to those with at least one citation; we anticipate that using both document and citation terms will be most effective.

Acknowledgements. The first author gratefully acknowledges the support of Microsoft Research through the European PhD Scholarship Programme.

References

1. Teufel, S., Siddharthan, A., Tidhar, D.: Automatic classification of citation function. In: Proceedings of Empirical Methods in Natural Language Processing, pp. 103–110 (2006)
2. Schwartz, A.S., Hearst, M.: Summarizing key concepts using citation sentences. In: Proceedings of the HLT-NAACL BioNLP Workshop on Linking Natural Language and Biology, pp. 134–135 (2006)
3. Schneider, J.: Verification of bibliometric methods' applicability for thesaurus construction. PhD thesis, Royal School of Library and Information Science (2004)
4. Strohman, T., Croft, W.B., Jensen, D.: Recommending citations for academic papers. In: Proceedings of the 30th annual international ACM SIGIR conference on Research and development in information retrieval (SIGIR), pp. 705–706 (2007)
5. Fujii, A.: Enhancing patent retrieval by citation analysis. In: Proceedings of the 30th annual international ACM SIGIR conference on Research and development in information retrieval (SIGIR), pp. 793–794 (2007)
6. Meij, E., de Rijke, M.: Using prior information derived from citations in literature search. In: Proceedings of the International Conference on Recherche d'Information Assistée par Ordinateur (RIAO) (2007)
7. McBryan, O.: GENVL and WWWW: Tools for taming the web. In: Proceedings of the World Wide Web Conference (WWW) (1994)
8. Hawking, D., Craswell, N.: The very large collection and web tracks. In: Voorhees, E.M., Harman, D.K. (eds.) TREC: Experiment and Evaluation in Information Retrieval, MIT Press, Cambridge (2005)

9. Brin, S., Page, L.: The anatomy of a large-scale hypertextual Web search engine. Computer Networks and ISDN Systems 30, 107–117 (1998)
10. Davison, B.D.: Topical locality in the web. In: Proceedings of Research and Development in Information Retrieval (SIGIR), pp. 272–279 (2000)
11. Bradshaw, S.: Reference directed indexing: Redeeming relevance for subject search in citation indexes. In: Koch, T., Sølvberg, I.T. (eds.) ECDL 2003. LNCS, vol. 2769, pp. 499–510. Springer, Heidelberg (2003)
12. Dunlop, M.D., van Rijsbergen, C.J.: Hypermedia and free text retrieval. Information Processing and Management 29(3), 287–298 (1993)
13. Pitkow, J., Pirolli, P.: Life, death, and lawfulness on the electronic frontier. In: Proceedings of the Conference on Human Factors in Computing Systems (1997)
14. O'Connor, J.: Citing statements: Computer recognition and use to improve retrieval. Information Processing and Management 18(3), 125–131 (1982)
15. Ritchie, A., Teufel, S., Robertson, S.: Creating a test collection for citation-based IR experiments. In: Proceedings of Human Language Technology conference and the North American Chapter of the Association for Computational Linguistics (HLT-NAACL) (2006)
16. Ritchie, A., Robertson, S., Teufel, S.: Creating a test collection: Relevance judgements of cited & non-cited papers. In: Proceedings of the International Conference on Recherche d'Information Assistée par Ordinateur (RIAO) (2007)
17. Kluck, M.: The GIRT data in the evaluation of CLIR systems - from 1997 until 2003. In: Peters, C., Gonzalo, J., Braschler, M., Kluck, M. (eds.) CLEF 2003. LNCS, vol. 3237, pp. 376–390. Springer, Heidelberg (2004)
18. Hersh, W., Bhupatiraju, R.T.: Trec genomics track overview. In: Proceedings of the Text REtrieval Conference (TREC), pp. 14–23 (2003)
19. Hersh, W., Bhupatiraju, R.T., Ross, L., Johnson, P., Cohen, A.M., Kraemer, D.F.: Trec 2004 genomics track overview. In: Proceedings of the Text REtrieval Conference (TREC) (2004)
20. Hersh, W., Cohen, A.M., Roberts, P., Rekapilli, H.K.: Trec 2006 genomics track overview. In: Proceedings of the Text REtrieval Conference (TREC) (2006)
21. Cleverdon, C., Mills, J., Keen, M.: Factors determining the performance of indexing sytems, vol. 1, design. Technical report, ASLIB Cranfield Project, (1966)
22. Strohman, T., Metzler, D., Turtle, H., Croft, W.B.: Indri: a language-model based search engine for complex queries. Technical report, University of Massachusetts (2005)
23. Powley, B., Dale, R.: Evidence-based information extraction for high accuracy citation and author name identification. In: Proceedings of the International Conference on Recherche d'Information Assistée par Ordinateur (RIAO) (2007)
24. Briscoe, E., Carroll, J.: Robust accurate statistical annotation of general text. In: Proceedings of the Conference on Language Resources and Evaluation (LREC), pp. 1499–1504 (2002)
25. Ritchie, A., Teufel, S., Robertson, S.: How to find better index terms through citations. In: Proceedings of COLING/ACL Workshop on How Can Computational Linguistics Improve Information Retrieval? (2006)
26. Zaragoza, H., Craswell, N., Taylor, M., Saria, S., Robertson, S.: Microsoft Cambridge at TREC-13: Web and HARD tracks. In: Proceedings of the Thirteenth Text REtrieval Conference (TREC) (2004)

Automatic Extraction of Domain-Specific Stopwords from Labeled Documents

Masoud Makrehchi[1] and Mohamed S. Kamel[1]

Pattern Analysis and Machine Intelligence Lab,
Department of Electrical and Computer Engineering,
University of Waterloo, Waterloo, Ontario N2L3G1, Canada
{mkamel,makrechi}@pami.uwaterloo.ca

Abstract. Automatic extraction of domain-specific stopword list from a large labeled corpus is discussed. Most researches remove the stopwords using a standard stopword list, and high and low document frequencies. In this paper, a new approach for stopword extraction based on the notion of backward filter level performance and sparsity measure of training data, is proposed. First, we discuss the motivation for updating existing lists or building new ones. Second, based on the proposed backward filter-level performance, we examine the effectiveness of high document frequency filtering for stopword reduction. Finally, a new method for building general and domain-specific stopwords is proposed. The method assumes that a set of candidate stopwords must have minimum information content and prediction capacity, which can be estimated by a classifier performance. The proposed approach is extensively compared with other methods including inverse document frequency and information gain. According to the comparative study, the proposed approach offers more promising results, which guarantee minimum information loss by filtering out most stopwords.

1   Introduction

Stopwords or so-called common words, noise words or negative dictionary are considered as non-predictive and non-discriminating words. They carry low information content, and cause a low retrieval rate and prediction results. In addition, stopwords make up a large portion of the textual data in text mining tasks, where dimensionality is a critical issue.

Stopwords are grouped into two categories: general and domain-specific. The first category includes those standard stopwords, which are available in the public domain or non-standard stopwords which are generated inside information retrieval or text categorization systems[12,20,1,15]. In the second category, domain-specific stopwords are recognized as a set of words which have no discriminant value within a specific domain or context. Domain-specific stopwords differ from one domain to another. For example, the term *learning*, can be a stopword in the domain of *education*, but a keyword in *computer science*. The result of removing these terms, are similar to those of removing general stopwords that is an

C. Macdonald et al. (Eds.): ECIR 2008, LNCS 4956, pp. 222–233, 2008.

improved performance of the retrieval and categorization tasks. Domain-specific stopwords have already been employed in areas such as physics, human resource management[3], bioinformatics, and gene ontology[19].

Stopword reduction is achieved by a standard stoplist, a high Document Frequency (high-DF) filtering, a term ranking scheme, or a combination of all three methods[5]. Inspired by Zipf's law, where the number of documents in the data base is sufficiently high, the terms with high document frequencies are treated as stopwords. However, this rule fails in most cases, including the one where the documents are not uniformly distributed across the categories.

In this paper, a newly developed method is described for generating nonstandard and domain-specific stopwords. The drawback of previous methods, which are based on document frequency, are outlined. Conventional term ranking-based methods are evaluated by a proposed evaluation approach called backward filter-level performance. Using this evaluation, we illustrate that the behavior of term ranking measures for scoring relevant terms differs from that for scoring the stopwords. According to the experimental results that are obtained by applying the proposed method to six data sets, choosing the proper term ranking measure for building stoplists depends on the data set characteristics, including the sparsity measure.

The paper consists of six sections. Following the introduction, related works are briefly reviewed in Section 2. In Section 3, stopword reduction methods is discussed. Stoplist evaluation using backward filter-level performance is described in Section 4. Section 5 includes experimental results and the discussion. Conclusions are presented in Section 6.

2   Related Works

Although general stoplists, mostly for the English language, are available, the need for automatically constructing stoplists has not been obvious to researchers in text mining. The first initiative in stopword extraction has been accredited to Van Rijsbergen in 1979 [22]. His stoplist is one of the most used ones in NLP and information retrieval applications.

In [12], information gain ranking has been applied for stopword extraction. The results have been compared with those of several document and term frequency measures. Additionally, the extracted stopwords are evaluated in an information retrieval query processing task by using TREC collections. As we show in this paper, the information gain ranking is not comprehensive enough for stopword extraction. Choosing the proper ranking measure depends significantly on the data set characteristics.

To generate stoplists for web-specific documents, word entropy was employed in [21]. Since the method is unsupervised, the generated stoplist is evaluated by a web clustering scheme. In [20], the stoplist, generated by word entropy, is optimized via a k-means clustering and stochastic search algorithm. In [8], an association algorithm for producing stoplists, based on Receiver Operating Characteristics (ROC) analysis, has been suggested.

In [12,20], outdated stoplists and web influences are mentioned as the two major motivations for the automatic extraction of the stoplists, there are also some other reasons to develop stopword extraction algorithms.

Outdated Stoplists: Standard stoplists become quickly outdated. For example, the first English stoplist was published in the 1970's[22]. Obviously, over time the usage of some popular words have changed, depending on social factors such as technological and cultural shifts, media, and education. It is not surprising that revising, updating and optimizing current stoplists are crucial[20].

Web Impact: Influenced by new technologies such as the Web and new communication tools such as chat, and email, some new words have become more common in daily English, for example, *email, contacts, URL*, and *link*[12].

Stoplist for non-English Text Retrieval: Most of the research on natural language processing, information retrieval, and text categorization has been on the English language text. Recently, stopword lists have been published for other European languages. In [18], a general stoplist has been developed for French text. There are still some languages without standard stoplists.

Domain-Specific Stopwords: For the automatic extraction of domain-specific text mining and retrieval applications, we always need stoplists from the local vocabulary of a corpus. Domain-specific stopwords have been manually extracted in areas such as physics, human resource management[3], bioinformatics and gene ontology[19].

Formal Language Text Mining: Recently, text mining and statistical machine learning have been applied to formal language texts such as software source codes[13,6]. For instance, in software clustering[6], each function or procedure is represented by a bag of words including reserved words, constants, variables, and function calls. The first two elements are the noise and stopwords to be removed. The identification of stopwords for various formal languages requires an automatic extraction system.

Building Ontology: Hub words have been introduced in [9]. They are related to many other words. Since one characteristic of stopwords is their correlation with other terms, hub words are viewed as a subset of domain-specific stopwords. Extracting hub words and building sub-domain vocabularies, also known as terminology, is a baseline for learning ontologies.

3   Stopword Reduction Methods

Most text categorization systems perform a low and high document frequency filtering (low-DF and high-DF). According to Zipf's law, in a corpus of natural language text, the usage frequency of any word is considered to be inversely proportional to its frequency rank. In other words, in a vocabulary, only a few words are very frequent, whereas the majority of words occur only once. Both high frequent and low frequent groups carry some linguistic content, and typically facilitate the understanding of the meaning of the text. For text mining purposes both groups are removed, because they do not relevant enough to contribute to the learning process.

Document Frequency (DF) is a measure that reflects the contribution of a term in a document collection. It is assumed that all the terms in the vocabulary have the same importance. This assumption does not always work, because from a pragmatic point of view, the importance of the terms across the collection and its categories varies. The second assumption in estimating DF is that all the terms are uniformly distributed over the categories[11]. In other words, the DF is biased to uniformly distributed terms across the categories, which means DF can be potentially employed in stopword reduction. Since DF ignores the labels and class information of the documents, it is an unsupervised scoring measure that is widely used in text clustering. Each term is assigned a measure, representing the number of documents, containing the term.

In the majority of text classification research, low-DF terms are removed from the vocabulary. The threshold used for low-DF filtering varies from one to more than ten, depending on the data sets. Those low-DF terms, which occur only once in the collection, are called *singletons*. The singletons are sometimes considered as stopwords. Removing low-DF terms dramatically reduces the vocabulary size, but slightly reduces the sparsity measure. Low-DF terms include rare terms or phrases, spelling errors, and those have no significant contribution in the discriminating process of document category.

Inverse Document Frequency (IDF), which is one of the variants of the DF ranking measure, is calculated by different formulations such as

$$IDF(t_j) = \log \frac{n - n_{t_j} + 0.5}{n_{t_j} + 0.5} \qquad (1)$$

where n is the number of documents in the training data set, and n_{t_j} is the number of documents containing the term t_j. IDF is widely used in removing high frequent words, which are potentially considered as stopwords. For example, in [4] stopword removal is performed by removing all the terms with $DF \geq n/2$. This rule fails in the case of domain-specific stopword reduction and data sets with high class skew[4].

One alternative approach to extracting stopwords is to employ term ranking methods, which have been widely used in text categorization[23]. All term ranking methods are based on the following three components: *(i)* Calculating the scoring measure as a merit index for each term; *(ii)* Sorting all the features in decreasing order, according to their merit measures; and *(iii)* Applying the threshold to the sorted list of the features. Some well-known term ranking measures employed in this paper are as follows:

Information Gain (IG)[23]:

$$IG(t_j) = -\sum_{k=1}^{C} P(c_k).\log P(c_k) + P(t_j) \sum_{k=1}^{C} P(c_k|t_j).\log P(c_k|t_j) + \qquad (2)$$

$$P(\bar{t}_j) \sum_{k=1}^{C} P(c_k|\bar{t}_j).\log P(c_k|\bar{t}_j)$$

The χ^2 Statistic[17]:

$$\chi^2(t_i; c_k) = \frac{(ad - bc)^2}{(a + b)(c + d)(a + c)(b + d)} \tag{3}$$

Odds Ratio (OR) [16]:

$$OR(t_j, c_k) = \frac{odds(t_j|c_k)}{odds(t_j|\bar{c}_k)} = \frac{ad}{bc} \tag{4}$$

F-measure Feature Ranking [5]:

$$F(t_j, c_k) = \frac{2P(t_j, c_k).R(t_j, c_k)}{P(t_j, c_k) + R(t_j, c_k)} = \frac{2a}{2a + b + c} \tag{5}$$

where $a = P(t_j, c_k)$, $b = P(t_j, \bar{c}_k)$, $c = P(\bar{t}_j, c_k)$, and $d = P(\bar{t}_j, \bar{c}_k)$.

4   Evaluating Stoplists Using Filter-Level Performance

In previous stoplist generations [12,20], the information retrieval systems have been used to evaluate the generated stoplist. In this paper, we use a text classifier to evaluate the extracted stoplists. In addition, this framework is employed to compare various extraction algorithms. The text classifier, employed for the stoplist evaluation, must be a weak classifier, sensitive to the noise and stopwords, and scalable and inexpensive as much as possible. The Rocchio classifier, which is used in this research, can meet these requirements. For estimating the text classifier performance, the F-measure is employed. Typically, F-measure is used for binary classification problems. In the case of multiple class problems, micro-averaged and macro-averaged F-measures are recommended. Micro-averaged F-measure is the weighted average of the F-measure by class distribution. In macro-averaged F-measure, classes have no weights and all are similarly treated. In uniformly distributed classes, both averages are the same. Otherwise the macro-averaged F-measure is less than the micro-averaged F-measure[4,5]. Due to this fact, a macro-averaged F-measure is adopted to estimate the classifier performance.

Let $V = \{v_1, v_2, ..., v_m\}$ and $\Lambda = \{v_m, v_{m-1}, ..., v_1\}$ be the descendingly and ascendingly sorted list of terms in the vocabulary according to a ranking measure such as IDF such that v_1 is the best and v_m is the worst terms. Term selection, based on term ranking, employs the *best-k*, $k \leq m$, in which the first k terms of V are retained and the others are removed. Obviously, for stopword extraction, we can follow the *worst-k* rule that keeps the first k terms of Λ and rejects the rest. By scaling k by m, which is the number of terms in the vocabulary, *filter-level* $q = k/m$, $0 < q \leq 1$ is defined. The estimated performance of a text classifier versus increasing levels of filtering is called the *Filter-Level Performance (FLP)*. Depending on employing whether *best-k* or *worst-k* rule, FLP is attributed to a *Forward* or *Backward* FLP, respectively. In Forward FLP, the classifier

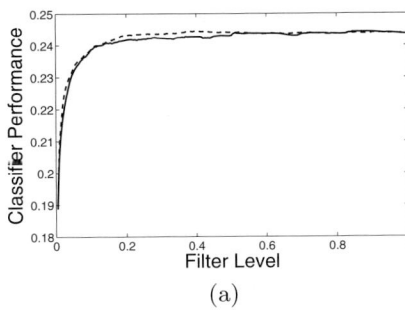 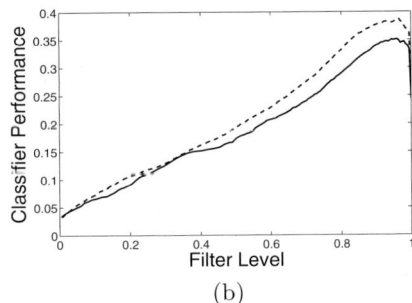

(a)   (b)

Fig. 1. Two examples of FLP characteristic for WebKB data set, Mean(χ^2) and IDF depicted by solid and dashed lines, respectively, (a) forward, (b) backward FLP

performance is a function of filter levels when employing the vector V. On the other hand, Backward FLP describes the classifier performance as a function of filter levels using the vector Λ.

From the text categorization viewpoint, a set of terms F_1 is more relevant than a set of F_2, if it offers a better classification performance. In the opposite direction, a set of stopwords S_1 is better than that of S_2, if S_1 offers less prediction capacity or classifier performance. To express these two statements into an unified expression, the Area Under FLP (AUF) is defined as follows:

$$A_\tau^+ = \sum_{0<q\leq\tau} J(V_q), \quad A_\tau^- = \sum_{0<q\leq\tau} J(\Lambda_q) \qquad (6)$$

where q is the variable filter-level, τ is the maximum threshold, and V_q (Λ_q) is the best (worst) $\lfloor q.m \rfloor$ terms of ordered set V (Λ) according to the forward (backward) FLP. A^+ (A^-) is defined as forward (backward) AUF. Let $J(v_q)$ be the estimated classifier performance by applying the set of terms V_q. By using AUF, two term selection scheme can be easily compared. A better term selection measure should posses larger AUF, whereas for a stoplist extraction, the best ranking measure should have the least AUF. Figure 1 illustrates the forward and backward FLPs of two ranking measures. According to Figure 1-a, we have:

$$AUF^+(IDF) > AUF^+(Mean(\chi^2)) \qquad (7)$$

then IDF is slightly better than Mean(χ^2) for choosing the most relevant terms. On the other hand, in Figure 1-b we have:

$$AUF^-(Mean(\chi^2)) < AUF^-(IDF) \qquad (8)$$

which means Mean(χ^2) acts better than IDF while selecting stopwords.

With the FLP approach, the behavior of a ranking measure is analyzed and compared with that of other measures. The key disadvantage of this approach is that it requires a text classifier for estimating the FLP response. Although the Rocchio classifier, employed in this paper, is simple and inexpensive, especially for large vocabularies, obtaining FLP characteristics is challenging.

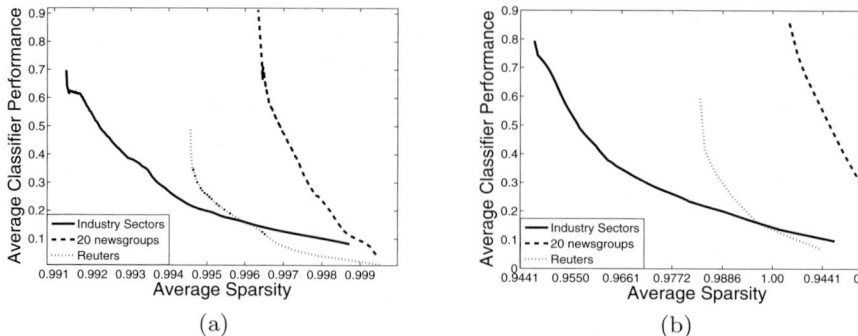

Fig. 2. Classifier performance as a function of sparsity: (a) Rocchio, and (b) SVM

One alternative approach is to approximately estimate the classifier performance without performing any classification task. Our idea is to analyze the data set characteristics and to estimate the performance trend. The data set sparsity is one of the appropriate data set characteristics, which, approximately, describes the trend of classifier performance[5].

$$S_d = 1 - \frac{\sum_{i=1}^{n} N_i}{m.n} \qquad (9)$$

where N_i is the number of distinct words in the ith document or the number of non-zero entries in i^{th} row of the document-term matrix. n and m are the number of training data and the number of terms (the size of vocabulary), respectively.

Figure 2-a depicts the correlation between the backward FLP and sparsity for Industry Sectors, 20 Newsgroup, and Reuters data sets. Each graph is the average of 9,000 experiments, including nine different ranking measures, 200 filter levels, and 5 different distributions of the training data (by 5-fold cross validation). In all the experiments, the Rocchio text classifier is used. These experiments are performed with a different classifier to investigate the impact of the classifier model on the correlation between the classifier performance and sparsity index. Figure 2-b illustrates the correlation between the backward FLP and sparsity for the data sets by the Support Vector Machine (SVM). Each graph is the average of 1,800 experiments, including 9 ranking methods, 40 filter levels, and 5 different distributions of the training data. To reduce the computation time, we use a smaller number of filter levels for the SVM-based FLP.

According to Figure 2, the FLP response and sparsity measure are strongly correlated. As a result, for the FLP analysis, instead of obtaining the classifier performance, which can be expensive when sophisticated classifiers such as the SVM are used, we can estimate backward FLP by adopting the sparsity of the document-term matrix, associated with the training data, when the same filter-level as FLP is applied as follows:

$$h(V_q) \approx -log(S_d(V_q)), \qquad (10)$$

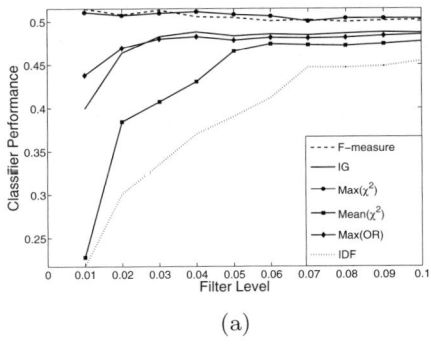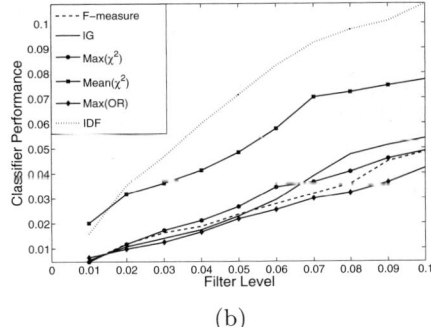

(a)   (b)

Fig. 3. Filter level performance of the ranking methods for Reuters data set: (a) forward, and (b) backward

where $S_d(V_q)$ is the sparsity of the filtered document-term matrix in which filter-level q is applied to reduce the number of terms.

5   Experimental Results

To evaluate the proposed framework, seven feature ranking measures are applied to six data sets. The set of feature ranking methods includes: *F-measure*, *IG*, *Max(χ^2)*, *Mean(χ^2)*, *Max(OR)*, *IDF*, and *Random*. All classification experiments and sparsity estimations are validated by a 5-fold cross validation schema.

Six document data sets have been used in this paper. Among them, four data sets are well-known benchmark collections including Industry Sectors[14], 20 Newsgroups[7], Reuters[10], and WebKB[2]. The remaining two data sets including Learning Object Metadata (LO Metadata) and Computer Science Abstracts, have been created by the authors. The Learning Object Metadata (LO Metadata) have been collected from the Schoolnet Canada[1]. The collection contains $1,525$ learning object metadata classified into 31 categories. Computer Science Abstracts have been also collected by the authors from CiteSeer Computer Science Directory[2]. The collection contains $2,912$ documents in 17 categories. All data sets are preprocessed by Porter stemmer.

Figure 3 illustrates forward and backward FLPs for Reuters data set with $\tau = 0.1$ and 200 filter levels. With respect to the backward FLP responses (Figure 3-b), it is evident that IDF offers the worst results compared to the other measures and it cannot be the best ranking for stopword extraction. This experiment challenges the use of high-DF filtering for stopword reduction. In addition, the experiment indicates we cannot rely on IDF to filter out the stopwords, since, in most cases, its backward FLP is usually higher than that of others. It is implied, by IDF ranking among terms with lower ranks, there are still some relevant and informative terms.

---

[1] http://www.schoolnet.ca/home/e/resources/
[2] http://citeseer.ist.psu.edu/directory.html

Table 1. The area under FLP of the term ranking methods (forward-backward)

	IG	Max(χ^2)	Mean(χ^2)	Max(OR)	F-measure	IDF	Random
Industry Sectors	6.61-1.50	6.51-1.32	6.30-1.45	6.41-1.07	6.50-1.26	6.09-1.84	2.98-2.98
LO Metadata	6.17-0.23	6.63-0.39	5.57-0.39	4.17-0.42	6.61-0.41	4.53-0.45	0.99-0.97
20 Newsgroups	9.09-2.31	9.09-1.27	9.04-1.35	8.12-0.83	9.09-1.31	8.99-3.21	2.69-2.17
Reuters	4.75-0.29	5.07-0.29	4.29-0.53	4.76-0.23	5.05-2.26	3.82-0.71	1.66-1.50
WebKB	2.27-0.51	2.32-0.49	2.29-0.52	2.36-0.49	2.33-0.49	2.310.56	1.64-1.47
CS Abstracts	4.381-1.50	4.21-1.20	4.23-1.15	4.28-0.97	4.23-1.20	4.11-2.25	2.32-2.43

The second point is the inconsistency of the behavior of the term ranking measures in forward and backward filtering. In Figure 3-a the forward FLPs are presented for all the ranking measures. The results in this figure are not consistent with those in Figure 3-b, which exhibits the backward FLPs of the ranking measures. In other words, a good ranking is supposed to assign not only higher ranks to the relevant words but also lower ranks to the stopwords. Table 1 summarizes the results for all data sets, describing the behavior of each ranking measure by the forward and backward AUF measures (A^+ and A^-). In Table 2, the rank of the ranking methods for each data set, with respect to their forward and backward AUF, is presented. According to Table 2, the best ranking measures for term selection (according to their Forward FLPs) are IG, $Max(\chi^2)$, F-measure, $Max(\chi^2)$, $Max(OR)$, and IG, for the six data sets, respectively. On the contrary, the best ranking measures to extract and remove stopwords are not exactly similar to those for term selection. The best ranking, in relation to the data sets are $Max(OR)$, IG, $Max(OR)$, $Max(OR)$, F-measure, and $Max(OR)$ which is different from term ranking results. This finding addresses the pitfall of using term ranking methods to remove noise and stopwords.

To compare the proposed approach, which is based on an estimated backward filter-level performance by using the sparsity measure, with other automatically generated stoplists, the following experiment is set up. First, the stoplist, extracted by backward FLP characteristic, is considered as the baseline and optimum list. It should be noted that for all methods, 10% of most irrelevant terms are selected as stopwords. The stopword list by using the IDF ranking, a classical approach, is also derived. Two well-appreciated term ranking measures, IG and $Max(\chi^2)$ [12] are also considered in the experiment. The optimum term ranking measures, which are obtained by forward FLP, are also examined.

Finally, the stoplist is extracted by the estimated backward FLP by using sparsity. All the lists are compared with the baseline stoplist by using the F-measure.

Table 2. The data sets and the rank of term ranking measures (forward, backward)

	IG	Max(χ^2)	Mean(χ^2)	Max(OR)	F-measure	IDF	Random
Industry Sectors	1,5	2,3	5,4	4,1	3,2	6,6	7,7
LO Metadata	3,1	1,2	4,3	6,5	2,4	5,6	7,7
20 Newsgroups	3,6	2,2	4,4	6,1	1,3	5,7	7,5
Reuters	4,4	1,3	5,5	3,1	2,2	6,6	7,7
WebKB	6,4	3,2	5,5	1,3	2,1	4,6	7,7
CS Abstracts	1,5	5,4	3,2	2,1	4,3	6,6	7,7

Table 3. F-measure of extracted stoplists compared to the baseline stoplist

	Sparsity Best	Forward FLP	IDF	IG	Max(χ^2)
Industry Sectors	0.9170	0.9170	0.5034	1.0000	0.7823
LO Metadata	1.0000	1.0000	0.5416	0.6891	0.3750
20 Newsgroups	1.0000	1.0000	0.7529	0.9831	0.6092
Reuters	0.6494	0.6494	0.6381	0.9736	0.6494
WebKB	1.0000	0.7724	0.5241	0.9983	0.8589
CS Abstracts	1.0000	0.8751	0.6749	0.8751	0.5458

Table 4. Extracted stoplists using the proposed method

Industry Sectors	LO Metadata	20 Newsgroups	Reuters	WebKB	CS Abstracts
deem	archer	cantaloup	ryan	mead	bibtex
inaccuraci	clickabl	cmu	pesticid	tei	citat
older	colombia	cs	helm	inet	context
av	gord	date	huski	puma	copyright
everydai	manor	id	vista	interplai	document
foster	motor	messag	rehear	sarita	download
purpos	toi	newsgroup	atla	unawar	entri
vari	2y2	path	wallcov	horizon	rate
oppos	mississauga	srv	hyo	judi	relat
creation	murphi	subject	medco	ajit	researchindex
carl	0a9	biscuit	tele	michal	previou
proud	0m5	stylu	lube	cuni	varieti
attribut	albertan	bobcat	proprietari	athena	advanc
stress	alex	ocsmd	absb	rowspan	intend
break	biotech	preffer	initit	marginwidth	typic
10k	bridg	sunil	spun	bud	exist
relianc	burn	abba	weis	pierr	examin
geograph	candid	overtur	caremark	wharton	determin
overse	cariboo	gtri	dyneer	toledo	directli
est	chase	belvilad	dyr	surgic	lot
trail	chevi	mattel	craa	usl	grant
jpg	conflict	fervour	55p	22th	main
unlaw	cyberu	hasta	microfilm	ariane5rep	advantag
teamwork	dole	implor	hazard	asplos7	separ
smaller	downtown	nirvana	produkt	cmg	difficult

According to Table 3, which illustrates the results of the comparison, the sparsity-based estimation of the FLP provides the most stopwords that are similar to the baseline stoplist. The IDF offers poor results compared to the other methods. Adopting the best term selection method by the forward FLP characteristic, and IG perform better than IDF, but they cannot outperform the sparsity-based estimated backward FLP results. Table 4 presents a short list of the first domain-specific stopwords, which are extracted by the proposed approach by using sparsity measure.

6   Conclusion

Standard stoplists, which are used in information retrieval and text categorization, are outdated. Automatically building stoplists are also required in applications such as domain-specific text mining, ontology generation, non-English text processing, and formal language textual data mining. Conventional methods for stopword extraction are based on removing the terms with low and high document frequencies. In this paper, the risks of the document frequency approach is discussed. For supervised stopword extraction, which uses labeled training data, term ranking measures such as information gain and χ^2 are employed. According to the results in this paper, if a given term ranking can perform well for selecting good features, the selection of poor features (stopwords) in the opposite direction, are not guaranteed. The reason is that term rankings behave differently, whereas ranking relevant terms from scoring stopwords. This fact is studied by introducing a new evaluation model, called the area under backward filter-level performance (backward AUF).

Using the notion of backward FLP, we can identify the optimum term ranking measure, which minimizes the prediction capacity of selected terms as candidate stopwords. The novel optimum solution can extract the most irrelevant words so-called the baseline stoplist. The major disadvantage of this approach is that it employs a classifier to obtain filter-level performance. One alternative approach is to use training data characteristics to estimate the classifier performance. In this paper, we used the sparsity measure, after term selection, to predict the trend of text classifier performance. According to the experimental results, sparsity measure offers a good estimation of classifier performance. The result of sparsity based estimation is almost better than other term ranking measures, and entirely outperforms traditional inverse document frequency.

References

1. Chen, A., Gey, F.C.: Building an Arabic stemmer for information retrieval. In: TREC (2002)
2. Craven, M., DiPasquo, D., Freitag, D., McCallum, A., Mitchell, T., Nigam, K., Slattery, S.: Learning to extract symbolic knowledge from the world wide web. In: Proceedings of the 15th National Conference on Artificial Intelligence (AAAI 1998), pp. 509–516 (1998)
3. Crow, D., De Santo, J.: A hybrid approach to concept extraction and recognition-based matching in the domain of human resources. In: ICTAI, pp. 535–539 (2004)
4. Forman, G.: An extensive empirical study of feature selection metrics for text classification. Journal of Machine Learning Research 3, 1289–1305 (2003)
5. Forman, G.: A pitfall and solution in multi-class feature selection for text classification. In: Proceedings of ICML 2004, Twenty-first international conference on Machine learning, pp. 297–304 (2004)
6. Hayes, J.H., Dekhtyar, A., Sundaram, S.: Text mining for software engineering: how analyst feedback impacts final results. In: MSR 2005: Proceedings of the 2005 international workshop on Mining software repositories, pp. 1–5. ACM Press, New York (2005)
7. Joachims, T.: A probabilistic analysis of the Rocchio algorithm with TFIDF for text categorization. In: Fisher, D.H. (ed.) Proceedings of ICML 1997, 14th International Conference on Machine Learning, Nashville, US, pp. 143–151. Morgan Kaufmann Publishers, San Francisco (1997)

8. Kawahara, M., Kawano, H.: Mining association algorithm with threshold based on roc analysis. In: Proceedings of the 34th Annual Hawaii International Conference on System Sciences (HICSS-34), vol. 3, pp. 3010–3017. IEEE Computer Society, Los Alamitos (2001)

9. Koo, S.O., Lim, S.Y., Lee, S.-J.: Building an ontology based on hub words for information retrieval. In: Web Intelligence, pp. 466–469 (2003)

10. Lewis, D.D., Yang, Y., Rose, T.G., Li, F.: RCV1: A new benchmark collection for text categorization research. Journal of Machine Learning Research 5, 361–397 (2004)

11. Liu, T., Liu, S., Chen, Z., Ma, W.-Y.: An evaluation on feature selection for text clustering. In: Proceedings of ICML 2003, pp. 488–495 (2003)

12. Lo, R.T., He, B., Ounis, I.: Automatically building a stopword list for an information retrieval system. The Journal on Digital Information Management: special issue on the 5th Dutch-Belgian Information Retrieval Workshop (DIR 2005) 3(1), 3–8 (2005)

13. Maletic, J.I., Valluri, N.: Automatic software clustering via latent semantic analysis. In: Proceedings 14th IEEE International Conference on Automated Software Engineering (ASE 1999), Cocoa Beach Florida, October 1999, pp. 251–254 (1999)

14. McCallum, A.K., Rosenfeld, R., Mitchell, T.M., Ng, A.Y.: Improving text classification by shrinkage in a hierarchy of classes. In: Shavlik, J.W. (ed.) Proceedings of ICML 1998, 15th International Conference on Machine Learning, Madison, US, pp. 359–367. Morgan Kaufmann Publishers, San Francisco (1998)

15. Petras, V., Perelman, N., Gey, F.C.: UC berkeley at clef-2003 - Russian language experiments and domain-specific retrieval. In: Peters, C., Gonzalo, J., Braschler, M., Kluck, M. (eds.) CLEF 2003. LNCS, vol. 3237, pp. 401–411. Springer, Heidelberg (2004)

16. Rijsbergen, C.J., Harper, D.J., Porter, M.F.: The selection of good search terms. In: Information Processing and Management, pp. 77–91 (1981)

17. Rogati, M., Yang, Y.: High-performing feature selection for text classification. In: Proceedings of the eleventh international conference on Information and knowledge management, pp. 659–661 (2002)

18. Savoy, J.: A stemming procedure and stopword list for general French corpora. Journal of the American Society for Information Science, 944–952 (1999)

19. Seki, K., Mostafa, J.: An application of text categorization methods to gene ontology annotation. In: SIGIR, pp. 138–145 (2005)

20. Sinka, M.P., Corne, D.W.: Evolving better stoplists for document clustering and web intelligence. Design and application of hybrid intelligent systems, 1015–1023 (2003)

21. Sinka, M.P., Corne, D.W.: Towards modernised and web-specific stoplists for web document analysis. In: Proceedings of the IEEE/WIC International Conference on Web Intelligence, pp. 396–402. IEEE Computer Society, Los Alamitos (2003)

22. Van Rijsbergen, C.J.: Information Retrieval, 2nd edn., Dept. of Computer Science, University of Glasgow (1979)

23. Yang, Y., Pedersen, J.O.: A comparative study on feature selection in text categorization. In: Fisher, D.H. (ed.) Proceedings of ICML 1997, 14th International Conference on Machine Learning, Nashville, US, pp. 412–420. Morgan Kaufmann Publishers, San Francisco (1997)

Book Search Experiments: Investigating IR Methods for the Indexing and Retrieval of Books

Hengzhi Wu[1], Gabriella Kazai[2], and Michael Taylor[2]

[1] Department of Computer Science
Queen Mary, University of London
UK
`hzwoo@dcs.qmul.ac.uk`
[2] Microsoft Research
Cambridge
UK
`{gabkaz,mitaylor}@microsoft.com`

Abstract. Through mass-digitization projects and with the use of OCR technologies, digitized books are becoming available on the Web and in digital libraries. The unprecedented scale of these efforts, the unique characteristics of the digitized material as well as the unexplored possibilities of user interactions make full-text book search an exciting area of information retrieval (IR) research. Emerging research questions include: How appropriate and effective are traditional IR models when applied to books? What book specific features (e.g., back-of-book index) should receive special attention during the indexing and retrieval processes? How can we tackle scalability? In order to answer such questions, we developed an experimental platform to facilitate rapid prototyping of a book search system as well as to support large-scale tests. Using this system, we performed experiments on a collection of 10 000 books, evaluating the efficiency of a novel multi-field inverted index and the effectiveness of the BM25F retrieval model adapted to books, using book-specific fields.

Keywords: Book search, multi-field indexing, BM25F, efficiency, effectiveness.

1 Introduction

Libraries around the world as well as commercial companies like Amazon, Google and Microsoft are digitizing thousands of books in an effort to enable online access to these collections. For example, more than 100 000 19th Century books previously unavailable to the public will go online thanks to a digitization programme at the British Library. Online access to these old books, many of which are unknown as few were reprinted after first editions, is believed to be of great help to teachers and scholars. Other examples of digitization efforts include the Virginia eText project[1], which produces replacement copies of texts that are

---

[1] http://etext.lib.virginia.edu/ebooks/

C. Macdonald et al. (Eds.): ECIR 2008, LNCS 4956, pp. 234–245, 2008.

deteriorating; and Project Gutenberg[2], which has made around 20 000 books available for the wider public since 1971. The world's largest, university-based digital library of freely accessible books is being created by the Million Book project[3] [9] lead by (among others) Carnegie Mellon University in the United States. In December 2007, the project has completed the digitization of more than 1.5 million books, which are now available online. The Open Content Alliance (OCA)[4] is a major library initiative with 80 contributing libraries, which is aimed at making the digitized materials broadly available through the Internet Archive[5]. The pace of digitization has been greatly accelerated by Google's and Microsoft's mass-digitization projects, which aim to create online digital libraries of tens of millions of volumes, accessible via their respective book search services: Google Books[6] and Live Search Books[7].

Through digitization, a book is turned into a series of files representing the pages of the book. With the application of OCR technologies the text of a book is extracted, thus enabling full-text retrieval over collections of digitized books. The application of standard IR techniques to this new domain, however, requires closer examination.

The unique characteristics of the digitized material, combined with the specialised domain that books represent, raise a range of research questions. For example, the scanning and OCR processes raise questions on how well traditional IR techniques will cope with such noisy data. Other issues are raised by the sheer size of such collections, where each book in itself can span several hundreds of pages. Add to this the size of the page image files that search engines will need to return to a user. Questions regarding suitable infrastructure for supporting the parsing, indexing, storage and access of books within such collections need to be examined. It is also necessary to examine which book-specific features (such as back-of-book index) should receive special attention during indexing and retrieval. Consideration need to be given to the issues that surround user's interactions with search systems. These include questions on how users may search for books and how they may navigate inside a book. These are important questions as books are, by their nature, different from the traditional notions of documents in IR. For example, books can vary widely in their genre, size, and style. These factors will likely affect the way in which books are thought of and used. User intent may be dependent upon the type of books. Users may search for books to purchase or to locate information within and they are likely to treat textbooks differently from novels [12].

While there are many more questions, in this paper, we aim to investigate indexing and retrieval strategies that can, on the one hand, scale to the size of large collections of books and, on the other hand, can exploit book specific fea-

---

[2] http://www.gutenberg.org/
[3] http://www.ulib.org/
[4] http://www.opencontentalliance.org/
[5] http://www.archive.org/
[6] http://books.google.co.uk/
[7] In US: http://search.live.com/results.aspx?mkt=en-US&scope=books

tures. With this aim, we implemented a prototype system for book search, which can be used as a test bed for investigating and developing indexing and retrieval models for searching books. Based on this prototype system, we experimented with various indexing strategies, and database and inverted index based storage structures. We developed a novel inverted index structure, where multiple fields (e.g., bibliography, headers, etc.) can be stored and searched efficiently. We also studied the effectiveness of retrieval models such as BM25 and BM25F, investigating how the structure of books (e.g., table of contents, back-of-book-index) can be exploited to improve retrieval performance.

The paper is structured as follows. Section 2 gives a brief overview of our prototype system. Section 3 reviews related works on indexing structures and presents our novel multi-field inverted index. The retrieval models used in our experiments are detailed in Section 4. Experiments evaluating the efficiency of our index and the effectiveness of the BM25 and our book-specific BM25F retrieval models are discussed in Section 5. We close with conclusions in Section 6.

2   System Overview

The two basic functionalities of any search system are the indexing and retrieval of documents. Other essential components include a user interface (front-end) and a storage (back-end). Based on this modularisation, we designed our book search experimentation system as an assembly of modular components (see Figure 1(a)).

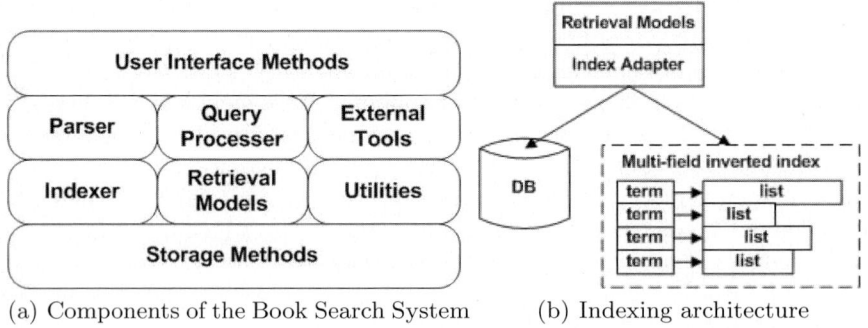

(a) Components of the Book Search System   (b) Indexing architecture

Fig. 1. Book Search System

The User Interface Methods module includes both command line tools for batch processing and graphical user interface tools for increased interaction and control. The Parser module is a collection of data source specific methods for processing the OCR input. During processing, the content of a book is separated out into six streams (fields): table of contents, back-of-book index, bibliography, headers, footers and body (any other content). The parser outputs the extracted information, such as keywords, in a predefined format based on a chosen index structure. A member of the Indexer class of components then takes this output and either creates an index in an SQL database or generates an inverted file through the Storage Methods component. The Query Processor module executes a ranking function, e.g., BM25 [10], implemented in Retrieval Models. The output is either passed to the User Interface or stored in a file as a set of feature vectors, which is then read by a neural net ranker via the External Tools layer. Utilities are common libraries used by other modules, e.g., matrix for storing feature vectors.

3   Indexing

The most widely used indexing mechanism in IR is an inverted index [15,8]. There are two main variants of inverted indexes, 1) a record-level inverted index contains a list of documents and within-document frequency *tf* pairs for each key (term),i.e., key $\rightarrow$ (doc_id,tf); 2) a word-level inverted index in addition also contains the positions of a key in the document, i.e., key $\rightarrow$ (doc_id,tf,{p1,..}). For efficient storage, compression methods are typically applied to reduce the size of the index. Challenges are presented due to the continuous growth of the data to be indexed, which calls for solutions to enable incremental updates [1]. Although an inverted index structure, representing a simple pair relation between terms and documents, is suitable for most IR applications, it needs to be extended for more complex retrieval scenarios, such as enterprise search (see [4]) as well as for book retrieval. In these domains, multiple sources of information need to be considered, for which separate collection statistics have to be calculated. The standard inverted index lacks the flexibility to cope with such applications.

Alternative mechanisms have been offered in the area of XML IR, which deals with the retrieval of XML documents, where the structure of documents is exploited to improve their retrieval. In order to support the retrieval of document parts (e.g., passages or XML elements inside a document), the notion of an "inverted document" has been extended to an "inverted element". The index in this case includes both content and structural information, for example, in the form of XPaths [2]. Solutions include configurable indexing (e.g., [7]) and the integration of inverted files and structure indexes (e.g., [6]). Similar studies have been conducted in Web retrieval and enterprise search [11,3].

In this work, we experimented with two different ways of indexing: 1) a middleware built on a database layer, and 2) a novel multi-field inverted index structure. Figure 1(b) shows the relationship between the different indexes and the query processor. The index adapter forms a logical layer between the physical data organisations and the retrieval models. It allows for switching between the database and the inverted index. In the current paper, we will use the database index when comparing the efficiency of our multi-field index, which is described next.

3.1   Multi-field Inverted Index

In order to store field-specific term frequency information for a number of different fields, we developed a novel multi-field inverted index by extending the traditional record-level index structure. An outline of the index structure is shown in Figure 2. Note that it is also possible to extend a word-level index structure the same way, which then enables phrase and proximity-based retrieval.

Similarly to the traditional inverted index, our multi-field index consists of two parts: 1) a key index, and 2) a posting file. The key index is based on a hash table, where a key unit stores an indexed term, the term's document frequency in the collection, and the offset of the corresponding posting list. For example, in Figure 2, the index term "Abraham" occurs in 123 documents in the collection, and the offset of its corresponding posting list is 12345. The key index is kept in memory during indexing and retrieval. A posting file contains a number of posting lists, where a posting list stores document id's and within-document term frequency (tf) pairs, i.e., (doc_id,tf), as well as field-specific tfs:

$$\langle doc_1, tf_{d1}, tf_{d1f1}, tf_{d1f2}, \cdots \rangle, \langle doc_2, tf_{d2}, tf_{d2f1}, tf_{d2f2}, \cdots \rangle, \cdots$$

Here, tf_{d1} is the term's tf in document $d1$, and tf_{d1fi} is the field-specific tf in document $d1$'s i-th field. For example, in Figure 2, the term "Abraham" in the posting list at the offset of 12345 has a tf value of 100 in the whole document, a tf of 80 in field 1, etc. The number in the header of a posting list indicates the number of posting units it contains.

In order to cater for any number of fields, an additional infrastructure called *fields-map* is added to the index header. A fields-map is a list of pairs:

$$\langle field\_name_1, field\_id_1 \rangle, \langle field\_name_2, field\_id_2 \rangle, \cdots$$

For example, for our book search application, we have $\langle body, 0 \rangle$ and $\langle toc, 1 \rangle$.

The number of fields is specified when a new index is created. At retrieval time, the fields-maps are loaded into memory and stored in a dictionary. A field-specific tf value is retrieved by looking up the dictionary using the field's name, which is then mapped to the field's id. The field id indicates the position of the field-specific tf value after the within-document tf value in the posting unit. For example, in the posting unit $\langle doc_1, 100, 80, 10, 2, 3, 4, 1 \rangle$, the field-specific tf value of the field with $id = 2$ is 2 (counts from 0).

The construction of multi-field index consists of two phases: Insertion and Merging. During the insertion phase, the posting units are created and appended to their respective posting lists, held in memory. If the length of a posting list reaches a predefined upper limit, the list is dumped to disk, and its address is added to the associated key index. To prevent the process from running out of memory, the posting lists are regularly dumped to disk. The key index is also dumped to disk when memory usage reaches an upper bound. In addition, malformed terms are also removed from the index. Once the insertion phase completed and all documents have been parsed, the resulting index may contain keys that are associated with many short posting lists. In order to optimize retrieval, these short lists are merged into a single long posting list during the merging phase. Figure 2(a) shows the configuration of the index before merging and figure 2(b) shows its state after merging. Once the index construction is completed, the in-memory key index is written to a file.

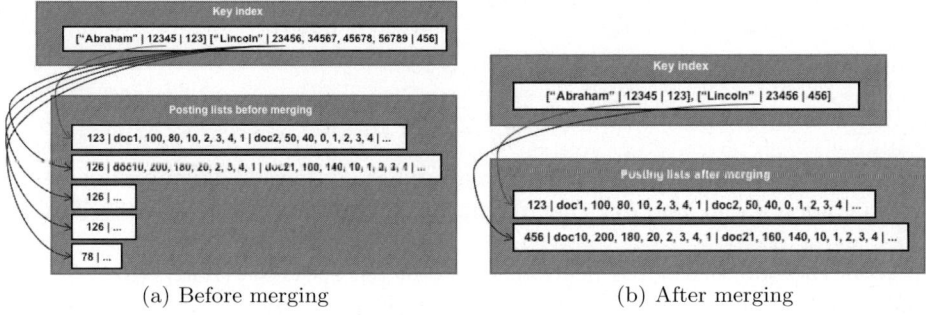

(a) Before merging   (b) After merging

Fig. 2. Multi-field inverted index

4   Retrieval Models

This section provides a brief overview of the BM25 and BM25F retrieval models that we experimented with for our book search application. We trained the parameters of both BM25 and BM25F by optimizing a smoothed approximation to the evaluation measure of normalised discounted cumulative gain (NDCG) [5] as used by LambdaRank [13] using gradient descent.

4.1   BM25

BM25 [10] is a 2-Poisson based retrieval function. The relevance status value (RSV) score of a document calculated by BM25 is given as:

$$RSV := \frac{tf}{k_1 B + tf} \cdot idf, \tag{1}$$

where tf is the within-document term frequency of the term, $B = ((1-b)+b\frac{dl}{avdl})$, k_1 and b are free parameters, dl is document length, $avdl$ is average document length across the collection, and $idf = \log \frac{N-df+0.5}{df+0.5}$ is the Robertson-Sparck-Jones inverse document frequency weighting formula.

4.2   BM25F

BM25F [11] is an extension of the BM25 retrieval function, where a document can be modeled as having a number of fields, where different fields may be of different importance. For example, the title of a document may be one such field. A term occurring in the title field then can be given higher importance to if the term occurred in the body of the document.

In BM25F different weights w_i are assigned to the different fields (i.e., reflecting importance). Although the parameter k_1 may also be chosen specifically for the different fields, the study of [11] has shown that field-specific b is more useful.

The definition of BM25F is given by (the subscript f indicates field-specific variables):

$$RSV := \frac{\sum_f w_f \frac{tf_f}{B_f}}{k_1 + \sum_f w_f \frac{tf_f}{B_f}} \cdot idf. \qquad (2)$$

In our system, we choose a fixed k_1 parameter, and used field-specific w and b parameters (i.e., w_f and b_f). The idf is the Robertson-Sparck-Jones weighting, same as for BM25.

From the books in our collection, we extracted six fields (body, table of contents, back-of-book index, header, footer and bibliography), thus for our experiments we used BM25F with 13 parameters.

5   Experiments

In this section we report on the efficiency of the our multi-field indexing strategy described in Section 3, and evaluate the retrieval effectiveness of the retrieval strategies introduced in Section 4. With respect to the latter we are interested to learn which of the features (fields) extracted from the books in our collection lead to improved performance.

5.1   Test Collection

For our evaluation we use a subset of the test collection used at the Book Search track of the INitiative for the Evaluation of XML Retrieval (INEX)[8], launched in 2007. This sub-collection consists of 10 000 out-of-copyright books (out of the over 42 000 books in the INEX collection), totaling around 80GB. Each book is an OCR file stored in djvu.xml format, and has an associated metadata file (.mrc), which contains publication (author, title, etc.) and classification information in MAchine-Readable Cataloging (MARC) record format. The average size of a book OCR file is 8MB. The basic XML structure of a book (djvu.xml) comprises the body of the book made up of individual pages, which include paragraphs (could be a header, a section or a footer paragraph), lines and words as XML tags.

For test queries, we used 211 queries extracted from the query log of a commercial book search engine. These queries are similar to the topic titles used at the TREC evaluation initiative [14]. The average query length is 2.1 terms.

Relevance judgements were again provided by the commercial search engine. These judgements were made over a much larger set of books and were filtered to only include relevance assessments for the 10 000 books included in our test set. The relevance assessments were made by human judges along a four point scale: Excellent, Good, Fair, and Non-relevant. There are a total of 3246 relevant books in the collection. The average number of relevant books per query is 18.5.

---

[8] http://inex.is.informatik.uni-duisburg.de/2007/bookSearch.html

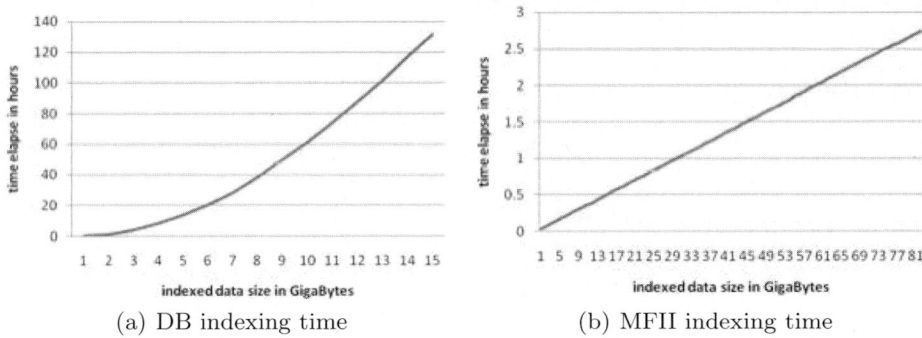

(a) DB indexing time   (b) MFII indexing time

Fig. 3. Indexing time

5.2   Efficiency

In this section, we test the efficiency of our multi-field inverted index and compare it to the performance of our database index.

We run these experiments on a Windows Server 2003, with 4GB memory, and a 3.6GHz Intel CPU. Figures 3(a) and 3(b) show the time required to build the database and our inverted index, repectively. We can see that after 130 hours, the database approach has only indexed 15GB of data, whereas the multi-field inverted index processed the whole collection (over 80GB) in 3 hours. The resulting inverted index size is 5Gb (of which 22MB is the key index). More importantly, however, the indexing time curve for the database appears to be polynomial, whereas it is linear for our inverted index.

Next, we look at the retrieval performance of both the database and inverted indexes. We randomly selected query terms with varying document frequencies from the range of 10 to 9 000. As we can see in Figure 4, retrieval time already reaches 20 seconds for $df = 1000$ for the database, while it only increases to 0.4 seconds at $df = 9000$ for our multi-field inverted index.

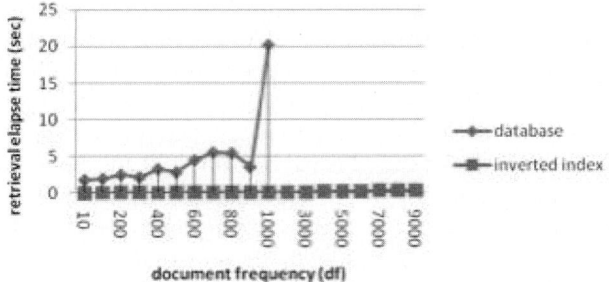

Fig. 4. Retrieval Time

Retrieval models	BM25, BM25F 1x Body, BM25F 1x TOC, BM25F 1x BOBI, BM25F 1x Header, BM25F 1x Footer, BM25F 1x Bib, and BM25F 6x
Query set bins	Random-1, Random-2, Sequential, Shuffle, Round-Robin
Neural Net input	Training: 3 bins, Validation: 1 bin, Test: 1 bin
Validation	5 folds, rotating bin allocation to training, validation and test sets

Fig. 5. Experimental Setup

5.3   Retrieval Effectiveness

Evaluation measure. We use as our primary metric the rank-based measure of normalised discounted cumulative gain (NDCG) [5], which makes use of the multi-level relevance judgements in our test collection. The measure works by summing up gains associated with relevant documents along the ranking while applying a discount function so that relevant documents retrieved later in the ranking achieve less gain. The obtained score is then compared to the best possible DCG value derived from an ideal ranking in which a more-relevant document always precedes a less-relevant document. The gains associated to the 4 grades of relevance are: Excellent (15), Good (7), Fair (3), and Not-relevant. The rank-based discount function is $\frac{1}{\log(1+j)}$ for rank j.

Evaluation Setup. For our baseline retrieval, we used BM25 to retrieve a ranked list of books estimated relevant to the 211 test queries. We then experimented with BM25F, where we considered evidence from six different book parts (fields): body, table of contents (TOC), back-of-book index (BOBI), header lines (Header), footer notes (Footer), and the bibliography (Bib). We investigate what retrieval performance can be achieved based on the different fields alone and when these are all combined.

As mentioned in Section 4, we train the parameters of BM25 (2 parameters) and BM25F (13 parameters) with the neural net ranker, LambdaRank [13]. As input, the neural net ranker requires a training set, a validation set, and a test set. To this end, we first split the 211 queries into 5 bins in 5 different ways: 2 random, 1 round robin, 1 shuffle (using hash value generated from the query string), and 1 sequential. From the 5 bins, we used 3 bins for the training set and the remaining 1-1 sets for validation and testing. We performed 5 fold validation, that is we repeated each run rotating the selection of the bins into the training, validation and test sets.

In total, we run 40 experiments: 5 runs each for the retrieval functions of BM25, BM25F with 6 of the individual fields and for all 6 fields combined. Repeating the experiments 5 times (using the different methods for splitting the 211 queries, e.g., shuffle) aimed at reducing the effect of outliers in the results (for example, caused by a cluster of badly performing queries). Figure 5 summarizes the experimental setup.

Results. Figure 6 shows the results of our experiments, averaged over the 5 runs (using the different methods for splitting the set of 211 queries) per retrieval method. The *NDCG* scores shown are for the queries in the test set, reporting

retrieval strategy	NDCG@1	NDCG@5	NDCG@10
BM25	55.22 (5.63)	55.1 (3.87)	54.57 (3.5)
BM25F 6x	**60.24** (5.62)	**57.68** (3.87)	**56.71** (3.46)
BM25F 1x, Body	42.39 (5.65)	43.23 (3.96)	43.22 (3.66)
BM25F 1x, TOC	33.36 (5.22)	32.16 (3.44)	30.57 (2.97)
BM25F 1x, BOBI	22.51 (4.76)	19.15 (3.09)	17.51 (2.56)
BM25F 1x, Header	19.33 (4.42)	15.97 (2.71)	14.5 (2.3)
BM25F 1x, Footer	1.62 (1.47)	1.02 (0.67)	0.84 (0.57)
BM25F 1x, Bib	57.78 (5.71)	50.35 (3.88)	47.28 (3.37)

Fig. 6. Results

performance at ranks 1, 5 and 10, respectively. The values in brackets reflect the error rate, where smaller numbers indicate more confidence in the reported NDCG scores.

Looking at the BM25 results, we can see that it achieves a respectable score based only on the full-text of the books without applying any special treatment. The BM25F model which combines evidence from all our 6 fields does, however, outperform BM25 in all our runs. This suggests that differentiating between different features (fields) and allowing these to be weighted differently does improve retrieval effectiveness. The difference in performance is especially visible at the top ranks. Some of the features considered here thus appear to have precision enhancing qualities.

Comparing the results obtained for the single fields with BM25F, we can see that the footer field is a rather poor indicator of relevance in this case (NDCG scores of 0.84-1.62%). The highest scoring field is the bibliography (with 47.28-57.78%). This would suggest that information contained in the bibliography section of a book is one of the most important sources of evidence for estimating the book's relevance. Although this sounds plausible and even intuitive, it is doubtful that most books in a general collection will contain a bibliography. While for most textbooks this may be true, for novels, it is most likely be not the case. Taking a closer look at the raw document and term frequency statistics extracted for the bibliography fields, we find a very sparse matrix with a few 'lucky' hits on relevant books. In addition, these hits get exaggerated by having a tf value that equals the length of the field. Aside from these handful of hits, the bibliography field also contains magnitudes more high impacting term occurrences in non-relevant books. Despite this, at NDCG@1, it actually outperforms our baseline BM25 model. Due to the sparseness of the data, we should be cautious to draw firm conclusions though. It is interesting however that while the sparseness issue also applies to the footer field, it affects performance rather differently. This issue will require further and more detailed investigations in the future. A further point to note is that bibliography has the largest variation in performance scores across ranks 1,5 and 10, changing over 10%.

The body, which contains the majority of a book's full text, achieves the second highest scores (42.39-43.23%). Interestingly, the body field is the only one where the NDCG score at rank 1 is worse than NDCG at ranks 5 or 10. This may suggest that, it is a less useful feature for improving performance at the very top ranks. The body field is followed by the table of contents (TOC) field, which achieves scores around 30-33%. This is somewhat of a surprise as we expected that the chapter titles which form the entries in a TOC would provide strong evidence for relevance and thus expected TOC to perform much better. A reason for the low score is that only about 10% of the books in the collection have a TOC. Genre-specific evaluation may thus yield better results in the future. The back-of-book index (BOBI) is very similar to the TOC, although it performs around 10% worse than the TOC. Just as with the TOC, only about 7% of the books contain a BOBI. Retrieval using the header field alone achieves scores of 14.5-19.33%, where the best results is obtained at the top rank. One reason why the header may not help as much as the TOC is that it quite often contains noisy information. A lot of OCR errors occur in this field as page numbers are often incorrectly recognised. Another reason is that the nature of the information printed in the header will differ from book to book. Some books just repeat the book title, sometimes alternating author names and the title on odd/even pages, while others may print the chapter titles here, and so on.

One important lesson we learnt here is that taking into account book features and combining these using BM25F appears to be a promising route to take to improve retrieval effectiveness for books. However, we should be careful with our choices of features that we build on and we may also need to consider genre-specific selections.

6   Conclusions

In this paper we reported on an experimental book search system that supports both database and IR style index structures, enabling fast prototyping and large-scale testing. We proposed a novel multi-field inverted index, which enables separate fields (contexts) to be indexed individually. For example, the table of contents or the back-of-book index of a book can be represented with their own set of term frequency statistics and can be weighted independently. These fields can then be combined using the BM25F retrieval model. We used a neural net ranker, LambdaRank, to tune the parameter weights of BM25F.

The evaluation of our indexing and retrieval strategies were based on a test collection of 10 000 books and 211 queries. Our novel multi-field inverted index completed the index of the whole 80GB collection in 3 hours, achieving a linear volume-time efficiency curve. The obtained retrieval effectiveness results showed that BM25F outperforms the baseline BM25 strategy. In addition, we found that the table of contents and the back-of-book index of books could be important features for the retrieval of a given genre of books (e.g., textbooks). Although our results showed that the bibliography field was the best performing feature, we will need to further investigate the reliability of this finding in our future work.

In conclusion, the contribution of this paper were two-fold: We introduced a novel multi-field inverted index and experimented with the BM25 family of retrieval models in a book retrieval task, where we identified a number of worthwhile features to investigate further. In particular, our findings indicate that fielded retrieval is a suitable strategy to apply to collections of books. Additional fields that we expect to work well include book title, author information and reviews.

Our future work will extend the range of studied retrieval models while also exploring further static and dynamic features, such as term occurrence frequencies extracted from related book reviews, metrics based on citation networks and authority scores based on publisher and author reputations. We also intend to study collections of books from a specific genre.

References

1. Büttcher, S., Clarke, C.L.A., Lushman, B.: Hybrid index maintenance for growing text collections. In: Efthimiadis, E.N., Dumais, S.T., Hawking, D., Järvelin, K. (eds.) SIGIR, pp. 356–363. ACM, New York (2006)
2. Clark, J., DeRose, S.: XML Path Language (XPath) version 1.0. W3C Recommendation. Technical Report REC-xpath-19991116, W3C (World Wide Web Consortium) (November 1999), http://www.w3.org/TR/xpath
3. Craswell, N., Zaragoza, H., Robertson, S.: Microsoft cambridge at TREC-14: Enterprise track
4. Hawking, D.: Challenges in enterprise search. In: Schewe, K.-D., Williams, H.E. (eds.) ADC. CRPIT, vol. 27, pp. 15–24. Australian Computer Society (2004)
5. Järvelin, K., Kekäläinen, J.: Cumulated gain-based evaluation of IR techniques. ACM Transactions on Information Systems (ACM TOIS) 20(4), 422–446 (2002)
6. Kaushik, R., Krishnamurthy, R., Naughton, J.F., Ramakrishnan, R.: On the integration of structure indexes and inverted lists. In: ICDE, p. 829 (2004)
7. Liu, S., Zou, Q., Chu, W.W.: Configurable indexing and ranking for XML information retrieval. In: Sanderson, M., Järvelin, K., Allan, J., Bruza, P. (eds.) SIGIR, pp. 88–95. ACM, New York (2004)
8. Moffat, A., Zobel, J.: Self-indexing inverted files for fast text retrieval. ACM Trans. Inf. Syst. 14(4), 349–379 (1996)
9. Reddy, R., StClair, G.: The million book digital library project
10. Robertson, S.E., Walker, S.: Some simple effective approximations to the 2-poisson model for probabilistic weighted retrieval. In: Croft, W.B., van Rijsbergen, C.J. (eds.) SIGIR, pp. 232–241. ACM/Springer (1994)
11. Robertson, S.E., Zaragoza, H., Taylor, M.J.: Simple BM25 extension to multiple weighted fields. In: Grossman, D., Gravano, L., Zhai, C., Herzog, O., Evans, D.A. (eds.) CIKM, pp. 42–49. ACM, New York (2004)
12. Sabine, Sabine: How people use books. Library Quarterly 56(4), 399–408 (1986)
13. Taylor, M.J., Guiver, J., Robertson, S., Minka, T.: Softrank: Optimising nonsmooth ranking metrics. In: First ACM International Conference on Web Search and Data Mining (WSDM), Stanford, California (2008)
14. Voorhees, E.M., Harman, D.K.: TREC: Experiment and Evaluation in Information Retrieval. MIT Press, Cambridge (2005)
15. Zobel, J., Moffat, A.: Inverted files for text search engines. ACM Comput. Surv. 38(2) (2006)

Using a Task-Based Approach in Evaluating the Usability of BoBIs in an E-book Environment

Noorhidawati Abdullah[1] and Forbes Gibb[2]

[1] Faculty of Computer Science and Information Technology, University of Malaya, Malaysia
[2] Department of Computer and Information Sciences University of Strathclyde, UK
noorhidawati@um.edu.my, Forbes.Gibb@cis.strath.ac.uk

Abstract. This paper reports on a usability evaluation of BoBIs (Back-of-the-book Indexes) as searching and browsing tools in an e-book environment. This study employed a task-based approach and within-subject design. The retrieval performance of a BoBI was compared with a ToC and Full-Text Search tool in terms of their respective effectiveness and efficiency for finding information in e-books. The results demonstrated that a BoBI was significantly more efficient (faster) and useful compared to a ToC or Full-Text Search tool for finding information in an e-book environment.

Keywords: E-book, Task-Based, BoBI, Book Index, Usability Evaluation.

1 Introduction

Typically, browsing a book or searching for specific content is accomplished via a ToC or a BoBI. In a digital or e-book environment these methods may be enhanced by hyperlink features and Full-Text Search tools. For instance, the California Digital Library Ebook Task Force [1] suggested that advanced search facilities (such as Boolean, truncation, proximity, etc.) should be incorporated in e-books. In general, a ToC provides information on the organisational structure of a book which may be skimmed to obtain some general ideas as to the book's content. A BoBI, on the other hand, provides more specific information on relevant sections of text by pointing to key concepts discussed in the book. However, although a ToC helps readers to browse through a book it relies on the reader's ability to interpret the section headings contained in the ToC. Some advantages of a BoBI compared to ToC are that it organises the information in the book into an alphabetical structure, groups together information that is scattered through the book, distinguishes important topics from random occurrences of information, and provides cross-references to indicate preferred and related terms.

In an e-book environment, we still need a BoBI even though we may already have a Full-Text Search tool. This is because a BoBI directly identifies significant topics in the book unlike a Full-Text Search tool which matches word strings specified by users. Previous research has found that users would like BoBIs to be incorporated in e-books [2-6] and this has been supported by the experiences of the Bureau of National Affairs in the United State. When the Bureau transferred its publications from paper to

C. Macdonald et al. (Eds.): ECIR 2008, LNCS 4956, pp. 246–257, 2008.
© Springer-Verlag Berlin Heidelberg 2008

CD-ROM they provided Full-Text Search tools for the electronic version of their publications but excluded BoBIs [7]. However users demanded BoBIs because they were very familiar with the tool and already knew how to use it, whereas a Full-Text-Search tool required users to have a certain level of skill and experience in order to be able to use features such as Boolean operators (AND, OR, NOT) effectively. Some other challenges raised by Full-Text Search tools are that they require users to specify search terms that match the terminology used in the text while allowing for variations in terms of spelling, hyphenation and synonyms; they retrieve and, usually, rank all instances of the occurrence of sought terms which the reader must evaluate in terms of the relevance of retrieved sections of text by browsing a list of titles.

The retrieval performance of BoBIs has been traditionally measured by researchers in terms of recall (the index finding ability) and precision (how well the index entries matched the text) [8]. With the emergence of the hypertext concept, and e-books that incorporated hyperlinks in BoBIs and ToCs, the issue of hypertext and e-book usability was introduced. Some studies such as [9-15] involved analysis of retrieval performance and user preferences between e-books or e-documents (with hypertext features) and printed books. However, the results of these studies are not directly comparable because of differences in the design of the usability tests (e.g. within subject-design or between subject design), the type of search tasks (e.g. fact finding or inference), the materials used in the evaluation (e.g. manuals, textbooks, or encyclopaedias), subject fields (e.g. chemistry or computer-based subjects), and participants (e.g. novice or computer experts) as well as interface and format issues (e.g. web or pdf versions). Therefore the study reported here may be considered significant based on the following factors:

i.   Testing was conducted using a within-subject design in which each participant was tested using each search tool (BoBI, ToC and Full-Text Search). Therefore each participant had experience of using every search tool. As a result they could provide more accurate responses on preferences and satisfaction through the interaction with the search tools. Other studies [12-14] only studied one of the search tools or let users choose the one with which they were most familiar.

ii. The choice of e-books and search tools was randomly selected for each participant. This method was used to minimise the likelihood of users becoming familiar with the e-book content.

iii. The study tested three different e-books from the information retrieval field. Previous studies normally used only one type of e-book for the evaluation and occasionally one book in several formats (such as [9, 12-14]).

iv. The study involved 45 participants (students in a UK university) which is a relatively large number of participants compared to most previous studies (e.g. [13], [14] and [16]).

2   Objectives

This study was conducted with the following two main purposes:

(i)   To evaluate whether a BoBI is more effective, efficient and useful compared to a ToC and Full-Text Search tool for finding information in an e-book environment.

(ii)   To measure users' attitudes with respect to a BoBI, ToC and Full-Text Search tool for finding information in an e-book environment in terms of their preferences, levels of satisfaction and ease of use.

There were six central hypotheses that ran through the usability evaluation:

> *H1.1*: A BoBI is more efficient compared to a ToC for finding information in an e-book environment.
> *H1.2*: A BoBI is more efficient compared to a Full-Text Search tool for finding information in an e-book environment.
> *H2.1*: A BoBI is more effective compared to a ToC for finding information in an e-book environment.
> *H2.2*: A BoBI is more effective compared to a Full-Text Search tool for finding information in an e-book environment.
> *H3.1*: A BoBI is more useful compared to a ToC for finding information in an e-book environment.
> *H3.2*: A BoBI is more useful compared to a Full-Text Search tool for finding information in an e-book environment.

3   Research Methodology

This evaluation was carried out with subject-specific users. This was because it was assumed that the target population must have some knowledge of the subject field covered by the test collections and that they would also have reasonable and similar levels of the computer skills necessary to be able to perform a search task. The target population was MSc and research students in the Department of Computer and Information Sciences, at the University of Strathclyde and involved a total of 45 respondents with 25 of them are male and 20 are female.

This study employed a task-based and within-subject approach its evaluation design as elaborated below:

3.1   Task-Based Approach

Vakkari [17] highlighted that to characterise a search task in task-based information searching the following factors should also be taken into consideration: (i) The selection of search terms and operators in query formulation, such as the use of narrower terms, synonyms, and the use of Boolean operators such as AND, and OR; (ii) The search tactics employed, such as browsing, initial general search, and final specific search; (iii) The use of search support tools, such as query expansion for refining the query; and (iv) Relevance and utility judgements of information found, such as degree of relevance, usefulness, and precision.

In addition, typical task that users would normally undertake when interacting with e-book are browsing, searching, analysing relevant contents, and so forth, depending on the purpose for consulting the book in the first place. This depends on factors such as: (i) The types of information users search for; for example, searching for individual facts, and searching for textual or non-textual information; and (ii) The selection of search terms, such as whether users are cued by the questions that contain words in the text or headings.

The task-based approach that was utilised in this study was based on the following characteristics:

i.   The types of information searched for. This involved searching for two types of information: specific facts and relevant content. Therefore two types of search task were involved: (a) Factual task. This was a straightforward question to find a specific piece of information in the e-book (e.g. "What are the definitions of the terms precision and recall as provided in this book?"); and (b) Analytical task. This was to identify relevant e-book content which would satisfy a query about an information need involving greater breadth. For example, "You are writing an essay about some methods and implementation of automatic classification. Which section(s) of this book discuss this topic? (Give the page numbers)."

ii. The selection of search terms in the query formulation. This study hypothesised that the appearance in the BoBI and ToC of a term in the target information probably affects users' search performance. Therefore three types of query were formulated: (a) A term in the target information appeared only in the BoBI; (b) A term in the target information appeared only in the ToC; and (c) A term in the target information appeared in both the BoBI and the ToC.

iii. The use of search support tools. In this case three types of search tools were used (the BoBI, the ToC and Full-Text Search) with the intent of evaluating whether one was superior to any of the others.

iv. Relevance judgements of information found. The relevance judgement of information is influenced by users' perceptions of past experience, their present situation, their knowledge and their search goals. In this study, relevance was constrained by the above factors and also by a predetermined correct answers set constructed by an expert in the chosen subject fields

3.2   Usability Evaluation Design

This study has employed a within-subjects approach in its usability evaluation design. The within-subject approach is also known as repeated measure where the same participants perform under all the possible combinations of conditions (in this study conditions are defined as types of search tools; i.e. BoBI, ToC and Full-Text Search). As subjects are tested under every condition, the problem of individual differences can be eliminated in which each participant acts as his/her own control. In addition, experience of using all the conditions could result in the respondents making a better judgement on each of the search tools. As repeated measures are always associated with an order effect (one task may be affected by the experience of having performed another task), or practice effect (participants' performance on the later task may have improved since the first tasks) a procedure of counterbalancing was taken to reduce these effects. In the counterbalancing the order of the condition in within-subjects designs was varied from participant to participant in order to balance out the effects across the conditions [18]. Some advantages of this approach are that it requires fewer participants and is hence less costly, it is suitable for evaluating a system where learning is involved, and it has less chance of effects from variation between participants. This approach was employed to minimise the number of participants that should be involved and to allow participants to interact with every search tool so that they could provide more accurate responses.

4 Results and Discussion

The data was analysed using the SPSS program. The performances of the three search tools were measured using two variables: (i) Speed (in minutes) of finding information in e-books and (ii) Count of success in finding information accurately in e-books. The data were analysed using analysis of variance (ANOVA) to test the statistical significance of the differences among the mean scores of two or more groups (in this study they were BoBI, ToC and Full-Text Search). Alternatively a Friedman Test was used to replace the ANOVA test for ordinal types of data (i.e. ranking of search tools usefulness).

4.1 Is a BoBI More Efficient Compared to a ToC and Full-Text Search Tool for Finding Information in an E-book Environment?

Fig. 1 below shows the average time (in minutes) of each search tool based on two types of search tasks. The chart shows that the BoBI outperformed the ToC and Full-Text Search for both factual and analytical tasks with an average of 3.04 and 2.44 minutes. The ToC was second best for answering factual tasks correctly (3.59 minutes) and third best for analytical tasks (4.21 minutes). The Full-Text Search on the other hand was worst for factual tasks (4.34 minutes) and second best (3.18 minutes) for analytical tasks. Fig. 1 shows that there were differences in speed performance between the search tools and search tasks. An ANOVA test was therefore performed to establish whether these differences were statistically significant.

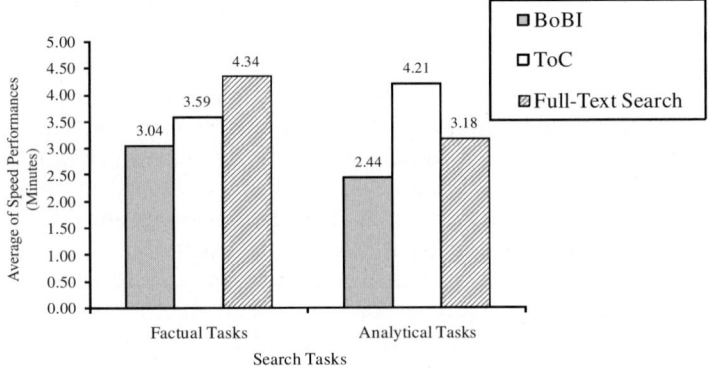

Fig. 1. Speed Performances in Answering Tasks Correctly

Table 1 shows that there was a significant difference in speed performance between the search tools: $F(2,86)=3.22; p=<0.04$[1]. The search tasks on the other hand exhibited no significant difference: $F(1,43)=0.55; p=<0.46$. There was also no significant difference in terms of the Search Tools*Search Tasks interaction: $F(2,86)=0.87; p=<0.39$. Hence, it can be concluded that the different types of search task did not significantly affect the speed performance of each of the search tools.

Table 1. ANOVA Test Result for Speed Performance

Independent Variables	Type III Sum of Squares	df	Mean Square	F
Search Tools	276395.73	2	138197.87	3.22
Error (Search Tools)	3689986.93	86	42906.82	
Search Tasks	41350.06	1	41350.06	0.55
Error (Search Tasks)	3215137.61	43	74770.64	
Search Tools * Search Tasks	109259.55	2	75702.73	0.87
Error (Search Tools*Search Tasks)	5386179.78	86	86788.96	

Measure: Speed
Significant at the 0.05 level.

As a significant difference existed in search tool performance, a multiple pairwise comparison test was performed to determine where the difference lay amongst the search tools (at a significance level of 0.05) by comparing it in a pair in order to judge which of the pair has a greater or lower amount of speed performance (minutes). This is shown in Table 2. The table indicated that BoBI and ToC and BoBI and Full-Text Search exhibited significant differences in the search performance ($p=<0.04$) in that the BoBI had performed better than the ToC and Full-Text Search. On the other hand, ToC and Full-Text Search showed no significant difference.

Table 2. Multiple Pairwise Comparisons for Search Tools Speed Performance

(I) Search Tools	(J) Search Tools	Mean Difference (I-J)	Std. Error	95% Confidence Interval for Difference	
				Lower Bound	Upper Bound
BoBI	ToC	-73.94	34.93	-144.40	-3.49
	Full-Text Search	-61.68	29.66	-121.50	-1.87
ToC	BoBI	73.94	34.93	3.49	144.40
	Full-Text Search	12.26	28.73	-45.67	70.20
Full-Text Search	BoBI	61.68	29.66	1.87	121.50
	ToC	-12.26	28.73	-70.20	45.67

Measure: Speed (Minutes)
The mean difference is significant at the 0.05 level.

---

[1] This formula is a standard way of writing an ANOVA test result based on the appropriate table that is referred to - in which "F(df value of the independent variable, df value of error of the independent variable)=F value; p=<Sig. (p-value))".

The relative speed differences between the search tools can be seen in Fig. 2 below, where on average the BoBI had the fastest performance at 5.48 minutes, followed by Full-Text Search at 7.52 minutes and finally the ToC at 8.19 minutes. In conclusion, hypotheses H1.1 and H1.2 were validated in that the BoBI was shown to be more efficient when compared to the ToC and Full-Text search tools for performing the search tasks. But there was no conclusive proof that the different search tasks had affected the performance of the search tools.

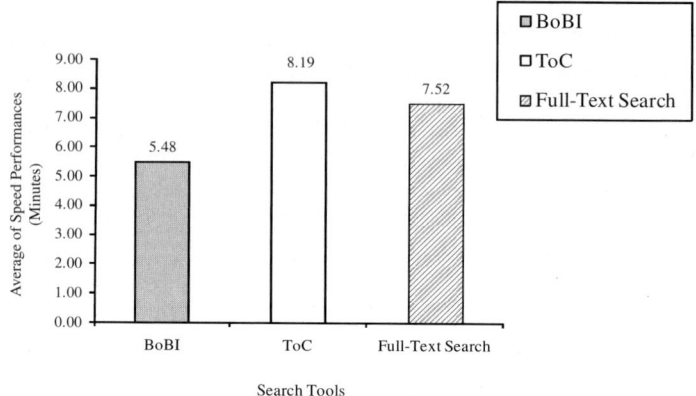

Fig. 2. Overall Speed Performances in Answering Tasks Correctly

4.2 Is a BoBI More Effective Compared to a ToC and Full-Text Search Tool for Finding Information in an E-book Environment?

Fig. 3 below shows the success in finding information accurately in e-books for each search tool based on two types of search tasks. The chart demonstrates that Full-Text Search outperformed the BoBI and ToC for finding information accurately for factual tasks with a success rate of 78%. This was followed by the BoBI (72%) and then the ToC (57%). The BoBI on the other hand, performed the best for accurately finding information in analytical tasks with a 76% success rate. Full-Text Search was second best (69% success) and finally the ToC (67% success). It can be seen from the graph that there were differences in success for finding information accurately amongst the search tools and search tasks in that Full-Text Search was more effective than the BoBI for factual tasks, whereas the BoBI was more effective than Full-Text Search for analytical tasks. These are interesting as a BoBI consists of selective and evaluated entries whereas as Full-Text Search undertakes string matching against a comprehensive index of terms (excluding stop words). An ANOVA test was therefore performed to establish whether these differences were statistically significant.

Table 3 below shows that there were no significant differences using the search tools variable: $F(2,86)=2.60;p=<0.08$. This was consistent with the search tasks variable: $F(1,43)=0.30;p=<0.59$ and Search Tools*Search Tasks interaction: $F(2,86)=2.14;p=<0.12$. Therefore types of search tools, search tasks and their interaction did not have significant differences in terms of the success in finding information accurately. Therefore hypotheses H2.1 and H2.2 were not validated.

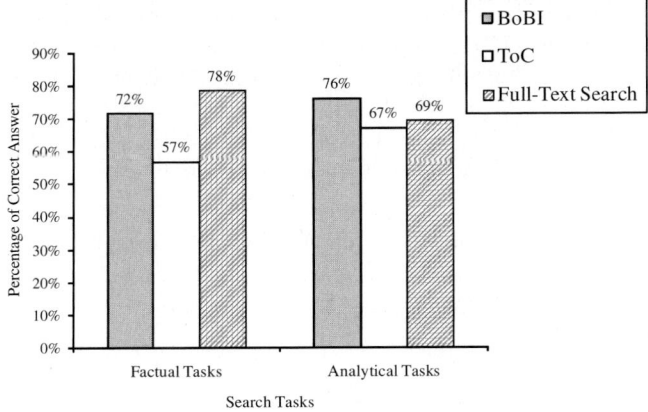

Fig. 3. Count of Success in Finding Information

Table 3. ANOVA Test Result for Success in Findings Information Accurately

Independent Variables	Type III Sum of Squares	df	Mean Square	F
Search Tools	3.34	2	1.67	2.60
Error(Search Tools)	55.33	86	0.64	
Search Tasks	0.09	1	0.09	0.30
Error(Search Tasks)	13.74	43	0.32	
Search Tools*Search Tasks	1.73	2	0.87	2.14
Error(Search Tools*Search Tasks)	34.93	86	0.41	

Measure: Count of Success
Significant test at the 0.05 level.

4.3   Is a BoBI More Useful Compared to a ToC and Full-Text Search Tool for Finding Information in an E-book Environment?

Table 4 below summarises the differences in users' ratings of how useful the three Search Tools were based on the average rank. A Friedman Test was then performed to validate if these differences were statistically significant at the level of 0.05.

The results of the Friedman test in Table 5 show that there were significant differences in students' ranking of Search Tools usefulness: $X^2(2) = 6.37$, p=<0.04[2]. A multiple comparison test was then conducted to find the relative differences amongst the Search Tools as shown in Table 6. The test indicated that only the ToC and BoBI had significant differences at the level of 0.05 (p=<0.02). It can be concluded that the BOBI had a higher ranking of usefulness (average rank = 2.27) compared to the ToC (average rank = 1.85).

---

[2] This formula is a standard of writing Friedman test result based on the appropriate table that is referred to - in which "X^2(df value of the independent variable)=Chi-Square value; p=<Sig. (p-value))".

Table 4. Users' Rating of the Usefulness of the Search Tools

Search Tools	Average Rank
BoBI	2.27
ToC	1.85
Full-Text Search	1.88

1-4 scale where 1 = useless to 4 = essential.

Table 5. Friedman Test of Search Tools Usefulness Rating

Friedman Test	
N	44
Chi-Square	6.37
df	2

* Significant at the 0.05 level.

Table 6. Multiple Comparisons for Search Tools Usefulness Rating

Sign Test	Toc - BoBI	Full Text Search - BoBI	Full Text Search - ToC
Z	-2.37346*	-1.54349	0

Measure: Count of Success
* Significant at the 0.05 level.

According to the statistical analysis above, it can be concluded that H3.1 was validated in that a BoBI was significantly more useful compared to a ToC for finding information in an e-book environment at 0.05 a significant level. While H3.2 was not validated in that there was not enough evidence to support that the BoBI was more useful compared to a Full-Text Search for finding information in an e-book environment.

4.4   Students' Attitudes Towards Search Tools for Finding Information in E-books

4.4.1   Students' Preferences with Respect to a BoBI, ToC and Full-Text Search

Fig. 4 below demonstrates students' ratings of Search Tools in percentages. It can be seen from the graph that most of the respondents rated the BoBI as 'most preferred' (53%) while the ToC was 'second most preferred' (46%) and Full-Text Search was 'least preferred' (34%).

4.4.2   Students' Satisfaction with Respect to a BoBI, ToC and Full-Text Search

Fig. 5 below shows students' rating of how satisfactory Search Tools were in use. It can be seen from the graph that most of the respondents were 'very satisfied' with the BoBI (42%), while most of them were 'satisfied' with the ToC and Full-Text Search at 44% and 38% respectively.

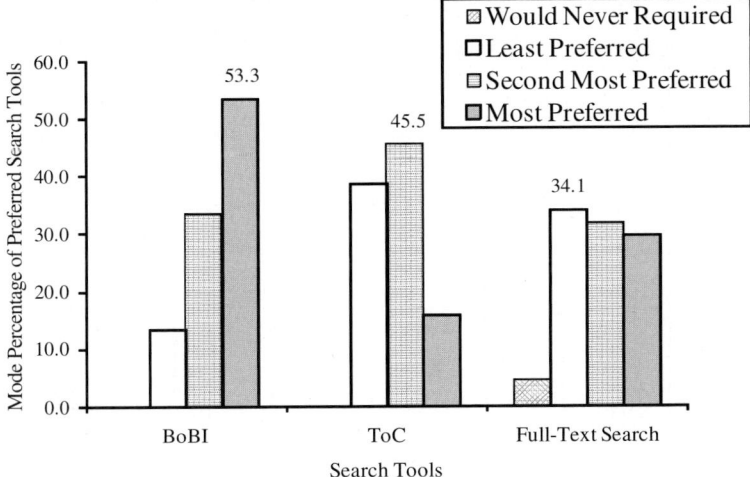

Fig. 4. Users' Rating of Preferred Search Tools

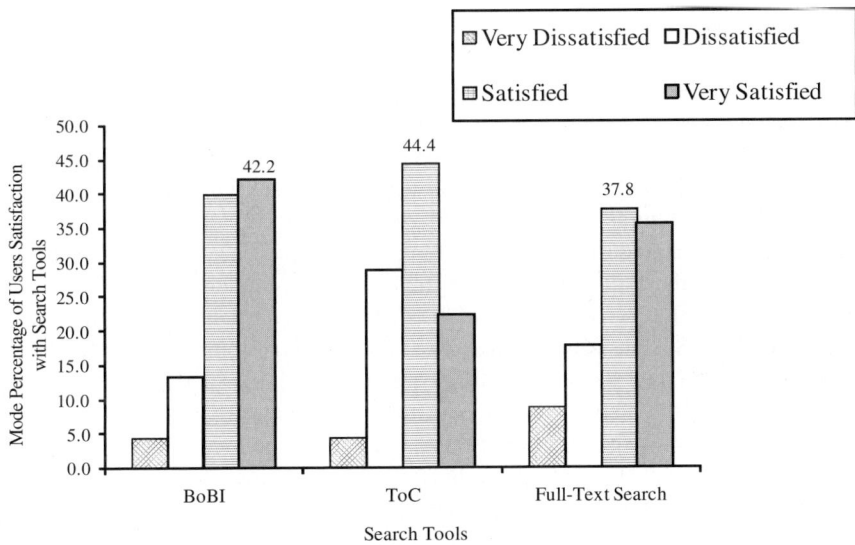

Fig. 5. Users' Rating on Satisfactory with the Search Tools

4.4.3 Students' Rating with Respect to Ease of Use with BoBI, ToC and Full Text Search

Fig. 6 below indicates students' rating of how easy or difficult they found it to use the Search Tools. As can be seen from the graph, most of the respondents found that all the search tools were 'easy' to use but at slightly different levels. The ToC had the highest mode percentage which was 49%, followed by the BoBI (42%), and finally Full-Text Search in 40%.

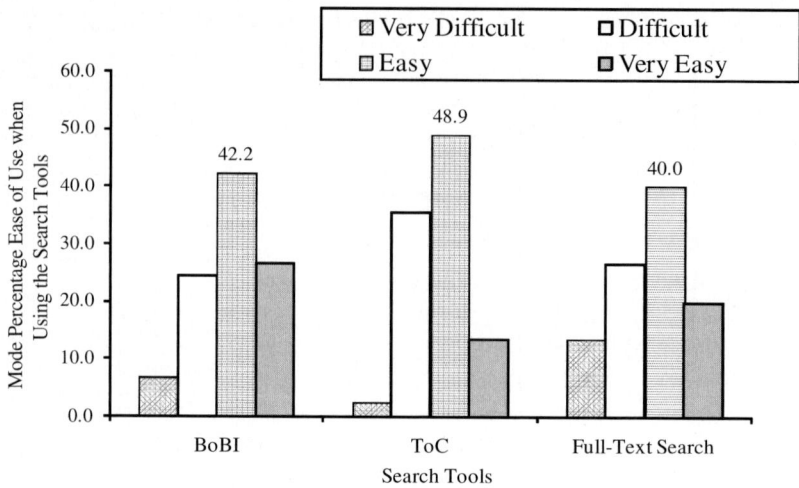

Fig. 6. Users' Rating of Ease of Use when Using the Search Tools

5   Conclusion

This study has demonstrated that a BoBI was significantly more efficient (faster) and useful compared to a ToC and Full-Text Search tool for finding information in an e-book environment. Preference ratings indicated that most students rated the BoBI as 'most preferred' with the ToC as 'second most preferred' and Full-Text Search as 'least preferred'. The search tools were rated similarly for satisfaction and ease of use, but the ToC had the highest mode percentage for satisfaction and ease of use with the BoBI second and finally the Full-Text Search. There was not enough evidence however to support that the different Search Tasks had effectively and efficiently affected the performance of the Search Tools. Although it is not directly comparable in terms of the evaluation design, in the main these findings are inline with [13] and [14].

As a conclusion, the e-book usability evaluation findings are important in gaining a better understanding of the retrieval performance of three search tools (BoBI, ToC and full text search) for browsing for relevant, and searching for specific information in e-books. This will be of value for designing better e-books and access systems.

It is important to acknowledge that the experiment presented here was constrained by subject-specific users and test materials (in information retrieval field) and therefore the generalisation of the results across other subject fields should be treated with caution.

References

1. Coyle, K., et al.: Report on California Digital Library Joint Steering Committee for shared collections ebook task force (2001), http://www.cdlib.org/ inside/ groups/ jsc/ebooks/
2. Landoni, M., Gibb, F.: The role of visual rhetoric in the design and production of electronic books: the visual book. The Electronic Library 18(3), 190–201 (2000)

3. Landoni, M., Wilson, R., Gibb, F.: From the Visual Book to the WEB Book: the importance of design. The Electronic Library 18(6), 407–419 (2000)
4. Landoni, M., Crestani, F., Melucci, M.: The Visual Book and the Hyper-TextBook: two electronic books, one lesson? In: Proceedings of the RIAO 2000 Conference, Paris, April 12-14, pp. 247–265. CID-CASIS, Paris (2000)
5. Landoni, M., Wilson, R., Gibb, F.: From the Visual Book to the WEB Book: the importance of design. The Electronic Library 18(6), 407–419 (2000)
6. Wilson, R., Landoni, M., Gibb, F.: A user-centred approach to e-book design. The Electronic Library 20(4), 322–330 (2002)
7. Brown, F.: Electronic media and the future of indexing (1995),
   http://www. allegrotechindexing.com/article01.htm
8. Bennion, B.C.: Performance testing of a book and its index as an information retrieval system. Journal of the American Society for Information Science 31(4), 264–270 (1980)
9. Egan, D.E., et al.: Formative Design-Evaluation of SuperBook. ACM Transactions on Information Systems 7(1), 30–57 (1989)
10. Egan, D.E., et al.: Hypertext for the electronic library? CORE sample results. In: Proceedings of the 3rd Annual ACM Conference on Hypertext, San Antonio, Texas, December 15-18, pp. 299–312. ACM Press, New York (1991)
11. Mynatt, B.T., et al.: Hypertext or book: which is better for answering questions? In: Proceedings of the SIGCHI conference on Human Factors in Computing Systems, Monterey, California, May 3-7, pp. 19–25. ACM Press, New York (1992)
12. Catenazzi, N., Sommaruga, L.: Hyperbook: an experience in designing and evaluating electronic books. Journal of Documentation and Text Management 2(2), 81–102 (1994)
13. Barnum, C., et al.: Index versus full-text search: a usability study of user preference and performance. Technical Communication 51(2), 185–206 (2004)
14. Henke, H.: An Empirical Design for eBooks. Chartula Press, Niwot (2003)
15. Wilson, R., Landoni, M.: Evaluating electronic textbooks: a methodology. In: Proceedings of the 5th European Conference on Research and Advanced Technology for Digital Libraries, Darmstadt, September 4-9, pp. 1–12. Springer, Berlin (2001)
16. Crestani, F., Ntioudis, S.P.: User centred valuation of an automatically constructed Hyper-Textbook. Journal of Educational Multimedia and Hypermedia 11(1), 3–19 (2001)
17. Vakkari, P.: Task-based information searching. Annual Review of Information Science and Technology 37, 413–464 (2003)
18. Kinnear, P.R., Gray, C.D.: SPSS 14 Made Simple. Psychology Press, East Sussex (2006)

Exploiting Locality of Wikipedia Links in Entity Ranking

Jovan Pehcevski[1], Anne-Marie Vercoustre[1], and James A. Thom[2]

[1] INRIA, Rocquencourt, France
{jovan.pehcevski,anne-marie.vercoustre}@inria.fr
[2] RMIT University, Melbourne, Australia
james.thom@rmit.edu.au

Abstract. Information retrieval from web and XML document collections is ever more focused on returning entities instead of web pages or XML elements. There are many research fields involving named entities; one such field is known as entity ranking, where one goal is to rank entities in response to a query supported with a short list of entity examples. In this paper, we describe our approach to ranking entities from the Wikipedia XML document collection. Our approach utilises the known categories and the link structure of Wikipedia, and more importantly, exploits link co-occurrences to improve the effectiveness of entity ranking. Using the broad context of a full Wikipedia page as a baseline, we evaluate two different algorithms for identifying narrow contexts around the entity examples: one that uses predefined types of elements such as paragraphs, lists and tables; and another that dynamically identifies the contexts by utilising the underlying XML document structure. Our experiments demonstrate that the locality of Wikipedia links can be exploited to significantly improve the effectiveness of entity ranking.

1 Introduction

The traditional entity extraction problem is to extract named entities from plain text using natural language processing techniques or statistical methods and intensive training from large collections. The primary goal is to tag those entities and use the tag names to support future information retrieval. *Entity ranking* has recently emerged as a research field that aims at retrieving entities as answers to a query. Here the goal is not to tag the names of the entities in documents but rather to get back a list of the relevant entity names. It is a generalisation of the expert search task explored by the TREC Enterprise track [14], except that instead of ranking people who are experts in the given topic, other types of entities such as organisations, countries, or locations can also be retrieved and ranked. For example, the query "European countries where I can pay with Euros" should return a list of entities representing relevant countries, and not a list of entities about the Euro and similar currencies.

The Initiative for the Evaluation of XML retrieval (INEX) has a new track on entity ranking, using Wikipedia as its XML document collection [7]. Two

C. Macdonald et al. (Eds.): ECIR 2008, LNCS 4956, pp. 258–269, 2008.

tasks are explored by the INEX 2007 entity ranking track: *entity ranking*, which aims at retrieving entities of a given category that satisfy a topic described in natural language text; and *list completion*, where given a topic text and a small number of entity examples, the aim is to complete this partial list of answers. The inclusion of the target category (in the first task) and entity examples (in the second task) makes these quite different tasks from the task of full-text retrieval, and the combination of the query and entity examples (in the second task) makes it quite different from the task addressed by an application such as Google Sets[1] where only entity examples are provided.

In this paper, we describe our approach to ranking entities from the Wikipedia XML document collection. Our approach is based on the following principles:

1. A good entity page is a page that answers the query (or a query extended with entity examples).
2. A good entity page is a page associated with categories close to the categories of the entity examples.
3. A good entity page is pointed to by a page answering the query; this is an adaptation of the HITS [10] algorithm to the problem of entity ranking.
4. A good entity page is pointed to by contexts with many occurrences of the entity examples. A broad context could be the full page that contains the entity examples, while smaller and more narrow contexts could be elements such as paragraphs, lists, or tables.

Specifically, we focus on whether the locality of Wikipedia links around entity examples can be exploited to improve the effectiveness of entity ranking.

2   Related Work

In this section, we review some related work on link analysis and entity disambiguation and extraction.

Link analysis. To calculate the similarity between a document and a query, most information retrieval (IR) systems use statistical information concerning the distribution of the query terms, both within the document and the collection as a whole. However, in hyperlinked environments, such as the World Wide Web and Wikipedia, link analysis is important. PageRank [3] and HITS [10] are two of the most popular algorithms that use link analysis to improve web search.

We use the idea behind PageRank and HITS in our approach; however, instead of counting every possible link referring to an entity page in the collection (as with PageRank), or building a neighbourhood graph (as with HITS), we only consider pages that are pointed to by a selected number of top-ranked pages for the query. This also makes our link ranking algorithm to be query-dependent (just like HITS), which allows for it to be dynamically calculated at query time.

---

[1] http://labs.google.com/sets

Cai et al. [4] recognise that most popular linkrank algorithms treat a web page as a single node, despite the fact that the page may contain multiple semantic blocks. Using the visual presentation of a page to extract the semantic structure, they adapted PageRank and HITS to deal with block nodes rather than full page nodes. Nie et al. [12] propose a topical link analysis model that formalises the idea of splitting the credit (the authority score) of a source page into different topics based on topical distribution. Our entity ranking approach is based on a similar idea, except that instead of using topics for discrimination we use list-like contexts around the entity examples.

Entity disambiguation and extraction. Kazama and Torisawa [9] explore the use of Wikipedia as external knowledge to improve named entity recognition, by using the first sentence of a Wikipedia page to infer the category of the entity attached to that page. These categories are then used as features in their named entity tagger. We do not use inferred categories in our approach; instead, we use categories that were explicitly associated with the entity page by Wikipedia authors. Cucerzan [6] also uses Wikipedia for entity disambiguation by exploiting (amongst other features) co-references in static contexts such as titles, links, paragraphs and lists. Callan and Mitamura [5] investigate if the entity extraction rules can be dynamically generated. Their rules are based on heuristics exploiting a few pre-defined HTML contexts such as lists and tables. The contexts are weighted according to the number of contained examples; the best contexts are then used to dynamically extract new data. We use pre-defined contexts in our entity ranking approach; however, we also develop a new algorithm that dynamically determines the contexts around entity examples.

ESTER [2] was recently proposed as a system for searching text, entities and relations. ESTER relies on the Wikipedia links to identify the entities and on the context of the links for disambiguation (using 20 words around the anchor text instead of just the anchor text). This approach primarily focuses on improving the efficiency of the proposed system, while we are more interested in improving the effectiveness of entity ranking.

3   The Wikipedia XML Document Collection

Wikipedia is a well known web-based, multilingual, free content encyclopedia written collaboratively by contributors from around the world. As it is fast growing and evolving it is not possible to use the actual online Wikipedia for experiments, and so we need a stable collection to do evaluation experiments that can be compared over time. Denoyer and Gallinari [8] have developed an XML-based corpus based on a snapshot of the Wikipedia, which has been used by various INEX tracks in 2006. It differs from the real Wikipedia in some respects (size, document format, category tables), but it is a very realistic approximation.

Entities in Wikipedia. In Wikipedia, an entity is generally associated with an article (a Wikipedia page) describing this entity. For example, there is a page

"The **euro** ... is the official currency of the Eurozone (also known as the Euro Area), which consists of the European states of Austria, Belgium, Finland, France, Germany, Greece, Ireland, Italy, Luxembourg, the Netherlands, Portugal, Slovenia and Spain, and will extend to include Cyprus and Malta from 1 January 2008."

Fig. 1. Extract from the Euro Wikipedia page

for every country, most famous people or organisations, places to visit, and so forth. In Wikipedia nearly everything can be seen as an entity with an associated page.

The entities have a name (the name of the corresponding page) and a unique ID in the collection. When mentioning such an entity in a new Wikipedia article, authors are encouraged to link every occurrence of the entity name to the page describing this entity. For example, in the Euro page (see Fig. 1), all the under-lined hypertext links can be seen as occurrences of entities that are each linked to their corresponding pages. In this figure, there are 18 entity references of which 15 are country names; more specifically, these countries are all "European Union member states", which brings us to the notion of category in Wikipedia.

Categories in Wikipedia. Wikipedia offers categories that authors can associate with Wikipedia pages. New categories can also be created by authors, although they have to follow Wikipedia recommendations in both creating new categories and associating them with pages. When searching for entities it is natural to take advantage of the Wikipedia categories since they would give a hint on whether the retrieved entities are of the expected type. For example, when looking for entities "authors", pages associated with the category "Novelist" may be more relevant than pages associated with the category "Book".

4   Our Entity Ranking Approach

We are addressing the task of ranking entities in answer to a query supplied with a few examples (task 2). However, our approach can also be used for entity ranking tasks where the category of the target entities is given and no examples are provided (task1).

Our approach to identifying and ranking entities combines: (1) the full-text similarity of the entity page with the query; (2) the similarity of the page's categories with the categories of the entity examples; and (3) the link contexts found in the top ranked pages returned by a search engine for the query.

4.1   Architecture

Our entity ranking approach involves several modules and functions that are used for processing a query, submitting it to the search engine, applying our entity ranking algorithms, and finally returning a ranked list of entities. We used Zettair[2] as our choice for a full-text search engine. Zettair is a full-text IR system developed by RMIT University, which returns pages ranked by their similarity scores to the query. Zettair is "one of the most complete engines" according to a recent comparison of open source search engines [11]. We used the Okapi BM25 similarity measure which was shown to be very effective on the INEX 2006 Wikipedia test collection [1].

The architecture of our approach is described as follows. The *topic* module takes an INEX topic as input and generates the corresponding Zettair query and the list of entity examples (as one option, the names of the entity examples may be added to the query). The *search* module sends the query to Zettair and returns a list of scored Wikipedia pages. The *link extraction* module extracts the links to target entities from a selected number of highly ranked pages, together with the information about the paths of the links (using an XPath notation). The *linkrank* module calculates a weight for a target entity based on (amongst other factors) the number of links to this entity and the number of entity examples that appear in the context of the link. The *category similarity* module calculates a weight for a target entity based on the similarity of its categories with that of the entity examples. The *full-text IR* module calculates a weight for a target entity based on its initial Zettair score. Finally, the global score for a target entity is calculated as a linear combination of three normalised scores coming out of the last three modules.

The above architecture provides a general framework for entity ranking which allows for replacing some modules by more advanced modules, or by providing a more efficient implementation of a module. It also uses an evaluation module to assist in tuning the modules by varying the parameters and to globally evaluate the entity ranking approach.

4.2   Score Functions and Parameters

The global score of an entity page is derived by combining three separate scores: a linkrank score, a category score, and a full-text similarity score.

LinkRank score. The linkrank function calculates a score for a page, based on the number of links to this page, from the first N pages returned by the search engine in response to the query. The parameter N has been kept to a relatively small value mainly for performance purposes, since Wikipedia pages contain many links that would need to be extracted. We carried out experiments with different values of the parameter N, by varying it between 5 and 100 with a step of 5, and found that N=20 was a good compromise between performance and discovering more potentially good entities.

The linkrank function can be implemented in a variety of ways; we have implemented a linkrank function that, for a target entity page t, takes into account the Zettair score of the referring page $z(p)$, the number of distinct entity

---

[2] http://www.seg.rmit.edu.au/zettair/

examples in the referring page $\#ent(p)$, and the locality of links around the entity examples:

$$S_L(t) = \sum_{r=1}^{N} \left(z(p_r) \cdot g(\#ent(p_r)) \cdot \sum_{l_t \in L(p_r,t)} f(l_t, c_r | c_r \in C(p_r)) \right) \quad (1)$$

where $g(x) = x + 0.5$ (we use 0.5 to allow for cases where there are no entity examples in the referring page); l_t is a link that belongs to the set of links $L(p_r, t)$ that point to the target entity t from the page p_r; c_r is a context around entity examples that belongs to a set of contexts $C(p_r)$ found for the page p_r; and $f(l_t, c_r)$ represents the weight associated to the link l_t that belongs to the context c_r. The contexts are explained in full detail in sub-section 4.3.

The weighting function $f(l_r, c_r)$ is represented as follows:

$$f(l_r, c_r) = \begin{cases} 1 & \text{if } c_r = p_r \text{ (the context is the full page)} \\ 1 + \#ent(c_r) & \text{if } c_r = e_r \text{ (the context is an XML element)} \end{cases}$$

Category similarity score. To calculate the category similarity score, we use a very basic similarity function that computes the ratio of common categories between the set of categories associated with the target page $\mathsf{cat}(t)$ and the set of the union of the categories associated with the entity examples $\mathsf{cat}(E)$:

$$S_C(t) = \frac{|\mathsf{cat}(t) \cap \mathsf{cat}(E)|}{|\mathsf{cat}(E)|} \quad (2)$$

Z score. The full-text (Z) score assigns the initial Zettair score to a target entity page. If the target entity does not appear among the initial ranked list of pages returned by Zettair, then its Z score is zero:

$$S_Z(t) = \begin{cases} z(t) & \text{if page } t \text{ was returned by Zettair} \\ 0 & \text{otherwise} \end{cases} \quad (3)$$

Global score. The global score $S(t)$ for a target entity page is calculated as a linear combination of three scores, the linkrank score $S_L(t)$, the category similarity score $S_C(t)$, and the Z score $S_Z(t)$:

$$S(t) = \alpha S_L(t) + \beta S_C(t) + (1 - \alpha - \beta) S_Z(t) \quad (4)$$

where α and β are two parameters that can be tuned differently depending on the entity retrieval task.

We consider some special cases that allow us to evaluate the effectiveness of each module: $\alpha = 1, \beta = 0$, which uses only the linkrank score; $\alpha = 0, \beta = 1$, which uses only the category score; and $\alpha = 0, \beta = 0$, which uses only the Z score.[3] More combinations of the two parameters are explored in the tuning phase of our approach (section 5).

4.3   Exploiting Locality of Links

The main assumption behind the idea of exploiting locality of links in entity ranking is that references to entities (links) located in close proximity to the entity examples, which typically appear in list-like contexts, are more likely to represent relevant entities than links that appear in other parts of the page. Here, the notion of *list* refers to grouping together objects of the same (or similar) nature. The aim is therefore to assign a bigger weight to links that co-occur with links to entity examples in such list-like contexts.

Consider the example of the Euro page shown in Fig. 1. Let us assume that the topic is "European countries where I can pay with Euros", and France, Germany and Spain are three entity examples. We see that the 15 countries that are members of the Eurozone are all listed in the same paragraph with the three entity examples. In fact, there are other contexts in this page where those 15 countries also co-occur together. By contrast, although there are a few references to the United Kingdom in the Euro page, it does not occur in the same context as the three examples (except for the page itself).

Statically defined contexts. We have identified three types of elements that correspond to list-like contexts in the Wikipedia XML document collection: paragraphs (tag `p`); lists (tags `normallist`, `numberlist`, and `definitionlist`); and tables (tag `table`). We design two algorithms for identifying the static contexts: one that identifies the context on the basis of the leftmost occurrence of the pre-defined tags (`StatL`), and another that uses the rightmost occurrence of the pre-defined tags to identify the context (`StatR`). We do this to investigate whether the recursive occurrences of the same tag, as often found in many XML documents in the INEX Wikipedia collection, has an impact on the ability to better identify relevant entities.

Consider Table 1, where the links to entity examples are identified by their absolute XPath notations. The three non-overlapping elements that will be identified by the `StatL` algorithm are the elements `p[1]`, `p[3]`, and `normallist[1]`, while with the `StatR` algorithm `p[5]` will be identified instead of `p[3]` in addition to also identifying the other two elements.

The main drawback of the static approach is that it requires a pre-defined list of element contexts which is totally dependent on the document collection. The advantage is that, once defined, the list-like contexts are easy to identify.

Dynamically defined contexts. To determine the contexts dynamically, we adapted the concept of *coherent retrieval elements* [13] initially used to identify

---

[3] This is not the same as the plain Zettair score, as apart from target entities corresponding to the highest N pages returned by Zettair, the remaining entities are all generated by extracting links from these pages, which may or may not correspond to the ranked pages returned by Zettair.

Table 1. List of links referring to entity examples (France, Germany, and Spain), extracted for the Euro topic

Page		Links		
ID	Name	XPath	ID	Name
9472	Euro	/article[1]/body[1]/p[1]/collectionlink[7]	10581	France
9472	Euro	/article[1]/body[1]/p[1]/collectionlink[8]	11867	Germany
9472	Euro	/article[1]/body[1]/p[1]/collectionlink[15]	26667	Spain
9472	Euro	/article[1]/body[1]/p[3]/p[5]/collectionlink[6]	11867	Germany
9472	Euro	/article[1]/body[1]/normallist[1]/item[4]/collectionlink[1]	10581	France
9472	Euro	/article[1]/body[1]/normallist[1]/item[5]/collectionlink[2]	11867	Germany
9472	Euro	/article[1]/body[1]/normallist[1]/item[7]/collectionlink[1]	26667	Spain
9472	Euro	/article[1]/body[1]/normallist[1]/item[8]/collectionlink[1]	26667	Spain

the appropriate granularity of elements to return as answers in XML retrieval.

For the list of extracted entities corresponding to entity examples, a *Coherent Retrieval Element* (CRE) is defined as an element that represents the *lowest common ancestor* (LCA) of at least two entity examples. To identify the CREs, we sequentially process the list of extracted entity examples by considering every pair of elements, starting from the first element down to the element preceding the last element in the list. For each pair of elements, their LCA is chosen to represent a dynamic context (a CRE). Starting from the first identified CRE, we filter the overlapping elements and end up with a final list of (one or more) non-overlapping CREs that represent the dynamically defined contexts for the page.[4] We refer to this algorithm as DynCRE.

For example, the two dynamic contexts that will be identified for the list of extracted entity examples shown in Table 1 are p[1] and normallist[1]. Although body[1] was also initially identified as a CRE, it was subsequently filtered from the final list since it overlaps with p[1] (the first identified CRE).

The main advantage of the dynamic approach is that it is independent of the document collection, and it does not require a pre-defined list of contexts. The possible drawback is that narrow contexts containing only one entity example (such as p[5] in Table 1) are never identified.

5   Experimental Results

We now present results that investigate the effectiveness of our entity ranking approach when using different types of contexts around the entity examples.

5.1   Test Collection

Since there was no existing set of topics with relevance assessments for entity ranking, we developed our own test collection, which we made available as a

---

[4] When the page contains *exactly one* entity example, the document element (article[1]) is chosen to represent a CRE.

training set for other participants in the INEX 2007 entity ranking track. So for these experiments we used our own test collection based on a selection of topics from the INEX 2006 ad hoc track, since most of these topics reflect real-life tasks represented by queries very similar to the short Web queries. We chose 27 topics that we considered were of an "entity ranking" nature, where for each page that had been assessed as containing relevant information, we reassessed whether or not it was an entity answer, and whether it *loosely* belonged to a category of entity we had *loosely* identified as being the target of the topic. If there were entity examples mentioned in the original topic these were usually used as entity examples in the entity topic. Otherwise, a selected number (typically 2 or 3) of entity examples were chosen somewhat arbitrarily from the relevance assessments. To this set of 27 topics we also added the Euro topic example that we had created by hand from the original INEX description of the entity ranking track [7], resulting in total of 28 entity ranking topics.

We use mean average precision (MAP) as our primary method of evaluation, but also report results using several alternative IR measures: mean of P[5] and P[10] (mean precision at top 5 or 10 entities returned), and mean R-precision. We remove the entity examples both from the list of returned answers and from the relevance assessments, as the task is to find entities other than those provided.

5.2   Full Page Context

We used the context of the full page to determine suitable values for the parameters α and β, and also to try out some minor variations to our entity ranking approach (such as whether or not to include the names of the entity examples in the query sent to Zettair).

We calculated MAP over the 28 topics in our test collection, as we varied α from 0 to 1 in steps of 0.1. For each value of α, we also varied β from 0 to $(1-\alpha)$ in steps of 0.1. We found that the highest MAP (0.3570) on this data set is achieved for $\alpha = 0.1$ and $\beta = 0.8$. We also trained using mean R-precision instead of MAP as our evaluation measure, but we also observed the same optimal values for the two parameters.

We used a selected number of runs to carry out a more detailed investigation of the performance achieved by each independent module and by the optimal module combination. We also investigated whether adding names of the entity examples to the query sent to Zettair would have a positive performance impact. The results of these investigations are shown in Tables 2(Q) and 2(QE).

Several observations can be drawn from these results. First, adding names of the entity examples to the query sent to Zettair generally performs worse for all but the linkrank module, for which we see a consistent performance improvement. Second, different optimal values are observed for the two parameters in the two tables, which suggest that adding the entity examples to the query can dramatically influence the retrieval performance. Third, we observe that the best entity ranking approaches are those that combine the ranking evidence from the three modules (runs $\alpha 0.1$–$\beta 0.8$ for Q and $\alpha 0.2$–$\beta 0.6$ for QE). With MAP, these two runs perform significantly better ($p < 0.05$) than the plain Zettair full-text

Table 2. Performance scores for runs using the context of the full page, obtained by different evaluation measures. Queries sent to Zettair include only terms from the topic title (Q), or terms from the topic title and the names of entity examples (QE). For each measure, the best performing score is shown in bold.

Run	P[r] 5	10	R-prec	MAP	Run	P[r] 5	10	R-prec	MAP
Zettair	0.2286	0.2321	0.2078	0.1718	Zettair	0.2000	0.1714	0.1574	0.1427
$\alpha 0.0$–$\beta 0.0$	0.2286	0.2321	0.2135	0.1780	$\alpha 0.0$–$\beta 0.0$	0.2000	0.1714	0.1775	0.1533
$\alpha 0.0$–$\beta 1.0$	0.3643	0.3071	0.3151	0.3089	$\alpha 0.0$–$\beta 1.0$	0.3357	0.2821	0.2749	0.2674
$\alpha 1.0$–$\beta 0.0$	0.1571	0.1571	0.1385	0.1314	$\alpha 1.0$–$\beta 0.0$	0.1857	0.1750	0.1587	0.1520
$\alpha 0.1$–$\beta 0.8$	**0.4714**	**0.3857**	**0.3902**	**0.3570**	$\alpha 0.1$–$\beta 0.8$	0.3357	0.3286	0.3109	0.3140
$\alpha 0.2$–$\beta 0.6$	0.4357	0.3786	0.3751	0.3453	$\alpha 0.2$–$\beta 0.6$	**0.3429**	**0.3357**	**0.3362**	**0.3242**
(Q) Topic title					(QE) Topic title and entity examples				

retrieval run, and they are also significantly better than any of the three runs representing each individual module in our entity ranking approach.

These results therefore show that the global score (the combination of the three individual scores), optimised in a way to give more weight on the category score, brings the best value in retrieving the relevant entities for the INEX Wikipedia document collection. However, the results also show that using only the linkrank module and the context of the full page results in a very poor entity ranking strategy, which is why below we also experiment with narrow contexts.

5.3 Static and Dynamic Contexts

We now investigate whether using smaller and more narrow contexts has a positive impact on the effectiveness of entity ranking. Tables 3(Q) and 3(QE) show the results of this investigation. These results reflect the case when only the linkrank module ($\alpha 1.0$–$\beta 0.0$) is used by our entity ranking approach.

As in the case with using the full page context, for all the four runs we observe a consistent performance improvement when names of the entity examples are added to the query sent to Zettair. Importantly, when compared to the baseline (the full page context), we observe a substantial increase in performance for the three runs that use smaller and more narrow contexts, irrespective of the type of query used. These increases in performance are all statistically significant ($p < 0.05$). However, the type of query sent to Zettair (Q or QE) seems to have an impact on the best performance that could be achieved by these three runs. Specifically, with MAP the `StatL` run performs best among the three runs when only the topic title is used as an input query (Q), while the `StatR` run is best when using terms from the topic title and the names of entity examples (QE). In both cases the `DynCRE` run achieves the best early precision but overall performs worst among the three runs, although the differences in performance between each of the three run pairs are not statistically significant.

Table 3. Performance scores for runs using different types of contexts in the linkrank module ($\alpha1.0$–$\beta0.0$), obtained by different evaluation measures. Queries sent to Zettair include only terms from the topic title (Q), or terms from the topic title and the names of entity examples (QE). For each measure, the best performing score is shown in bold.

Run	P[r] 5	10	R-prec	MAP
FullPage	0.1571	0.1571	0.1385	0.1314
StatL	0.2143	**0.2250**	**0.2285**	**0.1902**
StatR	**0.2214**	0.2143	0.2191	0.1853
DynCRE	**0.2214**	0.2107	0.2152	0.1828

(Q) Topic title

Run	P[r] 5	10	R-prec	MAP
FullPage	0.1857	0.1750	0.1587	0.1520
StatL	0.2429	0.2179	**0.2256**	0.2033
StatR	0.2429	**0.2214**	0.2248	**0.2042**
DynCRE	**0.2571**	0.2107	0.2207	0.1938

(QE) Topic title and entity examples

Implementing narrow contexts in our linkrank module allows for the locality of links to be exploited in entity ranking. By changing the context around entity examples, we would also expect the optimal values for the two combining parameters to change. We therefore varied the values for α and β and re-calculated MAP over the 28 topics in our test collection. For the three runs using narrow contexts we found that the optimal value for α has shifted from 0.1 to 0.2 (in the case of Q), while for the two static runs the optimal α value was 0.3 (in the case of QE). In both cases, the optimal value for β was found to be 0.6. The performances of the three optimal runs were very similar, and all of them substantially outperformed the optimal run using the full page context.

6   Conclusion and Future Work

We have presented our entity ranking approach for the INEX Wikipedia XML document collection which is based on exploiting the interesting structural and semantic properties of the collection. We have shown in our evaluations that the use of the categories and the locality of Wikipedia links around entity examples has a positive impact on the performance of entity ranking.

In the future, we plan to further improve our linkrank algorithm by varying the number of entity examples and incorporating relevance feedback that we expect would reveal other useful entities that could be used to identify better contexts. We also plan to carry out a detailed per-topic error analysis, which should allow us to determine the effect of the topic type on entity ranking. Finally, our active participation in the INEX entity ranking track will enable us to compare the performance of our entity ranking approach to those achieved by other state-of-the-art approaches.

Acknowledgements

Part of this work was completed while James Thom was visiting INRIA in 2007.

References

1. AwangIskandar, D., Pehcevski, J., Thom, J.A., Tahaghoghi, S.M.M.: Social media retrieval using image features and structured text. In: Fuhr, N., Lalmas, M., Trotman, A. (eds.) INEX 2006. LNCS, vol. 4518, pp. 358–372. Springer, Heidelberg (2007)
2. Bast, H., Chitea, A., Suchanek, F., Weber, I.: ESTER: efficient search on text, entities, and relations. In: Proceedings of the 30th ACM International Conference on Research and Development in Information Retrieval, Amsterdam, The Netherlands, pp. 671–678 (2007)
3. Brin, S., Page, L.: The anatomy of a large-scale hypertextual Web search engine. In: Proceedings of the 7th International Conference on World Wide Web, Brisbane, Australia, pp. 107–117 (1998)
4. Cai, D., He, X., Wen, J.-R., Ma, W.-Y.: Block-level link analysis. In: Proceedings of the 27th ACM International Conference on Research and Development in Information Retrieval, Sheffield, UK, pp. 440–447 (2004)
5. Callan, J., Mitamura, T.: Knowledge-based extraction of named entities. In: Proceedings of the 11th ACM Conference on Information and Knowledge Management, McLean, Virginia, pp. 532–537 (2002)
6. Cucerzan, S.: Large-scale named entity disambiguation based on Wikipedia data. In: Proceedings of the 2007 Joint Conference on EMNLP and CoNLL, Prague, The Czech Republic, pp. 708–716 (2007)
7. de Vries, A.P., Thom, J.A., Vercoustre, A.-M., Craswell, N., Lalmas, M.: INEX 2007 Entity ranking track guidelines. In: INEX 2006, pp. 481–486 (2007)
8. Denoyer, L., Gallinari, P.: The Wikipedia XML corpus. SIGIR Forum 40(1), 64–69 (2006)
9. Kazama, J., Torisawa, K.: Exploiting Wikipedia as external knowledge for named entity recognition. In: Proceedings of the 2007 Joint Conference on EMNLP and CoNLL, Prague, The Czech Republic, pp. 698–707 (2007)
10. Kleinberg, J.M.: Authoritative sources in hyperlinked environment. Journal of the ACM 46(5), 604–632 (1999)
11. Middleton, C.,Baeza-Yates, R.: A comparison of open source search engines. Technical report, Universitat Pompeu Fabra, Barcelona, Spain (2007), http://wrg.upf.edu/WRG/dctos/Middleton-Baeza.pdf
12. Nie, L., Davison, B.D., Qi, X.: Topical link analysis for web search. In: Proceedings of the 29th ACM International Conference on Research and Development in Information Retrieval, Seattle, Washington, pp. 91–98 (2006)
13. Pehcevski, J., Thom, J.A., Vercoustre, A.-M.: Hybrid XML retrieval: Combining information retrieval and a native XML database. Information Retrieval 8(4), 571–600 (2005)
14. Soboroff, I., de Vries, A.P., Craswell, N.: Overview of the TREC 2006 Enterprise track. In: Proceedings of the Fifteenth Text REtrieval Conference (TREC 2006), pp. 32–51 (2006)

The Importance of Link Evidence in Wikipedia

Jaap Kamps[1,2] and Marijn Koolen[1]

[1] Archives and Information Studies, University of Amsterdam, The Netherlands
[2] ISLA, University of Amsterdam, The Netherlands

Abstract. Wikipedia is one of the most popular information sources on the Web. The free encyclopedia is densely linked. The link structure in Wikipedia differs from the Web at large: internal links in Wikipedia are typically based on words naturally occurring in a page, and link to another semantically related entry. Our main aim is to find out if Wikipedia's link structure can be exploited to improve ad hoc information retrieval. We first analyse the relation between Wikipedia links and the relevance of pages. We then experiment with use of link evidence in the focused retrieval of Wikipedia content, based on the test collection of INEX 2006. Our main findings are: First, our analysis of the link structure reveals that the Wikipedia link structure is a (possibly weak) indicator of relevance. Second, our experiments on INEX ad hoc retrieval tasks reveal that if the link evidence is made sensitive to the local context we see a significant improvement of retrieval effectiveness. Hence, in contrast with earlier TREC experiments using crawled Web data, we have shown that Wikipedia's link structure can help improve the effectiveness of ad hoc retrieval.

1 Introduction

Wikipedia is a free Web-based encyclopedia, that is collaboratively edited by countless individuals around the globe [1]. As an encyclopedia, it consists of individual entries on a single subject (Wiki pages) that are densely hyperlinked to related content (using Wikilinks). Wikipedia's links are a special case of the general hyperlinks that connect the World Wide Web. On the Web, link structure has been exploited to improve information retrieval in algorithms like PageRank [2] and HITS [3]. In the simplest case of a link from a page A to a page B, we can count this as a vote by the author of page A for page B as being authoritative [3]. Pages with a high number of incoming links are considered important pages. Commercial Internet search engine companies have heralded the use of link structure as one of their key technologies.

Retrieval using Web data has been studied at TREC since TREC-8 in 1999. Despite high expectations, TREC experiments failed to establish the effectiveness of link evidence for general ad hoc retrieval [e.g., 4; 5]. As Hawking and Craswell [6, p.215] put it:

> Hyperlink and other web evidence is highly valuable for some types of search task, but not for others. Because binary judgements were employed and judges looked only at the text of the retrieved pages, the TREC-8 Small Web Task

C. Macdonald et al. (Eds.): ECIR 2008, LNCS 4956, pp. 270–282, 2008.
© Springer-Verlag Berlin Heidelberg 2008

and the TREC-9 Main Web Task did not accurately model typical Web search.
. . .

In prototypical TREC Ad Hoc methodology, the task presupposes a desire to
read text relevant to a fairly precisely defined topic, and documents are judged
on their own text content alone as either relevant or not relevant. By contrast,
Web searchers typically prefer the entry page of a well-known topical site to
an isolated piece of text, no matter how relevant. For example, the NASA
home page would be considered a more valuable answer to the query "space
exploration" than newswire articles about Jupiter probes or NASA funding
cuts.

These observations led to the definition of a range of Web-centric tasks, like
known-item (homepage, named-page) search and topic distillation. For these
special tasks, the URL-type, anchor text and link indegree are effective to improve retrieval performance [7; 8; 9].

Our conjecture is that the links in Wikipedia are different from links between
Web documents. Whereas in Web documents, an author can arbitrarily link his
page to any other page, whether there is a topical relation or not, in Wikipedia,
links tend to be semantic: a link from page A to page B shows that page B is
semantically related to (part of) the content of page A. Arguably, there will be
some fraction of links that do not denote an important topical relation between
pages, and not all links will be equally meaningful in all search contexts. For
example, Wikipedia "bots" may automatically insert links serving a particular
purpose (think of the year links). However, there is a clear mechanism in place
that results in links that are relevant to the context.[1] This immediately prompts
the question: what is the importance of link evidence in such a semantically
linked collection?

Given its encyclopedic content, Wikipedia is a particularly attractive resource
for informational search requests.[2] Hence, in this paper, our main aim is to find
out if Wikipedia's link structure can be exploited to improve the ad hoc retrieval
of relevant information. Our experimental evidence is based on the INEX 2006
test collection consisting of an XML version of Wikipedia containing over 650,000
articles, and a set of 114 ad hoc topics with judgments on the passage-level [11].
Specifically, we want to know:

- Can the link degree structure of a semantically linked document collection
  be used as evidence for the relevance of ad hoc retrieval results?

To answer this question, we analyse the link structure of Wikipedia pages, and
of Wikipedia pages relevant for a particular ad hoc retrieval topic. Furthermore,
link structure can be considered in various ways: on a global level, i.e., the
number of incoming links over the whole collection, or on a local level, i.e., the

---

[1] See, for example, `http://en.wikipedia.org/wiki/Wikipedia:Only_make_links_that_are_relevant_to_the_context`.

[2] That is, in terms of the Broder [10] taxonomy, Wikipedia seems a less suitable
resource for navigational queries (with the intent to reach a particular site), or for
transactional queries (with the intent to perform some web-mediated activity).

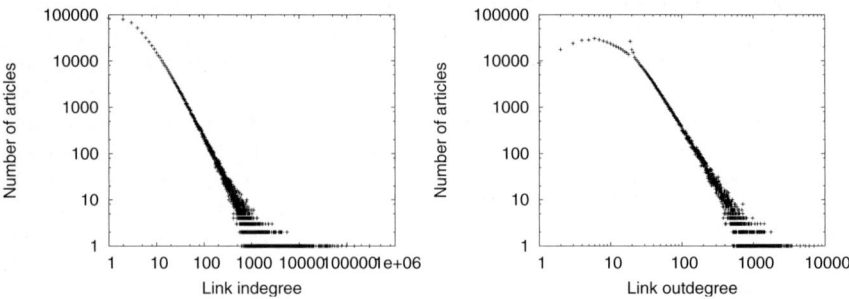

Fig. 1. Wikipedia collection link indegree (left) and outdegree (right) distribution over 659,304 pages

number of incoming links within the subset of articles retrieved as results for a given topic. So, more specifically, we want to know:

- Can global indegree, local indegree, or a combination of the two, be used effectively to rank results in general ad-hoc retrieval?

To answer this question, we investigate the effectiveness of a number of link degree priors for three different INEX ad hoc retrieval tasks.

The rest of this paper is structured as followed. Next, in Section 2, we analyse the link structure of Wikipedia, trying to establish its relation with relevance. Then, in Section 3, we detail on how link evidence can be incorporated into the scoring of a retrieval system. This is followed, in Section 4, by a range of retrieval experiments trying to establish the impact of link evidence on retrieval effectiveness. We end in Section 5 by drawing conclusions and discussing our findings.

2   Analysis of Wikipedia Link Structure

In this section, we analyse the link structure of Wikipedia. We use the XML'ified snapshot of the English Wikipedia (of early 2006) used at INEX [12]. We use the set of 114 ad hoc topics from INEX 2006, with their associated relevance judgments. In this section, we only consider Wikipedia pages and the internal links ignoring links to pages outside Wikpedia.

2.1   Degree Distribution

Is the link structure of Wikipedia different from the link structure of the Web? Recall from the above, links in Wikipedia are unlike generic Web links. Does the encyclopedic organization, where there is little redundant information, put a bound on the number of incoming links? Does the organization in mono-topical entries or lemmas restrict the number of outgoing links? We look at the number of different incoming links (indegree) and the number of outgoing links (outdegree). Figure 1 shows the degree distribution of Wikipedia. The indegree or number of

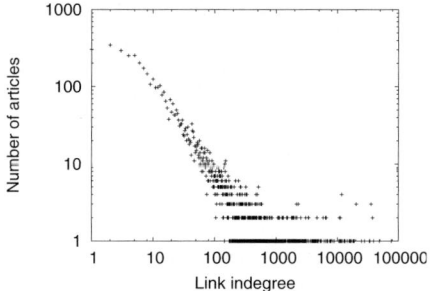

Fig. 2. Wikipedia collection link indegree distribution of 5,646 "relevant" pages

incoming links is shown on the left hand side, and the outdegree or number of outgoing links is shown on the right hand side. Both curves approximate straight lines on the log-log scale, suggesting a power-law distribution that is familiar for the Web at large [13]. In the rest of this paper, we focus on indegrees.

What is the degree distribution of relevant pages? The relevance assessments are at the passage level; we treat a Wikipedia page as relevant for a given topic if, and only if, it contains relevant information. Figure 2 plots the degree distribution for the subset of articles relevant for a INEX 2006 topic. Although there are far fewer data points, we see a similar distribution. There is no absolute evidence in the link indegree: both low indegree and high indegree pages can be relevant.

2.2   Local Degree Distribution

So far, we have looked at global evidence provided by the absolute number of links. We now zoom in on local evidence provided by the number of links among a subset of local pages. We used a standard retrieval system (discussed in detail in Section 4 below) to find the top 100 best matching Wikipedia articles for each of the INEX 2006 topics. We treat these pages as local context, and only consider links between pages in this subset and ignore all further links.

Recall again our conjecture that links in Wikipedia are semantic links. By restricting our view to the local context, a large fraction of these local links should relate to the topic at hand. Is this local structure different from the global link structure investigated above? The left hand side of Figure 3 shows the local degree distribution of pages in the local context of any of the INEX 2006 topics. Again, the plot suggests a power-law distribution. On the right hand side of Figure 3, we zoom in on only those articles which are relevant for any of the INEX topics. Also here we see a similar distribution. This also shows that local indegree is no absolute evidence of relevance: both low local indegree as well as high local indegree pages can be relevant.

2.3   Prior Probability of Relevance

Above, we saw that neither global nor local indegree provides absolute evidence of relevance. But can global or local indegree be used as a (possibly weak)

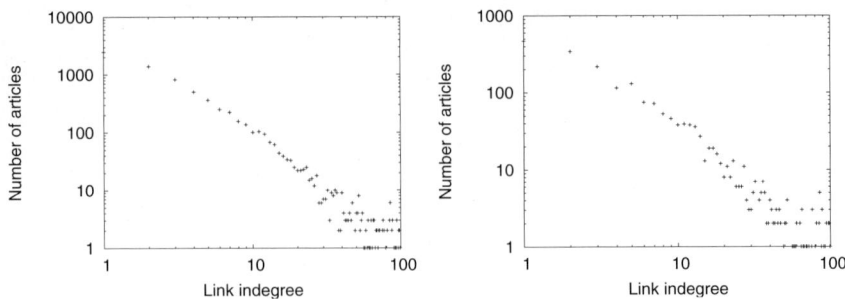

Fig. 3. Wikipedia local link indegree distribution of 11,339 local pages (left) and of 2,489 local relevant pages (right)

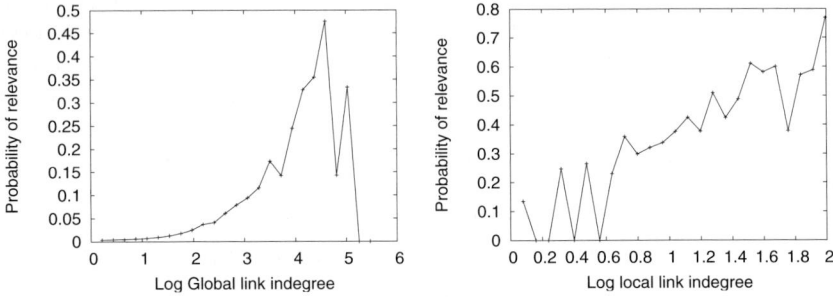

Fig. 4. Prior probability of relevance of Wikipedia global indegree (left) and local indegree (right)

indicator of relevance? That is, if we would know nothing more of a page than its global or local indegree, can we make an educated guess about the relevance of the page?

For a page of a given indegree, we can calculate the prior probability that it is relevant (with respect to at least one of the INEX topics). Specifically, we calculate the fraction of pages with that indegree that is relevant to any of the topics. To overcome data sparseness, we group the indegrees in bins for which we use an exponential scale. The left hand side of Figure 4 shows the prior probability of relevance of global indegree. We see a clear increase in the prior probability of relevance with increasing global indegree. Although there are more relevant pages with a low indegree (as was shown in Figure 2), this number is dwarfed by the total number of pages with a low indegree (as shown in Figure 1), leading to a relatively low prior probability of relevance. Conversely, although the number of relevant pages with a high indegree is modest, this is still a substantial fraction of all the pages with a high indegree—up to a third of these pages is relevant for at least one of the INEX topics.

We do the same analysis for the local indegree, shown on the right hand side of Figure 4. The prior probability of relevance also clearly increases with local

Table 1. Top 10 Wikipedia articles for topic 339 "Toy Story"

Title	Global indegree	Title	Local indegree
1990s	10,033	Toy Story	96
Screenwriter	762	Toy Story 2	34
Gnosticism	424	Pixar	34
Madeira Islands	339	Monsters, Inc.	25
Psychedelic music	339	Finding Nemo	22
1995 in film	314	Aladdin (1992 film)	14
Computer-generated imagery	310	Madeira Islands	12
Academy Award for Original Music Score	268	Computer-generated imagery	12
Tom Hanks	248	Buzz Lightyear	12
Debian	210	1990s	11

indegree. Again, although the absolute number of relevant pages with a low local indegree is higher (as shown in Figure 3), a larger fraction of pages with a high local indegree is relevant. The prior probability of relevance rises to well above 0.5 for pages with a high local indegree.

2.4 Naive Reranking

We selected one topic to look in detail at what happens to the top results when naively reranked by indegree. Topic 339 has title *Toy Story*, and is about the computer animated movie from 1995. We took the top 100 articles from the baseline run described in Section 3 below, and list the 10 articles with the highest global indegree (on the left) and the 10 highest local indegree (on the right) in Table 1.

The articles with the highest global indegree are at best slightly related to the *Toy Story* movie, but there are also infiltrations like *Screenwriter* and *Gnosticism*—pages that might be important or authoritative, but outside the scope of our topic at hand. The articles with the highest local indegree look much more promising, but still there are a lot of articles that are only weakly related to *Toy Story*. The qualitative analysis suggests that global and local indegree are weak indicators of relevance. Therefore, in reranking, their weight should be small compared to the weight of the content-based retrieval score.

Summarizing, our analysis of the link structure reveals that the Wikipedia link structure is a (possibly weak) indicator of relevance of Wiki pages. A naive reranking based on only global or local indegree is not effective: it leads to the infiltration of important but off-topic pages.

3 Incorporating Link Evidence

In this section, we discuss how link evidence can be incorporated in our retrieval model. Results at INEX can be arbitrary XML elements (such as paragraphs, sections, or the whole article), and we simply index each XML element as a separate document. Link structure is used at the article or Wiki page level, we simply associate each element with the link structure of the article it is part of.

3.1 Retrieval Model

We use a language model extension of Lucene [14], i.e., for a collection D, document d and query q:

$$P(d|q) = P(d) \cdot \prod_{t \in q} ((1 - \lambda) \cdot P(t|D) + \lambda \cdot P(t|d)), \tag{1}$$

where $P(t|d) = \frac{\text{freq}(t,d)}{|d|}$, $P(t|D) = \frac{\text{freq}(t,D)}{\sum_{d' \in D} |d|}$, and $P(d) = \frac{|d|}{\sum_{d' \in D} |d|}$. For our experiments we have used $\lambda = 0.15$ throughout. Our efficient implementation of the model calculates ranking-equivalent logs of the probabilities [15]. We take the exponent to get a score resembling a probability, and only then apply the length-prior.

3.2 Link Degree Priors

We incorporate link evidence by by multiplying the retrieval score with a further link degree prior:

$$Score = Score_{retrieved} \cdot Prior \tag{2}$$

We will, for convenience, refer to the link evidence as prior, even though we do not actually transform it into a probability distribution. Note that we can turn any prior into a probability distribution by multiplying it with a constant factor $\frac{1}{\sum_{d \in D} Prior(d)}$, leading to the same ranking.

Recall from above that we need to be careful when incorporating link evidence. We do not want to retrieve pages that only have a high link score, i.e., pages that may be important but unrelated to the topic of request. Hence, as a safe-guard, we apply the link priors only to the first 100 retrieved articles per topic. That is, we process the list of XML element results until we have encountered 100 articles, and only use the global or local indegree of these articles by simply treating the indegrees of lower ranked articles as zero. Note that elements deep down the result list may still be boosted if they belong to one of the top 100 articles.

3.3 Baseline

Our baseline is the retrieval model without using link evidence. To explain the impact of the link evidence, we look again in detail at Topic 339 and the effects of the priors on the top 10 articles. In the upper left corner of Table 2 the titles of the top 10 retrieved Wikipedia articles for the baseline run are given.

3.4 Global Indegree

The *global indegree prior* is proportional to the global degree of an article:

$$P_{\text{Glob}} \propto 1 + global \tag{3}$$

Table 2. Top 10 Wikipedia articles for topic 339 "Toy Story"

Baseline run	Global indegree prior
List of Disney animated features' titles in various languages	1990s
Toy Story	Screenwriter
Toy Story 2	Gnosticism
Buzz Lightyear	1995 in film
Toy Story 3	Computer-generated imagery
List of computer-animated films	Toy Story
100 Greatest Cartoons	Academy Award for Original Music Score
C64 Direct-to-TV	Tom Hanks
List of Capcom games	Pixar
Pixar	Debian
Local indegree prior	**Local/Global indegree prior**
Toy Story	Toy Story
Toy Story 2	List of Disney animated features' titles in various languages
Pixar	Toy Story 2
Monsters, Inc.	Toy Story 3
Buzz Lightyear	Buzz Lightyear
Finding Nemo	List of computer-animated films
Aladdin (1992 film)	100 Greatest Cartoons
Toy Story 3	Sheriff Woody
Computer-generated imagery	Timeline of CGI in film and television
1990s	Andrew Stanton

Alternatively, we use a conservative *log global indegree prior*:

$$P_{\mathsf{LogGlob}} \propto 1 + \log(1 + global) \qquad (4)$$

Inspecting our running example immediately confirms that we need to be careful when incorporating global link evidence. In the upper right corner of Table 2, the top 10 articles after reranking by global indegree are given. Although some top 10 results are retained, *Toy Story* at rank 6 and *Pixar* at rank 9, we see the infiltration of pages with a high global indegree—all of them are in Table 1—but with only a loose relation to the topic at hand.

3.5   Local Indegree

The *local indegree prior* is proportional to the local degree of an article:

$$P_{\mathsf{Loc}} \propto 1 + local \qquad (5)$$

Alternatively, we use a conservative *log local indegree prior*:

$$P_{\mathsf{LogLoc}} \propto 1 + \log(1 + local) \qquad (6)$$

The local indegree prior (shown in the lower left corner of Table 2) results in the *Toy Story* page at the top rank, thereby improving upon the baseline run. Also,

Table 3. Results of link evidence on three INEX 2006 ad hoc retrieval tasks. Best scores are in bold-face. Significance levels are 0.05 (*), 0.01 (**), and 0.001 (***).

Run ID	Thorough MAep,off		Focused nxCG@10,off		Relevant in Context MAgP	
Baseline	0.0353		0.3364		0.1545	
Global Indegree	0.0267	-24.40***	0.1979	-41.16***	0.1073	-30.57***
Log Global Indegree	0.0335	-4.99	0.3066	-8.87**	0.1352	-12.50***
Local Indegree	0.0405	+14.75*	0.3218	-4.34	0.1467	-5.02*
Log Local Indegree	0.0418	+18.46***	0.3460	+2.85	0.1515	-1.96
Local/Global Indegree	**0.0463**	+31.08***	**0.3629**	+7.88**	**0.1576**	+1.99*

Pixar moves up from rank 10 to 3. However, at rank 10 we see that *1990s*, the article with the highest local indegree in Table 1, has infiltrated the top results.

3.6   Local/Global Indegree

We can also weight the importance of observing local links by their number of global links—basically a *tf · idf* weighting of link evidence [16]. The combined *local/global indegree prior* is calculated as:

$$P_{\mathsf{LocGlob}} \propto 1 + \frac{local}{1 + global} \tag{7}$$

The combination prior (lower right corner of Table 2) improves further on the original top 10 by ranking *Toy Story* as the top articles and moving *Toy Story 3* from rank 5 to 4. Also, some unrelated articles like *C64 Direct-to-TV* and *List of Capcom games* are replaced by closer related articles, *Timeline of CGI in film and television* and *Andrew Stanton* (one of the writers of *Toy Story*).

Summarizing, we defined a number of ways—global, local, and combined local/global indegree—to incorporate link evidence into the retrieval model. The different indegree priors correspond to different levels of sensitivity to the local context of the topic of request.

4   Experimental Results

In this section, we discuss the results of applying the degree priors to three of the INEX 2006 Ad Hoc retrieval Tasks.

4.1   Baseline

Our baseline run is a standard language model run, using an index containing all the XML elements of the Wikipedia XML Collection [17]. The scores of the baseline run are in Table 3.

For the INEX 2006 *Thorough* task, where the aim is to estimate the relevance of individual XML elements, this run scores 0.0353 on the official MAep measure.[3] For the *Focused* task, no overlapping elements are allowed and we post-process the Thorough run using a top-down list-based removal of elements partially overlapping with earlier seen results. That is, we traverse the list top-down, and simply remove any element that is an ancestor or descendant of an element seen earlier in the list. The resulting run scores 0.3364 on the official nxCG@10 measure. Finally, for the *Relevant in Context* task, there is a further restriction that articles may not be interleaved and, again, we post-process the Focused run using a top-down list-based clustering of results per retrieved article. The resulting run scores 0.1545 on the official MAgP measure.

4.2 Global Indegree

Now, we turn our attention to global indegree. The results are negative: both the global indegree prior and the log global indegree prior lead to loss of performance for all three tasks. The decrease in performance is significant (bootstrap test, one-tailed) for all cases except for the log global indegree prior and the Thorough task. A plausible explanation is suggested by looking at Table 2. The original top 10 articles of the baseline run are infiltrated by non-relevant documents with high global indegrees—important pages, but off-topic.

4.3 Local Indegree

Next, we try the local link evidence, and use the (log) local indegree prior. The results are mixed. The local indegree prior leads to a significant gain in performance for Thorough (15%), but a loss for Focused (-4%) and for Relevant in Context (-5%). The more conservative log local indegree prior fares better and leads to a gain in performance for Thorough (18%, significant at $p < 0.001$) and for Focused (3%), but still a loss for Relevant in Context (-2%). Although the scores are much better than for the global indegree, there is still no overall improvement. This may still be due to the infiltration of non-relevant documents with high local indegrees. Since the local indegrees are generally much lower than global indegrees—with $N = 100$ the maximal local indegree is 99—the infiltration effect is also much smaller.

4.4 Local/Global Indegree

Finally, we experiment with the combined prior. Here the situation is quite different. For the Thorough task, we see an improvement of 31%. For the Focused Task, we see and improvement of 8%. For the Relevant in Context Task, we get an improvement of 2%. For all three tasks, the improvement is significant. The combined prior seems to effectively take the local context into account, and is effective for improving ad hoc retrieval.

---

[3] Mean average effort/precision (MAep) is a generalized MAP-like measure; normalized extended cumulative gain (nxCG@10) is resembling to precision at rank 10; mean average generalized precision (MAgP) is a version of MAP with partial scores per article. See [18] for details of the INEX 2006 measures.

Summarizing, we experimented with the use of global, local, and combined local/global link evidence, and found that only the combined local/global prior leads to a significant improvement of retrieval effectiveness for all tasks.

5   Discussion and Conclusions

In this paper, we investigated the importance of link evidence in Wikipedia ad hoc retrieval. The link structure of Wikipedia is an interesting special case of hyperlinking on the Web at large: the links in Wikipedia are semantic—they link to other pages relevant to the local context. Our main aim is to find out if Wikipedia's link structure can be exploited to improve the ad hoc retrieval of relevant information.

Our first research question was:

– Can the degree structure of a semantically linked document collection be used as evidence for the relevance of ad hoc retrieval results?

We analysed the degree structure of Wikipedia pages, and of Wikipedia pages relevant for a particular set of ad hoc retrieval topics. Our findings are that pages with a high global indegree are more likely to be relevant than pages with a low global indegree. Since global link evidence may lead to the retrieval of important but off-topic pages, we also looked at local indegree considering only links between pages retrieved in response to a search request. Also here we saw that pages with a high local indegree are more likely to be relevant than pages with a low local indegree. So the answer to our first research question is yes: the Wikipedia link structure is a (possibly weak) indicator of relevance.

Our second research question was:

– Can global indegree, local indegree, or a combination of the two, be used effectively to rerank results in general ad hoc retrieval?

In order to answer this question, we have to operationalize how to incorporate link evidence into our retrieval model, and then conduct experiments that try to establish its utility. The link topology in itself is not sensitive to the query or local context. Put differently, if we browse following a sequence of links, the similarity to the source page will water down quickly. This is especially true for the global link structure, which may lead quickly to a loss of focus on the topic at hand and allow for the infiltration of authoritative but off-topic pages. The local link structure ensures that only links within the local context are awarded, but the ranking may still suffer from a similar bias on authoritativeness over topicality (although the effect will be less strong). Hence, this leads to a third way in which the number of local links is normalized by the number of global links (basically a $tf \cdot idf$ weighting of link evidence). As it turns out, the use of global link evidence leads to a loss of performance, the use of local link evidence leads to mixed results, but the combined local/global link evidence leads to significant improvement of retrieval effectiveness. So, the answer to our second research question is also yes: if the link evidence is made sensitive to the local context we see an improvement of ad hoc retrieval effectiveness.

Earlier experiments at TREC using crawled Web data have failed to establish the utility of link evidence for ad hoc retrieval. In contrast with these TREC experiments, Wikipedia's link structure can help improve the effectiveness of ad hoc retrieval.

References

[1] Wikipedia: The free encyclopedia (2008), http://en.wikipedia.org/
[2] Page, L., Brin, S., Motwani, R., Winograd, T.: The pagerank citation ranking: Bringing order to the web. Technical report, Stanford Digital Library Technologies Project (1998)
[3] Kleinberg, J.M.: Authoritative structures in a hyperlinked environment. Journal of the ACM 46, 604–632 (1999)
[4] Hawking, D.: Overview of the TREC-9 web track. In: Ninth Text REtrieval Conference (TREC-9), National Institute for Standards and Technology, pp. 87–102. NIST Special Publication 500-249 (2001)
[5] Kraaij, W., Westerveld, T.: How different are web documents? In: Proceedings of the ninth Text Retrieval Conference, TREC-9, May 2001, NIST Special Publication (2001)
[6] Hawking, D., Craswell, N.: Very large scale retrieval and web search. In: TREC: Experiment and Evaluation in Information Retrieval, pp. 199–231. MIT Press, Cambridge (2005)
[7] Kraaij, W., Westerveld, T., Hiemstra, D.: The importance of prior probabilities for entry page search. In: Proceedings of the 25th Annual International ACM SIGIR Conference on Research and Development in Information Retrieval, pp. 27–34. ACM Press, New York (2002)
[8] Ogilvie, P., Callan, J.: Combining document representations for known-item search. In: Proceedings of the 26th Annual International ACM SIGIR Conference on Research and Development in Information Retrieval, pp. 143–150. ACM Press, New York (2003)
[9] Kamps, J.: Web-centric language models. In: CIKM 2005: Proceedings of the 14th ACM International Conference on Information and Knowledge Management, pp. 307–308. ACM Press, New York (2005)
[10] Broder, A.: A taxonomy of web search. SIGIR Forum 36(2), 3–10 (2002)
[11] INEX: INitiative for the Evaluation of XML retrieval (2007), http://inex.is.informatik.uni-duisburg.de/
[12] Denoyer, L., Gallinari, P.: The Wikipedia XML Corpus. SIGIR Forum 40, 64–69 (2006)
[13] Faloutsos, M., Faloutsos, P., Faloutsos, C.: On power-law relationships of the internet topology. In: SIGCOMM 1999: Proceedings of the conference on Applications, technologies, architectures, and protocols for computer communication, pp. 251–262. ACM Press, New York (1999)
[14] ILPS: The ILPS extension of the Lucene search engine (2008), http://ilps.science.uva.nl/Resources/
[15] Hiemstra, D.: Using Language Models for Information Retrieval. PhD thesis, Center for Telematics and Information Technology, University of Twente (2001)

[16] Sparck Jones, K.: A statistical interpretation of term specificity and its application in retrieval. Journal of Documentation 28, 11–21 (1972)

[17] Sigurbjörnsson, B., Kamps, J., de Rijke, M.: An element-based approach to XML retrieval. In: INEX 2003 Workshop Proceedings, pp. 19–26 (2004)

[18] Lalmas, M., Kazai, G., Kamps, J., Pehcevski, J., Piwowarski, B., Robertson, S.: INEX 2006 evaluation measures. In: Fuhr, N., Lalmas, M., Trotman, A. (eds.) INEX 2006. LNCS, vol. 4518, pp. 20–34. Springer, Heidelberg (2007)

High Quality Expertise Evidence for Expert Search

Craig Macdonald, David Hannah, and Iadh Ounis

University of Glasgow, G12 8QQ, UK
{craigm,hannahd,ounis}@dcs.gla.ac.uk

Abstract. In an Enterprise setting, an expert search system can assist users with their "expertise need" by suggesting people with relevant expertise to the topic of interest. These systems typically work by associating documentary evidence of expertise to each candidate expert, and then ranking the candidates by the extent to which the documents in their profile are about the query. There are three important factors that affect the retrieval performance of an expert search system - firstly, the selection of the candidate profiles (the documents associated with each candidate), secondly, how the topicality of the documents is measured, and thirdly how the evidence of expertise from the associated documents is combined. In this work, we investigate a new dimension to expert finding, namely whether some documents are better indicators of expertise than others in each candidate's profile. We apply five techniques to predict the quality documents in candidate profiles, which are likely to be good indicators of expertise. The techniques applied include the identification of possible candidate homepages, and of clustering the documents in each profile to determine the candidate's main areas of expertise. The proposed approaches are evaluated on three expert search task from recent TREC Enterprise tracks and provide conclusions.

1 Introduction

Modern expert search systems in Enterprise settings work by using documents to form the profile textual evidence of expertise for each candidate. The profiles represent the system's knowledge of the expertise of each candidate, and on receiving a user query, they are ranked by how well the documents in their profile are related to the query [1,2]. For example, the Voting Model for expert search [3] sees this as a voting process: documents in the collection are ranked in response to the query, and then each document retrieved that is associated with a candidate is seen as a vote for that candidate to be retrieved for the query.

The retrieval performance of an expert search system is very important. Indeed, expert search has been a retrieval task in the Enterprise tracks of the Text REtrieval Conferences (TREC) since 2005 [4], aiming to evaluate state-of-the-art expert search approaches. This effort has generated two test collections for expert search, namely the W3C collection, and the CERC collection.

Several important factors have been investigated that can impact the retrieval performance of an expert search system. Firstly, the manner in which the

C. Macdonald et al. (Eds.): ECIR 2008, LNCS 4956, pp. 283–295, 2008.

evidence of expertise in the associated documents of each candidate are combined has an impact on the retrieval performance of the expert search system [3,5]. Secondly, it has been shown several times that the retrieval performance of an expert search system can be improved if the means by which the topicality of a document to a query is improved [6,7,8,9]. The better the expert search system is able to identify only on-topic documents in the corpus, the more likely it is that the inferences of expertise that can be drawn from the documents will be correct - i.e. off-topic documents will not give erroneous votes to non-relevant candidates. Moreover, various past research has applied query expansion [6,8,9], document structure [6,7] and proximity of query terms in documents to improve the underlying document retrieval system [6,10]. Thirdly, various research in expert search has observed that the quality of the candidate profiles has a major impact on the retrieval performance of the expert search system [11,12]. In particular, if one or more documents about the query topic which should be associated to a relevant candidate are omitted, then retrieval performance can be impaired. Indeed, the principle of accumulation of evidence suggests it is better to obtain as much expertise evidence as possible for a candidate.

In the area of Web IR, documents usually have a notion of quality associated with them. For example, a document that is linked to by many other documents is considered to be more authoritative about a topic than another less linked document, or a document that has a short URL is likely to be a homepage which users prefer. Web IR retrieval systems often take such sources of evidence into account when ranking Web documents, to improve the retrieval performance of the search engine [13,14].

In a similar vein, the aim of this work is to investigate a new aspect of the expert search system, which is the identification of high-quality evidence in the candidate profiles. We believe that if a notion of high-quality expertise evidence for a candidate can be defined, then this evidence can be successfully taken into account when ranking candidate experts. For instance, a document which is the homepage of a candidate is more likely to contain useful evidence of expertise than the minutes of a meeting that the candidate attended. However, it is not necessarily safe to remove all meeting minutes from all the candidate profiles, as this could prevent a relevant candidate from being retrieved for a difficult query. Instead, it is safer to weight higher (i.e. give stronger votes) the documents in a profile that we believe bring more expertise evidence about the candidate.

In this paper, we propose five techniques to predict the quality documents in candidate profiles, which are likely to be good indicators of expertise. We carry out the experiments with integrating these technique using the Voting Model for expert search, because the voting paradigm provides a natural and flexible mechanism to incorporate such additional evidence into an expert search system. This paper is structured as follows: Section 2 reviews models for expert search, and defines the voting technique we apply in this work; Section 3 proposes the five techniques to determine the quality expertise evidence in candidate profiles; We detail the experimental setting, including the test collections used in Section 4;

Section 5 provides results and analysis of the proposed techniques; We make concluding remarks in Section 6.

2 Expert Search

There are two requirements for an expert search system: a list of candidate persons that can be retrieved by the system, and some textual evidence of the expertise of each candidate to include in their profile. In most Enterprise settings, a staff list is available and this list defines the candidate persons that can be retrieved by the system. Candidate profiles can be created either explicitly or implicitly: candidates may explicitly update their profile with an abstract or list of their skills and expertise; or alternatively, the expert search system can implicitly and automatically generate each profile from a corpus of documents. This documentary evidence can take many forms, such as intranet documents, documents or emails authored by the candidates, or even emails sent by the candidate or web pages visited by the candidate (see [3] for an overview). In this work, the profile of a candidate is considered to be the set of documents associated with the candidate.

Once a profile of evidence has been identified for every expert, these can then be used to rank candidates automatically in response to a query. Various expert search approaches were proposed by participants of the TREC 2005 and TREC 2006 Enterprise tracks. These include that of Balog et al., who proposed the use of language models in expert search [5]. They proposed two models for expert search, however the approach is limited to the use of language modelling to provide the estimates for the relevance of a document to the query. Similarly to Balog et al., Fang and Zhai [15] applied relevance language models to the expert search task. In contrast, the probabilistic approach proposed by Cao et al. [16] and the hierarchical language models proposed by Petkova and Croft [6] do not consider expertise evidence on a document level, but instead work on a more fine-grained approach using windowing.

Instead, this work uses the Voting Model for expert search proposed by Macdonald & Ounis in [3], which considers the problem of expert search as a voting process. Instead of directly ranking candidates, it considers the *ranking of documents*, with respect to the query Q, denoted by $R(Q)$. The ranking of candidates can then be modelled as a voting process, from the retrieved documents in $R(Q)$ to the profiles of candidates: every time a document is retrieved and is associated with a candidate, then this is a vote for that candidate to have relevant expertise to Q. The ranking of the candidate profiles can then be determined by aggregating the votes of the documents. Twelve voting techniques for ranking experts were defined in [3], each employing various sources of evidence that can be derived from the ranking of documents with respect to the query topic.

In this work, we only use the expCombMNZ voting technique [3], because it provides effective and robust results across several expert search test collections and document weighting models - for example experiments applying various voting techniques combined with BM25, PL2 and DLH13 showed that expCombMNZ is not only one of the best voting techniques, but that it is stable across different document weighting model [7]. expCombMNZ ranks candidates by considering the sum of the exponential of the relevance scores of the documents associated with each candidate's profile. Moreover, it includes a component which takes into account the number of documents in $R(Q)$ associated to each candidate, hence explicitly modelling the number of votes made by the documents for each candidate. Hence the relevance score of a candidate expert C with respect to a query Q, $score\_cand(C, Q)$, is:

$$score\_cand(C, Q) = \|R(Q) \cap profile(C)\|$$
$$\cdot \sum_{d \in R(Q) \cap profile(C)} exp(score(d, Q)) \tag{1}$$

where $profile(C)$ is the set of documents associated with candidate C, and $score(d, Q)$ is the relevance score of the document in the document ranking $R(Q)$. $\|R(Q) \cap profile(C)\|$ is the number of documents from the profile of candidate C that are in the ranking $R(Q)$, and $exp()$ is the exponential function. The exponential function boosts candidates that are associated to highly scored documents (strong votes).

Documents are ranked using the DLH13 document weighting model [17] from the Divergence from Randomness (DFR) framework. We chose to experiment using DLH13 because it has no term frequency normalisation parameter that requires tuning, as this is assumed to be inherent to the model. Hence, by applying DLH13, we remove the presence of any term frequency normalisation parameter in our experiments. Moreover, as mentioned above, it performs comparably to BM25 and PL2 when combined with expCombMNZ on this task [3,7].

3   Quality Evidence in Candidate Profiles

As described in the introduction, there are three factors that can have a major impact on the retrieval performance of an expert search system. Firstly, the technique used to generate the initial ranking of documents $R(Q)$ has an impact on the retrieval performance of the expert search system. Previous work has shown that applying various document retrieval enhancing techniques (such as query expansion) results in a better ranking of candidates [6,7,8,9].

Secondly, the technique used to aggregate the document votes into a ranking of candidates also has a bearing on the retrieval performance. Of the twelve voting techniques described in [3], only some techniques produce a good retrieval performance, of which we use expCombMNZ in this work for the reasons detailed in Section 2.

Lastly, the quality of the candidate profiles used in an expert search system can have a major impact on the retrieval performance of the system. Due to the ambiguity of names, obfuscation of email addresses etc., the authorship of a document is difficult to generically identify in a heterogeneous corpus. Hence, if an on-topic document is not associated with its author (say), then that candidate will not receive a vote from that document.

In [11], Balog et al. investigated how expertise evidence should be identified from the emails of the W3C corpus. Interestingly, it was found that being included in the CC field on an email was more important than being the author of an email, for use as expertise evidence. Similarly, in [12], the authors investigated the impact on retrieval performance of the method of identifying expertise evidence for each candidate. For instance, they compared the effectiveness of an expert search system when candidates were identified by their full names, by their emails or by their last-name alone in the documents. They found that the choice of identification method had a major impact on the performance of the expert search system, and that the most exact form of identification (full name) gave the best retrieval performance.

For this work, we aim not to investigate the identification of profile evidence for candidates, but instead to determine which part of the candidates profiles should be considered as quality expertise evidence. This is similar to the notion of quality documents that exists in the Web IR field, where techniques such as, to name but a few, link analysis and URL length can be used as measures of the quality of a document. As mentioned in Section 1, the central idea of this paper is to take into account a quality measure in assessing the documents within a candidate profile. In particular, we propose measures that predict the high quality expertise evidence in a candidate's profile. The central hypothesis of this paper is that by identifying and weighting quality expertise evidence in the candidate profiles, the retrieval performance of the expert search system will be improved. In this work we propose five different techniques for identifying high quality expertise evidence within a candidate profile. While all techniques depend on the document, some techniques take into account the query, and/or the name of the candidate. The techniques include Web IR techniques such as URL Length and document Inlinks, as well as techniques that examine the proximity of the query to occurrences of the candidate's name, attempt to identify each candidate's home page, and lastly determine if a document is about a central interest of a candidate by using clustering. These are detailed in Sections 3.1-3.4 below.

We can compute a score for each of the above sources of evidence of a quality document in a candidate profile, which is denoted as $Qscore(d, C, Q)$, and integrate it with the expCombMNZ voting technique as follows:

$$score\_cand(C, Q) = \|R(Q) \cap profile(C)\| \qquad (2)$$
$$\cdot \sum_{d \, \in \, R(Q) \cap profile(C)} exp(score(d, Q) + \omega \cdot Qscore(d, C, Q))$$

where ω is a parameter. Note that if $Qscore$ is 0, then the candidate still receives a vote equivalent to the relevance score of the document. In this way, no expertise evidence is removed and the principle of accumulation of evidence is upheld. Note also that Equation (2) is only one way in which the measures of quality could be integrated - other ways may exist that might improve the overall effectiveness of the expert search system, but for the purpose of this paper our main objective is to ascertain to which extent taking into account the quality evidence within a profile is important. In the remainder of this section, we detail each proposed technique for identifying quality documents, and explain how they can be weighted and the resultant *Qscore* integrated into the applied voting technique.

3.1   Candidate Homepages

Usually, the homepage of a person contains personalised information, particularly about professional interests and role in the organisation, while in a research environment, it may also contain the titles of their publications. If the corpus contains webpages that could be seen as the candidate's homepage, then we can assume that this page has good evidence of the candidate's expertise. We believe that this is a form of high quality evidence of expertise, which should be weighted higher if it matches an expert search query.

Both the TREC W3C and CERC collections pose a problem for the identification of candidate homepages, for various reasons. In the W3C collection, not all candidates are employed by the W3C and hence only some candidate have homepages within the w3c.org domain, even though the URL location of the homepages of the candidates that have them is fairly predictable. For the CERC collection, not all staff have homepages, and the form of the URL of these vary from person to person. Some employees have personal homepages that they maintain, while others have just database-managed pages detailing their research interests. However, the problem here is that these are difficult to identify from the URL structure, due to the compartmentalised nature of the CSIRO organisation (e.g. different research divisions), which is mirrored in the different URL hosts with different directory layouts in the corpus.

In this paper, we propose a general technique to identify homepages in both of the test collections used. It is based on the assumption that pages such as a candidate's homepage (or the candidate's research interests page) will often have anchor text linking to that page containing predominantly the candidate's name. To identify these homepages, we firstly build an index for all documents that consists only of the anchor text of the incoming hyperlinks to each document. Then, for each candidate, we construct a phrasal search query using the exact full name of the candidate. This query is then run on the anchor-text index, giving a ranking of predicted homepages for each candidate, and a score for the document as calculated by a document weighting model. For efficiency, this procedure can be done offline, before retrieval. During expert search, votes from the predicted homepage documents are strengthened.

We integrate this homepage evidence into the expCombMNZ voting technique (Equation (2)) by calculating *Qscore* as follows:

$$Qscore_{HP}(d, C, Q) = score_{Anchor}(name(C), d) \qquad (3)$$

where $score_{Anchor}(name(C), d)$ is the score calculated by the document weighting model on the anchor text only index, for document d and the query being the name of the candidate. To remain consistent with $score(d, Q)$, we use the DLH13 document weighting model to generate $score_{Anchor}(name(C), d)$.

3.2   Candidate-Name and Query Proximity

Some types of documents can have many topic areas and many occurrences of candidate names (for instance, the minutes of a meeting). In such documents, the closer a candidate's name occurrence is to the query terms, the more likely that the document is a high quality indicator of expertise for that candidate [6,16].

We define $Qscore_{prox}(d, C, Q)$ in terms of the DFR term proximity document weighting model [10]. The term proximity model is designed to measure the informativeness in a document of a pair of query terms occurring in close proximity. We adapt this to the expert search task and into the expCombMNZ voting technique (Equation (2)), by measuring the informativeness of a query term occurring in close proximity to a candidate's name, as follows:

$$Qscore_{prox}(d, C, Q) = \sum_{p=name(C) \times t \in Q} score(d, p) \qquad (4)$$

Here p is a tuple of a term t from the query and the full name of candidate C. $score(d, p)$ can be calculated using any DFR weighting model [10], however, for efficiency reasons, we use a model that does not consider the frequency of tuple p in the collection but only in the document:

$$
\begin{aligned}
score(d, p) = \frac{1}{pfn + 1} \cdot \Big(& - \log_2 (avg\_w - 1)! + \log_2 pfn! \\
& + \log_2(avg\_w - 1 - pfn)! \\
& - pfn \log_2(p_p) \\
& - (avg\_w - 1 - pfn) \log_2(p'_p) \Big)
\end{aligned}
\qquad (5)
$$

where $avg\_w = \frac{T-N(ws-1)}{N}$ is the average number of windows of size ws tokens in each document in the collection, N is the number of documents in the collection, and T is the total number of tokens in the collection. $p_p = \frac{1}{avg\_w - 1}$, $p'_p = 1 - p_p$, and pfn is the normalised frequency of the tuple p, as obtained using Normalisation 2 [10]: $pfn = pf \cdot \log_2(1 + c_p \cdot \frac{avg\_w - 1}{l - ws})$. In Normalisation 2, pf is the number of windows of size ws in document d in which the tuple p occurs. l is the length of the document in tokens and $c_p > 0$ is a hyper-parameter that controls the normalisation applied to pfn frequency against the number of windows in the document.

3.3   URL Length and Inlinks

In order to ascertain the high quality documents within a candidate profile, we apply sources of evidence inspired by work in the Web IR field about measuring the quality of a web page. In a Web IR setting, a document with many incoming links is likely to be of good quality, and indeed, link information within Enterprise settings has previously been found to be useful in intranet search [18,19].

In adapting this evidence to expert search, we assume that documents with shorter URLs are of higher importance and quality in the organisation, and that evidence of expertise obtained from them is of more importance. Similarly, documents with more inlinks are likely to be of good quality, and of more use in an expert search system. Note that most link analysis techniques (e.g. PageRank and Absorbing Model) have been shown to be strongly correlated to a simple count of the number of incoming hyperlinks (inlinks) to each document [20]. For this reason, in this paper we only use inlinks.

We follow Craswell et al. [14] by integrating URL path length and inlinks into the expCombMNZ voting technique (Equation (2)) using two saturation functions, respectively:

$$Qscore_{URL}(d, C, Q) = \frac{\kappa}{\kappa + URLPathLength(d)} \tag{6}$$

$$Qscore_{Inlinks}(d, C, Q) = \frac{\kappa \cdot \beta \cdot Inlinks(d)}{\kappa + \beta \cdot Inlinks(d)} \tag{7}$$

where $URLPathLength(d)$ is the number of characters in the path component of the URL of document d, κ is a parameter, $Inlinks(d)$ is the number of incoming hyperlinks to document d, and $\beta = \frac{N}{\sum_d Inlinks(d)}$, in which N is the number of documents in the collection. The purpose of β is to ensure that the mean of the inlinks distribution is 1.

3.4   Clustering of Candidate Profiles

Candidates can have many areas of expertise over the timespan of the organisation, and this can be measured as topic drift in their candidate profiles [9]. In this work, we use clustering to identify the main interests of each candidate, particularly for these prolific candidates. By clustering a candidate profile, the main expertise areas of the candidate should be reflected as the largest clusters. Votes for the candidate to be retrieved by documents that are about one of the candidate's main interests (i.e. one of the larger clusters) should be higher weighted.

In particular, in this paper we use a single-pass clustering algorithm to cluster the profiles of candidates who have more than θ documents in their profile. In the clustering, the cluster distance is defined as the Cosine between the average of each clusters. The clusters obtained are then ranked by the number of documents they contain, and we select the largest K clusters as representatives of the central interests of the expert. We integrate this evidence into the expCombMNZ voting technique (Equation (2)):

$$Qscore_{Cluster}(d, C, Q) = \begin{cases} \frac{1}{cluster(d,C)} & if \ cluster(d, C) \leq K \\ 0 & otherwise \end{cases} \tag{8}$$

where $cluster(d, C)$ is the rank of the cluster in which document d occurred for candidate C (largest cluster has rank 1). The above integration of cluster expertise evidence into the voting technique strengthens votes from documents which are found in larger clusters in the profile of candidate c, because the largest clusters are assumed to be the candidate's strongest expertise area. Note that if a document d does not occur in the top K clusters, then $Qscore_{Cluster}(d, C, Q)=0$, i.e. its vote is not strengthened further. Moreover, if no clustering has been applied for the candidate (i.e. they have less than θ documents in their profile), then $Qscore_{Cluster}(d, C, Q) = 0$.

In the remainder of this paper, we experiment with the proposed techniques for identifying quality evidence in the candidate profiles. In particular, we define the experimental setup of our experiments in the next section. Results and conclusions follow in Sections 5 and 6, respectively.

4   Experimental Setup

Our experiments are carried out in the setting of the Expert Search task of the TREC Enterprise tracks, namely 2005, 2006 and 2007. For TREC 2005 and 2006, the document collection used was a crawl of the World Wide Web Consortium (W3C), a virtual Internet organisation responsible for HTML, XML standards and the like. For TREC 2007, a different and more realistic corpus, known as CERC, was introduced, which is a crawl of the website of Commonwealth Scientific and Industrial Research Organisation (CSIRO). CSIRO is the national government body for scientific research in Australia. In terms of measuring retrieval performance, we use the Mean Average Precision (MAP) measure for all tasks. Moreover, for TREC 2005 and TREC 2006 for which there are generally more than 10 relevant candidates per-topic, we measure for Precision at 10 (P@10). In the CERC collection, in which there are typically less than 10 relevant candidates per topic, we measure Mean Reciprocal Rank (MRR).

Table 1. Statistics of the TREC W3C and CERC Enterprise research test collections

Statistic	W3C	CERC
# of Documents	331,037	370,715
# of Topics	99	50
# of Candidates	1,092	3,490
Average Profile Size (# of Documents)	913.2	217.7
Largest Profile Size (# of Documents)	88,080	62,285

The TREC W3C and CERC collections are indexed using Terrier [21], removing standard stopwords and applying the first two steps of Porters stemming algorithm. Moreover, we add onto each document, the anchor text of the incoming hyperlinks from other documents in the corpus. For the calculation of the clustering $Qscore$, we apply $K = 10$ and $\theta = 30$, because for prolific persons, 10 areas of expertise would seem intuitive for most people. The setting of all other $Qscore$ parameters is described in the following section. To identify the profile of documents to represent each candidate, we search for each candidate's full name in the corpus. For the CERC test collection, where no initial list of candidates is provided, candidates are initially identified by the presence of an email address in the form `firstname.lastname@csiro.au` in the corpus. Statistics of the W3C and CERC test collections are given in Table 1.

5   Experimental Results

In our experiments, we are not focused on the particular integration of the *Qscore* with expCombMNZ. Instead, we wish to see if any benefit is possible in applying that evidence. For this reason, we firstly train to maximise MAP on the set of topics being tested. Secondly, we use a more realistic setting, where for TREC 2006 we train using the TREC 2005 topics, and for TREC 2007, we train using the TREC 2005 and 2006 topics combined (even though it is not the same corpus). Appendix 1 details the obtained parameters for all settings. Table 2 presents the results of our experiments. On the first row, the median MAP is shown. Our baseline is the retrieval performance achieved by applying DLH13 with expCombMNZ. It can be seen that this baseline is markedly above the median performance of all participating groups (except MRR for TREC 2007). In particular, for TREC 2005 and TREC 2006, this baseline would have been ranked in the top three automatic title-only runs, and in the top four for TREC 2007 automatic title-only runs. The remainder of the table presents the retrieval performance of each proposed technique for identifying quality expertise. For the columns denoted '/test', the parameters have been trained on the test set, while '/train' denotes when the parameters were trained using a separate test set of topics, as detailed above.

Table 2. Results for TREC 2005, 2006 and 2007 expert search tasks, when trained on the test set. Significant increases over the baseline are denoted $>$ ($p < 0.05$) and $\gg$ ($p < 0.01$) respectively. '/test' and '/train' denote whether the parameters for the quality evidence techniques were trained using the test set or a separate training set.

TREC Year	2005/test		2006/test		2006/train		2007/test		2007/train	
	MAP	P@10	MAP	P@10	MAP	P@10	MAP	MRR	MAP	MRR
Median	0.1402	-	0.3412	-	0.3412	-	0.2468	0.5011	0.2468	0.5011
Baseline	0.2040	0.3100	0.5502	0.6837	0.5502	0.6837	0.3519	0.4730	0.3519	0.4730
+ Prox	0.2155	0.3200	**0.5621**>	0.6878	0.5427	0.6551	**0.4319**≫	**0.5742**≫	**0.3688**	0.4891
+ URL	0.2232≫	0.3300	0.5565	**0.7020**	**0.5657**	**0.7000**	0.3779>	0.5309>	0.3683	**0.5015**
+ Inlinks	0.2212≫	**0.3540**≫	0.5600	0.6857	0.5522	0.6755	0.3654	0.4847	0.3474	0.4778
+ Clusters	**0.2324**>	0.3420	0.5517	0.6816	0.4830	0.6020	0.3915>	0.5400	0.3584	0.4726
+ Homepage	0.2040	0.3100	0.5530	0.6837	0.5501	0.6837	0.3885	0.5334	0.3463	0.4569

On the optimal setting ('/test'), the Proximity quality evidence performs well, particularly on the CERC collection. URL and Inlinks evidence also appear to be reliable at discriminating between high and low quality expertise evidence in the candidate profiles. For the homepage, the results are mixed: it improves retrieval performance on the TREC 2007 collection (suggesting that many of the CSIRO experts do have homepages); for TREC 2005 and 2006, there are only minor differences in performance. By further examination of the W3C corpus, there are only 58 candidates from the 1092 in the collection are staff members of the W3C, therefore this evidence does not apply well in this case. Lastly, the clustering provides significant improvements for MAP on the TREC 2005 and TREC 2007 topic sets, while for TREC 2006 there is little change. For the plausible training ('/train'), Table 2 shows the performance is slightly less than the optimal training, the results are still similar. In particular, proximity and URL are the best indicators, followed by clustering. Again, the homepages and inlinks did not bring much difference in retrieval performance. The slightly lower performance of the clustering on TREC 2007 is explained by the fact that the combined TREC 2005 + 2006 topics are not a good training set for this quality evidence.

Overall, as mentioned above, our main aim was not to propose how to combine the quality evidences with the proposed voting technique. However, given that the retrieval performance could be improved in the future by better combinations and further training of parameters, some of the proposed quality evidences, such as proximity and clustering, seem very promising. In particular, the best setting for proximity on the TREC 2007 topics would have been ranked 2nd out of the submitted automatic title-only runs that year.

6   Conclusions

In this paper we have proposed five techniques to predict the quality of documents within a candidate's profile in the expert search task. We have thoroughly tested these techniques using two test collections and three TREC topic sets. The experiments show that among them, the novel clustering and proximity techniques seem very promising. However, in contrast to Web search settings, various Web IR features such as URL and Inlinks did not exhibit large increases in performance.

It is of interest that in the field of Web IR, it is natural to learn document features based on their occurrence in a set of relevance assessments. However, in the expert search task only the final outcome of the expert search system is evaluated. None of the three important performance-affecting factors described in this paper (see abstract, Sections 1 & 3) can be directly evaluated, making it particular difficult to have a complete overview of the performance of the system. While the initial steps taken in [22] work towards a more complete evaluation, perhaps in the future, the evaluation methodology can evolve to provide enough details such that a thorough failure analysis can be conducted and conclusions can be drawn about all components of an expert search system.

References

1. Craswell, N., Hawking, D., Vercoustre, A.M., Wilkins, P.: Panoptic expert: Searching for experts not just for documents. In: AusWeb Poster Proceedings, Queensland, Australia (2001)
2. Liu, X., Croft, W.B., Koll, M.: Finding experts in community-based question-answering services. In: Proceedings of CIKM 2005, pp. 315–316 (2005)
3. Macdonald, C., Ounis, I.: Voting for candidates: Adapting data fusion techniques for an expert search task. In: Proceedings of CIKM 2006 (2006)

4. Craswell, N., de Vries, A.P., Soboroff, I.: Overview of the TREC-2005 Enterprise Track. In: Proceedings of TREC-2005, pp. 199–204 (2006)
5. Balog, K., Azzopardi, L., de Rijke, M.: Formal models for expert finding in enterprise corpora. In: Proceedings of SIGIR 2006, pp. 43–50 (2006)
6. Petkova, D., Croft, W.B.: Hierarchical language models for expert finding in enterprise corpora. In: Proceedings of ICTAI 2006, pp. 599–608 (2006)
7. Macdonald, C., Ounis, I.: Voting Techniques for Expert Search. J. Knowledge and Information Systems (in press, 2008), DOI:10.1007/s10115-007-0105-3
8. Macdonald, C., Ounis, I.: Using Relevance Feedback in Expert Search. In: Amati, G., Carpineto, C., Romano, G. (eds.) ECIR 2007. LNCS, vol. 4425, pp. 431–443. Springer, Heidelberg (2007)
9. Macdonald, C., Ounis, I.: Expertise drift and query expansion in expert search. In: Proceedings of CIKM 2007 (in press, 2007)
10. Lioma, C., Macdonald, C., Plachouras, V., Peng, J., He, B., Ounis, I.: University of Glasgow at TREC 2006: Experiments in Terabyte and Enterprise Tracks with Terrier. In: Proceedings of TREC 2006 (2007)
11. Balog, K., de Rijke, M.: Finding experts and their details in e-mail corpora. In: Proceedings of WWW 2006 (2006)
12. Macdonald, C., Ounis, I.: Voting for Experts: The Voting Model for Expert Search. Special issue of the Computer Journal on Expertise Profiling (to appear, 2008)
13. Kraaij, W., Westerveld, T., Hiemstra, D.: The importance of prior probabilities for entry page search. In: Proceedings of SIGIR 2002, pp. 27–34 (2002)
14. Craswell, N., Robertson, S., Zaragoza, H., Taylor, M.: Relevance weighting for query independent evidence. In: Proceedings of SIGIR 2005, pp. 416–423 (2005)
15. Fang, H., Zhai, C.: Probabilistic models for Expert Finding. In: Amati, G., Carpineto, C., Romano, G. (eds.) ECIR 2007. LNCS, vol. 4425, pp. 418–430. Springer, Heidelberg (2007)
16. Cao, Y., Li, H., Liu, J., Bao, S.: Research on expert search at enterprise track of TREC 2005. In: Proceedings of TREC-2005 (2006)
17. Amati, G.: Frequentist and Bayesian approach to information retrieval. In: Lalmas, M., MacFarlane, A., Rüger, S.M., Tombros, A., Tsikrika, T., Yavlinsky, A. (eds.) ECIR 2006. LNCS, vol. 3936, pp. 13–24. Springer, Heidelberg (2006)
18. Hawking, D., Craswell, N., Crimmins, F., Upstill, T.: How valuable is external link evidence when searching enterprise webs? In: Proceedings of ADC 2004, pp. 77–84 (2004)
19. Fagin, R., Kumar, R., McCurley, K.S., Novak, J., Sivakumar, D., Tomlin, J.A., Williamson, D.P.: Searching the workplace web. In: Proceedings of WWW 2003, pp. 366–375 (2003)
20. Peng, J., Macdonald, C., He, B., Ounis, I.: Combination of Document Priors in Web Information Retrieval. In: Proceedings of RIAO 2007 (2007)
21. Ounis, I., Amati, G., Plachouras, V., He, B., Macdonald, C., Lioma, C.: Terrier: A high performance and scalable information retrieval platform. In: Proceedings of the OSIR Workshop 2006, pp. 18–25 (2006)
22. Macdonald, C., Ounis, I.: Expert Search Evaluation by Supporting Documents. In: Macdonald, C., et al. (eds.) ECIR 2008. LNCS, vol. 4956, pp. 555–563. Springer, Heidelberg (2008)

Appendix 1: Parameters

Table 3. Trained parameters, headings are as in Table 2: Proximity trained using manual scanning; other techniques trained using simulated annealing process for MAP

TREC Year	2005/test 2006/train	2006/test	2007/train	2007/test
Prox	$\omega = 1\ ws = 20\ c_p = 0.1$	$\omega = 1\ ws = 10\ c_p = 0.01$	$\omega = 1\ ws = 20\ c_p = 0.0001$	$\omega = 0.5\ ws = 200\ c_p = 1$
URL	$\omega = 14.12\ \kappa = 99.78$	$\omega = 12.22\ \kappa = 70.03$	$\omega = 8.27\ \kappa = 9.82$	$\omega = 18.41\ \kappa = 85.44$
Inlinks	$\omega = 5.88\ \kappa = 0.39$	$\omega = 3.04\ \kappa = 3.31$	$\omega = 4.55\ \kappa = 0.59$	$\omega = 5.74\ \kappa = 2.13$
Clusters	$\omega = 6.50$	$\omega = 0.80$	$\omega = 3.87$	$\omega = 1.74$
Homepage	$\omega = 0.004$	$\omega = 0.067$	$\omega = 0.03$	$\omega = 0.25$

Associating People and Documents

Krisztian Balog and Maarten de Rijke

ISLA, University of Amsterdam
{kbalog,mdr}@science.uva.nl

Abstract. Since the introduction of the Enterprise Track at TREC in 2005, the task of finding experts has generated a lot of interest within the research community. Numerous models have been proposed that rank candidates by their level of expertise with respect to some topic. Common to all approaches is a component that estimates the strength of the association between a document and a person. Forming such associations, then, is a key ingredient in expertise search models. In this paper we introduce and compare a number of methods for building document-people associations. Moreover, we make underlying assumptions explicit, and examine two in detail: (i) independence of candidates, and (ii) frequency is an indication of strength. We show that our refined ways of estimating the strength of associations between people and documents leads to significant improvements over the state-of-the-art in the end-to-end expert finding task.

1 Introduction

Since the launch of the TREC Enterprise track [4, 10] there has been a lot of work on models, algorithms, and evaluation methodology for the expert finding task, i.e., the task of returning a list of people within some given organization that are ranked by their expertise on some given topic. A feature shared by many of the models proposed for ranking people with respect to their expertise on a given topic is their reliance on *associations* between people and documents. E.g., if someone is strongly associated with an important document on a given topic, this person is more likely to be an expert on the topic than someone who is not associated with any documents on the topic.

Despite the important role of associations between candidate experts (from now on: "candidates") and documents for today's expert finding models, such associations have received relatively little attention in the research community. Various methods have been used for estimating the strength of associations, and these approaches come in two kinds: (i) *set-based*, where the candidate is associated with a set of documents (all with equal weights), in which (s)he is mentioned, and (ii) *frequency-based*, where the strength of the association is proportional to the number of times the candidate is mentioned in the document.

While a number of techniques have already been used to estimate the strength of association between a person and a document, these have never been compared. This gives rise to the research questions that we seek to answer in this paper: What is the impact of document-candidate associations on the end-to-end performance

C. Macdonald et al. (Eds.): ECIR 2008, LNCS 4956, pp. 296–308, 2008.

of expert finding models? What are effective ways of capturing the strength of these associations? How sensitive are expert finding models to different document-candidate association methods?

To answer our research questions, we use two principal expert search strategies (so-called candidate and document models), that cover most existing approaches developed for expert finding. Our models are based on generative language modeling techniques, which is a specific choice, but the need for estimating the strength of the association between document-candidate pairs is not specific to our models. Other approaches also include this component, not necessarily in terms of probabilities, but as a score or weight. Given these models, we study, and systematically compare, various association methods. To this end we first discuss the *boolean* model, which is a simple yet effective way of forming associations, and serves as our baseline approach to person-document associations.

Then we lift an assumption that underlies this method—the *independence of candidates*—, and use term weighting schemes familiar from Information Retrieval. The strategy we follow is this: we treat candidates as terms and view the problem of estimating the strength of association with a document as an importance estimation problem: how important is a candidate for a given document. Specifically, we consider TF, IDF, TFIDF, and language models.

As a next step, we examine a second assumptions underlying (at least some) document-person association methods: that *frequency is an indication of strength*. First, we consider *lean* document representations that contain only candidates, while all other terms are filtered out. We find that it seriously impacts the performance of some expert-finding models (esp. candidate models) while it affects others to a far lesser degree (esp. document models).

Then, to grasp the effect of using the frequency of a candidate, we propose a new person-document association approach, where instead of the candidate's frequency, the *semantic relatedness* of the document and the person is used. This is achieved by comparing the language model of the document with the candidate's profile. We find that frequencies succeed very well at capturing the semantics of person-document associations.

The remainder of the paper is organized as follows. In Section 2 we discuss related work. We describe our expert search models in Section 3 and our experimental setup in Section 4. We compare multiple people-document association methods, and report our results in Section 5. We conclude in Section 6.

2   Related Work

In this section we focus on expertise retrieval approaches developed and published since the launch of the TREC Enterprise Track in 2005. For an overview of expertise finding systems in organizations we refer the reader to [12].

There are two principal approaches to expert finding, which have been first formalized and extensively compared in [1], and are referred to as *candidate* and *document* models. Most systems that took part in the 2005 and 2006 editions of the Expert Finding task at TREC implemented (variations on) one of them; see

[4, 10]. Candidate-based approaches (also referred to as profile-based methods in [5] or query-independent approaches in [9]) build a textual (usually term-based) representation (profile) of candidate experts, and rank them based on the relevance of a query, using traditional ad-hoc retrieval models [1, 5, 6, 7, 8, 9]. Document-based models (called query-dependent approaches in [9]) first locate documents on the topic and then find the associated experts [1, 2, 3, 5].

Common to all approaches is that documents and candidates need to be linked, whether these associations are made explicit or encoded in the models. Association methods come two kinds: (i) *set-based*, where the candidate is associated with a set of documents (all with equal weights), in which (s)he occurs [7, 8], and (ii) *frequency-based*, where the strength of the association is proportional to the number of times the candidate occurs in the document [1, 5, 6, 9].

In [7, 8] candidate profiles are constructed based on a set of documents in which the person's name or email address occurs. The candidate's identifier(s) (name and/or e-mail address) are used as a query, and relevant documents contribute to this set of profile documents. These approaches do not quantify the strength of the document-candidate associations, thus use them implicitly. In our setting this corresponds to the *boolean* model of associations (Section 5.1), i.e., a person is either associated with a document or not.

Document-based expert finding models often employ language models (LMs) [1, 2, 3, 5, 9] and the strength of the association between candidate ca and document d is expressed as a probability (either $p(d|ca)$ or $p(ca|d)$). In [1], these probabilities are calculated using association scores between document-candidate pairs. The scores are computed based on the recognition of the candidate's name and e-mail address in documents. In [5, 9], $p(d|ca)$ is rewritten in terms of $p(ca|d)$, using Bayes' rule, then the candidate's representations are treated as a query given the document model. This corresponds to our LM approach in Section 5.2. The two-stage LM [2, 3] includes a co-occurrence model, $p(ca|d, q)$, which is calculated based on the co-occurrence of the person with one or more query terms in the document or in the same window of text. When co-occurrence is calculated based on the full body of the document, the query is not taken into account and document-candidate associations are estimated using LMs, where documents contain only candidate identifiers. This corresponds to our lean documents approach using LMs in Section 5.3.

The candidate-generation model in [5] covers the two-stage LM approach of [3], but it is assumed that the query q and candidate ca are independent given the document d, i.e., $p(ca|d, q) \approx p(ca|d)$. The document model in [1] (Model 2 in Section 3) makes the same assumption. That implies that we build associations on the document level only, and leave an exploration of candidate-"text snippet" associations (co-occurrence on the sub-document level) for future work.

3   Modeling Expert Search

In this section we briefly describe two models for expert finding, taken from [1]. These two models cover both expert search strategies; moreover, they are principled, and nicely demonstrate how the document-association component fits into the picture. We should point out that these models consider document-candidate associations on the document level only.

3.1   Model 1: Candidate Model

Model 1 builds a textual representation of a candidate ca using a multinomial unigram language model θ_{ca}. This model is used to predict how likely a candidate would produce a query q:

$$p(q|\theta_{ca}) = \prod_{t \in q} p(t|\theta_{ca})^{n(t,q)} \qquad (1)$$

where each term t in the query q is sampled identically and independently, and $n(t, q)$ is the number of times t occurs in q.

The candidate model is constructed as a linear interpolation of an empirical model $p(t|ca)$, and the background model $p(t)$ to ensure there are no zero probabilities:

$$p(t|\theta_{ca}) = (1 - \lambda) \cdot p(t|ca) + \lambda \cdot p(t), \qquad (2)$$

where parameter λ controls the amount of smoothing applied.

Using the associations between a candidate and a document, the probability $p(t|ca)$ can be approximated by $p(t|ca) = \sum_d p(t|d) \cdot p(d|ca)$, where $p(d|ca)$ is the probability that candidate ca generates supporting document d, and $p(t|d)$ is the probability of term t occurring in document d (calculated using the maximum-likelihood estimate of a term). The final estimation of Model 1 is:

$$p(q|\theta_{ca}) = \prod_{t \in q} \left\{ (1 - \lambda) \cdot \left(\sum_d p(t|d) \cdot p(d|ca) \right) + \lambda \cdot p(t) \right\}^{n(t,q)} \qquad (3)$$

3.2   Model 2: Document Model

Under this model, we can think of the process of finding an expert as follows. Given a collection of documents ranked according to the query, we examine each document and if relevant to our problem, we then see who is associated with that document. Conceptually, Model 2 differs from Model 1 because the candidate is not directly modeled. Instead, it assumes that q and ca are independent given d, the document acts like a "hidden" variable in the process which separates the query from the candidate. Formally, this can be expressed as

$$p(q|ca) = \sum_d p(q|d) \cdot p(d|ca) \qquad (4)$$

The probability of a query given a document $p(q|d)$ is estimated by inferring a document language model θ_d for each document d:

$$p(t|\theta_d) = (1 - \lambda) \cdot p(t|d) + \lambda \cdot p(t) \qquad (5)$$

where $p(t|d)$ is the probability of the term in the document. The probability of a query given the document model is:

$$p(q|\theta_d) = \prod_{t \in q} p(t|\theta_d)^{n(t,q)} \qquad (6)$$

The final estimate of Model 2, then, is:

$$p(q|ca) = \sum_d \left\{ \prod_{t \in q} \left((1 - \lambda) \cdot p(t|d) + \lambda \cdot p(t) \right)^{n(t,q)} \right\} \cdot p(d|ca) \qquad (7)$$

3.3   Document-Candidate Associations

In Model 1 and 2 the association between candidate ca and document d is expressed as $p(d|ca)$, the probability of the document given the candidate. We apply Bayes' rule to rewrite it:

$$p(d|ca) = \frac{p(ca|d) \cdot p(d)}{p(ca)} \qquad (8)$$

This allows us to incorporate document and candidate priors into the association component. We leave the estimation of document and candidate priors to future work and assume that $p(d)$ and $p(ca)$ are uniformly distributed. Hence, our task boils down to estimating of $p(ca|d)$. The reading of $p(ca|d)$ is different for the two models. For Model 1, it reflects the degree to which the candidate's expertise is described using this document. For Model 2, it provides a ranking of candidates associated with a given document d, based on their contribution made to d.

4   Experimental Setup

4.1   Test Collection

We use the test sets of the 2005 and 2006 editions of the TREC Enterprise track [4, 10]. The document collection used is the W3C corpus [11], a heterogenous document repository containing a mixture of document types crawled from the W3C site. We used the entire corpus, and handled all documents in the same way, as HTML documents. We did not resort to any special treatment of document types, nor did we exploit the internal document structure that may be present; instead, we represented all documents as plain text. We removed a standard list of stopwords, but did not apply stemming.

The TREC Enterprise 2005 topics (50) are names of working groups within the W3C. Members of a working group were regarded as experts on the corresponding topic. The 2006 topics (55) were contributed by TREC participants and assessed manually. We used only the titles of the topic descriptions.

We evaluate the methods with mean average precision (MAP), the official measure of the expert finding task at TREC.

4.2   Person Name Identification

In order to form document-candidate associations, we need to be able to recognize candidates' occurrences within documents. In the TREC setting, a list of possible candidates is given, where each person is described with a unique *person_id*, one or more *names*, and one or more *e-mail* addresses. While this is a specific way of identifying a person, and different choices are also possible (e.g., involving social security number instead of, or in addition to, the representations just listed), nothing in our modeling depends on *this* particular choice.

The recognition of candidate occurrences in documents (through one of these representations) is a restricted information extraction task. In [2], six match types (MT) of person occurrences are identified:

MT1 Full name (e.g., Ritu Raj Tiwari and Tiwari, Ritu Raj);
MT2 Email name (e.g., rtiwari@nuance.com);
MT3 Combined name (e.g., Tiwari, Ritu R and R R Tiwari);
MT4 Abbreviated name (e.g., Ritu Raj and Ritu);
MT5 Short name (e.g., RRT);
MT6 Alias, New Mail (e.g., Ritiwari and rtiwari@hotmail.com).

In [1], a similar approach is taken, and four types of matching are introduced; three attempt to identify candidates by name, and one uses email addresses. To facilitate comparison, we used the resources contributed by Bao et al. [2].[1]

Some of these matching methods create ambiguity, that is, a name may be shared by more than one person. To allow us to measure, how this noise introduced affects overall performance, we identify a group of matching methods, including MT1, MT2, and MT6, where ambiguity is insignificant, and refer to this set as STRICT matching methods. Using all matching methods is referred as ALL.

We replaced all candidate occurrences (name and email address) with a unique candidate identifier, which was then treated as a term in the document.

5   Establishing Document-Candidate Associations

In this section we address the problem of estimating $p(ca|d)$, the strength of the association between a document and a candidate.

5.1   The Boolean Model of Associations

Under the boolean model, associations are binary decisions; they exist if the candidate occurs in the document, irrespective of the number of times the person or other candidates are mentioned in that document. We simply set

$$p(ca|d) = \begin{cases} 1, & n(ca, d) > 0 \\ 0, & \text{otherwise,} \end{cases} \qquad (9)$$

where $n(ca, d)$ denotes the number of times the candidate's identifier appears in the document. It can be viewed as a set-based approach, analogously to [7], where a candidate is associated with a set of documents: $D_{ca} = \{d : n(ca, d) > 0\}$.

The boolean model is the simplest way of forming document-candidate associations. Simplicity comes at the price of two potentially unrealistic assumptions:

---

[1] URL: http://ir.nist.gov/w3c/contrib/

- **Candidate independence.** Candidates occurring in the document are independent of each other, and are all equally important given the document. The model does not differentiate between people that occur in its text.
- **Position independence.** The strength of the association between a candidate and a document is independent of the candidate's position within the document. Positional independence is equivalent to adopting the bag of words representation: the exact ordering of candidates within a document is ignored, only the number of occurrences is stored.

Common sense tells us that not all candidates mentioned in the document are equally important. Similarly, not all documents, in which a candidate occurs, describe the person's expertise equally well. For example, a person who is listed as an author of the document should be more strongly associated with the document, than someone who is only referred to in the body of the document. This goes against the candidate independence assumption. If we take into account that authors are also listed at the top or bottom of documents, the previous example also provides evidence against the position independence assumption.

In this paper, we stick with the position independence assumption, and leave the examination of that to further work. However, intuitively, candidate independence may be too strong an assumption. Therefore, we drop it as our next step, and discuss ways of estimating a candidate's importance given a document. In other words, our aim is a non-binary estimation of $p(ca|d)$.

5.2   Modeling Candidate Frequencies

Our goal is to formulate $p(ca|d)$ in such a way that it indicates the strength of the association between candidate ca and document d. The number of times a person occurs in a document seems to be the most natural evidence supporting the candidate being strongly associated with that document. This leads us to a new assumption: the strength of the association is proportional to the number of times the candidate is mentioned in the document.

A commonly employed technique for building document-candidate associations is to use the candidate's identifiers as a query to retrieve documents. The strength of the association is then estimated using the documents' relevance scores [5, 9]. This way, both the recognition of candidates' occurrences and the association's strength estimation is performed in one step. Our approach is similar, but limited to the estimation aspect, and assumes that the matching of candidate occurrences is taken care of by a separate extraction component.

We treat candidate identifiers as terms in a document, and view the problem of estimating the strength of association with a document as an importance estimation problem: how important is a candidate for a given document? We approach it by using term weighting schemes familiar from IR. Specifically, we consider TF, IDF, and TFIDF weighting schemes from the vector space model, and also language models. In the following, we briefly discuss these methods and the rationale behind them.

Table 1. Candidate mentions are treated as any other term in the document. For each year-model combination the best scores are in boldface.

Method	ALL MatchTypes				STRICT MatchTypes			
	TREC 2005		TREC 2006		TREC 2005		TREC 2006	
	Model 1	Model 2	Model 1	Model 2	Model 1	Model 2	Model 1	Model 2
Boolean	**.1742**	.2172	.2809	.4511	**.1858**	.2196	**.3075**	.4704
TF	$.0684^{(3)}$	$.2014^{(3)}$	$.1726^{(3)}$	.4408	$.0640^{(3)}$	$.2038^{(2)}$	$.1601^{(3)}$	$.4485^{(1)}$
IDF	.1676	$\mathbf{.2480}^{(3)}$	$.2488^{(3)}$	.4488	.1845	$\mathbf{.2512}^{(3)}$	$.2736^{(3)}$	.4670
TFIDF	$.1408^{(1)}$	.2227	**.2913**	.4465	$.1374^{(2)}$	.2266	.2828	.4514
LM	$.0676^{(3)}$	$.2013^{(3)}$	$.1619^{(3)}$	.4397	$.0642^{(3)}$	$.2031^{(2)}$	$.1586^{(3)}$	$.4470^{(1)}$

TF. The importance of the candidate within the particular document is proportional to the candidate's frequency (against all terms in the document): $p(ca|d) \propto TF(ca, d)$

IDF. It models the general importance of a candidate:

$$p(ca|d) \propto \begin{cases} IDF(ca), & n(ca, d) > 0 \\ 0, & \text{otherwise.} \end{cases} \tag{10}$$

Candidates that are mentioned in many documents, will receive lower values, while those who occur only in a handful of documents will be compensated with higher values. This, however is independent of the document itself.

TFIDF. A combination of the candidate's importance within the particular document, and in general is expected to give the best results.

Language Modeling. We employ a standard LM setting to document retrieval, using Equation 5. We set $p(ca|d) = p(t = ca|\theta_d)$, which is identical to the approach in [5, 9]. Our motivation for using language models is twofold: (i) expert finding models are also using LMs (pragmatic reason), and more importantly, and (ii) smoothing in language modeling has an IDF effect [13]. Tuning the value of λ allows us to control the background effect (general importance of the candidate), which is not possible using TFIDF. Here, we follow standard settings and use $\lambda = 0.1$ [13].

Table 1 presents the MAP scores for Model 1 and 2, using the TREC 2005 and 2006 topics. We report on two sets of experiments, using all (columns 2–5) and only the unambiguous (columns 6–9) matching methods. The first row corresponds to the boolean model of associations (Eq. 10), while additional rows correspond to frequency-based methods.

For significance testing we use a two-tailed, matched pairs Student's t-test, and look for improvements at significance levels [1] 0.95, [2] 0.99, and [3] 0.999. The boolean method is considered as the baseline, against which frequency-based methods are compared.

Our findings are as follows. First, there is a substantial difference between the performance on the TREC 2005 and 2006 topic sets. As pointed out in [5], this is due to the fact that judgments were made differently in these two years. In 2005, judgments are independent of the document collection, and were obtained artificially, while topics in 2006 were developed and assessed manually. Second, it is more beneficial to use rigid patterns for person name matching; the noise introduced by name ambiguity hurts performance. Hence, from now on we use the STRICT matching methods. Third, Model 2 performs considerably better than Model 1. This confirms the findings reported in [1].

As to the association methods, we find that the simple boolean model delivers excellent performance. The best results (using Model 2 and STRICT matching) are 0.2196 and 0.4704 for TREC 2005 and 2006, respectively; this beats the corresponding scores of 0.204 and 0.465 scores of (author?) [5]. However, in [5] candidate priors are used, and parameters of the models are tuned, while we use baseline settings. Compared with the official results of the TREC Enterprise track [4, 10], results produced by our boolean model would be in the top 3 for 2005 and top 10 for 2006. Top performing systems tend to use various kinds of heuristics, manual topic expansion, and sub-document models.

Surprisingly, in most cases the boolean model performed better than the frequency-based weighting schemes. The only noticeable difference is for Model 2 using the 2005 topics, where the IDF weighting achieves up to 0.25 MAP. The explanation of this behavior, again, lies in the nature of the 2005 topic set. Relevant experts in TREC 2006 are more popular in the collection compared to those identified in TREC 2005 [5], which means that penalizing popular candidates, which is indeed what IDF does, is beneficial for TREC 2005. Importantly, Model 1 shows much more variance in accuracy than Model 2. In case of the more realistic 2006 topic set, the use of various methods for Model 2 indicate hardly any difference. To explain this effect, we need to consider the inner workings of these two strategies. In case of the candidate model (Model 1), document-candidate associations determine the degree to which a document contributes to the person's profile. If the candidate is a "regular term" in the document, shorter documents contribute more to the profile than longer ones. E.g., if the person is an author of a document and appears only at the top of the page, a shorter document influences her profile more than a longer one. Intuitively, a length normalization effect would be desired to account for this. The boolean approach adds all documents with the same weight to the profile, and as such, does not suffer from this effect. On the other hand, this simplification may be inaccurate, since all documents are handled as if authored by the candidate.

For the document model (Model 2), we can observe the same length normalization effect. E.g., if two documents d_1, d_2 contain the same candidates, but have $|d_1| = 1000$ and $|d_2| = 250$, while the relevance scores of these documents are 1 and 0.5, respectively, then d_2 will add twice as much as d_1 to the final expertise score, even though its relevance is lower.

5.3   Using Lean Documents

To overcome the length normalization problem, we propose a *lean document representation*, where documents contain only candidate identifiers, and all other

Table 2. Lean document representation. For each year-model combination the best scores are in boldface.

Method	TREC 2005		TREC 2006	
	Model 1	Model 2	Model 1	Model 2
Boolean	.1858	.2196	.3075	.4704
TF	.2141$^{(3)}$ (+234%)	.1934 (-5.1%)	.3724$^{(3)}$ (+132%)	.4654 (+3.7%)
IDF	.1845	**.2512**	.2736	.4670
TFIDF	**.2304**$^{(3)}$ (+67.6%)	.2176 (-3.9%)	.3380$^{(2)}$ (+19.5%)	**.4728** (+4.7%)
LM	.2102$^{(3)}$ (+227%)	.1932 (-4.8%)	**.3763**$^{(3)}$ (+137%)	.4627 (+3.5%)

terms are filtered out. This can be viewed as "extreme stopwording," where all terms except candidate identifiers are stopwords. Given this representation, the same weighting schemes are used as before. Calculating TF on lean documents is identical to the candidate-centric way of forming associations proposed in [1]. IDF values remain the same, as they rely only on the number of documents in which the candidate occurs, which is unchanged.

For language models, the association's strength is calculated using

$$p(ca|d) = (1 - \lambda) \cdot \frac{n(ca, d)}{|d|} + \lambda \cdot \frac{n(ca)}{\sum_{d'} |d'|}, \qquad (11)$$

where $|d|$ denotes the length of d (total number of candidate occurrences in d), and $n(ca) = \sum_{d'} n(ca, d')$. Essentially, this is the same as the so-called document-based co-occurrence model of Cao et al. [3].

Table 2 presents the results. Significance is tested against the normal document representation (corresponding rows of Table 1, STRICT MatchTypes). The numbers in brackets denote the relative changes in performance.

For Model 1, using the lean document representation shows improvements of up to 227% compared to the standard document representation, and up to 24% compared to the boolean approach (differences are statistically significant). This shows the need of the length normalization effect for candidate-based approaches, such as Model 1, and makes frequency-based weighting schemes using lean documents a preferred alternative over the boolean method.

As to Model 2, the results are mixed. Using the lean document representation instead of the standard one hurts for the TREC 2005 topics, and shows moderate improvement (up to 4.7%) on the 2006 topics. For the document-based expert retrieval strategy the relative ranking of candidates for a fixed document is unchanged, and the length normalization effect is apparently of less importance than for the candidate-based model. Compared to the boolean association method, there is no significant improvement in performance (except the IDF weighting for 2005, which we have discussed earlier).

5.4 Semantic Relatedness

So far, we have used the number of times a candidate occurs in a document as an indication of its importance for the document. We will now re-visit this

Table 3. Comparing frequency-based associations using lean representations (FREQ) and semantic-relatedness of documents and candidates (SEM)

Method	TREC 2005						TREC 2006					
	Model 1			Model 2			Model 1			Model 2		
	FREQ	SEM	τ	FREQ	SEM	τ	FREQ	SEM	τ	FREQ	SEM	τ
TF	.2141	.2128	.750	.1934	.2012	.816	.3724	.3585	.761	.4654	.4590	.841
IDF	.1845	.1836	.982	.2512	.2541	.964	.2736	.2732	.986	.4670	.4586	.971
TFIDF	.2304	.2335	.748	.2176	.2269	.809	.3380	.3352	.771	.4728	.4602	.827
LM	.2102	.2117	.756	.1932	.2009	.816	.3763	.3671	.761	.4627	.4576	.841

assumption. We propose an alternative way of measuring the candidate's weight in the document—semantic relatedness. We use the lean document representation, but a candidate is represented by its semantic relatedness to the given document, instead of its actual frequency. We use $n'(ca, d)$ instead of $n(ca, d)$, where

$$n'(ca, d) = \begin{cases} \text{KLDIV}(\theta_{ca} \| \theta_d), & n(ca, d) > 0 \\ 0, & \text{otherwise.} \end{cases} \tag{12}$$

That is, if the candidate is mentioned in the document, his weight will be the distance between the candidate's and the document's language models, where the document's language model is calculated using Eq. 5 and the candidate's language model is calculated using Model 1, Eq. 3.

The absolute performance of the association method based on semantic relatedness is in the same general range as the frequency-based association method listed alongside it. Columns 4, 7, 10, 13 provide the Kendall tau rank correlation scores for the two columns that precede them—which are very high indeed. These correlation scores suggest that frequency-based associations based on lean documents are capable of capturing the semantics of the associations.

6   Discussion and Conclusions

As a retrieval task, expert finding has attracted much attention since the launch of the Enterprise Track at TREC in 2005. Two clusters of methods emerged, so-called candidate and document models. Common to these approaches is a component that estimates the strength of the association between a document and a person. Forming such associations is a key ingredient, yet this aspect has not been addressed as a research topic. In this paper we introduced and systematically compared a number of methods for building document-people associations. We made explicit a number of assumptions underlying various association methods and analyzed two of them in detail: (i) independency of candidates, and (ii) frequency is an indication of strength.

We gained insights in the inner workings of the two main expert search strategies, and found that these behave quite differently with respect to document-people associations. Candidate-based models are sensitive to associations. Lifting the candidate independence assumption and moving from boolean to frequency based methods can improve performance by up to 24%. However, the standard document representation (where candidate occurrences are treated as regular terms) suffers from length normalization problems, therefore, a lean document representation (that contains only candidates, while all other terms are filtered out) should be used.

On the other hand document-based models are less dependent on associations, and the boolean model turned out to be a very strong baseline. Only a moderate (up to 4.7%) improvement can be gained by moving to frequency-based associations. Absolute scores of the document-based model are substantially higher than of the candidate-based one, which makes it the preferred strategy.

To assess the *frequency is an indication of strength* assumption we proposed a new people-document association approach, based on the semantic relatedness of the document and the person. We find that frequencies succeed very well at capturing the semantics of person-document associations.

This study suggest that this is how far we can get by capturing expertise at the document level. For further improvements we seem to need sub-document models as well corpus-specific methods but in a non-heuristic way.

Acknowledgments

This research was supported by the Netherlands Organization for Scientific Research (NWO) by grants under project numbers 017.001.190, 220-80-001, 264-70-050, 354-20-005, 600.065.120, 612-13-001, 612.000.106, 612.066.302, 612.069.006, 640.001.501, 640.002.501, and by the E.U. IST programme of the 6th FP for RTD under project MultiMATCH contract IST-033104.

References

[1] Balog, K., Azzopardi, L., de Rijke, M.: Formal models for expert finding in enterprise corpora. In: SIGIR 2006, New York, NY, USA, pp. 43–50 (2006)

[2] Bao, S., Duan, H., Zhou, Q., Xiong, M., Cao, Y., Yu, Y.: Research on Expert Search at Enterprise Track of TREC 2006. In: The Fifteenth Text REtrieval Conference Proceedings (TREC 2006) (2007)

[3] Cao, Y., Liu, J., Bao, S., Li, H.: Research on Expert Search at Enterprise Track of TREC 2005. In: The Fourteenth Text REtrieval Conference Proceedings (TREC 2005) (2006)

[4] Craswell, N., de Vries, A.P., Soboroff, I.: Overview of the TREC-2005 Enterprise Track. In: The Fourteenth Text REtrieval Conference Proceedings (TREC 2005) (2006)

[5] Fang, H., Zhai, C.: Probabilistic Models for Expert Finding. In: Amati, G., Carpineto, C., Romano, G. (eds.) ECIR 2007. LNCS, vol. 4425, pp. 418–430. Springer, Heidelberg (2007)

[6] Fu, Y., Yu, W., Li, Y., Liu, Y., Zhang, M.: THUIR at TREC 2005: Enterprise Track. In: The Fourteenth Text REtrieval Conference Proceedings (TREC 2005) (2006)

[7] Macdonald, C., Ounis, I.: Voting for candidates: adapting data fusion techniques for an expert search task. In: CIKM 2006, pp. 387–396 (2006)

[8] Macdonald, C., Plachouras, V., He, B., Ounis, I.: University of Glasgow at TREC2005: Experiments in Terabyte and Enterprise tracks with Terrier. In: Proceedings of the 14th Text REtrieval Conference (TREC 2005) (2005)

[9] Petkova, D., Croft, W.B.: Hierarchical language models for expert finding in enterprise corpora. In: ICTAI 2006, pp. 599–608 (2006)

[10] Soboroff, I., de Vries, A.P., Craswell, N.: Overview of the TREC 2006 Enterprise Track. In: TREC 2006 Working Notes (2006)

[11] W3C. The W3C test collection (2005),
http://research.microsoft.com/users/nickcr/w3c-summary.html

[12] Yimam-Seid, D., Kobsa, A.: Expert finding systems for organizations: Problem and domain analysis and the demoir approach. Journal of Organizational Computing and Electronic Commerce 13(1), 1–24 (2003)

[13] Zhai, C., Lafferty, J.: A study of smoothing methods for language models applied to ad hoc information retrieval. In: SIGIR 2001, pp. 334–342 (2001)

Modeling Documents as Mixtures of Persons for Expert Finding

Pavel Serdyukov and Djoerd Hiemstra

Database Group, University of Twente,
PO Box 217, 7500 AE, Enschede, The Netherlands
{serdyukovpv,hiemstra}@cs.utwente.nl

Abstract. In this paper we address the problem of searching for knowledgeable persons within the enterprise, known as the expert finding (or expert search) task. We present a probabilistic algorithm using the assumption that terms in documents are produced by people who are mentioned in them. We represent documents retrieved to a query as mixtures of candidate experts language models. Two methods of personal language models extraction are proposed, as well as the way of combining them with other evidences of expertise. Experiments conducted with the TREC Enterprise collection demonstrate the superiority of our approach in comparison with the best one among existing solutions.

1 Introduction

In enterprises or in common web search settings users often experience the need not only for getting information, but for getting into the contact with those who could be the source of this information. The opportunity of interaction with a knowledgeable person is sometimes appreciated much higher than the access to a very relevant and clearly written document on the search topic [20]. An expert finding system helps to find individuals or even working groups possessing certain expertise and skills within an organization [6]. Quite like a typical document retrieval system, it uses a short user query and documents stored on personal desktops or within centralized databases as the input. The prediction of a personal expertise is made through the analysis of textual content of documents the person has relation to. The proof of relation can be authorship, simple occurrence in the text or just a fact that the document is stored locally at the PC (e.g. in the browser cache). For ensuring traceability, the system must return not only the ranking of people, but also the list of those documents that appeared to be the best indicators of expertness.

Apart from causing the boom on the enterprise search systems market [19], expert finding systems also compelled close attention of the IR research community. The expert search task is included into the Enterprise track of the Text REtrieval Conference (TREC) since 2005 [5]. The TREC community provided the experimental dataset and set up the standards for the evaluation.

The fundamental principle of state-of-art methods for expert finding is to infer personal expertise by studying the co-occurrence of personal identifiers (names,

C. Macdonald et al. (Eds.): ECIR 2008, LNCS 4956, pp. 309–320, 2008.
© Springer-Verlag Berlin Heidelberg 2008

email addresses etc.) and query terms in the scope of documents. The more often a person is detected in the documents containing many words describing the topic, the more likely we may rely on this person as an expert in this topic. However, all methods also consider that persons as well as terms occur in the document independently and do not influence the appearance of each other. Although, the assumption about independence among terms is a de facto standard in IR [7], the independence of terms from persons seems not quite adequate.

In this paper we consider that the occurrence of terms in the document can not be considered independent from the presence of a candidate expert. We propose a ranking model in which people are regarded as generators of the document's content. Our generative modeling method combines the features of both so-called profile- and document-centric approaches: it ranks candidates using their language models built from the retrieved documents and takes the frequency of appearance of a candidate in the top ranked documents as an additional evidence of his proficiency in a search topic.

The paper is organized as follows. In the next section we give a more detailed description of existing approaches to expert finding. In Section 3, we show how to utilize the assumption that persons mentioned in the document determine which terms it consists of. In Section 4, we explain how personal language models can be mined from retrieved documents and used further to build a good predictor of personal expertise. Experimental results supporting our assumptions are given in Section 5. Discussion of the paper outcome and a brief outline of potential future work can be found in Section 6.

2   Related Work

Existing approaches to candidate experts modeling and ranking are basically variations of two kinds. The first approach is *profile-centric* [15,21]. All documents related to a candidate expert are merged into a single personal profile prior to retrieval time. The personal profiles are ranked w.r.t a query as single documents using standard retrieval measures and corresponding best candidates are returned to the user. The second approach, *document-centric*, is based on the analysis of individual documents. It runs a query against all documents and ranks candidates by summarized scores of associated documents [1,9,17] or text windows surrounding the person's mentioning [16]. It is also suggested not to finish propagation of scores on the level of directly related persons and to propagate the scores further through reciprocal document-candidate links [23]. Document-centric approaches are claimed to be much more effective than profile-centric [1], probably due to the fact that they estimate the relevance of the text content related to a person on the much lower and hence less ambiguous level.

A subfamily of document-centric methods exploits the social network built using links among persons extracted from top documents (e.g. by utilizing *from* and *to* fields of emails). The persons are ranked by popular centrality measures calculated on the acquired network. Campbell et al. [4] proposed the use of HITS algorithm [13] which performed better than just ranking by candidate's in-degree (related documents number). However, Chen et al. [10] found that a document-centric approach is still better than HITS based ranking. Query-independent experts discovery using links acquired from posts and replies at specialized forums was studied recently by Zhang et al. [27].

A number of advanced pseudo relevance feedback techniques are consistently applied to the expert search task. Query expansion from the top retrieved documents performed quite well [2,21]. Macdonald and Ounis also successfully experimented with different numbers of expansion terms received from the top ranked candidate profiles [18]. Serdyukov et al. [22] applied massive query expansion using the mixture of two pseudo-relevance language models: built on top ranked documents and top ranked profiles.

Expert finding is only a subcase of the entity ranking task. Generalization of search for other entity classes in the Web (*countries, cities, dates etc.*) is made by Zaragoza et al. [25] and Tsikrika et al. [24].

3   Person-Centric Expert Finding

The key approaches to expert finding discussed in the previous section state that the level of personal expertise can be determined by the aggregation of document scores related to a person. As we show further, their intuition is based on measuring the co-occurrence degree of the query terms and personal id within the context of a document. In probabilistic terms, they suppose that our task comes to the estimation of the joint probability $P(e, q_1, ..., q_k)$ of observing the candidate expert e together with query terms $q_1...q_k$ in the documents ranked by the query. The methods which we describe here are graphically represented in Figure 1. We see that while in the typical document-centric method shown on the left (see Section 3.1), the document is responsible for producing terms, in our method, shown on the right (see Section 3.2), the document requests a person to generate its terms. Below, we define these models formally.

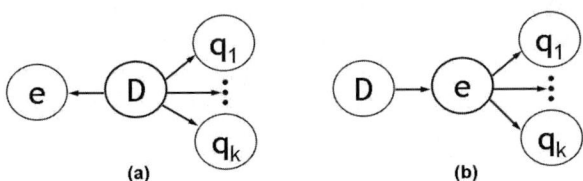

Fig. 1. Dependence networks for two methods of estimating $P(e, q_1, ..., q_k)$

3.1   Baseline Approach

Let's take a look at the well-known document-centric model by Balog and De Rijke (their Model 2) [1] using the principle shown in Figure 1a. According to their approach, we have the following formulas for the total joint probability $P(e, q_1, ..., q_k)$ over ranked documents set R:

$$P(e, q_1, ..., q_k) = \sum_{D \in R} P(D)P(e, q_1, ..., q_k|D) \tag{1}$$

$$P(e, q_1, ..., q_k|D) = P(e|D) \prod_{i=1}^{k} P(q_i|D) \tag{2}$$

where $P(D)$ is a document prior, which is uniform. $P(e|D)$ is the probability of relation between person e and document D, calculated as:

$$P(e|D) = \frac{a(e, D)}{\sum_{i=1}^{m} a(e_i, D)}, \tag{3}$$

where m is a number of candidate experts in the system and $a(e, D)$ is a nonnormalized association degree between the person and the document, which may depend on various factors: on the importance of the document part containing the person, on the number of occurrences of the personal identifier in the document, or on our confidence that a certain personal identifier found in the document matches person e.

The right part of the Equation (2) is a score of a document according to the language model based ranking principle [11], in which:

$$P(w|D) = (1 - \lambda_G)\frac{c(w, D)}{|D|} + \lambda_G P(w|G), \tag{4}$$

where $c(w, D)$ is the count of term w in document D, $|D|$ is its length, λ_G is the probability that term w will be generated from the global language model. $P(w|G)$ is the global language model estimated over the whole document collection.

As we may notice, this approach considers a candidate and the query terms to be conditionally independent given a ranked document (see Figure 1a). It is also similar to the popular query expansion method by Lavrenko [14] if only we consider the candidate expert as an expansion term. Since it's not only the most representative, but also the one of the most effective expert finding methods [1], it serves as a baseline in our experiments.

3.2   Putting Persons in the Middle

The person-centric method, which is the contribution of our paper, can be viewed as a hybrid method combining the features of both document- and profile-centric methods. It builds its prediction by analysing the top retrieved documents and summarizing the expertise evidence over them. However, the estimation of a personal language model (see next section) becomes a crucial step in this prediction.

Our approach is based on the assumption of dependency between the query terms and a candidate. We suppose that candidates are actually responsible for the generation of terms within retrieved documents. According to the model presented in Figure 1b, we calculate the required joint probability as follows:

$$P(e, q_1, .., q_k) = \sum_{D \in R} P(q_1, .., q_k | e) P(e|D) P(D) = P(q_1, .., q_k | e) \sum_{D \in R} P(e|D) P(D) \quad (5)$$

where $P(q_1, ..., q_k | e)$ is the probability of generating the query from the personal language model. It reflects the amount of relevant knowledge the candidate has. The sum in the right part of this formula can be considered as a person's prior $P(e)$:

$$P(e) = \sum_{D \in R} P(e|D) P(D), \quad (6)$$

which measures the influence/activity of the candidate in the topic area. It is proportional to the frequency of appearance of the candidate in the topical documents. We take a ranked document prior to be inversely dependent on the document rank: $P(D) = 1/rank(D)$ in order to distinguish the importance of a document in covering the aspects of the query topic. In our experiments we also show the performance with uniformly distributed $P(e) = 1/m$, where m is a number of candidate experts in the system.

We also consider that query terms occur independently given a candidate experts, what results in:

$$P(q_1, ..., q_k | e) = \prod_{i=1}^{k} P(q_i | e) \quad (7)$$

Now we present our algorithm of mining for personal language models from the top retrieved documents.

4   Mining for Personal Language Models

As we see, the personal query term generation probabilities $P(q_i|e)$ is the only part we miss so far. Of course, we can get them in the way similar to the one which profile-centric methods use: merge those retrieved documents that relate to the person e into one and calculate corresponding term frequencies. However, it would be justifiable if only there was only one person per document. Since we have already postulated that all candidates may be responsible for generating query terms in the documents they are mentioned in, such approach would give us only very rough approximation of a personal language model in most cases. Guided by these considerations, we represent a document as a mixture of personal models and the global language model. In formal terms, we define the likelihood of the top retrieved documents set R as:

$$P(R) = \prod_{D \in R} \prod_{w \in D} ((1 - \lambda_G)(\sum_{i=1}^{m} P(e_i|D) P(w|e_i)) + \lambda_G P(w|G))^{c(w,D)} \quad (8)$$

Here $e_1, ..., e_m$ are the persons occurring in the documents from R, $c(w, D)$ is the count of term w in document D, $(1 - \lambda_G)$ is a probability that a term will be generated from one of the personal models and not from the global language model. λ_G controls the ability of the algorithm to build personal models which are discriminative only for the terms which are topic-specific. Those terms which have high probability in the collection in total will get low generation probabilities over all persons.

Our approach to candidate experts modeling is based on the similar hypothesis with one used in pseudo-relevance feedback method for document retrieval by Zhai and Lafferty [26]. It also considers that the topical model of a user query can be mined from the top retrieved documents. The significant difference is that we define this model as a mixture of models of candidate experts. They are those who actually hold and share the knowledge which can meet the user information need. To say the truth, the personal language model which we get from top documents is only one of many the person uses. If we analyze the whole collection in the same way, we could get much more detailed personal term distribution. However, it would be much more difficult to distinguish candidate experts because the ambiguity of their expertise would increase dramatically in this case. Since we are interested only in the language model the person uses while generating documents that cover the query topic to some extent, it is reasonable to get it dynamically: at query execution time from retrieved documents. Our approach also shows some resemblance with Probabilistic Latent Semantic Indexing [12] with a distinction that our semantic topics are not 'latent', but personified and hence 'visible' in documents.

4.1   Using Fixed Personal Contribution Probabilities

Considering that all parameters, including $P(e_i|D)$ are given, we are able to calculate the maximum likelihood estimates of term generation probabilities from personal language models $P(w|e_i)$. In order to do that, we apply the EM algorithm [8], traditionally used to estimate unknown parameters. We propose the following formulas updating likelihood of the document set R (see Equation (8)) to be used recursively for its maximization:

E-step:

$$P(e|w, D) = \frac{(1 - \lambda_G)P(e|D)P(w|e)}{(1 - \lambda_G)(\sum_{i=1}^{m} P(e_i|D)P(w|e_i)) + \lambda_G P(w|G)} \tag{9}$$

M-step:

$$P(w|e) = \frac{\sum_{D \in R} c(w, D)P(e|w, D)}{\sum_w \sum_{D \in R} c(w, D)P(e|w, D)} \tag{10}$$

4.2   Measuring Personal Contribution

So far we relied on the assumption that probabilities $P(e|D)$ are fully determined by a person-document type of association. This practically means that if we have some document with probability distribution $P(e|D)$, then for some another document with the same set of persons having the same kind of associations with it, the probability $P(e|D)$ will be distributed likewise. However, in our method we extract not only personal language models, but also probability distributions $P(e|w, D)$, which show who is the most probable generator of the term w in the document D. It allows us to estimate the probability of contribution for each person solely based on the document's content.

For that purpose, we no more fix probabilities $P(e|D)$ and calculate them at every M-step of EM algorithm presented in Section 4.1 as follows:

$$P(e|D) = \frac{1 + \sum_{w \in D} c(w, D) P(e|w, D)}{m + \sum_{i=1}^{m} \sum_{w \in D} c(w, D) P(e_i|w, D)}, \tag{11}$$

where m is the number of candidate experts extracted from the retrieved documents in total, used here for the purposes of Laplace smoothing.

5   Experiments

5.1   Experimental Setup

For the evaluation we utilize the W3C corpus - the data from the expert search task in the Enterprise track of the TREC used in 2005 and 2006 - and its largest (1.8 GB) 'lists' part containing discussions within the W3C consortium. We focus our experiments on this part of the collection for several reasons. At first, this data has a standardized format (emails of average length 450 words) what means that its properties should not change significantly across different enterprises. Moreover, it allows to accomplish persons recognition using unique email addresses and hence to avoid uncertainty in determining person-document relations. Since these email addresses always occur in a specific email field, we are able to differentiate the types of person-document relations as well. The data is parsed and indexed using Java and the Lucene open-search engine.

TREC also provided a list of 1092 candidate experts with supplemented full names, email addresses and unique ids. Experiments were conducted by considering only these candidates as person entities. We also tested inclusion of other person entities by taking any unique email found in the collection as a new person id. This caused only small degradation of performance, probably due to the rapid increase of noisy features with each new document retrieved, so we do not report these results here.

We provide results separately for two sets of TREC queries with relevance judgments: used in 2005 (50 queries) and in 2006 (49 queries). These query sets are somewhat different in nature. In 2005 queries were made up using names of working groups in W3C as titles and members of these groups as experts on the query topic. In 2006 the the TREC community manually judged each candidate for each query using the provided list of documents where a person id occurred. While queries from 2006 allow to reproduce a classic expert search scenario, queries from 2005 partly simulate the search for sub-groups within organization (a search for any person in the group working on the query topic problem).

5.2   Results Discussion

First of all, we do candidates recognition by finding their email addresses in *from*, *to*, *cc* and *body* email fields. We additionally search for candidates in *body* fields using their full names. Association scores are $a(e, D^{from}) = 1.5$, $a(e, D^{to}) = 1.0$, $a(e, D^{cc}) = 2.5$ and $a(e, D^{body}) = 1.0$, what is the best combination according to recent studies of W3C 'lists' subcollection [3]. If a person appears in several fields, the highest association score is taken. The standard language model based IR approach, as defined in Equations (2) and (4), was used for the retrieval of documents.

We analyze the performance using the classic IR evaluation measures: Mean Average Precision (MAP), Mean Reciprocal Rank (MRR) and precision at top 5 ranked candidates (P@5). In our opinion, P@5 is more relevant to our task than precisions at greater ranks. The cost of a false recommendation in expert search is much higher than in document search: a conversation with an ignorant person or even reading all documents supporting the incorrect system's judgment takes much longer time than reading one irrelevant document. If we consider that the user can be satisfied with any single expert on the topic, than MRR becomes a decisive measure: it shows the ability of the system to present an expert as soon as possible if to go down by person's ranking one by one.

In order to demonstrate the quality of the mined personal language models (see Section 4), we start from presenting the performance of our methods considering that person's priors $P(e)$ are uniformly distributed and then using non-uniform priors, as in Equation (5), with the best of them. So, the following methods are evaluated:

- **Baseline:** the baseline document-centric method (see Section 3.1),
- **PCFix:** the person-centric method using fixed person-document association scores and uniform personal priors (see Sections 3.2 and 4.1),
- **PCUnf:** the person-centric method using unfixed dynamically calculated association scores and uniform personal priors (see Sections 3.2 and 4.2),
- **PCUnfNonUniPriors:** the person-centric method using unfixed dynamically calculated association scores and non-uniform personal priors (see Sections 3.2 and 4.2).

We have only two parameters in all models including the baseline model: λ_G, used in Equations (4) and (8), and the number of retrieved documents. Different values for λ_G between 0.1 and 0.9 showed negligible differences in performance, but 0.8 was slightly better than others. The second parameter was much more influential. It is always rather unclear how many top documents describe each query topic to the sufficient extent. So, a good algorithm should be robust to the size of a query result set. We vary its size from 1000 to 6000 of top ranked documents. We show MAP, P@5 and MRR values for both sets of queries in Figures 2, 3 and 4 respectively.

We see that the **PCFix** method performs similarly to the **Baseline** in average, except that it is notably better on P@5 for queries from 2005 (see Figure 3b). For other measures/queries, although it's better in half cases, it is worse in another half too.

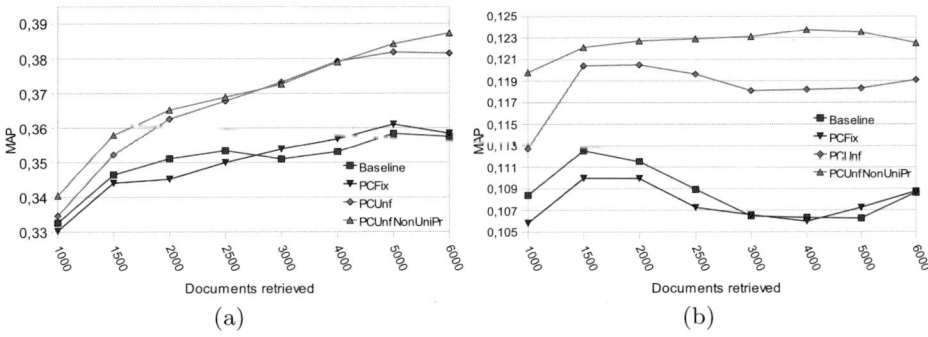

Fig. 2. MAP over different numbers of documents retrieved, for the queries from 2006 (a) and for the queries from 2005 (b)

However, the **PCUnf** method shows notably better performance than both the **Baseline** and the **PCFix** methods on all measures/queries, especially for MRR measure. It demonstrates that query-specific and purely content-based estimation of personal contribution to the document is crucial in personal language modeling.

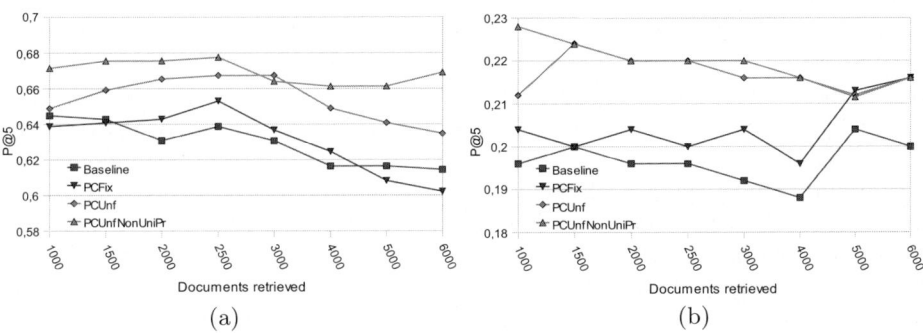

Fig. 3. Precision at 5 over different numbers of documents retrieved, for the queries from 2006 (a) and for the queries from 2005 (b)

Moreover, using non-uniform priors $P(e)$, as in Equation 6, with the **PCUnf** method (the method **PCUnfNonUniPriors**) improves performance further for all MAP and P@5 measures at almost all numbers of retrieved documents. The frequency of participation in discussions on the topic is of course a significant evidence of personal expertise. However, from a statistical point of view, this prior penalizes the score of those candidates whose models are built using insufficient amount of training data, i.e. related documents. Both effects in total prevent incidental persons from getting high scores. However, using non-uniform priors spoils the performance of the **PCUnf** in case of MRR measure. So, if the

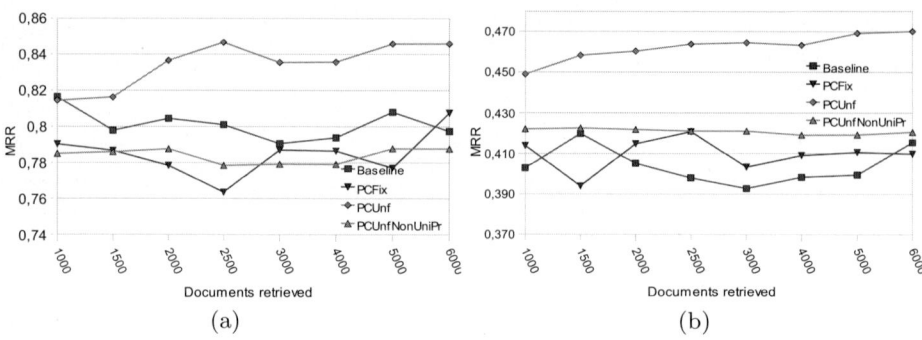

Fig. 4. MRR over different numbers of documents retrieved, for the queries from 2006 (a) and for the queries from 2005 (b)

user information need can be effectively satisfied with only one expert (and she is always available for requests), then the **PCUnf** is more preferable.

To sum things up, the presented results imply that our person-centric model is built on more realistic assumptions than the baseline document-centric model.

6 Conclusions and Further Work

We have presented the method for expert finding based on modeling of retrieved documents as mixtures of personal language models. Our approach assumed that terms in documents are generated by those persons who are mentioned in them. For the final ranking it combined two evidences of personal expertise: the probability of generation of the query by the personal language model and a prior probability of candidate experts expressing her level of activity in the important discussions on the query topic. We proposed two ways of personal models extraction from top ranked documents. In one case, we considered that person-document relation probabilities are fixed and fully depend on the field of a document where the person appeared. In another case, we obtained these probabilities dynamically by predicting the real contribution of persons to a document considering their intermediately calculated language models. When our method used this second way of modeling, it outperformed one of the best state-of-art approaches which we used as a baseline.

Several directions of improvement can be followed in the future. Certainly, the core person's modeling part can be extended up to higher complexity. We may imagine that a person is a mixture of sub-persons representing different fields of her expertise. These inside experts can be used differently across documents and their probability of use may even depend on the set of other persons appearing in the document. A document can be also represented not only as a mixture of persons, but also as a mixture of global latent topics, which in turn appear to be mixtures of persons, accumulating knowledge in the corresponding fields.

Or we can even suppose that terms and persons are independent given such latent topic, which generates both these kinds of entities.

It is also reasonable to find more use of specific data properties. Particularly, we can consider that persons in the email document appear non-independently: the occurrence of persons in the *to* and the *cc* fields depends on the email sender in the *from* field, who is selecting them for communication. It is promising to take document links into account: for instance, by regarding emails relating to one thread as a single document.

References

1. Balog, K., Azzopardi, L., de Rijke, M.: Formal models for expert finding in enterprise corpora. In: SIGIR 2006: Proceedings of the 29th annual international ACM SIGIR conference on Research and development in information retrieval, pp. 43–50 (2006)
2. Balog, K., Bogers, T., Azzopardi, L., de Rijke, M., van den Bosch, A.: Broad expertise retrieval in sparse data environments. In: SIGIR 2007: Proceedings of the 30th annual international ACM SIGIR conference on Research and development in information retrieval, pp. 551–558. ACM Press, New York (2007)
3. Balog, K., de Rijke, M.: Finding experts and their details in e-mail corpora. In: 15th International World Wide Web Conference (WWW 2006) (2006)
4. Campbell, C.S., Maglio, P.P., Cozzi, A., Dom, B.: Expertise identification using email communications. In: CIKM 2003: Proceedings of the twelfth international conference on Information and knowledge management, pp. 528–531. ACM Press, New York (2003)
5. Craswell, N., de Vries, A., Soboroff, I.: Overview of the trec-2005 enterprise track. In: Proceedings of TREC-2005, Gaithersburg, MD (2005)
6. Craswell, N., Hawking, D., Vercoustre, A.-M., Wilkins, P.: Panoptic expert: Searching for experts not just for documents. In: Ausweb Poster Proceedings, Queensland, Australia (2001)
7. Crestani, F., Lalmas, M., Rijsbergen, C.J.V., Campbell, I.: "Is this document relevant?: Probably": a survey of probabilistic models in information retrieval. ACM Comput. Surv. 30(4), 528–552 (1998)
8. Dempster, A., Laird, N.M., Rubin, D.B.: Maximum likelihood from incomplete data via the EM algorithm. Journal of the Royal Statistical Society, Series B 39(1), 1–38 (1977)
9. Fang, H., Zhai, C.: Probabilistic models for expert finding. In: Amati, G., Carpineto, C., Romano, G. (eds.) ECIR 2007. LNCS, vol. 4425, pp. 418–430. Springer, Heidelberg (2007)
10. Xiong, J., Tan, S., Chen, H., Shen, H., Cheng, X.: Social Network Structure behind the Mailing Lists: ICT-IIIS at TREC 2006 Expert Finding Track. In: Proceeddings of the 15th Text REtrieval Conference (TREC 2006) (2006)
11. Hiemstra, D., de Jong, F.M.G.: Statistical language models and information retrieval: Natural language processing really meets retrieval. Glot international 5(8), 288–293 (2001)
12. Hofmann, T.: Probabilistic latent semantic indexing. In: SIGIR 1999: Proceedings of the 22nd annual international ACM SIGIR conference on Research and development in information retrieval, pp. 50–57. ACM Press, New York (1999)

13. Kleinberg, J.M.: Authoritative sources in a hyperlinked environment. J. ACM 46(5), 604–632 (1999)
14. Lavrenko, V., Croft, W.B.: Relevance based language models. In: SIGIR 2001: Proceedings of the 24th annual international ACM SIGIR conference on Research and development in information retrieval, pp. 120–127. ACM Press, New York (2002)
15. Liu, X., Croft, W.B., Koll, M.: Finding experts in community-based question-answering services. In: CIKM 2005: Proceedings of the 14th ACM international conference on Information and knowledge management, pp. 315–316. ACM Press, New York (2005)
16. Lu, W., Robertson, S., Macfarlane, A., Zhao, H.: Window-based Enterprise Expert Search. In: Proceeddings of the 15th Text REtrieval Conference (TREC 2006) (2006)
17. Macdonald, C., Ounis, I.: Voting for candidates: adapting data fusion techniques for an expert search task. In: CIKM 2006: Proceedings of the 15th ACM international conference on Information and knowledge management, pp. 387–396. ACM Press, New York (2006)
18. Macdonald, C., Ounis, I.: Using relevance feedback in expert search. In: Amati, G., Carpineto, C., Romano, G. (eds.) ECIR 2007. LNCS, vol. 4425, pp. 431–443. Springer, Heidelberg (2007)
19. Maybury, M.T.: Expert finding systems. Technical Report MTR06B000040, MITRE Corporation (2006)
20. McDonald, D.W., Ackerman, M.S.: Just talk to me: a field study of expertise location. In: CSCW 1998: Proceedings of the 1998 ACM conference on Computer supported cooperative work, pp. 315–324. ACM Press, New York (1998)
21. Petkova, D., Croft, W.B.: Hierarchical language models for expert finding in enterprise corpora. In: ICTAI 2006: Proceedings of the 18th IEEE International Conference on Tools with Artificial Intelligence, pp. 599–608. IEEE Computer Society, Los Alamitos (2006)
22. Serdyukov, P., Chernov, S., Nejdl, W.: Enhancing expert search through query modeling. In: Amati, G., Carpineto, C., Romano, G. (eds.) ECIR 2007. LNCS, vol. 4425, pp. 737–740. Springer, Heidelberg (2007)
23. Serdyukov, P., Rode, H., Hiemstra, D.: University of Twente at the TREC 2007 Enterprise Track: Modeling relevance propagation for the expert search task. In: Proceeddings of the 16th Text REtrieval Conference (TREC 2007) (2007)
24. Tsikrika, T., Serdyukov, P., Rode, H., Westerveld, T., Aly, R., Hiemstra, D., de Vries, A.: Structured Document Retrieval, Multimedia Retrieval, and Entity Ranking Using PF/Tijah. In: Fuhr, N., Lalmas, M., Trotman, A. (eds.) INEX 2006. LNCS, vol. 4518, Springer, Heidelberg (2007)
25. Zaragoza, H., Rode, H., Mika, P., Atserias, J., Ciaramita, M., Attardi, G.: Ranking very many typed entities on wikipedia. In: CIKM 2007: Proceedings of the sixteenth ACM conference on Conference on information and knowledge management, pp. 1015–1018. ACM Press, New York (2007)
26. Zhai, C., Lafferty, J.: Model-based feedback in the language modeling approach to information retrieval. In: CIKM 2001: Proceedings of the tenth international conference on Information and knowledge management, pp. 403–410 (2001)
27. Zhang, J., Ackerman, M.S., Adamic, L.: Expertise networks in online communities: structure and algorithms. In: WWW 2007: Proceedings of the 16th international conference on World Wide Web, pp. 221–230. ACM Press, New York (2007)

Ranking Users for Intelligent Message Addressing

Vitor R. Carvalho[1] and William W. Cohen[1,2]

Language Technologies Institute[1] and Machine Learning Department[2]
Carnegie Mellon University, Pittsburgh, PA 15218 USA

Abstract. Finding persons who are knowledgeable on a given topic (i.e. Expert Search) has become an active area of recent research [1,2,3]. In this paper we investigate the related task of *Intelligent Message Addressing*, i.e., finding persons who are potential recipients of a message under composition given its current contents, its previously-specified recipients or a few initial letters of the intended recipient contact (intelligent auto-completion). We begin by providing quantitative evidence, from a very large corpus, of how frequently email users are subject to message addressing problems. We then propose several techniques for this task, including adaptations of well-known formal models of Expert Search. Surprisingly, a simple model based on the K-Nearest-Neighbors algorithm consistently outperformed all other methods. We also investigated combinations of the proposed methods using fusion techniques, which leaded to significant performance improvements over the base-lines models. In auto-completion experiments, the proposed models also outperformed all standard baselines. Overall, the proposed techniques showed ranking performance of more than 0.5 in MRR over 5202 queries from 36 different email users, suggesting intelligent message addressing can be a welcome addition to email.

1   Introduction

Expert search, the task of finding persons who are knowledgeable on a given topic, has been an active area of research recently [1,2,3]. Here we explore the related task of *Intelligent message addressing*, i.e., finding persons who are potential recipients for a message under composition given its current contents, its previously-specified recipients or a few initial letters of the intended recipient contact. This task can be a valuable addition to email clients, particularly in large corporations, where negotiations are frequently handled via email and the cost of errors in task management is very high. Intelligent Message Addressing can prevent a user from forgetting to add an important collaborator or manager as recipient, preventing costly misunderstandings, communication delays and missed opportunities. Below we present empirical evidence that such errors are very common in the corporate environment. The same technique can also potentially provide assistance in identifying people who have previously worked on specific topics or have relevant skills.

C. Macdonald et al. (Eds.): ECIR 2008, LNCS 4956, pp. 321–333, 2008.

In this paper we formalize two variants of this intelligent email addressing as user-ranking, i.e., finding a ranked list of email addresses that are likely to be intended recipients of a given message. We propose several methods for this task, including classification-based models and adaptations of successful Expert Search formal models [1]. Extensive experiments over 36 different users indicate that a simple model based on the K-Nearest Neighbors algorithm generally outperforms all other methods, including more refined Expert Search models. In a second set of experiments, we explore how to combine the rankings of different baseline models using rank-based data fusion techniques. Experiments clearly indicate that combined models can significantly outperform all base models on all prediction tasks.

Intelligent message addressing techniques can also be naturally adapted to improve email address auto-completion, i.e., suggesting the most likely addresses based on a few initial letters of the intended contact. Email auto-completion is an extremely useful and popular feature, but in spite of it, little is publicly known on how addresses are ranked in the most popular email clients, and we are not aware of any study comparing different techniques on this particular message addressing problem. In this paper we evaluate several ranking baselines for this problem — including alphabetical, frequency and recency ordering — in a large collection of users. Results clearly indicate that the proposed intelligent addressing models outperform all baselines, significantly improving suggestions in email auto-completion.

Overall we show that intelligent message addressing techniques are able to visibly improve email auto-completion, as well as to provide valuable assistance for users when composing messages. Results suggest it can be a desired addition to most email clients — for instance, on the task of predicting all email recipients, our methods reached 0.47 in MAP and more than 0.5 in MRR. Another advantage is that the best performing methods are computationally efficient and can be easily adapted to large-scale email systems, with no changes in the email server side.

2   Frequency of Message Addressing Problems

Although email is ubiquitous, large, public and realistic email corpora are not easy to find. The limited availability is largely due to privacy issues. For instance, in most US academic institutions, an email collection can only be distributed to researchers if all senders of the collection also provided explicit written consent. One of the few datasets available is the Enron Email Corpus, a large collection of real email messages from managers and employees of the Enron Corporation. This collection was originally made public by the US Federal Energy Regulatory Commission during the investigation of the Enron accounting fraud. The collection has approximately half a million messages from 150 users' inboxes.

By searching for messages containing the terms *sorry*, *forgot* or *accident* in the entire corpus, and then manually filtering the results[1], we found that at least 9.27% of the users have forgotten to add a desired email recipient in at least one sent

---

[1] Finding messages containing sentences such as "Oops, I forgot to send it to Vince." or "Sorry....missed your name on the cc: list!".

message, while at least 20.52% of the users were not included as recipients (even though they were intended recipients) in at least one received message[2]. These surprisingly high numbers clearly suggest that these problems are very common and that email users can benefit from an intelligent message addressing assistant.

3 Data Preprocessing and Task Definition

As expected, real email data have several inconsistencies. To help mitigate some of these problems, we used the Enron dataset version compiled by Jitesh and Adibi [5], in which a large number of repeated messages were removed. In addition, some users in the corpus used multiple email addresses. We partially addressed this issue by mapping "raw" email addresses into normalized email addresses for some users[3].

In this paper, we describe two possible settings for the recipient prediction task. The first setting is called the *TO+CC+BCC or primary prediction*, where we attempt to predict all recipients of an email given its message contents. It relates to a scenario where the message is composed, but no recipients have been added to the recipient list. The second setting is called *CC+BCC or secondary prediction*, in which message contents as well as the TO-addresses were previously specified, and the task is to rank additional addresses for the CC and BCC fields of the message.

We randomly selected 36 Enron users, and for each user we chronologically sorted their *sent collection* (i.e., all messages sent by this particular user) and then split the collection in two parts: the oldest messages were placed into *sent_train* and most recent ones into *sent_test*. Message counts statistics for the 36 randomly chosen Enron users are shown in Table 1. More specifically, *sent_test* collection was selected to contain at least 20 "valid-CC" messages, i.e., at least 20 messages with valid email addresses in both TO and CC (or both TO and BCC) fields. This particular subset of *sent_test*, with approximately 20 "valid-CC" messages, is called *sent_test**. The main idea is that TO+CC+BCC prediction will be tested on *sent_test*, and the CC+BCC prediction will be tested on the *sent_test** collection (a subset of *sent_test* in which all messages have a valid CC or BCC address).

This chronological split was necessary to guarantee a minimum number of test messages for secondary prediction task and to simulate a typical scenario in a user's desktop — where the user already has several sent messages, and the goal is to predict the recipients of the next sent messages. We also constructed, for each user, an address book set *AB* which is the set of all recipient addresses in the user's *sent_train* collection. A complete analysis of this data preparation over the different users can be found in an extended version of this paper [4].

---

[2] This is a lower bound since not all errors will be noticed by users and not all error-notification emails would be found by our search. A detailed analysis of these results can be found in an extended version of this paper [4].

[3] This mapping (author-normalized-author.txt) was produced by Andres Corrada-Emmanuel, and is currently available from the Enron Email webpage [6].

Table 1. Number of Email Messages in the Different Collections of the 36 selected Enron users. $|AB|$ is the Address Book size, i.e., the number of different recipients that were addressed in the messages of the *sent_train* collection. **Sent_test*** contains only messages having valid addresses in both TO and CC fields. User specific numbers can be found in [4].

	$|AB|$	sent_train	sent_test	sent_test*
Mean	377.67	1266.69	144.50	20.50
StDev	263.24	1099.05	116.79	0.69
Median	325	1025	109	20
Max	1262	4730	519	23
Min	36	99	26	19

4   Models

In this section we describe models and baselines to be used for recipient prediction. In all cases, we followed this terminology. The symbol *ca* refers to *candidate email address* and *t* refers to *terms* in documents or queries. A document *doc* refers to *documents* in the training set, i.e., email messages previously sent by the same Enron user. A *query q* refers to a message in the test set. Both documents and queries are modeled as distributions over (lowercased) terms found in the "body" of the respective email messages.

We also define other useful functions. The number of times a term *t* occurs in a query *q* or a document *doc* is, respectively, $n(t, q)$ or $n(t, doc)$. The *recipient function Recip(doc)* returns the set of all recipients of message *doc*. The *association function a(doc, ca)* returns 1 if and only if *ca* is one of the recipients (TO, CC or BCC) of message *doc*, otherwise it returns zero. $D(ca)$ is defined as the set of training documents in which *ca* is a recipient, i.e, $D(ca) = \{doc | a(doc, ca) = 1\}$.

4.1   Models

Expert Search Model 1. Predicting recipients (candidates) of a message under composition (query) is a very similar task to Expert Search, the task of predicting experts (candidates) associated with a particular topic (query). The analogy works so well that we can easily adapt many recently proposed Expert Search formal models to the task of recipient prediction.

The first recipient prediction model considered here is the *Model 1* proposed for Expert Search by Balog et al. [1]. In this model, the final candidate ranking for each query *q* is given by the probability of this query being generated by a smoothed candidate language model θ_{ca}[4]:

$$p(q|\theta_{ca}) = \prod_{t \in q} \left\{ (1 - \lambda) \left(\sum_{doc} p(t|doc) f(doc, ca) \right) + \lambda p(t) \right\}^{n(t,q)} \tag{1}$$

---

[4] Please refer to the original reference [1] for further details.

where λ is the Jelinek-Mercer smoothing parameter, $p(t)$ is the background model probability of term t (maximum likelihood estimates of term in *sent_train* collection), $p(t|doc)$ is the maximum likelihood estimate of the term in the document *doc*, and $f(doc|ca)$ is the document-candidate association function. Similarly to Balog et al. [1], we estimated the document-candidate association functions in two different ways:

$$f(doc, ca) = \begin{cases} \frac{a(doc,ca)}{\sum_{doc'} a(doc',ca)} & \text{, in } document\ centric\ \text{(DC) mode;} \\ \frac{a(doc,ca)}{\sum_{ca'} a(doc,ca')} & \text{, in } user\ centric\ \text{(UC) mode.} \end{cases} \tag{2}$$

Expert Search Model 2. The second recipient prediction model considered is the *Model 2* proposed by Balog et al. [1]. Basically, the final candidate ranking for each query q is given by the expression:

$$p(q|ca) = \sum_{doc} \left\{ \prod_{t \in q} [(1-\lambda)p(t|doc) + \lambda p(t)]^{n(t,q)} \right\} f(doc, ca) \tag{3}$$

where λ, $p(t|doc)$ and $p(t)$ are defined in the same way as in Section 4.1. Similarly, the two possible views of the document-candidate function $f(doc, ca)$ are defined according to equation 2. Please refer to [1] for further details.

TFIDF Classifier. The recipient recommendation problem can naturally be framed as a multi-class classification problem, with each candidate email ca representing a class ranked by some notion of classification confidence. Here we propose using the Rocchio algorithm with TFIDF (Term Frequency-Inverse Document Frequency) [7,8] weights as a classifier for recipient recommendation problems. For each candidate, a centroid vector-based representation is created:

$$\overrightarrow{centroid}(ca) = \frac{\alpha}{|D(ca)|} \sum_{doc \in D(ca)} \overrightarrow{tfidf}(doc) + \frac{\beta}{|sent\_train| - |D(ca)|} \sum_{doc \notin D(ca)} \overrightarrow{tfidf}(doc) \tag{4}$$

where $\overrightarrow{tfidf}(doc)$ is the TFIDF vector representation[5]. The final ranking score for each candidate ca is produced by computing the cosine similarity between the centroid vector and the TFIDF representation of the query, i.e., $score(ca, q) = \cos\left(\overrightarrow{tfidf}(q), \overrightarrow{centroid}(ca)\right)$.

K-Nearest Neighbors. We also adapted another multi-class classification algorithm, K-Nearest Neighbors as described by Yang & Liu [9], to the recipient prediction problem. Given a query q, the algorithm finds $N(q)$, i.e., the K most similar messages (or neighbors) in the training set. The notion of similarity here is also defined as the cosine distance between the TF-IDF query vector $\overrightarrow{tfidf}(q)$ and the TFIDF document vector $\overrightarrow{tfidf}(doc)$.

---

[5] For each term t in document doc, the value $tfidf(t) = log(n(t, doc) + 1)log(\frac{|sent\_train|}{DF(t)})$, where $DF(t)$ is the document frequency of t.

The final ranking is computed as the weighted sum of the query-document similarities (in which ca was a recipient):

$$score(ca, q) = \sum_{doc \in N(q)} a(doc, ca) \cos\left(\vec{tfidf}(q), \vec{tfidf}(doc)\right) \qquad (4)$$

Other Baselines: Frequency and Recency. For comparison, we also implemented two simple baseline models: one based on the frequency of the candidates in the training set, and another based on recently sent messages in the training set. The first method ranks candidates according to the number of messages in the training set in which they were a recipient: in other words, for any query q the *Frequency* model will present the following ranking of candidates:

$$Frequency(ca) = \sum_{doc} a(doc, ca) \qquad (5)$$

Compared to *Frequency*, the *Recency* model ranks candidates in a similar way, but attributes more weight to recent messages according to an exponential decay function. In other words, for any query q the *Recency* model will present the following ranking:

$$Recency(ca) = \sum_{doc} a(ca, doc)e^{\left(\frac{-timeRank(doc)}{\beta}\right)} \qquad (6)$$

where $timeRank(doc)$ is the rank of doc in a chronologically sorted list of messages in *sent_train* (the most recent message will have rank 1).

4.2   Effect of Threading

Threading information is expected to be a very important piece of evidence for recipient prediction tasks, but unfortunately it cannot be directly exploited here because the Enron dataset does not provide it explicitly. To approximately reconstruct message threads, we used a simple heuristic based on the approach adopted by Klimt & Yang [10].

For each test message q, we construct a set with all messages on the same thread as q (or $MTS(q)$, *Message Thread Set*) by searching for all messages satisfying two conditions. First, the message is among the last P messages sent previous to q. Second, the message must have the same "subject" information[6] as q. While small values of P may not be enough to find all previous messages on the same thread, larger values are expected to introduce more noise in the thread reconstruction process. In preliminary experiments, however, we observed that on average larger values of P did not degrade prediction performance, so only the second condition was imposed on the construction of $MTS(q)$.

In order to exploit thread information in all previously proposed models, we used the following backoff-driven procedure:

$$threaded\_model_i(q) = \begin{cases} MTS\_model(q) \text{ , if } \|MTS(q)\| \geq 1; \\ model_i(q) \qquad \text{ , otherwise.} \end{cases}$$

---

[6] Or subjects differing only in terms of reply-to (RE:) or forward (FWD:) markers.

where
$$MTS\_model(q) = \begin{cases} 1.0 \text{ , if } ca \in \bigcup_{d \in MTS(q)} Recip(d); \\ 0.0 \text{ , otherwise.} \end{cases}$$

That is, if q has no previous messages in its thread, predictions from the threaded version of $model_i$ will be made based on the original model $model_i$ (for instance, Frequency, Knn, TFIDF, Expert Model 1, etc.). Otherwise, if the thread of q contains at least one message ($\|MTS(q)\| \geq 1$), predictions are dictated by $MTS\_model(q)$ — a model that assigns weight 1.0 to all recipients found in the messages in $MTS(q)$ and weight 0.0 to all other candidates[7].

5   Results

5.1   Initial Results

In this section we present recipient prediction experiments using the models introduced in Section 4. All those models can be naturally applied to both primary and secondary recipient prediction tasks: the only difference is that, for obvious reasons, in the secondary prediction task, a post-processing step removes all TO-addresses from the final rank, and the test set contains only messages having at least one CC or BCC address.

Similarly to Balog et al. [1], in our experiments both Expert Model 1 and 2 used a smoothing parameter $\lambda = 0.5$. The TFIDF Classifier model had $\alpha = 1$ and $\beta = 0$, creating a centroid of positive examples for each candidate ca. We set $K = 30$ in the Knn Model and $\beta = 100$ in the Recency model, values that delivered the best results in preliminary tests.

Table 2 shows Mean Average Precision (MAP) results for all models presented in Section 4. *T-only* refers to *Thread Only* — the prediction based only on detecting threads, i.e., if no thread is detected, candidates are chosen randomly. *Freq* refers to the Frequency model, while *Rec* refers to the Recency model. The symbol *TFIDF* refers to the TFIDF Classifier model. Expert models one and two are referred as *M1* and *M2*, with the candidate-document association indicated by *-uc* (user centric) or *-dc* (document centric). *Thread* refers to models with thread processing (Section 4.2). Two-tailed paired t-test were used for statistical significance tests. Results in Table 2 clearly indicate that the best recipient prediction performance is typically reached by the Knn model, followed by TFIDF. It also reveals that Recency is typically a stronger baseline for this task than the Frequency model. Overall, the expert models M1 and M2 presented statistically significant inferior results when compared to Knn. It is also interesting that the best Expert Search-based model was consistently M2-uc, the same behavior observed by Balog et al. [1] on the TREC-2005 Expert Search task. The use of thread information clearly provided considerable performance gains for all models and tasks. These gains are somewhat expected because, in many cases, email users are simply using the "reply-to" or "reply-all" buttons to select recipients. These improvements are consequently a strong indication that the thread

---

[7] In all models of this paper, candidates with the same scores were ranked randomly.

Table 2. MAP recipient prediction results averaged over 36 users. Statistical significance relative to the best model results (in bold) is indicated with the symbols ** $(p < 0.01)$ and * $(p < 0.05)$.

	T-only	Freq	Rec	M1-dc	M1-uc	M2-dc	M2-uc	TFIDF	Knn
TOCCBCC	0.221**	0.203**	0.260**	0.279**	0.275**	0.279**	0.313**	**0.365**	0.361
CCBCC	0.261**	0.228**	0.309	0.262**	0.272**	0.236**	0.278**	0.301*	**0.332**
TOCCBCC (thread)	N/A	0.331**	0.363**	0.393**	0.385**	0.384**	0.408**	0.440	**0.441**
CCBCC (thread)	N/A	0.379**	0.424*	0.402**	0.407**	0.391**	0.425**	0.429*	**0.459**

Table 3. Recipient prediction results for the best model (Knn) averaged over 36 users

	MAP	MRR	R-Prec	P@5	P@10
TOCCBCC	0.361	0.440	0.294	0.182	0.135
CCBCC	0.332	0.405	0.266	0.177	0.126
TOCCBCC (threaded)	0.441	0.516	0.398	0.225	0.157
CCBCC (threaded)	0.459	0.540	0.425	0.239	0.156

reconstruction algorithm is working reasonably well in this dataset and also the fact that a large proportion of the test messages was found to have a non-empty Message Thread Set $MTS(q)$. In fact, 29% of the test messages in the primary prediction task had non-empty $MTS(q)$, while the same number for secondary predictions was 35%.

To give a complete picture of the best results, Table 3 shows the Knn performance metrics in terms of other metrics, such as Mean Reciprocal Rank (MRR), R-Precision (R-Prec), and Precision at Rank 5 and 10 (P@5 and P@10) [11]. Overall, the average performance over the 36 Enron users had MRR of more than 0.5, a very good result for such a large prediction task (5202 queries from 36 different users). A closer look in the numbers revealed a much larger variation in performance over different users than over different models. For the primary prediction (threaded), over the 36 users sample, the maximum MAP was 0.76, the minimum was 0.186, with a standard deviation of 0.101.

Based on this variability, we measured the Pearson's correlation coefficient R (quotient of the covariance of the two variables by the product of their standard deviations) between variables that might influence performance. First, the correlation between training set size ($|sent\_train|$) and the number of classes or ranked entities (address book size) is 0.636 — a clear indication that users who send more messages tend to have larger address books. More surprising, perhaps, was the fact that the Pearson's correlation between performance and training set size, as well as the one between performance and Address Book size, was smaller than 0.2 in absolute values — suggesting there is no apparent strong correlation between these variables[8]. One possible explanation is that these two variables contribute inversely to the performance (while recipient prediction is

---

[8] Similar results were observed for different models on both for primary and secondary predictions.

Table 4. MAP values for model aggregations with Reciprocal Rank. The $*$ and $**$ symbols indicate statistically significant results over the Knn baseline.

Task		Freq	Recency	TFIDF	M2-uc
TOCCBCC	Knn ⊙	0.417**	0.432	**0.457****	0.444
	Knn ⊙ TFIDF ⊙	0.455**	**0.464****	—	0.461**
Baseline: Knn	Knn ⊙ TFIDF ⊙ Rec ⊙	0.451**	—	—	**0.470****
MAP = 0.441	Knn ⊙ TFIDF ⊙ Rec ⊙ M2-uc ⊙	**0.464****	—	—	—
CCBCC	Knn ⊙	0.455	0.470	0.462	**0.474***
	Knn ⊙ M2-uc ⊙	0.476**	**0.491****	0.482**	—
Baseline: Knn	Knn ⊙ M2-uc ⊙ Rec ⊙	0.491**	—	**0.494****	—
MAP = 0.458	Knn ⊙ M2-uc ⊙ Rec⊙ TFIDF ⊙	**0.501****	—	—	—

certainly easier with smaller Address Book sizes, it is certainly harder with less training data) and the overall effect is hence weak.

5.2   Combining Evidence with Data Fusion Methods

Ranking results can be potentially improved by combining the results of two or more rankings to produce a better one. One set of the techniques commonly applied to rank combination is *Data Fusion* [12], whose methods have been successfully applied to many areas, including Expert Search [3] and Known Item Search [13].

Because not all ranking scores of the proposed methods in Section 4 are normalized, it is not reasonable to use score-based fusion techniques such as *Comb-SUM* and *CombMNZ* [3]. Instead, we utilized *Reciprocal Rank* [3] (or RR), a rank-based fusion techniques in which the aggregated score of a document is the sum of inverse ranks of this document in the rankings, i.e., the sum of one over the rank of the document across all rankings.

Table 4 shows experimental results on aggregating recipient recommendation techniques with rank-based Fusion methods. The symbol ⊙ represents the aggregation operation over different models (all threaded). On each line, the best performing model (in bold face) is selected to be part of the base aggregation in the following line. For instance, the second line displays aggregation results when Knn is combined with the best model in the previous line (TFIDF) and all other three remaining methods. The initial baseline model is threaded Knn. Results clearly show noticeable performance improvements over the baseline. MAP gains up to 0.042 in the secondary prediction task, and close to 0.03 on primary predictions. In most cases, the gains over the Knn baseline are statistically significant[9]. In a second set of experiments, we used a weighted version of RR, where the weights for each base ranking were determined by the performance obtained by the respective model in a development set. More specifically,

---

[9] We also experimented with the Borda Fuse [3] aggregation method, but it presented consistently worse results when compared to RR. A similar observation can be drawn from other rank aggregation tasks [3,13]

this development set was constructed using the 20% most recent messages in *sent_train*, and used as test after training the models in the remaining 80%. Overall, results were statistically significantly better than the Knn baseline, but not statistically significantly better than the unweighted results in Table 4.

5.3   Auto-completion Experiments

Email address auto-completion is the feature in email clients that provide a list of email addresses after the user typed a few initial letters of the intended contact address. Typically email clients allow users the option to turn on or off the auto-completion feature, but rarely are users allowed pick how the suggested addresses should be ranked. In this section we analyze different strategies for email auto-completion ranking.

In order to test different strategies and models for email auto-completion, we used the following experimental procedure. For each query message q, we extracted all its recipient $Recip(q)$, and for each recipient in $Recip(q)$, we extract its V initial letters[10]. Then these V initial letters are used to filter out candidates ranked by the recommendation model. Table 5 presents performance values in terms of MRR* for different values of V and different recommendation models. Notice that for each query q, $|Recip(q)|$ different auto-completion rankings are created, one for each member of $Recip(q)$ (each ranking contains a single relevant recipient and all other recipients in the Address Book who share the same initial letters). MRR* is the mean value of MRR over these rankings.

When $V = 0$, no initial letter of the email contact is known, just like in previous Sections 5.1 and 5.2. As V increases, more is known about the intended recipient and consequently prediction performance becomes better. In addition to the threaded versions of *Knn*, *Recency*(Rec) and *Frequency*(Freq), Table 5 shows results for when recipients are presented in alphabetical order (Alpha). It also contains a model called *All-Fusion* (Fus), displaying results with the aggregated rankings from all models in Table 4 (i.e., using rankings produced by the combinations indicated in the 4th and 8th lines of that Table).

In general, Table 5 indicates that Knn performs slightly better than Recency, which in turn performs better than Frequency. This difference is more noticeable for small values of V — exactly where most email users will benefit the most from auto-completion. When $V = 2$ or $V = 3$ the different between Knn and Recency is not statistically significant. The *All-Fusion* model shows the best auto-completion results overall, significantly outperforming all other models for all values of V. Table 5 also displays the relative performance gains between Knn and Recency, All-Fusion and Recency as well as All-Fusion and Knn.

Compared to any of the other models, auto-completion based only on the alphabetical order presents a rather low performance on both primary and secondary

---

[10] In a general case, initial letters from the contact's email address, last name, first name and nickname can be used. We used only email addresses because those were the only contact information consistently available in the Enron corpus; but results can be extended for the general case.

Table 5. Auto-completion Experiments. Performance values for different models and V values. Statistical significance relative to the previous column value is indicated with the symbols ** ($p < 0.01$) and * ($p < 0.05$).

		Primary Prediction (TOCCBCC)						
V	Alpha	Freq	Rec	Knn	Fus	Δ(Knn-Rec)	Δ(Fus-Rec)	Δ(Fus-Knn)
0	0.022	0.274**	0.300**	0.377**	0.394**	25.542%	31.124%	4.447%
1	0.250	0.620**	0.653**	0.690**	0.731**	5.753%	11.893%	5.806%
2	0.557	0.846**	0.857	0.858	0.895**	0.078%	4.412%	4.331%
3	0.737	0.911**	0.923*	0.917	0.942**	-0.683%	2.001%	2.702%
		Secondary Prediction (CCBCC)						
0	0.025	0.329**	0.364**	0.398*	0.436**	9.526%	19.927%	9.496%
1	0.265	0.668**	0.718**	0.717	0.777**	-0.125%	8.289%	8.424%
2	0.549	0.858**	0.875	0.865	0.910**	-1.189%	3.928%	5.178%
3	0.729	0.915**	0.929	0.915	0.946**	-1.558%	1.811%	3.423%

prediction tasks. All other methods can provide significant gains in performance when compared to it. It is surprising that some email clients still provide auto-completion based on this method, given that simple baselines such as Frequency or Recency can provide visible gains in recommendation ranking.

6 Related Work

The email recipient prediction problem is closely related to the *expert search* task. In the former, the task is to retrieve the most likely recipients of a message under composition, while in the latter the task is to retrieve the most likely experts in a topic specified by a textual query. In fact, it is easy to find similarities between recipient prediction and early expert search work using enterprise email data [14,15,16]. Recently, interesting models for Expert Search have been motivated by the TREC Enterprise Search, where different types of documents are taken as evidence in the process of finding experts. Because of the similarity between these tasks, many ideas in this paper were motivated by expert search models recently proposed by Balog et al. [1], Fang & Zhai [2] and Macdonald & Ounis [3].

Though relatively similar, expert search and email recipient prediction have some fundamental differences. First, the latter is focused on a single email user, while the former is typically focused in an organization or group. The former is explicitly trying to find expertise in narrow areas of knowledge (queries with a small number of words), while the latter is not necessarily trying to find expertise — instead, it is trying to recommend users related to one or more indiscriminate "topic(s)" (i.e., a message query that may have up to a few hundred words).

In a related work, Pal & McCallum [17] described what they called the CC Prediction problem. In their short paper, two machine learning models were used to predict email recipients in the personal collection of a single user. However their modeling assumptions is substantively different from ours: they assume that all recipients but one are given and the task is to predict the final missing recipient. Performance was evaluated in terms of the probability of having "recall at rank 5" larger than zero, i.e., the probability of having at least one correct guess in the top 5 entries of the rank. They report performance values around 44% for this metric on their single private email collection. For comparison, our best system achieves 64.8% and 70.6% on the same metric for primary and secondary predictions, respectively, averaged over the 36 different Enron users.

The recipient prediction task is also related to *email leak prediction* [18]. The goal of this task is preventing information leaks by detecting when a message is accidentally addressed to non-desired recipients. In some sense, the recipient prediction task can be seen as the negative counterpart of the email leak prediction task: in the former, we want to find the intended recipients of email messages, whereas in the latter we want to find the unintended recipients or email-leaks.

7   Conclusions

In this work we addressed the the problem of recommending recipients for messages under composition, a task relatively similar to Expert Search. Evidence from a very large real email corpus (Enron corpus) revealed that at least 9% of the users forgot to address an intended recipient at least once, while more than 20% of the users have been accidentally "forgotten" as intended recipients. We proposed several possible models for this task, and evaluated their predictive performance on 36 different users from the Enron corpus. Experiments showed that a simple model based on the K-Nearest Neighbors algorithm generally outperformed all other methods, including frequency or recency based models, and more refined formal models previously proposed for Expert Search.

We also investigate how to combine the rankings of different models using rank-based data fusion techniques, such as sum of Reciprocal Ranks. Experiments clearly indicated that aggregated models can generally outperform all base models, both on primary and secondary recipient prediction tasks. We then applied the proposed ideas to the email auto-completion problem, where the initial letters of the email contact are typed by the user. Results clearly indicate that the proposed models can provide intelligent email auto-completion, outperforming auto-completion based on alphabetical ordering (currently used by some email clients).

Acknowledgments. We would like to thank Jonathan Elsas for helpful comments. This material is based upon work supported by the DARPA under Contract Numbers NBCHD030010. Any opinions, findings and conclusions expressed in this material do not necessarily reflect the views of DARPA.

References

1. Balog, K., Azzopardi, L., de Rijke, M.: Formal models for expert finding in enterprise corpora. In: SIGIR 2006 (2006)
2. Fang, H., Zhai, C.: Probabilistic models for expert finding. In: Amati, G., Carpineto, C., Romano, G. (eds.) ECiR 2007. LNCS, vol. 4425, pp. 418–430. Springer, Heidelberg (2007)

3. Macdonald, M., Ounis, I.: Voting for candidates: Adapting data fusion techniques for an expert search task. In: CIKM, Arlington, USA, November 6-11 (2006)
4. Carvalho, V.R., Cohen, W.W.: Predicting recipients in the enron email corpus. Technical Report CMU-LTI-07-005 (2007)
5. Shetty, J., Adibi, J.: Enron email dataset. Technical report, USC Information Sciences Institute (2004), http://www.isi.edu/~adibi/Enron/Enron.htm
6. Cohen, W.W.: Enron Email Dataset Webpage, http://www.cs.cmu.odu/~onron/
7. Joachims, T.: A probabilistic analysis of the rocchio algorithm with TFIDF for text categorization. In: Proceedings of the ICML 1997 (1997)
8. Salton, G., Buckley, C.: Term weighting approaches in automatic text retrieval. Information Processing and Management 24(5), 513–523 (1988)
9. Yang, Y., Liu, X.: A re-examination of text categorization methods. In: 22nd Annual International SIGIR, August 1999, pp. 42–49 (1999)
10. Klimt, B., Yang, Y.: The enron corpus: A new dataset for email classification research. In: Boulicaut, J.-F., Esposito, F., Giannotti, F., Pedreschi, D. (eds.) ECML 2004. LNCS (LNAI), vol. 3201, pp. 217–226. Springer, Heidelberg (2004)
11. Baeza-Yates, R., Ribeiro-Neto, B.: Modern Information Retrieval. Addison-Wesley, Reading (1999)
12. Aslam, J.A., Montague, M.: Models for metasearch. In: Proceedings of ACM SIGIR, pp. 276–284 (2001)
13. Ogilvie, P., Callan, J.P.: Combining document representation for known item search. In: ACM SIGIR (2003)
14. Dom, B., Eiron, I., Cozzi, A., Zhang, Y.: Graph-based ranking algorithms for e-mail expertise analysis. In: Data Mining and Knowledge Discovery Workshop (DMKD2003) in ACM SIGMOD (2003)
15. Campbell, C.S., Maglio, P.P., Cozzi, A., Dom, B.: Expertise identification using email communications. In: CIKM (2003)
16. Sihn, W., Heeren, F.: Expert finding within specified subject areas through analysis of e-mail communication. In: Proceedings of the Euromedia 2001 (2001)
17. Pal, C., McCallum, A.: Cc prediction with graphical models. In: CEAS (2006)
18. Carvalho, V.R., Cohen, W.W.: Preventing information leaks in email. In: Proceedings of SIAM International Conference on Data Mining (SDM 2007), Minneapolis, MN (2007)

Facilitating Query Decomposition in Query Language Modeling by Association Rule Mining Using Multiple Sliding Windows

Dawei Song[1], Qiang Huang[1], Stefan Rüger[1], and Peter Bruza[2]

[1] Knowledge Media Institute, The Open Univeristy, Milton Keynes, UK
[2] Queensland University of Technology, Australia
{d.song,q.huang,s.rueger}@open.ac.uk, p.bruza@qut.edu.au

Abstract. This paper presents a novel framework to further advance the recent trend of using query decomposition and high-order term relationships in query language modeling, which takes into account terms implicitly associated with different subsets of query terms. Existing approaches, most remarkably the language model based on the Information Flow method are however unable to capture multiple levels of associations and also suffer from a high computational overhead. In this paper, we propose to compute association rules from pseudo feedback documents that are segmented into variable length chunks via multiple sliding windows of different sizes. Extensive experiments have been conducted on various TREC collections and our approach significantly outperforms a baseline Query Likelihood language model, the Relevance Model and the Information Flow model.

Keywords: Association Rule, Term Relationship, Query Expansion, Document Segmentation.

1 Introduction

Recent studies in language modeling (LM) have tried to exploit relevance feedback documents to establish an improved query model via a model-based approach [9, 10, 16]. One example is the Relevance Model (RM) [10], which estimates the joint probability of observing a term w in the vocabulary together with query topic $Q = \{q_1, \cdots, q_{|Q|}\}$. The assumption of independence among query terms has been made to reduce the complexity of computation. This, however, neglects the effect of the relationships between terms in determining the query language model.

More recent research [3, 5, 12, 13, 15] tries to incorporate term relationships or dependencies, for example, grammatical links [6], co-occurrence and Word-Net relations [5], in LM. It also has been shown in [5] that combining multiple types of term relationships, e.g., co-occurrences and WordNet relations, leads to improvement on average precision over the use of different types individually.

Furthermore, there has been a trend of decomposing a query into different combinations (subsets) of query terms, and using term relationships derived from

C. Macdonald et al. (Eds.): ECIR 2008, LNCS 4956, pp. 334–345, 2008.

the subsets of query terms rather than traditional pairwise term co-occurrences. These automatically derived term relationships are in higher order, in the sense that the information is flowing from a set of terms to another term (e.g. "(java, computer) $\rightarrow$ programming"). "java" and "computer" are combined to form a context-dependent premise for the derivation of "programming". Song and Bruza [15] propose an information flow model to capture the relationships, and in [13], Pickens and MacFarlane build a term context model based on a maximum entropy algorithm to estimate the co-occurrence of terms in documents with the query topic. The work in [12] expands the approach used in [13], and decomposes the query topic into "latent" concepts, which consist of the combinations of query terms. In [3], high-order inferential term relationships extracted by the information flow approach [15] have been employed in a LM framework combining the effects of information flows from different subsets of query terms.

Essentially, the Information Flow approach [15] is based on a lexical semantic space model, namely Hyperspace Analogue to Language (HAL). The HAL space is constructed by moving a fixed length sliding window over the corpus by a one term increment. All terms within the window are considered as co-occurring with each other with strengths inversely proportional to the distance between them. After traversing the corpus, numeric vectors representing the concepts (terms) are produced. Arbitrary terms (e.g., "Java" and "computer") that are related to each other (but not necessarily the syntactically valid phrases) can be combined to form a new concept, also represented as a vector, by a weighted addition of the underlying vectors of the terms. The information flow between two concepts is then computed by comparing their underlying vectors.

Despite its good performance, re-loading and manipulating vectors in the pre-computed HAL space, which is normally very large, for each query session may potentially lead to a high computational overhead. In particular, for query decomposition, the expensive information flow computation process (sequential scan of the vocabulary to compare each vector in the HAL space with the vector representing a subset of query terms) has to be performed for $2^{|Q|}$ times, i.e., for each of the subsets of query terms. Indeed, as a consequence, in both [15] and [3] the query decomposition was not actually performed. It was instead approximated by computing information flows only once from the whole set of query terms only. Moreover, the fixed sized sliding window approach used in HAL is less flexible to encode various levels of associations between terms.

This paper aims to further advance the trend of using query decomposition and high-order term relationships in query language modeling, by developing a novel method to overcome the two aforesaid limitations.

Firstly, we propose to use association rule mining which is a popular and well researched method for discovering interesting relations between variables in large databases. Compared with the information flow approach, the association rule mining shows a strong ability to capture the high-order term relationships from different subsets of query terms in one go, thus truly realizing the idea of query decomposition. It also does not need any training data (as in the MRF method) or a pre-computed semantic space.

Secondly, we propose dividing the documents into variable length segments through multiple sliding windows of different sizes to perform association rule mining. Using shorter segments instead of the whole documents will reduce the computational load of association rule mining. On the other hand, using viable length windows for document segmentation enables different levels of term associations generated from different sized segments to be taken into account in a mixture model. The segmentation-based approaches [2, 4, 11] have had a proven track record particularly in passage and sentence retrieval. However, to our knowledge, there has not been an approach to the use of multiple-sized sliding windows for query language modeling.

In this paper, we build a novel framework that integrates the advantages of association rule mining, multiple window segmentation and query decomposition to derive higher-order term relationships for query language modeling. Fig. 1 shows our overall approach, with pseudo feedback documents and query topic as input, and the generated query model as output.

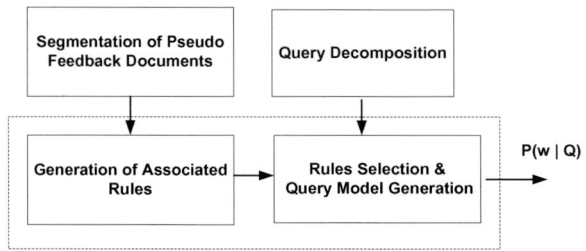

Fig. 1. Structure of Theory

Extensive empirical evaluation has been conducted to compare the effectiveness of our approach with a baseline language model, the Relevance Model and the Information Flow approach. Our approach has demonstrated a better performance than these existing models.

2   Generation of Association Rules from Documents

Mining association rules is an important technique for discovering meaningful patterns in transaction databases. Formally, the problem can be formulated as follows [1]. Let $I = \{i_1, i_2, \cdots, i_n\}$ be a set of n binary attributes called *items*. Let $D = \{t_1, t_2, \cdots, t_m\}$ be a set of transactions called the *database*. Each transaction in D has a unique transaction ID and contains a subset of the items in I [7]. An association rule is a rule of the form $X \Rightarrow Y$, where $X, Y \subseteq I$, and X, Y are two disjoint sets of items. It means that if all the items in X are found in a transaction then it is likely that the items in Y are also contained in the transaction. The sets of items X and Y are respectively called the *antecedent* and *consequent* of the rule [7]. To select interesting rules from the set of all possible rules, constraints on various measures of significance and strength can be used. The best-known constraints are minimum thresholds on support and confidence.

$$supp(X \Rightarrow Y) = supp(X \cup Y) = \frac{C_{XY}}{M} \qquad (1)$$

$$conf(X \Rightarrow Y) = \frac{supp(X \cup Y)}{supp(X)}, \qquad (2)$$

where C_{XY} is the number of transactions which contain all the items in X and Y, and M is the number of transactions in the database.

Support, in Equation 1, is defined as the fraction of transactions in the database which contain all items in a specific rule [14]. *Confidence*, in Equation 2, is an estimate of the conditional probability $P(E_Y|E_X)$, where E_X (E_Y) is the event that X (Y) occurs in a transaction [8].

According to the above definitions, by considering a term as an item and a text fragment, e.g., a sentence, a paragraph, a document, or a fixed sized window, as a transaction, we can easily apply association rule mining to the discovery of high-order term associations (i.e., the association rules between any subset (Q_j) of query terms $\{q_{j_1}, \cdots, q_{j_m}\}$ ($m = |Q_j|$) and the terms from vocabulary.

anti-missil $\Rightarrow$ *direct* (0.076, 36.67)
reform welfar $\Rightarrow$ *competit* (0.312, 177.16)
initi research defens $\Rightarrow$ *contract* (0.162, 1788)
high-combustion fuels create $\Rightarrow$ *laser* (0.03265, 61.11)

Fig. 2. Association Rules

Figure 2 shows some example association rules between various combinations of query terms on the left hand side and a word on the right hand side. The Porter stemmer has been used on these words . The numbers in the bracket following each rule are its *Support* and *Confidence* values respectively.

For association rules, a minimum support threshold is used to select the most frequent item combinations called frequent item sets. The computational complexity of finding these frequent item sets, in the worst case, can be exponential with respect to the number of items. Obviously, using the whole documents as transactions implies high computation cost. Segmenting a long document into shorter chunks can reduce such cost. There has been work carried out in passage and sentence retrieval. However, passages, which are often still quite long (e.g., minimum 50 terms [11]), may contain "noisy" information and lead to computational overhead. On the other hand, the use of sentences may miss some useful relationships between terms. In addition, using passages or sentences directly as transactions may also lead to the data sparseness problem. In this paper, we overcome these problems by segmenting documents into chunks using multiple sliding windows.

3   Document Segmentation

In this paper, we propose the use of multiple sliding windows. Each window slides over the documents with $1/3$ overlapping of the length of the window. Eventually, we segment the pseudo feedback documents into chunks of variable lengths.

Documents: □□□□□□□□□□□□□□□□□□□□□□□□□□□□□□□□ $\cdots$ □□□□□□□□□□
Chunk₁: □□□□□□□□□□
Chunk₂: □□□□□□□□□□
Chunk₃: □□□□□□□□□□
   $\cdots$ $\cdots$
Chunk$_M$:   □□□□□□□□□□

Fig. 3. Segmentation of Document

Fig. 3 illustrates the process of document segmentation by dividing the whole document into overlapped chunks using a sliding window. The generation of the overlapped chunks obviously increases the number of segments extracted from the pseudo feedback documents, which can reduce the problem of data sparsity to some extent, compared with the use of non-overlapping passages and sentences for retrieval. To alleviate the potential noise carried by longer windows and the potential missing information caused by shorter windows, we apply multiple windows to generating chunks in different lengthes. In our experiments, we tested 7 windows ranging from 15 to 45 terms with a 5 term increment.

4   Rule Selection and Query Model Generation

Association rule mining is then applied in the segmented chunks of the pseudo relevance feedback documents to discover terms associated with different combinations of query terms.

Instead of deriving association rules from a query as a whole, the query is first decomposed into all the possible combinations of query terms. Consequently, an example query $Q = \{q_1, q_2\}$ can be expanded to a list of subsets of query terms, $Q' = \{\{q_1\}, \{q_2\}, \{q_1, q_2\}\}$. A concrete example is shown in figure 4.

Based on the query decomposition, we refine the association rule mining process by selecting those rules derived from any subset of query terms. The process brings two advantages. One one hand, it collects the rules related to any portion

$$Query \quad \rightarrow \quad decomposed\ Query$$
$$\{theory, derivation\} \ \rightarrow \ \{\{theory\}, \{derivation\}, \{theory, derivation\}\}$$

Fig. 4. Example of the query decomposition

of the query instead of the whole query only. On the other hand, as the association rule mining is incremental, there is no additional computational costs for generating the rules from subsets other than the whole query.

Fig. 5 shows some refined rules from different subsets of query terms. In general, the *Confidence* value can effectively represent how good a captured rule is. As shown in Figure 5, the term combination "*high-combustion fuels hydrogen*" implies "*laser*" in a higher confidence than the others.

high-combustion fuel create ⇒ *laser* (0.03265, 61.11)
high-combustion fuel hydrogen ⇒ *laser* (0.01929, 100.00)
high-combustion fuel ⇒ *laser* (0.01484, 47.62)
high-combustion fuel energy ⇒ *laser* (0.02671, 56.25)
high-combustion create hydrogen ⇒ *laser* (0.01336, 100.00)

Fig. 5. Association Rules

After the associated rules derivable from all the subsets of query terms are obtained, the query model $P(w|Q)$ can be generated as follows.

$$P(w|Q) = \sum_{Q_j \in Q'} P(w|Q_j, Q)P(Q_j|Q) \tag{3}$$

Equation 3 shows a model by mixing the probability of term w given a specific subset of query terms Q_j and Q, weighted by a prior $P(Q_j|Q)$. By assuming $P(w|Q_j, Q) \approx P(w|Q_j)$, we can obtain a simplified version which has been used in [3]:

$$P(w|Q) = \sum_{Q_j \in Q'} P(w|Q_j)P(Q_j|Q) \tag{4}$$

Note that, in [3], the effect of Q_j was not actually implemented, due to the time consuming information flow computations for all the Q_j. Instead, Equation 4 was approximated by the information flows from the whole query Q only.

In addition, we have tested a number of ways for determining the prior distribution $P(Q_j|Q)$, e.g., based on the length of Q_j and the average IDF value of the terms in Q_j. However, our prior experiments show that they are not much of an improvement over the simple uniform distribution. Therefore, in our experiments, $P(Q_j|Q)$ is assumed to be uniform:

$$P(Q_j|Q) = \frac{1}{|Q'|} \tag{5}$$

Equation 4 is then rewritten as:

$$P(w|Q) = \sum_{Q_j \in Q'} P(w|Q_j)/|Q'| \tag{6}$$

Table 1. Test Collections and Query Topics

Coll.	Description	Size (MB)	# Doc.	Vocab.	Query	Q.fields	Q.length (words)
AP89	Associated Press (1989) Disk 1	254	84,678	137,728	1–50	title	3.2
AP88–89	Associated Press (1988–1989) Disk 1,2	492	164,597	254,872	101–150	title	3.6
					151–200	title	4.3
WSJ90–92	Associated Press (1990–1992) Disk 2	242	74,520	121,944	201–250	desc.	8
SJM	San Jose Mercury News (1991) Disk3	287	90,257	146,512	51–100	title & desc.	12.2

In the process of computing the conditional probability $P(w|Q_j)$, we propose using the *Confidence* values of those associated rules from Q_j.

$$P_{AR}(w|Q_j) = \frac{Conf(Q_j \Rightarrow w)}{\sum_{w'} Conf(Q_j \Rightarrow w')} \qquad (7)$$

To derive the "new" smoothed model, a linear mixture can typically be used. In this paper, we also use the method to mix the derived query model from association rule mining with an original query model $P_O(q_i|Q)$, where q_i is a term in the original query.

$$P_O(q_i|Q) = \frac{QTF * IDF(q_i)}{\sum_{j \in 1 \cdots |Q|} QTF * IDF(q_j)} \qquad (8)$$

where QTF is the number of q_i occurring in the query.

The smoothed query model can then be derived:

$$P_{NEW}(w|Q) = \lambda P(w|Q) + (1 - \lambda)P_O(w|Q) \qquad (9)$$

5 Empirical Evaluation

5.1 Data

The experiments are conducted using various TREC collections and query topics shown in Table 1. Different fields of the five topic sets are used in different experiments to verify the robustness of our method with respect to different average query lengths. All documents have been pre-processed in a standard manner: terms are stemmed and stop words are removed.

5.2 Experimental Setup

In our experiments, the Lemur Toolkit was used to construct the baseline. For association rule mining, the Apriori algorithm implemented in the WEKA toolkit was adapted with the granularity of transactions set to be at the chunk level. We

Table 2. Comparison between QL, RM, IF and AR

(a) Experimental results on AP89 collection for queries 1–50 (title)

	QL	RM	IF	AR	AvgPr change (% over QL)	AvgPr change (% over RM)	AvgPr change (% over IF)
AvgPr	0.1970	0.2270	0.2664	0.2731	+38.6**	+20.3**	+2.5
Recall	1702	2312	2372	2367			

(b) Experimental results on AP88-89 collection for queries 101–150 (title)

	QL	RM	IF	AR	AvgPr change (% over QL)	AvgPr change (% over RM)	Avgr change (% over IF)
AvgPr	0.2338	0.3069	0.3185	0.3287	+40.6**	+7.1*	+3.2*
Recall	3160	3910	3900	3935			

(c) Experimental results on AP88-89 collection for queries 151–200 (title)

	QL	RM	IF	AR	AvgPr change (% over QL)	AvgPr change (% over RM)	Avgr change (% over IF)
AvgPr	0.3063	0.3471	0.3942	0.4081	+33.2**	+17.6**	+3.5*
Recall	3319	3566	3841	3793			

(d) Experimental results on WSJ90-92 collection for queries 201–250 (description)

	QL	RM	IF	AR	AvgPr change (% over QL)	AvgPr change (% over RM)	Avgr change (% over IF)
AvgPr	0.2366	0.2403	0.2673	0.2846	+20.3**	+18.43**	+6.5*
Recall	978	990	1015	1038			

(e) Experimental results on SJM collection for queries 51–100 (title & description)

	QL	RM	IF	AR	AvgPr change (% over QL)	AvgPr change (% over RM)	Avgr change (% over IF)
AvgPr	0.2105	0.2154	0.2201	0.2372	+12.7*	+10.12*	+7.8*
Recall	1460	1486	1488	1498			

* indicates the difference is statistically significant at the level of $p-$value < 0.05.
** indicates the difference is statistically significant at the level of $p-$value < 0.01.

use the top 35 documents as pseudo feedback documents, and the top 100 terms from the new query model are selected. Our experiments show little variation in performance when λ is more than 0.9.

Our method (AR) is compared with a baseline language model, namely the Query Likelihood (QL) model, the Relevance Model (RM), and the language model based on Information Flow (IF). The effectiveness indicators are the standard non-interpolated average precision (AvgP) and recall, which are calculated based on 1000 retrieved documents for each query. We also perform the t-test to measure the statistical significance of performance improvements.

5.3 Result Analysis

Tables 2(a), 2(b) and 2(c) show the retrieval performance of the four models under comparison on the AP89 and AP8889 collections using the title field of three query sets (average query length: 4). Our approach (AR) shows statistically significant improvements over the Query Likelihood model (QL) by more than 30% (38.6%, 40.6% and 33.2%), and over the Relevance Model by at least 9% (23.4%, 9.3% and 19.9%). Our approach also improves recall over the QL and RM. As shown in Figures 6(a), 6(b) and 6(c), our approach generates better precision than QL and RM at almost all the recall points.

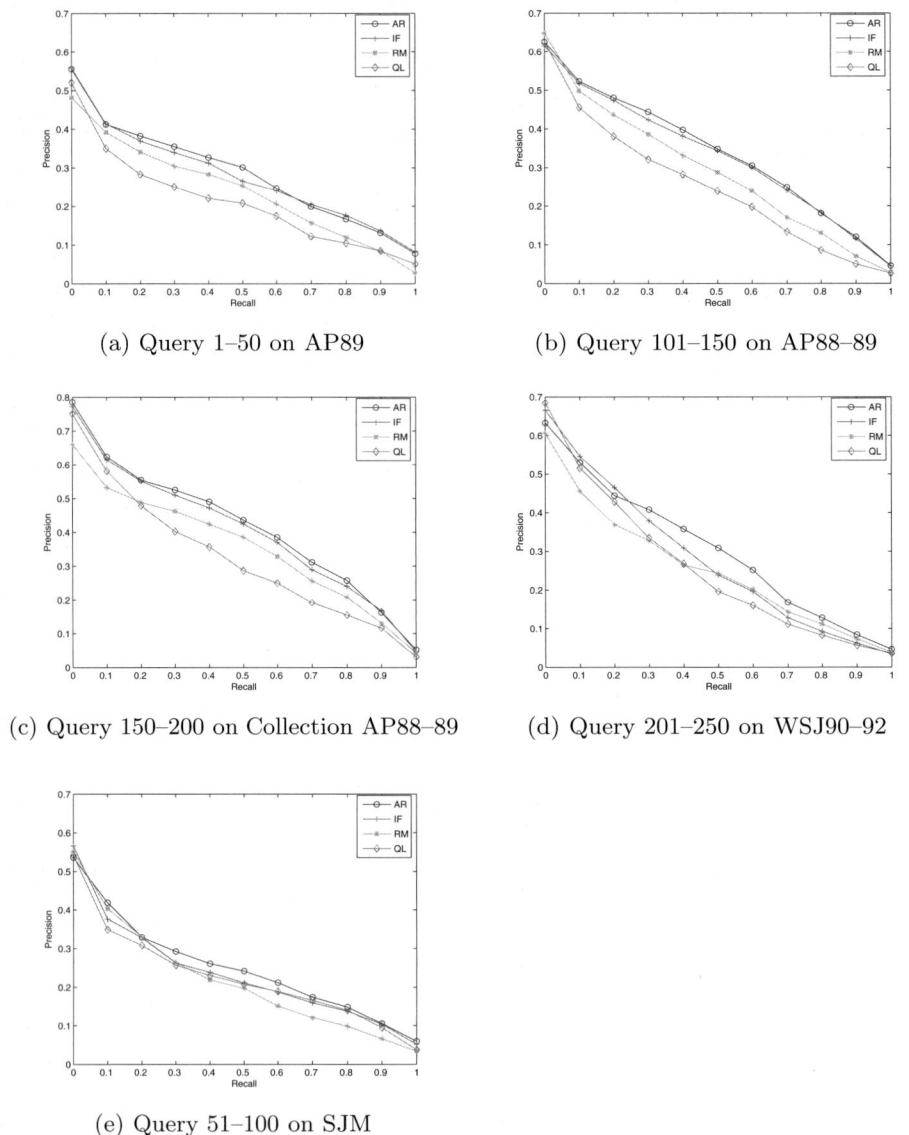

Fig. 6. Precision-recall Curves

Table 2(d) and Fig. 6(d) list the results on the WSJ collection using the description field of topics 201-250 (average query length: 8). Our approach also shows significant improvements in the average precision over the QL and RM, by 20.3% and 17.8% respectively.

Further, the experimental results on the SJM collection using a longer query set (title and description fields of topics 51-100, average query length: 12.2) are shown in Table 2(e) and Fig. 6(e). Again, significant improvements (12.7% and 10.38%) in average precision have been achieved, although not as much as the improvements obtained for the shorter queries. This is due to the query length. In general, the longer the query is, the more useful information it may have contained. Thus, it is reasonable that, for longer queries, a better baseline was also obtained. However, even in this case, our approach still shows its strong ability to capture the relationships between query terms and words in pseudo feedback documents.

Our approach has also shown improvements over the Information Flow based language model in all the experiments. Moreover, more improvements are obtained on longer queries. This reflects the effects of the query decomposition, i.e., the consideration of the contributions from any parts of the query will lead to improvement of retrieval effectiveness.

Fig. 7 compares the retrieval performance with the use of sentences as chunks as well as the use of individual sliding windows. For the former, we split the pseudo feedback documents into sentences based on the punctuation, such as full stops. The performance of the use of sentences is only slightly better than the 15-sized window, but lower than the others. The use of multiple-length chunks proves to be more effective than the use of individual fixed-sized windows.

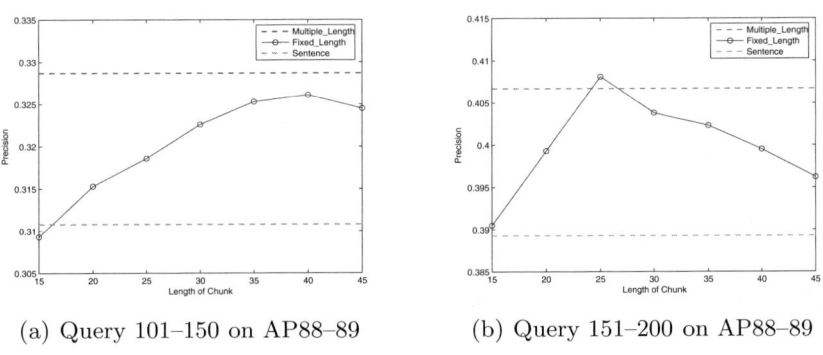

(a) Query 101–150 on AP88–89   (b) Query 151–200 on AP88–89

Fig. 7. Effects of Multiple Windows

Remark 1: We also test the effect of the mixture model (linear combination) of information flows and association rules. In all runs, the mixture model performs slightly better than the use of IF and AR individually (consistently by around +5% over the *IF* and less than +1% over the *AR*). This suggests that combining the different types of term relationships does not have to significantly improve retrieval performance. It also further explains the consistently better performance of association rule mining (AR) over information flow (IF). AR apparently produces better coverage of useful relationships. Therefore, using AR alone would seem to provide the basis of an effective solution.

Remark 2: The elapsed time for building the query model using AR and IF are also roughly compared. The AR (less than 1 second per query) is about three times faster than the IF (3.1 seconds per query). This further verifies our discussion about the computational issue in Section 1.

6   Conclusions and Future Work

We have proposed a novel approach, which integrates association rule mining, multiple window segmentation and query decomposition, to derive "higher-order" term relationships for query language modeling. Our framework takes into account inferences from any subset of query terms, facilitated by automatically derived multiple levels of "higher-order" term associations from the pseudo relevance feedback documents that are segmented into multiple sized chunks. A substantial suite of experiments have been conducted on various TREC collections and our approach outperforms a baseline language model, the Relevance Model and the Information Flow model. Based on experimental results, we can draw the following conclusions:

- The approach used in our paper considers the contributions from the different combinations of query words, i.e., the Q_j in Equation 4. This demonstrates that the incorporation of query decomposition to take into account all possible and partial inferences from the query is beneficial to retrieval performance.
- The multiple length document segmentation with overlapping sliding windows brings its benefits of avoiding the problem of data sparseness and the missing useful associations.
- The use of high-order terms relationships derived via association rule mining from segmented documents has proved more effective and efficient than the Information Flow model. This is, in our opinion, a significant step forward for developing operational query language models.

In the future, we will investigate more effective weighting function $P(Q_j|Q)$ to generate further improvement. The consideration of the importance of documents in high-order term relationship discovery could also be useful to improve the effectiveness of our system. In addition, other types of term relationships such as those from the WordNet, will be incorporated. Finally, some state-of-the-art algorithms, such as Metzler's MRF based method in [12], will be explored and compared in our future work.

Acknowledgement

This work is funded by the UK's Engineering Physical Sciences Research Council (EPSRC), grant number EP/E002145/1.

References

1. Agrawal, R., Imielinski, T., Swami, A.: Mining association rules between sets of items in large database. In: Proceedings of SIGMOD 1993, pp. 207–216 (1993)
2. Allan, J., Wade, C., Bolivar, A.: Retrieval and novelty detection at the sentence level. In: Proceedings of SIGIR 2003, Canada, August 2003, pp. 314–321 (2003)
3. Bai, J., Song, D., Bruza, P., Nie, J.-Y., Cao, G.: Query expansion using term relationships language models for information retrieval. In: Proceedings of CIKM 2005, Bremen, Germany (November 2005)
4. Balasubramanian, N., Allan, J., Croft, W.B.: A comparison of sentence retrieval techniques. In: Proceedings of SIGIR Poster Sessions, Amsterdam (July 2007)
5. Cao, G., Nie, J., Bai, J.: Integrating term relationships into language models. In: Proceedings of SIGIR 2005, pp. 298–305 (2005)
6. Gao, J., Nie, J., Wu, G., Cao, G.: Dependence language model for information retrieval. In: Proceedings of SIGIR 2004, pp. 170–177 (2004)
7. Hahsler, M., Buchta, C., Hornik, K.: Selective association rule generation. Computational Statistics 23 (2008)
8. Hipp, J., Guntzer, U., Nakhaeizadeh, G.: Algorithms for association rule mining - a general survey and comparison. In: SIGKDD Exploration, pp. 1–58 (2000)
9. Lafferty, J., Zhai, C.: Document Language Models, Query Models, and Risk Minimization for Information Retrieval. In: Proceedings of SIGIR 2001, pp. 111–119. ACM Press, New York (2001)
10. Lavrenko, V., Croft, W.B.: Relevance-based language models. In: Proceedings of SIGIR 2001, New York, pp. 120–127 (2001)
11. Liu, X., Croft, W.B.: Passage retrieval based on language models. In: Proceedings of CIKM 2002 (November 2002)
12. Metzler, D., Croft, W.B.: Latent concept expansion using markov random fields. In: Proceedings of SIGIR 2007 (July 2007)
13. Pickens, J., MacFarlane, A.: Term context models for information retrieval. In: Proceedings of CIKM 2006, November 2006, pp. 559–566 (2006)
14. Agrawal, R., Imielinski, T., Swami, A.: Mining association rules between sets of items in large databases. In: Proceedings of the SIGMOD 1993, Washington D.C., May 1993, pp. 207–216 (1993)
15. Song, D., Bruza, P.D.: Towards context sensitive information inference. Journal of the American Society for Information Science and Tecnology 54(3), 321–334 (2003)
16. Zhai, C., Lafferty, J.: A study of smoothing methods for language models applied to ad hoc information retrieval. In: Proceedings of SIGIR 2001, pp. 334–342 (2001)

Viewing Term Proximity from a Different Perspective

Ruihua Song[1,2], Michael J. Taylor[3], Ji-Rong Wen[2], Hsiao-Wuen Hon[2], and Yong Yu[1]

[1] Dept. of Computer Science and Engineer, Shanghai Jiao Tong University,
Shanghai, 200240 China
[2] Microsoft Research Asia, No.49 Zhichun Road,
Beijing, 100080, China
[3] Microsoft Research Ltd, 7 JJ Thomson Avenue,
Cambridge CB3 0FB, England
`{rsong,mitaylor,jrwen,hon}@microsoft.com, yyu@sjtu.edu.cn`

Abstract. This paper extends the state-of-the-art probabilistic model BM25 to utilize term proximity from a new perspective. Most previous work only consider dependencies between pairs of terms, and regard phrases as additional independent evidence. It is difficult to estimate the importance of a phrase and its extra contribution to a relevance score, as the phrase actually overlaps with the component terms. This paper proposes a new approach. First, query terms are grouped locally into non-overlapping phrases that may contain one or more query terms. Second, these phrases are not scored independently but are instead treated as providing a context for the component query terms. The relevance contribution of a term occurrence is measured by how many query terms occur in the context phrase and how compact they are. Third, we replace term frequency by the accumulated relevance contribution. Consequently, term proximity is easily integrated into the probabilistic model. Experimental results on TREC-10 and TREC-11 collections show stable improvements in terms of average precision and significant improvements in terms of top precisions.

1 Introduction

A document is usually represented as a bag of words in information retrieval theory in order to make both the development of retrieval models easier and the retrieval operation tractable. People often observe that the independence assumption does not hold in textual data and there has always been the feeling that term dependencies, if used correctly, should improve the retrieval quality. Consequently, there has been much research on incorporating term dependence into retrieval models over the last few decades.

Some recent work [10][17] on language models shows promising results by modeling dependencies on large web collections. However, most work on probabilistic models has not achieved consistent improvements. This paper aims to extend the state-of-the-art probabilistic model BM25 to take advantage of term proximity.

By surveying the literature, we find two problems in previous work. First, it is difficult to estimate the importance of phrases because they are different in nature from words. It may be not appropriate to apply the same weighting schemes for them.

C. Macdonald et al. (Eds.): ECIR 2008, LNCS 4956, pp. 346–357, 2008.

Second, a naïve linear combination of scores of words and those of phrases may break the non-linear property of term frequency. In probabilistic models, the non-linear term frequency is desirable because of the statistical dependence of term occurrences: the information gained on observing a term the first time is greater than the information gained on subsequently seeing the same term. As phrases are not independent from the component words, a linear combination of these two parts of scores is likely to be inappropriate.

To solve this problem, we take a brand new perspective in this paper. First, neighboring query terms are grouped into non-overlapping phrases that contain as many query terms as possible under some constraints. Second, these phrases are regarded as providing a context for the component query terms. The relevance contribution of a component query term occurrence is measured by how many query terms occur in the context phrase and how compact they are. Finally, we accumulate the relevance contribution from each query term. By replacing term frequency with this accumulated relevance contribution, term proximity is built into the probabilistic model. Experimental results on TREC-10 and TREC-11 collections show stable improvements in terms of average precision and significant improvements in terms of top precisions.

In the next section, previous work on term proximity is reviewed. Section 3 describes the details of the new approach that builds term proximity into the probabilistic model. Experimental results are presented in Section 4. Section 5 concludes and discusses future work.

2 Related Work

There has been some work dealing with term proximity or dependency in several information retrieval models, including vector space models [8], probabilistic models [6][12][16][28][29][22], language models [11][13][25][27][10][19][18][17], and inference network models [5] .

Fagan's thesis [8] is one of the most comprehensive studies on automatic indexing using phrases. He identifies both syntactic phrases by using linguistic evidence and statistical phrases by using factors, such as the number of times the phrase occurs in a collection and the proximity between phrase terms. Fagan regards the phrases as new words in scoring a document. On weighting a phrase, he associates the average of the *tf-idf* weights of the component words with the phrase. His results suggest that even though the statistical method does not consistently perform well, it yields improvements that are competitive with some syntactic methods.

Croft et. al. [5] proposes a different way to use phrases in inference network models. They identify phrases in a natural language query and then use the phrases to construct a structured query. The results on CACM collection indicate that this approach slightly outperforms statistical phrases, but the long natural language queries are not typical on the web today.

Several studies have examined term dependency in the language modeling framework [11][13][25][27][10][19][18][17]. Some models only consider dependencies among pairs of terms. Most approaches replace the unigram in the language model by bigrams or bi-terms that are two adjacent words. However, term dependency does not only exist between adjacent words. It may also occur between more distant words.

Thus Gao et. al. [10] extends the bigram or biterms models previously proposed to a more general one, in which word dependencies are not only restricted to adjacent words. Given a sentence, only the strongest word dependencies are considered to reduce estimation errors. The model shows consistent improvements over a baseline query likelihood system on a number of TREC collections. Metzler and Croft [17] develop a general term dependency model via Markov random fields. Sequential dependency and full dependency models are compared with the independent model. In their work, query terms within a window N terms are identified, so the number of query terms in a phrase may be more than two. Ordered and unordered phrases are assigned different weights in scoring. The experimental results show significant improvements, especially on the larger Web collections.

Only a few approaches take account of term proximity in ranking directly. In the 1995 TREC conference, both University of Waterloo and Australian National University adopted relevance measure based on term proximity [3][4][14][15]. Similarly to our approach, they detect spans that cover all concepts in a query and then calculate a score based on those spans. The score is proportional to the reciprocal of the length of span in [3][4], or the inverse square root of the length in [14][15]. These approaches are quite simple, but do not consider partial match of a query, i.e. a span covers a part of the query. Brin and Page described the usage of proximity in Google [1], but no detail or performance has been discussed.

Rasolofo and Savoy [22] demonstrate that one can improve retrieval effectiveness by combining a kind of simple term proximity based on word pairs into the BM25 ranking function,. As this work is most related to ours, we choose it as a competitor in our experiments. We now introduce the baseline approach BM25 and Rasolofo and Savoy's approach in detail.

BM25 is the state-of-the-art probabilistic model [24]. It combines term frequency, inverse document frequency and document length together to measure relevance. The formula is showed as below:

$$\sum_{t \in Q} w^{(1)} \frac{(k_1 + 1) \cdot tf}{K + tf} \tag{1}$$

where K is defined as

$$K = k_1 \cdot [(1 - b) + b \cdot \frac{l}{avdl}] \tag{2}$$

where, t denotes a word in the query Q; tf is term frequency of t; l is document length, and $avdl$ is average document length; k_1 and b are parameters. $w^{(1)}$ is the Robertson/Sparck-Jones weight [23], and one of its simple variants is:

$$w^{(1)} = \log \frac{N - n + 0.5}{n + 0.5} \tag{3}$$

where N is the sum of documents within a collection; n is the number of documents containing the term t within the collection.

In [22], for a query $q = (t_i, t_j, t_k)$, the following set S of term pairs is obtained: $\{(t_i, t_j), (t_i, t_k), (t_j, t_k)\}$. Rasolofo and Savoy extend formula (1) by adding an analogous function for term pairs.

$$\sum_{t \in Q} w^{(1)} \frac{(k_1 + 1) \cdot tf}{K + tf} + \sum_{(t_i, t_j) \in S} \min(w_i^{(1)}, w_j^{(1)}) \cdot \frac{(k_1 + 1) \cdot \sum_{occ(t_i, t_j)} tpi(t_i, t_j)}{K + \sum_{occ(t_i, t_j)} tpi(t_i, t_j)} \tag{4}$$

Here, $occ(t_i, t_j)$ denotes a term pair occurrence and the distance between t_i and t_j is no more than five in [22]. They calculate a term pair instance (*tpi*) weight as follows:

$$tpi(t_i, t_j) = \frac{1.0}{d(t_i, t_j)^2} \tag{5}$$

where $d(t_i, t_j)$ is the distance expressed in number of words. Based on formula (5), the highest value is 1.0, corresponding to a distance of one (the terms are adjacent), whereas the lowest value is 0.04, corresponding to a distance of 5.

Büttcher et al [2] propose a score function that is similar to Rasolofo and Savoy's function. The difference is that in Büttcher et al's approach only neighboring query terms can affect each other and the estimated weight of a term pair is smaller because it is limited to 1.

These approaches still suffer the problems of weighting phrases and over-scoring a term pair given the component words are already scored. Our work will try to solve these problems.

3   The Probabilistic Model with Term Proximity Built-In

Here we propose a way to go beyond the simple linear combination of relevance scores of single terms and phrases. Although we agree that the contribution of phrases is greater than that of an isolated query term, we do not treat phrases as separate objects. When a term occurs within a phrase, its contribution to relevance is boosted by factors related to its context, such as the density of the phrase and the number of unique terms that occur together with it. In other words, the contribution of a phrase is distributed to each term that composes the phrase. As a consequence, term frequency in existing ranking functions is replaced by the accumulation of single relevance contribution. While term proximity is easily plugged in, such approach could also take advantage of well-defined inverse document frequency and document length normalization parts, and preserve non-linear property of term frequency as well. In this section, we will mainly address the following two problems:

1. How to detect phrases without overlapping, called expanded spans.
2. How to measure relevance contribution of a single term by considering its context.

3.1 Expanded Span

Each document d is treated as an ordered sequence of terms

$$t_1, t_2, ..., t_{|d|}.$$

$|d|$ is the document length in terms of words. Given a query Q that is a set of query terms, all query term occurrences compose a chain of ordered hits:

$$t_{p_1}, t_{p_2}, ..., t_{p_m}.$$

Here, $t_{p_i} \in Q, \forall i \in \{1, 2..m\}$, and p_i is the position where t_{p_i} occurs. A hit means a term occurrence.

We wish to group the most related query terms together into an expanded span. To avoid counting a term twice, there is no overlap terms between any two expanded spans. Based upon such a strategy, we observe that an expanded span will have the following two properties.

First, two adjacent query terms are far enough away from each other, we regard them as unrelated query terms.

Second, given two different query terms a and b, if b is surrounded by more than one a, we would connect b with the closest a. For an instance, "$a\ x\ x\ x\ x\ a\ x\ b$" is a segment of a document where "x" represents any term not in the query. It is more natural that we combine the second a, instead of the first one, with b as an expanded span because they are closer. For another example, if another a follows b as "$a\ x\ x\ x\ a\ x\ b\ a$," the third a may be more related to b than the second one now because it is nearer to b.

Based upon these observations, we propose an algorithm to segment a chain of hits into a set of expanded spans. First, hits in the chain are scanned one by one from the head to the tail. Second, for the current hit, if the next hit exists, four possible cases are to be processed respectively:

(1) The distance between the current and the next is bigger than a threshold $d_{\max}$: Separate the chain between these two hits;
(2) The current and the next hit are identical: Separate the chain between these two hits;
(3) The current and next hits are different but the next hit is identical to a previous hit: Compare the distance between the current and the next hits, and the distance between the previous and the current hits. Then separate the chain at the bigger gap.
(4) Otherwise: Go to the next hit.

Finally, when the last hit has been processed, the chain is segmented into several spans called expanded spans like ($t_{p_i} ... t_{p_j}$).

For a chain composed of m hits, the worst time complexity of expanded span detection algorithm is $O(m \cdot |Q|)$. For each hit t_{q_i}, in case 1 and 2, no further comparison with hits in the current sub-chain is conducted, while in case 3, there are $|Q|$ comparisons at most to search identical hits. The width of an expanded span of $(t_{p_i} \dots t_{p_j})$ is defined as:

$$Width(t_{p_i} \dots t_{p_j}) = \begin{cases} p_i - p_j + 1 & \text{if } p_i \neq p_j, \\ d_{max} & \text{otherwise.} \end{cases} \tag{6}$$

To illustrate expanded spans, we will use the short document [21] shown in Fig. 1, which was quoted by Clarke et al. in [4].

Erosion[1]

It[2] took[3] the[4] sea[5] a[6] thousand[7] years,[8]
A[9] thousand[10] years[11] to[12] trace[13]
The[14] granite[15] features[16] of[17] this[18] cliff,[19]
In[20] crag[21] and[22] scarp[23] and[24] base.[25]

It[26] took[27] the[28] sea[29] an[30] hour[31] one[32] night,[33]
An[34] hour[35] of[36] storm[37] to[38] place[39]
The[40] sculpture[41] of[42] these[43] granite[44] seams,[45]
Upon[46] a[47] woman[48]'s[49] face.[50]

—E.[51] J.[52] Pratt[53] (1882[54]–1964)[55]

Fig. 1. A short document with position labels

Superscripts indicate term positions. Suppose that a query is "**sea thousand years**" and d_{max} is set as 10. In this document, the chain of ordered hits is:

sea[5], thousand[7], years[8], thousand[10], years[11], sea[29]

According to the segmentation algorithm, scanning starts from sea[5]. For sea[5] and thousand[7], the fourth case is applied. For years[8], the next hit, i.e. thousand[10], is identical to thousand[7], so the third case is applied. As thousand[7] is nearer to years[8] than thousand[10], the chain is separated before thousand[10]. Thus, *(sea[5] ... years[8])* forms the first expanded span. Next, the fourth case is applied for thousand[10]. When years[11] is scanned, the distance between sea[29] and years[11] is further than d_{max}. By applying the first case, we get *(thousand[10] years[11])* as a new expanded span. Finally, *(sea[29])* becomes an expanded span with a single query term only. The set of expanded spans for the document is:

{(sea[5] ... years[8]), (thousand[10] years[11]), (sea[29])}

The corresponding span widths are as follows:

{4, 2, 10}

Please note that the width (or length) of *(sea[29])* is not 1 but 10 that is d_{max}.

In summary, an expanded span has the following properties:

(1) There is no overlapping between any two expanded spans
(2) Each expanded span contains as many unique hits as possible while its length is minimized.

3.2 Viewing Term Proximity from a Different Perspective

When a document is viewed as a bag of words, term frequency accumulates the contribution to relevance of each term, independent of any individual term occurrence's proximity to other query terms. However, when term position information is taken into account, the relevance contribution of each term occurrence could vary with its context. For instance, sea^5 contributes more than sea^{29} as sea^5 is near to other two query terms while sea^{29} is isolated in Fig. 1.

Intuitively, the relevance contribution is related to several factors. First, the relevance contribution is inversely proportional to a function of the width of the expanded span. For example, *(thousand^{10} years^{11})*, whose width is 2, looks more relevant than *(thousand^{56} ... years^{65})*, whose width is 10. Second, the number of unique query terms within an expanded span also boosts relevance contribution. For example, though the width of *(thousand^{10} years^{11})* is less than that of *(sea^5 ... thousand^7 years^8)*, the latter matches all the query terms so that implies more relevance to the query of "sea thousand year."

Therefore, we propose a function to represent relevance contribution of one term occurrence:

$$f(t, espan_i) = \frac{n_i^{\lambda}}{Width(espan_i)^{\gamma}} \tag{7}$$

Where:

t is a query term,

$espan_i$ is an expanded span that covers t ,

n_i is the number of query terms that occur in $espan_i$,

$Width(espan_i)$ is the width of $espan_i$,

λ and γ are two parameters.

This function increases with increasing n_i and decreasing width of the expanded span. For example, when $\lambda = 2$ and $\gamma = 1$,

$$f(sea, (sea^5 ... year^8)) = \frac{3 \times 3}{4} = 2.25$$

$$f(year, (thousand^{10} year^{11})) = \frac{2 \times 2}{2} = 2$$

$$f(sea, (sea^{29})) = \frac{1 \times 1}{10} = 0.1$$

So far, $f(t, espan_i)$ has no factor for the specific term, thus we will get the same values for any term t in $espan_i$, e.g. $f(sea, (sea^5...year^8)) = f(year, (sea^5...year^8))$.

Next, we accumulate relevance contribution (rc) of all occurrences for a term t:

$$rc = \sum_i f(t, espan_i) \tag{8}$$

The more expanded spans found containing t in a document, the higher the relevance score.

When tf in traditional ranking functions is replaced by rc, a term proximity factor is naturally achieved. In our experiments, rc is built in Okapi's ranking function (1) [24]:

$$\sum_{t \in Q} w^{(1)} \frac{(k_1 + 1) \cdot rc}{K + rc} \tag{9}$$

Even if two terms in the same expanded span have the same relevance contribution, term weights preserved in the ranking function could distinguish them easily now.

4   Experiments

Experiments are conducted on the ad hoc task of TREC-9, TREC-10 and the web task of TREC-11. Each task has 50 queries. TREC-9 and TREC-10 use the dataset of WT10g [7], while TREC-11 uses .GOV [7]. Both the data and queries are processed by Fox's stop list [9] and Porter stemmer [20].

We tune parameters to optimize the baseline BM25/Okapi ranking function. With setting b as 0.3, k_1 as 0.4 for WT10g and b as 0.45, k_1 as 2.5 for .GOV, we achieve the baselines that are competitive with top-flight performance in TREC-9, 10, and 11. In addition, we train three parameters of our new ranking function on TREC-9. As a result, we set d_{max} as 45, λ as 0.55 and γ as 0.25. With such settings, given a query of two terms a and b, the span "$a\ b$" that exactly matches the query contributes 1.23 to rc, three times as big as that of a separated "a" and "b".

We test our proposed approach on TREC-10 and TREC-11. We show evaluation results of the baseline (Okapi) and our approach (newTP) in Table 1. It shows that our approach consistently outperforms the baseline in terms of average precision and top precisions. In particular, the approach with term proximity built-in significantly improves BM25/Okapi ranking function in terms of precision at 5 and precision at 10.

Table 1. Retrieval Results

Collections	AvePre	P@5	P@10
TREC-10-Okapi	0.2026	0.3640	0.3240
TREC-10-newTP	0.2237	0.3960	0.3480
TREC-10-Diff	**+10.4%***	**+8.8%***	**+7.4%***
TREC-11-Okapi	0.1776	0.2776	0.2408
TREC-11-newTP	0.1855	0.3143	0.2653
TREC-11-Diff	**+4.4%**	**+13.2%***	**+10.2%***

Note: * indicates statistic significance (p<0.05 with a two-tailed paired t-test).

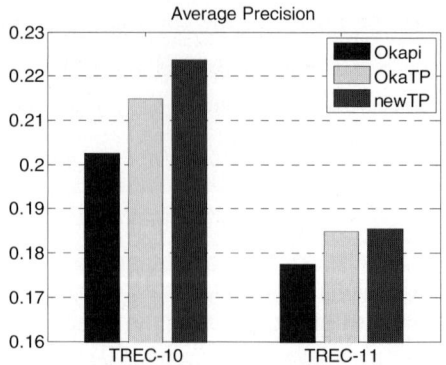

Fig. 2. Average precision

We implement Rasolofo and Savoy's approach (OkaTP) to do a comparison with ours (newTP). The results are shown in Fig. 2, 3, and 4.

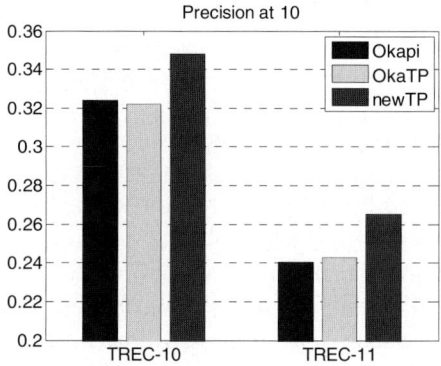

Fig. 3. Precision at 10

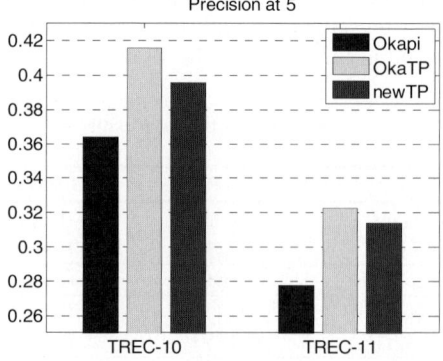

Fig. 4. Precision at 5

It indicates that our approach outperforms their approach in terms of average precision and especially precision at 10. In Fig.3 OkaTP does not achieve improvements over the baseline, but it performs even better than our approach in terms of precision at 5 (see Fig. 4).

It seems that OkaTP is more aggressive in bringing some documents with very close term pairs to the top, whereas our newTP is more stable in improving retrieval performance because we eliminate overlaps between a phrase and the component words. In addition, our newTP handles some documents with relatively distant query terms well, which may be why newTP brings more relevant documents into the top ten.

5   Conclusion and Future Work

This paper proposes a brand new approach that integrates term proximity into the state-of-the-art probabilistic model BM25. First, under some constraints, the chain of all the query term occurrences is segmented into spans. Second, the relevance contribution of a span is measured by the number of component words and its length. Third, term frequency in the probabilistic model is replaced by accumulated relevance contribution of the span that contains the term.

As a result, the BM25 formula is simply extended to a ranking function with term proximity built in. Experimental results on TREC-10, TREC-11 collections show stable improvements over the baseline in terms of average precision and significant improvements in terms of top precisions. When comparing the proposed approach with a previous approach, we find our approach is more stable and handles dependency between more distant terms better.

We have two directions of future work. On the one hand, there is room to improve the way of adding relevance contributions for a certain term. Though the relevance contribution of a span with a single query term is small, it is still possible to overwhelm the contribution of a compact span with more query terms, if the number of single-word spans is much bigger. Thus, we would try some other combination methods to solve the problem. On the other hand, evaluating the influence of term proximity in terms of perceived relevance is a part of our future research work. Sometimes, web search users appreciate returning a document with most or all of the query terms rather than a document that contains fewer, even if the later is more relevant [26]. Therefore, perceived relevance may be more appropriate to measure the impact of term proximity.

References

1. Brin, S., Page, L.: The anatomy of a large-scale hypertextual Web search engine. In: Proceedings of the 7th International World Wide Web Conference, Brisbane, Australia (1998)
2. Büttcher, S., Clarke, C.L.A., Lushman, B.: Term proximity scoring for ad-hoc retrieval on very large text collections. In: Proceedings of 29th Ann. Intl. ACM SIGIR Conf. on Research and Development in Information Retrieval (2006)
3. Clarke, C.L.A., Cormack, G.V., Burkowski, F.J.: Shortest substring ranking (multitext experiments for TREC-4). In: Proceedings of TREC-4 (1995)

4. Clarke, C.L.A., Cormack, G.V., Tudhope, E.A.: Relevance ranking for one to three term queries. Information Processing & Management 36(2), 291–311 (2000)
5. Croft, W.B., Turtle, H.R., Lewis, D.D.: The use of phrases and structured queries in information retrieval. In: Proceedings of 14th Ann. Intl. ACM SIGIR Conf. on Research and Development in Information Retrieval, pp. 32–45 (1991)
6. Croft, W.B.: Boolean queries and term dependencies in probabilistic retrieval models. JASIS 37(2), 71–77 (1986)
7. CSIRO, TREC Web Tracks home page, http://www.ted.cmis.csiro.au/TRECWeb/
8. Fagan, J.L.: Automatic phrase indexing for document retrieval: An examination of syntactic and non-syntactic methods. In: Proceedings of 10th Ann. Intl. ACM SIGIR Conf. on Research and Development in Information Retrieval, pp. 91–101 (1987)
9. Fox, C.: A stop list for general text. In: SIGIR Forum, December 1990, vol. 24(4), pp. 19–35. ACM Press, New York (1990)
10. Gao, J., Nie, J.-Y., Wu, G., Cao, G.: Dependence language model for information retrieval. In: Proceedings of 27th Ann. Intl. ACM SIGIR Conf. on Research and Development in Information Retrieval, pp. 170–177 (2004)
11. Harman, D.K.: Overview of the fourth Text Retrieval Conference (TREC-4). In: Proceedings of TREC-4, pp. 1–24
12. Harper, D.J., van Rijsbergen, C.J.: An evaluation of feedback in document retrieval using co-occurrence data. Journal of Documentation 34, 189–216
13. Harper, D.J., van Rijsbergen, C.J.: An evaluation of feedback in document retrieval using co-occurrence data. Journal of Documentation 34, 189–216
14. Hawking, D., Thistlewaite, P.: Proximity operators - So near and yet so far. In: Proceedings of TREC-4, pp. 131–143 (1995)
15. Hawking, D., Thistlewaite, P.: Relevance weighting using distance between term occurrences. Computer Science Technical Report TR-CS-96-08, Australian National University (August 1996)
16. Losee Jr., R.M.: Term dependence: truncating the Bahadur Lazarsfeld expansion. Information Processing and Management 30, 293–303 (1994)
17. Metzler, D., Croft, W.B.: A Markov random field model for term dependencies. In: Proceedings of 28th Ann. Intl. ACM SIGIR Conf. on Research and Development in Information Retrieval, pp. 472–479 (2005)
18. Mishne, G., de Rijke, M.: Boosting web retrieval through query operations. In: Losada, D.E., Fernández-Luna, J.M. (eds.) ECIR 2005. LNCS, vol. 3408, pp. 502–516. Springer, Heidelberg (2005)
19. Nallapati, R., Allan, J.: Capturing term dependencies using a language model on sentence trees. In: Proceedings of the 2002 ACM CIKM Intl. Conf. on Information and Knowledge Management, pp. 383–390 (2002)
20. Porter, M.: An algorithm for suffix stripping. Program 14(3), 130–137 (1980)
21. Pratt, E.J.: Complete poems. University of Toronto Press (1989)
22. Rasolofo, Y., Savoy, J.: Term proximity scoring for keyword-based retrieval systems. In: Sebastiani, F. (ed.) ECIR 2003. LNCS, vol. 2633, pp. 207–218. Springer, Heidelberg (2003)
23. Robertson, S.E., Spark Jones, K.: Relevance weighting for search terms. Journal of the American Society for Information Science 27(3), 129–146 (1976)
24. Robertson, S.E., Walker, S., Beaulieu, M.: Experimentation as a way of life: Okapi at TREC. Information Processing & Management 36(1), 95–108 (2000)
25. Song, F., Croft, W.B.: A general language model for information retrieval. In: Proceedings of CIKM 1999, pp. 316–321 (1999)

26. Spink, A., Wolfram, D., Jansen, B.J., Saracevic, T.: Searching the Web: The public and their queries. Journal of the American Society for Information Science and Technology 52(3), 226–234 (2001)
27. Srikanth, M., Srikanth, R.: Biterm language models for document retrieval. In: Proceedings of SIGIR 2002, pp. 425–426 (2002)
28. van Rijsbergen, C.J.: A theoretical basis for the use of cooccurrence data in retrieval. Journal of Documentation 33(2), 106–119 (1977)
29. Yu, C.T., Buckley, C., Lam, K., Salton, G.: A generalized term dependence in information retrieval. Technical report (1983)

Extending Probabilistic Data Fusion Using Sliding Windows

David Lillis[1], Fergus Toolan[2], Rem Collier[1], and John Dunnion[1]

[1] School of Computer Science and Informatics
University College Dublin
{david.lillis,rem.collier,john.dunnion}@ucd.ie
[2] Department of Computing Science
Griffith College Dublin
fergus.toolan@gcd.ie

Abstract. Recent developments in the field of data fusion have seen a focus on techniques that use training queries to estimate the probability that various documents are relevant to a given query and use that information to assign scores to those documents on which they are subsequently ranked. This paper introduces SlideFuse, which builds on these techniques, introducing a sliding window in order to compensate for situations where little relevance information is available to aid in the estimation of probabilities.

SlideFuse is shown to perform favourably in comparison with CombMNZ, ProbFuse and SegFuse. CombMNZ is the standard baseline technique against which data fusion algorithms are compared whereas ProbFuse and SegFuse represent the state-of-the-art for probabilistic data fusion methods.

1 Introduction

The aim of any Information Retrieval (IR) system is the identification of documents that best satisfy a user's information need, typically expressed in terms of a textual query. Traditional approaches to IR employ algorithms responsible for analysing the contents of the documents themselves in order to return those that most closely relate to the query provided.

More recently, there is a growing body of research focused on combining the output of several such systems with the aim of creating a single set of results that will have greater relevance than the output of any individual system [1,2,3]. Algorithms to perform this type of combination vary according to the situations in which they are intended to be used. This paper concentrates on the "data fusion" family of algorithms, which are intended for use in cases where each input system has access to the same document collections [4]. This is distinct from "collection fusion" [5], where the document collections are disjoint, or cases where only partial overlap exists between collections.

The principal difference between these situations is that data fusion algorithms may consider the presence of a document in multiple result sets as evidence of

C. Macdonald et al. (Eds.): ECIR 2008, LNCS 4956, pp. 358–369, 2008.

relevance, since a document's absence in a result set can only be as a result of it not being considered relevant by the corresponding input system. In contrast, where the overlap between document collections is not complete, the absence of a document from a result set may merely reflect its absence from the underlying document collection and so is not necessarily a reliable indication that the document has been considered to be nonrelevant.

This paper introduces SlideFuse, a novel probabilistic data fusion algorithm that uses the past performance of its underlying input systems as an indication of the probability that certain documents will be relevant to future queries. This assumption has been previously demonstrated to achieve favourable results [6,7]. It is robust in the face of incomplete training data by utilising information about a document's neighbours as evidence of its likelihood of relevance. It does this while avoiding some of the shortfalls of existing probabilistic methods.

Section 2 gives a brief outline of previous research in the area of data fusion. In Section 3, we present an overview of how SlideFuse operates, followed by a formal definition of the algorithm in Section 4. Section 5 details the setup of the experiments that were run to evaluate the SlideFuse algorithm, the results of which are presented in Section 6. This includes a comparison with the CombMNZ algorithm, which is a standard baseline frequently used in data fusion research, as well as ProbFuse and SegFuse, two recent probabilistic data fusion techniques. Finally, conclusions and future work are discussed in Section 7.

2   Data Fusion

Traditionally, data fusion techniques fall into two broad categories: score-based fusion and rank-based fusion. Score-based techniques make use of the scores each input system uses to rank the documents in its result set. This typically necessitates the use of some form of score normalisation [8], in order to ensure that the results cannot be skewed by the use of different methods of allocating scores (e.g. one input system may score documents on a scale of 0-100 whereas another may use a scale of 0-1).

A popular approach to score-based fusion is the use of a Linear Combination [1,9,3]. Here, weights are attached to each input system, which are multipled by the ranking scores assigned each document. The final score for each document is the sum of these. Normalised scores have also been used in this context [10].

An important suite of data fusion techniques based on normalised scores was proposed in [8]. Of these, CombMNZ has become the standard data fusion technique against which new algorithms are compared [2,11]. Here, the final score assigned to each document is the sum of the normalised scores it is given in each input result set, multiplied by the number of input systems that returned it. Significant work was carried out by Lee to demonstrate CombMNZ's effectiveness [12].

Interleaving is perhaps the simplest rank-based fusion technique [5]. This involves removing the top document from each input result set in turn and adding it to the fused set to be returned. Weighted variations on this have also been proposed so as to benefit input systems that have achieved superior performance in the past [13]. Two voting-based techniques based on document ranks were proposed by Aslam and Montague [11,14]. These used the analogy of the input systems representing few electors and the documents representing many candidates to be ranked.

An algorithm making use of the textual contents of the documents was presented in [15,16]. Another relies on the input systems providing metadata relating to the documents they return, which can be used in the fusion process [17].

In recent times, a variation of rank-based fusion has emerged, whereby result sets are divided into segments and documents are assigned a score based on the segments in which they appear, rather than their exact rank within the result set. The ProbFuse algorithm [6,18] divides each result set into equal length segments and uses training data to estimate the probability that a document returned in a particular segment by a particular input system is relevant. This is done by calculating the proportion of documents returned in each segment by each input system that are relevant to the training queries, compared to nonrelevant documents.

A similar approach is taken with SegFuse [7], with the major exception being that the segments are not of equal length, but rather increase in size exponentially later in the result set. As relevant documents are most likely to occur in the early part of a result set, maintaining small segment sizes in early positions advantages these early documents, as they are less likely to be grouped with less relevant documents occurring later on. SegFuse also takes normalised scores into account.

3   SlideFuse: Introduction

Existing segment-based data fusion techniques ProbFuse and SegFuse use the probability that a document is relevant to assign a score on which it is eventually ranked in the final result set. This probability is estimated by analysing the results of a number of training queries for which relevance judgments are available. Relevance judgments are typically included with IR test collections, and specify which documents in the collection have been judged to be relevant, or nonrelevant, to test queries. However, with large document collections these judgments tend to be incomplete, meaning that only relatively few documents have been judged for each query, leaving the majority unjudged. This incompleteness causes difficulty in analysing training data, as there may be positions in result sets in which a document that is known to be relevant is never returned, though this does not necessarily entail that a relevant document is never located at that rank.

For this reason, calculating probabilities at the individual rank level results in an extremely jagged probability distribution. For instance, with the Web Track from the TREC-2004 conference (which is the document collection used in the experiments presented in Section 5), calculating the probability for each position results in the graph presented in Figure 1. In that figure, the probability value used in each position is the number of relevant documents returned in that

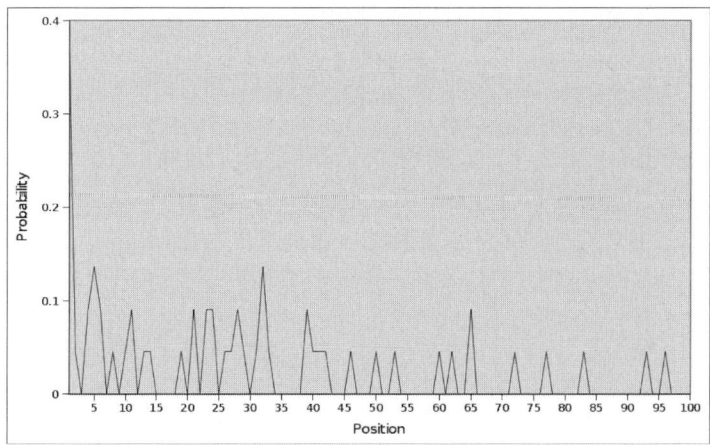

Fig. 1. Probability Distribution using Individual Positions

postion over all the training queries, divided by the total number of training queries that returned a document in that position (i.e. a result set of only 100 documents in length will not have returned a document in position 101).

One motivation behind segmenting result sets is to counter this effect, by not estimating the probability of relevance of a document returned at a particular rank solely based on documents returned at that exact rank for the training queries. Instead, relevant documents returned at other positions within the same segment are also taken into account, so smoothing the distribution of probability scores.

One consequence of this approach is that it is possible for a significant drop in probability score to occur at the boundary between segments. This effect is illustrated in Figure 2. For example, in a result set divided into segments of 40 documents each, the probability associated with the document returned in position 40 is likely to be much higher than that of the document returned in position 41. This is because the probability for the segment containing position 40 is calculated using positions 1 through 40, whereas the segment containing position 41 ranges from position 41 to position 80. As the former encompasses documents much higher in the result set (that are more likely to be relevant than documents further down the result set), position 40 is given an artificial advantage over position 41. This is easily demonstrated by plotting a graph of probability score against position. Unlike ProbFuse, SegFuse changes the size of each segment in different areas of the result set, with the smallest segments being at the beginning. This has the effect of reducing the distance between such segment boundaries at the beginning of the result set, where relevant documents are most likely to appear and consequently reducing the occurrence of sudden changes in probability scores.

In order to address the problem of the sudden drops in the probability scores associated with segmentation, and the problem of incomplete relevance judgments, we introduce the concept of a window surrounding each rank, where the

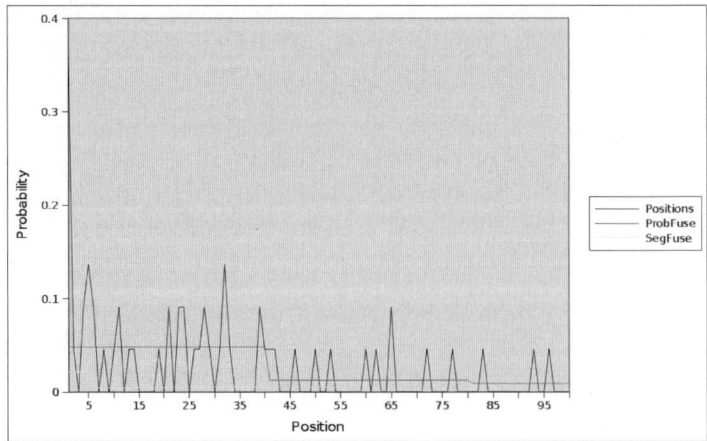

Fig. 2. Probability Distribution using ProbFuse and SegFuse

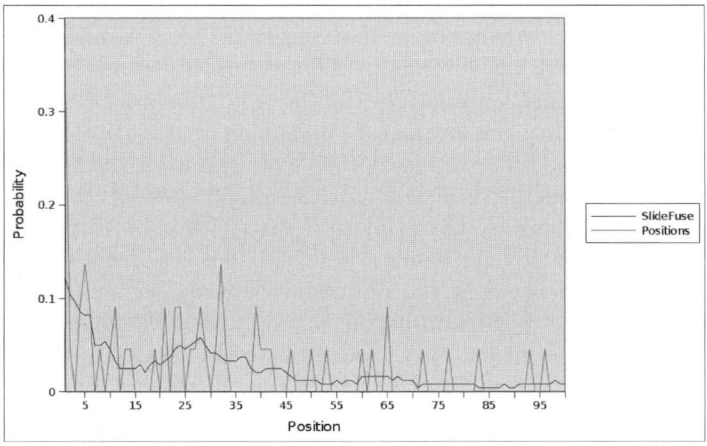

Fig. 3. Probability Distribution using SlideFuse

probability assigned to that rank is based on the proportion of relevant documents located in its surrounding window during the training phase. For example, if we define the size of the sliding window to extend to 5 documents on either side of the relevant position, the window for rank 40 will extend from position 35 to position 45 inclusive. Similarly, the sliding window for position 41 extends from rank 36 to rank 46. With this approach, the problem of the location of segment boundaries is eliminated, as it is the closest neighbouring positions that are always taken into account. The effect on probability distribution is shown in Figure 3. As a special case, SlideFuse ensures that a sliding window cannot extend beyond the boundaries of the result set.

The use of training data entails that the scores that are ultimately assigned to each document are based on the past performance of each input system, thus encompassing an implied weighting system wereby documents returned by input systems with a prior record of greater effectiveness will receive a higher score. A formal mathematical description is presented in Section 4.

4   SlideFuse: Description

In common with other probabilistic data fusion techniques, SlideFuse requires both a training phase and a fusion phase. In the training phase, relevance information is gleaned from result sets returned in response to training queries for which relevance judgments are available. Later, this training data is used to fuse result sets produced by the same input systems relating to other queries.

4.1   Training Phase: Rank Probability

The training phase consists of estimating for each input system the probability that a document returned in any given rank in that system's result set is relevant.

Formally, $P(d_p|s)$, the probability that a document d returned in position p of a result set is relevant, given that is has been returned by input system s is given by

$$P(d_p|s) = \frac{\sum_{q \in Q_p} R_{d_p,q}}{Q_p} \tag{1}$$

where Q_p is the set of all training queries for which at least p documents were returned by the input system and $R_{d_p,q}$ is the relevance of the document d_p to query q (1 if the document is relevant, 0 if not). This is calculated for each input system to be used in the fusion phase.

4.2   Fusion Phase: Window Boundaries

As noted in Section 3, using the probability at each rank leads to inconsistent results on document collections with incomplete relevance judgments, due to the high number of documents in each result set that have not been judged relevant (and are therefore assumed not to be relevant). In order to achieve more useful probability values, we construct a window around each position, so as to make use of relevance information about near neighbours when assigning probabilities to individual ranks.

The start and end points (a and b respectively) of the sliding window surrounding each result set position p are given by

$$a = \begin{cases} p - w & p - w >= 0 \\ 0 & p - w < 0 \end{cases} \tag{2}$$

$$b = \begin{cases} p + w & p + w < N \\ N - 1 & p + w >= N \end{cases} \tag{3}$$

where w is a parameter that indicates how many positions on either side of p should be included in the window and N is the total number of documents in the result set. In effect, the above definitions of a and b ensure that the window cannot begin before the first document in the result set and also cannot extend beyond the last document.

4.3   Fusion Phase: Assigning Probabilities to Windows

Once the window boundaries have been set around each position of each of the result sets that are to be fused, the next stage in the fusion process is to assign a probability score to each position based on those positions contained in the window surrounding it.

$P(d_{p,w}|s)$, the probability of relevance of document d in position p using a window size of w documents either side of p, given that it has been returned by input system s is given by

$$P(d_{p,w}|s) = \frac{\sum_{i=a}^{b} P(d_i|s)}{b - a + 1} \qquad (4)$$

The use of the sliding window results in a smoother decrease in the probabilities later in the result set, when compared with using probabilities based on data available at each position alone.

4.4   Fusion Phase: Ranking Score

Once the above stages have been completed, the final step is to assign a score to each document. R_d, the final ranking score given to document d is given by

$$R_d = \sum_{s \in S} P(d_{p,w}|s) \qquad (5)$$

where S is the set of all input systems used and p is the position in which document d was returned by input system s. Using the sum of the probability scores makes use of the "Chorus Effect", which argues that multiple input systems agreeing on the relevance of a document is evidence that the document in question is actually relevant [3]. The "Skimming Effect" is also important in the context of data fusion [3]. This states that since relevant documents are most likely to be located in early positions in a result set, weighting highly-ranked documents heavily is beneficial when performing fusion. Although there is no explicit consideration of this effect made in the definition of SlideFuse, the probability distribution in Figure 3 shows that this increased likelihood of relevance in early positions automatically benefits these highly-ranked documents.

5   Experiment Setup

The document collection used for evaluation is the Web Track from the TREC-2004 conference [19]. A feature of this document collection is that the relevance judgments are extremely incomplete. The available data includes 74 topfiles (each containing result sets produced by a single input system in response to each of 225 queries). A number of measures were taken in order to reduce the possibility of any bias being introduced by either the selection of input systems or the ordering of the queries.

Five runs of the experiment were performed. For each run, six topfiles were selected and the result sets from those topfiles were fused using SlideFuse, ProbFuse, SegFuse and CombMNZ. No topfile was used in more than one experimental run, the result of which being that of the 64 topfiles available, 30 were used for the purposes of this experiment. So as to eliminate the possibility of the ordering of the queries introducing any sort of bias, each run was performed five times, with the queries being shuffled each time. After shuffling, the first 10% of queries were used for the purposes of training SlideFuse, ProbFuse and SegFuse. As the CombMNZ algorithm does not require a training phase, these training queries were ignored for that technique. The evaluation results presented below for each run are the average evaluation results from all of the various query orderings.

When running ProbFuse, each result set was divided into 25 segments, as in [6]. For the purposes of SlideFuse, the value of the w parameter was set to 5 (i.e. 5 documents on both sides of each position were included in the window). It is desirable to use a small value for w, in order that the probabilities at each position are only influenced by positions that are close by. However, initial experiments showed that using windows that are too small failed to fully address the problem outlined in Section 3, as there were still positions for which probabilities could not be calculated due to a lack of available relevance judgments.

When performing the evaluation of the four data fusion techniques, three evaluation measures were used: *Mean Average Precision (MAP)* is the mean of the precision scores obtained after each relevant document has been retrieved. Relevant documents that are not included in the result set are given a precision of zero. MAP assumes that documents that have not been judged are nonrelevant. The *bpref* measure evaluates the relative position of relevant and nonrelevant documents, ignoring documents that are unjudged. It was proposed by Buckley and Voorhees to cater for situations where relevance judgments are incomplete [20]. *P10* measures the precision after 10 documents have been returned. Research has demonstrated that the vast majority of users of IR systems only examine the top 10 documents presented to them [21]. Thus, the P10 measure places emphasis on documents returned in those positions where they are likely to be of use to the user.

Table 1 illustrates the results of initial experiments aimed at choosing an appropriate training set size. Fusion was performed using each algorithm, with the training set sizes set to 10%, 20%, 30%, 40% and 50% of available queries in turn. The performance of each algorithm was evaluated for each training set size using MAP. The Coefficient of Variation relating to these scores was then calculated for each algorithm for each run. This reflects the degree to which fusion performance is affected by changing the training set size. As Table 1 illustrates, altering the number of training queries did not have any substantial effect on the performance of any of the fusion algorithms. Similar results were obtained for the bpref and P10 evaluation measures. Using only 10% of the available queries for training thus reduces the amount of training data without adversely affecting performance.

Table 1. Coefficient of Variation for MAP scores using training set sizes of 10%, 20%, 30%, 40%, 50%

	CombMNZ	ProbFuse	SegFuse	SlideFuse
first	0.0033	0.0131	0.0056	0.0056
second	0.0056	0.1188	0.0581	0.0104
third	0.0380	0.0168	0.0120	0.0103
fourth	0.0034	0.0143	0.0111	0.0019
fifth	0.0246	0.0229	0.0491	0.0179

6   Analysis of Results

The results of comparing SlideFuse with CombMNZ, ProbFuse and SegFuse are shown in Tables 2, 3 and 4. Each table presents the results from each of the five runs, along with the average result for each fusion technique. The "vs. Best" column displays the percentage difference between SlideFuse and the best of the other techniques (which is highlighted in bold in each case). The average in that column is the percentage difference between the average SlideFuse score and the best average score amongst the other algorithms. Values marked with "*" are statistically significant for a significance level of 5%, using a paired t-test. Entries marked with "**" are significant for a significance level of 1%.

Table 2. TREC-2004 performance of five individual runs evaluated with MAP

	CombMNZ	ProbFuse	SegFuse	SlideFuse	vs. Best
first	0.1598	**0.4045**	0.1789	0.4977	23.05% **
second	0.0783	**0.2809**	0.1493	0.4905	74.58% **
third	0.0426	0.2454	**0.4946**	0.5103	3.17% **
fourth	0.2454	0.2505	**0.4995**	0.5025	0.61%
fifth	0.1334	0.2892	**0.3348**	0.3849	14.98% **
average	0.1319	0.2941	**0.3314**	0.4772	43.99%

Of the baseline techniques, ProbFuse performs best on the "first" and "second" runs, with SegFuse achieving superior performance on the others, with one exception in the bpref data. Overall, SlideFuse achieves the highest evaluation scores on average for all evaluation measures, with the single exception of the bpref score for the "fourth" run where the difference is 0.58%, although this difference is not significant. Tests show that the performance improvements are

Table 3. TREC-2004 performance of five individual runs evaluated with bpref

	CombMNZ	ProbFuse	SegFuse	SlideFuse	vs. Best
first	0.2176	**0.2997**	0.2547	0.4009	33.75% **
second	0.3155	0.1877	**0.3529**	0.4085	15.75% **
third	0.1665	0.1281	**0.4228**	0.4331	2.44% *
fourth	0.4015	0.1375	**0.4155**	0.4131	-0.58%
fifth	0.1945	0.1968	**0.2971**	0.2996	0.83%
average	0.2591	0.1900	**0.3486**	0.3910	12.17%

Table 4. TREC-2004 performance of five individual runs evaluated with P10

	CombMNZ	ProbFuse	SegFuse	SlideFuse	vs. Best
first	0.1123	**0.1344**	0.1195	0.1413	5.15% **
second	0.0349	**0.1023**	0.0800	0.1436	40.39% **
third	0.0257	0.1164	**0.1401**	0.1445	3.15% **
fourth	0.1101	0.1124	**0.1381**	0.1408	1.96% *
fifth	0.0561	0.1070	**0.1113**	0.1189	6.83% **
average	0.0678	0.1145	**0.1178**	0.1378	16.99%

statistically significant in most cases, with the exceptions being the "fourth" run when evaluated using MAP and the "fifth" run when evaluated with bpref.

Additionally, SlideFuse outperforms the best other technique in all runs using all three evaluation measures. When compared on an overall basis against any individual technique, the improvement is over 12% in all cases, and is above 40% when measured using MAP.

7   Conclusions and Future Work

This paper describes SlideFuse, a probabilistic data fusion algorithm that addresses some of the limitations of existing segment-based probabilistic techniques. On experiments using the TREC-2004 Web Track dataset, SlideFuse was shown to outperform the CombMNZ, ProbFuse and SegFuse data fusion techniques when evaluated using MAP, bpref and P10. Despite the fact that the training data available for the dataset is incomplete, SlideFuse was still capable of outperforming two algorithms that use the same training data (ProbFuse and SegFuse) and one that does not rely on training data (CombMNZ).

This was achieved by using a sliding window to use the probable relevance of a document's neighbours to estimate the probability that a document itself is relevant.

At present, SlideFuse assumes that each result set returned by an input system is of the same quality, as the probabilities used for fusion will be same in each case. In the future, we aim to investigate methods of weighting a particular result set according to its quality. This could possibly involve the use of the scores assigned to each document as a measure of an input system's confidence in its own results. Another approach to weighting would be to introduce weights within the sliding windows themselves, so as to place more emphasis on those documents that are closest to the rank around which the window is centred. Finally, a minor drawback of SlideFuse is that documents returned in positions beyond the length of the training sets will not be taken into account when fusing. We aim to address this situation in a more satisfactory fashion.

References

1. Bartell, B.T., Cottrell, G.W., Belew, R.K.: Automatic combination of multiple ranked retrieval systems. In: SIGIR 1994: Proceedings of the 17th annual international ACM SIGIR conference on Research and development in information retrieval, pp. 173–181. Springer, New York (1994) Reference to show that it has long been demonstrated that fusion improves results
2. Beitzel, S.M., Jensen, E.C., Chowdhury, A., Grossman, D., Frieder, O., Goharian, N.: Fusion of effective retrieval strategies in the same information retrieval system. J. Am. Soc. Inf. Sci. Technol. 55(10), 859–868 (2004)
3. Vogt, C.C., Cottrell, G.W.: Fusion via a linear combination of scores. Information Retrieval 1(3), 151–173 (1999)
4. Aslam, J.A., Montague, M.: Bayes optimal metasearch: a probabilistic model for combining the results of multiple retrieval systems. In: SIGIR 2000: Proceedings of the 23rd annual international ACM SIGIR conference on Research and development in information retrieval, pp. 379–381. ACM Press, New York (2000)
5. Voorhees, E.M., Gupta, N.K., Johnson-Laird, B.: The collection fusion problem. In: Proceedings of the Third Text REtrieval Conference (TREC-3), pp. 95–104 (1994)
6. Lillis, D., Toolan, F., Collier, R., Dunnion, J.: ProbFuse: a probabilistic approach to data fusion. In: Proceedings of the 29th annual international ACM SIGIR conference on Research and development in information retrieval, pp. 139–146. ACM Press, New York (2006)
7. Shokouhi, M.: Segmentation of search engine results for effective data-fusion. In: Amati, G., Carpineto, C., Romano, G. (eds.) ECiR 2007. LNCS, vol. 4425, Springer, Heidelberg (2007)
8. Fox, E.A., Shaw, J.A.: Combination of multiple searches. In: Proceedings of the 2nd Text REtrieval Conference (TREC-2), National Institute of Standards and Technology Special Publication 500-215, pp. 243–252 (1994)
9. Callan, J.P., Lu, Z., Croft, W.B.: Searching distributed collections with inference networks. In: SIGIR 1995: Proceedings of the 18th annual international ACM SIGIR conference on Research and development in information retrieval, pp. 21–28. ACM Press, New York (1995)
10. Si, L., Callan, J.: Using sampled data and regression to merge search engine results. In: SIGIR 2002: Proceedings of the 25th annual international ACM SIGIR conference on Research and development in information retrieval, pp. 19–26. ACM Press, New York (2002)
11. Montague, M., Aslam, J.A.: Condorcet fusion for improved retrieval. In: CIKM 2002: Proceedings of the eleventh international conference on Information and knowledge management, pp. 538–548. ACM Press, New York (2002)
12. Lee, J.H.: Analyses of multiple evidence combination. SIGIR Forum 31(SI), 267–276 (1997)

13. Voorhees, E.M., Gupta, N.K., Johnson-Laird, B.: Learning collection fusion strategies. In: SIGIR 1995: Proceedings of the 18th annual international ACM SIGIR conference on Research and development in information retrieval, pp. 172–179. ACM Press, New York (1995)
14. Aslam, J.A., Montague, M.: Models for metasearch. In: SIGIR 2001: Proceedings of the 24th Annual International ACM SIGIR Conference on Research and Development in Information Retrieval, pp. 276–284. ACM Press, New York (2001)
15. Craswell, N., Hawking, D., Thistlewaite, P.B.: Merging results from isolated search engines. In: Australasian Database Conference, Auckland, New Zealand, pp. 189–200 (1999)
16. Lawrence, S., Giles, C.L.: Inquirus, the NECI meta search engine. In: Seventh International World Wide Web Conference, Brisbane, Australia, pp. 95–105. Elsevier, Amsterdam (1998)
17. Gravano, L., Chang, K., Garcia-Molina, H., Paepcke, A.: Starts: Stanford protocol proposal for internet retrieval and search. Technical report, Stanford, CA, USA (1997)
18. Lillis, D., Toolan, F., Collier, R., Dunnion, J.: Probabilistic data fusion on a large document collection. In: Proceedings of the 17th Irish Conference on Artificial Intelligence and Cognitive Science (AICS 2006), Belfast, Northern Ireland, Queen's University Belfast (2006)
19. Craswell, N., Hawking, D.: Overview of the TREC-2004 web track. In: Proceedings of the Thirteenth Text REtrieval Conference (TREC-2004) (2004)
20. Buckley, C., Voorhees, E.M.: Retrieval evaluation with incomplete information. In: SIGIR 2004: Proceedings of the 27th annual international ACM SIGIR conference on Research and development in information retrieval, pp. 25–32. ACM Press, New York (2004)
21. Silverstein, C., Henzinger, M., Marais, H., Moricz, M.: Analysis of a Very Large AltaVista Query Log. Technical Report 1998-014, Digital SRC (1998), http://gatekeeper.dec.com/pub/DEC/SRC/technical-notes/abstracts/src-tn-1998-014.html

Semi-supervised Document Classification with a Mislabeling Error Model

Anastasia Krithara[1], Massih R. Amini[2],
Jean-Michel Renders[1], and Cyril Goutte[3]

[1] Xerox Research Centre Europe, chemin de Maupertuis, F-38240, Meylan, France
Anastasia.Krithara@xrce.xerox.com
[2] University Pierre et Marie Curie, 104, avenue du President Kennedy,
75016 Paris, France
amini@poleia.lip6.fr
[3] National Research Council Canada, 283, boulevard Alexandre-Taché,
Gatineau, QC J8X 3X7, Canada
Cyril.Goutte@nrc-cnrc.gc.ca

Abstract. This paper investigates a new extension of the Probabilistic Latent Semantic Analysis (PLSA) model [6] for text classification where the training set is partially labeled. The proposed approach iteratively labels the unlabeled documents and estimates the probabilities of its labeling errors. These probabilities are then taken into account in the estimation of the new model parameters before the next round. Our approach outperforms an earlier semi-supervised extension of PLSA introduced by [9] which is based on the use of *fake labels*. However, it maintains its simplicity and ability to solve multiclass problems. In addition, it gives valuable information about the most uncertain and difficult classes to label. We perform experiments over the 20Newsgroups, WebKB and Reuters document collections and show the effectiveness of our approach over two other semi-supervised algorithms applied to these text classification problems.

1 Introduction

In this paper we present a new semi-supervised variant of the Probabilistic Latent Semantic Analysis (PLSA) algorithm [6] for text classification in which a mislabeling error model is incorporated.

Semi-supervised learning (SSL) algorithms have widely been studied since the 1990s mostly thanks to Information Access (IA) and Natural Language Processing (NLP) applications. In these applications unlabeled data are significantly easier to come by than labeled examples which generally require expert knowledge for correct and consistent annotation [3,4,13,16,11,1]. The underlying assumption of SSL algorithms is, if two points are *close* then they should be labeled similarly, resulting in that the search of a decision boundary should take place in low-density regions. This assumption does not imply that classes are formed from single compact clusters, only that objects from two distinct classes are not likely to be in the same cluster. This *cluster assumption* has first been expressed by [12] who proposed a mixture model to estimate the generation probability of examples by using both the labeled and unlabeled data. Prediction (the classification of new examples) is done by applying Bayes rule. Many practical algorithms have been implemented within this generative framework and successfully been applied to text classification [13].

Following the cluster assumption, we propose a new algorithm that iteratively computes class labels for unlabeled data and estimates the class labeling error using a mislabeling error model.

The parameters of this mislabeling error model are estimated within a semi-supervised PLSA (ssPLSA) model by maximizing the data log-likelihood, taking into account the class labels and their corresponding error estimates over the unlabeled examples. This work generalizes the study in [2], where a mislabeling error model was also proposed for SSL of discriminative models in the case of binary classification problems. We further show why the generative assumption leading to the ssPLSA we propose is more likely to hold than the one which serves to develop the semi-supervised Naive Bayes (ssNB) model. The empirical results we obtained confirm the effectiveness of our approach on 20Newsgroups, WebKB and Reuters document collections over the ssNB [13], the transductive Support Vector Machine (SVM) [8] and a previously developed ssPLSA model [9] in which fake labels are assigned to unlabeled examples.

In the remainder of the paper, we first briefly describe in section 2.2, the ssNB model proposed by [13] for text classification. Then in section 2.3, we present our extension of the aspect PLSA model for semi-supervised learning, in which we incorporate a mislabeling error. The previously developed ssPLSA model with fake labels is presented in the same section. The experiments we conducted are described in section 3. Finally, in section 4, we discuss the outcomes of this study and we also draw some pointers for the continuation of this research.

2 Semi-supervised Generative Models for Document Classification

This section presents two probabilistic frameworks for modeling the nature of documents in the case where a partially labeled training set is available. Each framework defines a generative model for documents and encompasses different probabilistic assumptions for their generation and their labeling. The ultimate aim of each framework is to assign a label to an unseen document.

2.1 Notations

We assume that the training set is a collection of partially labeled documents $\mathcal{D} = \{d_1, \ldots, d_{N_d}\}$ containing words from the vocabulary $\mathcal{W} = \{w_1, \ldots, w_{N_w}\}$. D_l and D_u denote respectively the set of labeled and unlabeled documents in $\mathcal{D}$. All documents from D_l have a class label $y \in \mathcal{C} = \{y_1, \ldots, y_K\}$ and each document $d \in \mathcal{D}$ is represented by the vector of word frequencies $\boldsymbol{d} = < n(w, d) >_{w \in \mathcal{W}}$.

2.2   Naive Bayes Model

In this framework each document is assumed to be generated by a mixture model:

$$p(\boldsymbol{d}, \Theta) = \sum_{k=1}^{K} p(y_k \mid \Theta)p(\boldsymbol{d} \mid y_k, \Theta) \tag{1}$$

We further assume that there is an univocal correspondence between each class $y \in \mathcal{C}$ and each mixture component. A document d is therefore generated by first selecting a mixture component according to the prior class probabilities $p(y_k \mid \Theta)$, and then generating the document from the selected mixture component, with probability $p(\boldsymbol{d} \mid y_k, \Theta)$ (Figure 1 (a)).

The probability of a new document is the sum over all mixture components as the true class to which the document belongs to is unknown.

In the Naive Bayes model the co-occurrence of words within each document is assumed to be independent; this essentially corresponds to the *bag-of-words* assumption. From this assumption the probability of a document d given the class y_k can be expressed as

$$p(\boldsymbol{d} \mid y_k, \Theta) \propto \prod_{j=1}^{N_w} p_{jk}^{n(w_j, d)} \tag{2}$$

Where, p_{jk} is the probability of generating word w_j in class y_k. The complete set of model parameters consists of multinomial parameters for the class priors $p(y_k)$ and word generation probabilities p_{jk}:

$$\Theta = \{p(y_k) : y_k \in \mathcal{C}; p_{jk} : w_j \in \mathcal{W}, y_k \in \mathcal{C}\}$$

[13] propose to estimate Θ by maximizing the complete data log-likelihood using an Expectation-Maximization (EM) algorithm, and modulating the influence of unlabeled documents in the estimation of the log-likelihood using a weighting parameter λ. The algorithm we used in our experiments may be sketched out as follows (refer to [13] for further details). The initial set of Naive Bayes parameters $\Theta^{(0)}$ is obtained by maximizing the likelihood over the set of labeled documents $D_l \subset \mathcal{D}$. We then iteratively estimate the probability that each mixture component $y_k \in \mathcal{C}$ generates each document $d \in \mathcal{D}$ using the current parameters $\Theta^{(j)}$, and update the Naive Bayes parameters $\Theta^{(j+1)}$ by maximizing the complete-data log-likelihood in which the effect of unlabeled documents are moderated via a parameter $\lambda \in [0, 1]$. The complexity of this algorithm is $O(K \times M)$, where $M = \#\{(w, d) | n(w, d) > 0\}$.

2.3   Probabilistic Latent Semantic Analysis

The PLSA model introduced by Hoffmann [6] is a probabilistic model which characterizes each word in a document as a sample from a mixture model, where mixture components are conditionally-independent multinomial distributions.

This model, also known as the aspect model [14], associates an unobserved latent variable (called aspect, topic or component) $\alpha \in A = \{\alpha_1, ..., \alpha_L\}$ to each observation corresponding to the occurrence of a word $w \in \mathcal{W}$ within a document $d \in \mathcal{D}$. One component or topic can coincide with one class or, in another setting, a class can be associated to more than one component. Although originally proposed in an unsupervised setting, this latent variable model is easily extended to classification with the following underlying generation process:

- Pick a document d with probability $p(d)$,
- Choose a latent variable α according to its conditional probability $p(\alpha \mid d)$
- Generate a word w with probability $p(w \mid \alpha)$
- Generate the document class y according to the probability $p(y \mid \alpha)$

The final result of this generation process is the document class $y \in \mathcal{C}$ as well as words $w \in \mathcal{W}$ within it, while the latent variable α is discarded. Figure 1 depicts the generation processes for the aspect models and the Naive Bayes model introduced earlier.

The generation of a word w within a document d can then be translated by the following joint probability model:

$$P(w, d) = p(d) \sum_{\alpha \in A} p(w \mid \alpha) P(\alpha \mid d) \tag{3}$$

for unlabeled data and, for labeled data:

$$P(w, d, y) = p(d) \sum_{\alpha \in A} p(w \mid \alpha) P(\alpha \mid d) P(y \mid \alpha) \tag{4}$$

This model overcomes some simplifying assumptions of Naive Bayes in two important ways. First, it relaxes the assumption that a class y is associated to a single topic. In PLSA, the number of topics $|A|$ may be larger than the number of classes K. The second and crucial difference is that in Naive Bayes, all words must be generated from the same topic (eq. 2). This requires the use of clever smoothing strategies to counter the fact that some words that are unrelated to a topic may appear by coincidence in a document from that topic. On the other hand, in PLSA, a topic is drawn independently from $p(\alpha \mid d)$ each time a new word is generated in a document. This provides a much more natural way to handle unusual words or multi-topicality.

Semi-supervised PLSA with Fake Labels. As the aspect PLSA model characterizes the generation of the co-occurrence between a word w and a document d, for learning the semi-supervised models we have to form two other labeled $\mathcal{Z}_l$ and unlabeled $\mathcal{X}_u$ training sets from D_l and D_u. We consider now each observation as a pair $x = (w, d)$ such that observations in $\mathcal{Z}_l$ are assigned to the same class label than the document d they contain.

We recall that we still characterize the data using a mixture model with L latent topic variables α, under the graphical assumption of aspect models (that

d and w are independent conditionally to a latent topic variable α). In this case the model parameters are

$$\Lambda = \{p(\alpha \mid d), p(y \mid \alpha), p(w \mid \alpha), p(d) : \alpha \in A, d \in \mathcal{D}, w \in \mathcal{W}\}$$

Krithara et al. [9], introduced a semi-supervised variant of PLSA, following the work of [5], where additional *fake* labels were introduced for the unlabeled data. The motivation for the latter was to try to solve the problem of the unlabeled components (components which contain only unlabeled examples). The lack of labeled examples in these components can lead to arbitrary class probabilities, and as a result, to arbitrary classification decisions. So all labeled examples in $\mathcal{Z}_l$ are kept with their real class labels and all unlabeled examples in $\mathcal{X}_u$ are assigned a new *fake* label $y = 0$.

The model parameters Λ are obtained by maximizing the complete data log-likelihood,

$$\mathcal{L}_1 = \sum_{x \in \mathcal{Z}_l \cup \mathcal{X}_u} \log p(x, y) = \sum_{x \in \mathcal{Z}_l \cup \mathcal{X}_u} \log p(w, d, y) \tag{5}$$

using the Expectation-Maximization algorithm. [9] showed how the EM iterations could be implemented via a single multiplicative update.

Once the model parameters are obtained, each new document d_{new} must first be "folded in" the model, by maximizing the likelihood on the new document using EM, in order to obtain the posteriors $P(\alpha|d_{new})$. We then need to distribute the probability associated with the *fake* label $y = 0$, on the "true" labels:

$$\forall y \neq 0, \ P(y|d_{new}) \propto \sum_{\alpha} P(\alpha|d_{new})P(y|\alpha) + \mu \sum_{\alpha} P(\alpha|x)P(y{=}0|\alpha) \tag{6}$$

with $\mu << 1$. This model corresponds to the graphical model in figure 1(b). A new document d is then assigned the class with maximum posterior probability. The complexity of this algorithm is $O(2 \times |A| \times M)$ where, as before, $M = \#\{(w, d)|n(w, d) > 0\}$.

Semi-supervised PLSA with a Mislabeling Error Model. In this section we present a new version of a semi-supervised PLSA model in which a misclassification error is incorporated. We assume that the labeling errors made by the generative model for unlabeled data come from a stochastic process and that these errors are inherent to semi-supervised learning algorithms. The idea here is to characterize this stochastic process in order to reduce the labeling errors computed by the classifier for unlabeled documents in the training set.

We assume that for each unlabeled example $d \in D_u$, there exists a perfect, true label y, and an imperfect label $\tilde{y}$, estimated by the classifier. We model the stochastic nature of the labeling by the following probabilities:

$$\forall (k, h) \in \mathcal{C} \times \mathcal{C}, \beta_{kh} = p(\tilde{y} = k|y = h) \tag{7}$$

with the constraint that $\forall h, \sum_k \beta_{kh} = 1$.

In this case, the new extension of the aspect model to unlabeled documents can be expressed by the graphical model represented in figure 1(c).

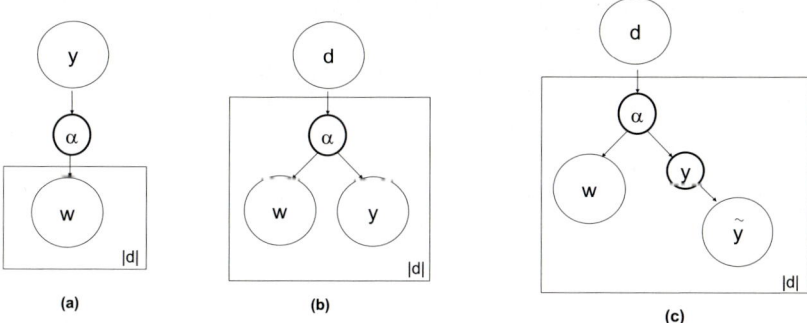

Fig. 1. Graphical model representation of the Naive Bayes model (a), PLSA/aspect models for labeled (b) and unlabeled (c) documents. The "plates" indicate repeated sampling of the enclosed variables.

The underlying generation process associated to this second latent variable model for unlabeled documents is:

- Pick a document d with probability $p(d)$,
- Choose a latent variable α according to its conditional probability $p(\alpha \mid d)$
- Generate a word w with probability $p(w \mid \alpha)$
- Generate the *latent* document class y according to the probability $p(y \mid \alpha)$
- The imperfect class label $\tilde{y}$ is generated with probability $\beta_{\tilde{y}|y} = p(\tilde{y} \mid y)$

With this new graphical model, the joint probability between an unlabeled example $x \in \mathcal{X}_u$ and its imperfect class label estimated by the classifier can be expressed as

$$\forall x \in \mathcal{X}_u, p(w, d, \tilde{y}) = p(d) \sum_{\alpha \in A} p(w|\alpha)p(\alpha|d) \sum_{y \in C} \beta_{\tilde{y}|y} p(y|\alpha)$$

With this formulation it becomes apparent that for each unlabeled document, the imperfect class probabilities estimated by the classifier is weighted over all possible true classes (i.e. $p(\tilde{y} \mid \alpha) = \sum_y p(\tilde{y} \mid y)p(y|\alpha)$). This lessens the possibility that the classifier makes a mistake over the document class as it aggregates the estimates over all true classes.

The model parameters

$$\Phi = \{p(\alpha \mid d), p(w \mid \alpha), p(d), \beta_{\tilde{y}|y} : d \in \mathcal{D}, w \in \mathcal{W}, \alpha \in A, y \in \mathcal{C}, \tilde{y} \in \mathcal{C}\}$$

are estimated by maximizing the log-likelihood

$$\begin{aligned}
\mathcal{L}_2 = &\sum_{d \in D_l} \sum_w n(w, d) \log \sum_\alpha p(d)p(w|\alpha)p(\alpha|d)p(y|\alpha) \\
&+ \sum_{d \in D_u} \sum_w n(w, d) \log \sum_\alpha p(d)p(w|\alpha)p(\alpha|d) \sum_y p(\tilde{y}|y)p(y|\alpha)
\end{aligned} \qquad (8)$$

Algorithm 1. Semi-Supervised PLSA with mislabeling error model

Input :
- A set of partially labeled documents $\mathcal{D} = D_l \cup D_u$,
- Training sets $\mathcal{Z}_l$ and $\mathcal{X}_u$ formed from D_l and D_u,
- Random initial model parameters $\Phi^{(0)}$.
- $j \leftarrow 0$

repeat

  - Re-estimate model parameters using multiplicative update rules (9–11)
  - $j \leftarrow j + 1$

until *convergence of* $\mathcal{L}_2$ *(eq. 8)* ;

Output : A generative classifier with parameters $\Phi^{(j)}$

---

using an EM-type algorithm. Joining the E and M steps in a single multiplicative update, we get:

$$p^{(j+1)}(w|\alpha) = p^{(j)}(w|\alpha) \times \left[\sum_{d \in D_l} n(w,d) \frac{p^{(j)}(\alpha|d)p(y|\alpha)}{p^{(j)}(w,y|d)} \right. \tag{9}$$

$$\left. + \sum_{d \in D_u} n(w,d) \frac{p^{(j)}(\alpha|d) \sum_y p(y|\alpha)\beta_{\tilde{y}|y}^{(j)}}{p^{(j)}(w,\tilde{y}|d)} \right]$$

$$p^{(j+1)}(\alpha|d) = p^{(j)}(\alpha|d) \sum_w n(w,d)p^{(j)}(w|\alpha) \times \begin{cases} \frac{p(y|\alpha)}{p^{(j)}(w,y|d)}, \forall d \in D_l \\ \\ \frac{\sum_y p(y|\alpha)\beta_{\tilde{y}|y}^{(j)}}{p^{(j)}(w,\tilde{y}|d)}, \forall d \in D_u \end{cases} \tag{10}$$

$$\beta_{\tilde{y}|y}^{(j+1)} = \beta_{\tilde{y}|y}^{(j)} \sum_w \sum_{d \in D_u} n(w,d) \frac{p^{(j)}(w,y|d)}{p^{(j)}(w,\tilde{y}|d)} \tag{11}$$

where $p^{(j)}(w,y|d) = \sum_\alpha p^{(j)}(\alpha|d)p^{(j)}(w|\alpha)p(y|\alpha)$, and
$p^{(j)}(w,\tilde{y}|d) = \sum_\alpha p^{(j)}(\alpha|d)p^{(j)}(w|\alpha) \sum_y p(y|\alpha)\beta_{\tilde{y}|y}^{(j)}$.

Note that the mislabeling probabilities are estimated over the unlabeled set.

In this new version of the semi-supervised PLSA algorithm, $P(y|\alpha)$ is fixed, and its values depend on the value of latent topic variable α. The overall number of topics, $|A|$, is given, and in addition, the number of latent topics α per class is also known. During initialization, we set $P(y|\alpha) = 0$ for all latent topic variables α which do not belong to the particular class y. This algorithm (Algorithm 1, above) is also an EM-like algorithm, and the iterative use of equations 9, 10 and 11 corresponds to alternating the E-step and M-step. Convergence is therefore guaranteed to a local maximum of the likelihood.

The complexity of this algorithm is $O(|A| \times M \times K)$, which is comparable with the previous algorithms, as the number of latent variables $|A|$ is generally set to a relatively low value.

3   Experiments

In our experiments we used two collections from the CMU World Wide Knowledge Base project, WebKB and 20Newsgroups[1], and the widely used text classification collection Reuters $-$ 21578. For each dataset, we ran 4 algorithms: the two flavours of semi-supervised PLSA presented above (with mislabeling error model and with fake labels), as well as the semi-supervised Naive Bayes and the transductive Support Vector Machine (TSVM) algorithm [7]. For the latter, we performed a one class vs. all TSVM for all existing classes using the SVM-light package of Joachims [7]. We used the linear kernel and we have optimized, for each of the different ratio of labeled-unlabeled documents in the training set, the cost parameter C by cross-validation. All performance reported below were averaged over 10 randomly chosen labeled, unlabeled and test sets.

The 20Newsgroups dataset is a commonly used document classification collection. It contains 20000 messages collected from 20 different Usenet newsgroups. The WebKB dataset contains web pages gathered from 4 different university computer science departments. The pages are divided into seven categories. In this paper, we focus on the four most often used categories: student, faculty, course and project, all together containing 4196 pages. Finally, the Reuters dataset consists of 21578 articles and 90 topic categories from the Reuters newswire. We selected the documents which belong only to one class, and in addition we only kept the classes which contain at least 100 documents. This gave us a base of 4381 documents belonging to 7 different classes.

All datasets were pre-processed by removing the email tags and other numeric terms, discarding the tokens which appear in less than 5 documents, and by removing a total of 608 stopwords from the CACM stoplist[2]. We used the microaverage F-score measure to compare the effectiveness of the semi-supervised algorithms. To this end, for each generative classifier, $\mathcal{G}_f$, we first compute its microaverage precision P and recall R by summing over all the individual decisions it made on the test set:

$$R(\mathcal{G}_f) = \frac{\sum_{k=1}^{K} \theta(k, \mathcal{G}_f)}{\sum_{k=1}^{K}(\theta(k, \mathcal{G}_f) + \psi(k, \mathcal{G}_f))}$$

$$P(\mathcal{G}_f) = \frac{\sum_{k=1}^{K} \theta(k, \mathcal{G}_f)}{\sum_{k=1}^{K}(\theta(k, \mathcal{G}_f) + \phi(k, \mathcal{G}_f))}$$

Where, $\theta(k, \mathcal{G}_f)$, $\phi(k, \mathcal{G}_f)$ and $\psi(k, \mathcal{G}_f)$ respectively denote the true positive, false positive and false negative documents in class k found by $\mathcal{G}_f$. The F-score measure is then defined as [10]:

$$F(\mathcal{G}_f) = \frac{2P(\mathcal{G}_f)R(\mathcal{G}_f)}{P(\mathcal{G}_f) + R(\mathcal{G}_f)}$$

---

[1] http://www.cs.cmu.edu/~webkb/
[2] http://ir.dcs.gla.ac.uk/resources/test_collections/cacm/

3.1   Results

We first compare the systems in a fully supervised way, that is where 100% of the documents in the training set have their true labels and are used for training the classifiers. As there are no unlabeled training document to consider here there are no fakes or mislabeling errors to characterize, both semi-supervised PLSA models behave identically. This comparison hence gives an upper bound on the performance of each generative approach and also provides a first comparison between these frameworks. We also compared our results with the TSVM model using the SVM-light package [8]. The number of latent class variables we used in the PLSA model, $|A|$, was found by cross-validation on each data set. Table 1 sums up these results. As we can notice, in all 3 datasets, the performance of PLSA is slightly better than the Naive Bayes and SVM classifiers. These results corroborate with the intuition that the generative hypothesis, which leads to the construction of the PLSA model, is more efficient than the Naive Bayes document generation assumption (section 2.3).

Table 1. Comparison of the F-score measures between the Naive Bayes and PLSA generative models as well as the SVM classifier on `20Newsgroups`, `WebKB` and `Reuters` test sets. All classifiers are trained in a fully supervised way.

System	20Newsgroups F-score (%)	WebKB F-score (%)	Reuters F-score (%)
Naive Bayes	88.23	84.32	93.89
PLSA	$|A| = 40$ **89.72**	$|A| = 16$ **85.54**	$|A| = 14$ **94.29**
SVM	88.98	85.15	89.50

Figures 2 and 3 (left) show the F-score measured over the test sets on all three data collections for semi-supervised learning at different ratio of labeled-unlabeled documents in the training set. 5% in the x-axis means that 5% of the training documents were labeled ($|D_l|$), the remaining 95% being used as unlabeled training documents ($|D_u|$). The ssPLSA with mislabeling consistently outperforms the three other models on these datasets. With only 5% of labeled documents in the training set, the F-score of the ssPLSA with mislabeling algorithm is about 15% higher than that of the ssPLSA with fake labels, on the `Reuters` dataset. Labeling only 10% of the documents allows to reach 93% F-score on `Reuters` while the 90% remaining labeled documents allows to reach the maximum performance level. The semi-supervised Naive Bayes model outperforms in the other hand the ssPLSA with fake labels on both datasets. This might be due to the fact that fake label parameterization makes it inappropriate to apply PLSA over both labeled and unlabeled documents.

The bad results of the TSVM in these experiments can be explained by the fact that the model was initially designed for 2-class classification problems and the one vs. all strategy does not give adequate recognition of classes.

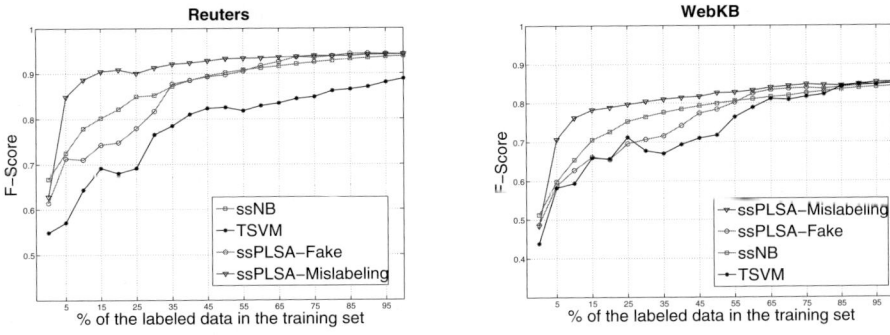

Fig. 2. F-Score (y-axis) vs. percentage of labeled training examples (x-axis), for the four algorithms on `Reuters` (left, $|A| = 14$) and `WebKB` (right, $|A| = 16$)

In order to evaluate empirically the effect of unlabeled documents for training the models we have also trained the PLSA model in a supervised manner using only the percentage of labeled documents in the training set. Figure 3 (right) shows these results on `20Newsgroups`. We can see that semi-supervised algorithms are able to take advantage from unlabeled data. For example, with 5% labeled data (corresponding to approximately 800 labeled documents with 40 documents per class), the fully supervised PLSA reaches 52.5% F-score accuracy while semi-supervised Naive Bayes and ssPLSA with fake labels achieve 63% and ssPLSA with mislabeling achieves 72%. This represents a 32% gain in F-score for the two former models.

Table 2. F-score for varying proportions of labeled-unlabeled training data, for semi-supervised Naive Bayes (ssNB), TSVM as well as semi-supervised PLSA with either the fake label (ssPLSA-f) or the mislabeling error model (ssPLSA-mem), and different numbers of latent topics $|A|$. Bold indicates statistically significantly better results, measured using a t-test at the 5% significance level.

		20Newsgroups				
		1%	5%	20%	40%	80%
	ssNB	51.45 ± 3.45	66.45 ± 0.67	75.65 ± 0.91	83.46 ± 0.46	87.98 ± 0.82
$|A| = 20$	ssPLSA-mem	53.69 ± 5.49	75.520 ± 0.22	81.59 ± 0.6	84.54 ± 0.3	87.76 ± 1.115
	ssPLSA-f	54.67 ± 4.11	75.48 ± 0.75	80.45 ± 1.09	78.86 ± 0.39	84.11 ± 0.93
$|A| = 40$	ssPLSA-mem	$\mathbf{53.52 \pm 6.46}$	$\mathbf{77.18 \pm 0.66}$	$\mathbf{82.89 \pm 0.73}$	$\mathbf{85.9 \pm 0.85}$	$\mathbf{89.04 \pm 0.75}$
	ssPLSA-f	$\mathbf{54.04 \pm 6.98}$	64.65 ± 3.54	67.61 ± 1.69	79.59 ± 0.28	$\mathbf{88.96 \pm 0.64}$
	TSVM	50.64 ± 1.79	54.37 ± 0.55	65.21 ± 0.75	71.31 ± 0.85	82.37 ± 1.03

One interesting aspect of our experimental results is that the behavior of the two ssPLSA variants is very different when the number of latent variables per class increases (Table 2).

For the *fake* label approach, the performance tends to decrease when more components are added to the model, and the variability of the results increases. Overall, this approach yields consistently lower performance than the "Mislabeling"

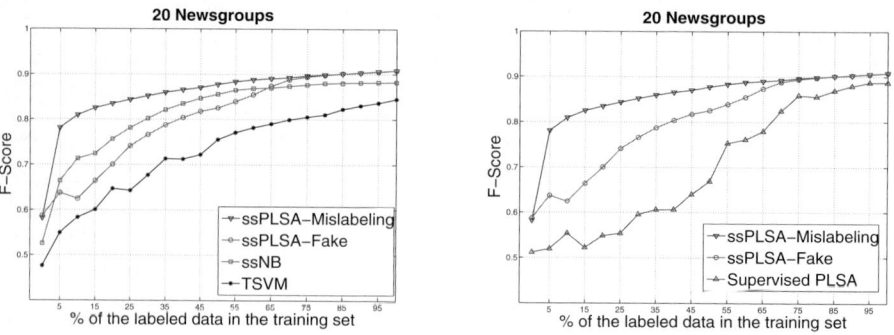

Fig. 3. Comparison of the ssPLSA models with the fully supervised PLSA (right) and with the other algorithms (left) for the 20Newsgroups dataset ($|A| = 40$)

approach, which in addition seems less sensitive to varying numbers of components. Notice in Table 2 how, when the number of components per class is increased from 1 to 2 - corresponding respectively to $|A| = 20$ and $|A| = 40$ (20Newsgroups), the performance of the mislabeling approach increases slightly, but consistently. In addition, the variability of the results is mostly well contained and generally smaller than for the "fake label" approach. The results are similar for the other two datasets.

4   Conclusion

We have presented a new version of the semi-supervised PLSA algorithm, where a mislabeling error model is incorporated in the generative aspect model. Our model has been compared to two state-of-the-art semi-supervised Naive Bayes and TSVM models as well as a previously designed ssPLSA algorithm. Performances on the 20Newsgroups, WebKB and Reuters datasets have shown promising results indicating decreases in the number of labeled documents used for training needed to achieve good accuracy, if an unlabeled document set is available. One of the advantages of our model is that it can be used directly to perform multiclass classification tasks, and as a result it is easily applicable to real world problems. A next step would be to try to combine the two presented variants of PLSA, that is the 'fake' label and the mislabeling error models. This combination would benefit from the advantages of each of the two versions and would hopefully improve the performance of our classifier. However, further experimental observations would be required to fully understand the behavior and the performance of these models.

Acknowledgments

The authors thank Nicolas Usunier for helpful comments. This work was supported in part by the IST Programme of the European Community, under the

PASCAL Network of Excellence, IST-2002-506778. This publication only reflects the authors' views.

References

1. Amini, M.R., Gallinari, P.: The use of unlabeled data to improve supervised learning for text summarization. In: SIGIR, pp. 105–112 (2002)
2. Amini, M.R., Gallinari, P.: Semi-supervised learning with an explicit label-error model for misclassified data. In: Proceedings of the 18th IJCAI, pp. 555–560 (2003)
3. Blum, A., Mitchell, T.M.: Combining labeled and unlabeled data with co-training. In: COLT 1998, pp. 92–100 (1998)
4. Collins, M., Singer, Y.: Unsupervised models for named entity classification. In: EMNLP/VLC (1999)
5. Gaussier, E., Goutte, C.: Learning from partially labelled data - with confidence. In: Learning from Partially Classified Training Data - Proceedings of the ICML 2005 workshop (2005)
6. Hofmann, T.: Probabilistic latent semantic indexing. In: Proceedings of the 22nd ACM SIGIR, pp. 50–57 (1999)
7. Joachims, T.: Transductive inference for text classification using support vector machines. In: Proceedings of ICML 1999, 16th International Conference on Machine Learning, pp. 200–209. Morgan Kaufmann Publishers, San Francisco (1999)
8. Joachims, T.: Text Categorization with Support Vector Machines: Learning with Many Relevant Features. In: Proceedings of the European Conference on Machine Learning (1998)
9. Krithara, A., Goutte, C., Renders, J.M., Amini, M.R.: Reducing the annotation burden in text classification. In: Proceedings of the 1st International Conference on Multidisciplinary Information Sciences and Technologies (InSciT 2006), Merida, Spain (October 2006)
10. Lewis, D., Ringuette, M.: A comparison of two learning algorithms for text categorization. In: Proceedings of SDAIR 1994, 3rd Annual Symposium on Document Analysis and Information Retrieval, Las Vegas, US, pp. 81–93 (1994)
11. McLernon, B., Kushmerick, N.: Transductive pattern learning for information extraction. In: Proc. Workshop Adaptive Text Extraction and Mining (2006), Conf. European Association for Computational Linguistics
12. Miller, D.J., Uyar, H.S.: A mixture of experts classifier with learning based on both labelled and unlabeled data. In: Proc. of NIPS-(1997)
13. Nigam, K., McCallum, K.A., Thrun, S., Mitchell, T.M.: Text classification from labeled and unlabeled documents using EM. Machine Learning 39(2/3), 103–134 (2000)
14. Saul, L., Pereira, F.: Aggregate and mixed-order Markov models for statistical language processing. In: Proc of 2nd ICEMNLP (1997)
15. Si, L., Callan, J.: A semi-supervised learning method to merge search engine results. ACM Transactions on Information Systems 24(4), 457–491 (2003)
16. Slonim, N., Friedman, N., Tishby, N.: Usupervised Document Classification Using Sequentiel Information Maximization. In: SIGIR, pp. 129–136 (2002)
17. Zhang, T.: The value of unlabeled data for classification problems. In: ICML (2000)

Improving Term Frequency Normalization for Multi-topical Documents and Application to Language Modeling Approaches

Seung-Hoon Na[1], In-Su Kang[2], and Jong-Hyeok Lee[1]

[1] POSTECH, Pohang, South Korea
{nsh1979,jhlee}@postech.ac.kr
[2] KISTI, Daejeon, South Korea
dbaisk@kisti.re.kr

Abstract. Term frequency normalization is a serious issue since lengths of documents are various. Generally, documents become long due to two different reasons - verbosity and multi-topicality. First, verbosity means that the same topic is repeatedly mentioned by terms related to the topic, so that term frequency is more increased than the well-summarized one. Second, multi-topicality indicates that a document has a broad discussion of multi-topics, rather than single topic. Although these document characteristics should be differently handled, all previous methods of term frequency normalization have ignored these differences and have used a simplified length-driven approach which decreases the term frequency by only the length of a document, causing an unreasonable penalization. To attack this problem, we propose a novel TF normalization method which is a type of partially-axiomatic approach. We first formulate two formal constraints that the retrieval model should satisfy for documents having verbose and multi-topicality characteristic, respectively. Then, we modify language modeling approaches to better satisfy these two constraints, and derive novel smoothing methods. Experimental results show that the proposed method increases significantly the precision for keyword queries, and substantially improves MAP (Mean Average Precision) for verbose queries.

1 Introduction

The highly-performed retrieval models rely on two different factors - TF (term frequency) and IDF (inverse document frequency). Among them, TF factor becomes a non-trivial, since long-length documents may increase term frequency, different to short-length ones, so that the naive estimation of term frequency would not be successful. Thus, term frequency of long-length documents should be seriously considered. Regarding this, Singhal observed the following two different types of reasons for making the length of a document long [1][1].

1. High term frequency: The same term repeatedly occurs in a long-length document. As a result, the term frequency factors may be large for long documents, increasing the average contribution of its terms towards the query-document similarity.

---

[1] Robertson and Walker mentioned two types of reasons as scope hypothesis and verbosity hypothesis, respectively [2].

C. Macdonald et al. (Eds.): ECIR 2008, LNCS 4956, pp. 382–393, 2008.
© Springer-Verlag Berlin Heidelberg 2008

2. More terms: Long-length document has large size of vocabulary. This increases the number of matches between a query and a long document, increasing the query-document similarity, and the chances of retrieval of long documents in preference over shorter documents.

Without loss of meaning, we can conceptualize these two reasons as verbosity and multi-topicality. First, verbosity means that the same topic is repeatedly mentioned by terms related to the topic, making term frequencies high. Second, multi-topicality indicates that a document has a broad discussion of multi-topics, rather than single topic, making more terms. Using these concepts, we divide long-length documents into two different ideal types - verbose documents and multi-topical documents. Verbose document is the document which becomes long mainly due to verbosity, rather than multi-topicality, while multi-topical document is the document which follows typical characteristics of multi-topicality, rather than verbosity.

Singhal pre-assumed that long-length documents should be penalized regardless of whether or not their types are verbosity (or multi-topicality) [1]. Basically, their approach belongs to a simplified length-driven method which decreases the term frequency of all long-length documents according to documents' length factor only. However, we insist that this Singhal's pre-assumption would be failed. We argue that the penalization should be applied to verbose document only, not to multi-topical document. As a main reason, terms in a multi-topical document are less repeated than ones in a verbose document, since the length of the multi-topical document is increased due to its broad topics. However, Singhal missed this point that these types of documents should be differently handled. Therefore, the retrieval function adopting Singhal's penalization will make multi-topical documents unreasonably less-preferred, causing an unfair retrieval ranking.

To clearly support our argument for verbose document and multi-topical document, we will exemplify two different situations to discuss different tendencies of term frequencies in verbose document and multi-topical document. First, let us examine the situation by considering two different document samples of D_1 and D_2 which have the same term frequency ratio.

D_1: Language modeling approach	D_2: Language modeling approach Language modeling approach

D_2 is twice the concatenation of D_1. Suppose that a query is given by "language modeling approach". Then, a question arises as "which one of D_1 and D_2 is more relevant?". By comparing the contained information, we know that two documents have the exactly same contents, although the length of D_2 is twice than that of D_1. Thus, D_1 and D_2 should have the same relevance score. However, the absolute term frequency of D_2 is twice than that of D_1, thus, the naive TF · IDF prefers D_2 to D_1. To avoid this unfair comparison, we should introduce a TF normalization. To this end, suppose that l is the length of documents, and tf is the term frequency of a query term. Then, one reasonable strategy of TF normalization is to use $tfn = tf/l$, instead of tf. Then, the modified TF · IDF produces the same score for D_1 and D_2. Note that Singhal's pivoted length normalization will also well-work since tfn can be well-reflected in Singhal's original formula. Remark that D_2 is a verbose document, not a multi-topical document, which is the main reason for the success of the normalization. Now, we examine the second situation by considering a multi-topical document sample D_3, which contains all topics of D_1 and D_2 as a subpart.

> D_3: Information retrieval model
> Language modeling approach

Here, D_3 describes a broad topic - "information retrieval model", and contain "language modeling approach" as a subtopic. Again, suppose that the same query of "language modeling approach" is given. Consider the question about "what relevance score should assigned to D_3 be, compared with D_1 and D_2?". D_3 contains all contents of D_1 and D_2, although D_3 is different from D_1 and D_2. In this case, if user sees D_3, he or she would think that D_3 is also relevant, because all relevant content - D_1 - is embedded to D_3. From this viewpoint, D_3 should have the same score as D_1 and D_2 (due to a partial relevance). However, if we apply the previous version of TF-normalization (i.e. $tfn = tf/l$) to D_3, then D_3 is much-less preferred to D_1 and D_2, since its term frequency of a query term is the same as D_1 but its length is twice than that of D_1. Of course, Singhal's method will assign less-score to D_3 than D_1 and D_2. The mean reason of this failure is that D_3 is not a verbose document but a multi-topical document. This result means that TF normalization problem is more complex, at least requiring the different strategies according to types of long-length documents. To avoid the unreasonable penalization for multi-topical ones, TF normalization problem should be more deeply re-investigated by discriminating multi-topical documents from verbose documents.

To obtain a more accurate TF normalization, we propose a novel TF normalization method which is a type of axiomatic approach. We try to modify language modeling approach as a case study without the loss of its elegance and principle. To this end, we first formulate two constraints that the retrieval scoring functions should satisfy for verbose and multi-topical documents, respectively. Then, we present the analysis result that previous language modeling approaches do not sufficiently satisfy these constraints. After that, we modify the language modeling approaches such that better satisfy these two constraints, derive a novel smoothing methods, and evaluate the proposed ones.

2   Formal Constraints of New TF Normalization, and Analysis of Previous Language Modeling Approaches

2.1   Constraints

From now on, we assume that $\tau(D)$ is a measurement for calculating the number of topics in document D. We define *K-verbosity* and *N-topicality* as follows.

Definition (K-verbosity): Suppose that D_1 and D_2 are given. Let $tf_1(w)$ and $tf_2(w)$ be the term frequency of term w in D_1 and D_2, respectively. For all term w, if $tf_2(w) = K \cdot tf_1(w)$ and $\tau(D_1) = \tau(D_2)$, then D_2 has K-verbosity to D_1 or D_2 is *K-verbose* to D_1.

Definition (N-topicality): Suppose that D_1 and D_2 are given as $\tau(D_2) = N \cdot \tau(D_1)$. Let l_1 and l_2 be the length of D_1 and D_2, respectively. If for all term w in D_1, $tf_2(w)/l_2 = tf_1(w)/l_1/N$, then D_2 has N-topicality to D_1 and D_2 is *N-topical* to D_1.

In our three samples from the introduction, D_2 has 2-verbosity to D_1, and D_3 has 2-topicality to D_1. Remind that we have argued that D_1, D_2 and D_3 should have the same relevance score. This argument can be re-formulated to following two constraints - VNC and TNC which the retrieval function should satisfy for two cases when one document has K-verbosity and N-topicality to another document, respectively. Let $score(Q,D)$ be a similarity function between a document D and a query Q.

VNC (Verbosity Normalization Constraint): Suppose a pair of D_1 and D_2. If D_2 is K-verbose to D_1, then $score(Q,D_1) = score(Q,D_2)$.

TNC (Topicality Normalization Constraint): Suppose a pair of D_1 and D_2. If D_2 is N-topicality to D_1, then $score(Q,D_1) = score(Q,D_2)$.

These constraints can be directly utilized to derive a new class of retrieval function as Fang's exploration [3]. Originally, Fang formulated two constraints related to term frequency - LNC1 and LNC2 [3]. Among them, LNC2 is highly relevant to VNC, where VNC is a more specific constraint - VNC entails LNC2, not vice versa. TNC is a new constraint which is not connected to Fang's any constraint. Note that our exploration of a retrieval function is different from Fang's one. We focus on only few constraints related to our issue, without identifying all constraints. Then, we select as the backbone model one among a previous well-performed retrieval model, and modify it to better satisfy the focused few constraints, without losing the elegance and the principle of the original model. In this regard, our exploration method belongs to the *partially-axiomatic* approach - 1) using partial constraints rather than full constraints, 2) using the restricted functional space which the backbone retrieval model can allows, rather than relying on full functional space. In contrast, Fang's approach is the *fully-axiomatic* approach [3,4]. In Fang's approach, full constraints are completely identified as well as the focused constraints. A new class of retrieval function is explored as one in separate functional space which is not related to previous retrieval models. However, the fully-axiomatic approach such as Fang's exploration approach requires un-principled heuristics which are not derived from a well-designed retrieval model. A partially-axiomatic approach doesn't need to discard the well-founded retrieval model such as language modeling approach, enabling us to pursue a more elaborated retrieval model, without losing its mathematical elegance and principles.

2.2   Analysis of Language Modeling Approaches

We selected the language modeling approaches as the backbone retrieval model [5]. Our goal is to modify the language modeling approaches such that better satisfies the proposed two constraints - VNC and TNC. We investigate two popular smoothing methods - Jelinek-Mercer smoothing (JM) and Dirichlet-prior smoothing (Dir) [6]. Before modifying them, we begin by discussing whether or not each smoothing method satisfies VNC and TNC in this subsection. Notations used in this paper are summarized as follows:

Q	A given query
$tf_D(w)$	Term frequency of w in document D
l_D	Length of document D
$tf_C(w)$	Term frequency of w of collection
l_C	Total term frequency of collection
θ_D	Smoothed document language model of D
$\hat{\theta}_D$	Unsmoothed document language model of D (MLE)
θ_C	Collection language model (MLE)

Analysis of Jelinek-Mercer Smoothing. In JM (Jeliner-Mercer Smoothing), a smoothed document model is obtained by the interpolation of MLE (Maximum Likelihood Estimation) of a document model and the collection model as follows [6]:

$$P(w|\theta_D) = (1-\lambda)P(w|\hat{\theta}_D) + \lambda P(w|\theta_C) \tag{1}$$

where λ is a smoothing parameter. By using JM, $score(Q,D)$, the similarity score of document D for query Q can be written by using only query-matching terms as follows:

$$score(Q,D) = \sum_{w \in Q} \log\left(\frac{1-\lambda}{\lambda}\frac{P(w|\hat{\theta}_D)}{P(w|\theta_C)} + 1\right) = \sum_{w \in Q} log\left(\frac{1-\lambda}{\lambda}\frac{tf_D(w)}{l_D}\frac{l_C}{tf_C(w)} + 1\right) \tag{2}$$

Our analysis of whether or not JM satisfies VNC and TNC is given as follows:

1. JM satisfies VNC: Suppose that D_2 is K-verbose to D_1. Then, MLEs of two document models are the same, resulting in the same scores.
2. JM does not satisfy TNC: Generally, JM prefers normal documents to multi-topical documents, regardless of our definition of topicality measurement τ. This proof is skipped.

Analysis of Dirichlet-Prior Smoothing. In Dir (Dirichlet-prior smoothing), a smoothed document model is estimated as posterior model when taking $\mu P(w|\theta_C)$ as a prior probability of term w as follows [6]:

$$P(w|\theta_D) = \frac{tf_D(w) + \mu P(w|\theta_C)}{l_D + \mu} \tag{3}$$

The equation is rewritten by

$$P(w|\theta_D) = \frac{l_D}{l_D + \mu}P(w|\hat{\theta}_D) + \frac{\mu}{l_D + \mu}P(w|\theta_C) \tag{4}$$

If we set λ_D by $\mu/(l_D+\mu)$, then Dir is equivalent to JM-style smoothing using document-specific smoothing parameter λ_D. $score(D,Q)$ based on Dir is formulated as follows:

$$score(D,Q) = \sum_{w \in Q} \log\left((1-\lambda_D)\frac{P(w|\hat{\theta}_D)}{P(w|\theta_C)} + \lambda_D\right)$$

The analysis on whether or not Dir satisfies VNC and TNC is somewhat complicated, due to its document-specific smoothing parameter. We can easily show that Dir does not satisfy VNC and TNC. The following lists up the analysis result.

1. Dir doesn't satisfy VNC: Generally, Dir makes inconsistent preferences according to whether or not a query term is topical. For a topical query term, Dir assigns the more score for verbose documents than normal documents. For a non-topical query terms, Dir assigns the less score for verbose documents than normal documents. The detailed proof is skipped.
2. Dir doesn't satisfy TNC: The detailed proof is skipped.

3   Modification of Previous Retrieval Models

In the previous section, we have shown that two different smoothing methods do not satisfy two constraints well. In this section, we introduce the measurement of the number of topics, and modify the previous retrieval model such that it better satisfies VNC and TNC.

3.1   Measurement of the Number of Topics

To figure out which measurement $\tau(D)$ is acceptable to calculate the number of topics in document D, we propose two simple measurements for $\tau(D)$ - The first one is *vocabulary size*, and the second one is *information quantity*.

Vocabulary Size: Generally, as there are more terms, a given document has more topics. Based on this idea, we can use the vocabulary size - $v(D)$ - which indicates the number of unique terms in a given document, as a measurement for the number of topics.

Information Quantity: Even though the vocabulary size is simple and reasonable, it cannot discriminate the mainly topical terms from the causally-occurred terms. When using the vocabulary size, the number of topics may be unreasonably increased due to causally occurred terms. As for an alternative measurement, we consider the entropy-driven value. Remind that entropy means the uncertainty of a generated sample. Entropy has the following positive properties for resolving the limitation of the vocabulary size. 1) As the number of possible events increases, entropy becomes larger. Here, events correspond to terms, hence the more terms are, the larger the entropy is likely to be. Thus, when a document has more topics, the content of the document can be described in more various ways, resulting in a larger entropy value. 2) Term generative probability of a document is used as the weight for calculating entropy value. As a term has more large probability, it makes more contribution to the final-entropy value. This property allows us to differentiate the effects of mainly topical terms and causally occurred terms.

The information quantity - $\varepsilon(D)$ - is defined as an exponential function of entropy of a document as follows:

$$\tau(D) = \varepsilon(D) = \exp\left(-\sum_w P(w|\theta_D)\log P(w|\theta_D)\right)$$

Some Useful Definitions: We define some useful notations. Let us define the normalized measurement of the number of topics - $\tau'(D)$ -, and define the *informative verbosity* - $\omega(D)$ - as follows:

$$\tau'(D) = \tau(D)/\tilde{\tau}, \qquad \omega(D) = l_D/\tau(D)$$

where $\tilde{\tau}$ is the mean of $\tau(D)$ for all documents in a given test collection. Note that the informative verbosity indicates the average term frequency per unit information.

3.2   Modification of JM

First Modification of JM. Since JM exactly satisfies VNC, we would try to modify JM to additionally support TNC. The core idea of the modification of JM smoothing is a pseudo document. The pseudo document mainly consists of relevant parts to a query, which is constructed by extracting relevant parts from non-relevant parts. Then, the score of a document is calculated by using the pseudo document model, instead of original document model.

Thus, the pseudo-document makes us take a dynamic viewpoint of document representation where a document is dynamically changed according to a query. Note that a pseudo document is an imaginary concept, which is not really constructed at real time. All we require is generative probabilities for query terms from the pseudo document model.

To estimate probability of query terms in a pseudo document, we simplify the estimation problem by using probability in original document. In other words, for terms in the pseudo document having non-zero probabilities, their probabilities are assumed to be proportional to the probabilities of terms in the original document. As a result, the estimation problem is completed only if we determine the length of the pseudo document from the original length l_D.

Intuitively, the length of the pseudo document will be smaller, as topics are more. This intuition makes the length of the pseudo document proportional to $l_D/\tau(D)$. Thus, if $\theta_{Pseudo(D)}$ is the language model of pseudo document, then the probability of pseudo document model is

$$P(w|\theta_{Pseudo(D)}) \propto tf_D(w)/l_D/\tau(D) = tf_D(w) \cdot \tau(D)/l_D$$

It is rewritten by using $\tau'(D)$ instead of $\tau(D)$, and the constant K as follows:

$$P(w|\theta_{Pseudo(D)}) = K \cdot tf_D(w) \cdot \tau'(D)/l_D$$

If we assume that the constant K is independent to any document and query, then K is not a tuning parameter since it can be included in smoothing parameter λ.

Let us derive a modified JM by substituting the original document model to this pseudo document model in Eq. (2). Then, $score(Q,D)$ is reformulated as follows:

$$score(Q,D) = \sum_{w \in Q} \log \left(\frac{1-\lambda_0}{\lambda_0} \frac{K \cdot \tau'(D) \cdot tf_D(w)}{l_D} \frac{l_C}{tf_C(w)} + 1 \right) \qquad (5)$$

where λ_0 is another smoothing parameter for the pseudo document model. Since K is independent to any document and query, we can select λ such that $(1 - \lambda_0)K : \lambda_0$ is $(1 - \lambda) : \lambda$, in order to eliminate constant K. Then, Eq. (5) is re-written by

$$score(Q,D) = \sum_{w \in Q} \log \left(\frac{1-\lambda}{\lambda} \frac{\tau'(D) \cdot tf_D(w)}{l_D} \frac{l_C}{tf_C(w)} + 1 \right) \qquad (6)$$

By using MLE of the original document model $P(w|\hat{\theta}_D)$, Eq. (6) is rewritten by

$$score(Q,D) = \sum_{w \in Q} \log \left(\frac{1-\lambda}{\lambda} \tau'(D) P(w|\hat{\theta}_D) \frac{l_C}{tf_C(w)} + 1 \right) \qquad (7)$$

Eq. (7) is the final modified JM, which is called JMV. JMV satisfies both of VNC and TNC.

1. JMV satisfies VNC: Let D_2 be K-verbose to D_1. Then, $\tau(D_1) = \tau(D_2)$ and $P(w|D_1) = P(w|D_2)$. Thus, $score(Q,D_1) = score(Q,D_2)$.
2. JMV satisfies TNC: Let D_3 be N-topical to D_1. Then, $\tau(D_3) = N\tau(D_2)$ and $P(w|D_1) = NP(w|D_3)$. It makes that $\tau(D_3)P(w|D_3) = \tau(D_1)P(w|D_1)$. Therefore, $score(Q,D_1) = score(Q,D_3)$.

Second Modification of JM. In our preliminary experiments, we found that JMV performs well for keyword queries (i.e. title query), but is not reliable for verbose queries (i.e. description query), by showing serious sensitivity according to smoothing parameter λ. To discuss the reason of this result, we focus on the main differences of keyword query and verbose query. First, there are common terms in a verbose query. Different from topical terms, common terms can be shared by all topics. A common term always verbosely acts regardless of verbose documents and multi-topical documents. Thus, the previous TF normalization would prefer multi-topical documents for queries including common terms. Second, verbose queries often contain noise terms such as "relevant", "find" and "documents". When a document has more topics, it will increase the chance of existence of such noise terms. However, when our previous TF normalization is applied, noise term becomes very serious, because the number of topics is further multiplied to the normalized term frequency. Thus, the previous TF normalization would increase the scores of multi-topical documents for noise queries. These two differences may be the reason why Singhal et. al. penalized even multi-topical documents, as well as verbose documents [1]. However, we already discussed that their approach is not acceptable to topical terms.

To handle the problems of verbose-type queries, our TF normalization should be restricted to only document-specific terms, not to noise terms or common terms. As a query term is more topical term in a given document, we hope to perform more TF normalization, and vice versa. To this end, we define $s(w,D)$ as term specificity of w in document D. As for $s(w,D)$ this paper uses a probabilistic metric $P(D|w)$ which is defined as follows:

$$s(w,D) = P(D|w) = \frac{\lambda_s P(w|\hat{\theta}_D)}{\lambda_s P(w|\hat{\theta}_D) + (1-\lambda_s)P(w|\theta_C)}$$

where λ_s is an additional smoothing parameter, which has 0.25 as the default value. By using the term specificity $s(w,D)$, we newly modify the pseudo document model as follows:

$$P(w|\theta_{Pseudo(D)}) = K \cdot tf_D(w) \cdot \tau'(D)^{P(D|w)}/l_D \tag{8}$$

Since $P(D|w)$ is between 0 and 1, the normalization is perfectly reflected when $P(D|w)$ is 1, while it is weaken as $P(D|w)$ is close to 0. One problem arises when $\tau'(D)$ is smaller than 1. In this case, as $P(D|w)$ is larger, the effect of normalization becomes weaker. To resolve this problem, we considered the exceptional TF normalization, making the normalization proportional to $P(D|w)$ even when $\tau'(D)$ is smaller than 1. In preliminary experiments, we found that the final retrieval performance is almost not changed, even after the exceptional TF normalization is applied. Thus, we select Eq. (8) for second modification. We call it JMV2.

4   Modification of Dir

Our goal for Dir modification is to provide VNC. We introduce the concept of pseudo document model to modify Dir. Different from the pseudo document for JM modification that consists of query-relevant parts only, the pseudo document for Dir modification consists of all topics in the original document, but has a different length from the original length. Note that the change of the length only makes different models, since the smoothed model - $P(w|\theta_D)$ - is different according to the document length. In fact, the length-dependence was the main reason why Dir does not satisfy VNC.

We assume that the pseudo document model is proportional to original MLE document model. In addition, we set the length of the pseudo document by $\tau(D)$. Remind that informative verbosity - $\omega(D)$ - is defined as $l_D/\tau(D)$. That is, the pseudo document with length of $\tau(D)$ compacts the original document with length l_D by $\omega(D)$ time. Therefore, each term w of document D has the following term frequency in the pseudo document.

$$tf_{Pseudo(D)}(w) = tf_D(w)/\omega(D) \tag{9}$$

As a result, the pseudo document model becomes length-independent model, even though MLE of pseudo document model is the same as the original document model. By using pseudo document model, Dir produces the following smoothed model.

$$P(w|\theta_{Pseudo(D)}) = \frac{tf_{Pseudo(D)}(w) + \mu P(w|\theta_C)}{\tau(D) + \mu} \tag{10}$$

By substituting Eq. (9) to Eq. (10), Eq. (10) becomes

$$P(w|\theta_{Pseudo(D)}) = \frac{\tau(D)}{\tau(D) + \mu} P(w|\hat{\theta}_D) + \frac{\mu}{\tau(D) + \mu} P(w|\theta_C) \tag{11}$$

This final modified model can be viewed as JM-style smoothing using document-specific smoothing paramter λ_D with $\mu/(\tau(D) + \mu)$, which is not dependent to the length any more. We call this modification DirV. We can easily prove that DirV additionally satisfies VNC.

1. DirV satisfies VNC: Let D_2 be K-verbose to D_1. Then, two MLE models are equal (i.e $P(w|\theta_{D_1}) = P(w|\theta_{D_2})$). λ_{D_1} is λ_{D_2} since $\tau(D_1)$ and $\tau(D_2)$ are the same. Thus, DirV gives the same score for D_1 and D_2.
2. DirV does not satisfy TNC: For DirV, we do not have a special consideration for supporting TNC.

5   Experimentation

5.1   Experimental Setting

For evaluation, we used five TREC test collections. The standard method was applied to extract index terms; We first separated words based on space character, eliminated stop-words, and then applied Porter's stemming. Table 1 summarizes the basic information of each test collection. In columns, #Q, *Topics*, #R, #Doc, *avglen*, and #*Terms* are the number of topics, corresponding query topic IDs, the number of relevant documents, the number of documents, the average length of documents, and the number of terms, respectively.

Table 1. Collection summaries

Collection	# Q	Topics	# R	# Doc	avglen	# Term
TREC7	50	350-400	4,674	528,155	154.6	970,977
TREC8	50	401-450	4,728			
WT2G	50	401-450	2,279	247,491	254.99	2,585,383
TREC9	50	451-500	2,617	1,692,096	165.16	13,018,003
TREC10	50	501-550	3,363			

According to Zhai's work [6], we used the following three different types of queries:

1) **Short keyword (SK):** Using only the title of the topic description.
2) **Short Verbose (SV):** Using only the description field (usually one sentence).
3) **Long Verbose (LV):** Using the title, description and the narrative field (more than 50 words on average).

As for retrieval evaluation, we used MAP (Mean Average Precision), Pr@5 (Precision at 5 documents), and Pr@10 (Precision at 10 documents).

5.2   Experimental Results

Table 2 shows the best performances (MAP, Pr@5, Pr@10) of DirV and JMV2, compared with Dir. As for topic measurement $\tau(D)$, we selected the information quantity $(\varepsilon(D))$ since JMV2 and DirV using the information quantity is better than those using vocabulary size. We used MLE (Maximum Likelihood Estimation) for $P(w|\theta_D)$ to calculate the information quantity without any smoothing. We selected Dir as the baseline

Table 2. Performances of Dir, DirV and JMV2 (MAP, Pr@5, Pr@10). Bold faced numbers indicate runs showing significant improvement over Dir.

MAP	Dir			DirV			JMV2		
	SK	SV	LV	SK	SV	LV	SK	SV	LV
TREC7	0.1786	0.1790	0.2209	0.1835	**0.1967‡**	**0.2348‡**	0.1825	**0.1926†**	0.2250
TREC8	0.2481	0.2294	0.2598	0.2492	**0.2393‡**	**0.2621‡**	**0.2505†**	**0.2354†**	0.2500
WT2G	0.3101	0.2854	0.2863	0.3125	**0.3103‡**	**0.3267‡**	**0.3278‡**	**0.3112‡**	**0.3263‡**
TREC9	0.2038	0.1990	0.2468	0.2040	**0.2336‡**	**0.2581‡**	**0.2068**	**0.2245‡**	0.2494
TREC10	0.1950	0.1865	0.2347	**0.2049†**	0.2248	0.2640	0.2091	**0.2133†**	0.2555

Pr@5	Dir			DirV			JMV2		
	SK	SV	LV	SK	SV	LV	SK	SV	LV
TREC7	0.4400	0.4280	0.5240	0.4560	**0.4840†**	**0.5680†**	0.4680	**0.4920†**	**0.5800†**
TREC8	0.4920	0.4320	0.5120	0.5120	**0.5040†**	0.5360	**0.5240‡**	0.4880	0.5280
WT2G	0.5160	0.5120	0.5280	0.5360	0.5520	**0.5720†**	0.5400	0.5560	**0.5920†**
TREC9	0.3000	0.3480	0.4160	0.3320	**0.4240†**	0.4320	0.3440	0.3720	0.3880
TREC10	0.3520	0.4040	0.4720	0.3840	0.4520	0.4920	0.3800	0.4200	0.4880

Pr@10	Dir			DirV			JMV2		
	SK	SV	LV	SK	SV	LV	SK	SV	LV
TREC7	0.3980	0.4120	0.4420	**0.4180†**	0.4420	**0.4720†**	0.4100	0.4440	**0.4800†**
TREC8	0.4460	0.4120	0.4660	**0.4740†**	0.4380	0.4780	**0.4700†**	0.4400	0.4480
WT2G	0.4660	0.4220	0.4240	0.4840	**0.4840†**	**0.4800‡**	0.4920	**0.4900‡**	**0.4820‡**
TREC9	0.2560	0.2860	0.3160	0.2780	**0.3260‡**	**0.3540‡**	0.2780	**0.3160†**	0.3220
TREC10	0.3060	0.3500	0.4040	0.3300	0.3820	0.4340	0.3300	0.3700	0.4340

due to its superiority over JM in all test collections. To obtain the best performance of each run, we searched 20 different values between 0.01 and 0.99 for λ, and 22 values between 100 and 30,000 for μ. To check whether or not the proposed method (DirV and JMV2) significantly improves the baseline, we performed the Wilcoxon sign ranked test to examine at 95% and 99% confidence levels. We attached † and ‡ to the performance number of each cell in the table when the test passes at 95% and 99% confidence level, respectively. The results are summarized as follows:

1. DirV significantly improves MAP of Dir for verbose type of query (SV and LV). Exceptionally, TREC10 did not show an improvement for verbose type of query.
2. DirV does not significantly improve MAP of Dir for keyword type of query (SK), but improves precisions (Pr@5 or Pr@10). Especially, on TREC7 and TREC8, Pr@10 is significantly improved over Dir. Although other test collections do not statistically show a significant improvement, there is large portion of the numerical increase.
3. DirV or JMV2 show improvement on a specific test collection even for keyword type of query. For DirV, TREC10 is such a collection by showing a significant improvement of MAP. For JMV2, WT2G is such a test collection.
4. Overall, DirV is slightly better than JMV2 in most of test collections. WT2G is an exceptional collection to show that JMV2 significantly improves DirV.

6   Conclusion

This paper introduced a new issue for TF normalization by considering two different types of long-length documents - verbose documents and multi-topical documents. We proposed a novel TF normalization method which uses a partially-axiomatic approach. To this end, we formulated two desirable constraints, which the retrieval function should satisfy, and showed that previous language modeling approaches do not satisfy these constraints well. Then, we derived novel smoothing methods for language modeling approaches, without losing basic principles, and showed that the proposed methods satisfies these constraints more effectively. Experimental results on five standard TREC collections show that the proposed methods are better than previous smoothing methods, especially for verbose type of query. JMV2 significantly improved JM for all type of queries, and DirV eliminated the limitation of Dir by providing the robustness of performances for verbose type of query, as well as improving precisions (Pr@5 or Pr@10) for keyword type of query. This is comparable to recent results using more complicated query-specific smoothing based on Poisson language model [7].

To handle long-length documents, passage-based retrieval could be applied [8]. However, passage-based retrieval has a burden of decreasing efficiency, since it requires additional process such as indexing of position information, pre-segmenting individual passages, and more importantly the additional overhead at online retrieval time. Contrast to the complicated method such as the passage retrieval, this paper handles multi-topical documents in a simplified manner by investigating a more accurate TF normalization without additional cost of efficiency.

Acknowledgement. This work was supported by the Korea Science and Engineering Foundation (KOSEF) through the Advanced Information Technology Research Center (AITrc), also in part by the BK 21 Project and MIC & IITA through IT Leading R&D Support Project in 2007.

References

1. Singhal, A., Buckley, C., Mitra, M.: Pivoted document length normalization. In: SIGIR 1996, pp. 21–29 (1996)
2. Robertson, S.E., Walker, S.: Some simple effective approximations to the 2-poisson model for probabilistic weighted retrieval. In: SIGIR 1994, pp. 232–241 (1994)
3. Fang, H., Tao, T., Zhai, C.: A formal study of information retrieval heuristics. In: SIGIR 2004, pp. 49–56 (2004)
4. Fang, H., Zhai, C.: An exploration of axiomatic approaches to information retrieval. In: SIGIR 2005 (2005)
5. Ponte, J.M., Croft, W.B.: A language modeling approach to information retrieval. In: SIGIR 1998, pp. 275–281 (1998)
6. Zhai, C., Lafferty, J.: A study of smoothing methods for language models applied to ad hoc information retrieval. In: SIGIR 2001, pp. 334–342 (2001)
7. Mei, Q., Fang, H., Zhai, C.: A study of poisson query generation model for information retrieval. In: SIGIR 2007, pp. 319–326 (2007)
8. Kaszkiel, M., Zobel, J.: Effective ranking with arbitrary passages. Journal of the American Society for Information Science and Technology (JASIST) 52(4), 344–364 (2001)

Probabilistic Document Length Priors for Language Models

Roi Blanco and Alvaro Barreiro

IRLab. Computer Science Department
University of A Coruña, Spain
{rblanco,barreiro}@udc.es

Abstract. This paper addresses the issue of devising a new document prior for the language modeling (LM) approach for Information Retrieval. The prior is based on term statistics, derived in a probabilistic fashion and portrays a novel way of considering document length. Furthermore, we developed a new way of combining document length priors with the query likelihood estimation based on the risk of accepting the latter as a score. This prior has been combined with a document retrieval language model that uses Jelinek-Mercer (JM), a smoothing technique which does not take into account document length. The combination of the prior boosts the retrieval performance, so that it outperforms a LM with a document length dependent smoothing component (Dirichlet prior) and other state of the art high-performing scoring function (BM25). Improvements are significant, robust across different collections and query sizes.

1   Introduction

Information retrieval (IR) systems aim to retrieve relevant documents in response to a user need, which is usually expressed as a query. The retrieved documents are returned to the user in decreasing order of relevance. Most retrieval models use term statistics, such as term frequency, to assign weights to individual terms, which represent the contribution of the term to the document content. These term weights are then used to estimate the score of relevance of a document for a query [14].

In addition to term statistics, IR models are often extended with further evidence that can improve retrieval performance, e.g. using the term frequency in specific fields of structured documents (e.g. title, abstract) [11], or integrating query-independent evidence in the retrieval model in the form of prior probabilities for a document [3,6] ('prior' because they are known before the query). In short, when determining the relevance between a query and a document, most IR models use primarily query-dependent term statistics, and sometimes also add query-independent evidence to further enhance retrieval performance. In this paper, we propose a new form of prior for documents, which we combine with IR models from the language modeling (LM) approach [8] .

C. Macdonald et al. (Eds.): ECIR 2008, LNCS 4956, pp. 394–405, 2008.

Language models for IR view documents as models and queries as segments of text generated or sampled from those models. Documents are ranked according to the probability of each query text string being generated from the respective document model. Although traditionally language models abandoned the explicit notion of document and query *relevance*, the work in [7] connects the notion of relevance and generative language models

The LM framework models the relevance of documents to queries by estimating two probabilities (namely, query likelihood and document prior). Considering a multinomial generation of events [17], documents are ranked against queries according to those estimations. The query likelihood component is a query dependent feature representing the probability of the query being generated by the language model of a document and the document prior is a query-independent feature representing the probability of seeing the document. Typically, this probability is assumed to be the same for any document, hence the document prior is taken to be uniform [17]. Alternatively, the document prior is useful for representing and incorporating other sources of information to the retrieval process; this is currently an active area of research. For instance, document priors can be derived from the link structure of Web pages. In fact, this is a popular source for priors: [16] introduced the number of incoming links (inlinks) count, which was subsequently used in various Web retrieval tasks of the Text REtrieval Conference (TREC [15]) repeatedly and with success. Another type of evidence from which document priors are derived is URL depth, also introduced in [16]. These two priors were further explored by the work in [6]. Other URL-derived information and also the Pagerank [1] algorithm for ranking Web documents according to their popularity, have been used to derive document priors [13].

Overall, incorporating prior knowledge on documents into retrieval has been particularly effective on Web retrieval, namely *homepage* and *named page finding*. Homepage and named page finding refer to the retrieval of a single Web page; on the contrary, *ad-hoc* retrieval refers to the more general application of retrieving as much relevant information to the query as possible.

In this paper, we revisit the idea of deriving a high-quality document prior based on document length and term statistics. Most retrieval models include a document length normalisation component, so that longer documents do not have an unfair advantage over shorter documents of being retrieved. This normalisation is fairly critical and some successful models of retrieval are based in part on document length models, like BM25 [10]. We show that it is possible to encode document length information as a prior probability and improve significantly retrieval effectiveness of a simple language model that uses Jelinek-Mercer (JM) smoothing. In particular, we experiment with two length-based priors: one prior estimated proportionally to document length (we call this prior *linear* [6],[16]), and a document length based prior which is not computed directly from the number of tokens in the document but estimated in a probabilistic fashion from term statistics, which are typically used by retrieval models. To our knowledge, deriving a document prior from these term statistics is a novel approach.

Generally, priors are combined with the retrieval model either using heuristics or handtuned parameters [3]. In this work, we combine our proposed priors with the LM in two different ways: using a standard logarithmic combination, and proposing a novel combination that considers the prior as a measure of the risk of accepting the score given by the query likelihood estimation of the LM. A thorough experimentation on four TREC collections of different size and domain, and 450 short and long queries show that our proposed prior benefits retrieval performance significantly, and in a robust way. Specifically, we find that it is possible to boost the performance of a retrieval model based on JM smoothing up to values comparable to state of the art retrieval models, and further outperform retrieval models traditionally considered to be more effective in previous literature [16].

This paper is organised as follows: section 2 describes in detail the formulation used for the document priors in the rest of the paper and related work; section 3 presents a simple well-known linear document prior and a novel way of approximating document length in a probabilistic fashion; section 4 explores new ways for combining the document prior with the query likelihood, and section 5 describes our experimental findings.

2   Document Priors in the Language Modeling Approach

The language modeling framework allows a mathematically elegant way of incorporating query-independent features, i.e. just related to a document *without seeing a query*. Next, it follows a derivation of the LM retrieval model where the probability of relevance $p(r|Q,D)$, given a query and a document is estimated indirectly by invoking Bayes' rule. For the formal connection between language models and the probabilistic model of retrieval refer to [7].

Let the random variables D and Q denote a document and a query, respectively. Let the binary random variable R stand for relevance r, $p(r) = p(R = 1)$ and non-relevance $\overline{r}$, $p(\overline{r}) = p(R = 0)$.

$$p(r|Q,D) = \frac{p(D,Q|r)\,p(r)}{p(D,Q)} \tag{1}$$

$$= p(Q|D,r)\,p(D|r)\,\frac{p(r)}{p(D,Q)} \tag{2}$$

$$= p(Q|D,r)\,p(r|D)\frac{p(D)}{p(D,Q)} \tag{3}$$

Assuming independence between queries and documents $p(D,Q) = p(D)p(Q)$, and given that $p(Q)$ does not affect the ranking (it is document-independent), equation 3 becomes

$$p(r|Q,D) = \frac{p(Q|D,r)\,p(r|D)}{P(Q)} \stackrel{rank}{=} p(Q|D,r)\,p(r|D)\,, \tag{4}$$

where $p(Q|D,r)$ is the query likelihood and $p(r|D)$ is the document prior. In equation 4, we took a strong independence assumption to get a final formulation with dependence on $p(r|D)$. The derivation presented in [7] took a more reasonable assumption, Q and D are independent under $\bar{r}$, and starting from the odds-ratio of relevance the final relevance score is dependent on $p(r|D)/(1 - p(r|D))$.

It is usual to decompose the query into its query terms $Q = \{q_1, q_2, \ldots, q_n\}$ and assume that, given relevance and the document, they are independent of each other and generated by a multinomial distribution.

$$p(Q|D, r) = \prod_{i=1}^{n} p(q_i|D, r) \tag{5}$$

In order to rule out zero probabilities for non-seen terms in a document, this estimate has to be *smoothed*, which eventually leads to different language models-based scoring functions. Most smoothing methods employ two distributions, one for words occurring in the document (p_s) and one for *unseen* words (p_u). Taking logs (refer to [17] for a complete derivation) it can be shown that equation 6 suffices to provide a document rank using sums of logarithms, equivalent to the one that equation 5 would yield.

$$\log p(Q|D, r) \stackrel{rank}{=} \sum_{i \backslash tf(q_i, D) > 0} \log \frac{p_s(q_i|D)}{\alpha_d p(q_i|\mathcal{C})} + n \cdot \log \alpha_d, \tag{6}$$

where $tf(q_i, D)$ stands for the frequency of term q_i in document D, α_d is a parameter and $p(q_i|\mathcal{C})$ is the collection language model.

The smoothing technique we considered as our baseline in this study is Jelinek-Mercer (JM) (also known as linear interpolation):

$$p_s(q_i|D) = (1 - \lambda)p_{mle}(q_i|D) + \lambda\, p(q_i|\mathcal{C}), \ \lambda \in [0, 1]\,, \alpha_d = \lambda \tag{7}$$

where $|D| = \sum_{w_i \in D} tf(w_i, D)$ (the document length), p_{mle} is the maximum likelihood estimator for a term q_i given a document d, $p_{mle}(q_i|D) = \frac{tf(q_i, d)}{|D|}$ and λ is a parameter controlling the amount of mass distribution assigned to the *document* and *collection*.

Another popular and effective smoothing technique is Dirichlet prior smoothing:

$$p_s(q_i|D) = \frac{tf(q_i, D) + \mu p(q_i|\mathcal{C})}{|D| + \mu}, \ \alpha_d = \frac{\mu}{|D| + \mu}, \tag{8}$$

where μ is a parameter.

In most cases $p(r|D)$ is taken to be uniform [17]. However, there have been several studies where the document length and link structure have been encoded as a prior probability, for ad-hoc and some non ad hoc tasks [6], [16].

Most weighting models include document length as a part of their core query-dependent retrieval model and that might be one of the reasons for traditionally not being considered a document static feature. For most retrieval models, the amount of normalisation contributed by document length is controlled by a parameter. This is not the case for JM smoothing, but it can be seen that the μ parameter in Dirichlet prior smoothing is playing the length normalisation role. The weight for a matched query term q_i in JM smoothing is $\log(1+(1-\lambda)p_{mle}(q_i, D)/\lambda p(q_i|\mathcal{C})))$ and for Dirichlet prior smoothing is $\log(1 + |D|p_{mle}(q_i, D)/(\mu p(q_i|\mathcal{C})))$. Clearly, $|D|/\mu$ and $(1-\lambda)/\lambda$ play the same role, with the difference that the former is document-dependent while the latter is document-independent [17]. It is assumed from past studies [17],[16], that Dirichlet prior smoothing outperforms JM smoothing, especially for short queries. In our opinion, this is due to the fact that Dirichlet prior smoothing includes document length normalisation as a part of the query likelihood estimation.

Although JM smoothing does not comprise document dependent length normalisation notions, it has the advantage of "explaining" the common words of the query. This is the reason JM behaves better with long queries: these kind of queries are usually more verbose. Experiments using short-verbose queries in [17] confirmed the query-modeling role of JM smoothing. Otherwise, it is assumed that Dirichlet prior smoothing has an effect of improving the accuracy of the estimated document language model. Incorporating a good document length prior into LM-JM would hopefully result in a model that will embody both roles mentioned before.

3   Length-Based Document Priors

3.1   Linear Prior

Previous studies [12,6], tried to establish a connection between the likelihood of relevance/retrieval and document length. In particular, [12] compared the results of a set of queries and tried to obtain a relevance versus retrieval pattern (of a particular scoring function) to see how they deviate from each other. The relevance pattern happened to follow a linear dependence on document length. The results presented in [6] on a another testbed, further confirmed that hypothesis. Then, our first document length based prior is proportional to document length. The intuition behind this prior is that longer documents span more topics and are more likely to be relevant *if no query has been seen* (denoted as *scope* hypothesis in [9]). It has been reported that this prior increases the retrieval performance [6] on the WT10G collection up to 0.03 on an absolute scale.

The linear document prior is given by:

$$p(r|D) \approx \frac{|D|}{\sum_{d_i \in \mathcal{C}} |D_i|} = C \cdot |D| ,\qquad (9)$$

where C is a constant that can be dropped out from the scoring function since it does not affect the ranking of documents.

3.2   Probabilistic Prior

We propose other prior indirectly based on document length by extending the idea of estimating the document prior as a function depending on the statistics of the terms it contains.

To estimate the conditional probability $p(r|D)$ we compute the expectation over the universe of terms $\{w_i\}$. Also, in 10 we make the additional assumption that r is independent of D once we picked a term w_i.

$$p(r|D) \approx \sum_{w_i} p(r|w_i)p(w_i|D) \tag{10}$$

$$= \sum_{w_i \in D} p(r|w_i)p(w_i|D) + \sum_{w_i \notin D} p(r|w_i)p(w_i|D) \tag{11}$$

$$\approx \sum_{w_i \in D} p(r|w_i)p(w_i|D) = \sum_{w_i \in D} (1 - p(\overline{r}|w_i))\, p(w_i|D) \tag{12}$$

$$= \sum_{w_i \in D} \left(1 - p(w_i|\overline{r})\frac{p(\overline{r})}{p(w_i)}\right) p(w_i|D) \approx \sum_{w_i \in D} \left(1 - p(w_i|\mathcal{C})\frac{p(\overline{r})}{p(w_i)}\right) p(w_i|D) \tag{13}$$

$$\approx \sum_{w_i \in D} p(w_i|D) \tag{14}$$

In the derivation we made the following assumptions, in order to obtain a simple model for the prior. In 11, $p(w_i|D) \approx 0$ if $w_i \notin D$. In 12 $p(w_i|\overline{r}) \approx p(w_i|\mathcal{C})$; this assumes the collection to be a model of *non-relevance*, which goes accordingly to the hypothesis taken in [4], that every document is non-relevant (and eventually leading to the inverse document frequency formula as we know it). Lastly, in 13, it is assumed for convenience that $p(\overline{r})p(w_i|\mathcal{C}) << p(w_i)$.

The final form of this prior comes from the distribution for the terms on a document, by smoothing the maximum likelihood estimator as follows:

$$p(r|D) \approx \sum_{w_i \in D} p(w_i|D) \tag{15}$$

$$= \sum_{w_i \in D} \left[(1 - \lambda')\frac{tf(w_i, d)}{|D|} + \lambda'p(w_i|\mathcal{C})\right] \tag{16}$$

$$= (1 - \lambda') + \lambda' \cdot \sum_{w_i \in D} p(w_i|\mathcal{C}) \tag{17}$$

In this work, it is not required that the document model employed in the prior and the document model used to compute the query likelihood be the same. The former, has a parameter, $\lambda' \in [0, 1]$, coming out from the JM smoothing formula.

The result of this derivation results in a prior obtained from the sum of the individual contributions of each term occurring in the document. The linear document length-based prior (equation 9) has a similar form: it is a sum over the document terms frequencies, floored by a constant:

$$p(r|D) \approx \frac{1}{\sum_{D_i \in \mathcal{C}} |D_i|} \cdot \sum_{w_i \in D} tf(w_i, D) \tag{18}$$

The probabilistic prior is higher for documents with common terms than for documents with many rare terms, which may seem counter-intuitive. Note that the probabilistic prior counts the contribution of a term only once, despite of its document frequency. Hence, documents with many *different common* words will receive a higher prior value. Very common stopwords are likely to appear in every document, and therefore their effect is the same for every document. However, in heterogeneous collections, there may be a number of keywords describing generally its different topics or clusters. Keywords are likely to be frequent (at least inside the clusters), and documents containing many of those terms will be promoted in the rank list by the prior. This goes accordingly to the *scope* hypothesis [9]: documents covering many topics are more likely to be relevant.

4   Combination of the Prior and the Query Likelihood

In order to evaluate both priors we combine them with the query likelihood $p(Q|D, r)$ component in two different ways: a *standard* logarithmic sum and a novel method presented below. If we follow a log sum derivation from equation 4 then, the standard way of combining the document prior with the query likelihood estimation in order to produce a document score would be:

$$score(D, Q) = \log p(Q|D, r) + \log p(r|D) \qquad (19)$$

We further devised a new prior-query likelihood combination, taking into account the fact that probability estimates for longer documents are more reliable than for shorter ones. We modeled this fact by considering the risk of accepting a certain score s, $\hat{R}(s) \in [0, 1]$. It is possible to bias s and calculate a new score for the document and query $score(Q, D)$ as

$$score(Q, D) = s^{1 - \hat{R}(s)} \qquad (20)$$

Taking into account the fact that *longer* documents may provide a better estimate of $p(Q|D, r)$, it is reasonable to associate the document prior $p(r|D)$ with $1 - \hat{R}(s)$, resulting in

$$score(Q, D) = scoreLM(Q, D)^{\hat{p}(r|D)} \qquad (21)$$

or in logarithmic notation

$$score(Q, D) \stackrel{rank}{=} \hat{p}(r|D) * \log(scoreLM(Q, D)), \qquad (22)$$

where $scoreLM(Q, D)$ stands for the score a language model assigns to document D under a query Q. We combined both priors (linear and probabilistic) with the query-likelihood using this new approach. However, for the risk-based combination and linear prior, we modified the document length with a logarithmic transformation given that the probability of relevance versus logarithm of document length curve seems to be approximately linear in some ranges [12]:

$$score(Q, D) = \log(|D|) * \log(scoreLM(Q, D)) \qquad (23)$$

5   Experiments and Results

The main goal of these experiments is to evaluate the effectiveness of both priors and combinations proposed before, and assess their effect on retrieval. To evaluate the new priors and combinations, we plug them into a LM with Jelinek-Mercer smoothing (equation 7). This scoring function without the prior serves as the baseline.

The TREC datasets used are described in table 1. The collections differ in size and domain, hence they represent a broad and varied experimental dataset. We experiment with short (title-only) and long (title, description and narrative) queries. We apply the standard Porter stemming algorithm, and we skip any stopwords removal, in order to avoid any bias by any choice of stoplist[1]. For all the retrieval experiments we use the Terrier IR platform[2].

The metrics used are Mean Average Precision (MAP), precision at top ten retrieved documents (P@10) and binary preference (BPref [2]). The value of the λ parameter in JM smoothing (with and without priors) has been optimised for every measure in every collection by using increasing values of 0.05 in the range (0,1]. We performed a preliminary tuning for the λ' parameter in some datasets (values increasing in 0.1 steps), and decided to set it to 0.7 for every collection. We report that it is possible to obtain marginal gains if λ' is tuned specifically for a given collection, but that step is omitted to prove the robustness of the technique.

Table 1. Collections and Topics

Collection	size	Topics	# queries
LATimes	450 MB	401-450	50
TREC disks 4&5	2G	301-450+601-700	250
WT2g	2G	401-450	50
WT10g	10G	451-550	100

The experiments presented next, compare separately the best scores produced by the two priors and two ways of combining them with the query-likelihood, with the best scores the LM-JM baseline produces. Finally, the best performing prior and combination is compared against two state of the art retrieval models (Dirichlet prior and BM25).

Table 2 presents the results for all the priors and combinations. The first column is the type of prior and combination used. JM is the baseline (without any prior). PL C1 stands for the model that uses the standard log sum combination (equation 19) and a linear prior (equation 9), and this is the only prior-combination form out of the four presented that can be found in previous studies [6]. PP C1 stands for the probabilistic prior (equation 14) and the log sum

---

[1] We repeated these experiments using a standard stop-word list and the conclusions derived from this experimentation are the same.

[2] http://ir.dcs.gla.ac.uk/terrier/

Table 2. Optimal performance comparison of JM with the different priors and combinations for short(left) and long(right) queries. Best values are bold. Significant MAP differences according to the Wilcoxon test ($p < 0.05$) are bold and starred.

	LATimes						LATimes				
Model	MAP	P@10	Bpref	Δ%	p-value	Model	MAP	P@10	Bpref	Δ%	p-value
JM	0.2322	0.2711	0.2275	–	–	JM	**0.3010**	**0.3067**	**0.2865**	–	–
PL C1	0.2560	0.2889	0.2398	10.24	0.323	PL C1	0.2696	0.2978	0.2527	-10.43	0.059
PP C1	0.2332	0.2680	0.2256	0.43	0.642	PP C1	0.2937	**0.3200**	0.2848	-2.43	0.259
PL C2	0.2591	0.2784	0.2370	11.58	0.149	PL C2	0.2856	0.2978	0.2511	-5.01	0.669
PP C2	**0.2685**	**0.2889**	**0.2507**	**15.63**	**0.043***	PP C2	0.2996	0.3044	0.2861	-0.46	0.986

	Disks 4&5						Disks 4&5				
Model	MAP	P@10	Bpref	Δ%	p-value	Model	MAP	P@10	Bpref	Δ%	p-value
JM	0.2333	0.3908	0.2395	–	–	JM	0.2844	0.4791	0.2838	–	–
PL C1	0.2544	0.4313	0.2583	9.04	≈0*	PL C1	0.2731	0.4514	0.2741	-3.97	**0.019***
PP C1	0.2377	0.3996	0.2479	1.72	≈0*	PP C1	0.2849	0.4847	0.2876	0.18	0.337
PL C2	0.2535	0.4307	0.2570	8.65	≈0*	PL C2	0.2822	0.4711	0.2783	-0.77	0.537
PP C2	**0.2639**	**0.4454**	**0.2651**	**13.11**	≈0*	PP C2	**0.2967**	**0.4984**	**0.2992**	**4.32**	≈0*

	WT2g						WT2g				
Model	MAP	P@10	Bpref	Δ%	p-value	Model	MAP	P@10	Bpref	Δ%	p-value
JM	0.2495	0.3480	0.2407	–	–	JM	0.2678	0.4300	0.2748	–	–
PL C1	0.3110	0.4660	0.2946	24.64	≈0*	PL C1	0.2871	0.4660	0.2925	7.20	0.184
PP C1	0.2572	0.3760	0.2507	3.09	0.013	PP C1	0.2750	0.4280	0.2796	2.69	0.120
PL C2	0.3123	0.4640	0.2998	25.17	≈0*	PL C2	0.3017	0.4580	0.3010	12.65	0.112
PP C2	**0.3335**	**0.4820**	**0.3182**	**33.66**	≈0*	PP C2	**0.3145**	**0.4840**	**0.3138**	**17.43**	≈0*

	WT10g						WT10g				
Model	MAP	P@10	Bpref	Δ%	p-value	Model	MAP	P@10	Bpref	Δ%	p-value
JM	0.1479	0.2469	0.1474	–	–	JM	0.2274	0.3850	0.2202	–	–
PL C1	0.1926	0.2959	0.1889	30.22	≈0*	PL C1	0.2298	0.3730	0.2338	1.05	0.592
PP C1	0.1574	0.2582	0.1597	6.42	≈0*	PP C1	0.2312	0.3890	0.2291	1.67	**0.005***
PL C2	0.1939	0.3153	0.1928	31.10	≈0*	PL C2	0.2366	0.3810	0.2297	4.04	0.291
PP C2	**0.1984**	**0.3316**	**0.1956**	**34.14**	≈0*	PP C2	**0.2509**	**0.4020**	**0.2351**	**10.33**	≈0*

combination. PL C2 stands for the linear prior and the new risk-based query likelihood combination (equation 23). Finally, PP C2 denotes the new probabilistic prior and risk-based combination (equation 22). The Δ% column stands for the MAP difference between the row value and the baseline. The p-value reported in the last column is obtained from the standard Wilcoxon-paired ranks sign test for the MAP results of the prior in that row and the baseline. Significant values ($p < 0.05$) are bold and starred. The best values in each column for the three measures used are bold.

Results show that under the linear combination C1, the linear prior P1 performs better for short queries whereas the probabilistic prior P2 is slightly better with long queries (in three out of four collections). Overall, improvements respect to the baseline are significant with short queries and not significant with long queries under combination C1. The risk-based combination C2 is able to improve the performance of both priors in almost every case. The behaviour of the priors changed in this case, and P2 performed better than P1 with queries of any size. In any case, the probabilistic prior under this combination always yielded the best performance among all combinations and methods tested, with some impressive improvements. Effectiveness gains are higher with shorter queries, which may be due to the fact that JM smoothing performs better for longer queries, and reduces the importance of the length normalisation step in those cases.

Table 3. Optimal performance comparison between JM+probabilistic prior, Dirichlet prior smoothing and BM25 on different collections for short(left) and long(right) queries. Best values are bold. Significant MAP differences according to the Wilcoxon test ($p < 0.05$) are bold and starred.

	LATimes						LATimes				
Model	MAP	P@10	Bpref	Δ%	p-value	Model	MAP	P@10	Bpref	Δ%	p-value
JM-Prior	**0.2685**	0.2889	**0.2507**	-	-	JM-Prior	0.2996	0.3044	0.2861	-	-
BM25	0.2586	**0.2978**	0.2398	3.82	**0.041***	BM25	0.3022	0.3044	0.2870	-0.86	0.450
Dirichlet	0.2572	0.2889	0.2355	4.39	**0.017***	Dirichlet	**0.3061**	**0.3111**	**0.2970**	-2.12	0.604
	Disks 4&5						Disks 4&5				
Model	MAP	P@10	Bpref	Δ%	p-value	Model	MAP	P@10	Bpref	Δ%	p-value
JM-Prior	**0.2639**	**0.4454**	**0.2651**	-	-	JM-Prior	**0.2967**	**0.4984**	**0.2992**	-	-
BM25	0.2548	0.4402	0.2565	3.57	**0.047***	BM25	0.2825	0.4896	0.2814	5.03	**0.004***
Dirichlet	0.2559	0.4329	0.2569	3.12	**≈0***	Dirichlet	0.2743	0.4667	0.2737	8.27	**≈0***
	WT2g						WT2g				
Model	MAP	P@10	Bpref	Δ%	p-value	Model	MAP	P@10	Bpref	Δ%	p-value
JM-Prior	**0.3335**	**0.4820**	**0.3182**	-	-	JM-Prior	**0.3145**	**0.4840**	**0.3138**	-	-
BM25	0.3205	0.3560	0.3039	4.05	0.250	BM25	0.2833	0.4600	0.2910	11.01	**0.060***
Dirichlet	0.3087	0.4500	0.2924	8.03	**0.002***	Dirichlet	0.2906	0.4280	0.2805	8.22	**0.012***
	WT10g						WT10g				
Model	MAP	P@10	Bpref	Δ%	p-value	Model	MAP	P@10	Bpref	Δ%	p-value
JM-Prior	**0.1984**	**0.3316**	**0.1956**	-	-	JM-Prior	**0.2509**	**0.4020**	**0.2351**	-	-
BM25	0.1954	0.3102	0.1872	1.53	0.45	BM25	0.2319	0.3940	0.2295	8.19	**0.012***
Dirichlet	0.1932	0.2898	0.1887	2.69	**0.035***	Dirichlet	0.2435	0.3910	0.2223	3.03	0.2708

One possible explanation for the different behaviour of both prior combinations may be due to the contribution of the prior with respect to the contribution of the query likelihood. The linear combination C1 sums the logarithm of query likelihood and prior; as the query likelihood increases (by adding more query terms) the prior contribution (query independent) diminishes. The probabilistic prior contribution does not affect much the final results when combined this particular way. A high query likelihood score is not so dominant with the risk-based combination C2: the prior is still important for the final score because the combination multiplies the prior by the query likelihood logarithm. Another result is that the effect of the prior is not very sensitive to query length with the C2 combination.

A second batch of experiments compared the new prior and combination developed in this work, probabilistic prior with the risk combination, with LM and Dirichlet prior smoothing and also against BM25. The comparison is fair, as this two matching functions already incorporate a document-dependent normalisation factor. Dirichlet prior smoothing is presented in equation 8. The μ parameter chosen is the one that optimised the performance for each metric in every collection, picked up from a reasonable set of possible choices[3]. The second weighting function considered was the probabilistic Okapi's Best Match25 (BM25) [10] which has proved to be robust, high-performing and stable in many IR studies. The behaviour of the BM25 scores is governed by three parameters, namely k_1, k_3, and b. Some studies ([5]) have shown that both k_1 and k_3 have little impact on retrieval performance, so for the rest of the paper they are set as constant to the values recommended in [10] ($k_1 = 1.2$, $k_3 = 1000$). The b parameter

---

[3] $\mu \in \{100, 500, 800, 1000, 2000, 3000, 4000, 5000, 8000, 10000\}$.

controls the document length normalisation factor and it has been optimised in the same way as λ for JM (parameter exploration in the $(0, 1]$ range with 0.05 steps), independently for each metric and collection. The p-values and $\Delta\%$ differences reported in table 3 are calculated considering the Dirichlet prior/BM25 run as a baseline and compared to the JM+prior (PP C2) values.

This second set of results is presented in table 3. These results prove that the PP C2 combination is able to outperform significantly high-performing retrieval matching functions in most cases (again, LATimes and long queries being the exception). We can conclude that by including a high-quality length prior, JM smoothing outperforms Dirichlet prior smoothing, which was considered superior, and also well-tuned BM25.

6   Conclusions

We developed a new document prior that takes into account term statistics and give a probabilistic derivation for it. The effect of the priors in retrieval is also dependent on the way they are combined with the query likelihood. Hence, we also demonstrated the effectiveness of a new way of combining document-length based priors with the query likelihood, that leverages the effect of the prior and likelihood components. The prior boosts the performance of a LM based on JM smoothing significantly, with robust and stable results across collections of different nature and topics of different sizes. The retrieval effectiveness of JM with the new prior is also able to outperform LM using Dirichlet prior smoothing and BM25, when the optimal parameters are used for all of them, and on the basis of three different effectiveness measures. The excellent outcome in terms of retrieval effectiveness of the prior and risk-based combination opens ground for future research directions, for instance we will try to address the problem of using this new developed way of considering document length into other retrieval matching functions, and other retrieval tasks.

Acknowledgements. We thank the reviewers for their helpful comments on the paper. This work is co-funded by FEDER, SEUI and 'Xunta de Galicia' under projects TIN2005-08521-C02 and 07SIN005206PR.

References

1. Brin, S., Page, L.: The anatomy of a large-scale hypertextual web search engine. In: WWW7: Proceedings of the seventh international conference on World Wide Web 7, pp. 107–117 (1998)
2. Buckley, C., Voorhees, E.: Retrieval evaluation with incomplete information. In: SIGIR 2004: Proceedings of the 27th annual international ACM SIGIR conference on Research and development in information retrieval, pp. 25–32 (2004)
3. Craswell, N., Robertson, S., Zaragoza, H., Taylor, M.: Relevance weighting for query independent evidence. In: SIGIR 2005: Proceedings of the 28th annual international ACM SIGIR conference on Research and development in information retrieval, pp. 416–423 (2005)

4. Harper, D.J., Croft, W.B.: Using probabilistic models of document retrieval without relevance information. Journal Of Documentation 35(4), 285–295 (1979)
5. He, B., Ounis, I.: A study of parameter tuning for term frequency normalization. In: CIKM 2003 Proceedings of the twelfth international conference on Information and knowledge management, pp. 10–16 (2003)
6. Kraaij, W., Westerveld, T., Hiemstra, D.: The importance of prior probabilities for entry page search. In: SIGIR 2002: Proceedings of the 25th annual international ACM SIGIR conference on Research and development in information retrieval, pp. 27–34 (2002)
7. Lafferty, J., Zhai, C.: Probabilistic relevance models based on document and query generation. In: Croft, W.B., Lafferty, J. (eds.) Language Modeling and Information Retrieval. Kluwer International Series on Information Retrieval (2002)
8. Ponte, J.M., Croft, W.B.: A language modeling approach to information retrieval. In: SIGIR 1998: Proceedings of the 21st annual international ACM SIGIR conference on Research and development in information retrieval, pp. 275–281 (1998)
9. Robertson, S.E., Walker, S.: Some simple effective approximations to the 2-poisson model for probabilistic weighted retrieval. In: SIGIR 1994: Proceedings of the 17th annual international ACM SIGIR conference on Research and development in information retrieval, pp. 232–241 (1994)
10. Robertson, S.E., Walker, S., Jones, S., Hancock-Beaulieu, M.M., Gatford, M.: Okapi at TREC-3. In: Proceedings of the tenth Text Retrieval Conference (TREC-3) (1995)
11. Robertson, S., Zaragoza, H., Taylor, M.: Simple BM25 extension to multiple weighted fields. In: CIKM 2004: Proceedings of the thirteenth ACM international conference on Information and knowledge management, pp. 42–49 (2004)
12. Singhal, A., Buckley, C., Mitra, M.: Pivoted document length normalization. In: SIGIR 1996: Proceedings of the 19st annual international ACM SIGIR conference on Research and development in information retrieval, pp. 21–29 (1996)
13. Upstill, T., Craswell, N., Hawking, D.: Query-independent evidence in home page finding. ACM Transactions on Information Systems (TOIS) 21(3), 286–313 (2003)
14. van Rijsbergen, C.J.: Information Retrieval. Butterworths (1979)
15. Voorhees, E.M., Harman, D.K.: TREC: Experiment and Evaluation in Information Retrieval. The MIT Press, Cambridge (2005)
16. Westerveld, T., Kraaij, W., Hiemstra, D.: Retrieving web pages using content, links, urls and anchors. In: Proceedings of the tenth Text Retrieval Conference (TREC-10), pp. 663–672 (2002)
17. Zhai, C., Lafferty, J.: A study of smoothing methods for language models applied to information retrieval. ACM Transactions on Information Systems 22(2), 179–214 (2004)

Applying Maximum Entropy to Known-Item Email Retrieval

Sirvan Yahyaei and Christof Monz

Department of Computer Science
Queen Mary, University of London
London E1 4NS, UK
{sirvan,christof}@dcs.qmul.ac.uk

Abstract. It is becoming increasingly common in information retrieval to combine evidence from multiple resources to compute the retrieval status value of documents. Although this has led to considerable improvements in several retrieval tasks, one of the outstanding issues is estimation of the respective weights that should be associated with the different sources of evidence. In this paper we propose to use maximum entropy in combination with the limited memory LBFG algorithm to estimate feature weights. Examining the effectiveness of our approach with respect to the known-item finding task of enterprise track of TREC shows that it significantly outperforms a standard retrieval baseline and leads to competitive performance.

1 Introduction

In several information retrieval tasks, such as web retrieval [15], structured document retrieval [7] and email retrieval [3], a number of approaches combine evidence from multiple resources to compute the retrieval status values.

Typically the different sources of evidence include term frequencies within different fields of a document (e.g., body and anchor text), different ways to compute within-document and collection term frequencies or the combination of different document similarity functions as a whole. Zobel and Moffat [16] show that it is very difficult to find a similarity measure which is best in all cases, but at the same time they show that there is still a lot of room for improvement by varying retrieval strategies.

Unfortunately, most of the evidence formulas and combining functions that have been developed were tuned by heuristic approaches. Thus, an approach which combines evidence from different representations of the documents and automatically estimates the importance of each component in the retrieval ranking function can be very useful. For this purpose, we have adapted maximum entropy, a statistical machine learning method, to perform the retrieval task. This paper contains a description of the method and a number of experiments to verify its effectiveness.

The remainder of this paper is organized as follows: In the next section the problem of combining evidence and document representations is introduced. The

C. Macdonald et al. (Eds.): ECIR 2008, LNCS 4956, pp. 406–413, 2008.

maximum entropy method and its adaptation to IR are provided in sections 3 and 4. Related work is discussed in section 5. In section 6 we discuss the experimental set-up and results. Section 7 concludes the paper.

2   Combining Evidence

The problem of evidence combination can be re-formulated as the problem of finding a ranking function $W(\mathbf{d}, q, \mathbf{C})$, where collection $\mathbf{C}$ contains a set of documents $\mathbf{d}$ with k fields $\{f_1, f_2, ..., f_k\}$. As mentioned, most of the work in this area is in the form of combining scores, particularly, linear combination of scores. However, Ogilvie and Callan [11] have shown that their mixture language model approach outperforms various meta-search methods in almost all cases. Moreover, Robertson et al. [14] have discussed the dangers of linear combination of entire document similarity scores and criticized it in detail.

To deal with the problem of combining evidence, we propose a method that addresses most of the issues in previous approaches. This method, was designed to have following features:

1. Automatically learn different features from different sources of evidence
2. Learn complex features such as term proximity or user preferences in a manner similar to simple features
3. Do not make assumptions which are not realistic, for example, term independence assumption due to mathematical convenience by many methods
4. Deal with documents with both single and multiple representations in a unified manner

3   Maximum Entropy Modeling

Statistical modeling is used to build a model to predict the behavior of a process. A labeled training set is employed to learn a model predict future behavior of the process [1]. The first modeling task is feature selection and the second one is model selection. Firstly, a set of statistics is determined and then these statistics will be employed to construct an accurate model of the desired process.

One of the approaches to build that model is through maximum entropy modeling. The idea behind maximum entropy method is very simple: model all that is known and assume nothing about that which is unknown [1]. It means, choose a model consistent with all the facts, but otherwise as uniform as possible.

The probability distribution for the process based on maximum entropy has two characteristics: Firstly, it is in accordance with the constraints, secondly it is as uniform as possible.

It can be shown that there is a unique distribution that satisfies these constraints and it is always of the exponential form. Berger et al. [1] have shown that the solution has the following parametric form:

$$p_\lambda(y|x) = \frac{1}{Z_\lambda(x)} \exp \sum_{i=1}^{n} \lambda_i f_i(x, y) \qquad (1)$$

$$Z_\lambda(x) = \sum_y \exp \sum_{i=1}^{n} \lambda_i f_i(x, y) \tag{2}$$

where $Z_\lambda(x)$ is a constraint to satisfy the requirement that $\sum_y p_\lambda(y|x) = 1$ for all x, because it is a probability distribution.

Except for simple problems, equation 1 cannot be solved analytically and numerical methods have to be used to find the optimal weights of the features. We decided to use Nocedal's limited-memory BFGS optimization algorithm [10] which is a very efficient and robust method to solve large scale optimization problems and significantly outperforms the other two optimization approaches we experimented with.

4   Maximum Entropy in IR

By viewing the IR problem as a classification task, it is possible to apply discriminative classifiers to it, such as a classifier based on maximum entropy modeling. The retrieval process output values are $r \in \mathcal{R} = \{R, \bar{R}\}$ which are affected by the contextual information from the collection, documents, and queries. The parametric form of the distribution, which is mentioned in section 3, can be expressed as the conditional probability $p(r|d, q)$ as follows:

$$p(r|d, q) = \frac{1}{Z_\lambda(d, q)} \exp \sum_{i=1}^{n} \lambda_i f_i(d, q, r) \tag{3}$$

$$Z_\lambda(d, q) = \sum_{r \in \{R, \bar{R}\}} \exp \sum_{i=1}^{n} \lambda_i f_i(d, q, r) \tag{4}$$

There are two classes of features: Firstly, *atomic features* for documents with a single representation, which are functions of different term frequency statistics in the collection, documents and queries. Secondly, statistics for various *representations* of documents which in our case amounts to the different sections of the text. Representations are combined with atomic functions to have real-valued numbers as value. Table 1 shows some of the atomic and complex features. As we evaluate our approach in the context of email retrieval, our documents are e-mail message and each of the features is applied to the different fields of an e-mail: the subject, body and the body of the replied messages.

For training the maximum entropy model we normally use a set of queries for each of which we take a number of relevant and non-relevant documents. However, in the known-item finding task, there is exactly one relevant document for each query and the remainder of the collection is considered non-relevant with respect to this query. Therefore, we have to repeat the relevant constraints as much as non-relevant examples or choose a small portion of non-relevant examples. Due to the large number of documents we decided to under-sample a set of non-relevant documents, also repeating the relevant examples to balance out the training set.

Table 1. Functions used as features in our maximum-entropy retrieval approach

Name	Atomic Feature	Description
NTF	$\sum_{t \in Q \cap D} \log \left(1 + \frac{tf(D,t)}{|D|}\right)$	Normalized term frequency
IDF	$\sum_{t \in Q \cap D} \log \frac{N}{df(t)}$	Inverse document frequency
CT	$|q_i \in Q \cap D|$	Number of common terms
ICF	$\sum_{t \in Q \cap D} \log \frac{|C|}{tf(C,t)}$	Inverse collection frequency

Name	Complex Feature	Description
NTF-ICF	$\sum_{t \in Q \cap D} \log \left(1 + \frac{tf(D,t)}{|D|} \frac{C}{tf(C,t)}\right)$	Normalized $tf \times icf$
NTF-IDF	$\sum_{t \in Q \cap D} \log \left(1 + \frac{tf(D,t)}{|D|} \frac{N}{df(t)}\right)$	Normalized $tf \times idf$
BM25	$\sum_{t \in Q \cap D} \frac{(k1+1) \cdot tf(D,t)}{tf(D,t) + k1 \cdot (1 - b + b \cdot \frac{|D|}{avgdl(C)})}$ $\cdot \log \frac{N - df(t) + 0.5}{df(t) + 0.5}$	Okapi BM25
TP	$\sum_{t_i, t_j \in \mathcal{P}} \log(1 + (\min\{distance(t_i, t_j)\})^{-2}$ $\cdot \frac{|C|}{tf(C,t_i) tf(C,t_j)})$	Term proximity
FO	$\sum_{t \in Q \cap D} \log \left(1 + \frac{|D|}{firstOccurrencePos(t)} \frac{|C|}{tf(C,t)}\right)$	First occurrence position
TD	$\log \left(1 + \frac{4}{td(D)}\right)$	Depth in thread

5   Related Work

Ogilvie and Callan [11] compare the effectiveness of meta-search methods for combining document representations with their language modeling retrieval approach. In particular, they compared rank-based and score-based meta-searching with a mixture language model approach, showing that the latter slightly outperforms the best meta-search algorithms.

Their mixture method uses a unigram language model where the language model θ_D is specified by $p(w|\theta)$. During retrieval, documents are ranked by $p(Q|\theta_D) = \prod_{i=1}^{|Q|} p(q_i|\theta_D)$, where θ_D is the language model estimated for document D, $|.|$ is the length function and q_i is the ith term in the query Q.

Their approach is very similar to ours, except that we use an exponential model. Moreover, Ogilvie and Callan's mixture language model is based on unigram language modeling assuming term independence.

Robertson et al. [14] use BM25 as scoring function, but they have mentioned that this is a general method that can be used for many other scoring functions. The linear combination of frequencies method, similar to mixture language model by Ogilvie and Callan, uses a linear combination of a single scoring function over the representations. However, we have shown that our proposed maximum entropy method can combine any scoring function in a unified manner. On the other hand, using advanced features in a linear combination of frequencies will not be as easy as integrating them into a maximum entropy approach.

Another difference between maximum entropy and the above approaches is the estimation of the optimal weights or parameters of the ranking functions. Ogilvie and Callan [12] did not mention any optimization algorithms for finding the appropriate feature weights. Robertson et al. [14] used grid search for finding the parameters of their function. On the other hand, there are a number of well-studied optimization algorithms such as IIS, and L-BFGS for maximum entropy.

There have been a few attempts to explore maximum entropy in IR. Cooper [2] applied maximum entropy to information retrieval. Kantor and Lee [5] explored the application of maximum entropy, but more recently ([6]) they reported low performance on large document collections.

Greiff and Ponte [4] showed that ranking formulas of the Binary Independence Model (BIR) and Combination Match Model (CMM) can be derived from the maximum entropy principle with suitable features. Nallapati [9] explored discriminative models for IR and applied maximum entropy and support vector machines to several ad-hoc retrieval test sets. However, because of the rather discouraging results in these tasks, he did not examine maximum entropy in other tasks such as web or email retrieval.

6   Experiments

As mentioned before, one of the benefits of using maximum entropy is its ability to automatically learn arbitrary features. Thus to demonstrate the effectiveness of our approach in the context of information retrieval, we evaluate it with respect to email retrieval as defined in the Known-Item Finding Task of TREC 2005's Enterprise Track [3]. As emails are structured documents containing a number of fields (subject line, body, quoted text, etc.) this task is well-suited to evaluate the effectiveness of retrieval approaches that combine different sources of evidence. The collection is the W3C corpus which contains 174,311 documents. 25 queries are provided for training purposes and 125 additional queries are set aside for testing only. In our experiment, only the 25 training queries are used to generate the training data and learn the weights of the features.

There are three official measures for evaluating TREC's known-item finding task. The primary measure is the *Mean Reciprocal Rank (MRR)* and the other two are *success at 10* ($S@10$), indicating whether a relevant document is ranked among the top 10 retrieved documents, and *success at infinity* ($S@inf$) indicating whether a relevant document had been retrieved at any rank [3]. The Okapi BM25 ranking function has been used as one of the baselines for this experiment. The best results for the parameters after several attempts were $b = 0.25$ and $k1 = 1.2$. For this baseline, documents are treated as they have only one representation, i.e. all fields are merged and documents are indexed with one field which contains the whole text of the e-mail message.

6.1   Features

Three categories of features are used in this experiment: Firstly, features based on term frequencies of different representation of e-mail messages such as NTF-ICF-S$= \sum_{t \in Q \cap D_s} \log\left(1 + \frac{tf(D_s,t)}{|D_s|} \frac{C_s}{tf(C_s,t)}\right)$. Secondly, we use position based features such as term proximity, phrase match and first occurrence position features. Lastly, we use query independent features such as message depth in the thread.

There are many different methods to calculate term proximities [13,8]. Our term proximity feature computes the sum of minimum distances between term pairs. We chose this method of calculating proximity to avoid using features which carry similar information. For example, this term proximity metrics does not contain information about term frequency in the document.

$$\sum_{t_i,t_j \in \mathcal{P}} \log(1 + (\min\{distance(t_i,t_j)\})^{-2} \frac{|C|}{tf(C,t_i)tf(C,t_j)})$$ (5)

where, $distance(t_i,t_j)$ returns the set of distances between terms t_i and t_j, $tf(C,t_i)$ is the collection frequency of term i and $\mathcal{P}$ is the set of all possible pairs of query terms.

Similar to term proximity, the phrase match feature computes the maximum length of an exact match between the query and the document. Thus, a phrase match of 3 terms has a greater value than three matches of length 2.

The position of the first query term in the document is another feature that is used in the experiment:

$$\sum_{t \in Q \cap D} \log(1 + \frac{|D|}{firstOccurrencePos(t)} \frac{|C|}{tf(C,t)})$$ (6)

Thread depth feature is a query independent feature that is computed as follows:

$$\log 1 + \frac{4}{td(D)}$$ (7)

where, $td(D)$ is the depth of document D in the thread. The depth of emails is capped at level 4.

6.2   Results

Table 2 shows the best runs of the maximum entropy system, compared to the baseline systems. For statistical significance testing we used the two-sided Wilcoxon signed-rank test. All the fields are stemmed by the first two steps of Porter stemmer after stop-word removal. We use three baselines: a standard BM25 run, where all fields are merged (run 1), a maximum entropy run with a BM25 feature applied separately to the body and subject (run 2), and a maximum entropy run, using term frequency statistics only (run 3).

As the results show, the best maximum entropy based system significantly outperforms all baselines. Although the first occurrence position feature is somewhat unstable, it improved overall performance. In accordance with earlier approaches, our experiments show the importance of the subject field. Runs using only the subject field outperform runs using only the body and thread fields.

In general, the results show that our maximum entropy approach leads to strong results, substantially outperforming competitive baselines. Comparing runs 5 and 3 shows that the new, term-position based features, such as the term proximity and first occurrence features described above, lead to the best results.

Table 2. E-mail search results. $\cdot^*$ indicates whether the improvement with respect to each of the three baselines (runs 1–3) statistically significant at level $\alpha = 0.05$. S, B and T indicate the field that the function is applied to, which are subject, body and replies of the message in the thread, respectively.

Run	Features	MRR	S@10	S@inf
1 Baseline	BM25 on S+B+T, b=0.25, k1=1.2	$0.483^{-,*,-}$	0.696	**0.976**
2 Baseline	BM25-S, BM25-B	$\mathbf{0.557}^{*,-,*}$	**0.728**	0.968
3 Baseline	NTF-ICF-S, NTF-ICF-B, NTF-ICF-T	$0.520^{-,*,-}$	0.68	0.968
4 MaxEnt	NTF-ICF-S, NTF-ICF-B, FO-B, PM-B, TP-S, TP-B, TD	$0.603^{*,-,*}$	**0.816**	0.944
5 MaxEnt	NTF-ICF-S, NTF-ICF-B, FO-B, PM-B, TP-S, TP-B	$\mathbf{0.609}^{*,*,*}$	0.800	**0.976**
6 MaxEnt	BM25-S, BM25-B, BM25-T, FO-B, PM-B, TP-S, TP-B	$0.587^{*,-,*}$	0.76	0.960
7 MaxEnt	BM25-S, BM25-B, FO-B, PM-B, TP-S, TP-B	$0.603^{*,-,*}$	0.808	**0.976**
8 MaxEnt	NTF-ICF-S, NTF-ICF-B, PM-B, TP-S, TP-B	$0.565^{-,-,*}$	0.768	**0.976**
9 MaxEnt	NTF-S, NTF-B, IDF-S, IDF-B, FO-B, PM-B, TP-S, TP-B	$0.554^{-,-,-}$	0.736	0.952

7   Conclusions

We have shown that maximum entropy can be applied successfully to known-item email retrieval leading to statistically significant improvements over various baselines.

The advantages of using maximum entropy are twofold: Firstly, it is easy to integrate and experiment with additional features that are more tailored towards the retrieval task at hand. Here, we used three additional term-position based feature functions, term proximity, first term occurrence, and phrase matching. Using these additional features resulted in the highest performance, and led to statistically significant improvements over a maximum entropy based retrieval system that did not use these features.

The second advantage of maximum entropy over a number of related evidence combination approaches concerns the problem of estimating the appropriate feature weights. Earlier work often estimated the feature ways in a rather ad-hoc way by just experimenting with a number of weight combinations or by applying grid search. Both approaches are likely to miss the optimal weights and therefore leading to sub-optimal performance. On the other hand, there are a number of well-established and well-studied feature weight optimization algorithms for maximum entropy.

Acknowledgments. This work has been supported, in part, by the Nuffield Foundation, Grant No. NAL/32720.

References

1. Berger, A.L., Della Pietra, V.J., Della Pietra, S.A.: A maximum entropy approach to natural language processing. Comput. Linguist. 22(1), 39–71 (1996)
2. Cooper, W.S.: Exploiting the maximum entropy principle to increase retrieval effectiveness. Journal of the American Society for Information Science 34(1), 31–39 (1983)
3. Craswell, N., de Vries, A., Soboroff, I.: Overview of the trec-2005 enterprise track. In: Proceedings of the 14th Text REtrieval Conference (2006)
4. Greiff, W.R., Ponte, J.M.: The maximum entropy approach and probabilistic ir models. ACM Trans. Inf. Syst. 18(3), 246–287 (2000)
5. Kantor, P.B., Lee, J.J.: The maximum entropy principle in information retrieval. In: SIGIR 1986: Proceedings of the 9th annual international ACM SIGIR conference on Research and development in information retrieval, pp. 269–274. ACM Press, New York (1986)
6. Kantor, P.B., Lee, J.J.: Testing the maximum entropy principle for information retrieval. J. Am. Soc. Inf. Sci. 49(6), 557–566 (1998)
7. Lalmas, M.: Uniform representation of content and structure for structured document retrieval. In: 20th SGES International Conference on Knowledge Based Systems and Applied Artificial Intelligence (2000)
8. Monz, C.: From Document Retrieval to Question Answering. PhD thesis, University of Amsterdam (2003)
9. Nallapati, R.: Discriminative models for information retrieval. In: Proceedings of the 27th Annual International ACM SIGIR Conference on Research and Development in Information Retrieval, pp. 64–71 (2004)
10. Nocedal, J.: Updating quasi-newton matrices with limited storage. Mathematics of Computation 35, 773–782 (1980)
11. Ogilvie, P., Callan, J.: Combining document representations for known-item search. In: SIGIR 2003: Proceedings of the 26th annual international ACM SIGIR conference on Research and development in information retrieval, pp. 143–150. ACM Press, New York (2003)
12. Ogilvie, P., Callan, J.: Experiments with language models for known-item finding of e-mail messages. In: Proceedings of the Fourteenth Text Retrieval Conference (TREC-14) (2005)
13. Rasolofo, Y., Savoy, J.: Term proximity scoring for keyword-based retrieval systems. In: Sebastiani, F. (ed.) ECIR 2003. LNCS, vol. 2633, p. 79. Springer, Heidelberg (2003)
14. Robertson, S., Zaragoza, H., Taylor, M.: Simple bm25 extension to multiple weighted fields. In: CIKM 2004: Proceedings of the thirteenth ACM international conference on Information and knowledge management, pp. 42–49. ACM Press, New York (2004)
15. Tsikrika, T., Lalmas, M.: Combining evidence from web retrieval using the inference network model - an experimental study. Information Processing & Management, Special Issue in Bayesian Networks and Information Retrieval 40(5), 751–772 (2004)
16. Zobel, J., Moffat, A.: Exploring the similarity space. SIGIR Forum 32(1), 18–34 (1998)

Computing Information Retrieval Performance Measures Efficiently in the Presence of Tied Scores

Frank McSherry and Marc Najork

Microsoft Research, Mountain View CA 94043, USA

Abstract. The Information Retrieval community uses a variety of performance measures to evaluate the effectiveness of scoring functions. In this paper, we show how to adapt six popular measures — precision, recall, F1, average precision, reciprocal rank, and normalized discounted cumulative gain — to cope with scoring functions that are likely to assign many tied scores to the results of a search. Tied scores impose only a partial ordering on the results, meaning that there are multiple possible orderings of the result set, each one performing differently. One approach to cope with ties would be to average the performance values across all possible result orderings; but unfortunately, generating result permutations requires super-exponential time. The approach presented in this paper computes precisely the same performance value as the approach of averaging over all permutations, but does so as efficiently as the original, tie-oblivious measures.

1 Introduction

One of the fundamental problems in Information Retrieval is the ranking problem: Ordering the results of a query such that the most relevant results show up first. Ranking algorithms employ scoring functions that assign scores to each result of a query, where the score is an estimate of the result's relevance to the query at hand. So, ranking the results of a query consists of assigning a score to each result and then sorting the results by score, from highest to lowest.

Ranking algorithms are typically evaluated against a test collection consisting of a set of queries. Each query in the test collection has a set of results, and the results were arranged into a (partial or total) order by a human judge. In order to evaluate a ranking algorithm, the algorithm is applied to the result set of each query, the distance between the computed ranking and the "optimal" ordering determined by the judge is measured, and the distances are averaged over the entire test collection. Coming up with suitable distance metrics (or performance measures) has been the subject of considerable research, and there are numerous such metrics.

Typically these performance measures assume that a ranking algorithm arranges the results of a query into a total ordering, *i.e.* that no two results to a query have the same score. This assumption is reasonable for scoring functions that map a rich set of features of the result document to a real-valued score, but

C. Macdonald et al. (Eds.): ECIR 2008, LNCS 4956, pp. 414–421, 2008.

it is less warranted for evaluating the performance of a single discrete feature, *e.g.* page in-degree, click count, and page visits.

We stress that we are not concerned with evaluating the final results of a ranking system, which will almost certainly combine many features into a real-valued score, but rather the internals of such a system, where the performance of individual, potentially discrete features needs to be assessed. A typical modern search engine will use hundreds of features and combine the evidence provided by these features using *e.g.* a neural network. An important first step in training is feature selection, to restrict the number of inputs and avoid overfitting. A natural approach is to treat each feature as a scoring function in its own right and assess its performance under various IR metrics.

Despite the fact that performance evaluation of IR systems is a mature field and that much thought has gone into devising appropriate measures to compare different ranking systems, not much work has been done on adapting these measures to evaluate the performance of ranking algorithms that impose only a partial ordering on the result set. The impact of ties in ranked result sets on measures of retrieval effectiveness was first considered by Cooper [2], who proposed *expected search length* as a performance measure robust to ties. Raghavan and Jung [5] investigated the problem of ties in the context of precision and recall. Their focus is on precision at varying levels of recall, and they develop approaches that are sensible in the presence of ties. In contrast, our approaches focus on a larger space of evaluation measures, including F1, RR, AP, and NDCG, but are aimed at settings with fixed document cut-off values.

The simplest approach to dealing with the problem (which is in fact the approach that was taken by TREC competitions; see for example [4]), is to arbitrarily pick one of the valid (that is, well-sorted) orderings of a result vector and evaluate it. However, in our own experiments on large-scale test sets for web collections, and using scoring functions prone to produce tied results, we found that different well-sorted permutations of result vectors can have appreciably different performance values, large enough to affect the relative ordering of several of the scoring functions we compared.

A more disciplined approach is to average over all possible orderings. A naive realization of this approach to dealing with tied scores would entail generating the possible orderings of a result set by generating all permutations of each subset of tied results. This approach, while straightforward to implement, is computationally very expensive; its time complexity is super-exponential to the number of tied results. This is especially troublesome in cases where the distance metric is the cost function of an optimization algorithm, such as a dynamic optimization program to determine optimal parameters of a scoring function.

In this paper, we show how to compute six of the most popular performance measures efficiently in the presence of tied scores. Our approach is arguably superior to the two aforementioned approaches: It is completely deterministic just like the second approach (and in fact produces precisely the same results), and at the same time not substantially more expense to compute than the first approach of computing the traditional measures for a single permutation. We have implemented all of the tie-aware measures described, and are routinely using them as cost functions (*i.e.* the inner loop) of dynamic optimization systems.

The remainder of the paper is structured as follows: section 2 adapts six well-established (and tie-oblivious) performance measures to handle result vectors with tied scores in a robust and efficient manner. In section 3 we assess the performance impact of tie-awareness. Finally, section 4 offers concluding remarks.

2   Dealing with Tied Scores

We will now show how six standard performance measures can be adapted to deal with ties in a robust, deterministic matter, while still being about as efficient as the standard tie-oblivious definitions. The approach we take is to consider all possible consistent orderings, but rather than explicitly producing these orderings we analytically derive their average, which in each case we can easily compute.

We will first introduce some mathematical notation to help us describe performance measures, and then develop the tie-aware versions of these measures.

A typical ranking algorithm applies a *scoring function* s to all result documents retrieved in response to a query q, and sorts the results by decreasing score. This produces a result vector $V = \langle v_1, \cdots, v_n \rangle$, where $s(v_i) \geq s(v_{i+1})$ for all $1 \leq i < n$. Document v_1 is the highest-scoring result.

Scoring functions are evaluated using test collections of queries and associated result sets, where the results have been labeled by human judges as to their relevance. Labels can be binary (*e.g. relevant* or *irrelevant*), drawn from a small range (*e.g. excellent, good, fair, bad*), or fine-grained enough to impose a total ordering on the results (*e.g. best, second-best, etc.*). Five of the six performance measures described in this paper assume that the judges have used a binary labeling scheme, *i.e.* have marked the results in the test collection as relevant or irrelevant to their associated query. We write $rel(v_i)$ to denote the relevance of document v_i in a result set; $rel(v_i)$ is 1 if v_i is relevant to the query and 0 otherwise. In order to evaluate the performance of a scoring function s, we iterate over the query/result-set pairs in a test collection, use s to rank the results associated with each query, use a performance measure to quantify how much the ranking imposed by the scoring function diverges from the assessment of the human judges, and average these quantities over all queries in the test collection. This approach is often called *macro-averaging*.

2.1   Precision

The *precision* measure [1] is based on the observation that users of an IR system tend to peruse only the first k results of a search. It measures what fraction of these k results are relevant to the query on average. The value k is commonly called the *document cut-off value*. For the rest of the paper we assume that $k \leq n$. The precision at k is defined as:

$$P@k(V) = \frac{1}{k} \sum_{i=1}^{k} rel(v_i)$$

In order to deal with ties, we introduce a *tie-vector* $T = \langle t_1, \cdots, t_{m+1} \rangle$ whose first element t_1 is 0 and whose remaining elements are the ending indices of the m equivalence classes in V, so that $V_i = \langle v_{t_i+1}, \cdots, v_{t_{i+1}} \rangle$ all have the same score. These classes need not have more than a single element, as in the case that there is not a tie. We use the notations r_i and n_i to reference the number of relevant and total elements in V_i, respectively. We use R_i to denote the number of relevant elements that precede V_i in V. Furthermore, we assume *w.l.o.g.* that the document cut-off occurs in sub-vector V_c, or putting it differently, that k is in the half-open interval $(t_c, t_{c+1}]$.

In order to compute the precision at k in the presence of ties, we sum up the expected number of relevant results contained in each sub-vector V_i of tied results. For any sub-vector preceding V_c (the sub-vector where the cut-off occurs), the contribution is exactly the same as in the tie-oblivious case, since any permutation of tied results does not change the number of relevant results in the sub-vector. For any sub-vector succeeding V_c, the contribution is 0, since per definition none of its results fall below the document cut-off. Finally, the sub-vector V_c contains n_c results, r_c of which are relevant, and it has $k - t_c$ "slots" within the document cut-off window. So, V_c contains on average $(k - t_c)\frac{r_c}{n_c}$ relevant results. This leads to the following tie-aware definition of precision at k:

$$P@k(V) = \frac{1}{k}\left(R_c + \frac{k - t_c}{n_c}r_c\right) = \frac{1}{k}\left(\sum_{i=1}^{t_c} rel(v_i) + \frac{k - t_c}{n_c}\sum_{i=t_c+1}^{t_{c+1}} rel(i)\right)$$

The time complexity for computing $P@k(V)$ is $O(k)$ in the tie-oblivious case, since we only need to examine the first k elements of the result vector V, and $O(t_{c+1})$ in the tie-aware case, since we need to scan the result vector up to and including the sub-vector V_c overlapping the document cut-off specified by k.

2.2   Recall

The *recall* measure [1] quantifies what fraction of all the relevant results was ranked to fall within the first k documents. The recall at k is defined as:

$$R@k(V) = \frac{\sum_{i=1}^{k} rel(v_i)}{\sum_{i=1}^{n} rel(v_i)}$$

Recall can be adapted to results with tied scores in much the same way as precision. Again, we sum up the contributions of each sub-vector V_i of tied results, *i.e.* the expected number of relevant results contained in each sub-vector. But while precision normalizes this sum by the document cut-off value k, recall normalizes it by the total number of relevant results in the entire result vector. Neither normalization factor is influenced by permutations of results with tied scores. So, we arrive at the following tie-aware definition of recall at k:

$$R@k(V) = \frac{R_c + (k - t_c)\frac{r_c}{n_c}}{\sum_{i=1}^{n} rel(v_i)} = \frac{\sum_{i=1}^{t_c} rel(v_i) + \frac{k-t_c}{n_c}\sum_{i=t_c+1}^{t_{c+1}} rel(i)}{\sum_{i=1}^{n} rel(v_i)}$$

The time complexity for computing $R@k(V)$ is $O(n)$ in both cases, since we need to iterate over the entire result vector V to find out how many relevant results there are in total. If the number of relevant results in V is known, $R@k(V)$ can be computed in $O(k)$ time in the tie-oblivious case and $O(t_{c+1})$ in the tie-aware case.

2.3   F1 Measure

A common combination of precision and recall is the F1 measure, defined as the harmonic mean of precision and recall, which can be rewritten so as to avoid division by zero in the absence of any relevant results:

$$F1@k(V) = \frac{2}{\frac{1}{P@k(V)} + \frac{1}{R@k(V)}} = \frac{2\sum_{i=1}^{k} rel(v_i)}{k + \sum_{i=1}^{n} rel(v_i)}$$

The denominator of the rewritten equation is independent of ties in V, and the numerator is exactly as we have seen in both precision and recall. We thus adapt F1 similarly, replacing $\sum_{i=1}^{k} rel(v_i)$ with $R_c + (k - t_c)\frac{r_c}{n_c}$, the average number of relevant documents over all possible ties.

$$F1@k(V) = \frac{2(R_c + (k - t_c)\frac{r_c}{n_c})}{k + \sum_{i=1}^{n} rel(v_i)}$$

The time complexity of computing the F1 measure is no more than precision or recall in either case; the numerator can be computed in $O(k)$ or $O(t_{c+1})$ time, and the denominator, if unavailable, can be computed in $O(n)$ time.

2.4   Average Precision

The *average precision* measure computes a precision for every relevant result in a result vector, and averages these precision values. More precisely, the *average precision* at k is defined to be:

$$AP@k(V) = \frac{\sum_{i=1}^{k} P@i(V)rel(v_i)}{\sum_{i=1}^{n} rel(v_i)} .$$

The numerator is the only quantity that has the opportunity to vary based on the choice of ordering given ties. To simplify presentation, when considering a position j, we will bind i to be the index of the tie that contains j; *i.e.* the value of i such that $t_i < j \leq t_{i+1}$. To analyze the average contribution of a position j in a tie V_i, we note that position j is relevant in a $\frac{r_i}{n_i}$ fraction of orderings. When an element is relevant, the average number of relevant documents preceding it in the tie is $\frac{r_i-1}{n_i-1}$ times the number of available slots in the tie, $j - t_i - 1$. We thus define the tie-aware variant of AP as:

$$AP@k(V) = \frac{\sum_{j=1}^{k} \frac{r_i}{n_i}\left(R_i + (j - t_i - 1)\frac{r_i-1}{n_i-1} + 1)\right)\frac{1}{j}}{\sum_{j=1}^{n} rel(v_j)}$$

Computing $P@i(V)$ and R_i for increasing values of i does not require two nested loops, but can be done in a single loop running from 1 to k. Hence, the time complexity to compute $AP@k(V)$ is $O(k)$ in the tie-oblivious case and $O(t_{c+1})$ in the tie-aware case, since all of the terms in the above formulas can be evaluated in a single pass through the first k or t_{c+1} elements.

2.5   Reciprocal Rank

The *reciprocal rank* measure [6] favors scoring functions that rank relevant results highly. The value of the measure is inversely proportional to how far a user has to go down the ranked list of results on average to find the first relevant result:

$$RR@k(V) = \begin{cases} \frac{1}{i} & \text{if } \exists i \leq k : rel(v_i) = 1 \wedge \forall j < i : rel(v_j) = 0 \\ 0 & \text{otherwise} \end{cases}$$

The tie-aware variant requires attention only in the case where the first relevant result in the partial order occurs in a tie with at least one other object. In this case, we must determine the average value of $\frac{1}{i}$ for the first relevant result in that set of ties. Additionally, if this set of tied results crosses the imposed document cut-off value of k, we must consider the possibility that all relevant results are ranked beyond k, yielding no score at all.

To compute the tie-aware reciprocal rank, we first identify the first group V_i containing a relevant result. For each of the values j from t_i+1 up to $\min(t_{i+1}, k)$, we compute the fraction of orderings in which the first relevant result occurs at exactly that position. Multiplying this fraction by $\frac{1}{j}$ and accumulating over j gives the correct answer.

We compute the fraction of orderings with the first relevant result at position t_i+x by computing for each t_i+x the fraction of orderings whose first x elements are irrelevant, and then computing the difference between adjacent fractions. Taking those orderings whose first x elements are relevant, minus those whose first $x+1$ elements are irrelevant, gives the fraction whose first relevant element is at x. The fraction $f(x, r, n)$ of the orderings of r out of n relevant elements for which the first x are irrelevant follows as simple recursive definition:

$$f(x, r, n) = \begin{cases} 1 - \frac{r}{n} & \text{if } x = 1 \\ (1 - \frac{r}{n-x+1})f(x-1, r, n) & \text{otherwise} \end{cases}$$

Intuitively, each ordering that contributes to $f(x-1, r, n)$ will contribute to $f(x, r, n)$ if the next element is irrelevant, which occurs when none of the r relevant results are chosen from the set of $n-x+1$ remaining results.

Letting V_i be the first group containing a relevant result,

$$RR@k(V) = \sum_{j=t_i+1}^{\min(t_{i+1},k)} \frac{f(j-t_i, r_i, n_i)}{j}$$

The time complexity for computing $RR@k(V)$ is $O(k)$ in the tie-oblivious case, since it requires a linear scan of at most the first k elements of the result vector V, and at most $O(t_{c+1})$ in the tie-aware case, as we need only scan as far as the end of the last possible tied group. Once the first relevant group V_i is identified, the dynamic program takes $O(n_i)$ time to compute the fractions and accumulate the weighted reciprocals, which is at most $O(t_{c+1})$.

2.6 Normalized Discounted Cumulative Gain

The *discounted cumulative gain* measure [3] assumes that judges have assigned labels to each result, and accumulates across the result vector a gain function G applied to the label of each result, scaled by a discount function D of the rank of the result. A common example uses integer labels, the gain function $G(l) = 2^l - 1$, and discount funtion $D(i) = \frac{1}{\log(1+i)}$. We define the *discounted cumulative gain* at k as follows:

$$DCG@k(V) = \sum_{i=1}^{k} G(label(v_i))D(i)$$

We normalize $DCG@k(V)$ into the range $[0,1]$ by dividing by the DCV of an "ideal" result vector I (produced by a hypothetical clairvoyant scoring function that maximizes $DCG@k(I)$):

$$NDCG@k(V) = \frac{DCG@k(V)}{DCG@k(I)}$$

NDCG can be adapted fairly easily to deal with ties, as the normalization requires no special attention, and discounted cumulative gain is a simple sum over the returned results. Notice that for each position in a tied group, the average gain at that position is the average of the gain function across tied elements. As the discount function is multiplicative, we need only multiply it by this average gain at each position:

$$DCG@k(V) = \sum_{i=1}^{m} \left(\left(\frac{1}{n_i} \sum_{j=t_i+1}^{t_{i+1}} G(label(v_j)) \right) \sum_{j=t_i+1}^{\min(t_{i+1},k)} D(j) \right)$$

As mentioned above, the normalization step is independent of ties in a candidate partial ordering, and is computed and applied as usual.

Computing $NDCG@k(V)$ involves computing $DCG@k(V)$ and computing $DCG@k(I)$. Computing $DCG@k(V)$ takes $O(k)$ time in the tie-oblivious case and $O(t_{c+1})$ in the tie-aware case, as the average gain can be computed for each V_i in time $O(n_i)$, with the discounted gain accumulated for that group in a similar amount of time. As $DCG@k(()I)$ is independent of the ordering of V, its computation time is unaffected.

3 Implementation

We have built C# implementations[1] of both tie-oblivious and tie-aware versions of the performance measures described in this paper. We compared the wall-clock

---

[1] Available at http://research.microsoft.com/research/sv/tie-aware-measures

running times of the tie-oblivious and tie-aware variants of each method. In order to account for the effects of disk latency, we first loaded the vector of ranks and scores for a test set of 28,043 queries into main memory. The scoring function used was the in-degree of the web page. We found that for most of the performance measures the overhead is negligible. The exception is reciprocal rank, where the overhead is roughly 25%. Reciprocal rank is the only measure we considered that can be computed without sorting the result vectors; consequently, the overhead of the tie-awareness is much more noticeable, as it is not drowned out by the cost of sorting.

4   Conclusions

This paper addressed the issue of defining deterministic performance measures for scoring functions that are prone to assign identical scores to many results in a result set. Our approach is inspired by the idea of evaluating the performance of all possible well-ordered permutations of the result set and averaging the performances, but it avoids the factorial time complexity that would go along with such an approach, despite the fact that it produces precisely the same performance values as averaging over all well-ordered permutations does. We have applied our approach to six well-established performance measures: recall, precision, F1, average precision, reciprocal rank, and normalized discounted cumulative again. For these six measures, computing the tie-aware measures is not appreciably slower than computing the standard, tie-oblivious performance measures.

References

1. Cleverdon, C.W., Mills, J.: The testing of index language devices. Aslib Proceedings 15(4), 106–130 (1963)
2. Cooper, W.: Expected Search Length: A Single Measure of Retrieval Effectiveness Based on the Weak Ordering Action of Retrieval Systems. American Documentation 19(1), 30–41 (1968)
3. Järvelin, K., Kekäläinen, J.: Cumulated gain-based evaluation of IR techniques. ACM Transactions on Information Systems 20(4), 422–446 (2002)
4. National Institute of Standards and Technology. TREC 2005 Robust Track Guidelines (2005), http://trec.nist.gov/data/robust/05/05.guidelines.html
5. Raghavan, V., Jung, G.: A Critical Investigation of Recall and Precision as Measures of Retrieval System Performance. ACM Transactions on Information Systems 7(3), 205–229 (1989)
6. Voorhees, E.M., Harman, D.K.: TREC: Experiment and Evaluation in Information Retrieval. MIT Press, Cambridge (2005)

Towards Characterization of Actor Evolution and Interactions in News Corpora

Rohan Choudhary[1,*], Sameep Mehta[2], Amitabha Bagchi[1], and Rahul Balakrishnan[1]

[1] Indian Institute of Technology, New Delhi, India
[2] IBM India Research Lab, New Delhi, India

Abstract. The natural way to model a news corpus is as a directed graph where stories are linked to one another through a variety of relationships. We formalize this notion by viewing each news story as a set of actors, and by viewing links between stories as transformations these actors go through. We propose and model a simple and comprehensive set of transformations: *create, merge, split, continue, and cease*. These transformations capture evolution of a single actor and interactions among multiple actors. We present algorithms to rank each transformation and show how ranking helps us to infer important relationships between actors and stories in a corpus. We demonstrate the effectiveness of our notions by experimenting on large news corpora.

1   Introduction

Browsing news websites and searching for relevant news forms a major portion of a user's interaction with the web. With the presence of efficient and accurate search engines, it has become extremely simple for a user to find news of interest. However, the amount of online news data available makes it difficult and time consuming for the user to logically arrange and read the news. Therefore, there is a strong need to organize the data in a manner that allows the user to extract meaningful information quickly. Moreover the user must be presented news items in a manner which captures the interrelated nature of news items in an evolving news corpus. Simply arranging news items in order of their timestamps is not enough.

> *Kerry says President would cut retiree payouts:* " That's up to $500 a month less for food, for clothing, for the occasional gift for a grandchildren." *Kerry* warned on Sunday as he addressed elderly and middle-aged worshipers at a black church in Columbus, Ohio, bringing to the fore a major issue in the election that he has rarely touched on. Kerry's comments on *social security* came as he headed to Florida for a voter turnout push timed to Monday's start of early voting.

News Story 1

The Topic Detection and Tracking (TDT) [1] research initiative was formed in 1998 to address such issues in news organization. A topic is defined as a cluster of news stories connected by a seminal event. For example, the US elections 2004 is a topic and all the news stories connected with it are labeled as being inside the topic. Nallapati et.al. [5] presented an algorithm to discover dependencies between news stories by

---

* rohan@cse.iitd.ernet.in

C. Macdonald et al. (Eds.): ECIR 2008, LNCS 4956, pp. 422–429, 2008.

taking into account the content of the news. For example, in US Elections 2004 topic, stories about Bush are related to each other and stories about Kerry are related to each other. The news items can now be arranged as a graph such that each node represents one news item and each edge captures both kinds of dependencies between two news stories: textual and temporal.

Same-Sex Marriages: Bush Backs Ban in Constitution Pres *Bush* backs constitutional amendment to ban *same-sex* marriages; holds marriage cannot be separated from its 'cultural, religious and natural roots' without weakening society

---

News Story 2

These algorithms were based on the key assumption that *a single theme is associated with each news item*. However, this assumption does not hold true in many cases. For example, a news item discussing Bush's health care policy indeed has two themes/actors *Bush* and *Health Care*. Going beyond just a simple multiplicity of actors is the fact that the interrelationship between actors is major feature of a news corpus, and it is a feature that users look for, implicitly or explicitly. Keeping this in view our key contention is this: *Actors interact and these interactions provide valuable cues which can be used to discover useful parts, patterns and properties of the news corpus.* We define five key types of evolutions/transformations which actors can undergo. These are *create, merge, split, continue,* and *cease.* Some of the transformations inside a news corpus are more important than others. Based on this idea, we provide quantitative metrics to measure importance of any transformation. The usefulness of these transformations is demonstrated by the empirical observation that top ranked transformations in-fact, correspond to important events, stories and actors in a news corpus.

The Final Debate: The mission of Wednesdays night presidential debate was to engage *George W. Bush* and *John Kerry* in a discussion of domestic issues. True, both men hewed to their talking points and tried harder to score cheap shots than to offer clear explanations. But its hard to believe that anyone who watched with attention didn't come away with a good handle on who John Kerry and George W. Bush are, what they believe, and how they would approach running the country

---

News Story 3

The focus of this article is to characterize the interactions among actors and propose quantifying measures for them. We do not approach the problem of identifying actors, instead we depend on the algorithms proposed by Mei et al. [3] to identify actors/themes. To reiterate,the key contributions of this paper are: i) We present an actor based view of news corpora and posit an interaction graph of actors as the appropriate organizational framework for these corpora. ii) We define, discover and rank the key transformations that capture the evolution of a single actor and its interactions with other actors. We also empirically show how the ranking aids a user in retrieving important and interesting aspects of the news corpus.

We have also proposed an automatic interaction graph generation algorithm. The algorithm enforces the top transformations that are mined from the news corpus. Due to shortage of space, we have not detailed the algorithm in this paper. Interested Readers are directed to [2] for details of the algorithm.

2   An Actor Based View of News Corpora

In this section we present and develop an actor based view of news corpora. We first define interaction graphs which form the basic structure of a news corpus organized by actors, then we study the transformations these actors undergo in the interaction graphs.

The Interaction Graph

The basic structure we proceed with is an *interaction graph* which is a major improvement on the structure proposed by Nallapati et. al. [5]. In our interaction graph each story is represented by a node. The actors present in a story are enumerated inside the node. Links may be established between news stories having common actors. Edges connecting two stories are annotated with actors common to the two stories. It is our contention that this is a natural and satisfactory way of organizing a news corpus being presented to a human user. For expository purposes consider a news corpus consisting of three news items: S1, S2 and S3(temporally ordered). Relevant actors in each news item are identified and marked. An interaction graph of these stories is shown in Figure 1. Stories 2 and 3 are linked because of the presence of a common actor, i.e., *Bush*. The actors *Bush* and *Kerry* both are present in Story 3. The presence of edges from Story 1 and Story 2 to Story 3 implies that previously non co-occurring actors appeared together in story 3. We call such a transformation a *merge* of two actors. Similar definitions hold for other transformations. Once all the transformations have been discovered we score them to ascertain their significance using a scoring procedure that takes into account stories in the temporal neighborhood.

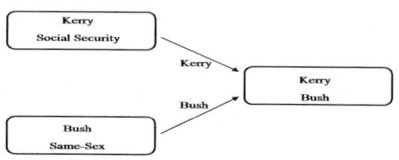

Fig. 1. Interaction Graph for the three stories

We would like to clarify again that we have not discussed the interaction graph generation algorithm in this paper. For details of the algorithm, the reader is encouraged to look at our technical report. [2].

Basic Notations

Given a news corpus consisting of D news items with respective time stamps $\{t_1, t_2, \ldots, t_D\}$, where $t_i \leq t_{i+1}$ D_i represents i^{th} news item with a time stamp of t_i. Associated with each news item D_i is an actor vector K_i of length n_i, $\{K_i^1, K_i^2, \ldots K_i^{n_i}\}$. A word or a phrase appearing in the news corpus is considered an actor if it occurs repeatedly in a time period. This vector can be derived by using the theme extraction algorithms proposed by Mei and Zhai [3]. These actors are subsets of salient themes across a topic. $G^l = (V^l, E^l)$ denotes a news graph till time t_l. Whenever there is no ambiguity we denote the graph simply by G. Each node represents a unique news item, i.e., $|V_l|$ is same as the number of news items collected till t_l and vertex V_i represents news item D_i. A direction edge $e_{(i,j)}$ from node(news items) V_i to V_j implies that $t_i < t_j$ and there is overlap between the corresponding actor vectors, i.e., $K_i \cap K_j \neq \phi$ We maintain the list of actors associated with such an edge in $K_{i,j}$. Also let $C_{t_j}^{t_l} = \cup_{i=j}^{l} K_i$ represent the set of all the actors discovered in the time window $[t_j, t_l]$.

Actor transformations

We now develop a framework for extracting information from news corpora:*actor transformations*. We contend that the interaction between news stories can be modeled as one of five fundamental transformations that one or more actors involved in those news stories undergo. These five transformations are *create, merge, split, continue* and *cease*. We assume that G^l and other variables, as defined above are available to us. The definitions below then serve as a way of mining the transformations at news story D_i. We now formally define these transformations. Figure 2 shows a sample interaction graph for US election 2004. The numbers inside the node establish a temporal order (not continuous dates) and the annotation on the edge represents the common actors. We will require the following functions for this formalization and the other measures we define in later sections:

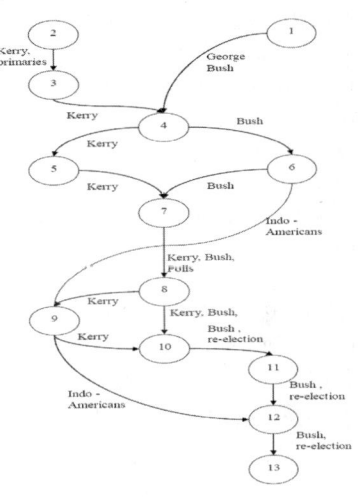

Membership Testing Function: The declaration of this function is **BOOL IsMember(List, A)**. The function returns TRUE if $A \in List$ else it returns FALSE.

Set Intersection Function: The declaration of this function is **List SetIntersect (List$_1$, List$_2$)**. This function returns a list of actors common in both $List_1$ and $List_2$.

Set Union Function: The declaration of this function is **List SetUnion(List$_1$, List$_2$)**. This function returns a list of actors present in either $List_1$ or $List_2$.

Fig. 2. Example of an Interaction Graph

Merge Actors $\underline{A}$ and $\underline{B}$ are marked as merged at D_i if the following conditions hold:

Condition 1- $\underline{A}$ and $\underline{B}$ are present in K_i.
Test: IsMember(K$_i$,A) = T $\wedge$ IsMember(K$_i$,B) = T.
Condition 2- Both $\underline{A}$ and $\underline{B}$ never co-occur in an edge to this news story D_i.
Test: $\nexists$ j $< (i)$ IsMember($K_{j,i}$,A) = T $\wedge$ IsMember($K_{j,i}$,B) =T
In figure 2 actors in node 1(*Bush*) and 3 (*Kerry*) merge at node 4.

Split Actors $\underline{A}$ and $\underline{B}$ are marked as split at D_i if:
Condition 1- $\underline{A}$ and $\underline{B}$ co-occur at t_i
Test: (IsMember(K_i,A) = T $\wedge$ IsMember(K_i,B) =T)
Condition 2- There is a news story D_k such that there is an edge from Story D_i to D_k and only actor $\underline{B}$ is present in the Story D_k.
Test: $\exists$ k $> i$ IsMember($K_{i,k}$, A) = F $\wedge$ IsMember($K_{i,k}$, B) = T.
Condition 3- There is a news story D_j such that there is an edge from Story D_i to D_j and only actor $\underline{A}$ is present in the Story D_j.
Test: $\exists$ j $> i$ IsMember($K_{i,j}$, A) = T $\wedge$ IsMember($K_{i,j}$, B) = F.

Condition 4- There is no news news story D_j such that there is an edge from Story D_i to D_j and both actor $\underline{A}$ and $\underline{B}$ are present in the Story D_j.
Test: $\nexists$ j $> i$ IsMember($K_{i,j}$, A) = T $\wedge$ IsMember($K_{i,j}$, B) = T.

Please note that swapping A and B in the above conditions also constitutes a valid split. An example of split can be seen at node 4 because the co-occurring actors now occur individually at node 5 and node 6.

Create An actor A is marked as created at D_i if:

Condition 1 A is present in K_i

Test: IsMember(K_i,A) = T

Condition 2 There is no news story D_j such that there is an edge from D_j to D_i and A is present in K_j

Test: $\nexists \, j < (i)$ IsMember($K_{j,i}$,A) = T

Indo-Americans and *Polls* was created at node 6 and node 7 respectively.

Continue An actor A is marked as continued at D_i if:

Condition 1 A is present in K_i

Test: IsMember(K_i,A) = T

Condition 2 There is a news story D_j such that A is present in K_j and there is an edge from D_j to D_i

Test: $\exists \, j < (i)$,IsMember($K_{j,i}$,A) = T

Polls continued at (7,8) whereas *Bush* was present at (1,4,6,7,8,10,11,12,13).

Cease An actor A is marked as ceased at D_i if:

Condition 1 A is present in in K_i

Test: IsMember(K_i,A) = T

Condition 2 There is no news story D_j such that A is present in K_j and there is an edge from D_i to D_j

Test: $\nexists \, j > (i)$,IsMember($K_{i,j}$,A) = T

Indo-Americans and *Polls* ceased to exist after node 12 and node 8 respectively.

We would like to emphasize that each news story or an actor can be involved in multiple transformations.For example between node 11 and node 12 in Figure 2 *Bush* is continuing as well as merging with *Indo-Americans*.

3   Ranking Transformations

The actor transformations described in the previous section can be used to gain insights into the data and extract useful information about the structure, evolution, key events, and storylines of a topic. However, in a typical large news corpus, we expect to discover a number of key transformations. To extract useful information, the user would have to iterate through all the transformations and find the important ones. This iterative process will soon become cumbersome and error prone. Therefore, one major challenge is to rank the discovered transformations.

In this section we, first, define a set of metrics to quantify the importance of an actor and co-occurrences of two or more actors. These metrics are then used to rank the transformations. Recall that $List_A^{[t_1,t_2]}$ denotes list of all the news stories in the time interval $[t_1, t_2]$ containing actor A and $N^{[t_1,t_2]}$ represents total stories in the interval $[t_1, t_2]$.

Strength: Strength of A during time interval $[t_1, t_2]$ is: $Strength_A^{[t_1,t_2]} = \frac{|List_A^{[t_1,t_2]}|}{N^{[t_1,t_2]}}$

This metric captures the fraction of news stories in which an actor appears during a given time interval. $Strength_A^{[t_1,t_2]} = 1$ implies that all the news stories contain A

and therefore $\underline{A}$ is regarded as a very important actor in the specified time period. This metric is used to rank individual actors. The definition can be extended to calculate the collective strength of a set of L actors as: $Strength^{[t_1,t_2])}_{(A_1,A_2...A_L)} = \frac{|\cap^{L}_{i=1} List^{[t_1,t_2]}_{A_i}|}{N^{[t_1,t_2]}}$

Coupling: Coupling between $\underline{A}$ and $\underline{B}$ during time interval $[t_1, t_2]$ is given by:

$Coupling^{[t_1,t_2]}_{(A,B)} = \frac{|List^{[t_1,t_2]}_A \cap List^{[t_1,t_2]}_B|}{|List^{[t_1,t_2]}_A \cup List^{[t_1,t_2]}_B|}$

This metric measures co-occurrence of $\underline{A}$ and $\underline{B}$ in the given time period, i.e, how many news stories contain both $\underline{A}$ and $\underline{B}$. $Coupling^{[t_1,t_2]}_{(A,B)} = 1$ implies that all the news stories in the given time period which contain A (B) also contains B (A) which implies a high and therefore an important coupling.

Next, we discuss how these metrics are used to rank the transformation. In this discussion we will be using P to denote a retrospective window i.e. P is the number of previous time steps (news stories) that are taken into account. Similarly, F denotes a future window i.e. F is the number of subsequent time steps (news stories) that are taken into account. The reader is encouraged to read our technical report [2] for the motivation behind the measures.

Importance of Split Transformation: A split transformation between $\underline{A}$ and $\underline{B}$ at time t is considered important if i) $Strength^{[t-P,t]}_{(A,B)}$ is high and ii) $Coupling^{[t,t+F]}_{(A,B)}$ is low.

Using these two conditions, score of a split is given as: $\frac{e^{Strength^{[t-P,t]}_{(A,B)}}}{e^{Coupling^{[t,t+F]}_{(A,B)}}}$

Importance of Merge Transformation: A merge transformation between $\underline{A}$ and $\underline{B}$ at time t is considered important if i) $Strength^{[t-P,t]}_A$ and $Strength^{[t-P,t]}_B$ is high and ii) $Coupling^{[t-P,t]}_{(A,B)}$ is low. Using these two conditions, the score of a merge is given as:

$\frac{e^{Strength^{[t-P,t]}_B} \times e^{Strength^{[t-P,t]}_A}}{e^{Coupling^{[t-P,t]}_{(A,B)}}}$

Importance of Continue Transformation: : Continuation of concept vector $K_{i,j}$ from story D_i to D_j is important if $Strength^{[t-H,t+F]}_{K_{i,j}}$ is high. The score simply is collective strength of $K_{i,j}$ in $[t-H, t+F]$.

Importance of Create Transformation: Creation of $\underline{A}$ at time t is considered important if $Strength^{[t,t+F]}_A$ is high. F denotes the number of future time steps (news stories) which should be considered to ascertain the quality of create transformation. The score is simply its strength in $[t, t+F]$.

Importance of Cease Transformation: Cessation of $\underline{A}$ at time t is considered important if $Strength^{[t-P,t]}_A$ is high. The score is simply the strength in $[t-P, t]$.

4 Experimental Results

Due to lack of space we provide detailed experiments only on FIFA World Cup, 2006 and US Presidential Elections, 2004. In our technical report [2], we have discussed experiments on other datasets and more interesting inferences.

FIFA World Cup, 2006: The first dataset FIFA World Cup 2006, consists of 459 news stories published between 02/06/2006 and 15/07/2006 by www.rediff.com.The main actors of the topic are the teams and some of the well reported players. We mined the transformations from the complete FIFA dataset. Next, we assigned scores to the transformations and picked the top 14 merges. The stories associated with these 14 transformations are shown in Table 1. The first column shows the stories according to their rank (in decreasing order) and the second column shows the same transformation arranged by time (decreasing). As evident from the list all the major stories received high score. These results strengthen our belief that the ranking procedure is indeed useful and that the user can be provided top stories based on score. The user can then explore any of these stories in more detail. Similarly, the top two creations discovered are: *Zidane's Head Butt* and *Fan Clashes*.

Table 1. Top ranked Merges in FIFA

Germany v/s Portugal	Italy v/s France
Portugal v/s England	Germany v/s Portugal
Italy v/s France	France v/s Portugal
Italy v/s Germany	Italy v/s Germany
Argentina v/s Germany	Brazil v/s France
Brazil v/s France	England v/s Portugal
France v/s Portugal	Argentina v/s Germany
England v/s Ecuador	Italy v/s Ukraine
England v/s Sweden	Spain v/s France
Sweden v/s Germany	Brazil v/s Ghana
Spain v/s France	Italy v/s Australia
Brazil v/s Ghana	Germany v/s Sweden
Italy v/s Australia	England v/s Sweden
Italy v/s Ukraine	Germany v/s Ecuador

US Elections 2004: This dataset consists of 389 news stories published between 02/02/04 and 15/11/04 by nytimes.com. The key actors of the topic are *Bush*, *Kerry* and important election issues like *abortion* and *social security*. We again mined the top transformations from the corpus and ranked them using our measures. Table 2 shows the abstract of new stories where top 8 creations occurred in this corpus. The actual actors are also noted in the table. The size of future window F is taken as 8 days. We can see that the top creations actually correspond to the major stories and events inside the topic.

5   Related Work

Topic detection and tracking has been a popular research topic in the areas of text mining, information retrieval and organization. Interested readers are pointed to an excellent survey in [1] The need for having a temporal structure within a topic was identified by Nallapati et al. [5]. The authors proposed a directed acyclic graph where each node represented an event and each edge represented a

Table 2. Synopsis of the top ranked creations in US Election 2004 corpus

Date	Story and Creation of Actor
30/08	Republican Convention kicks off (convention)
13/04	Iraq issue starts coming up (Iraq)
06/07	Kerry chooses Edwards as running mate (Edwards, running mate)
14/05	Issue of same-sex marriage (same-sex marriage)
28/07	Issue of economy during democratic convention (economy)
28/07	Issue of global terrorism at democratic convention (terror)
01/08	Republicans challenge Kerry's Vietnam records (Vietnam)
13/05	Ralph Nader wins endorsement of Reforms Party (Nader)

dependency between the two nodes. Although we also work on directed acyclic graph, the nodes in our graph are the individual news stores. Also, in their work, the focus was on generating the graph. In this paper, we use properties of the graph to draw interesting inferences about the topic.

The problem of discovering evolutionary theme patterns from text was first identified by Mei and Zhai [3,4]. The authors defined notion of theme across a time period and salient themes across the whole topic. The evolution of a theme was captured, however, the interaction between themes was not accounted for. The algorithms proposed in [3] can be used for detection of the major actors of a topic. Mei and Zhai [4] also demonstrated that a document can belong to multiple contexts. This is very similar to our modeling of each news story as an interaction of major actors which belong to that story. In their seminal work, Silver and Wang [6] enumerated the key transformations which a three dimensional scientific feature can undergo. Recently Spiliopoulou et al. [7] presented similar transformations to capture and monitor evolving clusters. Both these algorithm defined a customized overlap (intersect) function to derive the relationships. Our algorithms use set intersection algorithm.

6 Conclusions

In this article we presented definitions and algorithms for discovering the key transformations which actors in a news corpus can undergo. The intuition behind our approach is that each news story encompasses multiple themes/actors. Each individual actor evolves over time and simultaneously interacts with other actors. These interactions point to interesting and important parts of a news corpus. To reduce the number of transformation which the user has to evaluate, we outlined a scoring procedure to rank the transformations. We empirically showed that the transformations with high score typically point to the important stories in the corpus by discussing the results on two large datasets.

References

1. Allan, J., Carbonell, J., Doddington, G., Yamron, J., Yang, Y.: Topic detection and tracking pilot study: Final report. In: DARPA Broadcast News Transcription and Understanding Workshop, pp. 194–218 (2006)
2. Choudhary, R., Mehta, S., Bagchi, A., Balakrishna, R.: A framework for exploring news corpora by actor evolution and interaction. IBM Research Report- RI07004 (2007)
3. Mei, Q., Zhai, C.: Discovering evolutionary theme patterns from text: an exploration of temporal text mining. In: KDD 2005: 11th ACM SIGKDD international conference on Knowledge Discovery and data mining, pp. 198–207 (2005)
4. Mei, Q., Zhai, C.: A mixture model for contextual text mining. In: KDD 2006: 12th ACM SIGKDD international conference on Knowledge Discovery and data mining, pp. 649–655 (2006)
5. Nallapati, R., Feng, A., Peng, F., Allan, J.: Event threading within news topics. In: CIKM 2004: 13th ACM International Conference on Information and Knowledge Management, pp. 446–453 (2004)
6. Silver, D., Wang, X.: Volume tracking. In: VIS 1996: 7th conference on Visualization, pp. 157–164 (1996)
7. Spiliopoulou, M., Ntoutsi, I., Theodoridis, Y., Schult, R.: Monic: modeling and monitoring cluster transitions. In: KDD 2006: 12th ACM SIGKDD international conference on Knowledge Discovery and data mining, pp. 706–711 (2006)

The Impact of Semantic Class Identification and Semantic Role Labeling on Natural Language Answer Extraction

Bahadorreza Ofoghi, John Yearwood, and Liping Ma

Centre for Informatics and Applied Optimization, University of Ballarat
PO Box 663, Ballarat VIC 3353, Australia
bofoghi@students.ballarat.edu.au,
{j.yearwood,l.ma}@ballarat.edu.au

Abstract. In satisfying an information need by a Question Answering (QA) system, there are text understanding approaches which can enhance the performance of final answer extraction. Exploiting the FrameNet lexical resource in this process inspires analysis of the levels of semantic representation in the automated practice where the task of semantic class and role labeling takes place. In this paper, we analyze the impact of different levels of semantic parsing on answer extraction with respect to the individual sub-tasks of frame evocation and frame element assignment.

Keywords: Shallow Semantic Parsing, FrameNet, Question Answering.

1 Introduction

The main advantage of the exploitation of semantic information in the process of QA is realized by the semantic classes and roles added to the texts. These develop more comprehensive understanding of the texts by the automated system and lead to a more meaning-aware QA process. The task of shallow semantic parsing to add such information to texts mainly consists of two phases: i) sense disambiguation of the predicative target word to identify the semantic class, and ii) role assignment to the arguments of the predicate with regard to its specific sense [1]. In the context of FrameNet, which is a lexical resource for English [2] relying on Frame Semantics [3] [4], the class is the specific semantic *frame* which is evoked in the true sense of the context of the sentence [5], while the arguments are the different participant roles known as *frame elements* (FEs). An example of the process is shown in Figure 1.

The task of semantic class and role labeling in shallow semantic parsers has not been studied in the context of question answering by many researchers. Narayanan and Harabagiu first introduced the importance of the task in question answering when articulated in Coordinated Probabilistic Relational Models (CPRMs) [6]. Similar methods of answer extraction have been implemented in [7], [8], and [9]. Kaisser proposes the use of semantic frames to overcome the paraphrasing phenomenon by question reformulations [10]. One of the most recent efforts in this context is the work by Shen and Lapata [11] which formulates the usage of semantic role labeling via

C. Macdonald et al. (Eds.): ECIR 2008, LNCS 4956, pp. 430–437, 2008.
© Springer-Verlag Berlin Heidelberg 2008

bipartite graph optimization and matching for answer extraction using the FrameNet frames and FEs. They have not, however, studied the impact of different levels of semantic class identification and semantic role labeling on QA individually.

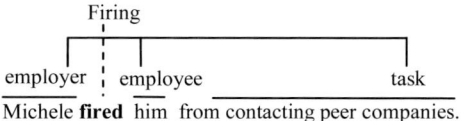

Fig. 1. Shallow semantic analysis of an example sentence evoking the frame "Firing"

In this paper, we investigate the significance of the individual tasks of semantic class and role labeling and their benefits for the problem of answer extraction in a baseline QA system at different levels of parsing. Before empirical studies, it is not obvious that with higher levels of annotation the QA performance will always increase as there are barriers such as frame redundancy in exhaustively annotated texts which may interfere with the practical task of correct answer identification.

2   Configuration of Experiments

We have developed an experimental configuration to measure the importance of each sub-task of shallow semantic parsing with respect to the FrameNet frames and FEs in the task of answer extraction in a QA system. To this end, we have i) implemented a baseline QA system without any frame semantics involved, ii) used a baseline shallow semantic parser to add the FrameNet elements to texts, and iii) defined a few levels of semantic representation that can be realized in the sense of how perfectly a semantic parser can identify the semantic classes and assign corresponding semantic roles.

2.1   Experimental QA System

The pipelined architecture of the implemented QA system is depicted in Figure 2. We perform the question classification task using our shallow rule-base containing about 130 rules to categorize the focus of the questions into different classes described in [12]. Then, passages are retrieved from the TREC-reported documents per target using a modified version of the MultiText passage retrieval algorithm [13, 14].

The answer extraction module uses two answer extraction models: i) the *Entity-based model* (ENB), and ii) the *Frame semantic-based model* (FSB). The ENB model extracts the Named Entities (NEs) from the related passages, filters them with respect to the answer type, and finally ranks them according to the score of the answer-bearing passages. The NE tagger used in this model is LingPipe[1] which is capable of identifying *locations*, *persons*, and *organizations*. We have implemented a pattern-based *date* and *time* tagger to identify temporal references in this model as well as a short list of *definitional* adjectives.

---

[1] LingPipe: http://alias-i.com/lingpipe/

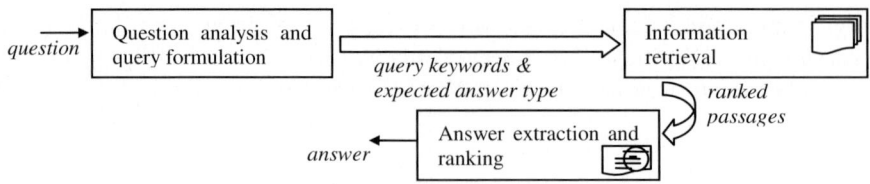

Fig. 2. The pipelined architecture of the baseline QA system

In the second model – FSB – both the question and potential answer bearing passages are annotated against the FrameNet frames and FEs. Having the vacant FE identified in the question, the process of answer extraction includes frame and FE alignment to instantiate the vacant FE in the question. Figure 3 shows an example answer extraction process in FSB. The baseline QA method, however, only uses ENB which is to be further elaborated by the FSB model in our study.

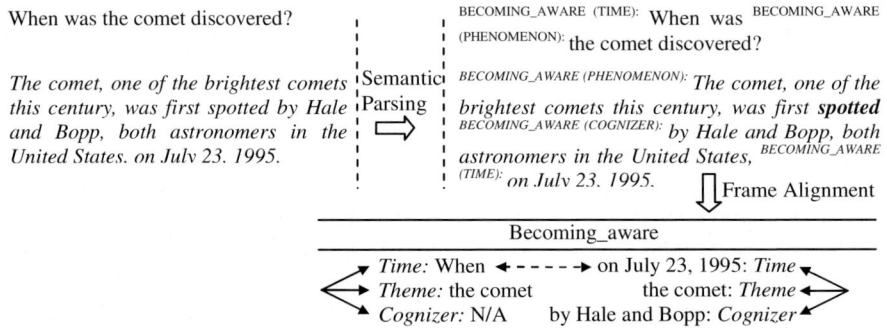

Fig. 3. Answer extraction via frame and frame element alignment in the FSB model

The baseline shallow semantic parser used to automatically assign frames and FEs to the text is the *Shalmaneser* parser [1]. Shalmaneser is a loosely coupled tool chain which can benefit from different tools at each processing step. Table 1 shows the setting used in our experiments.

Table 1. Shalmaneser settings at each processing step

Processing step	System	Version
POS-tagging	TNT	2
Lemmatization	TreeTagger	–
Syntactic parsing	Collins' Parser	1.0
Machine learning	Mallet	mallet 0.4

2.2 Levels of Semantic Parsing

With respect to the outputs of Shalmaneser, four levels of shallow semantic parsing have been defined as depicted in Figure 4. In L1 and L2, the frames are those evoked by Shalmaneser manually augmented in L2 with respect to the FE assignments. In L3, there are other frames manually evoked with their complete FE assignations. All miss-classifications of the word senses into wrong frames are rectified along with a complete FE assignment in L4. There is also a possibility of drawing the levels of semantic representation with respect to the parts-of-speech of the frame-evoking predicates. We conduct some experiments considering only verbal predicates as well.

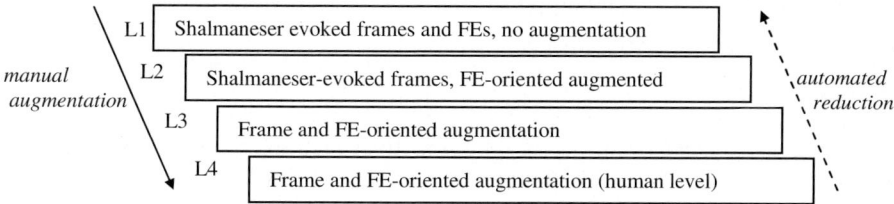

Fig. 4. Levels of shallow semantic parsing considering the baseline parser

The manual annotation task, after automated parsing by Shalmaneser, has been performed on the subset of the TREC 2004 factoid questions (143 questions and their top 10 passages) using SALTO (the SALSA annotation tool) [15]. The accuracy of the semantic parsing task at the different levels of annotation is shown in Table 2. The manual annotation has been performed using the FrameNet version 1.3 on the outputs of Shalmaneser trained on FrameNet 1.2. In this sense, the *italicized* numbers in Table 2 show the accuracies with respect to the FrameNet dataset 1.3 at the human level annotation and the regular measures are those relative to FrameNet 1.2.

Table 2. Average accuracy of semantic labeling task

Semantic parsing level	Top 10 passages		Question	
	Frame	FE	Frame	FE
Shalmaneser	41.765%	17.000%	59.207%	60.256%
	38.003%	*15.665%*	*57.109%*	*58.158%*
FE aug. on verb frames	17.480%	21.966%	36.713%	38.488%
	15.924%	*20.226%*	*34.615%*	*36.390%*
FE aug. on all Shalmaneser frames	41.765%	43.539%	59.207%	60.256%
	38.003%	*40.042%*	*57.109%*	*58.158%*
Human level aug.	100%	100%	100%	100%
	100%	*100%*	*100%*	*100%*

Equation 1 shows the formula for measuring the accuracies over the item (frame and FE) assignment at each level where n_{cai} is the number of items correctly assigned and N is the total number of items assigned at the human level annotation.

$$Acc_{item} = \frac{n_{cai}}{N} * 100 \tag{1}$$

The inter-annotator agreement over the frame and FE assignment tasks has also been measured. The overall agreement (for two annotators) on the frame assignment task is 0.684 (using the Kappa statistics) while in the FE assignment task the agreement ratio drops to ~23% by exact matches. We will make the automatically annotated data and their manually augmented version available in the near future. More details of the annotation task and the different method of measuring the inter-annotator agreement on frames and FEs can be found in [16].

3 Experiments

The different levels of semantic parsing and their contributions to the task of QA are experimented in this section.

3.1 Data

A subset of 143 TREC 2004 factoid questions for which the passage retrieval system retrieves the correct answers in the top 10 passages is considered as the experimental dataset; however, only 75 questions are regarded in the evaluations. The rest are not considered as, according to our error analysis after gold standard annotation, they come with the difficulties like answers being in non-predicate-argument structures, no frame evoking questions, and different answer and question frames. The answer resource is the AQUAINT collection[2].

3.2 Results and Discussion

The individual performance of each answer extraction model is shown along with the overall MRR of the system at each level in Table 3. The merging strategy to fuse the results of the two answer extraction models is based on the scores of the answers retrieved by each model. A single answer with the highest score from either model is reported. The answers in both models are scored based on their passage scores.

There are a few observations that can be made from the results in Table 3. First, the ENB performance has a decreasing trend as a result of FSB performing increasingly at higher levels of annotation. There are overlaps between the results of the two answer extraction models and since the answers from the FSB model obtain higher scores the FSB model dominates and the performance of ENB decreases. Second, the overall MRR of the system, being the sum of the MRRs of the models, after using Shalmaneser pure outputs is lower than that of the baseline system. This seems to be mainly due to the poor coverage of the semantic classes and roles and wrong answers obtained by the FSB model which damage the overall performance of the system. However, once the manual augmentation of the FEs is accomplished, an improvement in the performance of the answer extraction task from 0.400 to 0.440 is achieved. In addition, the performance of the FSB model rises from 0.000 to 0.227 showing how the FSB model can perform on more sophisticatedly annotated texts.

There is also a rise from 0.400 to 0.413 when filtering the frames to verb frames which are manually augmented with complete semantic roles. This supports the importance of semantic roles in QA even when other parts-of-speech predicates are not considered as frame-evoking target words. The fact that there is some improvement in the FSB and overall MRRs from the verb frames to all Shalmaneser frames augmented is indicative of the fact that other parts-of-speech predicates play an important role in QA as well. Having all frame evoking predicates in the texts

---

[2] http://www.ldc.upenn.edu/Catalog/docs/LDC2002T31/

annotated the FSB model performs higher again and the overall MRR reaches its highest measure, 0.507 for FrameNet version 1.2. The coverage shows its impact where this value rises to 0.520 with FrameNet 1.3.

Table 3. Results of the different QA runs on 75 questions – strict[3] evaluation

QA level	MRR		
	ENB	FSB	Total
Baseline (BL)	**0.400**	N/A	0.400
BL and Shalmaneser	0.347	0.000	0.347
BL and Shalmaneser – verb frames FE-augmented	0.253	0.160	0.413
BL and Shalmaneser – all frames FE-augmented	0.213	0.227	0.440
BL and Shalmaneser – human level augmentation – FN1.2	0.120	0.387	0.507
BL and Shalmaneser – human level augmentation – FN1.3	0.107	**0.413**	**0.520**

The failure for FSB in the cases where Shalmaneser pure outputs are used is due to three main reasons: i) semantic class assignment is incorrect with respect to the predicate sense, ii) semantic role assignment is partially or completely incorrect, despite the correct semantic class assignment, and iii) frame redundancy interferes where there is more than one single frame matching with that of the question.

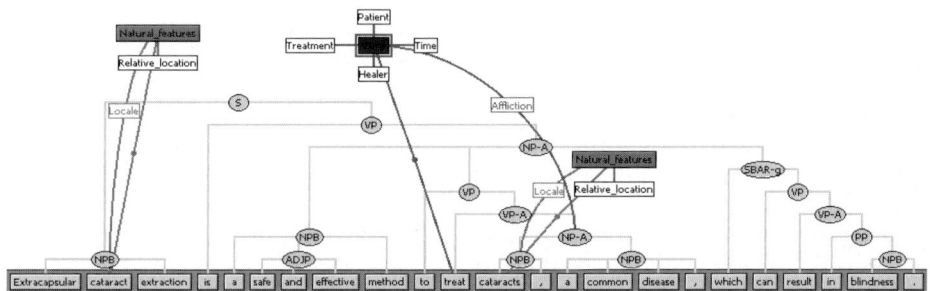

Fig. 5. Pure Shalmaneser output visualized in SALTO

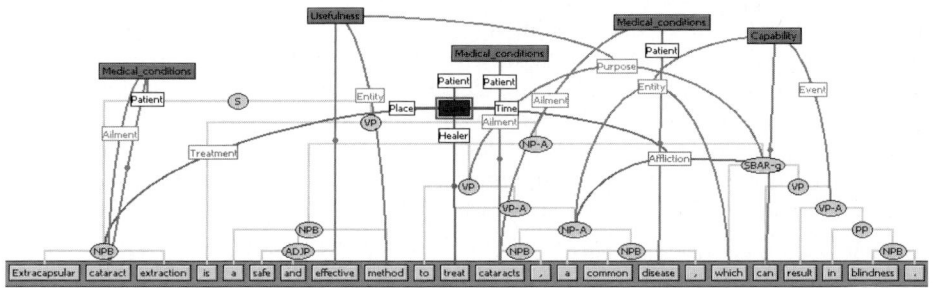

Fig. 6. Manually augmented Shalmaneser output visualized in SALTO

---

[3] The strict evaluation inherits from the TREC terminology. Lenient answers have the same trend and skipped due to space limitation.

When the semantic class assignment goes wrong, there is no chance for the system to identify the matching frame for alignment. Figure 5 and Figure 6 illustrate the problem with the task of semantic role assignment. In Figure 5 the pure output of Shalmaneser is shown with the main predicate "treat" evoking the correct frame "Cure" and incomplete FE assignment. The vacant FE "Treatment" in the question *"How are they treated?"* is not instantiated with the value of its corresponding FE in the answer-containing sentence annotated by Shalmaneser. However, this is alleviated after the manual augmentation of the output in Figure 6.

The third reason is a general barrier which interferes in all levels of semantic annotation. This is worse in the case of complete annotations with more frames. We are trying to overcome this problem with other strategies of answer candidate scoring.

4   Concluding Remarks

We have shown how the different levels of shallow semantic parsing can impact the performance of a QA system. In conjunction with the ENB model (the baseline model) the FSB model that extracts answers from annotated texts brings some improvement. The ratio of such improvement is proportional to the level of the shallow semantic parsing of the questions and their answer-bearing passages.

Our experimental results show that not only can a FSB model contribute to the answer extraction task, but also the two subtasks of semantic class and semantic role (frames and FEs) assignment may have different impacts on the same goal.

Our best-performing hybrid QA system was able to answer ~26% of the questions not answered by the best performing system in TREC 2004 in the strict method (~28% lenient) all of which correctly answered by FSB. This proves how effective the FSB model can be at the highest level of semantic parsing using a more sophisticated baseline QA system. However, it is still a direction of future study on how to best fuse the FSB model with any existing answer extraction model the way that the overall performance can reach to a maximal measure.

It is the goal of our future work to more effectively attack and overcome the problem of shallow semantic parsing. The subtask of semantic role assignment, in fact, is the more valuable part which requires further elaboration according to our experimental results. There are two reasons for this: i) the current state-of-the-art semantic parsers perform relatively higher in semantic class identification rather than semantic role assignment, and ii) slightly less improvement over the QA MRR has been achieved after frame-oriented augmentation of the baseline semantic parser's outputs. The second point can be observed from Table 3. Since the improvement achieved with frame-oriented augmentation is due to a complete FE assignment, in the presence of noises in the FE assignment of these frames, the improvement would have been lower which offers more emphasis on the task of semantic role labeling.

References

1. Erk, K., Pado, S.: Shalmaneser - A toolchain for shallow semantic parsing. In: LREC 2006 (2006)
2. Baker, C.F., Fillmore, C.J., Lowe, J.B.: The Berkeley FrameNet project. In: International Conference on Computational Linguistics (1998)

3. Fillmore, C.J.: Frame semantics and the nature of language. Annals of the New York Academy of Sciences: Conference on the Origin and Development of Language and Speech 280, 20–32 (1976)
4. Lowe, J.B., Baker, C.F., Fillmore, C.J.: A frame-semantic approach to semantic annotation. In: SIGLEX Workshop on Tagging Text with Lexical Semantics: Why, What, and How? (1997)
5. Erk, K.: Frame assignment as word sense disambiguation. In: 2006 IAENG International Workshop on Computer Science, IWCS 2006, Tilburg University, Tilburg, The Netherlands (2006)
6. Narayanan, S., Harabagiu, S.: Question answering based on semantic structures. In: International Conference on Computational Linguistics (COLING-2004), Switzerland (2004)
7. Ofoghi, B., Yearwood, J., Ghosh, R.: A hybrid question answering schema using encapsulated semantics in lexical resources. In: Sattar, A., Kang, B.-h. (eds.) AI 2006. LNCS (LNAI), vol. 4304, pp. 1276–1280. Springer, Heidelberg (2006)
8. Hickl, A., et al.: Question answering with LCC's CHAUCER at TREC 2006. In: 2006 Text Retrieval Conference (TREC 2006) (2006)
9. Sun, R.X., et al.: Using syntactic and semantic relation analysis in question answering. In: 2005 Text Retrieval Conference (TREC-2005) (2005)
10. Kaisser, M., Scheible, S., Webber, B.: Experiments at the University of Edinburgh for the TREC 2006 QA track. In: 2006 Text Retrieval Conference (TREC 2006) (2006)
11. Shen, D., Lapata, M.: Using semantic roles to improve question answering. In: 45th Annual Meeting of the Association for Computational Linguistics, Prague (2007)
12. Moldovan, D., et al.: The structure and performance of an open-domain question answering system. In: Conference of the Association for Computational Linguistics (ACL-2000), Hong Kong (2000)
13. Clarke, C., Cormack, G., Tudhope, E.: Relevance ranking for one to three term queries. In: 5th International Conference Recherche d'Information Assistee par Ordinateur, RIAO 1997 (1997)
14. Ofoghi, B., Yearwood, J., Ghosh, R.: A semantic approach to boost passage retrieval effectiveness for question answering. In: 29th Australian Computer Science Conference (2006)
15. Burchardt, A., et al.: SALTO – a versatile multi-level annotation tool. In: LREC 2006 (2006)
16. Ofoghi, B., Yearwood, J., Ma, L.: Two-step comprehensive open domain text annotation with frame semantics. In: Australasian Language Technology Workshop 2007, Melbourne, Australia (2007)

Improving Complex Interactive Question Answering with Wikipedia Anchor Text

Ian MacKinnon[1] and Olga Vechtomova[2]

[1] David R. Cheriton School of Computer Science
University of Waterloo
Waterloo, ON, Canada
imackinn@cs.uwaterloo.ca
[2] Department of Management Sciences
University of Waterloo
Waterloo, ON, Canada
ovechtom@engmail.uwaterloo.ca

Abstract. When the objective of an information retrieval task is to return a nugget rather than a document, query terms that exist in a document will often not be used in the most relevant information nugget in the document. In this paper, a new method of query expansion is proposed based on the Wikipedia link structure surrounding the most relevant articles selected automatically. Evaluated with the Nuggeteer automatic scoring software, an increase in the F-scores is found from the TREC Complex Interactive Question Answering task when integrating this expansion into an already high-performing baseline system.

1 Introduction

With the Complex Interactive Question Answering (CiQA) task introduced at TREC in 2006[1], the focus of evaluation is shifted from documents and facts to more elaborate nuggets. However, due to the concepts being sought having multiple terms to describe them, it becomes difficult to determine which sentences in the AQUAINT corpora of news articles contain the query terms being sought as they may be represented in the parent document by a variety of different phrases still making reference to the query term. For example, if the term "John McCain" was being sought, the phrase might appear in a document; however, the sentence which has the vital piece of information may simply contain "Senator McCain": an imperfect match.

In CiQA, templates are used with several bracketed items we call "facets" which are the basis for the information being sought. We can see from an example CiQA topic and answer key in Figure 1. A system must return text as a response which is then mapped to answer nuggets for scoring. Responses that correspond to 'vital' nuggets contribute to the score, 'okay' nuggets do not harm the score, and unassigned nuggets penalize the system for verbosity as a surrogate for precision[1].

Traditional query expansion of facets would introduce new terms which are related but do not necessarily mean the same as the original facet. This does not

C. Macdonald et al. (Eds.): ECIR 2008, LNCS 4956, pp. 438–445, 2008.

Qid 27: What evidence is there for transport of [drugs] from [Mexico] to [the U.S.]?

Topic	Number	Value	Nugget
27	1	vital	Mexico, Switzerland to cooperate on Salinas - Swiss seized over $114 million in bank accounts opened by Salinas
27	2	okay	Anti-drug police in Mexico confiscate 3.5 tons of marijuana
27	3	vital	Mexican heroin trafficking emerges - Mexican authorities discover a new organization smuggling heroin into the US
27	4	okay	Mexican navy seized 20 tons of cocaine off ships traveling Mexico's coast using technology and info supplied by American law enforcement
27	5	okay	Despite the often spectacular seizures and arrests, the bilateral structures to fight drugs put into effect by U.S. and Mexican governments... have been incapable of reducing the intensity of drug trafficking

Fig. 1. Templated query and answer key for a CiQA topic

always help the problem of query terms appearing in relevant documents but not within relevant sentences of the documents; it only introduces related terms which cannot be considered synonymous with the facet being retrieved.

Many of the CiQA facets are proper nouns and most thesauri, such as WordNet, do not contain entries for these. Thus, a new manner of finding synonyms must be found. In recent years, several new approaches have been proposed to use Wikipedia as a source of lexical information as it can be downloaded in its entirety and contains relatively high quality articles[2]. Wikipedia has previously been used in a lexical capacity to disambiguate named entities[3], explicitly compute semantic relatedness[4,5] and for word sense disambiguation[6].

As pointed out in previous work about creating an explicit semantic analysis engine based on Wikipedia[4], the anchor text which points to a Wikipedia article contains high quality terms which can be taken as synonyms for the articles which they link to. For example, the article "United States" will have frequent anchor texts such as "U.S.", "America", "American", "United States of America", or "USA".

While drawing potentially hundreds of articles becomes useful for semantic analysis, to find expansion terms we must first map facets to a small set of Wikipedia articles from which we can draw anchor texts to ascertain synonyms for the article title. Fortunately, by analyzing the whole Wikipedia corpus we can see the frequency of anchor text that links to articles.

We propose an algorithm to automatically select articles which best describe the facets of a CiQA topic in order to extract high quality phrases for expansion.

2   Wikipedia Article Selection for Facets

2.1   Automatic Article Selection Algorithm

We have devised a method of using the anchor text within Wikipedia links in order to resolve a small set of concepts which are represented in a candidate sentence.

Every article in Wikipedia represents a concept and all links from other articles to that article will have an anchor text associated with the link. We also know that there are Wikipedia guidelines for what the anchor text should be for a link, and that we can assume that, provided editors are following the rules, the anchor text of the link will be of high quality. As we can see from this excerpt from the Wikipedia manual of style[1]:

> "It is possible to link words that are not exactly the same as the linked article title, for example, [[English language—English]]. However, make sure that it is still clear what the link refers to without having to follow the link." -Wikipedia Manual of Style

The anchor texts which point to the article will contain other terms for the same concept which are necessary to get a better understanding of phrases that are used to describe the concept in the text. As we can see in Table 1, there are several different articles to the 'radio waves' anchor text of varying frequency.

Table 1. Frequency of links to articles that have "radio waves" as anchor text

Article Name	Anchor Text Frequency
radio waves	72
radio frequency	10
Electromagnetic radiation	3
radio	2
Radio Waves (album)	1

We define the algorithm to turn a facet into a list of concepts as follows:

1. Set window length to n.
2. For each possible position of window, check all anchor text in Wikipedia to see if the phrase or term is recognized. If it is, record the matching string and drop the words covered in the window from future consideration. See Fig 2.
3. Decrease the length of the window by one ($n = n - 1$). If the window length is 1, do not look up stopwords in term dictionary, simply ignore. Go to step 2 if window length is greater than 0.
4. For terms extracted from the query, look at the frequency of that term when linking to different articles. If an article has a majority of the links with that term as anchor text pointing to it, resolve that article to be the most relevant article for that multi-word unit. If no article has more than half the links with that anchor text pointing to it, drop the multi-word unit from consideration, as the term is ambiguous. However, if the frequency of anchor text linking to that article is less than 2, it is ignored.
5. If there are multiple articles resolved for the query, select whichever article has the highest number of incoming links from all other Wikipedia articles to be the most relevant Wikipedia article for the given facet. See Fig 3.

---

[1] http://en.wikipedia.org/wiki/Wikipedia:Manual_of_Style_%28links%29

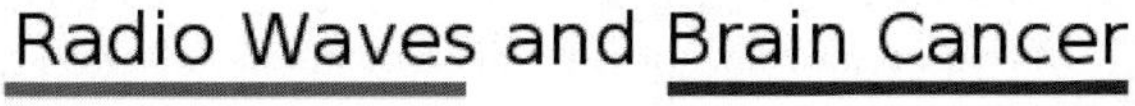

Radio Waves and Brain Cancer

Fig. 2. When a window recognizes a multi-word unit from the nugget, it saves it and drops the text from future consideration

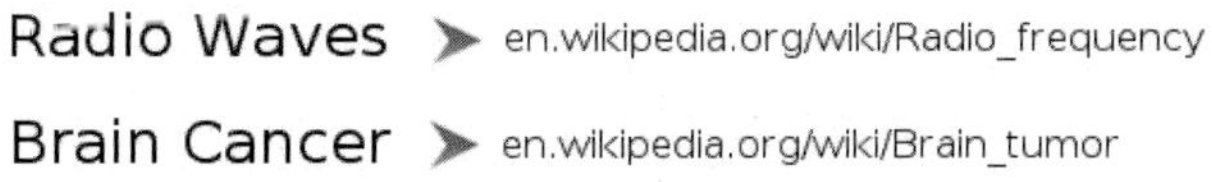

Radio Waves ➤ en.wikipedia.org/wiki/Radio_frequency

Brain Cancer ➤ en.wikipedia.org/wiki/Brain_tumor

Fig. 3. Multi-Word Units are resolved to whichever article has the most links with that anchor text

In our experiments, we initially set $n = 5$.

By running this algorithm on the CiQA 2006 and CiQA 2007 test topics, we get sets of articles for every facet in each topic. To compare these automatically retrieved articles with the consensus articles of 12 human assessors, we use Fleiss' Kappa. Looking at this agreement, we find there to be a 0.6206 agreement between the human consensus articles and the automatically retrieved articles for the CiQA 2006 topics, and an agreement of 0.6764 for the CiQA 2007 topics. Both of these coefficients would be considered "substantial agreement" using the informal interpretation given by Landis and Koch[7].

We see a greater degree of agreement among the CiQA 2007 data, possibly on account of the more time-relevant data in the AQUAINT-2 corpus for Wikipedia. AQUAINT has articles from 1998 to 2000; before Wikipedia was launched. AQUAINT-2 has articles from when Wikipedia was considerably more popular, meaning the coverage of named entities from the news articles is likely more complete.

2.2 Baseline System

We base our system on the one which yielded the highest F-scores for initial automatic runs[8] at the CiQA 2006 task at TREC. To gain an initial set of documents, the system parses out the initial topic to get the 2 or 3 facets from the test topics, performs a BM25 retrieval[2] using the facet words as query terms, and returns the top 50 documents from the AQUAINT newswire corpus.

Once a list of documents has been retrieved, every document is split into candidate sentences. Preserving the rankings provided to us by BM25, we keep the sentences in order in which their parent document occurred in the top 50 ranking. Afterwards, a score of 0,1,2, or 3 to a sentence depending on the number of facets which are represented in a candidate sentence. Each topic will have 2 or 3 facets containing a number of terms within them. For each facet, let us consider $\Gamma = \gamma_1...\gamma_n$ to be the set of non-stopword terms for a facet in a CiQA topic.

---

[2] Using default parameters k=1.2, b=0.75.

A score is assigned to a candidate sentence S, by iterating through all the γ_i in Γ, and determining if any of the non-stopword stems of the terms exist in the sentence. If at least one exists, a nugget is said to be represented in the facet. More formally:

$$score(S, \Gamma) = \begin{cases} 1 & \text{if at least one of } \gamma_i \in \Gamma \text{ exist in S} \\ 0 & \text{otherwise} \end{cases} \tag{1}$$

We get the total score for S by taking the sum of the $score(S, \Gamma)$ for each facet in the topic.

A sort is then performed on the list of sentences, but it is of great importance that the sort preserve the original ordering of the sentences with the same score. This allows for sentences which come from a document with a higher BM25 score to be ranked higher, given that they are likely more relevant to the test topic. The top $n = 30$ ranked sentences for each topic are output by the system.

3 Experiments

The only accurate way to judge a binary ("vital" or "okay") F-score for a CiQA run is to have human assessors assign system responses to answer key nuggets. However, this poses a problem for experimentation since the turnaround time for an assessment. For all experiments, the Nuggeteer system is used for determining the F-scores of the system responses as it has shown itself to have a highest correlation with human scores of all the automatic evaluation systems[9].

3.1 CiQA Runs

In order to test the ability of anchor text to improve ciQA retrieval, we must first introduce a method of using the articles we have selected for each facet to be integrated into the base CiQA system described earlier.

If, for a given facet Γ, we have corresponding Wikipedia articles which have anchor text linking to them, the set of anchor text phrases for that facet will be $A = \alpha_1\alpha_2...\alpha_n$, each α_i being an anchor text which links to one of the Wikipedia articles resolved for the facet, with a frequency across the Wikipedia corpus greater than 1. Ensuring that at least 2 articles link to the facet-corresponding one with the same anchor text will prevent potentially vandalized articles from introducing noise into the set of synonyms for the facet, A.

In the ideal situation, only one Wikipedia article is resolved for a facet, with not terms leftover from the facet. In this case, each α_i represents a high-quality phrase which multiple editors on Wikipedia have agreed is a reasonable referent for the concept being described in the linked article. Thus, we can use is it as a substitute for the facet being sought. However, we find that only 45 of the 72 facets, or 62.5%, of the CiQA 2006 facets fit this optimal case.

We modify the baseline system described earlier to incorporate the information from a facet's A set of anchor text in addition to the set of terms in the facet, Γ. A higher score is given to a candidate sentence, S, if it contains an anchor text term from A in it as opposed to simply a term from the facet. More formally:

$$score(S, \Gamma, A) = \begin{cases} 1.2 & \text{if at least one of } \alpha_i \in A \text{ exist in S} \\ 1 & \text{if at least one of } \gamma_i \in \Gamma \text{ exist in S, and no } \alpha_i \in A \text{ exist in S} \\ 0 & \text{otherwise} \end{cases}$$

$$(2)$$

The score of 1.2 is rather arbitrary. It just needed to be higher than 1, but low enough such that 2 matches from A would not be ranked higher than 3 from Γ. Experiments were conducted with various weighting techniques, but none garnered a significant change in scores.

Afterwards, sentences are sorted according to score as before. The only difference from the baseline system is the integration of the A terms from the anchor text. The remaining issue is what method is used to select the articles from Wikipedia for the given facet, for which we described an automatic method earlier.

To test this, we compare the baseline system F-score against the Wikipedia-enhanced system. The results of these runs can be seen in Table 2.

Table 2. F-Scores for CiQA runs using Nuggeteer

Run	F-score	Percent Improvement
2006 Baseline	0.3356	n/a
2007 Baseline	0.3388	n/a
2006 Wiki	0.3718	10.8%
2007 Wiki	0.3663	8.1%

From the table we can see a modest improvement in F-scores using the proposed Wikipedia method.

Looking at the individual results of the 30 2006 topics, we find that the automatic article selection improves F-scores in 8 of the topics, leaves 20 static (less than 2% change), and decreases 2.

Looking at the 2007 topics more closely, we see the most improved topics being "What evidence is there for transport of [automobiles] from [China] to [Russia]?" and "What effect does [glucosamine] have on [arthritis]?". The "automobiles" facet being expanded to include the term "car" being the probable cause for the former, and the expansion of "glucosamine" to include other marketed names for the drug for the later.

The most under-performing queries were "What evidence is there for transport of [illegal immigrants] from [Croatia] to [the European Union]?" and "What effect does [the Red Tide] have on [sea creatures]?". "The European Union" resolved to the political entity, thus the country names in the vital nugget were not contained within it. In this case, a "PART-OF" relation would need to be established. For example, "Italy" would be referenced in a vital nugget, so a system would have to recognize "Italy" as a potential substitute for "the European Union" since Italy is part of the European Union political entity; something that could be plausible by using category information. "Sea Creatures" resolved to the "Marine Biology" article, which was fairly general and caused query drift. This is on account of a small number of links pointing to that article with that anchor text.

4   Conclusions and Future Work

We proposed an algorithm to automatically select a small set of relevant Wikipedia articles. This method was found to have a substantial amount of agreement with the consensus of the human assessors.

Using Nuggeteer, we were able to show a modest improvement in F-scores for CiQA topics which used the Wikipedia anchor text method of query expansion.

It is likely that a few well selected articles were enhancing the retrieval, which we selected in both cases, while the poorly selected ones were noise that was not affecting the retrieval.

This line of research introduces several new directions involving Wikipedia, which has shown itself to be an up and coming source for lexical information. The first being the resolution of articles from a query. We showed that many previous approaches looked at the selection of a large array of articles for traditional latent semantic analysis. However, our approach is close to ones involving WordNet, in that a small set of lexical data is sought. When trying to resolve an article for a given phrase, there are many interesting questions, such as disambiguation of the potentially multiple articles with similar titles and whether a term is significant enough to warrant resolving to an article. We hope to improve our article resolution algorithm by incorporating a part-of-speech tagger and word sense disambiguation tools to more accurately select articles.

Further work could also be done to fine-tune the procedure for extracting synonyms for articles by looking at anchor text. The current method of only taking anchor text which labels a link to an article with a frequency higher than 1 was mostly done because a lack of CiQA datasets meant that there could be no effective training set. Once more sets become available, statistical models could be found to give the most appropriate synonyms based on the distribution of the anchor text.

In the future we hope to also begin looking at a connectionist model of Wikipedia articles, treating every link in the corpus as a semantic link between two concepts. Clearly, weights on the links would depend on the strength of the semantic bond between two concepts. Using this method it may be possible to retrieve a list of high-quality related terms which could also be used to aid in nugget retrieval. More importantly, it could be used to find intersections of related terms between two facets.

References

1. Kelly, D., Lin, J.: Overview of the TREC 2006 ciQA task. SIGIR Forum 41(1), 107–116 (2007)
2. Giles, J.: Internet encyclopaedias go head to head. Nature 438(7070), 900–901 (2005)

3. Bunescu, R., Pasca, M.: Using encyclopedic knowledge for named entity disambiguation. In: Proceedings of the 11th Conference of the European Chapter of the Association for Computational Linguistics (EACL 2006), Trento, Italy, April 2006, pp. 9–16 (2006)
4. Gabrilovich, E., Markovitch, S.: Computing semantic relatedness using wikipedia-based explicit semantic analysis. In: Proceedings of The Twentieth International Joint Conference for Artificial Intelligence, Hyderabad, India (2007)
5. Strube, M., Ponzetto, S.P.: Wikirelate! computing semantic relatedness using wikipedia. In: Proceedings of the Twenty-First National Conference on Artificial Intelligence, Boston, Mass, July 2006, pp. 1419–1424 (2006)
6. Mihalcea, R.: Using Wikipedia for automatic word sense disambiguation. In: Human Language Technologies 2007: The Conference of the North American Chapter of the Association for Computational Linguistics; Proceedings of the Main Conference, Rochester, New York, Association for Computational Linguistics, April 2007, pp. 196–203 (2007)
7. Landis, J.R., Koch, G.G.: The measurement of observer agreement for categorical data. Biometrics 33, 159–174 (1977)
8. Vechtomova, O., Karamuftuoglu, M.: Identifying relationships between entities in text for complex interactive question answering task. In: TREC (2006)
9. Marton, G., Radul, A.: Nuggeteer: Automatic nugget-based evaluation using descriptions and judgements. In: Proceedings of NAACL/HLT (2006)

A Cluster-Sensitive Graph Model for Query-Oriented Multi-document Summarization

Furu Wei[1,2], Wenjie Li[2], Qin Lu[2], and Yanxiang He[1]

[1] Department of Computer Science and Technology
Wuhan University, China
{frwei,yxhe}@whu.edu.cn
[2] Department of Computing
The Hong Kong Polytechnic University, Hong Kong
{csfwei,cswjli,csluqin}@comp.polyu.edu.hk

Abstract. In this paper, we develop a novel cluster-sensitive graph model for query-oriented multi-document summarization. Upon it, an iterative algorithm, namely QoCsR, is built. As there is existence of natural clusters in the graph in the case that a document comprises a collection of sentences, we suggest distinguishing intra- and inter-document sentence relations in order to take into consideration the influence of cluster (i.e. document) global information on local sentence evaluation. In our model, five kinds of relations are involved among the three objects, i.e. document, sentence and query. Three of them are new and normally ignored in previous graph-based models. All these relations are then appropriately formulated in the QoCsR algorithm though in different ways. ROUGE evaluations shows that QoCsR can outperform the best DUC 2005 participating systems.

Keywords: Query-Oriented Summarization, Multi-document Summarization, Graph Model and Ranking Algorithm.

1 Introduction

Graph models have been drawn considerable attention from the document summarization community in the past few years [3, 4, 5, 6, 8, 9, 10, 12, 13, 14]. Normally, they model the documents as a graph constructed by taking the text unit such as the term [5, 6, 12, 14], or the sentence [3, 4, 8, 9] as the vertex and the similarity or the association between the text units as the edge. The importance of a vertex in a graph can then be estimated by graph-based ranking algorithms, which normally take into account global information recursively computed from the entire graph, rather than only rely on local information. Accordingly, the sentences in the documents are evaluated according to the computed vertex importance and the most salient ones are extracted into the summary. The most popular graph-based ranking algorithm applied in document summarization is Google's PageRank [1]. It has also been extended to the topic-sensitive version [10, 13] to accommodate the new challenge of query-oriented summarization promoted by DUC evaluations [2], which aims to produce a summary from a set of relevant documents to a given topic expressed by a short description of user's information need.

C. Macdonald et al. (Eds.): ECIR 2008, LNCS 4956, pp. 446–453, 2008.

When dealing with the task of query-oriented multi-document summarization, a graph model can actually characterize five different kinds of information and utilize them to build the corresponding ranking algorithm. They are the relevance of a sentence or a document to the query, the similarity between the sentences in the same or different documents and the similarity (or diversity) among the documents. Notice that there exist natural clusters in the graph reflecting the fact that a collection of sentences constitute a document. Therefore, there is a need to make a difference between the edges of inter-document and intra-document sentences. This thereby requires a ranking algorithm consider them discriminatingly. Also because of the existence of the clusters, when a sentence is ranked, the influence from the document which contains it could be integrated so that the ranking algorithm could be enhanced by imposing document global information to local sentence evaluation. Unfortunately, these considerations are ignored by almost all the previous graph-based ranking algorithms, which equally look on all the edges and the vertices. These considerations motivate us to study how to make full use of the information provided in a graph in the context of query-oriented multi-document summarization.

The remainder of this paper is organized as follows. Section 2 introduces the proposed cluster-sensitive graph model and the corresponding iterative ranking algorithm. Section 3 then present experiments and evaluations. Finally, Section 4 concludes the paper.

2 Cluster-Sensitive Graph Model for Query-Oriented Multi-document Summarization

2.1 Model Description

Three objects are involved in our query-oriented multi-document summarization graph model. They are document (d), sentence (s) and query (q). There exist five kinds of meaningful relations (r) among them, including s-d (inclusion), s-s (similarity), d-d (similarity), s-q (relevance) and d-q (relevance). All of them are concerned in the model though in different ways.

Similar to the previous work (e.g. Otterbacher et al in [10]), the document set $D=\{ d_1, d_2, ..., d_n \}$ is represented as a simple weighted undirected graph G (called similarity graph) by taking sentences in D as vertices and adding edges to connect vertices if the sentences concerned are similar enough, i.e. the s-s similarity is used to measure the strength of the edge. Meanwhile, the s-q relevance is used to assess the importance of each individual sentence vertex.

Different from the prior models, the other three relations, which have been ignored in the literature, are also implied in our model by introducing the concept of cluster. A cluster here is defined as a document containing a collection of sentences. It realizes the s-d inclusion relation and, most important, can further differentiate two different types of s-s edges, i.e. the one within the cluster (called intra-document s-s) and one spanning over two clusters (called inter- document s-s). Then, the d-q relevance and the d-d similarity can be represented by the edges from cluster to query or from cluster to cluster, as illustrated in Figure 1.

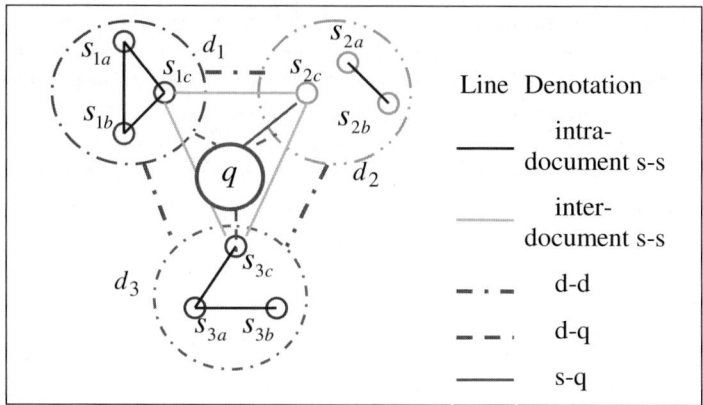

Fig. 1. Illustration of similarity graph

Notice that they are not used directly to evaluate the vertices and the edges in the graph, but to adjust the weight or the strength of them. These have been motivated by the observed evidences that the sentence in a more relevant document should be ranked higher and two inter-document sentences are more close to each other if the two documents are more similar. The following figure illustrates the objects d, s and q together with the relations among them.

2.2 Graph Modeling

In our graph model, the document set D is represented as the similarity graph $G = \left(S, D, E^S, E^D, \alpha, \beta, \phi, \varphi\right)$, where S and D are sentence and document vertex sets, $E^S \subseteq S \times S$ and $E^D \subseteq D \times D$ are sentence and document edge sets. $\alpha : S \to \Re_+^*$ and $\beta : D \to \Re_+^*$ are functions labeling sentence and document vertices, while $\phi : E^S \to \Re_+^*$ and $\varphi : E^D \to \Re_+^*$ are functions labeling sentence and document edges. A document in G corresponds to a cluster.

The sentence vertex function is calculated by the relevance $rel(s_i \mid q)$ of s_i with respect to the query q, which is defined as:

$$\alpha(s_i) = rel(s_i \mid q) = \frac{sim(s_i, q)}{\sum_{s_k \in D} sim(s_k, q)} \tag{1}$$

where, $sim(s_i, q) = \frac{\vec{s}_i \bullet \vec{q}}{\|\vec{s}_i\| * \|\vec{q}\|}$.

And, the sentence edge function is formulated as the similarity $sim(s_i, s_j)$ between the two sentences s_i and s_j, which is defined as:

$$\phi(s_i, s_j) = sim(s_i, s_j) = \frac{\vec{s}_i \bullet \vec{s}_j}{\|\vec{s}_i\| * \|\vec{s}_j\|} (i \neq j) \tag{2}$$

A sentence edge is inserted to link a pair of sentences only when the cosine similarity between them exceeds a given threshold. The edges linking the sentences from the same document are referred to as the intra-document edges, while the edges linking the sentences from the different documents are referred to as the inter-document edges. They are assigned different weights by the ranking algorithm introduced in Section 3.3 according to how similar the documents they reside in are.

Similarly, the document vertex d_i and edge (d_i, d_j) functions are defined as the relevance of d_i to q and the normalized similarity between the two documents d_i and d_j, i.e.

$$\beta(d_i) = rel(d_i \mid q) = \frac{sim(d_i, q)}{\sum_{d_k \in D} sim(d_k, q)} \tag{3}$$

where $rel(d_i, q) = \dfrac{\vec{d_i} \bullet \vec{q}}{\|\vec{d_i}\| * \|\vec{q}\|}$, and

$$\varphi(d_i, d_j) = sim\_norm(d_i, d_j) = \frac{sim(d_i, d_j)}{\sum_{d_k \in D \cap k \neq i} sim(d_i, d_k)} (i \neq j) \tag{4}$$

where $sim(d_i, d_j) = \dfrac{\vec{d_i} \bullet \vec{d_j}}{\|\vec{d_i}\| * \|\vec{d_j}\|}$. These two document- level functions are used as weighing functions for the sentence-level vertices and edges during sentence ranking.

2.3 Graph-Based Ranking Algorithm

Based on the above enhanced similarity graph, we develop a new iterative graph-based sentence ranking algorithm extended from query-oriented PageRank proposed in [10], which however calculates the graph only in sentence-level but neglects the differentiation of the documents and the influence from them. In the remaining parts of this subsection, we will first introduce the query-sensitive ranking algorithm and then explain our cluster-sensitive extension.

2.3.1 Query-Oriented Ranking (QoR) Algorithm

Let $R(s_i)$ denote the rank of the sentence s_i, we have the following iterative ranking algorithm:

$$R(s_i) = (1-d)rel(s_i \mid q) + d * \sum_{s_j \in N(s_i)} \left(R(s_j) * \frac{sim(s_i, s_j)}{\sum_{s_k \in N(s_j)} sim(s_j, s_k)} \right) \tag{5}$$

where $N(s)$ denotes the set of neighboring sentence vertices on s and d is the PageRank damping factor.

As we can see in QoR algorithm, the edges are processed without considering the information carried by the documents that the edges connect. Moreover, the importance of the sentences from different documents of different relevance with respect to the query is not differentiated. It will be shown in Section 3 later that these useful aspects can be effectively contributed to the improved performance.

2.3.2 Query-Oriented Cluster-Sensitive Ranking (QoCsR) Algorithm

The idea of this novel algorithm is inspired by the work in [11], where a weighted inter-cluster edge ranking (WICER) was proposed for clustered graphs. The major contributions of this work are to weight edges based on whether it is an inter-cluster or intra-cluster edges and to weight the vertex based on the number of clusters it connects in the case of a graph containing natural clusters. But, WICER is query-independent. We borrow the spirit of it and integrate it into our model so that both the query and the document (i.e. the cluster of sentences) perspectives can be involved in sentence ranking process. The new ranking algorithm is defined as follows:

$$R(s_i) = (1-d)rel(s_i \mid q) + d * \left(1 + \frac{|D(s_i)|}{|D|}\right) *$$

$$\left(\sum_{d_j \in D(s_i)} \left[(1 + rel(d_j \mid q)) * \sum_{s_j \in N(s_i) \cap (s_j \in S(d_j))} \left(R(s_j) * \frac{sim(s_i, s_j)}{\sum_{s_k \in N(s_i)} sim(s_j, s_k)}\right) * (1 + sim\_norm(d_i, d_j))\right]\right)$$

$$(6)$$

where, $|D|$ is the number of the documents to be summarized in the document set D. $D(s_i)$ is the set of documents that have connections to s_i. $|D(s_i)|$ is the size of $D(s_i)$. $S(d_j)$ represents the set of sentences in the document d_j.

Intuitively, the edges between the sentences implement and reflect the concept of recommendation among them. We believe that the sentences with more inter-document connections should have higher ranks. This can be understood as that the recommendation from more varying communities is supposed to be more trustful. This assumption is formulated by the fraction of $|D(s_i)|/|D|$ in formula (6).

Furthermore, we hold the assumption that a sentence from the document with higher importance should be ranked higher. This can be explained as that a recommendation from a reputable person should be more important. Accordingly, the relevance of the document to the query is taken as a weight and adopted as the product in the algorithm. Finally, the different sentences edges are differentiated corresponding to the fact that the recommendations exist in two familiar communities are more credible. So the sentence edges are then weighted by the similarity between the two documents they connect in formula (6). All these assumptions will be validated by a series of experiments conducted in the following sections.

3 Experimental Studies

The experiments are set up on DUC 2005 50 document sets. Each set of documents is given a query description which represents a user information need. All the documents and the queries are pre-processed by sentence segmentation and word splitting. Words are stemmed by Porter Stemmer[1] and stop-words are removed[2].

According to task definitions, system-generated summaries are strictly limited to 250 English words in length. After sentence ranking, we simply select the highest ranked sentences from the original documents into the summary until the word limitation is reached. Duplicate sentences are prohibited in the generated summary. But more complex post-processing, such as sentence compression, is not performed in this study. Since DUC 2005 officially adopted ROUGE toolkit[3] [7] as the automatic evaluation method. Therefore, like other researchers, it is taken as the evaluation means in this work.

3.1 Comparison of Component Contributions

In our model, three new components are introduced for the first time into the previous graph-based methods and ranking algorithms. They are the relevance of the document with respect to the query ("A"), the different weigh treatment for the edges combining different documents ("B"), and the fraction of the number of the documents from which there exists an edge to the sentence ("C"). Notice that, it's the QoR algorithm [10] when A, B and C are not considered (Θ), while it's the QoCsR algorithm when A, B and C are all taken into account ($A \cup B \cup C$).

The aim of the first set of experiments is to examine the contributions of these three new components (i.e. A, B and C) and their combinations. Table 1 below shows the results of the average recall scores of ROUGE-1, ROUGE-2 and ROUGE-SU4.

Table 1. Evaluations of components and combinations

	ROUGE-1	ROUGE-2	ROUGE-SU4
Θ / QoR	0.37228	0.07257	0.13064
A	0.37303	0.07338	0.13068
B	0.37309	0.07354	0.13073
C	0.37447	0.07493	0.13308
$A \cup B$	0.37298	0.07363	0.13086
$B \cup C$	0.37507	0.07545	0.13294
$A \cup C$	0.37501	0.07545	0.13281
$A \cup B \cup C$ / QoCsR	0.37545	0.07605	0.13339

In this set of experiments, sentences, queries, and documents are all represented as vectors of words, and the similarity or relevance is calculated by the cosine similarity based on the word co-occurrence. The threshold used as the criteria for determining when an edge is inserted is set to 0.25. Moreover, the damping factor is set to 0.75.

---

[1] http://www.tartarus.org/~martin/PorterStemmer.
[2] A list of 199 words is used to filter stop words.
[3] ROUGE 1.5.5 is used, and the ROUGE parameters are "-n 2 -x -m -2 4 -u -c 95 -r 1000 -f A -p 0.5 -t 0", according to the DUC task definition.

As we can see in Table 1, our implementation of the QoR algorithm has achieved very competing performance. As a matter of fact, the ROUGEs are significantly lower in traditional PageRank implementation, where the weights of words are measured by the classical $tf \cdot idf$ method from IR community. In our preliminary experiments, the ROUGE-2 score is only 0.06465 with this approach. We argue that it's not reasonable to adopt the inverse document frequency (IDF), because here we are working on sentence-level retrieval, other than document-level retrieval as it is defined in IR.

As seen, there are still improvements over our basic approach when the three new components are added step by step. The best result is achieved when all are considered. These results strongly validate our extension to the cluster-sensitive query-oriented graph model and the corresponding ranking algorithm.

3.2 Comparison with DUC Systems

Table 2 shows the comparison of our model with DUC 2005 participating systems, where S15, S17 and S10 are the top three performing systems. It is happy to see that the proposed model outperforms significantly over the best systems in DUC 2005, i.e. 0.43% on ROUGE-2, 4.89% on ROUGE-2 and 1.34% on ROUGE-SU4 over S15.

Table 2. Comparison with DUC top-3 systems

	ROUGE-1	ROUGE-2	ROUGE-SU4
QoCsR	0.37545	0.07605	0.13339
S15	0.37383	0.07251	0.13163
S17	0.36901	0.07174	0.12972
S10	0.36640	0.07089	0.12649
NIST Baseline	0.30217	0.04947	0.09788

4   Conclusion

In this paper, a cluster-sensitive graph model and the corresponding iterative algorithm, namely QoCsR, are proposed for query-oriented multi-document summarization. The main contributions of our work are to introduce the concept of cluster into the graph model, differentiate intra- and inter-document sentence relations and consider the influence of entire document context on individual sentence evaluation. ROUGE evaluations on DUC 2005 dataset shows that QoCsR can improve the results by integrating these new ideas into the previous graph-based models. In addition, the implemented QoCsR outperforms the best participating systems in DUC 2005.

Acknowledgments

The work described in this paper was supported by a grant from the Research Grants Council of the Hong Kong Special Administrative Region, China (Project No. PolyU5211/05E), a grant from National Natural Science Foundation of China (Project

No. F020303), and an internal grant from the Hong Kong Polytechnic University (Project No. A-PA6L).

References

1. Brin, S., Page, L.: The Anatomy of a Large-scale Hypertextual Web Search Engine. Computer Networks and ISDN Systems 30(1-7), 107–117 (1998)
2. DUC, http://www-nlpir.nist.gov/projects/duc/pubs.html
3. Erkan, G., Radev, D.R.: LexPageRank: Prestige in Multi-Document Text Summarization. In: Proceedings of EMNLP, pp. 365–371 (2004)
4. Erkan, G., Radev, D.R.: LexRank: Graph-based Centrality as Salience in Text Summarization. Journal of Artificial Intelligence Research 22, 457–479 (2004)
5. Leskovec, J., Grobelnik, M., Milic-Frayling, N.: Learning Sub-structures of Document Semantic Graphs for Document Summarization. In: Proceedings of LinkKDD Workshop, pp. 133–138 (2004)
6. Li, W., Wu, M., Lu, Q., Xu, W., Yuan, C.: Extractive Summarization using Intra- and Inter-Event Relevance. In: Proceedings of ACL/COLING, pp. 369–376 (2006)
7. Lin, C.-Y., Hovy, E.: Automatic Evaluation of Summaries Using N-gram Co-occurrence Statistics. In: Proceedings of HLT-NAACL, pp. 71–78 (2003)
8. Mihalcea, R., Tarau, P.: TextRank – Bringing Order into Text. In: Proceedings of EMNLP, pp. 404–411 (2004)
9. Mihalcea, R.: Graph-based Ranking Algorithms for Sentence Extraction, Applied to Text Summarization. In: Proceedings of ACL (Companion Volume) (2004)
10. OtterBacher, J., Erkan, G., Radev, D.R.: Using Random Walks for Question-focused Sentence Retrieval. In: Proceedings of HLT/EMNLP, pp. 915–922 (2005)
11. Padmanabhan, D., Desikan, P., Srivastava, J., Riaz, K.: WICER: A Weighted Inter-Cluster Edge Ranking for Clustered Graphs. In: Proceedings of 2005 IEEE/WIC/ACM International Conference on Web Intelligence, pp. 522–528 (2005)
12. Vanderwende, L., Banko, M., Menezes, A.: Event-Centric Summary Generation. In: Working Notes of DUC 2004 (2004)
13. Wan, X., Yang, J., Xiao, J.: Using Cross-Document Random Walks for Topic-Focused Multi-Document Summarization. In: Proceedings of the 2006 IEEE/WIC/ACM International Conference on Web Intelligence, pp. 1012–1018 (2006)
14. Yoshioka, M., Haraguchi, M.: Multiple News Articles Summarization based on Event Reference Information. In: Working Notes of NTCIR-4 (2004)

Evaluating Text Representations for Retrieval of the Best Group of Documents

Xiaoyong Liu and W. Bruce Croft

CIIR, Computer Science Department, 140 Governors Drive,
University of Massachusetts, Amherst, MA 01003, USA
{xliu,croft}@cs.umass.edu

Abstract. Cluster retrieval assumes that the probability of relevance of a document should depend on the relevance of other similar documents to the same query. The goal is to find the best *group* of documents. Many studies have examined the effectiveness of this approach, by employing different retrieval methods or clustering algorithms, but few have investigated text representations. This paper revisits the problem of retrieving the best group of documents, from the language-modeling perspective. We analyze the advantages and disadvantages of a range of representation techniques, derive features that characterize the good document groups, and experiment with a new probabilistic representation as a first step toward incorporating these features. Empirical evaluation demonstrates that the relationship between documents can be leveraged in retrieval when a good representation technique is available, and that retrieving the best group of documents can be more effective than retrieving individual documents.

Keywords: Text Representation, Document Retrieval, Cluster Retrieval, Cluster Representation, Representation Techniques.

1 Introduction

The standard approach to document retrieval has been based on the Probability Ranking Principle [13]. It assumes that the relevance of documents could be assessed independently. The fact that a document is relevant does not contribute to predicting the relevance of a closely-related document. Cluster retrieval, on the other hand, assumes that the probability of relevance of a document should depend on the relevance of other similar documents to the same query [17]. Document groups are usually formed by utilizing some clustering algorithms, and the system's goal is to find the best *group* of documents [4]. Jardine and van Rijsbergen, and others [4, 3, 15] studied the performance of the ideal retrieval strategy that infallibly finds the best group (they call it an "optimal" cluster), and showed that effectiveness would be far better than a search based on individual documents.

Many studies have examined the effectiveness of cluster retrieval, by employing different retrieval methods or clustering algorithms [1, 2, 4, 7, 15, 16, 18, 20]. The findings have been inconclusive as to whether a real retrieval strategy is able to retrieve the good document groups in the top ranks. Except for precision-oriented

C. Macdonald et al. (Eds.): ECIR 2008, LNCS 4956, pp. 454–462, 2008.
© Springer-Verlag Berlin Heidelberg 2008

searches on very small data sets [1, 4], retrieving individual documents is found to be generally more effective [2, 18, 20, 8]. However, most studies represented document groups either by concatenating the documents within each group [1, 6, 8] or a centroid vector [18], and only a couple of studies [7, 10] have compared different representations. The number of representations examined is small. There has been a resurgence of research in cluster-based retrieval in the past few years [8, 6, 14]. The general approach is to use clusters as a form of document smoothing. The IR system's goal is still directly ranking individual documents, not clusters. The issue of how to identify good document groups remains unaddressed. In this paper, we revisit the problem of retrieving the best group of documents, from the language-modeling perspective. We aim to provide an extensive evaluation of existing and new representation techniques. We argue that whether good document groups could be successfully identified by an IR system largely depends on how they are represented.

In this work, document groups are generated by a clustering algorithm. It is possible to have other types of document groups (e.g. user-generated discussion groups) in other applications. For simplicity of discussion and to avoid possible confusion in this paper, we use "cluster" and "document group" interchangeably. We distinguish between cluster retrieval and cluster smoothing. Cluster retrieval directly ranks groups of documents (clusters) whereas cluster smoothing ranks documents but uses clusters to smooth the document probabilities. We will use "good cluster" instead of "optimal cluster" or "the best groups of documents" in our discussions. An optimal cluster is one that gives the best precision out of all clusters [4] and a good cluster is a relaxed definition of optimal cluster. It refers to any cluster that gives better precision than document retrieval with the same cutoff from the top of the result list.

2   Cluster Representations

To use the language modeling (LM) approach for retrieving clusters, we first need to derive language models from cluster representations and then apply retrieval models. Let's take the query likelihood (QL) retrieval model for example. Clusters are ranked based on their likelihood of generating the query, i.e. $P(Q|Cluster)$. It can be estimated by equation (1) where Q is the query, q_i is the ith term in the query, and $P(q_i|Cluster)$ is the cluster language model (computed using equation (2)). $P_{ML}(w|Cluster)$ and $P_{ML}(w|Coll)$ are the maximum likelihood estimates of word w in the document and the collection, $tf(w, Cluster)$ and $tf(w, Coll)$ are the term frequencies of w in the cluster and the collection, V is the vocabulary, and λ is a general symbol for smoothing which takes different forms when different smoothing methods are used [8].

1. Concatenating documents. The standard approach to representing clusters is to treat them as if they were big documents formed by concatenating their member documents. Thus, $tf(w, Cluster)$ is computed by equation (3) where $Cluster = \{D_1, ..., D_k\}$ and k is the number of documents in a cluster. Clusters are ranked by equation (1) with components estimated from equations (2) and (3).

This representation, while being simple and intuitive, may have a number of problems. For example, if cluster A has a document that is very long and has many occurrences of the query terms while other member documents are short with only few query terms appearing, then simply concatenating these documents would result in a representation that is largely biased by one particular document [9]. In contrast, a cluster B has more relevant documents but do not have as many occurrences of the query terms when combined. Cluster A will be ranked higher because of the probability estimates. This is what we want to avoid because the quality of clusters is usually judged by the total number of relevant documents they contain rather than how good one of the documents is [15]. Clusters with more relevant documents are considered better. The problem with this representation is that the differences of query term frequencies in documents with a higher QL mask the differences in term frequencies in the documents with a lower QL. A lesson learned from this is that a good representation should offset the bias toward documents with a higher QL, and one way to achieve this is to put more emphasis on documents with a lower QL.

2. *Centroid vector.* Clusters can also be represented by a centroid vector, or the document that is the most similar to the actual centroid, as in e.g. [7]. The representation can be formulated as equation (4). Clusters are ranked by equation (1)

$$P(Q \mid Cluster) = \prod_{i=1}^{m} P(q_i \mid Cluster) \tag{1}$$

$$
\begin{aligned}
P(w \mid Cluster) &= \lambda P_{ML}(w \mid Cluster) + (1-\lambda) P_{ML}(w \mid Coll) \\
&= \lambda \frac{tf(w, Cluster)}{\sum_{w' \in cluster} tf(w', Cluster)} + (1-\lambda) \frac{tf(w, Coll)}{\sum_{w' \in V} tf(w', Coll)}
\end{aligned}
\tag{2}
$$

$$tf(w, Cluster) = \sum_{i=1}^{k} tf(w, D_i) \tag{3}$$

$$tf(w, Cluster) = \frac{\sum_{i=1}^{k} tf(w, D_i)}{k} \tag{4}$$

$$
\begin{aligned}
P(Q \mid Cluster) &= \max_{D_i \in Cluster} P(Q \mid D_i) \\
&= \max_{D_i \in Cluster} [\lambda P_{ML}(w \mid D_i) + (1-\lambda) P_{ML}(w \mid Coll)]
\end{aligned}
\tag{5}
$$

$$
\begin{aligned}
P(Q \mid Cluster) &= \min_{D_i \in Cluster} P(Q \mid D_i) \\
&= \min_{D_i \in Cluster} [\lambda P_{ML}(w \mid D_i) + (1-\lambda) P_{ML}(w \mid Coll)]
\end{aligned}
\tag{6}
$$

$$tf(w, Cluster) = \sum_{i=1}^{k} (\alpha_i * tf(w, D_i)) \qquad \text{where} \quad \sum_{i=1}^{k} \alpha_i = 1 \tag{7}$$

$$P(w \mid Cluster) = \sum_{i=1}^{k} [\beta_i * (\lambda P_{ML}(w \mid D_i) + (1-\lambda) P_{ML}(w \mid Coll))] \quad \text{where} \quad \sum_{i=1}^{k} \beta_i = 1 \tag{8}$$

$$P(w \mid Cluster) = \left(\prod_{0 \le i \le k} P(w \mid D_i) \right)^{\frac{1}{k}} \tag{9}$$

Fig. 1. Lanugage model formulations for different representations

with components estimated from equations (2) and (4). Similar to concatenating documents, this method may also suffer from bias introduced by some member documents. It is possible that each member document only contributes largely to the estimate associated with one query term but different document contributes to different terms. So even if the individual documents are not relevant, the centroid vector may still look good.

3. Best document. [7] used the highest ranked document (e.g. by QL model in document retrieval) in a cluster as the representative. The hypothesis is that if this document is non-relevant then the rest of the cluster is very likely non-relevant. Clusters are ranked according to equation (5). The problem with this approach is not difficult to see with an example. Suppose we have two clusters, one with five relevant documents and the other with one relevant and four non-relevant documents. If the relevant document in the second cluster has a better QL score than any of those in the first cluster, then the retrieval model will rank the second cluster higher. But in reality, the first cluster is better.

4. Worst document. The lowest ranked document in a cluster was also used as the cluster representative in [7]. The hypothesis is that if that document is relevant then it is very likely that the rest of the cluster is also relevant. Clusters are ranked by equation (6). Again, we illustrate the problem with an example. Suppose we have two clusters, one with five non-relevant documents and the other with four relevant and one non-relevant document. If the non-relevant document in the second cluster has a lower QL than any of the non-relevant documents in the first cluster, the retrieval model will rank the first cluster higher, but in fact the second cluster is better.

5. TF mixture. [10] proposes a weighted mixture of term frequencies from member documents for representation, i.e. equation (7), where α is a weighting parameter between 0 and 1. Clusters are ranked by equation (1) with components estimated from equations (2) and (7). α in equations (7) is estimated by the first-stage retrieval log QL score of each document divided by the sum of log QL scores of all member documents in a cluster. Note that the log QL scores are negative. Setting α this way penalizes clusters with documents that match the query poorly.

   The advantage of this approach lies in that it explicitly considers the contribution of individual documents to the cluster model. The disadvantage is that the α weight is difficult to determine. The current way of setting the weight may not be optimal as the performance of this representation does not vary much from the centroid vector representation discussed earlier (see section 4). We have experimented with several other ways of determining the weight but have not found a setting that will perform better than document retrieval.

6. DM mixture. The second method proposed by [10] is to build language models for individual member documents and the cluster language model is a weighted mixture of these member document models, i.e. equation (8). Again, λ is a general symbol for smoothing, and β is a weighting parameter between 0 and 1. β is estimated in the same way as α for the TF mixture method. Clusters are ranked by equation (1) with components estimated from equation (8).

Similar to TF mixture, this method has the advantage of explicitly modeling contributions from member documents. But again, it suffers from the difficulty of setting the β weight. Empirically, using the current way of setting the weight, this representation performs slightly better than TF mixture (see section 4).

7. Geometric mean. As we can see from previous representations, especially concatenating documents and centroid vector, the problem with summing up or averaging the query term frequencies in member documents is that differences in term frequencies in documents with higher QL mask the differences in term frequencies in documents with lower QL. We analyzed the ideal and real results of cluster retrieval using the QL model and the representation of concatenating documents in [9]. We found that, despite that there are plenty of good clusters per query, those clusters are typically not retrieved in the top ranks. We further identified the following features that characterize good clusters: a) a cluster model with good query likelihood, b) member document models with good query likelihood, and c) low variability in document model estimates. The existing representations don't account for these features and thus often fail to assign top ranks to good clusters.

These observations suggest a non-linear rescaling of the individual documents' language model estimates before averaging over the cluster as a way of emphasizing the documents with low QL. We experimented with a new representation that is based on the geometric mean of document model estimates. It is formulated as equation (9). We first derive the member document models P(w|D) and compute their geometric mean. Clusters are ranked by equation (1) combined with (9). The geometric mean is equivalent to taking the log of individual documents' estimates, computing the arithmetic mean of the logs, and exponentiating back for the final geometric mean score. This representation has the desired effect of emphasizing estimates close to 0 (documents with low QL) while minimizing differences between larger estimates. There is no need for additional parameter tuning other than the smoothing parameters associated with the document models. Theoretically, the geometric mean estimates need to be renormalized so that they still qualify as probability estimates. We found in our experiments, however, that the normalization significantly increases the computer processing time while being less effective in ranking clusters than the un-normalized method. We evaluate this representation and present the results using the un-normalized geometric mean in section 4. Geometric mean has been used in the geometric mean average precision measure introduced in TREC 2004 Robust track to account for a similar phenomenon observed with evaluating topic sets that contain poorly performed topics [19].

3   Experimental Setup

The data sets used in the experiments and analysis come from the TREC collections: Wall Street Journal (WSJ) 1987-92 with topics 51-100, Associated Press newswire (AP) 1988-90 with topics 101-150, TREC disks 1 & 2 (TREC12) with topics 151-200, and TREC disks 4 & 5 (TREC45) with topics 301-400. The queries are taken from the "title" field of TREC topics. The query sets are determined such that different collections do not share the same queries. Both queries and documents are stemmed with K-stem [5], and stopwords are removed based on the standard INQUERY list of 418 words. The WSJ data set is used as the training collection if parameter tuning is needed.

We use query-specific clustering in this work. Document retrieval using the query likelihood retrieval model [12, 11] is first performed with Dirichlet smoothing at 1000. The top 1000 retrieved documents are then clustered using the K Nearest Neighbor method (KNN) [6]. K is set to 5 (i.e. each cluster has five documents). The cosine similarity measure is used to determine the similarity between documents. Once we have the clusters, we represent and rank them using one of the methods described in section 2. As we mentioned previously, for cluster retrieval, the system's goal is to retrieve the best group of documents. Theoretically, only one cluster should be displayed. However, since the system has a ranked list of clusters, it is also a common practice to display some or all of them. This work focuses on the top retrieved cluster and the precision at 5 documents (PREC-5) is used for evaluation.

4   Experimental Results

There are four experimental questions that we would like to address. The first question is to compare the performance of the geometric-mean representation with the performance of the standard approach of concatenating documents. The results are given in table 1. The percentage improvement is given in parentheses. We observe that there is a large difference in effectiveness between these two representations. The geometric-mean approach gives at least a 9.9% improvement on any evaluation set over the standard approach.

The second experimental question is to compare the performance of the geometric-mean representation with that of document retrieval. Table 2 shows the results for precision at 5, 10, 15, and 20 documents. We observe that, except for precision at 15 and 20 on the AP collection, the geometric-mean representation for cluster retrieval consistently outperforms document retrieval across different data sets and at varying precision levels. If we focus on the first retrieved cluster, large performance gain (over 9%) is obtained on both WSJ and TREC45 collections while smaller improvements are observed on AP and TREC12 collections.

In order to gain a better understanding as to why the new representation works better on some of the collections than the others, we analyzed the queries and the intermediate and final outputs of the system. We found that the geometric-mean approach works well for queries that have four or fewer index terms. All queries on the TREC45 collection have fewer than 5 index terms, so most of the queries benefited from cluster retrieval with only 9 out of 100 queries that were slightly hurt by this

Table 1. Comparing cluster representations: geometric mean and concatenating docs

Collection	Prec. At 5 docs	
	Concatenating docs	Geometric mean
WSJ	0.4400	0.5040 (+ 14.5%)
AP	0.4040	0.4440 (+ 9.9 %)
TREC12	0.4360	0.6000 (+ 37.6 %)
TREC45	0.3240	0.4520 (+ 39.5 %)

Table 2. Comparing cluster (geometric-mean representation) and document retrieval

Eval. Metric	WSJ		AP		TREC12		TREC45	
	Doc	Cluster	Doc	Cluster	Doc	Cluster	Doc	Cluster
Prec. @ 5	0.4600	0.5040 (+9.6%)	0.4240	0.4440 (+4.7%)	0.5920	0.6000 (+1.4%)	0.4140	0.4520 (+9.2%)
Prec. @ 10	0.4320	0.4760 (+10.2%)	0.4040	0.4080 (+1.0%)	0.5460	0.5960 (+9.2%)	0.3820	0.4060 (+6.3%)
Prec. @ 15	0.4173	0.4587 (+9.9%)	0.3867	0.3813 (-1.4%)	0.5427	0.5747 (+5.9%)	0.3553	0.3700 (+4.1%)
Prec. @ 20	0.3950	0.4350 (+10.1%)	0.3880	0.3780 (-2.6%)	0.5210	0.5450 (+4.6%)	0.3385	0.3410 (+0.7%)

technique. For queries that are longer, however, the proposed representation seems to lose its advantage. One possible reason is that, for shorter queries, good clusters tend to have all query terms but for longer queries it is rarely the case. Both relevant and non-relevant documents contribute to only some of the query terms, and at often times good clusters can contain fewer unique query terms than bad clusters. As the geometric mean is based on a product of query term probabilities in documents and clusters, if a term doesn't occur in a cluster, its collection probability is used instead, which will result in smaller overall probability estimate for that cluster. Good clusters can receive lower ranks because of this. This type of queries is also difficult for document retrieval due to similar problems. Shorter queries do not have this problem because there are at least some good clusters that contain all the query terms, and bad clusters will not have more unique query terms than them.

The next experimental question is the comparison of seven different cluster representations (described in section 2). The results are presented in table 3. We can see that the geometric-mean representation is consistently better than all others on all four data sets. DM mix, TF mix, and Centroid methods are very similar to each other in performance. Except for the geometric-mean method, the performance of all the other representations is typically lower than that of document retrieval. If we order the representations from best to worst, we have this list: Geometric mean, DM mix, TF mix, Centroid, Concatenating documents, Worst Doc, Best Doc. We noticed that some of the representations are not stable and can perform well on some data sets but badly on others. For example, TF mix outperforms document retrieval on WSJ but does poorly on TREC45. Best Doc performs poorly on WSJ and AP but gives one of the best results on TREC12 and TREC45. Centroid, TF mix, DM mix, and Concatenating documents all seem to perform poorly on TREC45. Compared to these, the geometric-mean approach seems to be most stable.

Table 3. Comparing different cluster representations. Prec @ 5 is used for evaluation.

Coll.	Doc Ret.	Cluster Smoothing	Cluster Retrieval					
			Concat.	Best Doc	Worst Doc	Centroid	TF mix	DM mix	Geometric
WSJ	0.4600	0.4480	0.4400	0.3840	0.4080	0.4800	0.4800	0.4920	0.5040
AP	0.4240	0.4440	0.4040	0.3600	0.3760	0.3800	0.3860	0.4240	0.4440
TREC12	0.5920	0.5440	0.4360	0.5080	0.4680	0.4400	0.4180	0.4120	0.6000
TREC45	0.4140	0.4140	0.3240	0.4120	0.4060	0.2940	0.3020	0.3960	0.4520

The last question is comparing the performance of cluster retrieval with cluster smoothing [8]. Cluster smoothing is implemented following [8] and with query-specific clusters (which is the same set of clusters for cluster retrieval). The results are shown in table 3. Cluster retrieval using the geometric mean representation is consistently better than cluster smoothing in retrieval effectiveness. The other representations are typically less effective than cluster smoothing.

5   Conclusions and Future Work

In this paper, we have revisited the problem of retrieving the best group of documents within the language modeling framework. We empirically evaluated and compared document retrieval, cluster smoothing, and cluster retrieval with seven different cluster representations, including a new approach based on geometric mean as a first step toward incorporating these features. Experimental results show that the geometric-mean representation is a relatively stable method, and performs consistently better than document retrieval, cluster smoothing, and cluster retrieval using other representations. This work demonstrates that, with a good representation method, we can leverage the relationship between documents, and the effectiveness of retrieving documents as a group can be consistently better than that of retrieving them individually, especially in the top rank positions. This work is in progress and we plan to look into other features that are likely to benefit cluster retrieval as well as feature combination.

Acknowledgments

This work was supported in part by the Center for Intelligent Information Retrieval and in part by NSF grant #CNS-0454018 and #CCF-005575.

References

[1] Croft, W.B.: A model of cluster searching based on classification. Information Systems 5, 189–195 (1980)
[2] Griffiths, A., Luckhurst, H.C., Willett, P.: Using interdocument similarity information in document retrieval systems. Journal of the American Society for Information Science 37, 3–11 (1986)
[3] Hearst, M.A., Pedersen, J.O.: Re-examining the cluster hypothesis: Scatter/Gather on retrieval results. In: SIGIR 1996, pp. 76–84 (1996)
[4] Jardine, N., van Rijsbergen, C.J.: The use of hierarchical clustering in information retrieval. Information Storage and Retrieval 7, 217–240 (1971)
[5] Krovetz, R.: Viewing Morphology as an Inference Process. In: SIGIR 1993, pp. 191–203 (1993)
[6] Kurland, O., Lee, L.: Corpus structure, language models, and ad hoc information retrieval. In: Proceedings of SIGIR 2004 conference, pp. 194–201 (2004)
[7] Leuski, A.: Evaluating Document Clustering for Interactive Information Retrieval. In: Proceedings of CIKM 2001 conference, pp. 33–40 (2001)

[8] Liu, X., Croft, W.B.: Cluster-based retrieval using language models. In: Proceedings of SIGIR 2004 conference, pp. 186–193 (2004)

[9] Liu, X.: Cluster-based retrieval from a language-modeling perspective. In: The Doctoral Consortium of SIGIR 2006 conference, pp. 737–738 (2006), Abstract in SIGIR 2006 Proceedings

[10] Liu, X., Croft, W.B.: Representing clusters for retrieval. In: Proceedings of SIGIR 2006 conference, pp. 671–672 (2006)

[11] Miller, D., Leek, T., Schwartz, R.: A hidden Markov model information retrieval system. In: SIGIR 1999, pp. 214–221 (1999)

[12] Ponte, J., Croft, W.B.: A language modeling approach to information retrieval. In: SIGIR 1998, pp. 275–281 (1998)

[13] Robertson, S.E.: The probability ranking principle in IR. Journal of Documentation 33, 294–304 (1977)

[14] Tao, T., Wang, X., Mei, Q., Zhai, C.: Language model information retrieval with document expansion. In: Proceedings of HLT/NAACL 2006 (2006)

[15] Tombros, A., Villa, R., Van Rijsbergen, C.J.: The effectiveness of query-specific hierarchic clustering in information retrieval. Information Processing and Management 38, 559–582 (2002)

[16] van Rijsbergen, C.J., Croft, W.B.: Document clustering: An evaluation of some experiments with the Cranfield 1400 collection. Information Processing & Management 11, 171–182 (1975)

[17] van Rijsbergen, C.J., Sparck Jones, K.: A test for the separation of relevant and non-relevant documents in experimental retrieval collections. Journal of Documentation 29, 251–257 (1973)

[18] Voorhees, E.M.: The cluster hypothesis revisited. In: SIGIR 1985, pp. 188–196 (1985)

[19] Voorhees, E.M.: The TREC robust retrieval track. SIGIR Forum 39(1) (2005)

[20] Willet, P.: Query specific automatic document classification. International Forum on Information and Documentation 10(2), 28–32 (1985)

Enhancing Relevance Models with Adaptive Passage Retrieval

Xiaoyan Li[1] and Zhigang Zhu[2]

[1] Department of Computer Science, Mount Holyoke College, South Hadley,
MA 01075, USA
xli@mtholyoke.edu
[2] Department of Computer Science, CUNY City College, New York, NY 10031, USA
zzhu@ccny.cuny.edu

Abstract. Passage retrieval and pseudo relevance feedback/query expansion
have been reported as two effective means for improving document retrieval in
literature. Relevance models, while improving retrieval in most cases, hurts
performance on some heterogeneous collections. Previous research has shown
that combining passage-level evidence with pseudo relevance feedback brings
added benefits. In this paper, we study passage retrieval with relevance models
in the language-modeling framework for document retrieval. An *adaptive
passage retrieval* approach is proposed to document ranking based on the best
passage of a document given a query. The proposed passage ranking method is
applied to two relevance-based language models: the Lavrenko-Croft relevance
model and our *robust relevance model*. Experiments are carried out with three
query sets on three different collections from TREC. Our experimental results
show that combining adaptive passage retrieval with relevance models
(particularly the robust relevance model) consistently outperforms solely
applying relevance models on full-length document retrieval.

Keywords: Relevance models, passage retrieval, language modeling.

1 Introduction

Language modeling approach is a successful alternative to traditional retrieval models
for text retrieval. The language modeling framework was first introduced by Ponte
and Croft [19], followed by many research activities related to this framework since
then [1, 3, 4, 8, 10-12, 14-18, 20, 21, 23]. For example, query expansion techniques
[3,11,12,17,18,21,23], pseudo-relevance feedback [4,11,12,17,18,21,23], parameter
estimation methods [10], multi-word features [20], passage segmentations [16] and
time constraints [14] have been proposed to improve the language modeling
frameworks. Among them, query expansion with pseudo feedback can increase
retrieval performance significantly [11,18,23]. It assumes a few top ranked documents
retrieved with the original query to be relevant and uses them to generate a richer
query model.

However, two major problems remain unsolved in the query expansion techniques.
First, the performance of a significant number of queries decreases when query

C. Macdonald et al. (Eds.): ECIR 2008, LNCS 4956, pp. 463–471, 2008.
© Springer-Verlag Berlin Heidelberg 2008

expansion techniques are applied on some collections. Second, existing query expansion techniques are very sensitive to the number of documents used for pseudo feedback. Most approaches usually achieved the best performance when about 30 documents are used for pseudo feedback. As the number of feedback documents increases beyond 30, retrieval performance drops quickly. In our recent work [15], a robust relevance model is proposed based on a study of features that affected retrieval performance. These features included key words from original queries, relevance ranks of documents from the first round retrieval, and common words in the background data collection. The robust relevance model seamlessly incorporated these features into the relevance-based language model in [11] and further improved the performance and robustness of the model. The three features were also used in a recent work by Tao and Zhai [22] with regularized mixture models.

The robust relevance model and the regularized mixture model greatly ease the second problem, i.e. sensitivity of the retrieval performance to the number of documents used for pseudo feedback. However, the solution to the first problem is only partially. As we have reported in [15], the performance of the robust relevance model outperformed the Lavrenko-Croft relevance model and the simple query likelihood model on four test query sets, but it underperformed the simple query likelihood model on a query set against a subset of the TREC terabyte collection.

Passage retrieval is another effective means to improve document retrieval [5,6,7,16]. Particularly in [16], it was incorporated into the language modeling framework via various approaches. However, a major concern of passage retrieval in the language modeling framework is that it hurts retrieval performance on some collections, although it can provide comparable results and sometimes significant improvements over full-length document retrieval on collection with long and multi-topic documents. Therefore, one important research issue for both relevance models and passage retrieval is when and how to apply relevance models and passage retrieval for better retrieval performance.

In this paper, an *adaptive* passage retrieval approach is proposed to document ranking based on the *best passage* of a document given a query. The best passage of a document is the passage with the highest relevance score with respect to the query. The size of the best passage varies from document to document and from query to query. The best passage of a document can be a passage of the smallest window size considered or the document itself depends on whether it has the highest relevance score among all available passages. This adaptive passage selection is applied to two relevance-based language models: the Lavrenko-Croft relevance model [11] and our robust relevance model [15]. Experiments are carried out with three query sets on three different collections from TREC, including the ones that caused under-performance in the robust relevance model [15] and the fixed-size passage retrieval approach [16]. Our experimental results show that combining adaptive passage retrieval with relevance models consistently outperforms solely applying relevance models on full-length document retrieval. It indicates that passage-level evidence, if used appropriately, can be incorporated in relevance models to achieve better performance in terms of mean average precision, especially in the case of the robust relevance model.

The rest of the paper is structured as follows. In Section 2, we give a brief overview of the two relevance-based language models used in this paper. Section 3 describes our approach to combining the adaptive passage retrieval with the relevance models. Section 4 provides experimental results, compared to baseline results of full-length document retrieval. Section 5 summarizes the paper with conclusions and some future work.

2   Relevance Models

2.1   The Lavrenko-Croft Relevance Model

Lavrenko and Croft's relevance-based language model [5] is a model-based query expansion approach in the language-modeling framework [18]. A relevance model is a distribution of words in the relevant class for a query. Both the query and its relevant documents are treated as random samples from an underlying relevance model R, as shown in Figure 1. Once the relevance model is estimated, the *KL-divergence* between the relevance model (of a query and its relevant documents) and the language model of a document can be used to rank the document. Documents with smaller divergence are considered more relevant thus have higher ranks. Equations (1) and (2) are the formulas used in [5] and in this paper for approximating a relevance model for a query:

$$P_o(w \mid R) \approx \frac{P(w, q_1...q_k)}{P(q_1...q_k)} \tag{1}$$

$$P(w, q_1...q_k) = \sum_{D \in M} P(D)P(w \mid D)\prod_{i=1}^{k} P(q_i \mid D) \tag{2}$$

where $P_o(w \mid R)$ stands for the relevance model of the query and its relevant documents, in which $P(w, q_1...q_k)$ stands for the total probability of observing the word w together with query words $q_1...q_k$. A number of top ranked documents (say N) returned with a query likelihood language model are used to estimate the relevance model. In Equation (2), M is the set of the N top ranked documents used for estimating the relevance model for a query (together with its relevant documents). $P(D)$ is the prior probability to select the corresponding document language model D for generating the total probability in Equation (2). In the original relevance model approach, a uniform distribution was used for the prior.

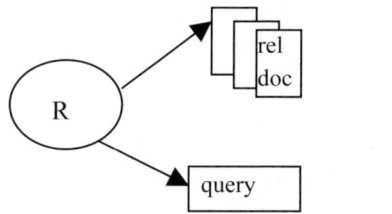

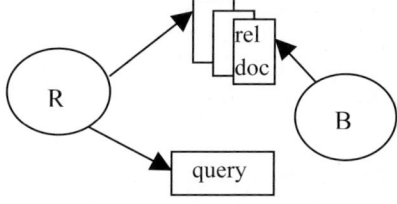

Fig. 1. The Lavrenko-Croft relevance model   **Fig. 2.** Our robust relevance model

2.2   Our Robust Relevance Model

Based on the Lavrenko-Croft relevance model approach, we have proposed a robust relevance model to further improve retrieval performance and robustness [15]. In the robust relevance model, Queries are random samples from the underlying relevance model R, and relevant documents are sampled from both the underlying relevance model R and a background language model B, as shown in Figure 2.

Three significant changes were made to the original relevance model in order to estimate a more accurate relevance model for a query: *treating the original query as a special document, introducing rank-related prior,* and *discounting common words*. The robust relevance model seamlessly incorporated these three features into the original relevance-based language model. Equations (3), (4) and (5) are the formulas used in [15] and also in this paper for approximating a relevance model for a query:

$$P_{new}(w, q_1...q_k) = \sum_{D \in S} P(D)P(w \mid D)\prod_{i=1}^{k} P(q_i \mid D) \tag{3}$$

$$P(D) = \frac{1}{Z_1} * \frac{\alpha + |D|}{\beta + Rank(D)}, \quad Z_1 = \sum_{D \in S} \frac{\alpha + |D|}{\beta + Rank(D)} \tag{4}$$

$$P_{new}(w \mid R) = \frac{1}{Z_2} \frac{P_{new}(w, q_1...q_k)}{\gamma + P(w \mid B)}, \quad Z_2 = \sum_{w \in V} \frac{P_{new}(w, q_1...q_k)}{\gamma + P(w \mid B)} \tag{5}$$

Unlike the set M including only top N documents' models in equation (2) for the Lavrenko-Croft relevance model, the robust relevance model treats the original query as a special document: the set S in equation (3) includes both the query model and the document models for the top N documents.

The robust relevance model also introduces a rank-related prior. In equation (4), $|D|$ denotes the length of document D or the length of the query – the special document. *Rank(D)* denotes the rank of document D in the ranked list of documents returned by using the basic query likelihood language model. The rank of the query is set to 0 so that it has the highest rank among all the documents used for relevance model approximation. Z_1 is the normalization factor that makes the sum of the priors to 1. Parameters α and β are used to control how much a document's length and its rank affect the prior of the document, respectively.

Finally, the robust relevance model discounts common words in the background data collection. In equation (5), $P_{new}(w \mid R)$ denotes the probability of word w in the new relevance model. $P(w \mid B)$ denotes the probability of word w in the background language model B. γ is the parameter for discounting the probability of a word in the new relevance model by its probability in the background language model. Z_2 is the normalization factor that makes the sum of the probabilities of words in the new relevance model to 1. The best values for parameters α, β and γ reported in [15] are used in our experiments in this paper.

3   Combining Passage Retrieval with Relevance Models

Passage retrieval can be applied in the language modeling framework. Various approaches were proposed in [16] to implement passage retrieval in the language modeling environment. In the context of relevance models, three different methods R1, R2 and R3 were developed in [16], and the method R1 was reported as the best candidate. Different types of passages including half-overlapped window and arbitrary passage with fixed or variable lengths were also tried for passage retrieval.

In our paper, as baselines, we use the method R1 with fixed-size half-overlapped windows to retrieve relevance documents. Half-overlapped windows of 150, 350 and 500 words are considered in our experiments.

Given a window size, documents are first broken into half-overlapped passages. A language model is then built for each passage. At query time, a simple query likelihood language model implemented in LEMUR [13] is used to retrieve top passages. In the case of the Lavrenko-Croft relevance model, the top retrieved passages are assumed relevant and used to build a relevance model for the query. In the case of our robust relevance model, both the top passages and the query itself are used to build a relevance model. Once the relevance model is built, a *KL divergence score* is computed between each passage model and the relevance model. The KL divergence score is then used for ranking passages. Documents finally are ranked based on the score of their best passage.

However, the problem with a fixed-size window approach is that the best performance is achieved with different window size on different collections. Therefore, it is not clear how to preset a window size at query time on a new collection that is previously unseen. To solve this problem, we propose an *adaptive* passage retrieval approach in this paper. Documents are ranked based on their *best passage*. The best query in this context is the passage that can represent a document better than other passages with respect to a query. We observe that the size of a best passage can vary from document to document and query from query, because documents may discuss multiple topics, have a different focus, and by authors with different writing styles. There are various approaches that can be used to locate the best passage of a document. In this paper, we choose a very simple but efficient way to find the best passage of a document to improve retrieval performance for a given query.

We simply take the retrieval results from relevance models on full-length documents retrieval, fixed-size passages on relevance models with several preset window sizes, for example 150, 350 and 500, respectively. With the four result files, each document has four scores: full-length document score, the highest score among all 150-word-long passages, the highest score among all 350-word-long passages, and the highest score among all 500-word-long passages. We take the highest score among the four scores as the score of the best passage of the document. Documents are then ranked according to the score of their best passages. Note that the size of the best passage varies from document to document. The best passage of a document can be a passage of the smallest window size considered (say 150 in this paper) or the full-length document itself depends on whether it has the highest relevance score among all available passages of variable size. Therefore, the adaptive passage retrieval approach combines the best of the full document and fixed-size passage retrieval results.

4   Experiments and Results

We have carried out experiments with three TREC query sets on three data collections. All experiments were performed with the Lemur toolkit [13]. The Krovetz stemmer [9] was used for stemming and the standard stopword list of LEMUR was used to remove about 420 common terms. Top 30 documents or passages are used to estimate a relevance model for a query when using relevance model approaches. Parameters α, β and γ in Equations (3)-(5) are set to the same values as used in [15].

4.1 Data

We used three query sets on three document collections in our experiments. (1). Queries 51 to 150 on a homogeneous collection AP88_90. AP88_90 includes newswires from Associated Express 1988, 1989 and 1990. It is a collection of short documents. This was used in [16] where fixed-size passage retrieval hurt the relevance retrieval performance. (2). Queries 101 to 150 on a heterogeneous collection AP&FR collection, which includes the Associated Press data set (AP88 and AP89) and the FR88&89 collection. We created this collection to test the performance of our approach to such a heterogeneous data collection. (3). Queries 701 to 750 on a sub-collection of the TREC Terabyte data set on which the robust relevance model [15] had some problem. To construct the subset, the top-ranked 10,000 documents for each of the 50 queries that were retrieved using the basic query likelihood language model were selected. The subset has 466,724 unique web documents and is about 2% of the entire terabyte collection [2]. This collection is by nature a more heterogeneous collection with web documents, blogs, emails as well as news articles. The statistics of AP88_89 collection, AP&FR collection, and the subset of terabyte collection are shown in Table 1. Table 2 summaries the information about the three sets of queries used in our experiments and relevant documents on the corresponding three document collections. The queries are taken from TREC topics and only title field are used in our experiments. The queries are on average 3 or 4 words long, and the number of relevant documents per query varies across collections.

Table 1. Statistics of the three document collections

Collection Statistics	AP88_90	AP&FR	Terabyte (GOV2)
# of documents	242,918	210,417	466,724
# of terms	61,975,608	83,936,199	958,740,730
# of unique terms	255,617	362,886	3,637,433
Average Length of documents	255	398	2,054
Average frequency of terms	242	231	264

Table 2. Information of the three query sets (N1: # of queries with relevant documents; N2: total # of relevant documents; N3: average # of relevant doc. per query)

Collections	Queries (title only)	N1	N2	N3
AP88_90	TREC topics 51-150	99	21829	220.5
AP&FR	TREC topics 101-150	50	5,211	104.2
Terabyte	TREC topics 701-750	49	10,617	216.7

Table 3. Performance comparison of passage retrieval + relevance models (RM: Lavrenko-Croft relevance model; RRM: our robust relevance model)

Datasets	Methods	FullDoc	P150	P350	P500	BestP
AP88_90	RM	0.2779	0.2677	0.2747	0.2771	0.2844
	RRM	0.2821	0.2655	0.2800	0.2822	0.2882
AP&FR	RM	0.2696	0.2799	0.2720	0.272	0.2761
	RRM	0.2724	0.3084	0.3093	0.3106	0.3113
Terabyte	RM	0.1872	0.2026	0.2119	0.2067	0.2202
	RRM	0.2361	0.2256	0.2448	0.2376	0.2528

4.2 Experimental Results

We have carried out experiments on three query sets. In each query set, the four baselines - full document retrieval (FullDoc) and three fixed-size passage retrieval baselines with three different window sizes (P150, P350 and P500), and the adaptive passage retrieval method (BestP), is applied to the Lavrenko-Croft relevance model (RM) and our robust relevance model (RRM), respectively. Mean average precision is used for performance evaluation. Three different window sizes in the fixed-size passage retrieval baselines are 150, 350 and 500. The "best" passage of a document in the proposed adaptive passage retrieval approach (BestP) is the passage with the highest KL divergence score among all passages of different sizes (150, 350, 500 and full-length document). The performance of three query sets in terms of mean average precision is given in Table 3. The following observations can be obtained based on the experimental results.

(1) Combining adaptive passage retrieval with the two relevance models consistently outperforms solely applying the relevance models on full-length document retrieval on all the three collections. Robust relevance model with fixed-size passages also gives better performance than full-length document retrieval on all three collections. But original relevance model with fixed-size passage achieves outperforms full-length document retrieval only on two of the collections.

(2) The adaptive passage retrieval consistently provides the best performance than the full-length document retrieval and the fixed-size passage retrieval, when using the two relevance models, on all three collections. The only exception is for queries 101-150 with the original relevance model, where the best performance was achieved when the passage size is fixed to 150. However, the adaptive passage retrieval method ranked the second best, and is very close to the first best.

(3) Better performance is achieved when the robust relevance model is used. This is true for all the four baselines as well as the adaptive passage retrieval approach. The performance is always the best when combining the adaptive passage retrieval with the robust relevance model.

5 Conclusions and Future Work

In this paper, we study how to better combine passage retrieval with relevance models in the language modeling framework for better retrieval performance. Three main conclusions have been drawn from the experimental results. First, combining passage retrieval with relevance models consistently outperforms relevance models on full-length document retrieval in terms of mean average precision on document retrieval. Second, the proposed adaptive passage retrieval approach for identifying best passage gives better performance than using passages of fixed sizes. Third, the robust relevance model uniformly outperforms the original relevance models, especially when combining with passage-level evidence.

In the current experiments, for testing the ideas of the adaptive passage sizes, we only used a few typical document sizes that have been tested empirically in literature. As a future work, the approach proposed by Jiang and Zhai [5] for identifying variable-length passages using HMMs could be used. As another future work, new approaches to query expansion techniques need to be developed for retrieval on heterogeneous collections (e.g., the Terabyte collection), which may include web documents, blogs, emails as well as news articles. In this case, incorporating selective query expansion techniques, such as Cronen-Townsend et al's work in [3], and features like metadata into relevance models may be helpful.

Acknowledgments. The first author is supported by Mount Holyoke College and second author by NSF (Grant No. CNS-0551598).

References

[1] Abdul-Jaleel, N., et al.: UMASS at TREC2004. In: Thirteen Text Retrieval Conference Notebook (2004)

[2] Clarke, C., Craswell, N., Soboroff, I.: Overview of the TREC 2004 terabyte track. In: Thirteen Text Retrieval Conference Notebook (2004)

[3] Cronen-Townsend, S., Zhou, Y., Croft, W.B.: A framework for selective query expansion. In: Proc. 13th Int. Conf. on Information and Knowledge Management, pp. 236–237 (2004)

[4] Hiemstra, D.: Using language models for information retrieval. PhD thesis, University of Twente (2001)

[5] Jiang, J., Zhai, C.: UIUC in HARD 2004 – Passage Retrieval using HMMs. In: Thirteen Text Retrieval Conference Notebook (2004)

[6] Kaszkiel, M., Zobel, J.: Passage retrieval revisited. In: Proc. 20th ACM-SIGIR Conf. on Research and Development in Information Retrieval, pp. 178–185 (1997)

[7] Kaszkiel, M., Zobel, J.: Effective ranking with arbitrary passages. Journal of the American Society for Information Science and Technology 52(4), 344–364 (2001)

[8] Kraaij, W., Westerveld, T., Hiemstra, D.: The importance of prior probabilities for entry page search. In: 25th ACM SIGIR Conf. on Research and Development in Information Retrieval, pp. 27–34 (2002)

[9] Krovetz, R.: Viewing morphology as an inference process. In: Proc. 16th ACM SIGIR Conf. on Research and Development in Information Retrieval, pp. 191–202 (1993)

[10] Lafferty, J., Zhai, C.: Document language models, query models, and risk minimization for information retrieval. In: 24th ACM SIGIR Conf. on Research and Development in Information Retrieval, pp. 111–119 (2001)

[11] Lavrenko, V., Croft, W.B.: Relevance-based language models. In: Proc. 24th ACM SIGIR Conf. on Research and Development in Information Retrieval, pp. 120–127 (2001)

[12] Lavrenko, V., Croft, W.B.: Relevance models in information retrieval. In: Croft, W.B., Lafferty, J. (eds.) Language modeling for information retrieval, pp. 11–56. Kluwer Academic Publishers, Dordrecht (2003)

[13] LEMUR, http://www-2.cs.cmu.edu/~lemur/3.1/doc.html

[14] Li, X., Croft, W.B.: Time-based language models. In: Proc. 12th Int. Conf. on Information and Knowledge Management, pp. 469–475 (2003)

[15] Li, X.: A new robust relevance model in the language model framework, Information Processing and Management (2007), doi:10.1016/j.ipm.2007.07.005

[16] Liu, X., Croft, W.B.: Passage retrieval based on language models. In: Proc. 11th Int. Conf. on Information and Knowledge Management, pp. 375–382 (2002)

[17] Miller, D.H., Leek, T., Schwartz, R.: A hidden Markov model information retrieval system. In: Proc. 22nd ACM SIGIR Conf. Research and Development in Information Retrieval, pp. 214–221 (1999)

[18] Ponte, J.: A Language modeling approach to information retrieval. PhD thesis, UMass-Amherst (1998)

[19] Ponte, J., Croft, W.B.: A language modeling approach to information retrieval. In: Proc. 21st ACM SIGIR Conf. Research and Development in Information Retrieval, pp. 275–281 (1998)

[20] Song, F., Croft, W.B.: A general language model for information retrieval. In: Proc. 22nd ACM SIGIR Conf. Research and Development in Information Retrieval, pp. 279 280 (1999)

[21] Tao, T., Zhai, C.: A two-stage mixture model for pseudo feedback. In: Proc.27th ACM SIGIR Conf. Research and Development in Information Retrieval, pp. 486–487 (2004)

[22] Tao, T., Zhai, C.: Regularized estimation of mixture models for robust pseudo-relevance feedback. In: Proc. 29th ACM SIGIR Conf. on Research and Development in Information Retrieval, pp. 162–169 (2006)

[23] Zhai, C., Lafferty, J.: Model-based feedback in the language modeling approach to information retrieval. In: Proc. 10th Int. Conf. on Information and Knowledge Management, pp. 403–410 (2001)

Ontology Matching Using Vector Space

Zahra Eidoon[1], Nasser Yazdani[1], and Farhad Oroumchian[2,3]

[1] ECE Department, University of Tehran, Tehran, Iran
[2] University of Wollongong , Dubai
[3] Control and Intelligent Processing Center of Excellence, Faculty of Eng.,
Univ. of Tehran,Tehran, Iran
{z.eidoon,yazdani}@ece.ue.ac.ir,
FarhadOroumchian@uowdubai.ac.ae

Abstract. Interoperability of heterogeneous systems on the Web will be achieved through an agreement between the underlying ontologies. Ontology matching is an operation that takes two ontologies and determines their semantic mapping. This paper presents a method of ontology matching which is based on modeling ontologies in a vector space and estimating their similarity degree by matching their concept vectors. The proposed method is successfully applied to the test suit of Ontology Alignment Evaluation Initiative 2005 [10] and compared to the results reported by other methods. In terms of precision and recall, the results look promising.

Keywords: semantic web, ontology matching, vector space.

1 Introduction

The current World Wide Web has over 22.47 billion pages [17], but the vast majority of them are in human readable format only. In order to allow software agents to understand and process the web information in a more intelligent way, researchers have created the Semantic Web vision [15], where data has structure. Like the Web, the semantic Web will necessarily be distributed and heterogeneous. Therefore, the integration of resources found on the semantic Web is a key issue. A standard approach to the resulting problem lies in the use of ontologies for data description. Ontologies allow users to organize information into taxonomies of concepts, each with their properties, and describe relationships between concepts [16]. However, the available ontologies could themselves introduce heterogeneity: given two ontologies, the same entity can be given different names in each of them or simply be defined in different ways, whereas both ontologies may express the same knowledge but in different languages. So, one of the key challenges of Semantic Web is to find semantic correspondences between ontologies.

The underlying problem, which we call the ontology matching (or alignment), is the operation of taking two distinct ontologies, finding a set of entities with similar relationships which exist in both ontologies and return the similar entities. Shvaiko et al. classifies ontology alignment techniques in two general categories: element-level techniques and structure-level techniques [5]. The former techniques concentrate just

C. Macdonald et al. (Eds.): ECIR 2008, LNCS 4956, pp. 472–481, 2008.

on individual elements while in latter approaches the structural arrangement of elements and their relation to each other is more of interest. The structural-level techniques involve Graph-based techniques which consider the input as labeled graph, Taxonomy-based techniques which consider only the specialization relation, Repository of structures which stores schemas/ontologies and their fragments together with pair-wise similarities (e.g., coefficients in the [0 1] range) between them and finally Model-based algorithms which handle the input based on its semantic interpretation (e.g., model-theoretic semantics). Furthermore, ontology matchers can be categorized into automatic and semi-automatic techniques. Automatic ontology matchers are those which perform their operation independent of human operator, while semi-automatic techniques are dependent on user preferences.

This paper presents an automatic taxonomy-based ontology alignment technique that is based on a vector matching method. Any ontology consists of a set of concepts and each concept is described by a set of properties. These concepts and properties define a space such that each distinct concept and property represents one dimension in that space. Modeling ontologies in multi-dimensional vector spaces will enable us to use vector matching methods for performing ontology alignment. An iterative approach has been employed to achieve convergence, in which vectors representing ontology concepts and properties are matched iteratively and their similarity degree is estimated. In order to model two ontologies in a vector space, RDF [1] and OWL [7] subclass predicates are utilized and concepts are described with respect to their ancestors and successors and properties. Properties are also described with respect to their domain and range concepts.

The rest of the paper is organized as follows. In section 2 we discuss state of the art of matching systems from the structured base ontology matching perspective. Our approach is presented in section 3. Experimental results are reported in Section 4. Finally, Section 5 contains some conclusions and future work.

2   Related Work

The Cupid system [9] implements a generic schema matching algorithm combining linguistic and structural schema matching techniques, and computes normalized similarity coefficients with the assistance of a precompiled thesaurus. The algorithm contains two phases. The first phase, called linguistic matching and the second one is the structural matching of schema elements based on the similarity of their contexts or vicinities. Finally the weighted similarity, a mean of the first and second phases results are calculated. Anchor-PROMPT [2] is another structure base algorithm. It takes as input a set of pairs of related terms—anchors—from the source ontologies. Either the user identifies the anchors manually or the system generates them automatically with the help of string-based techniques, or another matcher computing linguistic (dis)similarity between frame names (labels at nodes) [6]. Then it refines them based on the ontology structures and users' feedback. Anchor-PROMPT traverses the paths between the anchors in the corresponding ontologies. As it traverses the two paths, Anchor-PROMPT increases the similarity score for the pairs of terms in the same positions in the paths. It aggregates the similarity score from all the traversals to generate the final similarity score.

The compositional systems like [12],[4] consist of a set of elementary matchers based on rules, exploiting codified knowledge in ontologies, such as information about super- and sub-concepts, super- and sub-properties, etc. The approach described in [11] is relatively similar to our method. It uses vector characteristics and presents a semantic similarity measure based on a matrix representation of nodes from an RDF labeled directed graph. In this algorithm an entity is described with respect to how it relates to other entities using N-dimensional vectors, N being the number of selected external predicates. Similarities are computed using graph matching algorithm [13]. There are some other methods that benefit from structure of ontologies as well as other techniques such as ola[14], foam[8] and omap[18]. Vector Based Ontology Matching, which we present here, is another vector based model that providing another suggestions for possible matching terms.

3   Vector Based Ontology Matching (VBOM)

As mentioned before, the proposed method of ontology matching is based on vector similarity algorithms. Thus, the first step is to model source ontologies in vector notation and then apply a vector matching algorithm to estimate the degree of similarity among them. Similarity of the two vectors can be computed with cosine of angle between those vectors. If the cosine of the angle is 1, the two vectors are exactly the same. As the cosine approaches 0, the similarity degree reduces. Considering $\vec{A}$ and $\vec{B}$ as two vectors, the cosine of their angle can be computed using the following formula:

$$Cos\theta = \frac{\vec{A}.\vec{B}}{\|\vec{A}\|.\|\vec{B}\|} \tag{1}$$

$\vec{A}.\vec{B}$ represents the dot product of two vectors (sum of the product of their corresponding elements). $\|\vec{A}\|$ and $\|\vec{B}\|$ represent the size of the vectors $\vec{A}$ and $\vec{B}$ respectively.

3.1   Ontology Vectorization

Ontology Vectorization is the method of modeling two source ontologies (for which the matching problem is of interest) in a single multi dimensional vector space. Any ontology consists of a set of concepts and any concept may have a set of properties which describes that concept. Two types of properties are distinguished:

➤ *datatype properties*, relations between instances of classes and RDF literals and XML Schema datatypes.
➤ *object properties*, relations between instances of two classes.

The overall perspective of the method is to make a vector space that any of its dimensions represents a unique concept, property or the range of datatype property of the two source ontologies. The vector space must have certain characteristics to be appropriate for utilization in matching algorithm:

✓ Similar concepts, properties and the ranges of datatype properties of the source ontologies will not be duplicated in the vector space.
✓ The order of elements is not important. Thus the concepts, properties and the ranges of datatype properties can be arranged in any order for constructing the vector space.
✓ The vector space must fully cover all the distinct concepts, properties and the ranges of datatype properties which exist in the two ontologies.

As mentioned before, given a pair of ontologies, vector space is built by extracting all distinct concepts, properties and the ranges of datatype properties belonging to these two source ontologies as its dimensions. Then each of these elements is presented as a vector in this vector space.

Let us have a look at a simple example. Take the following ontologies O_A and O_B in figure 1(the left hand ontology is O_A and the right hand one is O_B). The distinct concepts of the two ontologies are: "Address", "Institution", "Publisher", "School", "Directions", "Organization", having "Publisher" and "School" as the subclasses (successors) of "Institution" in O_A and "Organization" in O_B. In other words, "Institution" and "Organization" are the ancestors of "School" and "Publisher" in O_A and O_B, respectively. The distinct properties are "country", "city", "name", "address" and "town". Each ontology contains 3 datatype properties and one object property (the values in the brackets show min and max cardinality of the property for that concept). Properties are defined in the following style:

property Name   #domain Name->#range Name.

Dimensions of our vector space are:{"Address", "Institution", "Publisher", "School", "Directions", "Organization", "country", "city", "name", "address", "town", "http://www.w3.org/2001/XMLSchema#string"}. As we mentioned earlier, there is no particular order among the dimensions in the vector space. (Hereafter for simplicity we use "string" instead of http://www.w3.org/2001/XMLSchema#string.)

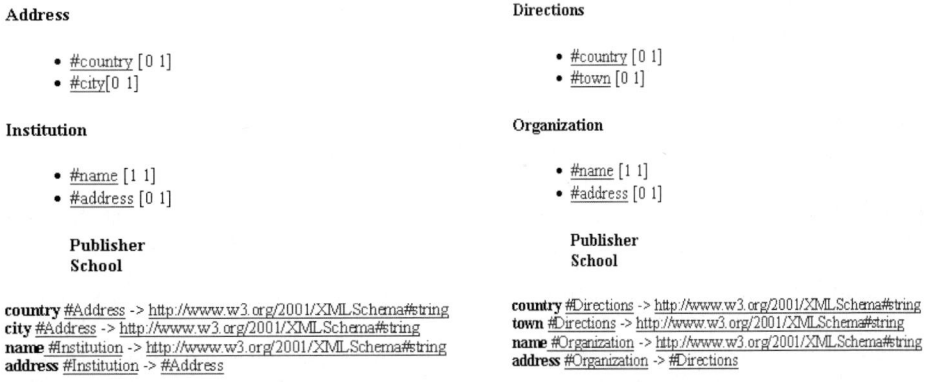

Fig. 1. O_A and O_B

Each concept is then described by a vector of weights for itself, all of its properties and ancestors and successors. Furthermore each property is described by a vector of nonzero weights for itself and all of its domain and range concepts.

3.2 Weighting Mechanism

3.2.1 Concept Vectors. The following shows the weight of each element in the concept vector.

$$W_C(X) = \begin{cases} \log(\dfrac{1}{d_X(c)+1}) & \text{if } d_X(c) > 0 \\ 1 & \text{if } d_X(c) = 0 \end{cases} \qquad (2)$$

Where $W_c(X)$ is the weight of concept c in the concept vector X, and $d_X(c)$ is the level of distance of concept c from X in its sub/super class chain. In fact the concept itself acts as a pivot and all of its super/sub classes would receive weights based on their distance from this pivot.

$$W_p(X) = \begin{cases} 1 & \text{if } X \quad has \quad p \\ 0 & \text{if } X \quad doesn't \quad have \quad p \end{cases} \qquad (3)$$

where $W_p(X)$ is the weight of property p in the concept vector X.

3.2.2 Property Vectors

$$W_C(x) = \begin{cases} 1 & \text{if } C \in \{xDomain, xRange\} \\ 0 & otherwise \end{cases} \qquad (4)$$

where $W_C(x)$ is the weight of concept c in the property vector x, $xDomain$ is a set of concepts which are the domain of property x, and $xRange$ is a concept which is the range of property x.

$$W_p(x) = \begin{cases} 1 & \text{if } p = x \\ 0 & otherwise \end{cases} \qquad (5)$$

where $W_p(x)$ is the weight of property p in the property vector x.

Consider we want to produce the "Institution" concept vector of O_A in figure 1. As we know "Institution" concept has 2 sub classes: "Publisher" and "School" and 2 properties: "name" and "address". Therefore its vector contains 5 none zero elements: "Institution" "Publisher", "School", "name" and "address". The weight of "Institution" will be 1, the weight of its 2 direct sub classes is $\log(\dfrac{1}{1+1})$ and the weight of its properties is 1. Thus, according to the vector space which is constructed above, the "Institution" concept vector of O_A is { 0, 1, log (1/2), log (1/2), 0, 0, 0, 0, 1, 1, 0, 0}. Some other concept vectors are: the "Address" concept vector of O_A: {1, 0, 0, 0, 0, 0, 1, 1, 0, 0, 0, 0}, the "Publisher" concept vector of O_B: {0, 0, 1, 0, 0, log (1/2),

0, 0, 0, 0, 0, 0} and so on. Property vectors are also produced. For example "country" property vector of O_A contains 3 none zero elements: its domain ("Address"), itself and its range ("string"). Thus "country" property vector of O_A equals {1,0,0,0,0,0,10,0,0,0,1}. Other vectors are constructed in the same way.

3.3 Matching Process

After vectorizing two source ontologies, finding similarities between two ontologies would be easy. As we mentioned in section 3 the correlation between two concept vectors in an N dimensional vector space can be calculated using the cosine of angle between them.

$$sim(\vec{X},\vec{Y}) = \frac{\vec{X}.\vec{Y}}{\| \vec{X} \|.\| \vec{Y} \|} \tag{6}$$

We compute the cosine of all the pairs of concept vectors between the two source ontologies. Then for each concept, we choose the most similar concept with the highest similarity score. This operation is repeated for all the pairs of property vectors.

VBOM is an iterative approach. In each iteration, it selects pairs of similar concepts and similar properties that each participates only in one similarity relation. Then it updates all the vectors of all concepts and properties by setting the weights of participating elements of each selected pair to their biggest non-zero weight. In this way, in each iteration, VBOM benefits from similarities that were discovered in previous iteration. These iterations continue until there are no new similar pairs.

4   Results

We carried out experiments on OAEI (Ontology Alignment Evaluation Initiative) 2005 test suite [10]. The evaluation organizers provide a systematic benchmark test suite with pairs of ontologies to align as well as expected (human-based) results. The ontologies are described in OWL-DL and serialized in the RDF/XML format. The expected alignments are provided in a standard format expressed in RDF/XML.

There are different groups of tests in this benchmark [10]:

Simple tests (tests 1xx). such as comparing the reference ontology with itself, with another irrelevant ontology or the same ontology in its generalization or restriction to OWL-Lite .

Systematic tests (tests 2xx). that are obtained by discarding some features of the reference ontology. (The considered features are names, comments, hierarchy, instances, relations, restrictions, etc.)

- Tests 201 to 210: focus on labels and comments of entities. Names of entities can be replaced by random strings, synonyms, names with different conventions, strings in a language other than English.
- Tests 221 to 247: for these tests the structure is changed. In fact hierarchy can be suppressed, expanded or flattened; properties can be suppressed or

their imposed restrictions on classes are discarded and classes can be expanded or flattened.

- Tests 248 to 266: for these tests, names of entities are replaced by random strings; hierarchy can be suppressed, expanded or flattened and properties can be suppressed.

Four real-life ontologies of bibliographic references (3xx). that were found on the web and left mostly untouched. These real world ontologies are a combination of complications of the previously mentioned tests.

Table 1. Ontologies with similar labels

test	Name	Precision	Recall
101	Reference	1	1
102	Irrelevant Ontology	-	-
103	Language Generalization	1	1
104	Language restriction	1	1
221	No specialization	1	1
222	Flattened hierarchy	0.9	0.9
223	Expanded hierarchy	1	1
224	No instance	1	1
225	No restrictions	1	1
228	No properties	1	1
230	Flattened classes	1	1
231	Expanded classes	1	1
232		1	1
233		1	1
236		1	1
237		0.9	0.9
238		0.91	0.91
239		1	1
240		1	1
241		1	1
246		1	1
247		1	1

We obtained 3 kinds of results in our experiments:

1) Excellent results from ontologies that have similar names (labels) (in tests 1xx, 221 to 247). Because similar names make vectors more similar to each other. In fact the labels are the most important feature to recognize alignments in this approach and if the labels denote an alignment, every thing else can be abandoned. As table 1 shows, both precisions (the number of correct alignments found, divided by the total number of alignments found) and recalls(the number of correct alignments found, divided by the total number of expected alignments) are equal to "1" except for 3 cases;

2) Good results in ontologies are those with similar structures but different naming conventions (in tests 201 to 210 and 249). However the labels are the most important feature in distinction of alignments, the structures of ontologies also play a key role in our approach. We obtained precisions and recalls in the range of 0.78 to 1 and 0.85 to 1 respectively (table 2);

3) Weak results in cases that the two source ontologies are different in both their naming conventions and structures (in tests 248, 250 to 266.). Especially the recall factor is affected more in these situations.(table 3)

Table 2. Ontologies with similar structures and different labels

test	Name	Precision	Recall
201	No names	0.89	0.94
202	No names, No comments	0.89	0.94
203	No comments	1	1
204	Naming conventions	0.94	0.97
205	Synonyms	0.89	0.94
206	Translation	0.78	0.85
207		0.89	0.94
208		0.94	0.97
209		0.89	0.94
210		0.89	0.94
249		0.89	0.94

Table 3. Ontologies with difference in both their labels and structures

test	Name	Precision	Recall
248		1	0.76
250		0.6	0.09
251		0.42	0.17
252		0.59	0.7
253		1	0.76
254		0	0
257		0.6	0.09
258		0.42	0.17
259		0.59	0.7
260		0.6	0.1
261		0.4	0.06
262		0	0
265		0.6	0.1
266		0.4	0.06
301	Real: BibTeX/MIT	0.73	0.53
302	Real: BibTeX/UMBC	0.57	0.62
303	Real: Karlsruhe	0.5	0.53
304	Real: INRIA	0.84	0.9

Table 4, depicts summarized results of the three groups of tests and comparison of our method with some other systems. The last row of the Table 1 shows the harmonic mean (H-mean) of three upper values.

Table 4. A comparison of VBOM with other systems on OAEI2005 test suit

algo	VBOM		foam		omap		ola	
test	Prec.	Rec.	Prec.	Rec.	Prec.	Rec	Prec.	Rec
1xx	1.00	1.00	0.98	0.65	0.96	1.00	1.00	1.00
2xx	0.81	0.74	0.89	0.69	0.31	0.68	0.80	0.73
3xx	0.66	0.65	0.92	0.69	0.93	0.65	0.50	0.48
H-means	0.80	0.77	0.93	0.68	0.56	0.75	0.71	0.67

Although VBOM only focuses on sub/super class chains and properties in ontologies, our experiments show that it is comparable with hybrid models like foam [8] and ola [14] and omap [18] that use linguistic and structural methods. Even in some cases VBOM worked better than the hybrid methods.

VBOM results show that in ontologies that include the sub/super predicate, it is possible to achieve reasonable results by focusing on this predicate and properties. This method is simple and efficient.

5   Conclusions

We have presented a structure-based semantic similarity measurement approach for mapping ontologies that can be directly applied to OWL ontologies. The work is based on the intuition that the similarity of two entities can be defined in terms of how these entities are similar with respect to their ancestors, successors and properties. We converted the source ontologies into a vector space of N dimensions. These dimensions represent distinct concepts, properties and ranges of datatype properties of two source ontologies. We mapped the concepts in the source ontologies into vectors containing nonzero weights in order to represent their properties and relationships with their ancestors and successors. Also properties are mapped into vectors containing nonzero weights in order to represent their domains and ranges. The results obtained from the tests performed over the Ontology Alignment Evaluation Initiative 2005 test suite are promising. Labels are very important in our approach. After that structures can help the alignment process. In future, we are going to use a dictionary to benefit more from the same labels.

References

1. Resource description framework, http://www.w3.org/RDF/
2. Noy, N., Musen, M.: Anchor-PROMPT: Using Non-Local Context for Semantic Matching. In: Conference on Artificial Intelligence (IJCAI) (2001)
3. Euzenat, J., Valtchev, P.: An integrative proximity measure for ontology alignment. In: Proceedings of Semantic Integration workshop at ISWC (2003)

4. Ehrig, M., Staab, S.: QOM-quick ontology mapping. In: Proc. 3rd ISWC, Hiroshima (JP) (November 2004)

5. Shvaiko, P., Euzenat, J.: A survey of schema-based matching approaches. Journal on Data Semantics IV (2005)

6. McGuinness, D.L., Fikes, R., Rice, J., Wilder, S.: An Environment for Merging and Testing Large Ontologies. In: Principles of Knowledge Representation and Reasoning: Proceedings of the Seventh International Conference (KR 2000) (2000)

7. Owl web ontology language overview. w3c recommendation, February 10 (2004), http://www.w3.org/TR/owl-features/

8. Ehrig, M., Sure, Y.: FOAM – Framework for Ontology Alignment and Mapping Results of the Ontology Alignment Evaluation Initiative. Results of the Ontology Alignment Evaluation Initiative. In: Integrating Ontologies (2005)

9. Madhavan, J., Bernstein, P.A., Rahm, E.: Generic Schema Matching using Cupid. In: VLDB (2001)

10. Ontology alignment evaluation initiative (2005), http://oaei inrialpes. fr/ 2005/

11. Tous, R., Delgado, J.: A Vector Space Model for Semantic Similarity Calculation and OWL Ontology Alignment. In: Bressan, S., Küng, J., Wagner, R. (eds.) DEXA 2006. LNCS, vol. 4080, pp. 307–316. Springer, Heidelberg (2006)

12. Ehrig, M., Sure, Y.: Ontology mapping - an integrated approach. In: Bussler, C.J., Davies, J., Fensel, D., Studer, R. (eds.) ESWS 2004. LNCS, vol. 3053, pp. 76–91. Springer, Heidelberg (2004)

13. Blondel, V.D., et al.: A measure of similarity between graph vertices: Applications to synonym extraction and web searching. SIAM Rev. 46(4), 647–666 (2004)

14. Euzenat, J., Loup, D., Touzani, M., Valtchev, P.: Ontology Alignment with OLA. In: 3rd EON Workshop on Evaluation of Ontology based Tools (EON), Hiroshima, Japan (2004)

15. Berners-Lee, T.: The semantic web. Scientific American 284(5), 35–43 (2001)

16. Doan, A., Madhavan, J., Domingos, P., Halevy, A.: Learning to Map Between Ontologies on the Semantic Web. In: Proceedings of the 11th international conference on World Wide Web, Honolulu, Hawaii, USA, May 07-11 (2002)

17. http://www.worldwidewebsize.com

18. Straccia, U., Troncy, R.: OMAP: Combining Classifiers for Aligning Automatically OWL Ontologies. In: Ngu, A.H.H., Kitsuregawa, M., Neuhold, E.J., Chung, J.-Y., Sheng, Q.Z. (eds.) WISE 2005. LNCS, vol. 3806, pp. 133–147. Springer, Heidelberg (2005)

Accessibility in Information Retrieval

Leif Azzopardi[1] and Vishwa Vinay[2]

[1] Dept. of Computing Science,
University of Glasgow,
Glasgow UK
leif@dcs.gla.ac.uk
[2] Microsoft Research Cambridge
Cambridge, UK
vvinay@microsoft.com

Abstract. This paper introduces the concept of accessibility from the field of transportation planning and adopts it within the context of Information Retrieval (IR). An analogy is drawn between the fields, which motivates the development of document accessibility measures for IR systems. Considering the accessibility of documents within a collection given an IR System provides a different perspective on the analysis and evaluation of such systems which could be used to inform the design, tuning and management of current and future IR systems.

1 Introduction

Information Retrieval is the area that deals with the storage, organization, management and retrieval of information, where the goal of continual research in the field is to find *better* methods of doing the same. In pursuit of this betterment evaluation has been instrumental in the development of IR Systems. While evaluation has typically focused on the effectiveness [12], or the efficiency [14] of the IR system, these are only two ways in which to assess the quality of an IR system. In this paper, we introduce a complementary view to evaluation which provides a higher level view of the IR system by focusing on the *accessibility* an IR system provides to the documents in a collection.

Accessibility is an abstract concept coined almost 50 years ago in the land use and transportation planning field [8], where it was defined as a measure of potential opportunities for interaction with resources like employment, schooling, shopping, dining, etc. Measuring the accessibility in this context enabled many studies (e.g. [8][10][13][5]) to be performed which examined, for example, how changes in the levels of accessibility to such opportunities affected the urban area (in terms of economic impact, social changes and so forth). The results of such studies provide valuable information to transportation planners and city designers in the development of land use and transportation systems. Before this, planners and designers would focus on measures which were based on the effectiveness and efficiency of the transportation system (such as, the travel time between particular locations). However, accessibility provided a different

C. Macdonald et al. (Eds.): ECIR 2008, LNCS 4956, pp. 482–489, 2008.

perspective, and while related to effectiveness and efficiency, it takes a more abstracted or high level view on the evaluation of transportation systems, considering more general concerns relating to access, instead of focusing on specific instances.

Their definition of accessibility[1] considers the accessibility of opportunities at locations in a physical space (such as a city). The transportation system is the means by which opportunities are made accessible (i.e., the road network and the bus, cycle path and a bicycle, etc). In this context, the main consideration in the design and management of the transportation system is to look beyond efficiency and effectiveness and to consider the accessibility of opportunities given a certain distance or the generalized cost the user is willing to incur to reach these opportunities and the desirability of these opportunities.

In the context of Information Retrieval, an analogy of accessibility can be made as follows. Instead of an actual physical space, in IR, we are predominately concerned with accessing information within a collection of documents (i.e., information space), and instead of a transportation system, we have an Information Access System (i.e., a means by which we can access the information in the collection, like a query mechanism, a browsing mechanism, etc). The accessibility of a document is indicative of the likelihood or opportunity of it being retrieved by the user in this information space given such a mechanism. For example, in a hyper-linked collection exposed by a browsing-based system, a page with no incoming or outgoing links will have no accessibility. Conversely, a page with thousands of incoming links would be very accessible. Here, we consider the accessibility of documents given an IR system, where documents are accessed by querying the system. Each query provides a different ordering in which to access the documents in the information space. Much like a particular bus taking a pre-defined route through a city. However, unlike in the physical space, in the information space, there is no constraint imposed by the user's current location (i.e., at a particular document) because the IR system facilitates access to the collection regardless of location. The IR system is like being at a bus stop where every possible bus route is available, (i.e., the universe of all possible queries), and we can select any route desired, at any time. While this makes every document potentially accessible, the choice of route and distance the user is willing to travel will affect just how accessible documents are in the information space.

In this paper, our main contribution is the introduction of the concept of accessibility and the proposal of how to measure accessibility in this context. To do so, we first describe the related research in Section 2 and draw upon the

---

[1] Accessibility is also a key concept in other areas but defined differently. For instance, the disability rights movement advocates equal access to social, political, and economic life which includes not only physical access but access to the same tools, services, organizations and facilities. Another example is the World Wide Web Consortium (W3C)'s Web Accessibility Initiative (WAI), which is aimed to improve the accessibility of the World Wide Web for people using a wide range of user agent devices, not just standard web browsers. However, accessibility in these contexts concentrates on the physical aspects of accessing the information, and even extends to issues regarding usability and mobility.

extensive body of work in transportation planning and land use to provide the basis in developing measures of accessibility for IR system. Then in Section 3, we propose two IR based accessibility measures that are analogous to those in the field of transportation. The introduction of accessibility presents many different possibilities and challenges which can not be fully addressed here, so we summarize this initial contribution in Section 4.

2   Related Work

In Hansen's seminal paper [8] on measuring accessibility in transportation planning and land use, he defines how accessibility could be measured:

> a measurement of the spatial distribution of activities about a point adjusted for the ability and the desire of people or firms to overcome spatial separation. More specifically, the formulation states that the accessibility at point 1 to a particular type of activity at area 2 (say employment) is directly proportional to the size of the activity at area 2 (number of jobs) and inversely proportional to some function of the distance separating point 1 from area 2. The total accessibility to employment at point 1 is the summation of the accessibility to each of the individual areas around point 1. Therefore, as more and more jobs are located nearer to point 1, the accessibility to employment at point 1 will increase.

Key to this definition is the notion that as opportunities become further away the less accessible they become, and that by considering all possibilities to opportunities subject to the cost function based on the distance apart, provides a measure of accessibility. Essentially, this measure quantifies the *potential of opportunities for interaction* [8]. In the context of IR, the opportunities are the documents in the information space, and we wish to capture the *potential of documents for retrieval*.

2.1   Measures of Accessibility in Transportation Planning

There are numerous measures of accessibility that have been proposed in the field of Transportation Planning; the simplest and most popular measures are the Cumulative Opportunity Measures and Gravity Based Measures.

Cumulative Opportunity Measures also known as Isochrone measures count the number of opportunities that can be reached within a given travel time, distance, or generalized cost [13]. An example application of the measure is "the total number of dining opportunities within 400 metres". The advantage of this measure is that it is intuitive and easy to compute. However, the measure is sensitive to the size of the range (around the point of interest) to be considered, and the representation of the opportunities.

First derived by [8], *Gravity Based Measures* provide a general method for measuring accessibility, which is widely used. They differ from cumulative based measures in that they include a cost function within the calculation. Generally, the cost function takes the form of a negative exponential function (as described by [8], above), such that opportunities that are further away will have a lower impact on the final accessibility value. By "further", it is meant in terms of time, distance or generalized cost.

While more sophisticated measures have been developed, such as *Utility Based Measures* [10] and *Activity Based Measures* [5], we shall only be considering the former two methods in this work as they are the most widely used and accepted measures in transportation and planning. Thus, it seems reasonable to use these as a starting point to determine if they can be useful and informative in IR, before developing more sophisticated measures.

2.2   Accessibility in Information Retrieval

Accessibility issues in IR have focused on restrictions (physical and virtual) to index or retrieve information, whether this is because of a physical impairment [6], restricted access due to security clearance [9], or the inability to crawl portions of the web [2]. In each case, documents are inaccessible to the user or the system because of some physical or virtual limitation. For instance, in the latter case, the inability to crawl a web site means that certain documents are not indexed by the IR system, and therefore are not accessible to the user via the IR system. Recently, it was posited that the "searchability" of a web site would be affected by how easily pages can be crawled and how well the search engine matches and ranks them [11]. Searchability and accessibility are therefore very similar concepts. However, we are concerned with the influence of the IR system on accessing documents. Others (e.g. [7][3]) have considered how documents are accessed from the index in the retrieval process to facilitate more efficient retrieval by considering processor, disk and memory constraints. For instance "access-ordered indices" [7] are where the documents which are more likely to be returned at higher ranks are placed before those that are not likely to be returned at higher ranks. Another example, is the caching of queries [1], in web search engines, where results pages are cached in response to popular queries in order to facilitate efficient access.

In essence, IR is all about *accessing information*, and *how the information is accessed*. Our work is focused on measuring the accessibility of documents in the collection given the IR system used to access these documents. This is different from past work, in that we are specifically examining the influence of the IR system to restrict or promote access to the information within the collection as opposed to other restrictions. This paper hopes to establish the idea of accessibility as an integral concept in the field by highlighting its potential in the practical task of developing, building, and optimizing IR Systems, as well as diagnostics and evaluation.

3   Measuring Document Accessibility

Given a collection $\mathbf{D}$, an IR system accepts a user query $\mathbf{q}$ and returns a ranking of documents $\mathbf{R_q}$, which are deemed to be relevant to $\mathbf{q}$ from within $\mathbf{D}$ by the IR system. We can consider the accessibility of a document as a system dependent factor that measures how retrievable it is, with respect to the collection $\mathbf{D}$ and the ranking function used by the IR system. Using the analogy of transportation, entering a query is like to choosing a particular bus, where the order of documents returned are like the order of destinations reached for that given bus route. Opportunities to interact with resources while traveling along the route are reflected by going through the documents returned in the ranking $\mathbf{R_q}$. The accessibility of the resources (i.e., documents) is dependant on the willingness of the user to travel a certain distance along the route (i.e., traverse down the ranked list) and all the queries that users are likely to travel along. So, by adapting the measures from transportation planning, we propose a general measure of the accessibility of a document, as:

$$A(\mathbf{d}) = \sum_{\mathbf{q} \in \mathbf{Q}} o_q \cdot f(c_{dq}, \theta) \tag{1}$$

where o_q denotes the likelihood of expressing query $\mathbf{q}$ from the universe of queries $\mathbf{Q}$ and $f(c_{dq}, \theta)$ is a generalized utility/cost function where c_{dq} is the distance associated with accessing $\mathbf{d}$ through $\mathbf{q}$ which is defined by the rank of the document, and θ is a parameter or set of parameters given the specific type of measure.

A cumulative based measure can then be defined as follows: $\theta = c$, where c denotes the maximum rank that a user is willing to proceed down the ranked list. The function $f(c_{dq}, c)$ returns a value of 1 if $c_{dq} \leq c$ (with the top-most position considered as rank 1), and 0 otherwise. So, if returning a document in response to a given query has a distance greater than c associated with it, then it is considered unaccessible (for this query). For another query however, the document may be accessible because the cost of accessing it is within the distance c. Alternatively, the document could be considered accessible for the same query but to a user who has a higher cost threshold. Since all the documents within the cutoff defined by c are equally weighted, this type of measure emphasizes the number of times the document can be retrieved within that cutoff over the set $\mathbf{Q}$.

A gravity based measure can also be defined by setting the function to reflect the effort of going further down the ranked list, such that the further down the ranking the less accessible a document becomes. There are numerous ways in which such a function could be determined. Here, we adopt the function suggested in [8], where the accessibility of the document is inversely proportional to the rank of the document, such that:

$$f(c_{dq}, \beta) = \frac{1}{(c_{dq})^\beta} \tag{2}$$

where, the set of parameters θ includes β which is a dampening factor that adjusts how accessible the document is in the ranking. Interestingly, if the β parameter is set to one, then accessibility of the document for the given query is equivalent to the reciprocal rank of the document, which is related to the (expected) search length [4]. When there is only one relevant document, the expected search length is equivalent to the reciprocal rank of the document. Intuitively, the expected search length (ESL) and accessibility of documents is related, because the expected search length corresponds to how many irrelevant documents have to be examined in order to find the relevant documents. The expected search length to a particular document is proportional to the accessibility of the document for a given query. However, what the accessibility measure captures is more general, i.e., how retrievable the document is given all possible/likely queries regardless of relevancy, but this link to ESL and reciprocal rank appears to provide a connection between accessibility and effectiveness. As we have previously mentioned this direction is left for future work.

Given either measure, $A(\mathbf{d})$ provides an indication of the opportunity of retrieving $\mathbf{d}$. This value can be obtained for each document $\mathbf{d} \in \mathbf{D}$ so that we can compare whether there is more opportunity to retrieve one document over another. Using this measure to compare groups of documents has potential to aid in the design, management and tuning of retrieval systems in a number of ways. Imagine that for a given collection of documents and a given IR system, the average $A(\mathbf{d})$ of a set of documents is extremely high, while for another set of documents the average $A(\mathbf{d})$ is very low. Perhaps, the first set of documents was a group of site entry pages, and our system has a prior towards such pages, thus we would expect these pages to have a higher $A(\mathbf{d})$. In this case, it is desirable that these documents are so accessible. On the other hand, if the set of highly accessible pages was composed of spam pages, because these pages have used "tricks" to artificially inflate the number of queries for which they are retrieved, then this is not desirable and the system needs to be adjusted. Alternatively, if there is a set of documents which are virtually inaccessible in the collection, then it is a management decision to decide whether these documents should be included in the index or not.

At a higher level, the measure $A(\mathbf{d})$ motivates questions regarding how accessible documents in the collection should be, and whether we are interested in trying to "hide" or "promote" certain documents within the collection. Or whether we should adopt an approach that ensures access to the information is free from bias, i.e. "universal access"[2] so that *any document is as accessible as any other document* in the collection. This provides a novel framework for measuring document accessibility, which enables the consideration of such questions and issues.

4   Conclusion and Future Work

The main contribution of this paper is the introduction of the concept of accessibility and quantifying the accessibility of documents in the collection given a particular IR system. Measures of accessibility are not performance measures like effectiveness or efficiency, but instead are measures of the *potential of documents*

---

[2] As previously mentioned, the disability rights movement advocates equal access and terms this notion as universal access.

for retrieval. This abstraction provides a novel way to quantify and detect different levels of accessibility within the collection imposed by the IR system. For a system administrator, this could prove to be very useful in designing, managing and tuning the IR system.

This work represents the initial step towards formalizing accessibility and developing accessibility measures for information spaces, in IR and more generally for any Information Access system. However, there are many open problems, challenges and issues which have arisen as a result of this work. Further research needs to be conducted in two main directions:

1. the calibration, computation and estimation of document accessibility measures, and
2. the application of document accessibility measures.

Acknowledgements. The authors would like to thank Stephen Robertson, Keith van Rijsbergen and Murat Yakici for their helpful and insightful comments and suggestions.

References

1. Baeza-Yates, R., Gionis, A., Junqueira, F., Murdock, V., Plachouras, V., Silvestri, F.: The impact of caching on search engines. In: Proceedings of the 30th ACM SIGIR conference, pp. 183–190 (2007)
2. Bailey, P., Craswell, N., Hawking, D.: Chart of darkness: Mapping a large intranet. Technical report, CSIRO Mathematical and Information Sciences (2000)
3. Buttcher, S., Clarke, C.L.A.: A document-centric approach to static index pruning in text retrieval systems. In: Proceedings of the 15th ACM Conference on Information and Knowledge Management (2006)
4. Cooper, W.S.: Expected search length: A single measure of retrieval effectiveness based on weak ordering action of retrieval systems. Journal of the American Society for Information Science 19(1), 30–41 (1968)
5. Dong, X., Ben-Akiva, M.E., Bowman, J.L., Walker, J.L.: Moving from trip-based to activity-based measures of accessibility. Transportation research. Part A, Policy and practice 40(2), 163–180 (2006)
6. Fajardo, I., Canas, J.J., Salmeron, L., Abascal, J.: Improving deaf users' accessibility in hypertext information retrieval: are graphical interfaces useful for them? Behaviour and Information Technology 25(6), 455–467 (2006)
7. Garcia, S., Williams, H.E., Cannane, A.: Access-ordered indexes. In: Twenty-Seveth Australasian Computer Science Conference (ACSC 2004), pp. 7–14 (2004)
8. Hansen, W.: How accessibility shape land use. Journal of the American Institute of Planners 25(2), 73–76 (1959)
9. Hawking, D.: Challenges in enterprise search. In: ADC 2004: Proceedings of the 15th Australasian database conference, Darlinghurst, Australia, pp. 15–24. Australian Computer Society, Inc. (2004)
10. Neuburger, H.: User benefit in the evaluation of transport and land use plans. Journal of Transport Economics and Policy 5, 52–75 (1971)
11. Upstill, T., Craswell, N., Hawking, D.: Buying bestsellers online: A case study in search & searchability. In: 7th Australasian Document Computing Symposium, Sydney, Australia (2002)

12. van Rijsbergen, C.J.: Information Retrieval, 2nd edn. Butterworths, London (1979)
13. Wachs, M., Kumagai, T.G.: Physiscal accessibility as a social indicator. Socioeconomic Planning Science 7, 327–456 (1973)
14. Witten, I.H., Moffat, A., Bell, T.C.: Managing Gigabytes: Compressing and Indexing Documents and Images, 2nd edn. Morgan Kaufmann Publishing, San Franciso (1999)

Semantic Relationships in Multi-modal Graphs for Automatic Image Annotation*

Vassilios Stathopoulos, Jana Urban, and Joemon Jose

Department of Computer Science, University of Glasgow,
17 Lilybank Gardens, Glasgow G12 8QQ, UK
{stathv,jj}@dcs.gla.ac.uk

Abstract. It is important to integrate contextual information in order to improve the inaccurate results of current approaches for automatic image annotation. Graph based representations allow incorporation of such information. However, their behaviour has not been studied in this context. We conduct extensive experiments to show the properties of such representations using semantic relationships as a type of contextual information. We also experimented with different similarity measures for semantic features and results are presented.

1   Introduction

Multimedia content, and especially image and video, is produced at highly increasing rates. This indicates the need for effective methodologies for storing and organising multimedia content in order to render it accessible and reusable. Early Content Based Image Retrieval (CBIR) systems were solely based on indexing low-level visual features. The success of such systems, however, was limited mainly due to the semantic gap [1]. A solution towards bridging the semantic gap is to index images using also semantic features, such as keywords, describing the content of the image. The majority of Automatic Image Annotation (AIA) systems incorporate statistical approaches for finding correlations between image visual features and words used to annotate images in a training set. The learnt correlations can then be used to annotate new images.

Often not all the keywords are distinguishable from the visual features alone. For example the concepts of 'meeting' and 'corporate leader' are two of the concepts used in the TrecVid 2006 evaluation campaign [2]. Contextual image information can be used to identify concepts non-distinguishable from visual features and improve object detection. Recently, it was shown that relationships between semantic features can be utilised to improve the annotation performance of existing algorithms [3]. Removing irrelevant terms and identifying others more relevant to be included in the annotation can significantly improve performance.

Graphs and graph learning algorithms provide an interesting alternative for the problem of inference using multi-modal representations of documents. Graph

---

* The research leading to this paper was supported by European Commission under contracts FP6-027026(K-Space) and FP6-027122(Salero).

C. Macdonald et al. (Eds.): ECIR 2008, LNCS 4956, pp. 490–497, 2008.

representations of image collections have been previously used for Automatic Image Annotation in [4] and Image Retrieval in [5]. In [4] only similarities between visual features are incorporated in the graph while, in [5] relationships incorporating image usage information are also considered. In this paper a graph representation is extended to incorporate semantic relationships and the effects in annotation performance as well as the properties of the correlation measure between graph nodes are investigated. Doing so we wish to study the potential of graph models and graph correlation measures for integrating contextual information in an ad-hoc manner.

2   Images and Their Captions as a Graph

An image can be represented by a number of low-level visual features which can be global (extracted from the whole image) or local (extracted from image regions after a segmentation algorithm and concatenated to a single vector describing the image region). In either case an image can be decomposed into a number of feature vectors. Images, their corresponding feature vectors and words can be represented as nodes in a graph $G =< V, E >$, where V is the set of all nodes and E is the set of all edges. A similar strategy with those in [4] and [5] is followed to construct the graph. Let W be the set of nodes representing unique words w used as captions for all the images in the collection. Also let F be the set of nodes representing all the feature vectors f extracted from all the images. Finally let I be the set of all nodes representing images i in the collection. Then the vertices of the Image Graph (IG) can be defined as $V = I \cup F \cup W$.

The relationship between images and their feature vectors can be encoded in the IG by a pair of edges (i_n, f_j) and (f_j, i_n) connecting image nodes i_n and their feature vectors f_j. In a similar way relationships between images and their caption words are encoded by a pair of edges (i_n, w_j) and (w_j, i_n). Now assume that a function $dist(f_i, f_j)$ returns a positive real value measuring the distance, or dissimilarity, between two feature vectors. This function can be the Euclidean distance or any other valid distance metric on feature vectors. Using this function, the k nearest neighbours of each feature vector f_i are selected and a pair of edges $\{(f_i, f_k), (f_k, f_i)\}$ for all the k nearest neighbors, is used to denote their similarity in IG. Similarly, assume a function $dist(w_i, w_j)$ or $sim(w_i, w_j)$ returning a positive real number quantifying the distance or similarity of two words. We will discuss these two functions in the next section. Again the k nearest neighbours of each w_i can be selected and a pair of edges $\{(w_i, w_k), (w_k, w_i)\}$, for all k, is included to the graph to represent semantic relationships between words.

2.1   Finding Correlations between Graph Nodes

One measure of correlation between nodes in a graph can be derived as follows. By performing a random walk on a graph the long term visit rate, or the stationary probability, of each node can be calculated. Random Walks with Restarts

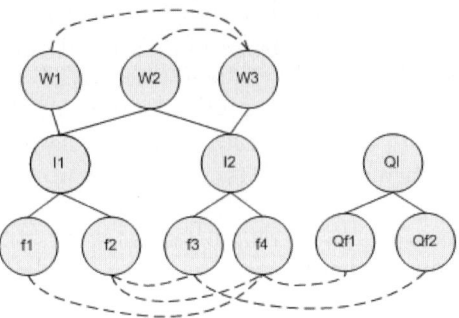

Fig. 1. Image graph with nodes corresponding to a new query image and its feature vectors, QI, Qf1 and Qf2. k = 1 for this graph

(RWR) [6] are based on the same principle but the stationary probabilities are biased towards a specific node, referred to as the restart node. Starting from a node s, a RWR is performed by randomly following a link to another node at each step with a probability a to restart at s. Let $\mathbf{x}^{(t)}$ be a row vector where $\mathbf{x}_i^{(t)}$ denotes the probability that the random walk at step t is at node i. $\mathbf{s}$ is a row vector of zeros with the element that corresponds to the starting node set to 1. Also let $\mathbf{A}$ be the row normalized adjacency matrix of the graph IG. In other words $\mathbf{A}$ is the transition probability table where the element $a_{i,j}$ gives the probability of j being the next state given that the current state is i. The next state is then distributed as

$$\mathbf{x}^{(t+1)} = (1 - a)\mathbf{x}^{(t)}\mathbf{A} + a\mathbf{s} \tag{1}$$

To annotate a new image it's feature vectors are calculated and the corresponding nodes are inserted to the graph, see Fig. 1. The starting node is set to the new node corresponding to the query image (QI in Fig. 1). The stationary probability of all words are calculated by recursively applying (1) until convergence. Words are sorted in decreasing order of their stationary probability and the top, say 5 words are selected to annotate the new image.

3   Semantic Relationships

In this study we adopt two approaches for calculating the semantic similarity of words used as captions of images. The first method exploits the co-occurrence of words in the WWW assuming a global meaning of words. The second exploits the co-occurrence of words in the training set in order to identify particular uses of words in the image collection.

3.1   Normalized Google Distance

Although the WWW is not the most reliable source of information, it does reflect the average interpretation of words' meaning globally. Cilibrasi and Vitanyi

[7] propose a method for estimating a distance between words by utilizing page counts returned from web search engines. The probability of a word w_i can be taken to be the relative frequency of pages containing w_i thus $p(w_i) = f(w_i)/N$, where f is a function which returns the number of pages containing w_i in the search engine's index and N is the number of web pages indexed by the search engine. Similarly the probability of a word w_j, $p(w_j)$ as well as the joint probability of the two words $p(w_i, w_j)$ can also be obtained using a web search engine. Therefore the conditional probability $p(w_i|w_j)$ can be estimated as $p(w_i|w_j) = p(w_i, w_j)/p(w_j)$. Since $p(w_i|w_j) \neq p(w_j|w_i)$, the minimum is taken in order to calculate a distance between w_j and w_i giving $dist(w_i, w_j) = min\{p(w_i|w_j), p(w_j|w_i)\}$.

Based on this simple measure the authors in [7] develop the Normalized Google Distance (NGD) that utilizes the Google search engine to estimate the meaning and similarities of words. NGD is expressed as

$$NGD(w_i, w_j) = \frac{\max(\log(1/f(w_i)), \log(1/f(w_j))) - \log f(w_i, w_j)}{\log N - \min(\log f(w_i), \log f(w_j))} \quad (2)$$

3.2 Automatic Local Analysis

Automatic Local Analysis (ALA) is mainly used for query expansion in traditional IR [8]. It utilises documents returned as a response to a user query in order to calculate co-occurrences of words. Then the query can be expanded with highly correlated keywords. The aim of the approach followed in this study is to calculate a similarity between words regardless of the query, in order to enhance the structure of the image graph with semantic relationships. Images can be considered as documents while the frequency of a word in an image is either 1 or 0.

Let H be an $N \times M$ matrix where N is the number of unique words used to annotate the image collection and M is the number of images in the collection. An element $H_{i,j}$ is equal to 1 if and only if word w_i is in the caption of image w_j and 0 otherwise. The co-occurrence correlation between w_i and w_j is then defined as $corr(w_i, w_j) = \sum_{t=1}^{M} H_{it} \times H_{jt}$

This measure gives the number of images where the two words appear together. Words that appear very often in the collection will tend to co-occur frequently with most of the words in the vocabulary and thus the score can be normalized to take into account the frequency of the words in the collection.

$$NormCorr(w_i, w_j) = \frac{corr(w_i, w_j)}{corr(w_i, w_i) + corr(w_j, w_j) - corr(w_i, w_j)} \quad (3)$$

Using (3) the neighborhood of a word w_i can be defined as a vector $\mathbf{s}_{w_i} = \{NormCorr(w_i, w_1), \ldots, NormCorr(w_i, w_N)\}$. Words having similar neighborhood frequently co-occur with a similar set of words and thus they have some synonymic relation. The semantic relationship between two words can then be calculated using the cosine of $\mathbf{s}_{w_i}$ and $\mathbf{s}_{w_j}$.

4   Experiments and Results

In this study a subset of the Corel image collection consisting of 5000 manually annotated images was used. The dataset is divided into a training set (4500 images) and a test set (500 images). For this dataset image regions are extracted using the NormalisedCuts algorithm and visual features extracted from each region are concatenated in a single vector. The visual features used are average and standard deviation of RGB and LUV values, mean oriented energy and 30 degrees increments, region and location of the region, region convexity, region angular mass and the region boundary length divided by the region's area. For more information about the features extraction and segmentation process refer to [9]. The dataset is available for download[1] and is extensively used in the literature [9,10,11].

For the first run (RWR) of the algorithm described in Section 2, edges between word nodes denoting semantic relationships are discarded while individual features in the region feature vectors are normalized to 0 mean and 1 variance. The values for the restart probability and the number of nearest neighbors for each region feature vector, as have been shown in [4], can be set empirically to $a = 0.65$ and $k = 3$ respectively. The second run (RWR+ALA) incorporates edges between word nodes indicated by the similarity of words as calculated by Automatic Local Analysis described in Section 3.2. Finally in the third run (RWR+NGD) the Normalized Google Distance is used to create edges between word nodes based on their semantic distances.

Results in Fig. 2(a) and Fig. 2(b) are reported using average Accuracy [4], Normalized Score[12] and average Precision-Recall [9,11,10] measures. Accuracy for an image is defined as the number of correctly annotated words divided by the number of the expected words for the particular image. The expected number of words is the number of words in the true annotation of the image taken from the test set. Normalized Score is defined as $NS = Accuracy - inc_i/(N_w - e_i)$ where inc_i is the number of incorrectly predicted words for the ith image, N_w is the number of words in the vocabulary and e_i is the number of expected words for the ith image. Averages are taken over all images in the test set. Precision and Recall are measured for each word in the vocabulary and are defined as follows. Precision is the number of correctly annotated images with a particular word divided by the number of images annotated by that word. Recall is the number of correctly annotated images divided by the number of relevant images in the test set. The relevant images are simply the images having the particular word in their true annotations. In this study we report average Precision Recall values over all the words in the vocabulary.

Despite the small increase in performance, the differences in the Accuracy and Normalized Score averages between the RWR and RWR+ALA runs are statistically significant using a paired t-test with a 0.05 threshold. On the other hand, the average differences between the RWR and RWR+NGD runs are not statistical significant. This indicates that semantic relationships calculated using

---

[1] http://kobus.ca/research/data/eccv_2002/

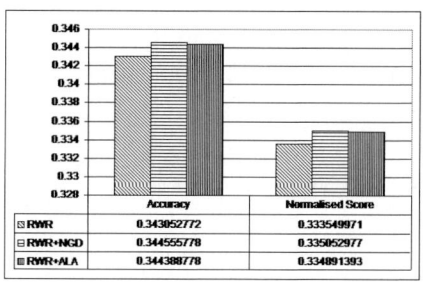

	Accuracy	Normalised Score
RWR	0.343052772	0.333549971
RWR+NGD	0.344555778	0.335052977
RWR+ALA	0.344388778	0.334891393

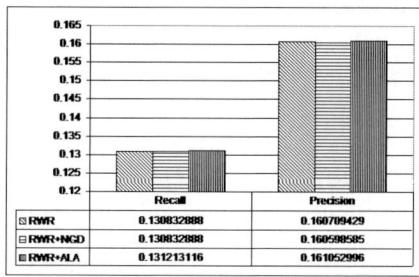

	Recall	Precision
RWR	0.130832888	0.160709429
RWR+NGD	0.130832888	0.160598585
RWR+ALA	0.131213116	0.161052996

(a) Avg. Accuracy and Normalised Score   (b) Avg. Precision and Recall

Fig. 2. Results obtained from the three runs of the algorithm. See text for description.

Automatic Local Analysis in the image graph can improve the annotation performance although the increase is not dramatic.

The not statistically significant improvement of results obtained by RWR+NGD can have two possible explanations. Firstly, in contrast to the assumption of the authors in [7], NGD is not symmetric $NGD(w_i, w_j) \neq NGD(w_j, w_i)$. In [7] is assumed that the Google search engine returns the same number of pages regardless of the order of the words in the query. Thus the second term in the numerator of (2) is assumed to be symmetric. However, during this study it was found that $f(w_i, w_j) \neq f(w_j, w_i)$ which was probably due to changes in the implementation of the Google search engine. Secondly, NGD reflects the co-occurrence of words in the WWW. While for some words these can be beneficial for image annotation, for others might lead to the opposite results.

The improvement in performance in the RWR+ALA run is due to the improved detection accuracy of particular words. Studying the raw results we found that only 5 words are affected by the semantic relationships and the corresponding Precision Recall values are given in Fig. 4. Studying the Precision Recall measures for each individual word we also found interesting properties of the stationary probability obtained by RWR. Firstly, we found that although for some words there are significantly more training images in the training set than for other words, most of the time the more frequent words are erroneously predicted. For example, for the word 'water' there are 1004 images in the training set. The Precision and Recall for this word is 0.269 and 0.931 indicating that this word is erroneously predicted mostly due to its frequency in the training set. On the other hand, for the word 'jet' the corresponding Precision and Recall values are 0.705 and 0.63 while there are only 147 images in the training set.

Secondly we found a relationship of the restart probability with the number of words having at least one image correctly annotated. As the restart probability increases the number of words with at least one image correctly annotated increases. For a small restart probability only the most frequently occurring words in the training set are predicted, while for a larger restart probability the stationary probability favours word nodes closer to the query image node. In this context the distance between nodes is the geodesic distance in the graph. In Fig. 4 we show how the number of words with positive recall behaves for different values of the restart probability. For this experiment we did not use semantic relationships; however the behavior is similar to when edges between word nodes are incorporated regardless of the method used (ALA or NGD).

	RWR		RWR+ALA	
Word	Precision	Recall	Precision	Recall
grass	0.22929	0.70588	0.23076	0.70588
rocks	0.16279	0.31818	0.16666	0.31818
ocean	0.35714	0.55555	0.38461	0.55555
tiger	0.62532	0.5	0.66666	0.6
window	0.33333	0.125	0.52356	0.125

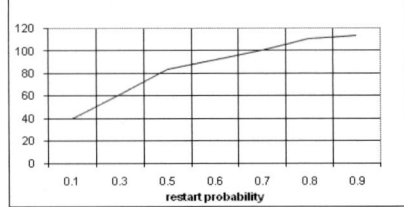

Fig. 3. Precision Recall values for the five words which are affected from semantic relationships

Fig. 4. Number of words with positive recall for different values of restart probability

These findings suggest that the stationary probability obtained by RWR is mostly affected by the frequency of occurrence of the words in the training collection. In other words, it is affected by the connectivity of the nodes in the graph. The notion of distance between nodes in the graph is reflected by the restart probability, although there is not an explicit relation. We conclude that both connectivity of nodes and the geodesic distances in the graph are important properties that must be explicitly considered in the correlation measure.

5   Discussion and Conclusion

Although we achieved a small statistically significant improvement in annotation performance, we have identified two drawbacks to the stationary probability as a correlation measure. First, in contrast to traditional machine learning techniques, the number of training samples for a particular word had negative effect on annotation performance. Second, although the geodesic distances of nodes in the graph are of significant importance in order to facilitate inference, the stationary probability does not define a distance in the graph. The notion of distance is encoded by the restart probability, but the relation is not clear.

One of the most successful applications of the stationary probability obtained by Random Walks with Restarts is the so called PageRank[13] measure of web page relevance. PageRank is an indicator of relevance based on the quality of web page citation measuring in-links of each web-page. In such application the edges in the graph denote attribute value relationships of the form "page A suggests/links to page B". For AIA, the majority of edges in the image graph denote similarities or distances between nodes. We conclude that for such type of edges a correlation measure based mostly on the connectivity of nodes in the graph is not appropriate, leading to the problems above mentioned.

The graph representation, however, provides an interesting approach for integrating contextual information which is vital for improving performance. There are number of different graph correlation measures that can be defined which might be more appropriate for the Automatic Image Annotation problem. We are currently experimenting with other measures such as the Average First Passage Time [14,15].

References

1. Smeulders, A.W.M., Worring, M., Santini, S., Gupta, A., Jain, R.: Content-based image retrieval at the end of the early years. IEEE TPAMI 22(12), 1349–1380 (2000)
2. Smeaton, A.F., Over, P., Kraaij, W.: Evaluation campaigns and trecvid. In: MIR 2006, Santa Barbara, CA, USA, pp. 321–330. ACM Press, New York (2006)
3. Wang, Y., Gong, S.: Refining image annotation using contextual relations between words. In: CIVR, Amsterdam, The Netherlands, ACM, pp. 425–432 (2007)
4. Pan, J.Y., Yang, H.J., Faloutsos, C., Duygulu, P.: Gcap: Graph-based automatic image captioning. In: CVPRW 2004, vol. 9, p. 146. IEEE Computer Society Press, Los Alamitos (2004)
5. Urban, J., Jose, J.M.: Adaptive image retrieval using a graph model for semantic feature integration. In: MIR 2006, Santa Barbara, CA, USA, pp. 117–126. ACM Press, New York (2006)
6. Lova'sz, L.: Random walks on graphs: A survey. In: Combinatronics Bolyai Society for Mathematical Studies, Budapest, vol. 2, pp. 353–397 (1996)
7. Cilibrasi, R., Vitanyi, P.M.: Automatic meaning discovery using google. In: Hutter, M., Merkle, W., Vitanyi, P.M. (eds.) Kolmogorov Complexity and Applications. Number 06051 in Dagstuhl Seminar Proceedings, Schloss Dagstuhl, Germany, IBFI (2006)
8. Baeza-Yates, R., Ribeiro-Neto, B.: Modern Information Retrieval. Addison Wesley, Reading (1999)
9. Duygulu, P., Barnard, K., de Freitas, J.F.G., Forsyth, D.A.: Object recognition as machine translation: Learning a lexicon for a fixed image vocabulary. In: Heyden, A., Sparr, G., Nielsen, M., Johansen, P. (eds.) ECCV 2002. LNCS, vol. 2350, pp. 97–112. Springer, Heidelberg (2002)
10. Yavlinsky, A., Schofield, E.J., Rüger, S.: Automated image annotation using global features and robust nonparametric density estimation. In: Leow, W.-K., Lew, M., Chua, T.-S., Ma, W.-Y., Chaisorn, L., Bakker, E.M. (eds.) CIVR 2005. LNCS, vol. 3568, pp. 507–517. Springer, Heidelberg (2005)
11. Jeon, J., Lavrenko, V., Manmatha, R.: Automatic image annotation and retrieval using cross-media relevance models. In: SIGIR 2003, Toronto, Canada, pp. 119–126. ACM Press, New York (2003)
12. Barnard, K., Duygulu, P., Forsyth, D., de Freitas, N., Blei, D.M., Jordan, M.I.: Matching words and pictures. JMLR 3, 1107–1135 (2003)
13. Page, L., Brin, S., Motwani, R., Winograd, T.: The pagerank citation ranking: Bringing order to the web. Technical report, Stanford Digital Library Technologies Project (1998), http://dbpubs.stanford.edu:8090/pub/1999-66
14. Kemeny, J.G., Snell, J.L.: Finite Markov Chains. Springer, New York (1960)
15. Gallager, R.G.: Discrete Stochastic Processes. Kluwer Academic, Boston (1996)

Conversation Detection in Email Systems

Shai Erera[1,2] and David Carmel[2]

[1] University of Haifa, 31905 Haifa, Israel
shaie@il.ibm.com
[2] IBM Research Lab in Haifa, 31905, Israel
carmel@il.ibm.com

Abstract. This work explores a novel approach for conversation detection in email mailboxes. This approach clusters messages into coherent conversations by using a similarity function among messages that takes into consideration all relevant email attributes, such as message subject, participants, date of submission, and message content. The detection algorithm is evaluated against a manual partition of two email mailboxes into conversations. Experimental results demonstrate the superiority of our detection algorithm over several other alternative approaches.

1 Introduction

Electronic mail (email) has become one of the most popular tools for handling conversations among people. In general, a typical user mailbox contains hundreds of conversations. Detecting these conversations has been identified a long ago as an important task [4]. Clustering the messages into coherent conversations has many useful applications, among them are allowing users to see a greater context of the messages they are reading and collating related messages automatically.

Several email clients deal with conversation detection by detecting *email threads*. The Internet message format, RFC 2822 [10], is a common standard that specifies a syntax for text messages within the framework of email systems. It does not provide a precise definition of an email thread, but instead defines how an email thread can be detected using structural attributes embedded in email messages, such as "In-Reply-To", "References" and "Message-ID".

Thread detection based on structural attributes has some challenges, starting with that not all email clients support the structural attributes required for detecting email threads [5,8]. In this work, we follow the path of conversation detection based on email attributes. We begin by sharpening the distinction between email threads and conversations. A thread is defined and detected according to the RFC 2822 standard, while a conversation is defined as *an exchange of messages among the "same" group of people on the "same" topic*[1]. We group messages into coherent conversations by using a similarity function that takes into consideration all relevant email attributes, such as message subject, participants, date of submission, and message content. We study the contribution of

---

[1] The notions of the *same* group of people and the *same* topic will be clarified in Section 3.

C. Macdonald et al. (Eds.): ECIR 2008, LNCS 4956, pp. 498–505, 2008.

the message attributes to conversation detection by experimental analysis using a set of email messages that were manually clustered into coherent conversations. We show experimentally that the detected conversations better suit the user's expectations (as reflected by the manually marked conversations) than structural-based email threads.

2   Related Work

Extensive research has been done over the past years on using email structure, and especially email threads, in several email-based applications [4,6]. Thread detection has also attracted significant attention [5,8,9]: Lewis and Knowels [8], and recently Aaron and Jen-Yuan [1], show that by applying text matching techniques to the textual portions of messages they are able to detect threads effectively.

While these methods are highly effective in detecting threads, they may fail to detect all conversations. Klimt and Yang [5] group messages that have the same subject attributes and are sent among the same group of people. Conversations, on the other hand, may span several threads with similar (but not exact) subject lines. In addition, a conversation may not include all the participants in all the messages.

Recently, Gabor et al. [2] developed an email client extension that clusters messages by topic. The similarity between two messages, as applied by the clustering algorithm, is based on the email message *subject*, *date*, *participants*, and *content* attributes. Their approach is the most similar to ours and will likely detect similar conversations to those observed by our method. However, as they noted, their clustering approach is focused on topic detection, hence messages belonging to different conversations on the same topic will be clustered together.

3   Conversation Detection

In this section we formally describe the process of detecting conversations from a collection of email messages. We begin by defining the basic concepts used in this work:

Email Thread: is a sequence of messages, ordered by their date of submission, that are related according to their structural attributes as defined by the Internet standard RFC 2822 [10].

Subject Thread: is a sequence of messages, ordered by their date of submission that are related according to their *subject* attribute regardless of their structural relationships. Two messages, e_1 and e_2, belong to the same subject thread if and only if their "core" subject is identical. A "core" subject is extracted from the full message subject by eliminating common prefixes such as "Re:", "Re: Re:", "Fwd:", etc., which are very common in email systems.

Sub-Conversation: is a sequence of messages, ordered by their date of submission, belonging to a subject thread that focus on the same topic and are among the same group of people. A subject thread may contain one or more sub-conversations. One breaks a subject thread by replying to a message while keeping the subject, however changing the topic or modifying the participants. Another indication for a new conversation in the same subject thread, as noted in [8], is a long time break between two consecutive messages. In addition, users may compose a new message on a different topic, but with the same subject appeared in other previous messages.

Conversation: is a sub-conversation, or a sequence of sub-conversations that focus on the same topic and are among the same group of people. The sequence of the sub-conversations is ordered by their *date* attribute.

3.1   Message Similarity

We measure the similarity between two messages as a linear combination of the similarity between their attributes.

Subject: The similarity of two subject attributes is determined by the subject words. Let S_i and S_j be the sets of words belonging to the core subject attributes of two messages, e_i and e_j, respectively. Then the subject similarity is defined by the Dice coefficient similarity: $subj(e_i, e_j) = \frac{2|S_i \cap S_j|}{|S_i| + |S_j|}$. For a conversation containing several sub-conversations, the subject attribute is determined by concatenating the subjects of all sub-conversations. Therefore, the same function will be used to measure subject similarity between conversations.

Date: Date attributes are highly important for detecting conversations. As Kalman and Rafaeli [3] discovered, a reasonable response time for an email message is five days after it has been sent while most messages are being replied in a matter of hours. Therefore, we use a *max date difference* threshold, *mdf*, above which the date similarity is zeroed. Let d_i and d_j be the date attributes of two messages, e_i and e_j respectively. The date similarity of two messages is defined as: $date(e_i, e_j) = 1 - \min(1, \frac{|d_i - d_j|}{mdf})$.

The date of a conversation is defined as a (d_s, d_e) pair where d_s and d_e are set to the earliest and latest dates of all messages in the conversation. Let (d_{s_i}, d_{e_i}) and (d_{s_j}, d_{e_j}) be the date attributes of two conversations, c_i and c_j respectively, and without loss of generality let $d_{s_i} \leq d_{s_j}$. The date similarity of two conversations is defined as:

- If $d_{e_j} \leq d_{e_i}$ then the time range of conversation c_j is fully contained in the time range of c_i and $date(c_i, c_j) = 1$.
- If $d_{e_i} \leq d_{s_j}$ then the conversations are disjoint. If $d_{e_i} > d_{s_j}$ then the conversations intersect. In both cases $date(c_i, c_j)$ is computed as $date(e_i, e_j)$, where $d_i = d_{e_i}$ and $d_j = d_{s_j}$.

Participants: A *participants* attribute is determined by an aggregation of all participants mentioned in the *From, To,* and *Cc* attributes of the message. In [2], the similarity between two participant attributes is calculated by the Dice coefficient similarity between two sets, while excluding the mailbox owner due to his/her participation in all the mailbox messages. One drawback of this approach is that there is no distinction between active and passive participants.

We define the participants similarity by using a variant of the Dice similarity, taking the activity role of participants into consideration. Let P_i and P_j be the sets of participants of messages e_i and e_j respectively, including the mailbox owner. Let $w(p, e)$ be the activity weight of participant p in message e. We associate a high activity weight for active participants (coming from the *From* and *To* message attributes), a lower weight for passive ones (from the *Cc* attribute), and a zero weight when p does not participate in message e. Since participants might have different activity roles in the two messages, we average the activity weights of the participants over the two messages: $aw(p, e_i, e_j) = \frac{w(p,e_i)+w(p,e_j)}{2}$. The participants similarity between two massages is defined as: $part(e_i, e_j) = \frac{\sum_{p \in P_i \cap P_j} aw(p,e_i,e_j)}{\sum_{p \in P_i \cup P_j} aw(p,e_i,e_j)}$.

The activity weight of a participant in a conversation is determined by averaging its activity weight over the sequence of messages. Hence, $part(c_i, c_j)$ can be computed as $part(e_i, e_j)$ while using the average weight of participants over the sequence of messages.

Content: While creating a reply message, many email clients automatically quote the previous message content in the reply. In a long thread of messages, the quoted part of some messages might be very long comparing to their unquoted part. We therefore recursively split the content of a message to its quoted and unquoted parts.

Given two messages e_i and e_j with their corresponding elements $e_{i_1}, \ldots, e_{i_m}$ and $e_{j_1}, \ldots, e_{j_n}$, as extracted from the message content. In addition, and without loss of generality, let $date(e_i) > date(e_j)$. We compute the textual similarity between each element in e_j to the unquoted part of e_i taking the maximum value as the similarity measure between the two messages.

The similarity between two elements is calculated using the well-known $tf - idf$ cosine similarity, $sim(e_i, e_j)$. Given el_{u_i}, the unquoted part of message e_i, and $el_{j_1} \ldots el_{j_k}$ the elements of message e_j, the content similarity between two messages (e_i, e_j) is defined as: $content(e_i, e_j) = \max_{1 \le t \le k} sim(el_{u_i}, el_{j_t})$.

For a sub-conversation S_i, the content attribute includes the unique elements from all its messages. The content similarity between a message e_k and sub-conversation S_i, $content(e_k, S_i)$, is calculated exactly the same, only we compare the unquoted part of e_k to all the elements of S_i. Given two sub-conversations S_i and S_j and their corresponding unique elements $S_{i_1}, \ldots, S_{i_m}$ and $S_{j_1}, \ldots, S_{j_n}$, we compute the maximal similarity between all elements of S_i to all elements of S_j.

Similarity Function. Finally, the similarity between two messages, e_i and e_j is computed by a linear combination of the similarities between their attributes:

$$Sim(e_i, e_j) = w_s \times subj(e_i, e_j) + w_d \times date(e_i, e_j) + \qquad (1)$$
$$w_p \times part(e_i, e_j) + w_b \times content(e_i, e_j)$$

Since all similarity functions of all attributes are also defined for conversation attributes, Equation 1 can also be used to measure the similarity between such instances.

3.2 The Conversation Detection Algorithm

Figure 1 illustrates the conversation detection process. It begins by grouping messages with an identical "core" subject into Subject threads. These threads are broken to sub-conversations, each includes all messages belonging to the same conversation according to their similarity matching (using Equation 1). Finally, similar sub-conversations are grouped together to form conversations (using Equation 1).

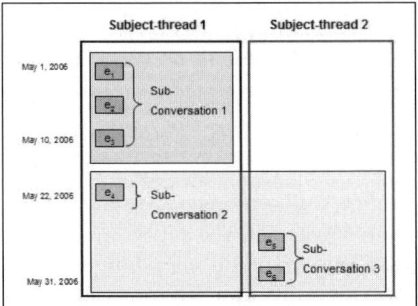

Fig. 1. The conversation detection process

The clustering algorithm can also be applied incrementally. Sorting messages and conversations by their *date* attributes allows us to efficiently compare a new arriving message to all the conversations detected so far. If a good candidate is found (if the similarity of the message to a conversation is higher than a certain threshold), we can add this message to that conversation. Otherwise we start a new one.

4   Experiments

To evaluate our approach, we need a set of manually marked messages of a single mailbox. One candidate is to use a mailbox from the publicly available Enron corpus [5]. However, marking conversations in a mailbox is a subjective task which strongly depends on the user's familiarity with the mailbox content. Our attempts to mark messages in this corpus proved almost infeasible - since often looking at a single message gave no obvious clues as to which conversation the message should belong. We therefore manually marked conversations in two mailboxes, each marked by its owner.

We used two manually marked mailboxes (mailbox A and B). The recent 448 messages of mailbox A were classified to 145 subject threads, that were manually split to 161 sub-conversations which were grouped to 147 conversations. The recent 500 messages of mailbox B were classified to 355 subject threads, that were manually split to 376 sub-conversations which were grouped to 334 conversations.

4.1   Evaluation Process

One way to compare manual and automatic detected conversations is to measure the distance between the two partitions of the same set of messages. We compare the two partitions by measuring the agreement between them. Given a set of messages, partitioned into n manual conversations $M_c = \{MC_1, \ldots, MC_n\}$, and k automatic conversations, $A_c = \{AC_1, \ldots, AC_k\}$. We use Tp to mark the number of message pairs, each belongs to one of the automatic conversations, AC_i, for which there exists a manual conversation MC_j containing both messages of the pair.

The number of all message pairs, N_p, in the manual conversations is: $N_p = \sum_{i=1}^{n} \frac{|MC_i|(|MC_i|-1)}{2}$. We define the similarity between the manual and automatic partitions to be $\frac{T_p}{N_p}$. Note that the similarity between two identical partitions is maximal (1.0), since any pair of messages belonging to an automatic conversation also belongs to the corresponding manual conversation.

4.2   Results

Sub-Conversation detection evaluation. To evaluate the applicability of using sub-conversations for conversation detection, we break subject threads to sub-conversations using the algorithm described in Section 3.2, experimenting with different coefficient weights of the similarity function defined in Equation 1. We then evaluate the similarity between the two partitions.

The results are given in Table 1 and demonstrate the significance of considering all message features by the similarity function. Column *All* shows that considering all the attributes gives better results than considering only a subset of them. The detection quality for both mailboxes was very high (0.99), which suggests that detecting sub-conversations in subject-threads is an easy task. In addition, Table 1 shows that both mailbox owners consider the *date* attribute as the most important one for detecting sub-conversations.

Table 1. Sub-conversation detection results, using different coefficient weights

Coeff.	Independent	Pairs	All
Date	1.00 0.00 0.00	0.50 0.50 0.00	**0.50**
Content	0.00 1.00 0.00	0.00 0.50 0.50	**0.20**
Part.	0.00 0.00 1.00	0.50 0.00 0.50	**0.30**
Score A	0.98 0.91 0.87	0.89 0.93 0.90	**0.99**
Score B	0.97 0.84 0.85	0.97 0.86 0.84	**0.99**

Table 2. Conversation detection results, using different coefficient weights

Coeff.	Independent	All 1	All 2
Date	1.00 0.00 0.00 0.00	**0.05**	**0.05**
Content	0.00 1.00 0.00 0.00	**0.40**	**0.60**
Part.	0.00 0.00 1.00 0.00	**0.25**	**0.10**
Subject	0.00 0.00 0.00 1.00	**0.30**	**0.25**
Score A	0.93 0.93 0.91 0.92	**0.96**	**0.92**
Score B	0.71 0.82 0.72 0.79	**0.85**	**0.90**

Conversation detection evaluation. To evaluate the applicability of our conversation detection process we group sub-conversations to conversations, applying the algorithm described in 3.2 with different coefficient weights of the similarity function. We then evaluate the similarity between the two partitions.

The results are given in Table 2 and show that the detection quality for both mailboxes was higher when considering all message attributes than only independent ones. In addition, the results show that to get higher-quality conversations for mailbox *B*, the *content* attribute should be weighted higher than for mailbox *A*. This shows that detecting conversations in ones mailbox is not an intuitive task and that different coefficient weights apply to different mailbox owners. However, Table 2 shows that both mailbox owners consider the *date* attribute as almost insignificant for the conversation detection task, and that the *content* attribute is more important than the *subject* attribute.

The coefficient weights in the *All* columns in Tables 1 and 2 were selected empirically after applying various weights combinations. The results show that the weights in column *All 1* in Table 2 are better for mailbox *A* than for mailbox *B* and the weights in column *All 2* produce better results for mailbox *B*. The weights somewhat reflect the mailbox owner's opinion on what constitutes a conversation. Therefore the algorithm should learn and adjust the proper weights for a mailbox over time, using machine learning approaches similar to those described in [7].

Figure 2 summarizes the evaluation scores of the different detection methods based on email threads, subject threads, sub-conversations and automatic detected conversations. It shows very clearly the drawbacks of email threads for this task. It also shows that subject threads encapsulate most conversations in the mailbox. Sub-conversations alone, as expected, are inferior to detecting full conversations as they are only part of the process. However, they are also inferior to subject threads. The reason is that some of the sub-conversations represent partial conversations, which are scored lower than the subject threads containing them, while the full conversations that include those partial sub-conversations are only detected by the final stage of the algorithm.

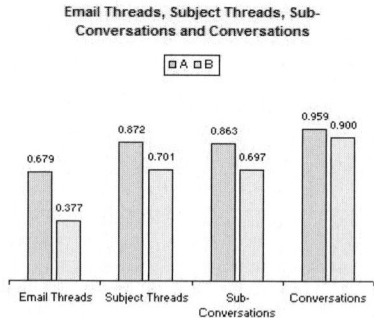

Fig. 2. A summary of the evaluation scores of the detection methods

5   Summary

This work explored a novel approach for detecting conversations in a mailbox. We defined a conversation as *an exchange of messages among the "same" group of people on the "same" topic*, and discussed the distinction between such conversations and traditional email threads that are usually used for conversation detection.

Our experiments (and their corresponding results) show that for detecting conversations in email messages, one should consider all message attributes. Using all the attributes for measuring similarity gave the highest results in both experiments of sub-conversation and conversation detection. In addition, we have shown the superiority of our algorithm to alternative approaches.

We believe that the type of conversations described in this work can benefit applications that rely on email threads from the end user's perspective. For example, the email summarization methods described in [6] may produce better summaries for the end user when coherent conversations are used rather than email threads. Similarly, the visualization systems described in [4] might better display conversations rather than email threads to the end user. We believe that users will benefit even more when provided with a view that displays coherent conversations rather than email threads.

References

1. Aaron, H., Jen-Yuan, Y.: Email thread reassembly using similarity matching. In: Proceedings of the Third Conference on Email and Anti-Spam (CEAS) (2006)
2. Gabor, C., Keno, A., Roger, W.: BuzzTrack: Topic Detection and Tracking in Email. In: Proceedings of the 12th international conference on Intelligent user interfaces IUI 2007, ACM Press, New York (2007)
3. Kalman, Y.M., Rafaeli, S.: Email Chronemics: Unobtrusive Profiling of Response Times. In: Proceedings of the 38th Annual Hawaii International Conference on System Sciences (HICSS 2005), vol. 04, pp. 108.2 (2005)
4. Kerr, B.: THREAD ARCS: An Email Thread Visualization. In: Proceedings of IEEE InfoVis, Seattle, WA, pp. 211–218 (2003)
5. Klimt, B., Yang, Y.: Introducing the Enron Corpus. In: Proceedings of the First Conference on Email and Anti-Spam (CEAS), Mountain View, CA (2004)
6. Lam, D., Rohall, S.L., Schmandt, C., Stern, M.K.: Exploiting e-mail structure to improve summarization. In: ACM 2002 Conference on Computer Supported Cooperative Work (CSCW2002), New Orlenes, LA (2002)
7. Lewis, D.D., Gale, A.W.: A Sequential Algorithm for Training Text Classifiers. In: Proceedings of the 17th annual international ACM SIGIR conference on Research and development in information retrieval, Dublin, Ireland, pp. 3–12 (1994)
8. Lewis, D.D., Knowels, K.A.: Threading Electronic Mail: a preliminary study. In Information Processing and Management 33(2), 209–217 (1997)
9. Rudy, I.A.: A Critical Review of Research on Electronic Mail. European Journal of Information Systems 4, 198–213 (1996)
10. The Internet Society. RFC 2822 – Internet Message Format (2001), http://www.faqs.org/rfcs/rfc2822.html

Efficient Multimedia Time Series Data Retrieval Under Uniform Scaling and Normalisation

Waiyawuth Euachongprasit and Chotirat Ann Ratanamahatana

Chulalongkorn University
Department of Computer Engineering
Phayathai Rd., Pathumwan, Bangkok 10330 Thailand
{g50wch,ann}@cp.eng.chula.ac.th

Abstract. As the world has shifted towards manipulation of information and its technology, we have been increasingly overwhelmed by the amount of available multimedia data while having higher expectations to fully exploit these data at hands. One of the attempts is to develop content-based multimedia information retrieval systems, which greatly facilitate us to intuitively search by its contents; a classic example is a Query-by-Humming system. Nevertheless, typical content-based search for multimedia data usually requires a large amount of storages and is computationally intensive. Recently, time series representation has been successfully applied to a wide variety of research, including multimedia retrieval due to the great reduction in time and space complexity. Besides, an enhancement, Uniform Scaling, has been proposed and applied prior to distance calculation, as well as it has been demonstrated that Uniform Scaling can outperform Euclidean distance. These previous work on Uniform Scaling, nonetheless, overlook the importance and effects of normalisation, which make their frameworks impractical for real world data. Therefore, in this paper, we justify this importance of normalisation in multimedia data and propose an efficient solution for searching multimedia time series data under Uniform Scaling and normalisation.

Keywords: Content-Based Multimedia Retrieval, Time Series, Uniform Scaling.

1 Introduction

As the world has shifted towards manipulation of information and its technology, multimedia data have played a crucial role in our daily lives. We feel much more comfortable in using multimedia data as a medium to communicate with each other, to present new ideas, or even to entertain ourselves. While the amount of multimedia data has been increasing dramatically and continually, our expectations of manipulating these data have as well been escalating. However, multimedia data manipulation typically requires a large amount of storages and is computationally intensive. Recently, time series representation has been proposed to help alleviate this burden, and it has been successfully applied in multi disciplines such as science, bioinformatics, economics, and multimedia [1].

C. Macdonald et al. (Eds.): ECIR 2008, LNCS 4956, pp. 506–513, 2008.
© Springer-Verlag Berlin Heidelberg 2008

Transformation from multimedia data into time series data is simple and straightforward, e.g., audio data in query by humming. We can just extract a sequence of pitch from a sung query and then use this time series as a query to retrieve an intended song from the database. For other types of multimedia data such as videos and images, several techniques have been proposed to convert them into time series data [2-5].

The main objective of the time series transformation of multimedia data is to achieve an efficient representation for data manipulation, including information retrieval. The heart of information retrieval, especially in time series retrieval, is similarity measurement, where Euclidean distance is prevalent. However, Euclidean distance seems to be impractical in several applications, particularly for multimedia applications, where shrinking and stretching of the data are very typical. For example, in query-by-humming system, users tend to sing queries slower or faster than the songs in the database. Hence, scaling of the data before distance calculation is very important. Uniform Scaling (US) was introduced to solve this problem [3]. Unfortunately, US comes at a cost and cannot scale well with large databases. Therefore, a lower bounding of US [3] was introduced to achieve significant speedup over the calculation by efficiently pruning a large number of unwanted sequences before costly distance calculation. Although this technique appears to be a practical solution for multimedia retrieval, unawareness of normalisation causes serious flaws in the previous framework.

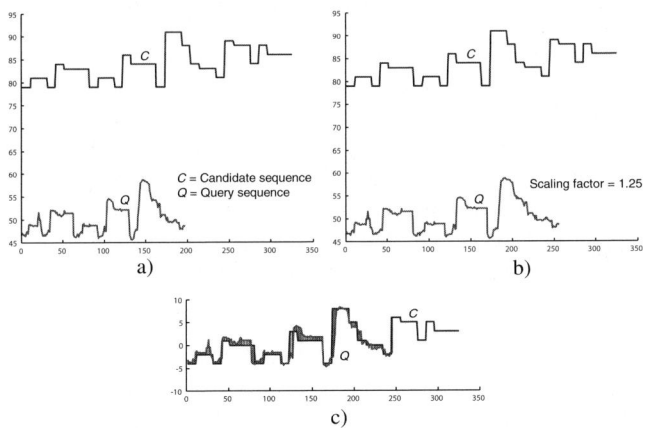

Fig. 1. a) A sequence of pitch extracted from a sung query sequence of a "Happy Birthday" song represented by a query sequence Q, and a MIDI pitch contour of the same song represented by a candidate sequence C. b) A rescaled query sequence with a scaling factor = 1.25 c) Both sequences after normalisation at the query's length. The shaded region shows their Euclidean distance.

Generally, normalisation is crucial for similarity measurement in time series data since it enables us to measure the underlying shapes of time series. For example, in query by humming, we want to search a song database for the one with a segment that is most similar to the sung query, regardless of the higher or lower key sung. However, to achieve this music key invariance, we must remove any existing offsets of both a query and a candidate sequence. If we do not normalise both sequences before distance calculation, this measurement will not be sensible. In Fig. 1 a), the shape of a query sequence extracted from a "Happy Birthday" sung query and the shape of a candidate sequence from the same part of the song are quite similar. Nevertheless, if we measure the similarity of these sequences by using only Euclidean distance without any pre-processing to the data, the distance will be excessive. Thus, rescaling and then normalising both sequences before distance calculation are crucial steps to achieve accurate and meaningful retrieval (see Fig. 1 b) and c)).

Though normalisation is one of the most important parts of similarity measurement, the current technique [3] has been developed while overlooking the importance of normalisation, which causes their framework to appear impractical for multimedia retrieval. Besides, their proposed lower-bounding function, which claims large pruning power, is invalid under normalisation condition. Hence, we propose a lower-bounding function that specifically deals with this normalisation problem efficiently and calculates a distance under the US, where no false dismissals are guaranteed.

2   Background

We begin with a formal problem definition and reviews of necessary background.

Problem definition. Suppose we have a query sequence Q of length m, where $Q = q_1,q_2,q_3,...,q_m$. It is scalable between lengths $sfmin*m$ and $sfmax*m$, where $sfmin$ and $sfmax$ are minimum and maximum scaling factors respectively, i.e., we can shrink or stretch a query sequence from any length $sfmin*m$ to $sfmax*m$, where $sfmax \geq 1$ and $0 < sfmin \leq 1$. In addition, each candidate sequence C of length n, $C = c_1,c_2,c_3,...,c_n$, is stored in a database D. For simplicity, here, we define $n \geq sfmax*m$. Finally, we want to find the most similar-shaped candidate sequence C in the database D to the query sequence Q, which is also scalable in arbitrary lengths between $sfmin*m$ and $sfmax*m$.

Definition 1. *Squared Euclidean distance*: We define a squared Euclidean distance measure in eq.(1), which calculates distance between two sequences of equal length m (query's length). Note that the square root from the original Euclidean distance has been removed for an optimization purpose [3].

$$D(Q,C) \equiv \sum_{i=1}^{m}(q_i - c_i)^2 \qquad (1)$$

Definition 2. *Uniform Scaling*: Uniform Scaling is a technique that uniformly stretches or shrinks a time series. In this approach, if we want to stretch a prefix of a candidate sequence C of length l to length m, we can use a scaling function in eq.(2); shrinking of a candidate sequence is done similarly to a stretching process.

We can formally define the US function as follows.

$$c_j = c_{\lfloor j*l/m \rfloor} \quad \text{where} \quad 1 \le j \le m \tag{2}$$

For the US distance calculation, prefixes of a candidate sequence C of length l, where $\lfloor sfmin * m \rfloor \le l \le \min(\lfloor sfmax * m \rfloor, n)$, are rescaled to length m (query's length). Then we use a squared Euclidean distance function to calculate distance between a query sequence and all rescaled prefix sequences in order to find a minimum distance value ranging from $sfmin$ to $sfmax$.

The formal definition of a Uniform Scaling distance function (US) is defined in eq.(3), where $RP(C,m,l)$ is a Rescaled Prefix function that returns a prefix of a candidate sequence of length l rescaled to length m.

$$US(Q, C, sfmin, sfmax) = \min_{l=\lfloor sfmin*m \rfloor}^{\min(\lfloor sfmax*m \rfloor, n)} D(RP(C, m, l), Q) \tag{3}$$

$$\text{where} \quad RP(C, m, l)_i = c_{\lfloor j*l/m \rfloor} \quad ; 1 \le i \le m \text{ and } 1 \le j \le m$$

Definition 3. *Lower bounding of Uniform Scaling* [3, 6]: Lower bounding of Uniform Scaling is a distance approximation function, which can quickly compute a lower-bounding distance between a query and a candidate sequences; however, this lower bound value must not exceed the true distance value in order to be a valid lower-bounding function. To illustrate the idea, two new sequences are created, an upper envelope sequence UY and a lower envelope sequence LY enclosing a candidate sequence. This envelope represents all scaled candidate sequences for a lower-bounding distance calculation.

UY and LY are formally defined in eq.(4). Note that a lower bounding distance can simply be a squared Euclidean distance between a query sequence and the candidate's envelope, as defined in eq.(5).

$$UY_i = \max(c_{\lfloor i*sfmin \rfloor}, \dots, c_{\lfloor i*sfmax \rfloor}) \tag{4}$$

$$LY_i = \min(c_{\lfloor i*sfmin \rfloor}, \dots, c_{\lfloor i*sfmax \rfloor})$$

$$LBY(Q, C) = \sum_{i=1}^{m} \begin{cases} (q_i - UY_i)^2 & \text{if } q_i > UY_i \\ (q_i - LY_i)^2 & \text{if } q_i < LY_i \\ 0 & \text{otherwise} \end{cases} \tag{5}$$

3 Our Proposed Method

As mentioned earlier, without realising an importance of normalisation, the distance measurement under US scheme becomes almost meaningless. Besides, the existing lower-bounding function cannot correctly calculate the distance for normalised sequences without false dismissals. For example, in Fig. 2, a query Q is a rescaled version of the candidate's prefix C; hence, the distance between these two sequences should be zero according to their shape similarity. However, in Fig. 2 b), it is apparent that the lower-bounding distance between a normalised query and the lower-bounding envelope is not zero as it should be. This phenomenon definitely violates the lower-bounding criteria, where the lower-bounding distance must guarantee not to exceed the true distance. Therefore, the existing lower-bounding function may cause some false dismissals. In an attempt to correct this flaw, we propose the US under *normalisation* condition, together with a corresponding lower-bounding function, which guarantees not to cause false dismissals, as shown in Fig. 2 c).

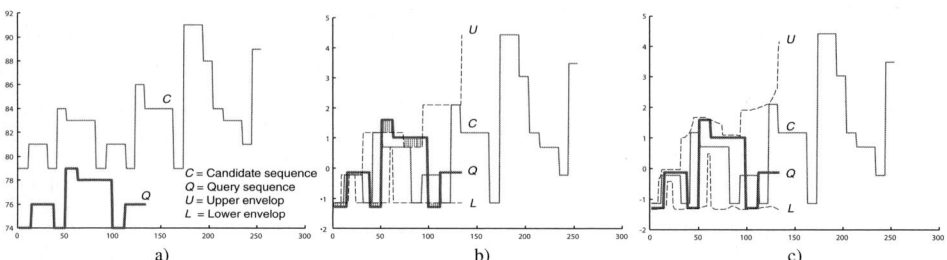

Fig. 2. a) Raw pitch contours are extracted from a "Happy Birthday" song sequence. C represents the candidate song, and Q is the query sung in a slower tempo (scaling factor = 1.2). b) and c) the query and the candidate sequences are z-normalised within the query's length enclosed by different lower-bounding envelopes with the scaling factors ranging from 0.7 to 1.3. b) The previous lower bounding of US function [3]. c) Our proposed lower bounding of US that is guaranteed not to cause any false dismissals.

Definition 4. *Uniform Scaling with Normalisation*: The formal definition of $\overline{\text{US}}$ with normalisation is shown in eq.(6), where Q' is a normalised query, and $\overline{c_{1...l}}$ and $SD(c_{1...l})$ are mean and standard deviation values of a candidate's prefix of length l respectively.

$$US_{norm}(Q',C,sfmin,sfmax) = \min_{l=\lfloor sfmin*m \rfloor}^{\min(\lfloor sfmax*m \rfloor,n)} D(RP_{norm}(C,m,l),Q')$$

$$\text{where} \quad RP_{norm}(C,m,l)_i = \frac{c_{\lfloor j*l/m \rfloor} - \overline{c_{1...l}}}{SD(c_{1...l})} \quad ; 1 \le i \le m \text{ and } 1 \le j \le m \tag{6}$$

Definition 5. *Lower bounding of Uniform Scaling with Normalisation*: We develop a bounding envelope as shown in eq.(7), where UZ'_i and LZ'_i are an upper envelope and a lower envelope respectively. The distance calculation function is shown in eq.(8).

$$UZ'_i = \max(\frac{c_{sfmin*i} - \overline{c_{1...sfmin*m}}}{SD(c_{1...sfmin*m})}, \ldots, \frac{c_{sfmax*i} - \overline{c_{1...sfmax*m}}}{SD(c_{1...sfmax*m})})$$

$$= \max_{j=0}^{\lfloor m*(sfmax-sfmin) \rfloor} (\frac{c_{\lfloor a \rfloor} - \overline{c_{1...\lfloor b \rfloor+j}}}{SD(c_{1...\lfloor b \rfloor+j})}, \frac{c_{\lceil a \rceil} - \overline{c_{1...\lfloor b \rfloor+j}}}{SD(c_{1...\lfloor b \rfloor+j})}, \frac{c_{\lfloor a \rfloor} - \overline{c_{1...\lceil b \rceil+j}}}{SD(c_{1...\lceil b \rceil+j})}, \frac{c_{\lceil a \rceil} - \overline{c_{1...\lceil b \rceil+j}}}{SD(c_{1...\lceil b \rceil+j})}) \tag{7}$$

$$LZ'_i = \min(\frac{c_{sfmin*i} - \overline{c_{1...sfmin*m}}}{SD(c_{1...sfmin*m})}, ..., \frac{c_{sfmax*i} - \overline{c_{1...sfmax*m}}}{SD(c_{1...sfmax*m})})$$

$$= \min_{j=0}^{\lfloor m*(sfmax-sfmin) \rfloor} (\frac{c_{\lfloor a \rfloor} - \overline{c_{1...\lfloor b \rfloor + j}}}{SD(c_{1...\lfloor b \rfloor + j})}, \frac{c_{\lceil a \rceil} - \overline{c_{1...\lfloor b \rfloor + j}}}{SD(c_{1...\lfloor b \rfloor + j})}, \frac{c_{\lfloor a \rfloor} - \overline{c_{1...\lceil b \rceil + j}}}{SD(c_{1...\lceil b \rceil + j})}, \frac{c_{\lceil a \rceil} - \overline{c_{1...\lceil b \rceil + j}}}{SD(c_{1...\lceil b \rceil + j})})$$

$$\text{where} \quad a = sfmin*i + j*\frac{i}{m}, b = sfmin*m$$

$$\text{LBZ}(Q', C) = \sum_{i=1}^{m} \begin{cases} (q_i - UZ'_i)^2 & \text{if } q_i > UZ'_i \\ (q_i - LZ'_i)^2 & \text{if } q_i < LZ'_i \\ 0 & \text{otherwise} \end{cases} \qquad (8)$$

4 Experiment

In this section, we conduct an experiment to reconfirm that our approach is able to efficiently search through large multimedia databases by evaluating pruning power of the proposed lower-bounding function. The pruning power directly reflects the quality of the lower-bounding function. It is defined as the ratio of the candidate objects that can be disregarded from further calculations to the total number of candidates [6, 7], as shown in eq.(9).

$$Pruning\ Power = \frac{The\ number\ of\ pruned\ candidates}{The\ total\ number\ of\ candidate\ sequences} \qquad (9)$$

In our experiments, we use 55 sung queries collected from 12 subjects of both genders. Then we extract sequences of pitch from the sung queries by using autocorrelation algorithm [8]. To observe the effect of different sequence lengths, we select sung queries that are sufficiently long, and crop their prefixes to lengths 75, 100, and 125 data points. The candidate sequences in databases are generated from the subsequences extracted from the MIDI files using a sliding window size of 125*sfmax data points, where maximum scaling factor is 1.4.

To demonstrate the quality and utilities of our lower-bounding function, we test our proposed lower-bounding function on a simple query-by-humming system in two aspects using 1-nearest-neighbour algorithm. First, we inspect the pruning power using different lengths and different ranges of scaling as shown in Fig. 3, where the database contains 22441 subsequences. Second, we conduct an experiment to observe the effect of database sizes on pruning power by varying the number of subsequences; we use 22441, 55595, 107993 and 220378 subsequences from 100, 250, 500 and 1000 songs respectively to construct the databases. Note that queries' length is 100 data points, and the scaling factor ranges from 0.7 to 1.3. The result is shown in Fig. 4.

According to the experiment result in Fig. 3, the pruning power of our proposed method is quite impressive since the proposed lower-bounding function can prune a

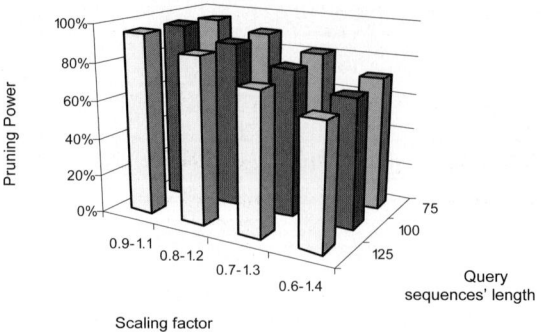

Fig. 3. The pruning power of different length and different ranges of scaling factors

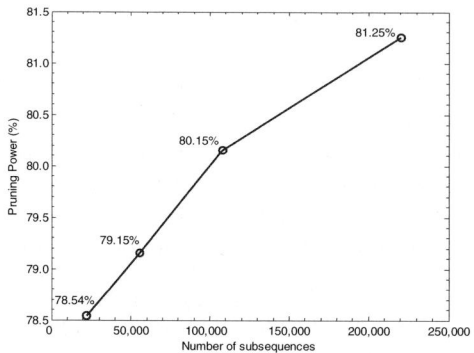

Fig. 4. The pruning powers in different-sized databases

large number of candidate sequences in database in every scaling. However, the wider the range of scaling factors are, the smaller the pruning power will become because the increasing range of scaling factors will also increase the size of the lower-bounding envelope. In addition, longer sequences appear to have smaller pruning power than that of the shorter sequences.

Fig. 4 demonstrates that pruning power will increase as the size of database increases. This is one of the most desirable properties for lower-bounding functions. However, the normalisation also affects the pruning power because the distance between the normalised query and the normalised candidate sequences is greatly reduced comparing with the distance between unnormalised sequences. Nonetheless, we would like to reemphasize the importance and necessity of normalisation, especially in multimedia applications.

5   Conclusions

We have shown that a lower-bounding function of US for *normalised* time series can efficiently prune a large number of candidate sequences in the database and significantly reduce the time complexity in the retrieval, especially for multimedia data. Although this lower-bounding function can achieve dramatic speedups by pruning almost all the candidate sequences before finer calculations, scalability to a truly massive database is still a challenge for future research directions.

Acknowledgments. We would like to thank Dr. Boonserm Kijsirikul for valuable comments and enlightening discussions. This work is partially supported by the Thailand Research Fund (Grant No. MRG5080246).

References

1. Sakurai, Y., Yoshikawa, M., Faloutsos, C.: FTW: Fast Similarity Search under the Time Warping Distance. In: Proceedings of 24th ACM SIGMOD-SIGACT-SIGART symposium on Principles of database systems, pp. 326–337. ACM Press, Baltimore, Maryland (2005)
2. Keogh, E., Palpanas, T., Zordan, V.B., Gunopulos, D., Cardle, M.: Indexing Large Human-Motion Databases. In: Proceedings of 30th VLDB Conference, Toronto, Canada (2004)
3. Keogh, E.: Efficiently Finding Arbitrarily Scaled Patterns in Massive Time Series Databases. In: Lavrač, N., Gamberger, D., Todorovski, L., Blockeel, H. (eds.) PKDD 2003. LNCS (LNAI), vol. 2838, pp. 253–265. Springer, Heidelberg (2003)
4. Keogh, E., Celly, B., Ratanamahatana, C.A., Zordan, V.B.: A novel technique for indexing video surveillance data. In: 1st ACM SIGMM international workshop on Video surveillance, pp. 98–106. ACM Press, Berkeley, California, USA (2003)
5. Shu, S., Narayanan, S., Kuo, C.-C.J.: Efficient Rotation Invariant Retrieval of Shapes using Dynamic Time Warping with Applications in Medical Databases. In: Proceedings of 19th IEEE International Symposium on Computer-Based Medical Systems (CBMS), pp. 673–678 (2006)
6. Fu, A., W.-c., K.E., Lau, L.Y.H., Ratanamahatana, C.A.: Scaling and time warping in time series querying. In: Proceedings of 31st international conference on Very large data bases, VLDB Endowment, Trondheim, Norway, pp. 649–660 (2005)
7. Keogh, E., Ratanamahatana, C.A.: Exact indexing of dynamic time warping. Knowledge and Information Systems: An International Journal (KAIS) 7, 358–386 (2004)
8. Boersma, P., Weenink, D.: Praat: doing phonetics by computer (version 4.4.13) (2005), http://www.praat.org/

Integrating Structure and Meaning: A New Method for Encoding Structure for Text Classification

Jonathan M. Fishbein[1] and Chris Eliasmith[1,2]

[1]Department of Systems Design Engineering
[2]Department of Philosophy
Centre for Theoretical Neuroscience
University of Waterloo
200 University Avenue West
Waterloo, Canada
{jfishbei,celiasmith}@uwaterloo.ca

Abstract. Current representation schemes for automatic text classification treat documents as syntactically unstructured collections of words or 'concepts'. Past attempts to encode syntactic structure have treated part-of-speech information as another word-like feature, but have been shown to be less effective than non-structural approaches. We propose a new representation scheme using Holographic Reduced Representations (HRRs) as a technique to encode both semantic and syntactic structure. This method improves on previous attempts in the literature by encoding the structure across all features of the document vector while preserving text semantics. Our method does not increase the dimensionality of the document vectors, allowing for efficient computation and storage. We present classification results of our HRR text representations versus Bag-of-Concepts representations and show that our method of including structure improves text classification results.

Keywords: Holographic Reduced Representations, Vector Space Model, Text Classification, Part of Speech Tagging, Random Indexing, Syntax, Semantics.

1 Introduction

Successful text classification is highly dependent on the representations used. A representation of a dataset that leaves out information regarding prominent features of the dataset will result in poor performance, no matter how good the classification algorithm may be. In the case of natural language text, there are many choices which must be made in converting the raw data to high-dimensional vectors for these algorithms to process. Currently, most approaches to text classification adopt the 'bag-of-words' document representation approach, in which the grammatical structure, and semantic relationship between words in a document is largely ignored, and their frequency of occurrence is considered as most

C. Macdonald et al. (Eds.): ECIR 2008, LNCS 4956, pp. 514–521, 2008.

important. This is largely because past approaches that have tried to include more complex structures or semantics have often been found lacking [1].

However, these negative conclusions are premature. More recent work that employs automatically generated semantics using Latent Semantic Analysis and Random Indexing have been shown to be more effective than bag-of-words approaches in some circumstances [2]. As a result, it seems more a matter of determining how best to represent semantics, than of whether or not semantics is useful for classification.

Here we demonstrate that the same is true of including syntactic structure. A recent comprehensive survey suggests that including parse information will not help classification [1]. However, the standard method for including syntactic information is simply to add the syntactic information as a completely new, independent feature of the document. In contrast, our method takes a very different approach to feature generation by distributing syntactic information across the document representation. This avoids limitations of past approaches.

2   Bag-of-Words and Bag-of-Concepts

One of the simplest and most common text representation is the Bag-of-Words (BoW) scheme, where a document is represented as a vector of weighted (typically term frequency-inverse document frequency) word frequency counts. The dimensionality of these document vectors is typically very high; however, they are also typically very sparse.

The Bag-of-Concepts (BoC) text representation is a more recent representation scheme [2] meant to address the deficiencies of the BoW representations by implicitly representing synonymy relations between document terms. BoC representations are based on the intuition that the meaning of a document can be considered as the union of the meanings of the terms in that document. BoC representations is often significantly less than the dimensionality of BoW representation yielding better computational efficiency for classification tasks.

There have been two approaches taken to define a 'context' in BoC representations. The first is to use the Latent Semantic Indexing (LSI) model [3], which uses the entire document as a single context and each term context vector is a vector of the weighted counts in which it occurs in each document. The second is the Hyperspace Analogue to Language model [4], which uses individual words as contexts and each term context vector is a vector of the weighted counts in which it co-occurs with other words as determined by passing a fixed-size sliding window over the document. In this paper, we investigate both approaches for our new method.

3   Context Vectors and Dimensionality Reduction

Reducing the dimensionality of document term frequency count vectors is a key component of BoC context vector generation. Exploiting the Johnson-Lindenstrauss lemma [5], which states that if we project points into a random subspace of sufficiently high dimensionality, we will approximately preserve the distances between the points, we can reduce the dimensionality of a large matrix in a more computationally efficient manner than using principal component analysis (PCA). Specifically for an $m \times n$ sparse matrix, the computational complexity of PCA using as singular value decomposition is $O(mnc)$ while the computational complexity of this random mapping is $O(nc \log m)$, where c is the number of non-zero entries per row (i.e., the average number of terms in a document). This random mapping dimensionality reduction is accomplished by multiplying a large $F_{m \times n}$ matrix by a random $R_{n \times k}$ matrix, with $k \ll n$ and where each row is constructed by randomly distributing a small number of +1s and -1s (usually around 1-2% of the matrix) and setting the rest of the elements to 0. The resulting context vector matrix FR is now $m \times k$, with the distance between every pair of rows approximately preserved from that in F.

However, performing this large matrix multiplication can be costly in terms of memory requirements, since the full $m \times n$ matrix F must be built. The random indexing technique [6], in contrast, assembles this lower dimensional matrix incrementally and avoids building this large matrix. In Random Indexing, we first create k-dimensional random index vectors for each dimension in our data, where k is significantly less than the total number of dimensions in the data. These random index vectors are created identically to the rows in the random projection matrix. Term context vectors are created by adding the context's random index vector to the term context vector every time a word occurs in a given context, The resulting term context vectors are equivalent to the ones created using the random mapping approach.

The advantage of random indexing is that it is an incremental approach, meaning that context vectors can start to be created without sampling all the data, while still avoiding the computationally costly singular value decomposition as utilized in LSI. But more importantly, random indexing avoids constructing the large context count matrix required in random mapping.

4   Limitations of Bag-of-Concepts

Sahlgren & Cöster [2] have shown that BoC has a classification advantage over BoW in certain situations. Nevertheless, the BoC scheme still ignores the large amount of syntactic data in the documents not captured implicitly through word context co-occurrences. For instance, although BoC representations can successfully model some synonymy relations, since different words with similar meaning will occur in the same contexts, it can not model polysemy relations. For example, consider the word "can". Even though the verb form (i.e., "I *can* perform that action.") and the noun form (i.e., "The soup is in the *can*.") of the word occur in different contexts, the generated term vector for "can" will be a combination of these two contexts in BoC. As a result, the representation will not be able to correctly model polysemy relations involving a word that can be used in different parts of speech.

5   Holographic Reduced Representations

In order to solve the problem of modeling certain polysemy relations in natural language text, we need a representation scheme that can encode both the semantics of documents, as well as the *syntax* of documents. Borrowing from a representation scheme introduced in cognitive science [7], Holographic Reduced Representations (HRRs), we can complement the BoC semantic modeling with parts of speech information to generate a more robust text representation. Eliasmith and Thagard [8] have previously shown that HRRs can be used to model both syntactic and semantic psychological data. As well, Eliasmith [9] has shown that HRRs can be successfully applied to language processing. The intuition behind this approach, is that we can "bind" part-of-speech information with a word's term context vector in order to encode both pieces of information in our representation.

HRRs use holographic transformations to encode and decode information in flat, constant dimension vectors. In order to encode the information contained within multiple vectors into a single vector, HRRs depend on circular convolution. Circular convolution binds two vectors $\underline{A} = (a_0, a_1, \ldots, a_{n-1})$ and $\underline{B} = (b_0, b_1, \ldots, b_{n-1})$ to give $\underline{C} = (c_0, c_1, \ldots, c_{n-1})$ where $\underline{C} = \underline{A} \otimes \underline{B}$ with $c_j = \sum_{k=0}^{n-1} a_k b_{j-k}$ for $j = 0, 1, \ldots, n - 1$. Circular convolution is efficiently computed in time $O(n \log n)$.

There are a number of properties of circular convolution that make it ideal to use as a binding operation. First, the expected similarity between a convolution and its constituents is zero. So, the same term acting as different parts of speech in similar contexts, such as the word *can* that can act as both a noun and a verb, would not be similar in their bound HRR representation (e.g., "He kicked the *can*." would be distinct from "He *can* kick"). Second, the dimensionality of the vectors are constant under HRR operations, so the number of vectors encoded in the structure does not affect the complexity of the representation. Third, similar semantic concepts bound to the same part-of-speech result in similar vectors. So, since similarity reflects the structure of the semantic space, these binding operations usefully preserve the relevant geometric relations of the original semantic space.

HRRs also need to be combined in a manner that assembles the parts of the desired structure while preserving the similarity of the final structure to its components. For this, superposition (i.e. vector addition) is used. So if $\underline{C} = \underline{A} + \underline{B}$, $\underline{C}$ is most likely more similar to $\underline{A}$ or $\underline{B}$ than to any other vector.

6   HRR Document Representation

Our natural language representation takes advantage of the ability of HRRs to encode a document's structure in a way that is non-destructive to the document's semantics. Using the circular convolution and superposition operations of HRRs, our representation scheme can augment the semantic modeling of the BoC representations with part-of-speech information to better disambiguate document classes for classification.

We first determine the term context vectors for the data by adopting the random indexing method, described earlier. We then use a part-of-speech tagger to extract some syntactic structure of the corpus documents. We collapse the set of possible part-of-speech tags returned by the tagger into the basic linguistic set (e.g.: nouns, verbs, pronouns, prepositions, adjective, adverbs, conjunctions, and interjections), and generate random HRR vectors of the same dimension as our term context vector for each possible tag.

To build the HRR document representation, we perform the following steps:

1. for each word in a document we take the term context vector of that word and bind it to the word's identified part-of-speech vector;
2. we take the $tf \times idf$-weighted sum of the resulting vectors in order to obtain a single HRR vector representing the document.

Like BoC document vectors, these HRR document vectors are normalized by dividing by the number of terms in the document in order to ensure that there is no classification bias to longer documents. But unlike BoC vectors, these HRR document vectors encode both semantic and syntactic information.

7   Experimental Setup

In the following sections we describe the setup for our text classification experiments. Specifically, we describe the text representations used for classification, and the classifiers and evaluation methodology used in the experiments.

7.1   Representations

We used the 20 Newsgroups corpus[1] as the natural language text data for our experiments. The purpose of these experiments was to compare the classification effectiveness of BoC and HRR text representations[2], not produce a top score for the 20 Newsgroups corpus.

The BoC representations were generated by first stemming all words in the corpus, using the Porter stemmer, to reduce the words to their root form. We then used Random Indexing to produce context vectors for the given text corpus. The dimensionality of the context vectors was fixed at 512 dimensions[3], which should be compared to the 118 673 unique stems within the corpus. We investigated the effects of both document-based context vectors and word-based context vectors. For word-based context vectors, we produced contexts using a sliding window extending 4 words in each direction from the focus word, where the term vector of the focus word was updated by adding to it the context vector of each word inside the sliding window weighted by $2^{(1-d)}$, where d is the

---

[1] Available at http://people.csail.mit.edu/jrennie/20Newsgroups/.

[2] We did not pursue comparison experiments with BoW representations as there are already published results (e.g. [2]) of BoW/BoC experiments in the literature.

[3] The dimensionality of the vectors has been chosen to be consistent with other work. There is as yet no systematic characterization of the effect of dimensionality on performance.

distance from the focus word. These context vectors were then $tf \times idf$-weighted and summed for each document.

The context vectors used in the HRR representations were generated in the exact same way as the BoC representations. The part-of-speech data was extracted using the Stanford Log-linear Part-of-Speech tagger[4] and random 512 dimensional HRR vectors were created for each tag in our collapsed tag set. This part-of-speech tag vector was then bound to its word's associated context vector by circular convolution, $tf \times idf$-weighted and summed for each document.

7.2   Classification and Evaluation

We performed Support Vector Machine (SVM) classification experiments[5] in order to investigate the classification effectiveness of the HRR and BoC representation. For the experiments in this paper, we used a linear SVM kernel function (with a slack parameter of 160.0). In these classification experiments, we used a one-against-all learning method employing 10-fold stratified cross validation[6]. The SVM classifier effectiveness was evaluated using the $\mathcal{F}_1$ measure.

8   Results

We only present the comparison results between the BoC text representations and HRR representations using document-based context vectors since the results for word-based context vectors showed the same comparison trends in the $\mathcal{F}_1$ scores, but produced lower total $\mathcal{F}_1$ scores.

The macro-averaged $\mathcal{F}_1$ showed that the HRR representations produced the best results, with a score of 58.19. The BoC representations produced a macro-averaged $\mathcal{F}_1$ score of 56.55. These results were calculated to be statistically significant under a 93.7% confidence interval.

Figure 1 shows the correlation between the macro-averaged SVM $\mathcal{F}_1$ scores of BoC and HRR text representations for each category in the 20 Newsgroups corpus. The graph shows that the HRR representations produce similar classification scores for some classes and significantly higher scores for other classes. This may be explained by noticing that the classes that the HRR representations outperform BoC representations are the classes in the corpus that are highly related to other classes in the corpus.

The learning curves for the representations are included in Figure 2. The graph shows that the HRR representations consistently produce better SVM classification when compared to BoC representation no matter how much of the class data is used for training. This result indicates that in situations where there is limited class data from which to learn a classification rule, the extra

---

[4] Available at http://nlp.stanford.edu/software/tagger.shtml.

[5] We used the SVM^{perf} implementation, which optimizes for $\mathcal{F}_1$ classification score, available at http://svmlight.joachims.org/svm_perf.html.

[6] This cross-validation scheme was chosen as it better reflects the statistical distribution of the documents, although produces lower $\mathcal{F}_1$ scores.

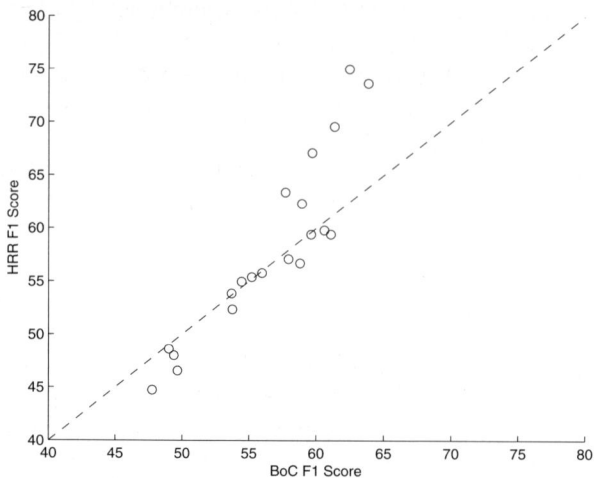

Fig. 1. Correlation between SVM $\mathcal{F}_1$ scores of BoC and HRR text representations for each corpus category

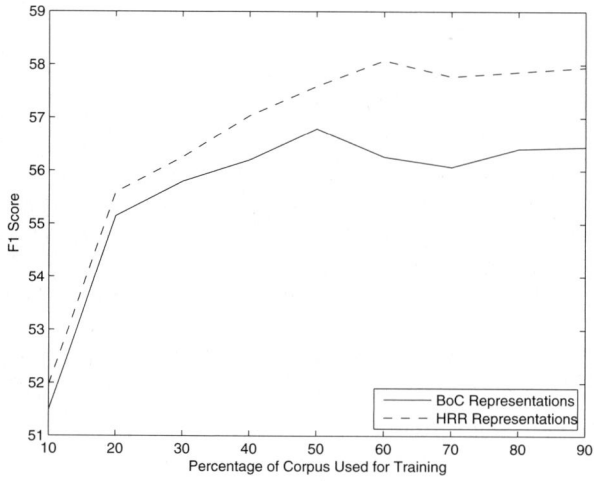

Fig. 2. Learning curves of SVM $\mathcal{F}_1$ scores of BoC and HRR text representations

part-of-speech information contained within the HRR representation assists in better classifying documents.

9   Conclusions and Future Research

Using HRRs, we have created a novel document representation scheme that encodes both the structure and the semantics of natural language text documents.

Our results show that including both the structure and semantics of natural language text in our HRR text representations can improve the text classification $\mathcal{F}_1$ score of SVM classifiers when compared to the BoC approach. We have also demonstrated the sustained superiority of the HRR representations when using various amounts of data to train the classifiers.

Our results suggest many areas of further research. We have only investigated a single natural language corpus in this paper and further investigations using different corpora should be undertaken to examine the effectiveness of HRR representations under different text domains. As well, the document vectors were fixed to 512 dimensions in the experiments, but it would be interesting to analyze the effects of the vector dimensionality on the classification results.

Acknowledgment. This research is supported in part by the Canada Foundation for Innovation, the Ontario Innovation Trust, the Natural Science and Engineering Research Council and the Open Text Corporation.

References

1. Moschitti, A., Basili, R.: Complex linguistic features for text classification: a comprehensive study. In: McDonald, S., Tait, J.I. (eds.) ECIR 2004. LNCS, vol. 2997, pp. 181–196. Springer, Heidelberg (2004)
2. Sahlgren, M., Cöster, R.: Using Bag-of-Concepts to Improve the Performance of Support Vector Machines in Text Categorization. In: Proceedings of the 20th International Conference on Computational Linguistics, pp. 487–493 (2004)
3. Deerwester, S.C., et al.: Indexing by latent semantic analysis. Journal of the American Society of Information Science 41(6), 391–407 (1990)
4. Lund, K., Burgess, C.: Producing high-dimensional semantic spaces from lexical co-occurrence. Behaviour Resrach Methods, Instrumentation and Computers 28(2), 203–208 (1996)
5. Johnson, W.B., Lindenstrauss, J.: Extensions to Lipshitz mapping into Hilbert space. Contemporary Mathematics 26 (1984)
6. Sahlgren, M.: An Introduction to Random Indexing. In: Methods and Applications of Semantic Indexing Workshop at the 7th International Conference on Terminology and Knowledge Engineering (2005)
7. Plate, T.A.: Holographic Reduced Representation: Distributed representation for cognitive structures. CSLI Publications (2003)
8. Eliasmith, C., Thagard, P.: Integrating structure and meaning: A distributed model of analogical mapping. Cognitive Science 25(2), 245–286 (2001)
9. Eliasmith, C.: Cognition with neurons: A large-scale, bilogically realistic model of the Wason task. In: Proceedings of the XXVII Annual Conference of the Cognitive Science Society (2005)

A Wikipedia-Based Multilingual Retrieval Model

Martin Potthast, Benno Stein, and Maik Anderka

Bauhaus University Weimar, Faculty of Media, 99421 Weimar, Germany
`<first name>.<last name>@medien.uni-weimar.de`

Abstract. This paper introduces CL-ESA, a new multilingual retrieval model for the analysis of cross-language similarity. The retrieval model exploits the multilingual alignment of Wikipedia: given a document d written in language L we construct a concept vector $\mathbf{d}$ for d, where each dimension i in $\mathbf{d}$ quantifies the similarity of d with respect to a document d_i^* chosen from the "L-subset" of Wikipedia. Likewise, for a second document d' written in language L', $L \neq L'$, we construct a concept vector $\mathbf{d}'$, using from the L'-subset of the Wikipedia the topic-aligned counterparts $d_i'^*$ of our previously chosen documents.

Since the two concept vectors $\mathbf{d}$ and $\mathbf{d}'$ are *collection-relative representations* of d and d' they are language-independent. I. e., their similarity can directly be computed with the cosine similarity measure, for instance.

We present results of an extensive analysis that demonstrates the power of this new retrieval model: for a query document d the topically most similar documents from a corpus in another language are properly ranked. Salient property of the new retrieval model is its robustness with respect to both the size and the quality of the index document collection.

1 Introduction

Retrieval models are used to assess the similarity between documents. For this purpose a retrieval model provides (*i*) the rationale for the construction of a particular document representation $\mathbf{d}$ given a real-world document d, and, (*ii*) a similarity measure φ to quantify the similarity between two representations $\mathbf{d}$ and $\mathbf{d}'$.

This paper deals with retrieval models that can be applied in a cross-language retrieval situation, i.e. when a document d is written in language L and we would like to assess its similarity to a document d' written in language L'. Our contributions are the following:

- *Multilingual Retrieval Model.* Section 2 introduces the new multilingual retrieval model CL-ESA, which overcomes many of the restrictions of previous approaches when quantifying cross-language similarity.
- *Evaluation.* Section 3 reports on experiments related to retrieval performance, dimensionality dependence, and runtime. In particular, with the so-called bilingual rank correlation a new measure for cross-lingual retrieval performance is proposed.
- *Comparable Corpus Wikipedia.* We use Wikipedia as a comparable corpus and demonstrate its usability for cross-lingual retrieval.

C. Macdonald et al. (Eds.): ECIR 2008, LNCS 4956, pp. 522–530, 2008.

Table 1. Assessment results for cross-language retrieval models with respect to six criteria: the number of languages, the computational complexity to compute the retrieval model, the number of documents which can be represented with reasonable effort, the availability of resources to construct the retrieval model, the achievable retrieval quality, and the specificity of the retrieval model to a (topic) domain.

Multilingual retrieval model	Multilinguality (# languages)	Computational complexity	Scalability (# documents)	Resource availability	Retrieval quality	Domain specificity
CL-VSM [1,13,6]	2	low	Web	medium	medium	none
Eurovoc-based [8,9]	21	medium	Web	poor	medium	medium
CL-LSI [2,10]	3	high	10^4	–	very good	total
CL-KCCA [15]	2	high	10^4	poor	very good	total
CL-RM [5]	2	medium	Web	good	good	medium
CL-ESA	**14**	**low**	**Web**	**good**	**good**	**low**

1.1 Comparison of Related Work

Cross-language retrieval models are generalizations of monolingual models such as the vector space model (VSM), latent semantic indexing (LSI), principal component analysis (PCA), language or relevance models (RM), or—as in the case of our approach—explicit semantic analysis (ESA). However, deductions that come at no expense in a monolingual retrieval situation are difficult to be achieved between two languages L and L': terms, named entities, time or currency expressions, etc. have to be identified and mapped from L to L', which entails the problem of translation ambiguity. Basically, there are two possibilities to bridge the language barrier: (*i*) dictionaries, gazetteers, rules, or thesauri versus (*ii*) parallel corpora or comparable corpora.[1] The former provide a means to translate words and concepts such as locations, dates, and number expressions directly from L to L', whereas the latter provide "aligned" documents from L and L' that are translations of each other or that cover the same topic, and which are utilized to translate arbitrary texts.

We have analyzed the existing approaches to overview their strong and weak points; the results are comprised in Table 1: the first group shows dictionary-based approaches and the second group corpus-based approaches. For large-scale retrieval tasks the computational complexity and the resource availability disqualify the Eurovoc-based approach, CL-LSI, and CL-KCCA. Among the remaining approaches CL-RM and CL-ESA have the advantage that no direct translation effort is necessary.

2 Explicit Semantic Analysis

This section introduces the principle of cross-language explicit semantic analysis, CL-ESA, a new multilingual retrieval model which does without automatic translation capabilities. Our starting point is a recently proposed monolingual retrieval model, the explicit semantic analysis, ESA [3,4].

---

[1] There has been much confusion concerning corpora termed "parallel" and "comparable"; the authors of [7] provide a consistent definition.

Let D^* denote a document collection of so-called index documents, and let φ denote the cosine similarity measure. Under ESA a document d is represented as an n-dimensional concept vector $\mathbf{d}$:

$$\mathbf{d} = (\varphi(\mathbf{v}, \mathbf{v}_1^*), \ldots, \varphi(\mathbf{v}, \mathbf{v}_n^*))^T,$$

where $\mathbf{v}$ is the vector space model representation of d, $\mathbf{v}_i^*$ is the vector space model representation of the ith index document in D^*, and n is the size of D^*. If $\varphi(\mathbf{v}, \mathbf{v}_i^*)$ is smaller than a noise threshold ε the respective entry is set to zero. Let $\mathbf{d}'$ be the concept representation of another document d'. Then the similarity between d and d' under ESA is defined as $\varphi(\mathbf{d}, \mathbf{d}')$.

Rationale of the ESA retrieval model is to encode the specific knowledge of d relative to the collection D^*. In this sense each document in D^* is used as a single concept to which the document d is compared, say, $\mathbf{d}$ can be understood as a projection of d into the concept space spanned by D^*. The authors in [4] achieved with ESA an average retrieval improvement of 20% compared to the vector space model.

To function as a generic retrieval model the index document collection D^* must be of a low domain specificity: D^* should contain documents from a broad range of domains, and each index document should be of "reasonable" length. A larger subset of the documents in Wikipedia fulfills both properties.

2.1   CL-ESA

Let $\mathcal{L} = \{L_1, \ldots, L_m\}$ denote a set of languages, and let $\mathcal{D}^* = \{D_1^*, \ldots, D_m^*\}$ denote a set of index document collections where each D_i^* contains index documents of language L_i. Moreover, let $C = \{c_1, \ldots, c_n\}$ denote a set of concept descriptors. $\mathcal{D}^*$ is called a concept-aligned comparable corpus if it has the property that the ith index document, d_i^*, of each index document collection $D^* \in \mathcal{D}^*$ describes c_i in its respective language.

A document d written in language $L \in \mathcal{L}$ is represented as ESA vector $\mathbf{d}$ by using that index document collection $D^* \in \mathcal{D}^*$ that corresponds to L. Likewise, a document d' from another language $L' \in \mathcal{L}$ is represented as $\mathbf{d}'$. The similarity between d and d' is quantified in the concept space, by computing the cosine similarity between $\mathbf{d}$ and $\mathbf{d}'$.

CL-ESA exploits the following understanding of a comparable corpus alignment: if all concepts in C are described "sufficiently exhaustive" for all languages in $\mathcal{L}$, the documents d and d' are represented in comparable concept spaces under ESA, using the associated index document collections D^* and D'^* in $\mathcal{D}^*$.

CL-ESA requires a comparable corpus $\mathcal{D}^*$, and each index document collection $D^* \in \mathcal{D}^*$ should meet the requirements for the monolingual explicit semantic analysis. Again, a larger subset of the documents in Wikipedia fulfills these properties.

3   Evaluation

To analyze the power of the CL-ESA retrieval model we implemented various experiments on a multilingual parallel and a multilingual comparable corpus. The results can be summarized as follows. (*i*) Given a document d in language L, CL-ESA ranks the aligned document d' in language L' with 91% probability on the first rank. (*ii*) Given a rank ordering in language L, CL-ESA is able to reproduce this ordering in language L' at a high fidelity. (*iii*) CL-ESA is insensitive with respect to the quality of the underlying index document collection. (*iv*) CL-ESA behaves robust with respect to a wide range of the concept space dimensionality.

Altogether CL-ESA is a viable retrieval model to assess the cross-language similarity of text documents. The remainder of this section describes the experiments in greater detail.

Multilingual Corpora. In our experiments we have employed the parallel corpus JRC-Acquis [14], and the comparable corpus Wikipedia. As one of the largest corpora of its kind the JRC-Acquis corpus contains 26 000 aligned law documents per language from the European Union in 22 languages. Wikipedia has not been considered as a comparable corpus by now. This fact is surprising since up to 100 000 aligned documents are available from diverse domains and languages, and the corpus is constantly extended by Wikipedia's editors. On the downside the aligned documents may be of less quality than those of custom-made comparable corpora.

Test Collections. Two test document collections comprising 3 000 documents each were selected from the German (D, L) and the English (D', L') parts of the multilingual corpora. Both collections contain 1 000 randomly selected translation-aligned documents from JRC-Acquis, 1 000 concept-aligned documents from Wikipedia, and 1 000 not aligned documents from Wikipedia. The latter have no language link from L to L' or vice versa. In particular, we assured that the distribution of monolingual similarities among the documents in D and D' corresponds to normal orders of magnitude.

The aligned index document collections D^* and D'^* were constructed from Wikipedia so that $D^* \cap D = \emptyset$ and $D'^* \cap D' = \emptyset$: no document is index document and test document at the same time. The size $n = |D^*| = |D'^*|$ of these collections corresponds to the dimensionality of the resulting document representations in the concept space.

3.1   Experiments

This subsection describes six selected experiments from our evaluation.

Experiment 1: Cross-Language Ranking. Given an aligned document $d \in D$, all documents in D' are ranked according to their cross-language similarity to d. Let $d' \in D'$ be the aligned document of $d \in D$, then the retrieval rank of d' is recorded. Ideally, d' should be on the first or at least on one of the top ranks. The experiment was repeated for all of the aligned documents in D. The first column of Table 2 shows the recall at ranks ranging from 1 to 50. The probability of finding a document's translation- or concept-aligned counterpart on the first rank is 91%, and the probability of finding it among the top ten ranks is > 99%.

Experiment 2: Bilingual Rank Correlation. To quantify the retrieval quality related *to a set* of retrieved documents we propose a new evaluation statistic. Starting point is a pair of aligned documents $d \in D$ and $d' \in D'$, whereas the documents from D'

are ranked twice: (*i*) with respect to their cross-language similarity to d using a cross-language retrieval model, and, (*ii*) with respect to their monolingual similarity to d' using the vector space model. The top 100 ranks of the two rankings are compared using a rank correlation coefficient, e. g. Spearman's ρ, which measures their disagreement or agreement as a value between -1 and 1 respectively.

The idea of this statistic relates to "diagonalization": a reference ranking under the vector space model is compared to a test ranking computed under the CL-ESA concept space representation. The experiment is conducted for each pair of aligned documents d and d' in the test collections, averaging the rank correlations. The second column of Table 2 shows a high correlation, provided a high dimensionality of the concept space. Note that this experiment is a generalization of Experiment 1 and that it has much more explanatory power.

Experiment 3: Cross-Language Similarity Distribution. This experiment contrasts the distribution of pairwise similarities of translation-aligned documents and concept-aligned documents. The results show that, on average, for both kinds of aligned documents high similarities are computed (cf. Table 2, third column), which demonstrates that CL-ESA is robust with respect to the quality of the aligned documents in the index document collections D^* and D'^*.

Experiment 4: Dimensionality. Both retrieval quality and runtime depend on the concept space dimension of CL-ESA, which in turn corresponds the size of a language's index document collections D^* and D'^*. The dimensionality of a retrieval model affects the runtime of all subsequently employed retrieval algorithms. Under CL-ESA, documents can be represented with a reasonable number of 1 000 to 10 000 dimensions while both retrieval quality and runtime are maintained (cf. Table 2 and Figure 1).

Experiment 5: Multilinguality. Starting with the two most prominent languages in Wikipedia, English and German, we study how many concepts are described in both languages, and how many are in the intersection set if more languages are considered. Currently, the Wikipedia corpus allows that documents from up to 14 languages are represented with CL-ESA (cf. Figure 1, left plot).

Experiment 6: Indexing Time. The time to index a document is between 10 to 100 milliseconds, which is comparable to the time to compute a vector space representation (cf. Figure 1, right plot). Employed hardware: Intel Core 2 Duo processor at 2 GHz and with 1 GB RAM.

3.2 Discussion

The evaluation of this section provides a framework for the adjustment of CL-ESA to the needs of a cross-language retrieval task. If, for example, a high retrieval quality is desired, documents should be represented as 10^5-dimensional concept vectors: ranking with respect to a particular query document will provide similar documents on the top ranks with high accuracy (cf. Table 2, first row). High retrieval quality comes at the price that with the current Wikipedia corpus only 2 languages can be represented at the same time, and that the time to index a document will be high (cf. Figure 1). If high retrieval

Table 2. Landscape of cross-language explicit semantic analysis: each row shows the results of three experiments, depending of the dimenionality n of the concept space

Experiment 1 Cross-Language Ranking	Experiment 2 Bilingual Rank Correlation JRC-Acquis Wikipedia		Experiment 3 CL Similarity Distribution	Dimension n
	0.81	0.72		10^5
	0.46	0.61		10^4
	0.20	0.44		10^3
	0.09	0.22		10^2
	0.04	0.07		10

speed or a high multilinguality is desired, documents should be represented as 1000-dimensional concept vectors. At a lower dimension the retrieval quality deteriorates significantly. A reasonable trade-off between retrieval quality and runtime is achieved for a concept space dimensionality between 1 000 and 10 000.

Concerning the multilinguality of CL-ESA the left plot in Figure 1 may not show the true picture: if the languages in Wikipedia are not considered by their document number but by geographical-, cultural-, or linguistic relations, there may be more intersecting concepts in the respective groups. And, if only two languages are considered, the number of shared concepts between a non-English Wikipedia and the English Wikipedia will be high in most cases.

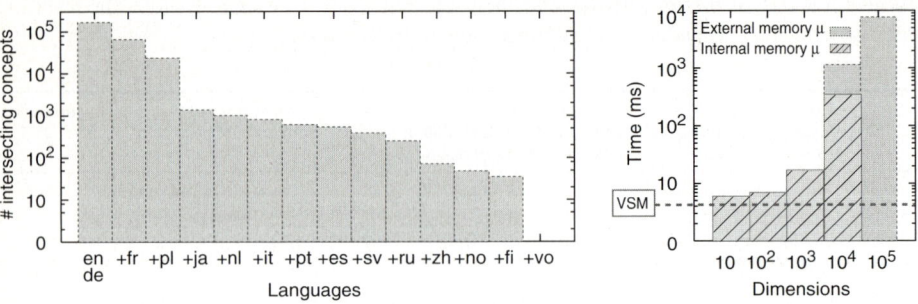

Fig. 1. Left: number of intersecting concepts among groups of languages in Wikipedia. The languages are organized in descending order wrt. the number of available documents. Right: average time to index a document under ESA, depending on the number of dimensions. We distinguish between indexes that fit in internal memory (μ), and external indexes.

4   Current Work

Our current work focuses on cross-language plagiarism detection. Plagiarism is the act of copying the work of another author and claiming it as own work. Though automatic plagiarism detection is an active field of research the particular case of cross-language plagiarism has not been addressed in detail so far.

The authors of [11] propose a three-step retrieval process to detect plagiarism, which can also be applied to detect cross-language plagiarism. Figure 2 illustrates the process. A suspicious document d of language L, which may contain a plagiarized section from a document d' in a reference corpus D' of language L', is analyzed as follows:

1. *Heuristic Retrieval.* A subset of D' is retrieved which contains candidate documents that are likely to be sources for plagiarism with respect to the content of d.
2. *Detailed Analysis.* The candidate documents are compared section-wise to d using CL-ESA for each pair of sections.

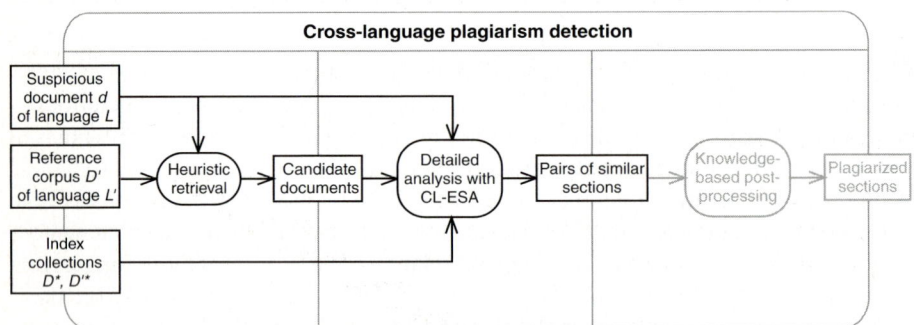

Fig. 2. A three-step process for cross-language plagiarism detection

3. *Knowledge-based Post-Processing.* Sections from the candidate documents that are similar to a section in d are processed in detail, for instance to filter cases of proper citation.

Cross-language explicit semantic analysis can be directly applied for Step 2, the detailed analysis, since it allows for a reliable assessment of cross-language similarity. However, for the preceding heuristic retrieval CL-ESA is not the best choice since a pairwise comparison of d to all documents from D' is required in order to cope with the high dimensionality of CL-ESA representations. To speed up this retrieval step we are investigating the following alternatives:

– Construction of a keyword index for D' which is queried with keywords extracted from d that are translated to L', and implementation of a focused keyword search.
– Construction of a keyword index for D' which is queried with keywords extracted from d', the machine translation of d to L', and, again, implementation of a focused search.
– Construction of a hash-based fingerprint index for D' which is queried with the fingerprint of d' [12].

The first two alternatives are based on keyword extraction as well as on cross-language keyword retrieval or machine translation technologies. The last alternative, which has the potential to outperform the retrieval recall of the first approaches, employs machine translation and similarity hashing technologies.

References

1. Ballesteros, L.: Resolving Ambiguity for Cross-Language Information Retrieval: A Dictionary Approach. PhD thesis, Director-W. Bruce Croft (2001)
2. Dumais, S., Letsche, T., Littman, M., Landauer, T.: Automatic cross-language retrieval using latent semantic indexing. In: AAAI 1997, Cross-Language, Text, and, Speech, Retrieval (1997)
3. Gabrilovich, E.: Feature Generation for Textual Information Retrieval Using World Knowledge. Phd thesis, Israel Institute of Technology (2006)
4. Gabrilovich, E., Markovitch, S.: Computing semantic relatedness using wikipedia-based explicit semantic analysis. In: IJCAI 2007, Hyderabad, India (2007)
5. Lavrenko, V., Choquette, M., Croft, W.: Cross-Lingual Relevance Models. In: SIGIR 2002, pp. 175–182. ACM Press, New York (2002)
6. Levow, G.-A., Oard, D., Resnik, P.: Dictionary-based techniques for cross-language information retrieval. Inf. Process. Manage. 41(3), 523–547 (2005)
7. McEnery, A., Xiao, R.: Parallel and comparable corpora: What are they up to? Incorporating Corpora: The Linguist and the Translator (2007)
8. Pouliquen, B., Steinberger, R., Ignat, C.: Automatic annotation of multilingual text collections with a conceptual thesaurus. In: OntoIE 2003 at EUROLAN 2003, pp. 9–28 (2003)
9. Pouliquen, B., Steinberger, R., Ignat, C.: Automatic identification of document translations in large multilingual document collections. In: RANLP 2003, pp. 401–408 (2003)
10. Rehder, B., Littman, M., Dumais, S., Landauer, T.: Automatic 3-language cross-language information retrieval with latent semantic indexing. In: TREC, pp. 233–239 (1997)
11. Stein, B., zu Eissen, S.M., Potthast, M.: Strategies for retrieving plagiarized documents. In: SIGIR 2007, pp. 825–826 (2007)

12. Stein, B.: Principles of hash-based text retrieval. In: SIGIR 2007, pp. 527–534 (2007)
13. Steinberger, R., Pouliquen, B., Ignat, C.: Exploiting multilingual nomenclatures and language-independent text features as an interlingua for cross-lingual text analysis applications. In: 4th Language Technology Conference at Information Society, Slovenia (2004)
14. Steinberger, R., Pouliquen, B., Widiger, A., Ignat, C., Erjavec, T., Tufis, D., Varga, D.: The JRC-Acquis:A multilingual aligned parallel corpus with 20+languages. In: LREC 2006 (2006)
15. Vinokourov, A., Shawe-Taylor, J., Cristianini, N.: Inferring a semantic representation of text via cross-language correlation analysis. In: NIPS 2002, pp. 1473–1480. MIT Press, Cambridge (2003)

Filaments of Meaning in Word Space

Jussi Karlgren, Anders Holst, and Magnus Sahlgren

Swedish Institute of Computer Science

Abstract. Word space models, in the sense of vector space models built on distributional data taken from texts, are used to model semantic relations between words. We argue that the high dimensionality of typical vector space models lead to unintuitive effects on modeling likeness of meaning and that the local structure of word spaces is where interesting semantic relations reside. We show that the local structure of word spaces has substantially different dimensionality and character than the global space and that this structure shows potential to be exploited for further semantic analysis using methods for local analysis of vector space structure rather than globally scoped methods typically in use today such as singular value decomposition or principal component analysis.

1 Vector Space Models

Vector space models are frequently used in information access, both for research experiments and as a building block for systems in practical use. There are numerous implementations of methods for modeling topical variation in text using vector spaces. These and related methods are used for information access or knowledge organisation of various levels of abstraction, all more or less based on quasi-geometric interpretations of distributional data of words in documents. Vector space models in various forms have been implicit in information retrieval practice at least since the early 1970's and their origin has usually been attributed to the work of Gerard Salton. His 1975 paper titled "A vector space model for automatic indexing" [1], often cited as the first vector space model, does not in fact make heavy use of vector spaces, but in his later publications the processing model was given more prominence as a convenient tool for topical modeling (see e.g. Dubin for a survey [2]). The vector space model has since become a staple in information retrieval experimentation and implementation.

Distributional data collected from observation of linguistic data can be modeled in many ways, yielding probabilistic language models as well as vector space models. Vector space models have attractive qualities: processing vector spaces is a manageable implementational framework, they are mathematically well-defined and understood, and they are intuitively appealing, conforming to everyday metaphors such as "near in meaning". In this way, vector spaces can be interpreted as a model of meaning, as semantic spaces. In this sense, the term "word space" is first introduced by Hinrich Schütze: "Vector similarity is the only information present in Word Space: semantically related words are close, unrelated words are distant" [3]. While there is some precedent to this definition

C. Macdonald et al. (Eds.): ECIR 2008, LNCS 4956, pp. 531–538, 2008.

in linguistic and philosophical literature, none of the classic claims in fact give license to construct spatial models of meaning: to do so, we first must examine how the model we build in fact preserves and represents the sense of meaning it sets out to capture.

1.1   How Many Dimensions?

Much of the theoretical debate on vector space models has to do with how many dimensions a semantic word space should have. The typical bare vector space of terms by contexts gives a word space of tens or hundreds of thousands or even millions of dimensions, a large number which typically, in most approaches, continues to grow when more data are added. Most every element of a typical word vector will be zero which seems a waste of dimensions, most words appear to be polysemous to some extent, and most concepts – taken on a suitably coarse-grained level of analysis – would seem to be representable by many different lexical items. Without further treatment of the data a bare model fails to generalise between terms with similar but non-identical distribution patterns. This calls for the informed reduction of dimensions to a smaller number, both for ease of processing and to be able to capture similarities.

What then seems to be an appropriate dimensionality? The word space research field frequently searches for a "latent", or intrinsic, dimensionality in the data, lower than the bare dimensionality resulting from the data collection. This intrinsic dimensionality can potentially be found by processing the data set in some informed way [4, e.g.]. Most efforts heretofore have used global measures such as singular value decomposition or principal component analysis to process, examine or reduce dimensions of the observed data. Some claims as to what this intrinsic dimensionality might be are $o(100)$ for data processed by singular value decomposition in the popular latent semantic analysis framework [5] and $o(1000)$ for data processed by us in previous work using the more recent random indexing approach [6]. These figures are obtained by reprocessing the data set with parameter variation based on trial-and-error experimentation, typically evaluated by synonym tests.

In this paper, in contrast with previous dimension reduction approaches, we argue that this appropriate dimensionality is determined locally, not globally, in the space. We base our argument on an inspection of the character of a typical vector space model built from textual data.

1.2   Are There Large Distances in Semantic Spaces?

There is a huge theoretical leap from the realisation that words are defined by their contexts to furnishing a whole vector space based on the postulated distances between words – however those distances are defined and however the distributional data are collected. What sort of information are the distances supposedly based on? It would seem there is very little purchase in the data to base *any* sort of distance between say "tensor" and "cardamom" or between "chilblain", "child-birth" and "chiliad". There is a limit to what questions one can expect a word space built by distributional data to answer. The intuitively attractive qualities of a semantic representation where meaning is distributed about a many-dimensional vector space leads us to forget that the *only* interest we ever will show the space is in its local neighbourhoods. Returning to the original discussion on word spaces cited above: "Vector similarity is the only information present in Word Space: semantically related words are close, unrelated words are distant" [3], we claim that "close" is interesting and "distant" is not, and that vector space models are overengineered to handle information that never is relevant in language modeling.

In this paper, the question we address is what sort of dimensionality one might need to model the context of a term – as opposed to how many dimensions would be necessary if one would wish to attempt to model the structure of an entire large sample of language in one contiguous and coherent space. We will investigate the *local character* of word spaces – the structure of the space within which semantically related words are expected to be found.

1.3   What Is a Typical Angle between Random Pairs of Words?

Human topological intuitions are based on our experiences in a two-to-three dimensional world. We live our lives more or less on a plane with occasional ventures or glances up or down from it. Many-dimensional spaces are in some important respects very different from two-to-three dimensional spaces. One such unintuitive feature of a high dimensional space is that two randomly picked points on e.g. a unit hypersphere are almost always at near orthogonal angles to each other with respect to the origin. The graphs displayed in Figure 1 show the probability distribution of the resulting angle between two randomly chosen points in 3-, 10-, and 1000-dimensional spaces respectively.

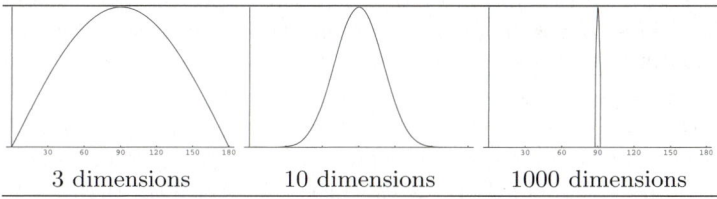

| 3 dimensions | 10 dimensions | 1000 dimensions |

Fig. 1. Probability distribution for angles between directions to randomly chosen points in many-dimensional spaces

The given distribution of points (and thus angles) has practical consequences for semantic models. In low dimensional spaces, given our experience of the physical world, we find it easy and intuitive to reason about distances. For example, if point A is close to point B, and point B is close to point C, then point A is fairly close to C. This makes it natural to imagine clusters of samples close to each other but separated from samples in other clusters.

However, in high dimensional spaces the transitivity of distance is not necessarily as obvious. It may well be that points A and C are both close to point B but still at considerable distance from each other. The triangle inequality still holds in high dimensional Euclidean spaces; the problem is that whereas an angle of $90°$ between two words always will mean that the words are completely unrelated, in a thousand-dimensional space an angle of $45°$ means a remarkable degree of similarity. This makes the notion of a cluster less useful in high dimensional spaces (a similar argument is given by Beyer et al. [7]).

1.4   What Does a Word Space Look Like?

For this experiment we have constructed a vector space, V_{text}, from textual data as provided in the TASA corpus[8], a corpus of short high-school level English texts on various factual subjects such as language, health, business and other general school curriculum related topics. The corpus consists of about 10 million words, 37 600 text samples, and 27 000 distinct terms.

To build the word space, we use the random indexing approach as described by Sahlgren [9]. We choose random indexing for two main reasons. First, its authors make strong claims about appropriate representational dimensionality, and indeed have set $d_{representation}$, the dimension of the representation, as a settable parameter for the algorithm. Second, where in most indexing approaches each distinct term encountered in the text is assigned a binary index vector with all elements zero and one element 1, random indexing assigns each distinct term a ternary index vector with most elements zero and some randomly assigned elements either 1 or -1. Using random indexing thus avoids overloading the positive section of the vector space: by the use of negative vector elements in the representation it has the resulting word space occupy the entire possible vector space. This is desirable for our experiment, since we want to be able to relate the observed distribution of words to the expected distribution over the entire hypersphere, without leaving large swathes of vector space vacant. (The fact that most vector space models only operate in the positive sector of the multi-dimensional space is usually never discussed.)

Random indexing is convenient for our purposes, and the validity of the results is spoken for by Johnson-Lindenstrauss' lemma [10], the basis of random projection approaches, which states that a vector space (in this case, the term-by-context matrix, which is of immense dimensionality) can be projected into a random subspace of appropriate dimensionality (in this case, V_{text} of dimension, $d_{representation}$) without corrupting the relative distances between points in the space.

In this experiment, V_{text} is built from occurrence data collected from a rolling $2+2$ window over the text segments, a setting which has previously been shown by us to provide consistent results in extrinsic evaluation schemes [11]. The dimensionality, $d_{representation}$, is set to 10 000, higher than most published experiments using random indexing, with five randomly positioned 1's and five randomly positioned -1's for the initial index vectors. The relatively high choice of $d_{representation}$ was chosen to ensure that the global dimensionality is high enough not to distort or constrain any local subspace structure.

A sample of 1 000 000 pairs of words were randomly selected from the material and the angles between them tabulated. The first neighbours in our test material appear at an angle of about 30°. Further neighbours are found at an rapidly increasing rate, and as expected, the distribution peaks around 90°. Figure 2 shows the distribution of angles between words in the sample from V_{text}. By comparing with the expected random distributions given in Figure 1 we find that the shape of the distribution for the observed data yields a global dimensionality of $o(10\,000)$, around the dimensionality of the representation. But the distribution in Figure 2 does not match the theoretical distributions in every detail. If we zoom in on the base, as shown in the right graph, we find structure in the left tail. This represents a non-homogenous distribution at smaller angles between word vectors than would have been expected if the words were homogeneously distributed in the 10 000-dimensional space. It is in this neighbourhoood we find the non-random, i.e. semantically interesting, word-word relations.

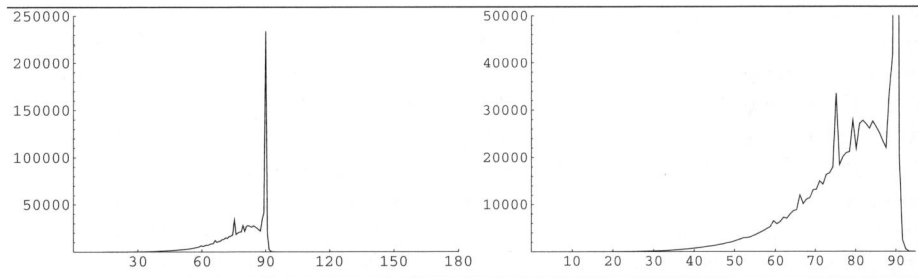

Fig. 2. Observed distance distributions; rightmost graph zoomes in on left tail

To analyse the intrinsic dimensionality of a local region of word space we use a method from the analysis of fractal dimensions. The method rests on the mathematically trivial observation that if we have samples homogeneously distributed in a $d-$dimensional space, the number of samples within a hypersphere of radius r increases proportionally to r^d. For example, if we double the radius of a circle in a two-dimensional space the number of samples within the circle will increase by a factor of four, and for a sphere in a three dimensional space with a factor of eight. To measure the intrinsic dimensionality of the word space we examine the neighbourhoood of a point within it: we begin by counting the number of samples within a sphere of some radius r, and then we double r and count again. If the number of samples increases at a rapid rate, this means a higher dimensionality. In detail, the dimensionality d in the span between two radii r_1 and r_2 is here computed by

$$d = \frac{log(n_{r_2}/n_{r_1})}{log(r_2/r_1)}$$

where n_{r_1} and n_{r_2} are the observed numbers of samples within those radii.

An advantage with this method to measure intrinsic dimensionality is that it can measure the dimensionality locally on the small scale, as well as medium scale, or large scale, as opposed to e.g. principal component analysis or singular value decomposition which can only be used to compute intrinsic dimensionality on the global scale. This enables us to compare the intrinsic dimensionality of local contexts with the global dimensinality.

Using the above method of analysis of the fractal dimension of the samples in the left tail we find a local intrinsic dimensionality somewhere just below 10, somewhat depending on where the tail is cut off, i.e. exactly how small scale structure we care to investigate. These findings support our claim that there indeed is a local neighbourhood for a typical vector: even allowing for a substantial margin of error due to the amount of noise in the data and the somewhat crude methodology, we find that the local dimensionality is several orders of magnitude less than that of most vector space models today, including those that use dimension reduction techniques.

1.5   Can We Model the Local Neighbourhood in Word Spaces?

In our experimental data and the word space V_{text}, we found, as shown in Figure 2, that the number of neigbours at small distances was greater than what would be expected from a theoretical completely homogeneous case. But we also saw that the small scale dimensionality was much less (< 10) than the large scale dimensionality ($\approx 10\ 000$). So, more neighbours, but lesser dimensionality. This gives us some clues to the local structure of the space. Consider the hand-made example graphs in Figure 3. The first graph shows a fairly homogeneous distribution of observations, both neighbours and dimensionality, in some subspace of a larger-dimensional space. If the observations instead were "clustered" or "lumpy" as in the second graph, then there is a higher number of small distances between samples than in the homogeneous case, but the small scale dimensionality of the subspace will be the same as the large scale dimensionality of the space. However, in the case shown in the third graph where the observations are found to occur in a filamentary structure, the number of small distances are also increased, but the small scale dimensionality (one-dimensional, along the filaments, in the figure) is smaller than the large scale dimensionality (two-dimensional, across the plane, in the figure).

This allows us to suggest that the word space is primarily neither homogeneous nor lumpy, but filamentary in its structure, and that these filaments, which are of much lower dimensionality than usually considered in semantic models is where the key to modelling similarity of meaning may reside.

1.6   How Much Data Do We Need to Train a Knowledge Representation?

It is often claimed that models of meaning based on distributional data need huge data sets for training. However, it is simple to observe that in actual language use only very few occurrences are in practice necessary to model the approximative

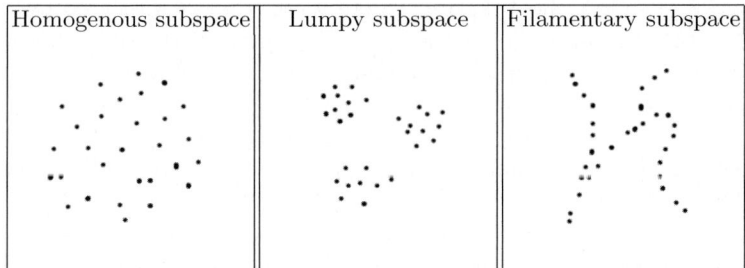

Fig. 3. Neighbourhoods of different character

meaning of a newly encountered term. How many occurrences of "Jarlsberg" do you need to figure out what manner of beast it is? Training data found by one of the more popular web search engines are given in Figure 4. This would seem to speak to the fact that semantic context in realistic knowledge representations in fact will be saturated rather rapidly even by a small set of sample observations, that semantic similarity in fact is modellable by few dimensions rather than a plentitude. The data collected for any single term can be fairly assumed to tell us a fair amount about the term in question – and of other terms that occur similarly. The distributional data for some term will, however, not tell us much about every other term in the language and relations from the observandum to them. This observation is borne out by the data analysis from our experiment.

> The famous Jarlsberg cheese is known for its distinctive sweet and nutty taste ...
>
> The largest producer of Jarlsberg today is the Tine BA factory in ...
>
> Within a few decades Jarlsberg has become one of Norway's greatest export successes ...
>
> Jarlsberg is the most popular Norwegian cheese in the UK. ...

Fig. 4. What does "Jarlsberg" mean?

1.7 Ramifications for Word Spaces

Our conclusions and claims are three-fold. Firstly, that the most interesting qualities of word spaces are found in their local structure rather than their global dimensionality, and that thus much of the discussion of representational dimensionality, latent semantic dimensionality, and of global methods for dimension reduction is of lesser theoretical and practical import. Secondly, since the high dimensionality of the global model saddles the practical system with tractability bottle-necks in processing, maintenance, and deployment, optimising the global character of the model is likely to provide respectable gains in efficiency. However, any claims of semantic relevance of such optimisations should be viewed with skepticism unless they expressly take local context into account. Thirdly, that elaborating the structure of the local filamentary structure further is likely to lead to less demanding models as regards size of training data sets. We do not claim to yet have a framework for how to realise such models.

We want to stress that these results are not meant to be part of the discussion of appropriate choice of dimensionality in vector space models. However, we believe that studying the local structure of vector space models will cast light on the structure of textual data, give insights into the design of future processing models, and provide new starting points for the informed design of semantic representations based on distributional data, whether in vector space models or not.

References

1. Salton, G., Wong, A., Yang, C.S.: A vector space model for automatic indexing. Commun. ACM 18(11), 613–620 (1975)
2. Dubin, D.: The most influential paper Gerard Salton never wrote. Library Trends 52(4), 748–764 (2004)
3. Schütze, H.: Word space. In: Proceedings of the 1993 Conference on Advances in Neural Information Processing Systems, NIPS 1993, pp. 895–902. Morgan Kaufmann Publishers Inc., San Francisco (1993)
4. Chávez, E., Navarro, G.: Measuring the dimensionality of general metric spaces. Technical Report TR/DCC-2000-1, Department of Computer Science, University of Chile (2000)
5. Deerwester, S., Dumais, S., Furnas, G., Landauer, T., Harshman, R.: Indexing by latent semantic analysis. Journal of the Society for Information Science 41(6), 391–407 (1990)
6. Kanerva, P., Kristofersson, J., Holst, A.: Random indexing of text samples for latent semantic analysis. In: Proceedings of the 22nd Annual Conference of the Cognitive Science Society, p. 1036. Erlbaum, Mahwah (2000)
7. Beyer, K., Goldstein, J., Ramakrishnan, R., Shaft, U.: When Is Nearest Neighbor Meaningful? In: Beeri, C., Bruneman, P. (eds.) ICDT 1999. LNCS, vol. 1540, pp. 217–235. Springer, Heidelberg (1998)
8. Landauer, T., Foltz, P., Laham, D.: Introduction to latent semantic analysis. Discourse Processes 25, 259–284 (1998)
9. Sahlgren, M.: An introduction to random indexing. In: Witschel, H. (ed.) Methods and Applications of Semantic Indexing Workshop at the 7th International Conference on Terminology and Knowledge Engineering. TermNet News: Newsletter of International Cooperation in Terminology, vol. 87 (2005)
10. Johnson, W., Lindenstrauss, J.: Extensions of lipshitz mapping into hilbert space. Contemporary Mathematics 26, 189–206 (1984)
11. Sahlgren, M.: The Word-Space Model: using distributional analysis to represent syntagmatic and paradigmatic relations between words in high-dimensional vector spaces. PhD thesis, Department of linguistics, Stockholm university (2006)

Finding the Best Picture:
Cross-Media Retrieval of Content

Koen Deschacht and Marie-Francine Moens

Katholieke Universiteit Leuven, Department of Computer Science,
Celestijnenlaan 200A, B-3001 Heverlee, Belgium
{Koen.Deschacht,Marie-Francine.Moens}@cs.kuleuven.be
http://www.cs.kuleuven.be/~liir/

Abstract. We query the pictures of Yahoo! News for persons and objects by using the accompanying news captions as an indexing annotation. Our aim is to find these pictures on top of the answer list in which the sought persons or objects are most prominently present. We demonstrate that an appearance or content model based on syntactic, semantic and discourse analysis of the short news text is only useful for finding the best picture of a person of object if the database contains photos each picturing many entities. In other circumstances a simpler bag-of-nouns representation has a good performance. The appearance models are tested in a probabilistic ranking function.

Keywords: Cross-media Retrieval, Information Extraction, Ranking, Image search.

1 Introduction

Repositories of multimedia content (e.g., provided by the World Wide Web) demand for effective means of retrieval without relying on manual annotations. In text-based image retrieval some form of textual description of the image contents is stored with the image, the image base is queried with a textual query, and correspondence is sought between the textual data when ranking the images. When people search for images, high precision on top of the answer list is very important. They might search for the best pictures of a person or object (in which the sought entities are most prominently present), or of a combination of them (e.g., picture of a meeting between Angela Merkel and George Bush), where a high recall of all best pictures is valuable, and a high recall of all images picturing the queried persons or objects is not.

Our goal is to find the best images of a person (persons) or object(s) in a database of photos (in our case found on the World Wide Web) that possibly picture many persons or objects and that have associated texts in the form of descriptive sentences. When a text describes an accompanying image, it is often the case that content described in the text is not present in the image and vice versa. In addition, retrieval of the images based on accompanying texts not always returns the best picture on top of the answer list. Our goal is to test

C. Macdonald et al. (Eds.): ECIR 2008, LNCS 4956, pp. 539–546, 2008.

several approaches with varying complexity of analysis of the caption texts for their capability of being discriminative indexing descriptions of the images.

Generative probabilistic retrieval models are suited for cross-media information retrieval. They rely on a content model generated from a document. We call the content model a language model when it represents the content of a text and an appearance model when it represents the content of an image. When retrieving images that have accompanying texts, one can design several content models that probabilistically model the textual and/or the visual content. In the research reported here we build an appearance model solely on the basis of the text in an attempt to capture persons and objects that are present in the image and to compute their degree of prominence in the image. This appearance model is used in a probabilistic ranking function for retrieval. We illustrate our approach by querying the pictures of Yahoo! News.

This article is organized as follows. First we give an overview for our methodology with focus on the construction of the appearance model and its integration into a probabilistic retrieval model. Then, we describe and discuss our experiments and conclude with related research and prospects for future research.

2   Methods

2.1   The Content Models

The most simple content model of the text is made by tokenization of the text into words, which gives us a bag-of-word representatation (BOW-representation). A more advanced model considers only the nouns (including proper nouns) because in the search for persons and objects only nouns are important (bag-of-noun representation or BON). Part-of-speech tagging detects the syntactic word class and we use here the LTChunk tool [8]. In advanced content models we rely on more sophisticated natural language processing techniques. We perform pronoun resolution[1], word sense disambiguation [6], named entity recognition (NER)[2] and consider the visualness and salience of a noun phrase.

2.2   Computation of the Visualness and Salience

When we build an appearance model, entities that are not visual do not play a role because they cannot be part of an image. We compute the visualness (value between zero and one) of each noun and proper noun, where visualness is defined as the degree that a noun entity can be perceived visually by humans or a camera. Proper nouns that were classified as persons by the NER tool receive a visualness of 1. We compute the visualness of a common noun based on knowledge of the visualness of a few seed words and their semantic distance with the target nouns in WordNet.

---

[1] http://www.alias-i.com/lingpipe/
[2] Adaptation of Lingpipe NER tool.

San Francisco Giants' Barry Bonds, right, holds a bat while sitting in the dugout with Omar Vizquel, left, of Venezuela in the ninth inning against the Florida Marlins Tuesday, May 30, 2006 at Dolphin Stadium in Miami. Bonds did not play as the Marlins defeated the Giants 5-3.

Barry Bonds 0.75	bat 0.259
dugout 0.254	Omar Vizquel 0.214
Dolphin 0.172	Stadium 0.084

Fig. 1. Image-text pair (source: AP Photo/Yahoo! News) with the probabilities that the text entities appear in the image

We also compute the salience (value between zero and one) of each noun and proper name, assuming that salient entities in texts that accompany images have a better chance of being present in the images. Computation of visualness and salience is described and evaluated in detail in [5].

2.3   Computation of the Appearance Model

We assume that entities found in a text T_j might be present in the accompanying image I_j, and that the probability of the occurrence of an entity e_i in the image, given a text T_j, $P(e_{i-im}|T_j)$, is proportional with the degree of visualness and salience of e_i in T_j. In our framework, $P(e_{i-im}|T_j)$ is computed as the product of the salience of the entity e_i and its visualness score, as we assume both scores to be independent, normalized by the sum of appearance scores of all entities in T_j. We have here used a very simple smooting method in order to counter errors in the named entity recognition, where we give all words which receive a zero score in the appearance model a fixed score of 0.01. $P(e_{i-im}|T_j)$ defines a ranking of the text's entities. Figure 1 gives an example of such a ranking generated from the text by our system.

In [5] the impact and a detailed error analysis of each step in the construction of the appearance model is given. It was shown that both the salience and visualness substantially contribute to an improved appearance model for describing and ranking the entities according to prominence in an accompanying image. We now want to find out how good this model is for discriminatively indexing the images in a cross-media retrieval task compared to more simpler models.

2.4   Integration in a Probabilistic Retrieval Model

Statistical language modeling has become a successful retrieval modeling approach [4]. A textual document is viewed as a model and a textual query as a string of text randomly sampled from that model. In case of our text-based image retrieval, the content model of image I_j is solely generated from the accompanying text T_j. Let the query be composed of one or more entities where the queried entity e_i is in the form of a person proper name or common noun representing a person or object. Our baseline appearance model considers a bag-of-words (BOW) representation of the text as content model of the image resulting in following retrieval model or ranking function:

$$P(e_1, ..., e_m|I_j) = \prod_{i=1}^{m} ((1 - \lambda)P(e_i|T_j) + \lambda P(e_i|C)) \tag{1}$$

where q_i is the ith query term in a query composed of m terms, and $P(e_i|T_j)$ is specified by the appearance model built from the text, and C represents the collection of documents. We estimate $P(e_i|T_j)$ by maximum likelihood estimation of the occurrence of the query term in the text. An intermediate model (BON) uses a bag-of-noun representation of the text and computes the probability that the text generates the entity $P(e_i|T_j)$ by maximum likelihood estimation of the occurrence of the query term in the text filtered by all words except nouns (including proper nouns). Both models do not take into account that an entity mentioned in the text can actually be shown in an image. In a limited way they consider salience as in longer texts the maximum likelihood of a term will be lower - especially when terms mostly occur only once - and thus the entities mentioned are likely to be less important.

We also integrate the appearance model (AP) described above:

$$P(e_i, ..., e_m|I_j) = \prod_{i=1}^{m} ((1 - \lambda)P(e_{i-im}|T_j) + \lambda P(e_i|C)) \tag{2}$$

Variations of this model only consider the factor salience (APS) or visualness (APV) when generating $P(e_{i-im}|T_j)$. We used the Lemur toolkit[3] and adapted it to suit our appearance models (AM). We used Jelinek-Mercer smoothing with the linear interpolation weight λ set to 0.1.

3   Experiments, Results and Discussion

3.1   The Data Collection, Queries and Ground-Truth Answer Lists

Because of the lack of a standard dataset that fits our tasks and hypotheses, we annotated our own ground truth corpus. Our dataset from the Yahoo! News website[4] is composed of 700 image-text pairs. Every image has an accompanying news text which describes the content of the image. This text will in general discuss one or more persons in the image, possibly one or more other objects, the location and the event for which the picture was taken. Not all persons or objects who are pictured in the photograph are necessarily described in the

---

[3] http://www.lemurproject.org/
[4] http://news.yahoo.com/

news text. The inverse is also true. The texts are short and contain maximum 3 sentences. On average the texts have a length of 40.98 words, and contain 21.10 words that refer to noun phrase entities of which 2.77 refer to distinct persons and objects present in the image (see table 1 for the distribution of visible entities in the documents).

Because of the many images with only one person pictured, we refer to this dataset as the EASYSET. From this set we select a subset of pictures where three or more persons or objects are shown, which varying degree of prominence in the image. We call this dataset that comprises 380 image-text pairs the DIF-FICULTSET (see example in figure 2). Tests on the latter set allows us to better understand the behavior of our different indexing methods when many persons or objects with varying degree of prominence are shown in the photographs.

Table 1. Number of image-text pairs for a given number of entities in the image

Entities	0	1	2	3	4	5	6	7	≥ 8
Documents	2	168	353	151	133	47	17	8	24

We annotated the images with the names of the persons and objects shown, and ranked these entities according to prominence in the image. Queries were randomly generated from the manual annotations of the images and were filtered in order to have images in which the queried persons or objects were present at several levels of prominence. In this way we obtained 53 queries that contain one name of a person or object, and 26 queries with two entities (23 queries with two person names and 3 queries with a person and object name). Larger queries do not seem to make sense, as people often search for a picture of one person, perhaps a person with an object (e.g., car, flag), or 2 persons meeting each other.

For each query we generated a list of images sorted according to relevancy for the query (ground truth answer list) where the prominence of the entities in the image is taken into account and where we give priority to images with fewer persons or objects, and take into account the centrality and size of the person(s) or object(s) of interest.

Note that all queries have at least one relevant image in the data sets, which makes a comparison among the methods for finding the best picture more transparent, and, most importantly, that multiple images can occupy the same relevance rank in the ground truth answer list.

3.2   Evaluation

Our aim is to retrieve the best pictures, i.e., the images on rank 1 in the ground truth answer list for a certain query, on top of the machine generated list. We use the mean R-precision or R-recall where R is defined as the number of relevant pictures on rank 1 in the ground truth answer list. This corresponds with precision@1 taking into account that the first position or rank might contain

U.S. President George W. Bush (2nd R) speaks to the press following a meeting with the Interagency Team on Iraq at Camp David in Maryland, June 12, 2006. Pictured with Bush are (L-R) Vice President Dick Cheney, Defense Secretary Donald Rumsfeld and Secretary of State Condoleezza Rice.

Fig. 2. Image-text pair (source: Reuters/Yahoo! News)

Table 2. Results in terms of Mean R-precision (MRP) and Mean Average Precision (MAP) for the ranking models based on the different text representations for the EASYSET and DIFFICULTSET where the query is composed of one entity

Content model	MRP EASYSET	MAP EASYSET	MRP DIFFICULTSET	MAP DIFFICULTSET
BOW	53.84 %	56.90 %	50.00 %	70.48 %
BON	69.23 %	61.25 %	58.00 %	74.14 %
AP	57.69 %	59.28 %	60.00 %	75.57 %
APS	57.69 %	57.14 %	56.00 %	74.07 %
APV	61.54 %	57.00 %	60.00 %	75.27 %

several best images. We also compute the classical average precision (AP) for the R relevant pictures. The above precision values are averaged over the queries and named in the tables below respectively as MRP and MAP.

3.3 Results

The results in terms of MRP and MAP are shown in tables 2 and 3. First, we see that the visualness measure in generally improves the retrieval model when the query is composed of one entity. This measure enables to determine how many entities in a given text are likely to appear in the image, and thus to create a more fine-grained ranking (since images with a small number of entities are prefered above images with a large number of entities). We see furthermore from tables 2 and 3 that this effect is most important when testing on the difficult set. This seems intuitive, since the difficult set contains only documents with large numbers of entities, for which it is important to determine what entities appear exactly in the image. The results also show that prominence is sufficiently captured by the maximum likelihood estimation of the term occurrence in the text. The longer the captions, the more content probably is shown in the image and the less important the individual entities in the image are. This simple heuristic yields good results when using captions for indexing images, while more advanced salience detection techniques are superfluous. When the queries

Table 3. Results in terms of Mean R-precision (MRP) and Mean Average Precision (MAP) for the ranking models based on the different text representations for the EASYSET and DIFFICULTSET where the query is composed of two entities

Content model	MRP EASYSET	MAP EASYSET	MRP DIFFICULTSET	MAP DIFFICULTSET
BOW	53.85 %	60.60 %	69.23%	68.08 %
BON	69.23 %	64.93 %	73.07 %	72.08 %
AP	57.69 %	59.28 %	57.70 %	63.83 %
APS	57.69 %	54.44 %	61.54 %	60.30 %
APV	53.85 %	52.81 %	61.54 %	62.78 %

contain more terms, the simpler bag-of-words or bag-of nouns models have better retrieval performance, possibly explained by the fact that short caption texts that contain the query entities retrieve the best pictures.

4   Related and Future Work

Since the early days of image retrieval, text-based approaches are common because users often express an information need in terms of a natural language utterance. Especially in a Web context text-based image retrieval is important given that users are acquainted with keyword searches. Recognizing content in the image that relies on descriptions of surrounding texts is researched, for instance, by [9,1]. [3] demonstrated the importance of content that surround the images on Web pages for their effective retrieval and have investigated how multiple evidence from selected content fields of HTML Web pages (e.g. meta tags, description tags, passages) contribute to a better indexing. Also [10] combine textual and visual evidence in Web image retrieval. The textual analysis in the above research does not go further than a bag-of-words representations scheme. The most interesting work here to mention is the work of Berg et al. [2] who also process the image-text pairs found in the Yahoo! news corpus. They consider pairs of person names recognized with named entity recognition (text) and faces (image) and use clustering with the Expectation Maximization algorithm to find all faces belonging to a certain person. Bayesian networks have been successfully used in image retrieval by [3] who integrate evidence of multiple fields of HTML Web pages. These authors found that a combination of description tags with a 40-term textual passage that most closely accompanies the image, provides best retrieval performance. However, still a bag-of-words approach is used. Our work can perfectly complement the above research as we provide a more accurate appearance model. The future lies in combining evidence from the different media relying on advanced technology for text and image analysis (cf. [7]). Our current and future work which combines visual and textual features goes in this direction. When we obtain evidence from different sources other probabilistic ranking models, such as inference or Bayesian networks are valuable [3].

5   Conclusions

We have built and tested several probabilistic appearance or content models from texts that accompany images. A simple bag-of-words approach is compared with a bag-of-nouns approach and with a more fine-grained identification of what content of the text can be visualized and of how prominent the content is in the image. The appearance models were integrated in a probabilistic retrieval function. The models based on more advanced text analysis taking into account syntactic, semantic and discourse analysis - although successful in automatically annotating the images - are not necessarily more discriminative for indexing purposes, except when querying a difficult data set for one person or object, where the images contain three or more persons or objects. A bag-of-nouns representation yielded overall the best results, especially when the query becomes more elaborated (with more entities), the overlap with the caption by using simpler representations is sufficient.

Acknowledgments

The work reported was supported by the CLASS project (EU-IST 027978). We thank Yves Gufflet (INRIA, France) for collecting the Yahoo! News dataset.

References

1. Barnard, K., Duygulu, P., de Freitas, N., Forsyth, D., Blei, D., Jordan, M.I.: Matching Words and Pictures. Journal of Machine Learning Research 3(6), 1107–1135 (2003)
2. Berg, T.L., Berg, A.C., Edwards, J., Forsyth, D.: Who's in the Picture? Neural Information Processing Systems, 137–144 (2004)
3. Coelho, T., Calado, P., Souza, L., Ribeiro-Neto, B.: Image Retrieval Using Multiple Evidence Ranking. Image 16(4), 408–417 (2004)
4. Croft, W.B., Lafferty, J.: Language Modeling for Information Retrieval. Kluwer Academic Publishers, Boston (2003)
5. Deschacht, K., Moens, M.: Text Analysis for Automatic Image Annotation. In: Proceedings of the 45th Annual Meeting of the Association of Computational Linguistics, pp. 1000–1007 (2007)
6. Deschacht, K., Moens, M.-F.: Efficient Hierarchical Entity Classification Using Conditional Random Fields. In: Proceedings of the 2nd Workshop on Ontology Learning and Population, Sydney, July 2006, pp. 33–40 (2006)
7. Hsu, W.H., Kennedy, L., Chang, S.-F.: Reranking methods for visual search. IEEE Multimedia Magazine 13(3) (2007)
8. Mikheev, A.: Automatic Rule Induction for Unknown-Word Guessing. Computational Linguistics 23(3), 405–423 (1997)
9. Mori, Y., Takahashi, H., Oka, R.: Automatic Word Assignment to Images Based on Image Division and Vector Quantization. In: RIAO 2000 Content-Based Multimedia Information Access, Paris, April 12-14 (2000)
10. Tollari, S., Glotin, H.: Web image retrieval on IMAGEVAL: Evidences on visualness and textualness concept dependency in fusion model. In: ACM International Conference on Image and Video Retrieval (CIVR) (July 2007)

Robust Query-Specific Pseudo Feedback Document Selection for Query Expansion

Qiang Huang, Dawei Song, and Stefan Rüger

Knowledge Media Institute,
The Open University, UK
{q.huang,d.song,s.rueger}@open.ac.uk

Abstract. In document retrieval using pseudo relevance feedback, after initial ranking, a fixed number of top-ranked documents are selected as feedback to build a new expansion query model. However, very little attention has been paid to an intuitive but critical fact that the retrieval performance for different queries is sensitive to the selection of different numbers of feedback documents. In this paper, we explore two approaches to incorporate the factor of query-specific feedback document selection in an automatic way. The first is to determine the "optimal" number of feedback documents with respect to a query by adopting the clarity score and cumulative gain. The other approach is that, instead of capturing the optimal number, we hope to weaken the effect of the numbers of feedback document, i.e., to improve the robustness of the pseudo relevance feedback process, by a mixture model. Our experimental results show that both approaches improve the overall retrieval performance.

1   Introduction

To document retrieval, the pseudo relevance feedback tries to build an expanded query language model using the top-selected documents according to the initial retrieval results. Naturally, the top-ranked documents are assumed to be relevant to the user's query. In the process of building an expanded query model, traditional methods tend to select a fixed number ($\leqslant 50$, typically) of top-ranked documents as feedback, regardless of different queries. However, an intuitive but critical fact has long been ignored: the retrieval performance for a specific query is often sensitive to the selected number of feedback documents.

Figure 1(a) and 1(b) show the effects of different numbers, $\{5,10,15,20,25,30,$ $35,40,45,50\}$, of feedback documents by testing TREC query topics 51-150 (only title field) on collection AP88-90. Figure 1(a) shows the manually identified "optimal" (i.e., best performing) number of documents for each query, which is obviously not a constant value for different queries. A comparison of the retrieval performances between the expanded query language model using the query-specific optimal numbers of feedback documents (based on Figure 1(a)) versus other four expanded query language models using a fixed number of top-N ($N \in$ $(5, 10, 30, 35)$) documents to all the queries. It turns out that the former can generate a large improvement in average precision over the others. Following

C. Macdonald et al. (Eds.): ECIR 2008, LNCS 4956, pp. 547–554, 2008.

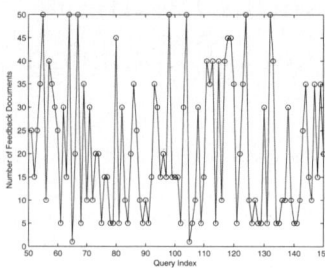

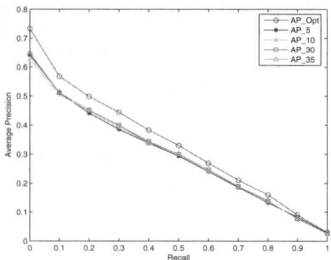

(a) The optimal number of feedback documents

(b) Precision-recall curves (TREC topics 51-150 on Collection AP8890)

Fig. 1. The optimal number of feedback documents

this preliminary experiments, the question we are concerned about is: Is there an automatic method of selecting the feedback documents with respect to individual query?

There can be three directions towards finding a solution to the problem. The first is to build a model by finding the truly relevant documents in the top-ranked documents [8,9] using a support vector machine (SVM) based semi-supervised method with the user's help. The second direction is to directly capture the optimal number of documents with respect to each query [1,3,11]. Some methods, such as computing a *clarity score* using Kullback-Leibler (KL) divergence [3] and using the maximum clarity score as the model-selection criterion [11]. The third direction is to build a mixture model combining several expanded query language models to weaken the effect of pseudo feedback document selection [6,2].

In summary, all the aforesaid attempts try to address the problem of document/model selection for generating a new query model. However, in order to build an optimal expanded query language model, a fully automatic method, for either pursuing a single optimal model or combining multiple models, still remains an open and attractive topic. In this paper, we explore novel approaches incorporating the factor of query-specific feedback document selection in a fully automatic way, and apply the existing *clarity score* (CS) and present two new approaches respectively based on *discount cumulative gain* (DCG) and *mixture model* (MM) for the document retrieval.

2   Determination of the Query-Specific Optimal Model

2.1   Clarity Score (CS)

In general, if the collection is large enough, it is often assumed that the distribution of words in the document collection is uniform. The model with uniform distribution is generally considered as the worst model for document retrieval because the importance of words to query can not be distinguished from each other. The clarity score defined here is the KL divergence of the expanded query language model M to the collection model M_{coll}, as shown in Equation 1.

$$clarity\ score = \sum_{w \in V} P(w|M) \log_2 \frac{P(w|M)}{P(w|M_{coll})} \tag{1}$$

where V is the word vocabulary for the collection. The smaller distance between the two models is assumed to imply a poor retrieval performance for the query. Based on this assumption, the clarity score can be used to predict the retrieval performance of an expanded query language model. The pseudo code below describes the application of the clarity score for selecting the optimal model from several query language models $M_i, (1 \leq i \leq m)$.

$$for\ i\ =\ 1\ :\ m,$$
$$CS_i = \sum_{w \in V}\ P(w|M_i) \log_2 \frac{P(w|M_i)}{P(w|M_{coll})}$$
$$end$$
$$M^* = max_{1 \leq i \leq m}\ CS_i$$

The model corresponding to the maximum clarity score is chosen as the optimal model. The clarity method has a clear advantage that it does not require doing the actual retrieval. However, it can not guarantee that the selected model is the truly best performing one. On one hand, the words in the collection model may not distribute uniformly. On the other hand, even if the collection model had the uniform distribution, the larger divergence between a query language model and the collection model does not necessarily mean the query language model closer to the best model we expect.

2.2   Discount Cumulative Gain (DCG)

Compared with the *clarity score* measure, *discount cumulative gain* (DCG) is a more complex approach to measure the possible highly relevant documents. Unlike the binary measure, by which queries are judged relevant or irrelevant with regard to the query, the cumulative gain generally uses multiple graded relevance judgments [10,4,5,7]. The cumulative gain based measure was summarized into two points: (1) highly relevant documents are more valuable than marginally relevant documents, (2) the lower the ranked position of a relevant document (of any relevant level), the less valuable it is for the user. The details are referred to [4]. In this paper, we apply the DCG to predicting the retrieval performance of a model. So we hope to select an "optimal" model by comparing the cumulative gains of each query language model. The cumulative gain is computed as below:

Collection: Given a query, collect the top 100 documents ranked after the initial retrieval.

Re-ranking: Re-rank the 100 documents based on 10 expanded query language models which is built by using $\{5, 10, 15, 20, 25, 30, 35, 40, 45, 50\}$ top-ranked documents, respectively. Simultaneously, 10 rank lists of the 100 documents are respectively obtained as well.

Identification: Compute the summation of the order of a document in the 10 rank lists, so there are 100 values of the summation corresponding to the 100 documents. Select 16 documents as "pseudo" highly relevant documents whose summation values are smaller.

Label: Label the 16 selected documents (16 is an experience value) with four grades of ranking (also called gain value in [7]), namely $R = [4, 4, 4, 4, 3, 3, 3, 3, 2, 2, 2, 2, 1, 1, 1, 1]$

Computation: Compute the cumulative gain:

$$DCG_{M_i} = \sum_{j=1}^{16} \frac{2^{label(j)} - 1}{\log_2(j+1)}$$

where $label(j)$ is the gain value associated with the label of the document at the jth position of the ranked list. $log_2(j+1)$ is a discounting function that reduces document's gain value as its rank increases [7].

In the process of computation, the relevance levels can be mapped to numerical values, with 4 corresponding to the highest level of relevance and 1 corresponding to the lowest level of relevance. The difference in gain values assigned to highly relevant and relevant documents changes the score of cumulative gain. The method of computing cumulative gain is almost same as that used in [7], in which a normalized discount cumulative gain (NDCG) averaged over the queries is used to evaluate the performance of the multiple nested ranker algorithm. In addition, the computation also means that the re-ranking is needed over all expanded query language models. The similar method using re-ranking over multiple models for model selection can also be found in [11], but our method only runs on the top 100 documents ranked by the initial retrieval rather than searching the whole collection of documents with each query model, as done in [11].

2.3   Mixture Models (MM)

The above two methods based on the *CS* and *DCG* aim to find the "optimal" model in the multiple models. In this section, we attempt to build a mixture model by combining all query language models rather than only selecting one. The application of mixture models is to bind all N models whatever the value of N is to a target model, aiming to smooth the effects from different models [6]. In the process of building a mixture model, the key step is to estimate the mixture weight of each model, as shown in Equation 2:

$$M_{opt} = \sum_j \lambda_j M_j \tag{2}$$

where $\sum_j \lambda_j = 1$. In [12,6], an approach based on Kullback-Leibler (KL) distance was used to optimize the weights for mixture models. Here we briefly describe the optimization procedure, and the details can be found in [6].

$$D = \sum_i T(w_i) log \frac{T(w_i)}{M_{opt}(w_i)} \tag{3}$$

Table 1. Test collection and test topics

Collection	Contents	# of docs	Size	Queries (topics)	# of Queries
AP88-90	Associated Press	242,918	0.73Gb	51-150	99
WSJ87-92	Wall Street Journal	173,252	0.51Gb	1-200	200
SJM91	San Jose Mercury News	90,257	0.29Gb	51-150	94

In equation 3, KL distance is computed between the target model T and the mixture model M_{opt}. In [6], a similar optimization was adopted as below:

$$H_\lambda(T|M_j) = -\sum_w T_w \log \sum_j (\lambda_j / \sum_j \lambda_j) M_{j,w} \qquad (4)$$

In order to find the maximum of Equation 4, a derivation on λ_k is taken, and the derivation is set to be zero.

$$\frac{\partial H_\lambda}{\partial \lambda_j} = -\sum_w \frac{T_w M_{j,w}}{\sum_j \lambda_j M_{j,w}} + \frac{1}{\sum_j \lambda_j} = 0 \qquad (5)$$

Suppose λ_k^n is the mixing weight of element k after n iterations of the algorithm. Then at the next iteration the weight should become:

$$\lambda_k^{n+1} \longleftarrow \sum_w \frac{T_w M_{j,w} \lambda_k^n}{\sum_j (\lambda_j^n / \sum_j \lambda_j^n) M_{j,w}} \qquad (6)$$

Here, the optimization of the weight to each model is to make the mixture model best approximate the target model, so the selection of the target model is actually key to the final results that the mixture models can achieve.

In [6], Lavrenko used the mixture model to weaken the effect of selecting the number of feedback documents. Here, we exploit this idea in two different ways. Firstly, we select the original words distribution on the top 50 documents as the target model instead of a known relevant document as used in [6]. The reason is that [6] needs a relevant document, which is generally selected manually, to build the target model. Secondly, the model built by using the top 50 documents could be the worst model compared with less documents being used because there are more irrelevant information being included. If the performance of MM based on the top 50 documents is good, then it could mean less documents used will generate better result.

We have presented three approaches to deal with the problem caused by selecting the number of feedback documents. The first two approaches, respectively based on *CS* and *DCG*, try to select the "optimal" model. The *MM* aims to smooth this factor. In the next section, we will test their performances with two TREC topic sets on three TREC collections.

3   Data and Experiments

The experiments are run by testing two query topics (only using the title field) on three standard TREC data sets, whose statistic are summarized in Table 1.

Table 2. Results (Average Precisions) of different models

| | The number of feedback documents | | | | | | | | | | |
	5	10	15	20	25	30	35	40	45	50	Optimal
AP8890	0.2829	0.2852	0.2863	0.2867	0.2886	0.2893	0.2888	0.2888	0.2862	0.2859	**0.3228**
SJM	0.2309	0.2303	0.2346	0.2339	0.2325	0.2335	0.2350	0.2342	0.2356	0.2346	**0.2727**
WSJ8792	0.3026	0.3065	0.3037	0.3026	0.3039	0.3042	0.3028	0.3031	0.3023	0.3021	**0.3356**

Table 3. The average precisions obtained by using three different approaches

	worst model	best model	Clarity score	Change over worst model(%)	Change over best model(%)
AP8890	0.2829	0.2893	0.2863	1.2	-1
SJM	0.2303	0.2356	0.2328	1.1	-1.2
WSJ8792	0.3021	0.3065	0.3028	0.2	-1.2

	worst model	best model	Cumulative Gain	Change over worst model(%)	Change over best model(%)
AP8890	0.2829	0.2893	0.2872	1.5	-0.7
SJM	0.2303	0.2356	0.2356	2.3	0
WSJ8792	0.3021	0.3065	0.3031	0.3	-1.4

	worst model	best model	Mixture Model	Change over worst model(%)	Change over best model(%)
AP8890	0.2829	0.2893	0.2889	2.1	-0.1
SJM	0.2303	0.2356	0.2402	4.3	1.9
WSJ8792	0.3021	0.3065	0.3087	2.1	0.7

In our system, each expanded query language model is built by using Jelinek-Mercer linear interpolation between a query language model and the collection model, in which the query language model is modeled using maximum likelihood with the top-N, ($N \in \{5, 10, 15, 20, 25, 30, 35, 40, 45, 50\}$) documents. The expansion is generated by running the Lemur toolkit. In this paper, we build 10 baseline expanded query language models $M_i, (1 \le i \le 10)$. For each model, the top 100 words are selected according to their distribution $P(w|M_i)$ to form the expanded query. The linear combination coefficient is set to be 0.9 and μ is set to be 1000 for the retrieval process.

In Table 2, the average precision obtained by using 10 baseline expansion models are listed in the order of increasing number of feedback documents used. At the most right-hand side, the optimal average precisions are listed, which are obtained by manually selecting the optimal model to each query. Naturally, the optimal performance is much better than the baseline expansion models generated by applying a fixed number of feedback documents to all queries, and can be considered as the upper bound of the retrieval performance. To show the different characteristics of the proposed automatic approaches, in the rest of this section, we use three performance measures, i.e. average precision, average precision @30 docs and robustness via query-by-query comparison.

Table 3 shows the retrieval performances using the three approaches for the different collections, The "worst expansion model" and "best expansion model" respectively represent the model with the lowest and highest average precision among the 10 baseline expansion models as shown in Table 2. All three approaches give higher average precision than the "worst expansion model". The average precisions of the CS and DCG are slightly lower than the "best

Table 4. The precisions @ 30 docs using three different approaches

	Best Expansion Model	Clarity Score	Cumulative Gain	Mixture Model
AP8890	0.4451	0.4411	0.4453	0.4455
SJM	0.2720	0.2626	0.2761	0.2762
WSJ8792	0.4002	0.3996	0.4062	0.4033

Table 5. Robust analysis to the retrieval performance on three collections

		vs. best expansion model			vs. worst expansion model		
		Better	Neutral	Worse	Better	Neutral	Worse
AP	Clarity Score	41	6	52	43	6	50
88-90	Cumulative Gain	48	6	45	52	5	42
	Mixture Model	50	3	46	57	2	40
WSJ	Clarity Score	43	3	48	45	2	47
87-92	Cumulative Gain	48	2	44	49	1	44
	Mixture Model	47	0	47	48	0	46
SJM	Clarity Score	96	6	98	98	5	97
91	Cumulative Gain	101	2	97	110	2	88
	Mixture Model	99	3	98	110	2	88

expansion model". The *CS* gives the lowest performance for all the three collections. The *MM* generates better results and even outperforms the "best expansion model" on two collections. As we discussed in Section 2.1, the *CS* simply measures the distance between a model and the collection model and it seems to fail in selecting the appropriate number of feedback document. On the other hand, the *MM* tries to combine the information from multiple models, which can help weaken the effect of the model selection.

In addition, we list the precisions @ 30 docs, where the *DCG* and *MM* perform better than the *CS*, and also outperform the best expansion model. This could be because the *DCG* takes into account the ranking of the relevant documents and MM combines the useful information from different models, and also smooth them by weighting scheme to weaken the effect of "noisy" information.

A robustness analysis is shown in Table 5. The baselines are the best expansion model and the worst expansion model with a fixed-number of feedback documents. We perform a comparison of the mean average precisions between each of the three methods and the two baseline models query by query. Here, the terms *better/neutral/worse* in Table 5 stand for the numbers of queries for which our approach gives a *better/neutral/worse* than the two baselines, respectively. We can observe the robustness of using *CS* is a little lower than the other two approaches. Furthermore, compared with the *CS*, both *DCG* and *MM* show more robust performance improvement. *DCG* improves the most number of queries' performance but hurts the least number of queries, thus is the most robust.

4 Conclusion

In this paper, we present three approaches to automatically determine the query-specific optimal number of pseudo feedback documents for query expansion. The *CS* and *DCG* are used to look for an optimal value to the number of feedback documents, and *MM* to reduce the effect of selecting the optimal number. The

MM can combine the multiple expansion models instead of trying to capture the best one. Its advantage is that it not only makes use of more useful information, but also smooths "noisy" information. It is verified by our experimental results: the *MM* shows better effectiveness (average precision and precision @30) than the other two. Using *DCG* also shows promising result, especially in the query by query robustness analysis. Both *DCG* and *MM* outperform the *CS* in terms of both effectiveness and robustness. There is still a big gap between the performance of our proposed approaches and the upper bound average precision generated by the manually selected optimal model (as shown in Table 2). This means there is a plenty of room for further performance improvement. In the future, we will not only take into account the effect of selecting documents, but also terms as well, which are kept constant in our experiments.

Acknowledgement

This work is funded by the UK's Engineering Physical Sciences Research Council (EPSRC), grant number EP/E002145/1.

References

1. Collins-Thompson, K., Callan, J.: Estimation and use of uncertainty in pseudo-relevance feedback. In: Proceedings of the SIGIR 2007, pp. 303–310 (2007)
2. Croft, W.B.: Combining approaches in information retrieval. In: Croft, W.B. (ed.) Advances in Information Retrieval: Recent Research from the CIIR, pp. 1–36. Academic Publishers, Boston (2000)
3. Cronen-Townsend, S., Zhou, Y., Croft, W.B.: A language modeling framework for selective query expansion. Technical report, CIIR (2004)
4. Jarvelin, K., Kekalainen, J.: Ir evaluation methods for retrieving highly relevant documents. In: Proceedings of SIGIR 2000, pp. 41–48 (2000)
5. Jarvelin, K., Kekalainen, J.: Cumulated gain-based evaluation of ir techniques. ACM Transaction on Information Systems 20, 422–446 (2002)
6. Lavrenko, V.: Optimal Mixture Models in IR. In: Crestani, F., Girolami, M., van Rijsbergen, C.J.K. (eds.) ECIR 2002. LNCS, vol. 2291, pp. 193–212. Springer, Heidelberg (2002)
7. Matveeva, I., Burges, C., Burkard, T., Laucius, A., Wong, L.: High accuracy retrieval with multiple nested ranker. In: Proceedings of SIGIR 2006 (2006)
8. Okabe, M., Umemura, K., Yamada, S.: Query Expansion with the Minimum Relevance Judgments. In: Lee, G.G., Yamada, A., Meng, H., Myaeng, S.-H. (eds.) AIRS 2005. LNCS, vol. 3689, pp. 31–42. Springer, Heidelberg (2005)
9. Onoda, T., Murata, H., Yamada, S.: One class classification methods based non-relevance feedback document retrieval. In: The International Workshop on Intelligent Web Interaction, Hong Kong, pp. 389–392 (2006)
10. Voohees, E.: Evaluating by highly relevant documents. In: Proceedings of SIGIR 2001, pp. 74–82 (2001)
11. Winaver, M., Kurland, O., Domshlak, C.: Towards robust query expansion: Model selection in the language modeling framework (poster). In: Proceedings of SIGIR 2007, Amsterdam (2007)
12. Yamron, J., Carp, I., Gillick, L., Lowe, S., van Mulbregt, P.: Topic tracking in a news stream. In: Proceedings of DARPA Broadcast News Workshop, pp. 133–136 (1999)

Expert Search Evaluation by Supporting Documents

Craig Macdonald and Iadh Ounis

Department of Computing Science,
University of Glasgow, Glasgow, G12 8QQ, UK
{craigm,ounis}@dcs.gla.ac.uk

Abstract. An expert search system assists users with their "expertise need" by suggesting people with relevant expertise to their query. Most systems work by ranking documents in response to the query, then ranking the candidates using information from this initial document ranking and known associations between documents and candidates. In this paper, we aim to determine whether we can approximate an evaluation of the expert search system using the underlying document ranking. We evaluate the accuracy of our document ranking evaluation by assessing how closely each measure correlates to the ground truth evaluation of the candidate ranking. Interestingly, we find that improving the underlying ranking of documents does not necessarily result in an improved candidate ranking.

1 Introduction

In large Enterprise settings with vast amounts of digitised information, an *expert search* system aids a user in their "expertise need" by identifying people with relevant expertise to the topic of interest. The retrieval performance of an expert search system is very important. If an expert search system suggests incorrect experts, then this could lead the user to contacting these people inappropriately. Similarly to document IR systems, the accuracy of an expert search system can be measured using the traditional IR evaluation measures such as precision and recall of the suggested candidates. Expert search has been a retrieval task in the Enterprise tracks of the Text REtrieval Conferences (TREC) since 2005 [1], aiming to evaluate expert search approaches.

Most of the existing models for expert search work by examining the set of documents ranked or scored with respect to the query, and then converting this into a ranking of candidates, based on some information about the associations between documents and candidates. However, while various studies have shown that applying known retrieval techniques to improve the quality of the document ranking lead to an improvement in the accuracy of the ranking of candidates [2,3,4], it has not been clear what characteristics in the improved document ranking have caused the increase of retrieval accuracy of the expert search system. This work attempts to approximate an evaluation of the underlying document ranking, to better understand how the document ranking can affect the retrieval accuracy of the expert search system.

C. Macdonald et al. (Eds.): ECIR 2008, LNCS 4956, pp. 555–563, 2008.

The objectives of our experiments are two-fold: Firstly, to assess whether the proposed methodology for evaluating the underlying document ranking can produce an accurate estimation of the final accuracy of the expert search system; Secondly, to examine which evaluation measures calculated on the document ranking exhibit the highest correlation with each evaluation measure calculated on the candidate ranking. In doing so, we gain an understanding into how various techniques for expert search behave when the underlying ranking is altered.

The remainder of this paper is as follows: Section 2 briefly reviews several models for expert search, and discusses the evaluation of expert search systems. In Section 3, we show how to approximate an evaluation of the document ranking of an expert search system, and investigate how the document ranking evaluation correlates with the ground truth evaluation of the ranking of candidates. Finally, in Section 4, we provide concluding remarks and points for future work.

2   Models for Expert Search

Given an input list of candidate experts, modern expert search systems work by using documents to form a profile of textual evidence of expertise for each candidate. This associated documentary evidence can take many forms, such as intranet documents, documents or emails authored by the candidates, or web pages visited by the candidate (see [2] for an overview). The candidate profiles can then be used to rank candidates automatically in response to a query.

The most successful models for expert search use an initial ranking or scoring of documents with respect to the query [2,4,5,6]. For instance, in Model 2 of the language models proposed by Balog et al. [5], the probability of a candidate is the sum of the probability of all retrieved documents, multiplied by the degree of association between each document and the candidate. Similarly, in the Voting Model for Expert Search [2], various voting techniques can be applied to aggregate the retrieval scores or ranks of all the retrieved documents associated to each candidate to form the final score for the candidate.

For all these techniques, there are three fundamental parameters that can impact the accuracy of the expert search system: Firstly, the technique(s) applied to generate the underlying ranking of documents impact the final ranking of candidates: various studies have shown that applying techniques (which normally improve a document IR system) improve the 'quality' of the document ranking results in increased accuracy of the candidate ranking [2,3,4]; Secondly, the quality of expertise evidence for each candidate (for instance how documents have been associated to each candidate) has a major impact on the performance of the system [5,7]; Lastly, the manner in which the document evidence is combined for each candidate impacts on how accurate the expert search system is [2].

This work is concerned with the document ranking experimental parameter. While it is possible to evaluate the final ranking of candidates, it has not been possible to determine the properties of a 'high quality' ranking of documents that produces an accurate ranking of candidates, because there has been no direct method of measuring this 'quality'. In the remainder of this section, we review how expert search system evaluation is normally performed, while the next section describes how we can approximate an evaluation of the document ranking.

2.1   Evaluation of Expert Search Systems

The evaluation of expert search systems presents more difficulties than that of a document retrieval system, primarily because a document assessor can read a document, and fairly easily make a judgement as to its relevance. However, an expert search system returns only a list of names, with nothing to allow an assessor to easily determine each person's expertise. To this end, using the TREC paradigm, there are essentially three strategies for expert search system evaluation, to generate relevance assessments for candidates:

Pre-Existing Ground Truth: In this method, queries and relevance assessments are built using a ground truth, which is not explicitly present in the corpus. For example, in the TREC 2005 expert search task, the queries were the names of working groups within the W3C, and participating systems were asked to predict the members of each working group [1]. The problem with this method of evaluation is that it relies on known grouping of candidates, and does not assess the systems for more difficult queries where the vocabulary of the query does not match the name of the working group. Moreover, candidates can have expertise in topics they are not members of working groups on.

Candidate Questionnaires: In this method, each candidate expert in the collection (or a person with suitable knowledge about the candidate experts' expertise areas), is asked if they have expertise in each query topic. While this process can be reduced in size by using pooling of the suggested candidates for each query, the process obviously does not scale to a large collection with hundreds or thousands of candidates. In particular, not all candidates are available to question, or assessors may not have knowledge of every candidates' interests. A derivative of this approach was used to assess the TREC 2007 expert search task in a medium-sized enterprise setting [9].

Supporting Evidence: This last method was proposed for the TREC 2006 expert search task [10]. In this method, each participating system is asked, for each suggested candidate, to provide a selection of ranked documents that supported that candidate's expertise. For evaluation, the top-ranked candidates suggested for each query are pooled, and then for each pooled candidate, the top-ranked supporting documents are pooled. Relevance assessment follows a two-stage process: assessors are asked to read and judge all the pooled supporting documents for a candidate, before making a judgement of his/her relevance to the query. Additionally, the pooled supporting documents which supported their judgement of expertise are marked. Figure 1 shows a section of the TREC 2006 relevance assessments, showing that candidate-0001 has relevant expertise to topic 52. Moreover, a selection of supporting documents are provided, which the relevance assessor used to support that judgement. In the final evaluation, only the

```
52 candidate-0001 2
   52 candidate-0001 lists-015-4893951 2
   52 candidate-0001 lists-015-4908781 2
   ....
52 candidate-0002 0
....
```

Fig. 1. Extract from the relevance assessments of the TREC 2006 expert search task (topic 52). candidate-0001 is judged relevant, with two positive supporting documents shown (lists-015-4893951 etc.). candidate-0002 is not judged relevant.

candidate relevant assessments are used to evaluate the accuracy of the expert search systems.

Once the (candidate) relevance assessments have been generated, using one of the methods described above, it is then simple to evaluate a ranking of candidates using standard retrieval evaluation measures, such as Mean Average Precision (MAP), etc. For clarity, we call these measures Candidate MAP, etc, as they are calculated on the ranking of candidates.

3   Document Ranking Evaluation

As noted above, the current effective models for expert search all take into account the notion of document relevance to the query topic, before ranking the associated candidates. We designate this underlying ranking of documents for the query as $R(Q)$. Because various studies have shown that improving $R(Q)$ has increased the accuracy of the candidate ranking, one could assume that the accuracy of the ranking of candidates is dependant on how well the underlying ranking of documents ranks highly documents related to the relevant candidates.

We aim to approximate an evaluation of the document ranking directly, to aid failure analysis of expert search systems. In doing so, we hope to gain new insights about the desirable characteristics of the document retrieval component of an expert search system, which will help to build more accurate expert search systems. To achieve this approximate evaluation, we use the supporting documents as relevance assessments: a document is assumed to be relevant to a query iff it was judged as a relevant supporting document for a relevant candidate of that query. Then to evaluate the document ranking, we use standard evaluation measures, applied using these supporting document relevance assessments. Mean Average Precision measured on the document ranking is denoted MAP of $R(Q)$.

This work has two central objectives: Firstly, we test if the evaluation using supporting documents of the underlying document ranking can approximate the evaluation of the final candidate ranking; Secondly, to determine which measures calculated on the document ranking best predict various measures calculated on the candidate ranking. For our experiments, we use the set of supporting documents for all relevant candidates from the TREC 2006 expert search task. In particular, 49 queries were assessed, for which there are on average 28.4 candidates with relevant expertise. For each relevant candidate, there is on average

9.8 supporting documents for that judgement, which over all candidates, gives a mean of 134.8 unique supporting documents per query.

In the following section, we use an expert search system to generate many document rankings and corresponding candidate rankings, and examine how changes in the document rankings are reflected in the candidate rankings. The following section details the experimental setting applied.

3.1   Experimental Setting

The TREC W3C collection is indexed using Terrier [8], removing standard stopwords and applying the first two steps of Porter's stemming algorithm. Documents in the initial ranking $R(Q)$ are ranked using the DLH13 document weighting model [2] from the Divergence from Randomness (DFR) framework. We chose to experiment using DLH13 because it has no term frequency normalisation parameter that requires tuning, and hence, by applying DLH13, we remove the presence of any term frequency normalisation parameter in our experiments. We then create many document rankings by varying the parameters of a document-centric query expansion technique. Next, we generate the profiles of documentary evidence of expertise for the candidates: for each candidate, documents which contain an exact match of the candidates full name are used as the profile of the candidate. The document candidate associations are not varied, however the applied associations have previously performed robustly on the same task [3].

For the combining of document ranking evidence into a ranking of candidates, we use three voting techniques from the Voting Model, namely CombSUM, CombMNZ and expCombMNZ [2], as these provide several distinct methods to transform a document ranking into a candidate ranking. Note that CombSUM is equivalent to the Model 2 approach of Balog et al [5], if a language modelling approach is used to generate $R(Q)$ [3]. For this reason, we do not experiment using the language modelling approach of Balog et al [5].

To assess how the document ranking evaluation correlates with the evaluation of the generated candidate ranking, we need to generate many alternative document rankings for each query, evaluate them, and see how these correlate to the final candidate evaluation measure. To this end, and as mentioned above, we use document-centric query expansion (DocQE) for expert search [3]. In document-centric QE, query expansion is applied on the document ranking, to identify some informative terms from the top-ranked documents (we use the Bo1 DFR term weighting model to measure the informativeness of terms [3]), which are added to the initial query. The expanded query is then re-run to give an enhanced document ranking, which should produce higher retrieval performance when transformed into a ranking of candidates [3]. The number of top retrieved documents to consider (exp_doc) and the number of terms to add to the query (exp_term) are parameters of the query expansion, and by varying these we can generate various initial ranking $R(Q)$ with varying retrieval performances. We vary $1 \leq exp\_term \leq 29$ and $3 \leq exp\_doc \leq 29$, giving 783 different parameter settings.

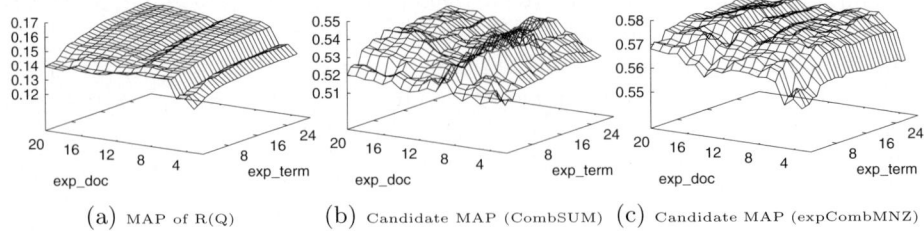

(a) MAP of R(Q)   (b) Candidate MAP (CombSUM)   (c) Candidate MAP (expCombMNZ)

Fig. 2. The effect of varying QE parameters (*exp_term* and *exp_doc*) on the various evaluation measures, i.e. MAP on the initial document ranking (denoted MAP of R(Q)), and final candidate MAP calculated on the candidate ranking produced by the CombSUM and expCombMNZ voting techniques.(Note different Z-axis scales).

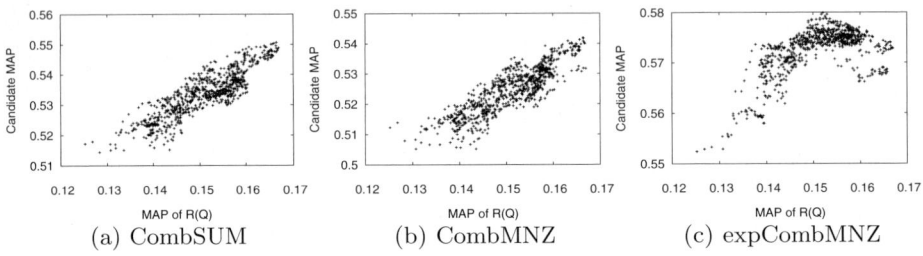

(a) CombSUM   (b) CombMNZ   (c) expCombMNZ

Fig. 3. Scatterplots showing the overall correlation between MAP of R(Q) and Candidate MAP, for three voting techniques.(Note different Y-axis scales).

3.2   Document Ranking Correlation

Figure 2(a) shows a surface plot for various settings of the *exp_term* and *exp_doc* QE parameters, evaluated using MAP of R(Q). Secondly, Figures 2(b) & (c) show the retrieval performance achieved when the ranking is aggregated into a ranking of candidates by CombSUM and expCombMNZ respectively[1]. Each point in Figure 2(a), (b) or (c) represents the MAP over the 49 topics in the TREC 2006 expert search task. Comparing these figures we can observe that while the outline of the surfaces between the MAP of R(Q) and candidate MAP plots are similar, the MAP of R(Q) plot is much smoother - this suggests that if the overall correlation trend between MAP of R(Q) and candidate MAP plots is similar, it may be easier for an automated training process (e.g. hill climber or simulated annealing) to train an expert search system on the smoother MAP of R(Q) surface.

Figures 3(a), (b) & (c) show scatterplots of the correlations between MAP of R(Q) vs Candidate MAP for the CombSUM, CombMNZ and expCombMNZ voting techniques respectively. From the figures, it is clear that the accuracy of the voting techniques is dependent on the accuracy of the underlying ranking of documents. In particular, we can quantify this by examining the overall

---

[1] The plot for CombMNZ is similar to CombSUM, and is hence omitted for brevity.

Table 1. Correlation between various document and candidate ranking evaluation measures, for three voting techniques. The best correlation for each Candidate ranking measure (column) and voting technique are emphasised, while correlations which are statistically different (using a Fisher Z-transform and the two-tailed significance test) from the best correlation ($p < 0.05$) in each column are denoted *.

| R(Q) | Candidate Measures | | | | | | | | | | | |
| | CombSUM | | | | CombMNZ | | | | expCombMNZ | | | |
	MAP	MRR	rPrec	P@10	MAP	MRR	rPrec	P@10	MAP	MRR	rPrec	P@10
MAP	**0.8552**	**0.7076**	**0.8192**	0.6461	**0.8561**	**0.6581**	**0.8260**	**0.6661**	**0.4898**	0.0942*	**0.3190**	-0.0397
MRR	0.3503*	0.5008*	0.2889*	0.0492*	0.3031*	0.4151*	0.2868*	0.2274*	0.0570*	-0.2701*	0.2092*	-0.0469
rPrec	0.8256*	0.5737*	0.8086	**0.6959**	0.8519	0.5942*	0.8034	0.6049*	0.4280	**0.2420**	0.1745*	-0.2041*
P@10	0.7225*	0.6008*	0.6955*	0.5300*	0.7340*	0.5378*	0.7206*	0.5929*	0.4235	0.0665*	0.0361*	**0.0361**

correlation between the ranking of settings by MAP of R(Q) and the Candidate MAP, using Spearman's ρ. In these cases, $\rho = 0.8552$ and $\rho = 0.8561$ over the 783 points each, for CombSUM and CombMNZ respectively. For expCombMNZ, which performs better overall, the correlation is lower ($\rho = 0.4898$), and interestingly a 'tail-off' in Candidate MAP can be observed for MAP of $R(Q) > 0.15$. Indeed, this technique exhibits a rather unexpected trait in the sense that improving the document ranking does not always result in an improved candidate ranking accuracy. We suspect that this is an example of a form of over-fitting of the QE technique to the document ranking evaluation. In general, we conclude that to improve the accuracy of an expert search system, we can apply techniques that are known to improve the accuracy of a standard document retrieval system, however, some techniques (e.g. expCombMNZ) can suffer when the document ranking is over-fitted to the R(Q) evaluation, and thus require further investigation to fully understand this phenomenon.

Next, we investigate which measures calculated on the initial document ranking predict best various evaluation measures for the candidate ranking. In doing so, we aim to understand what characteristics in the document ranking affect the generated candidate ranking. Table 1 presents the Spearman's ρ correlation between various evaluation measures on the document ranking (R(Q)), and the final ranking of candidates, for the CombSUM, CombMNZ and expCombMNZ voting techniques. The evaluation measures applied are MAP, precision at R documents (rPrec), reciprocal rank of first relevant document (MRR) and precision @10 (P@10).

From the results, we can draw the following conclusions: MAP and rPrec on the document ranking are good predictors for both the candidate MAP and rPrec measures. This is not surprising, given that rPrec is often the most correlated measure to MAP [11]. In general, for CombSUM and CombMNZ, MAP of R(Q) is the best predictor for any candidate ranking measure (an exception is CombSUM, where rPrec is a slightly better predictor for P@10). This is intuitive, as the voting techniques investigated here are recall orientated - i.e. they examine all the retrieved document associated with each candidate, so it makes sense that even small changes lower down the document ranking improve the overall effectiveness of the voting technique. In contrast, despite the higher retrieval performance of expCombMNZ technique, lower correlations are observed. In particular, MAP and rPrec on R(Q) are the best predictors for candidate MAP. P@10 is also a good predictor, due to the natural focus of expCombMNZ on the top of the document ranking. However, it appears to be impossible to predict the candidate P@10 measure for expCombMNZ, which is unexpected, because MAP and P@10 are normally strongly correlated [11].

Overall, while in general, we conclude that in order to improve an expert search system, it appears to be most effective to apply retrieval techniques that improve MAP, regardless of the evaluation measure that it is desired to improve.

4   Conclusions

The current effective expert search models all take into account, in some way, the relevance score of the documents with respect to the query, which are then converted into a ranking of candidates. Moreover, previous works on expert search show that somehow improving the quality of the underlying ranking of documents $(R(Q))$ results in a more accuracy expert search system.

In this work, we have proposed an approximate evaluation of $R(Q)$ using the supporting documents as relevance assessments. In our experiments, we examined how closely the R(Q) evaluation correlates to the final candidate ranking, using various evaluation measures, across various input document rankings of varying quality. Our experiments found that the document ranking could be evaluated using the proposed methodology. Furthermore, while various measures can be used to measure the quality of R(Q), for the voting techniques applied, MAP appears to be the most effective predictor of the candidate evaluation measures.

The initial step taken in this work towards the evaluation of expert search systems using the document ranking is important as the current evaluation is awkward due to its second-order nature. By showing that the accuracy of the ranking of candidates generated by an expert search system is indeed linked to the quality of the underlying document ranking, failure analysis becomes easier. Moreover, we are able to gain more insights into the characteristics of the document ranking which influence the generated candidate ranking.

In this paper, we did not evaluate the document ranking with real document relevance assessments, instead approximating these using the supporting document as relevance assessments. The newly available TREC 2007 Expert Search test collection [9] is the natural next step for this work, as it contains relevance assessments for candidates and documents on the same query topics. Additionally, using a more diverse source of document rankings than varying query expansion parameters would allow a fuller understanding of the evaluation methodology.

References

1. Craswell, N., de Vries, A.P., Soboroff, I.: Overview of the TREC 2005 Enterprise Track. In: Proceedings of TREC 2005, Gaithersburg, MD (2006)
2. Macdonald, C., Ounis, I.: Voting for candidates: Adapting Data Fusion techniques for an Expert Search task. In: Proceedings of ACM CIKM 2006, Arlington, VA (2006)

3. Macdonald, C., Ounis, I.: Using Relevance Feedback in Expert Search. In: Amati, G., Carpineto, C., Romano, G. (eds.) ECIR 2007. LNCS, vol. 4425, pp. 431–443. Springer, Heidelberg (2007)

4. Petkova, D., Croft, W.B.: Hierarchical language models for expert finding in enterprise corpora. In: Proceedings of ICTAI 2006, pp. 599–608 (2006)

5. Balog, K., Azzopardi, L., de Rijke, M.: Formal models for expert finding in enterprise corpora. In: Proceedings of ACM SIGIR 2006, Seattle, WA, pp. 43–50 (2006)

6. Cao, Y., Li, H., Liu, J., Bao, S.: Research on Expert Search at Enterprise Track of TREC 2005. In: Proceedings of TREC 2005, Gaithersburg, MD (2006)

7. Macdonald, C., Ounis, I.: High Quality Expertise Evidence for Expert Search. In: Macdonald, C., et al. (eds.) ECIR 2008. LNCS, vol. 4956, pp. 283–295. Springer, Heidelberg (2008)

8. Ounis, I., Amati, G., Plachouras, V., He, B., Macdonald, C., Lioma, C.: Terrier: A high performance and scalable information retrieval platform. In: Proceedings of OSIR Workshop 2006, Seattle, WA (2006)

9. Bailey, P., Craswell, N., de Vries, A.P., Soboroff, I.: Overview of the TREC-2007 Enterprise Track. In: Proceedings of TREC-2007, Gaithersburg, MD (2008)

10. Soboroff, I., de Vries, A.P., Craswell, N.: Overview of the TREC-2006 Enterprise Track. In: Proceedings of TREC 2006, Gaithersburg, MD (2007)

11. Buckley, C., Voorhees, E.M.: Retrieval evaluation with incomplete information. In: Proceedings of ACM SIGIR 2004, Sheffield, UK, pp. 25–32 (2004)

Ranking Categories for Web Search

Gianluca Demartini[1], Paul-Alexandru Chirita[2,⋆],
Ingo Brunkhorst[1], and Wolfgang Nejdl[1]

[1] L3S Research Center
Leibniz Universität Hannover
Appelstrasse 9a D-30167 Hannover, Germany
{demartini,brunkhorst,nejdl}@l3s.de
[2] Adobe Systems Incorporated
Anchor Plaza, Timisoara Blvd. # 26 Z
District 6, 061331 Bucharest, Romania
pchirita@adobe.com

Abstract. In the context of Web Search, clustering based engines are emerging as an alternative for the classical ones. In this paper we analyse different possible ranking algorithms for ordering clusters of documents within a search result. More specifically, we investigate approaches based on document rankings, on the similarities between the user query and the search results, on the quality of the produced clusters, as well as some document independent approaches. Even though we use a topic based hierarchy for categorizing the URLs, our metrics can be applied to other clusters as well. An empirical analysis with a group of 20 subjects showed that the average similarity between the user query and the documents within each category yields the best cluster ranking.

1 Introduction

When looking for information in the web, search engines are the place to start. However, in order to quickly find the sought answers, a high quality user interface is also necessary. For this purpose, categorizing output by assigning categories to URLs has been shown to perform much better than the classical ranked list interface [1]. In this paper we argue that an efficient, well-studied ordering of the search clusters is equally important for the search quality.

There are few works proposing methods to *rank* search clusters, and none of them analyses and compares the approaches available for this purpose. We distinguish two broad methods to cluster search results: (1) Unsupervised approaches, such as the seminal Scatter/Gather algorithm [2], which groups documents together based on their terms; and (2) Supervised approaches, such as the work of Zeng et al. [4], who ranked salient phrases using pre-learned regression models, and then formed clusters by assigning documents to the appropriate sentences.

⋆ This work was performed while the author was employed by the L3S Research Center, Hannover, Germany.

C. Macdonald et al. (Eds.): ECIR 2008, LNCS 4956, pp. 564–569, 2008.

While this paper studies the latter broad technique, its investigation could be easily carried out for unsupervised algorithms as well.

In this paper we analyse 10 different metrics for ordering search output clusters, grouped as follows: (1) Ranking by search engine scores, (2) Ranking by query to cluster similarity, (3) Ranking by intra-cluster similarity, and (4) Ranking with generic measures. An empirical investigation performed with 20 subjects indicated the similarity between the user query and the cluster documents as the best cluster ranking approach at a statistically significant difference.

2   Categories Ranking Algorithms

This section details the categories ranking algorithms we analysed, grouped into the previously mentioned 4 categories.

Ranking by Search Engine Scores. This set of metrics builds onto the importance score given by the search engine to each resource. It comprises the following approaches:

- *Average PageRank (AvgPR)* of resources in the category, higher values coming first:

$$\text{AvgPR}(C) = \frac{1}{n} \sum_{p=1}^{n} \text{PR}(p), \ \forall \text{ page } p \in C$$

  with n denoting the number of pages p contained in category C, and $PR(p)$ representing the actual search engine score[1] (i.e., PageRank) assigned to p.
- *Total PageRank (SumPR)* in the category:

$$\text{SumPR}(C) = \sum_{p=1}^{n} \text{PR}(p), \ \forall \text{ page } p \in C$$

  This scheme extends the previous one by incorporating cluster size, as categories containing more items will tend to have higher scores.
- *Average Rank (AvgRank)* of resources in the category, as opposed to score:

$$\text{AvgRank}(C) = \frac{1}{n} \sum_{p=1}^{n} \text{Rank}(p), \ \forall \text{ page } p \in C$$

- *Minimum Rank (MinRank)*, ranking by the best result in the category:

$$\text{MinRank}(C) = \min_{p} \text{Rank(p)}, \ \forall \text{ page } p \in C$$

Ranking by Query to Category Similarity. This measure is based on the similarity between the web query q and each page p within the category. It is called *Average Normalized Logarithmic Likelihood Ratio (NLLR)* [3]:

$$\text{NLLR}(C) = \frac{1}{n} \sum_{p=1}^{n} \sum_{t \in q} P(t|p) * \log \frac{(1-\lambda) \cdot P(t|p) + \lambda \cdot P(t|C)}{\lambda \cdot P(t|C)}, \ \forall \, p \in C$$

---

[1] In our experiments we approximated these values using a Power Law distribution.

where t are the terms in the query, n is the number of pages in the given category, λ is a constant set to 0.85, and C represents the category itself.

Ranking by Intra Category Similarity. This group exploits the quality of the resulting categories, as follows:

- *Average Intra Category Similarity (AvgValue)*, based on the average similarity between each URL and the category centroid. The metric favours clean clusters of pages, in which the results are closest to their specific category.
- *Maximum Intra Category Similarity (MaxValue)*, which considers only the maximum value of the above mentioned similarities. Thus, the focus here is only on the best matching document.

Other Ranking Approaches. Our last group covers metrics which seem to be used by commercial search engines, as well as a random approach, as baseline:

- *Order by Cluster Size (Size)*[2].
- *Alphabetical Order (AlphaBet)*[3].
- *Random Order (Random)*.

3   Experiments

Setup. We asked 20 PhD / PostDoc Students in Computer Science and Pedagogy to perform specific search tasks using our algorithms. We searched the web and we classified the search engine output using 2 methods, Supporting Vector Machines and Bayesian classifiers, in order to make sure that the performance of the ranking algorithms does not depend on the classification method. The classifiers were trained using the top-3 ODP category levels, using the 50,000 most frequent terms from their titles and web page descriptions. We discarded the non-English categories, applied Porter's stemmer, and removed the stopwords, obtaining in total 5,894 categories. Both algorithms performed similarly, SVM being slightly better. Note that although we experimented on textual data, most of our metrics also work on other types of input as well.

Each subject evaluated 20 queries, one for each [algorithm, classifier] pair, the category ranking being completely invisible. There were 12 queries randomly selected from the TREC Topic Distillation Task of the Web Track 2003 (TD), and 8 from the Web Track 2004 (WT), half of them randomly selected and the other half manually selected by us as ambiguous. One extra query at the beginning was performed for getting familiarized with the system and was not included in the evaluation. In total, 400 queries were evaluated. For each of them, the system performed web search using the Google API and retrieved the titles and snippets of the first 50 results.

For each category, we displayed the Top-3 results, including title and snippet, and a "More" button if there were more results in that category. The order

---

[2] Apparently a major component of Vivisimo's ranking.
[3] Used in some Faceted Search Engines, such as Flamenco.

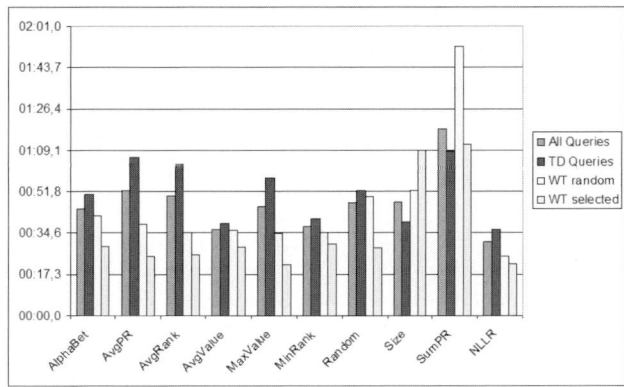

Fig. 1. Average time to find relevant results

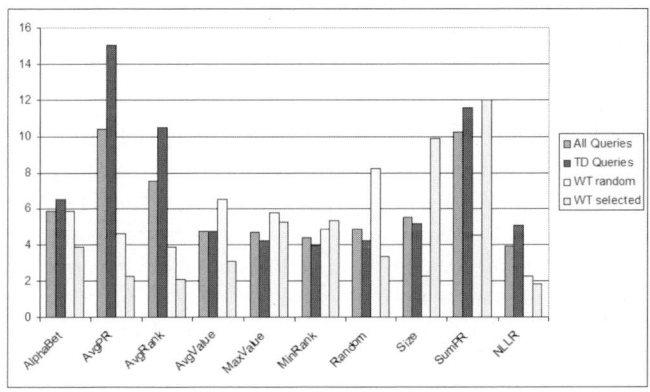

Fig. 2. Average Rank of the Results

inside a category was the one returned by Google. Each URL was classified into up to 3 categories. All output was cached to ensure that results from different participants were comparable and that each participant received the same URLs for the same query. Having the clustered output, the user had to select the first result relevant to the query.

Results. We depict in Figure 1 the average search time for our cluster ranking methods. The metrics using the similarity between the query and the results (NLLR) allowed for finding relevant results in the fastest way, with an average of 31s. In contrast, the performance of the widely used algorithms (e.g., Alpha-Bet, Size) was rather average, below the best 3 methods. The Topic Distillation queries were the most difficult, as they were selected to be more ambiguous, while

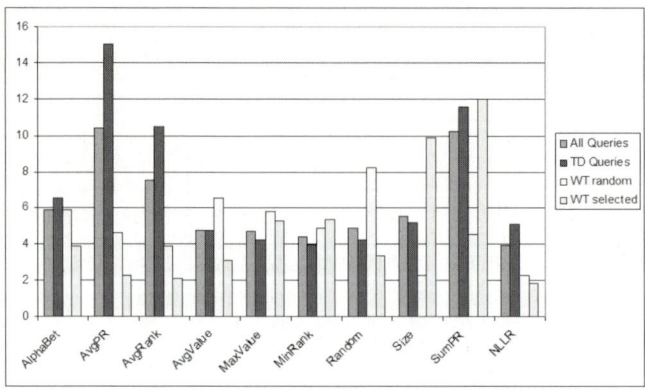

Fig. 3. Average Rank of the Categories

also having a specific task associated to them. The Web Track random queries were moderately difficult and the Web Track ambiguous ones were the easiest (this is correct, as they had no specific search task associated to them). For all cases, NLLR yielded the best performance. Performing an ANOVA analysis, the p-values versus Size and Alphabetical ordering were 0.03 and 0.04, respectively making the difference statistically significant. MinRank and AvgValue were also visibly better than the traditional methods, especially for TD queries. Their p-values were 0.04 and 0.03 compared to Size, as well as 0.17 and 0.09 compared to AlphaBet. Finally, since strong time differences between queries might mislead the results, we also calculated the ranking of each algorithm per user, the outcome being consistent with the conclusions drawn from the time analysis.

Another investigation related to the placement of the relevant results. Figure 2 presents the average rank of the first relevant URL and Figure 3 the average rank of the cluster containing it. Again, NLLR yields the best ranking, whereas AlphaBet and Size are mediocre. Only for the WT random queries did Size perform close to NLLR. We therefore conclude that NLLR is not only facilitating the fastest identification of relevant URLs, but it also generates the best overall ranking.

4   Conclusions and Future Work

We analysed 10 metrics for ranking web search clusters, grouped into 4 classes: (1) Ranking by search engine scores, (2) Ranking by query to cluster similarity, (3) Ranking by intra-cluster similarity, and (4) Ranking with generic measures. Our empirical results showed the similarity between the query and the cluster as the best ranking approach, significantly better than traditional methods such as ordering by size. While this paper focused on a supervised clustering method, in future work we will apply the same analysis onto unsupervised and semi-supervised techniques.

References

1. Chen, H., Dumais, S.T.: Bringing order to the web: automatically categorizing search results. In: ACM SIGCHI (2000)
2. Cutting, D.R., Pedersen, J.O., Karger, D.R., Tukey, J.W.: Scatter/gather: A cluster-based approach to browsing large document collections. In: ACM SIGIR (1992)
3. Rode, H., Hiemstra, D.: Using Query Profiles for Clarification. In: Lalmas, M., MacFarlane, A., Rüger, S.M., Tombros, A., Tsikrika, T., Yavlinsky, A. (eds.) ECIR 2006. LNCS, vol. 3936, pp. 205–216. Springer, Heidelberg (2006)
4. Zeng, H.-J., He, Q.-C., Chen, Z., Ma, W.-Y., Ma, J.: Learning to cluster web search results. In: ACM SIGIR (2004)

Key Design Issues with Visualising Images Using Google Earth

Paul Clough and Simon Read

Dept. of Information Studies, University of Sheffield, UK
p.d.clough@sheffield.ac.uk

Abstract. Using map visualisation tools and earth browsers to display images in a spatial context is integral to many photo-sharing sites and commercial image archives, yet little academic research has been conducted into the utility and functionality of such systems. In developing a prototype system to explore the use of Google Earth in the visualisation of news photos, we have elicited key design issues based on user evaluations of Panoramio and two custom-built spatio-temporal image browsing prototypes. We discuss the implications of these design issues, with particular emphasis on visualising news photos.

1 Introduction

The development of location-aware technology has increased the availability of media containing spatial and temporal information. This is particularly true for visual media such as photographs. Photo-sharing websites such as Flickr[1], Pikeo[2], Panoramio[3] and Woophy[4] now offer spatial (and temporal) image browsing facilities using generic map visualisation services such as Google Maps[5], and earth browsers such as Google Earth (GE)[6]. The use of such interfaces is interesting, as it seems to take advantage of the mind's inherent spatial-temporal reasoning capabilities (cf. Tversky's concept of cognitive collages [1] and Tomaszewski et al. [2]). However, little academic research has been conducted into their utility (what such a system is useful for) and functionality (how does such a system achieve this), two key design characteristics.

To address this, we chose to evaluate the Google Earth version of Panoramio, an existing spatial image browser run on a commercial basis by Google. Our key findings were then used to develop and evaluate two prototype spatio-temporal image browsers for news photos, media that is highly time and place specific. The prototypes are being developed for a major UK-based news agency, Press Association (PA) Photos[7], to investigate the use of new visualisation technologies, such as Google Earth, within their organisation and the news photo industry in general.

---

[1] http://www.flickr.com
[2] http://www.pikeo.com
[3] http://www.panoramio.com
[4] http://www.woophy.com
[5] http://maps.google.com
[6] http://earth.google.com
[7] http://www.paphotos.com

C. Macdonald et al. (Eds.): ECIR 2008, LNCS 4956, pp. 570–574, 2008.
© Springer-Verlag Berlin Heidelberg 2008

2 Background

Many online image databases (such as Flickr) currently use earth browsers to enable users to browse images. Research into the display of images in a geo-spatial context can be traced back through systems such as WING [6], which combined a 3-D map with location specific text and images in a way that made it a clear predecessor of Google Earth. More recent developments have occurred in the field of Geovisual Analytics (GA), which according to MacEachren [4] is an attempt to bridge the putative gap between traditional cartography and virtual reality. Google Earth is considered to be one of the most popular platforms for GA applications [2].

3 Methodology

The research was conducted as part of a user-centered, iterative design process; providing participants with hands-on access to image browsing software. Following an initial survey of four spatial image browsing facilities (provided by Flickr, Panoramio, Pikeo, and Woophy), the first iteration consisted of an explorative user evaluation of Panoramio (arguably the most successful of the four in its use of Google Earth) to identify key issues of utility and functionality (section 4). The codified data was extracted from contemporary notes of volunteers' behaviour and verbal feedback, together with a pre and post evaluation questionnaire completed by each volunteer. The second iteration explored these design issues by implementing two prototype systems, each displaying around 3,000 news images (from PA). The images were associated with metadata including geo-name (city and country), date, category, headline and caption. Each geo-name was manually associated with a spatial coordinate using the geonames.org online gazetteer. These two prototypes used the spatio-temporal browsing capabilities of Google Earth; the difference being that Prototype 2 also addressed the issue of clustering large numbers of spatially close images by providing single thumbnails linked to 'galleries' of similar (i.e. those sharing identical time, date and headline) images opened by a mouse click.

With no established user group within PA for such a system, non-employee volunteers were recruited. For consistency, five female and five male master's students were selected, all studying librarianship or information science. The number was not predetermined, but chosen when qualitative saturation was observed. For the first experiment, each volunteer completed a preliminary questionnaire to establish her/his Internet experience before carrying out four search tasks using a within-subjects design (tasks ordered using a Latin-Square arrangement). Thinking aloud was encouraged. To be representative of the widest variety of search tasks, these were chosen to correspond with Shneiderman's [5] four categories of search: specific fact-finding, extended fact-finding, open-ended browsing and exploration of availability. A ten minute limit was placed on each task. During the second iteration, seven volunteers (all prior participants) were invited to browse for images on their own choice of subject, but asked to consider their use of temporal selection and provide feedback on the two prototypes. User's comments, observations and preferences were recorded during all of the experiments.

4 Results

The initial survey of image browsers identified a number of key functional design issues related to (in particular) spatial visualisation, including how to deal with many images sharing spatial proximity (*clustering*) and what metadata to display alongside the image. Regarding clustering, this includes whether to display an icon for each image (Panoramio), have one icon representing a set of images (Woophy) or decompose into more specific icons when viewed at a higher magnification (Pikeo). In the case of one image representing a set, this includes whether to display a 'gallery' showing further images (Flickr/Pikeo) or an "exploding cluster" (GE 4.0).

Table 1. Panoramio utility issues, by frequency

Issue Code	Issue Description	Occurrence (% of volunteers)
U1	Preference for searching in places visited in real life	50%
U2	Expressed belief that local knowledge needed to search effectively	40%
U3	Tendency to browse pics not directly relevant to the task in hand	40%
U4	Found system highly engaging	30%
U5	Difficult to search for pics geographically unless linked to specific place	30%
U6	Would work better if combined with text based image search engine	30%
U7	Would make an excellent educational tool	20%
U8	Only being able to look at one location at a time makes searching inefficient	20%
U9	Hesitating at index search as can't think of words to enter	20%
U10	Index search hampered by spelling error	20%
U11	Tend to find interesting pics when not particularly looking for them	10%
U12	Geographical context of images is irrelevant	10%

Table 2. Panoramio functionality issues, by frequency

Issue Code	Issue Description	Occurrence (% of volunteers)
F1	Time lag before Panoramio thumbnails appear (at least 4 seconds after globe stops moving)	40%
F2	Not a big enough selection of pics available	40%
F3	Zoomed in on remote area (or body of water) and no pics available	40%
F4	Selected images vanish unexpectedly (due to network link updating)	30%
F5	Selecting nodes because thumbnails slow to appear	30%
F6	Not waiting long enough for Panoramio pics to appear before moving on	30%
F7	Delay in downloading full-size image	20%
F8	Not able to refind one or more pics previously selected	20%
F9	Not zooming in to see more images	10%
F10	Would be good to more text info in picture bubbles.	10%
F11	Would be good to see a selection of related thumbnails shown in bubble	10%
F12	Would be good if thumbnail significantly enlarged when mouse hovers over	10%

Table 3. Prototypes 1 & 2 temporal issues, by frequency

Issue Code	Issue Description	Occurrence (% of volunteers)
T1	Lack of text / numeric option for selecting dates	43%
T2	Selecting the most recent images contemporises the information available in Google Earth, as satellite images can be years out of date	14%
T3	Time slider is particularly useful for selecting news images	14%
T4	Slider is too small and fiddly	14%
T5	Good that it is bimodal (static or animated)	14%

Although it became apparent that Panoramio is not an efficient means of image retrieval, several volunteers found it highly engaging and continued using it after the session. No statistically significant relationship was found between the issues of utility and functionality identified (Tables 1 and 2) and the four search tasks. Table 3 shows temporal design issues elicited from evaluation of the two prototypes.

5 Discussion

Issues of utility are important because the technology is novel. Though not necessarily a time-efficient means of image retrieval, spatio-temporal image browsing seems to hold broad appeal. One explanation is offered by Tversky's [1] concept of *cognitive collages*; i.e. that spatial memory is not a single, coherent internal 'map' but rather a collage of diverse environmental sources (Hirtle & Sorrows [3] link this idea to WING). Combining news images and text with spatio-temporal information may be one way in which such collages could be built, thus companies like PA might use such a system to understand news images in a broader geo-spatial context, rather than simply filtering images by time and place. This corresponds with 30% of volunteers preferring a text based search to select images being viewed in GE (Table 1, U6), and 30% having difficulty searching for non location-specific images (U5).

If cognitive collages exist, then spatio-temporal display may reinforce existing collages as well as create new ones. Set in the context of PA, journalists might browse images from a familiar region to further enrich their understanding of it. A desire to reinforce existing collages could explain why 50% of volunteers gravitated towards places visited in real life (U1), 40% stated local knowledge is required for effective browsing (U2), and 40% diverted from set tasks to browse locations/images of personal interest (U3). The principle benefit that temporal image browsing adds is contemporisation (Table 3, T2), particularly useful if viewing images from a familiar (but not recently visited) place. Given that press photographers can now upload (in real time) photos embedded with GPS generated spatio-temporal information, there is strong potential for the spatio-temporal display of news images in a GA context.

Some of the functional issues identified can be applied to spatio-temporal image browsing in any context, not jut PA. Preloading a large set of thumbnails (rather than downloading thumbnails piecemeal) seems the best way to address the time lag in waiting for Panoramio thumbnails to appear (Table 6, F1) that 50% of volunteers mentioned. This reduces the likelihood of users giving up the search in a particular area (F6) or selecting other information providers in Google Earth (F5). It also prevents images disappearing unexpectedly (F4). However, preloading incurs a longer initial delay and may increase overall bandwidth usage. The selection of images (F2) and their uneven spatial distribution (F3) may be outside the control of the developer. The latter is especially problematic when images are provided with geo-names rather than user-specified geo-codes as thousands of images may share an identical location (e.g. "London"). GE (release 4.0 and higher) allows such clusters to spring apart when selected, but this is ineffective for very large clusters. Prototype 2's method of displaying one thumbnail per date/location/subject (and linking this to a 'gallery' containing the rest) provides a contingency; volunteers preferred this in most respects.

6   Conclusions

The increase in location-aware technologies is likely to cause more widespread use of spatial and temporal visualisation technologies. These are likely to appeal to users due to their inherent conceptual link with the mind's spatial-temporal reasoning capabilities. Organisations like PA Photos are able to exploit such technologies for novel interfaces and new business contexts. In this paper, we have identified some of the key general design issues with using Google Earth to provide spatial and temporal search and browse. We are using these design features to guide the development of prototype systems to explore the use of Google Earth in the news photo industry. In future work, we plan to test prototypes with staff from PA Photos, perhaps using a positivist study to strengthen or discount some of the key issues suggested by this first exploratory study, and to further explore the use of spatio-temporal visualisation.

Acknowledgments

We would like to thank PA Photos for the use of photographic materials from their archive and express particular thanks to Martin Stephens and Phil Lakin. This work was partially funded by the EU-funded MultiMatch project (IST-033104).

References

[1] Tversky, B.: Cognitive Maps, Cognitive Collages, and Spatial Mental Models. LNCS, vol. 716, pp. 14–24. Springer, Heidelberg (1993)
[2] Tomaszewski, B.M., et al.: Geovisual Analytics and Crisis Management. In: Proceedings of the 4th International ISCRAM Conference, Delft, the Netherlands, May 13-16, pp. 173–179. ISCRAM, Delft (2007)
[3] Hirtle, S.C., Sorrows, M.E.: Designing a Multi-modal Tool for Locating Buildings on a College Campus. Journal of Environmental Psychology 18(3), 265–276 (1998)
[4] MacEachren, A.M.: An evolving cognitive-semiotic approach to geographic visualization and knowledge construction. Information Design Journal [Online] 10(1) (2001)
[5] Shneiderman, B.: Designing the user interface: strategies for effective human-computer interaction Harlow. Addison-Wesley, Reading (1998)
[6] Masui, T., et al.: Multiple-view approach for smooth information retrieval. In: Proceedings of the 8th annual ACM symposium on User interface and software technology, Pittsburgh, Pennsylvania, United States, November 15-17, 1995, pp. 199–206. ACM Press, Pittsburgh (1995)

Methods for Augmenting Semantic Models with Structural Information for Text Classification

Jonathan M. Fishbein[1] and Chris Eliasmith[1,2]

[1] Department of Systems Design Engineering
[2] Department of Philosophy
Centre for Theoretical Neuroscience
University of Waterloo
200 University Avenue West
Waterloo, Canada
{jfishbei,celiasmith}@uwaterloo.ca

Abstract. Current representation schemes for automatic text classification treat documents as syntactically unstructured collections of words or 'concepts'. Past attempts to encode syntactic structure have treated part-of-speech information as another word-like feature, but have been shown to be less effective than non-structural approaches. Here, we investigate three methods to augment semantic modelling with syntactic structure, which encode the structure across all features of the document vector while preserving text semantics. We present classification results for these methods versus the Bag-of-Concepts semantic modelling representation to determine which method best improves classification scores.

Keywords: Vector Space Model, Text Classification, Parts of Speech Tagging, Syntactic Structure, Semantics.

1 Introduction

Successful text classification is highly dependent on the representations used. Currently, most approaches to text classification adopt the 'bag-of-words' document representation approach, where the frequency of occurrence of each word is considered as the most important feature. This is largely because past approaches that have tried to include more complex structures or semantics have often been found lacking [1], [2].

However, these negative conclusions are premature. Recent work that employs automatically generated semantics using Latent Semantic Analysis and Random Indexing have been shown to be more effective than bag-of-words approaches in some circumstances [3]. As a result, it seems more a matter of determining how best to represent semantics, than of whether or not semantics is useful for classification.

Here we demonstrate that the same is true of including syntactic structure. A recent comprehensive survey suggests that including parse information will not help classification [2]. However, the standard method for including syntactic information is simply to add the syntactic information as a completely new,

C. Macdonald et al. (Eds.): ECIR 2008, LNCS 4956, pp. 575–579, 2008.
© Springer-Verlag Berlin Heidelberg 2008

independent feature of the document. In contrast, the methods we investigate in this paper take a very different approach to feature generation by distributing syntactic information across the document representation, thus avoiding the limitations of past approaches.

2   Bag-of-Concepts and Context Vectors

The Bag-of-Concepts (BoC) [3] text representation is a recent text representation scheme meant to address the deficiencies of the Bag-of-Words (BoW) representations by implicitly representing synonymy relations between document terms. BoC representations are based on the intuition that the meaning of a document can be considered as the union of the meanings of the terms in that document. This is accomplished by generating term context vectors for each term within the document, and generating a document vector as the weighted sum of the term context vectors contained within that document.

Reducing the dimensionality of document term frequency count vectors is a key component of BoC context vector generation. We use the random indexing technique [4] to produce these context vectors in a more computationally efficient manner than using principal component analysis (PCA).

BoC representations still ignore the large amount of syntactic data in the documents not captured implicitly through word context co-occurrences. For instance, although BoC representations can successfully model some synonymy relations, since different words with similar meaning will occur in the same contexts, it can not model polysemy relations. For example, consider the word "can". Even though the verb form (i.e., "I *can* perform that action.") and the noun form (i.e., "The soup is in the *can*.") of the word occur in different contexts, the generated term vector for "can" will be a combination of these two contexts in BoC. As a result, the representation will not be able to correctly model polysemy relations involving a word that can be used in different parts of speech.

3   Methods for Syntactic Binding

To solve the problem of modeling certain polysemy relations in natural language text, we need a representation scheme that can encode both the semantics of documents, as well as the *syntax* of documents. We will limit syntactic information to a collapsed parts-of-speech (PoS) data set (e.g.: nouns, verbs, pronouns, prepositions, adjective, adverbs, conjunctions, and interjections), and look at three methods to augment BoC semantic modelling with this information.

3.1   Multiplicative Binding

The simplest method that we investigate is multiplicative binding. For each PoS tag in our collapsed set, we generate a unique random vector for the tag of the same dimensionality as the term context vectors. For each term context vector, we perform element-wise multiplication between that term's context vector and its identified PoS tag vector to obtain our combined representation for the term. Document vectors are then created by summing the the document's combined term vectors.

3.2   Circular Convolution

Combining vectors using circular convolution is motivated by Holographic Reduced Representations [5]. For each PoS tag in our collapsed set, we generate a unique random vector for the tag of the same dimensionality as the term context vectors. For each term context vector, we perform circular convolution, which binds two vectors $\underline{A} = (a_0, a_1, \ldots, a_{n-1})$ and $\underline{B} = (b_0, b_1, \ldots, b_{n-1})$ to give $\underline{C} = (c_0, c_1, \ldots, c_{n-1})$ where $\underline{C} = \underline{A} \otimes \underline{B}$ with $c_j = \sum_{k=0}^{n-1} a_k b_{j-k}$ for $j = 0, 1, \ldots, n-1$ (all subscripts are modulo-n). Document vectors are then created by summing the document's combined term vectors.

There are a number of properties of circular convolution that make it ideal to use as a binding operation. First, the expected similarity between a convolution and its constituents is zero, thus differentiating the same term acting as different parts of speech in similar contexts. As well, similar semantic concepts bound to the same part-of-speech will result in similar vectors; therefore, usefully preserving the original semantic model.

3.3   Text-Based Binding

Text-based binding combines a word with its PoS identifier before the semantic modelling is performed. This is accomplished by concatenating each term's identified PoS tag name with the term's text. Then, the concatenated text is used as the input for Random Indexing to determine the term's context vector. Document vectors are then created by summing the the document's term vectors.

4   Experimental Setup

We performed Support Vector Machine (SVM) classification experiments[1] in order to investigate the classification effectiveness of our syntactic binding methods compared against the standard BoC representation. For the experiments in this paper, we used a linear SVM kernel function (with a slack parameter of 160.0) and fix the dimensionality of all context vectors to 512 dimensions[2]. We used the 20 Newsgroups corpus[3] as the natural language text data for our experiments. In these classification experiments, we used a one-against-all learning method

---

[1] We used the SVM^{perf} implementation, which optimizes for $\mathcal{F}_1$ classification score, available at http://svmlight.joachims.org/svm_perf.html.

[2] The dimensionality of the vectors has been chosen to be consistent with other work. There is as yet no systematic characterization of the effect of dimensionality on performance.

[3] Available at http://people.csail.mit.edu/jrennie/20Newsgroups/.

employing 10-fold stratified cross validation[4]. The SVM classifier effectiveness was evaluated using the $\mathcal{F}_1$ measure. We present our aggregate results for the corpus as macro-averages[5] over each document category for each classifier.

5   Results

Table 1 shows the macro-averaged $\mathcal{F}_1$ scores for all our syntactic binding methods and the baseline BoC representation under SVM classification. All of the syntactic binding methods produced higher $\mathcal{F}_1$ scores than the BoC representation, thus showing that integrating PoS data with a text representation method is beneficial for classification. The circular convolution method produced the best score, with a macro-averaged $\mathcal{F}_1$ score of 58.19, and was calculated to be statistically significant under a 93.7% confidence interval.

Table 1. Macro-Averaged SVM $\mathcal{F}_1$ scores of all methods

Syntactic Binding Method	$\mathcal{F}_1$ Score
BoC (No Binding)	56.55
Multiplicative Binding	57.48
Circular Convolution	**58.19**
Text-based Binding	57.41

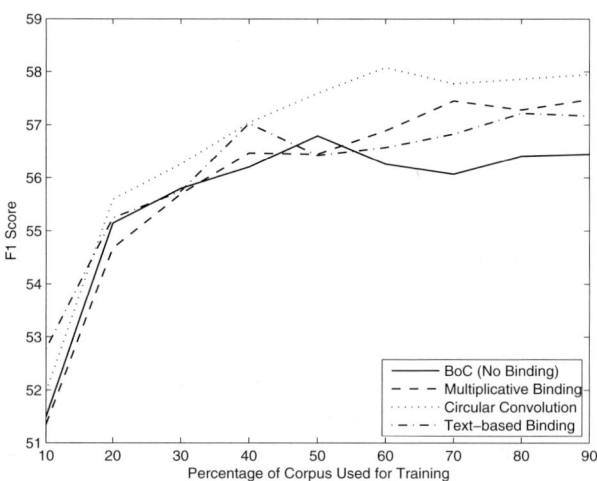

Fig. 1. Learning curves of SVM $\mathcal{F}_1$ scores of all methods

---

[4] This cross-validation scheme was chosen as it better reflects the statistical distribution of the documents, although produces lower $\mathcal{F}_1$ scores.

[5] Since the sizes of document categories are roughly the, same micro-averaging yields similar results and have been omitted for brevity.

The learning curves for the methods are included in Figure 2. The graph shows that circular convolution consistently produces better SVM classification results when compared to the other methods after 20% of the data is used for training. This result indicates that in situations where there is limited class data from which to learn a classification rule, combining the PoS data using circular convolution leads to the most efficient method to assist the classifier in better distinguishing the classes.

6   Conclusions and Future Research

Of all the methods investigated, the circular convolution method of binding a document's PoS information to its semantics was found to be the best. The circular convolution method had the best SVM $\mathcal{F}_1$ score and was superior using various amounts of data to train the classifiers.

Our results suggest areas of further research. One area of is to further investigate alternative binding schemes to augment text semantics, since all of the methods can bind more information than just PoS data. As well, further investigations using different corpora, such as the larger Reuters corpus, should be undertaken to examine the effectiveness of the syntactic binding methods under different text domains.

Acknowledgment. This research is supported in part by the Canada Foundation for Innovation, the Ontario Innovation Trust, the Natural Science and Engineering Research Council and the Open Text Corporation.

References

1. Kehagias, A., et al.: A comparison of word- and sense-based text categorization using several classification algorithms. Journal of Intelligent Information Systems 21(3), 227–247 (2003)
2. Moschitti, A., Basili, R.: Complex Linguistic Features for Text Classification: A Comprehensive Study. In: McDonald, S., Tait, J.I. (eds.) ECIR 2004. LNCS, vol. 2997, pp. 181–196. Springer, Heidelberg (2004)
3. Sahlgren, M., Cöster, R.: Using Bag-of-Concepts to Improve the Performance of Support Vector Machines in Text Categorization. In: Proceedings of the 20th International Conference on Computational Linguistics, pp. 487–493 (2004)
4. Sahlgren, M.: An Introduction to Random Indexing. In: Methods and Applications of Semantic Indexing Workshop at the 7th International Conference on Terminology and Knowledge Engineering (2005)
5. Plate, T.A.: Holographic Reduced Representation: Distributed representation for cognitive structures. CSLI Publications, Stanford (2003)

Use of Temporal Expressions in Web Search

Sérgio Nunes[1], Cristina Ribeiro[1,2], and Gabriel David[1,2]

[1] Faculdade de Engenharia da Universidade do Porto
[2] INESC-Porto
Rua Dr. Roberto Frias, s/n 4200-465 Porto Portugal
{sergio.nunes,mcr,gtd}@fe.up.pt

Abstract. While trying to understand and characterize users' behavior online, the temporal dimension has received little attention by the research community. This exploratory study uses two collections of web search queries to investigate the use of temporal information needs. Using state-of-the-art information extraction techniques we identify temporal expressions in these queries. We find that temporal expressions are rarely used (1.5% of queries) and, when used, they are related to current and past events. Also, there are specific topics where the use of temporal expressions is more visible.

1   Introduction

Query log analysis is currently an active topic in information retrieval. There is a significant and growing number of contributions to the understanding of online user behavior. However, work in this field has been somewhat limited due to the lack of real user data and the existence of important ethical issues [2]. The recent availability of large datasets has specially contributed to a growing interest in this topic. On the other hand, temporal information extraction has reached a point of significant maturity. Current algorithms and software tools are able to extract temporal expressions from free text with a high degree of accuracy. This paper contributes to the characterization of the use of temporal expressions in web search queries, by combining work from these two areas. Our main goal is to provide a better understanding of how users formulate their information needs using standard web search systems. Our focus is on a particular facet of this behavior, namely the use of temporal expressions.

In the following section is presented the experimental setup, including details about the datasets and software used. Section 3 describes the experiments and highlights the main results. An overview of related work is included in Section 4, followed by the conclusions in Section 5

2   Experimental Setup

We used two publicly available datasets containing web search queries. The first dataset includes a collection of manually classified web search queries collected

C. Macdonald et al. (Eds.): ECIR 2008, LNCS 4956, pp. 580–584, 2008.

from the AOL search engine [4]. Each one of the 23,781 queries has been manually classified using a set of predefined topics by a team of human editors. The classification breaks down as follows: Autos (2.9%), Business (5.1%), Computing (4.5%), Entertainment (10.6%), Games (2.0%), Health (5.0%), Holidays (1.4%), Home & Garden (3.2%), News & Society (4.9%), Organizations (3.7%), Personal Finances (1.4%), Places (5.2%), Porn (6.0%), Research (5.7%), Shopping (8.6%), Sports (2.8%), Travel (2.6%), URL (5.7%), *Misspellings* (5.5%) and *Other* (13.2%).

The second dataset is also from AOL [8] and includes more than 30 million (non-unique) web queries collected from more than 650,000 users over a three month period. This dataset is sorted by user ID and sequentially ordered. For each request there is also information about the time at which the query was issued and, when users follow a link, the rank and the URL of the link. An important feature of this second dataset is the availability of the query issuing time, making possible the positioning of temporal expressions. For instance, we are able to determine the specific date of a search for *"a week ago"* because we have access to this information. However, and unlike the first dataset, this one isn't classified.

Temporal expressions were extracted from each query using free, publicly available, Natural Language Processing (NLP) software. First, text was tagged using Aaron Coburn's `Lingua::EN::Tagger` [1], a Part of Speech tagger for English. Then, the output was redirected to TempEx [7], a text tagger that is able to identify a large number of temporal expressions. This tagger covers most of the types of time expressions contained in the 2001 TIMEX2 standard [5]. In Table 1 several examples of this process are presented, showing that TempEx is able to detect a wide range of temporal expressions (e.g. explicit dates, implicit dates, periods).

Table 1. Examples of Tagged Search Queries

olympics 2004
⇒ olympics <TIMEX2 TYPE="DATE" VAL="2004">2004</TIMEX2>
easter 2005
⇒ <TIMEX2 TYPE="DATE" ALT_VAL="20050327">easter 2005</TIMEX2>
monday night football
⇒ <TIMEX2 TYPE="DATE">monday night</TIMEX2> football
us weekly
⇒ us <TIMEX2 TYPE="DATE" SET="YES" PERIODICITY="F1W">weekly</TIMEX2>

3   Use of Temporal Expressions

First, we investigate how temporal expressions are distributed within distilled web queries. For this task we used the first dataset since it is manually annotated with classes. The topics containing the higher percentage of queries with

---

[1] http://search.cpan.org/~acoburn/Lingua-EN-Tagger

temporal expressions are: Autos (7.8%), Sports (5.2%), News & Society (3.9%) and Holidays (2.5%). Examples of queries containing temporal expressions are: *"1985 ford ranger engine head"* (Autos), *"chicago national slam 2003"* (Sports) and *"los angeles times newspaper april 1946"* (News & Society). Manual inspection reveals that the higher number of temporal expressions in the Autos class is mostly due to searches for vintage cars.

In a second experiment, we analyzed the overall distribution of temporal expressions in web search queries. Our first finding is that the use of these expressions is relatively rare. In the first AOL dataset the total number of queries including temporal expression was 347 (1.5%). Remarkably, on the second dataset, we found 532,989 temporal expressions resulting in an equal percentage of 1.5%. Removing duplicate queries results in a small increase in these percentages, specifically 1.6% for the first dataset and 1.9% for the second.

To evaluate the quality of the TempEx tagger when applied to web search queries, we manually classified a random subset of 1,000 queries from the large AOL corpus (including duplicates). We compared this classification with an automatic classification performed by the TempEx tagger. Standard IR measures were computed: accuracy (0.99), precision (0.92) and recall (0.63). The low recall value indicates that the tagger is being conservative, missing some temporal expressions. The non-parametric McNemar test was performed to evaluate the homogeneity of the two classifications. The test confirms that the automatic classification is equivalent to the human classification ($p > 0.05$).

Restricting our analysis to the second AOL dataset, we performed additional measurements. Since query issuing time is available, we are able to precisely date a large fraction of the temporal expressions found. Using this information we measured the number of expressions referencing past events, present events and future events. TempEx automatically detects some of these expressions. This module was able to identify generic references to the past (e.g. *"once"*, *"the past"*) (1.25% of the temporal expressions), to the present (e.g. *"now"*, *"current"*) (4%) and to the future (e.g. *"future"*) (0.82%). The vast majority of temporal expressions (94%) do not include explicit references like these.

We've extracted the year from all dated expressions and counted the occurrences. Taking into account that all queries were issued between March and May 2006, we see that the majority of temporal expressions are related to current events. The frequency distribution is positively skewed with a long tail toward past years. Summarizing, in all temporal expressions identified, 42.5% indicate a date from 2006, 49.9% are from dates prior to 2006, and 4.2% are from dates after 2006.

To better understand which temporal expressions were being used, we manually inspected a list of the top 100 more common expressions. These expressions account for slightly more than 80% of the queries containing temporal expressions. We grouped all references to a single year and to a single month in two generic expressions (i.e. $<Year>$ and $<Month>$). The top 10 expressions used in queries are: $<Year>$ (45.7%), *easter* (5.6%), *daily* (5.4%), $<Month>$ (4.6%), *now* (2.3%), *today* (2.1%), *mothers day* (1.8%), *current* (1.2%), *christmas* (1.1%) and *weekly* (0.8%). It is important to note that seasonal results (e.g. "Easter") are artificially inflated in the 3 month of data (March to May).

When starting this research one of our hypothesis was that temporal expressions were regularly used to improve initial queries. To investigate on this hypothesis we did a rough analysis on query refinements within the AOL dataset. Our algorithm is very simple and only identifies trivial query reformulations. In a nutshell, since the dataset is ordered by user and issuing time, we simply compare each query with the previous one to see if there is an expansion of the terms used. For instance, *"easter holidays"* is considered a reformulation of *"easter"*. With this algorithm we found 1,512,468 reformulated queries (4.2%). We then counted the presence of temporal expressions in this subset and verified that only 1.4% of these queries contained temporal expressions.

4   Related Work

We found no previous work on the specific topic of identifying and characterizing the use of temporal expressions in web search queries. Thus, the related work presented here is divided in the two parent topics: *temporal expression extraction* and *query log analysis*. In recent years, Temporal Information Extraction emerged from the broader field of Information Extraction [9]. Most work in this field is focused on the study of temporal expressions within semi-structured documents [6]. In our work we apply these techniques in processing short text segments that represent information needs.

Query log analysis has been the focus of increasing interest in recent years. Most work in this area has been devoted to the classification and characterization of queries. An example of a detailed work in this area is from Beitzel et al. [3]. In this work the authors perform a detailed characterization of web search queries through time using a large topically classified dataset. Our work differs from this since we are interested in how temporal expressions are used within queries.

5   Conclusions

Contrary to our initial expectations, the use of temporal expressions in web queries is relatively scarce. Using two different datasets, we've found that temporal expressions are used in approximately 1.5% of the queries. We speculate on three reasons that might explain this situation: (1) information needs of web users are mostly focused on current events; (2) users are generally happy with the results obtained using short text queries; (3) users resort to more advanced interfaces when they have dated information needs. Investigating these hypotheses is left for future work. Focusing on the small subset of temporal expressions extracted, we've found that most temporal expressions reference current dates (within the same year) and past dates (exhibiting a long tailed behavior). Future dates are rarely used. Finally, we've shown that these expressions are more frequently used in topics such as: Autos, Sports, News and Holidays.

Although temporal expressions appear in only a small fraction of all queries, the scale of the Web translates this percentage into a large number of users. Temporal expression extraction might be used in public search engines to improve ranking or result clustering. As the web grows older, and more content is accumulated in archives (e.g. Internet Archive), we think that the need for dated information will rise [1]. Search engine designers can respond to this challenge by incorporating temporal information extraction algorithms or by developing specialized search interfaces. As an example, Google has recently launched a prototype that provides date-based navigation in search results using timelines [2].

Acknowledgments. Sérgio Nunes was financially supported by the Fundação para a Ciência e a Tecnologia (FCT) and Fundo Social Europeu (FSE - III Quadro Comunitário de Apoio), under grant SFRH/BD/31043/2006. We would like to thank the ECIR 2008 reviewers, whose comments have contributed to improve the final version of this paper.

References

1. Alonso, O., Gertz, M., Baeza-Yates, R.: On the value of temporal information in information retrieval. ACM SIGIR Forum 41(2), 35–41 (2007)
2. Bar-Ilan, J.: Access to query logs - an academic researcher's point of view. In: Query Log Analysis: Social And Technological Challenges Workshop. 16th International World Wide Web Conference (WWW 2007) (May 2007)
3. Beitzel, S.M., Jensen, E.C., Chowdhury, A., Grossman, D., Frieder, O.: Hourly analysis of a very large topically categorized web query log. In: SIGIR 2004: Proceedings of the 27th annual international conference on Research and development in information retrieval, pp. 321–328. ACM Press, New York (2004)
4. Beitzel, S.M., Jensen, E.C., Frieder, O., Lewis, D.D., Chowdhury, A., Kolcz, A.: Improving automatic query classification via semi-supervised learning. In: ICDM 2005: Proceedings of the Fifth IEEE International Conference on Data Mining, Houston, Texas, USA, November 2005, pp. 42–49 (2005)
5. Ferro, L., Mani, I., Sundheim, B., Wilson, G.: TIDES temporal annotation guidelines (v. 1.0.2). Technical report, The MITRE Corporation, Virginia (June 2001)
6. Mani, I., Pustejovsky, J., Sundheim, B.: Introduction to the special issue on temporal information processing. ACM Transactions on Asian Language Information Processing (TALIP) 3(1), 1–10 (2004)
7. Mani, I., Wilson, G.: Robust temporal processing of news. In: Proceedings of the 38th Annual Meeting of the Association for Computational Linguistics (ACL2000), Hong Kong, pp. 69–76 (2000)
8. Pass, G., Chowdhury, A., Torgeson, C.: A picture of search. In: InfoScale 2006: Proceedings of the 1st international conference on Scalable information systems, ACM Press, New York (2006)
9. Wong, K.-F., Xia, Y., Li, W., Yuan, C.: An overview of temporal information extraction. International Journal of Computer Processing of Oriental Languages 18(2), 137–152 (2005)

---

[2] http://www.google.com/experimental

Towards an Automatically Generated Music Information System Via Web Content Mining

Markus Schedl[1,*], Peter Knees[1], Tim Pohle[1], and Gerhard Widmer[1,2]

[1] Department of Computational Perception
Johannes Kepler University
Linz, Austria
http://www.cp.jku.at
[2] Austrian Research Institute for Artificial Intelligence
Vienna, Austria
http://www.ofai.at

Abstract. This paper presents first steps towards building a music information system like *last.fm*, but with the major difference that the data is automatically retrieved from the WWW using web content mining techniques. We first review approaches to some major problems of music information retrieval (MIR), which are required to achieve the ultimate aim, and we illustrate how these approaches can be put together to create the *automatically generated music information system* (AGMIS). The problems addressed in this paper are *similar and prototypical artist detection, album cover retrieval, band member and instrumentation detection, automatic tagging of artists*, and *browsing/exploring web pages related to a music artist*. Finally, we elaborate on the currently ongoing work of evaluating the methods on a large dataset of more than $600,000$ music artists and on a first prototypical implementation of AGMIS.

1 Introduction and Context

Music information systems like *last.fm* [1] typically offer multimodal information about music artists, albums, and tracks (e.g. genre and style, similar artists, biographies, song samples, or images of album covers). In common music information systems, such information is usually gained and revised by experts (e.g. *All Music Guide* [2]), or relies on user participation (e.g. *last.fm*). In contrast, we are building such a system by automatically extracting the required information from the web.

Automatically retrieving descriptive information about music artists is an important task in music information retrieval (MIR) as it allows for enriching music players [13], for automatic biography generation [4], for enhancing user interfaces to browse music collections [7,6], or for defining similarity measures between artists, a key concept in MIR. Similarity measures enable, for example, creating relationship networks [10], building music recommender systems [15] or music search engines [5].

---

* markus.schedl@jku.at

C. Macdonald et al. (Eds.): ECIR 2008, LNCS 4956, pp. 585–590, 2008.

In the following, we first give a brief overview of the available techniques which we are refining at the moment. Hereafter, we present the currently ongoing work of combining these techniques to build the *automatically generated music information system* (AGMIS).

2 Mining the Web for Music Artist-Related Information

All of the applied methods rely on the availability of artist-related data in the WWW. Our principal approach to extracting such data is the following. Given only a list of artist names, we first query a search engine[1] to retrieve the URLs of up to 100 top-ranked search results for each artist. The content available at these URLs is extracted and stored for further processing. To overcome the problem of artist or band names that equal common speech words and to direct the search towards the desired information, we use task-specific query schemes, like *"band name"+music+members* to obtain data related to band members and instrumentation. Depending on the task to solve, we then create either a document-level inverted file index or a word-level index [16]. In some cases, we use a special dictionary of musically relevant terms to perform indexing. After having indexed the web pages, we gain artist-related information of various kinds as described in the following.

2.1 Relations between Artists

A key concept in music information retrieval and crucial part of any music information system are *similarity relations* between artists. To model such relations, we use an approach that is based on co-occurrence analysis [9]. More precisely, the similarity between two artists a and b is defined as the conditional probability that the artist name a occurs on a web page that was returned as response to the search query for the artist name b and vice versa, formally $\frac{1}{2} \cdot \left(\frac{df_{a,B}}{|B|} + \frac{df_{b,A}}{|A|} \right)$, where A represents the set of web pages returned for artist a and $df_{a,B}$ is the document frequency of the artist name a calculated on the set of web pages returned for artist b. Having calculated the similarity for each pair of artists in the artist list, we can output, for any artist, a list of most similar ones. Evaluation in an artist-to-genre classification task on a set of 224 artists from 14 genres yielded accuracy values of about 85%.

Co-occurrences of artist names on web pages (together with genre information) can also be used to derive information about the *prototypicality of an artist for a certain genre* [10,11]. To this end, we make use of the asymmetry of the co-occurrence-based similarity measure.[2] We developed an approach that is based on the forward link/backlink-ratio of two artists a and b from the same genre, where a backlink of a from b is defined as any occurrence of artist a on a web page that is known to contain artist b, whereas a forward link of a to b is defined

---

[1] We commonly used *Google* in our experiments, but also experimented with *exalead*.

[2] In general, $\frac{df_{a,B}}{|B|} \neq \frac{df_{b,A}}{|A|}$.

as any occurrence of b on a web page known to mention a. Relating the number of forward links to the number of backlinks for each pair of artists from the same genre, a ranking of the artist prototypicality for the genre under consideration is obtained. A more extensive description of the approach can be found in [11].

2.2   Band Member and Instrumentation Detection

Another type of information indispensible for a music information system is *band members and instrumentation*. In order to capture such aspects, we first apply a named entity detection approach that basically relies on extracting N-grams and on filtering w.r.t. capitalization and words contained in the *iSpell English Word Lists* [3]. The remaining N-grams are regarded as potential band members. Subsequently, we perform linguistic analysis to obtain the actual instrument(s) of each member. To this end, a set of seven patterns like *"M plays the I"*, where M is the potential member and I is the instrument, is applied to the N-grams (and the surrounding text as necessary). The document frequencies of the patterns are recorded and summed up over all seven patterns for each (M, I)-tuple. After having filtered out those (M, I)-pairs whose document frequency is below a dynamically adapted threshold in order to suppress uncertain information, the remaining (M, I)-tuples are predicted for the band under consideration. More details as well as an extensive evaluation can be found in [14].

2.3   Automatic Tagging of Artists

For automatically attributing textual descriptors to artists, we use a dictionary of about $1,500$ musically relevant terms to index the web pages. As for term weighting, three different measures (document frequency, term frequency, and $TF \times IDF$) were evaluated in a yet unpublished quantitative user study. This study showed, quite surprisingly, that the simple document frequency measure outperformed the well-established $TF \times IDF$ measure significantly (according to Friedman's non-parametric two-way analysis of variance). Thus, for the AGMIS, we will probably use this measure to automatically select the most appropriate tags for each artist.

2.4   Co-occurrence Browser

To easily access the top-ranked web pages of any artist, we designed a user interface, which we call the *Co-Occurrence Browser* (COB). Based on the dictionary used for automatic tagging, the COB groups the web pages of the artist under consideration w.r.t. the document frequencies of co-occurring terms. These groups are then visualized using the approach presented in [12]. Thus, the COB allows for browsing the artist's web pages by means of descriptive terms. Furthermore, the multimedia content present on the web pages is extracted and made available via the user interface.

2.5 Album Cover Retrieval

Preliminary attempts to automatically retrieve album cover artwork were made in [8]. We refined the methods presented in this paper and conducted experiments with content-based as well as context-based methods for detecting images of album covers. We found that using the text distance between album names and *img*-tags in the HTML file at character level gives a quite good indication whether an image is an album cover or not. The results could further be improved by rejecting images that have non-quadratic dimensions or appear to show a scanned disc (which happens quite often). On a challenging collection of about 3,000 albums, we estimated a precision of approximately 60%.

3 Building the AGMIS

Currently, our work is focusing on the large-scale retrieval of artist-related information and on building a prototypical implementation of the AGMIS user interface. As for retrieval, the search engine *exalead* was used to obtain a list of more than 26,000,000 URLs (for a total of 600,000 artists from 18 genres). We are fetching these URLs using a self-made, thread-based Java program that offers load balancing between the destination hosts. A file index of the retrieved web documents will be build subsequently.

As for the user interface, Figure 1 shows a sample web page created by a prototypical implementation of AGMIS (based on Java Servlet and Applet technologies). This prototype incorporates the information whose extraction and

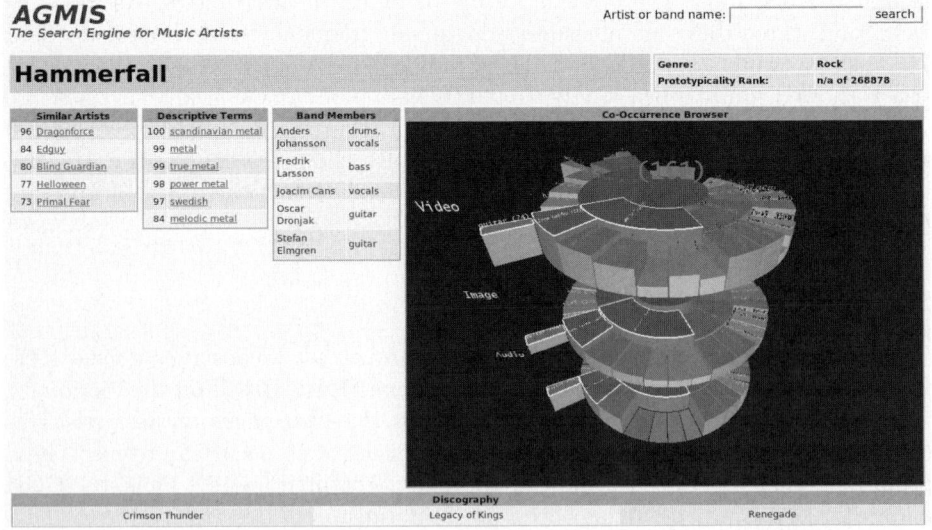

Fig. 1. Screenshot of a prototypical implementation of AGMIS

presentation was described in Section 2. On the left-hand side, textual information about the artist *Hammerfall* is offered to the user, whereas on the right, the user interface of the COB is embedded as a Java Applet. The page is further enriched by displaying images of album covers in its lower part (which are omitted in the screenshot due to copyright reasons).

4   Conclusions and Future Work

We presented a set of methods that address current problems in the field of web-based music information retrieval and showed how we will apply them to create an automatically generated music information system, which we call AGMIS.

Future work will mainly focus on evaluating the presented approaches on the large corpus which we are currently building. After having fetched them, we will look into efficient methods for high-speed indexing of the retrieved web pages and for organizing and storing the information extracted from the index via the approaches presented in Section 2. Finally, the user interface for accessing the music information will probably need some updates.

Acknowledgments

This research is supported by the Austrian Fonds zur Förderung der Wissenschaftlichen Forschung (FWF) under project number L112-N04.

References

1. http://last.fm, (2007) (last access: 2007-10-04)
2. http://www.allmusic.com, (2007) (last access: 2007-10-04)
3. http://wordlist.sourceforge.net, (2007) (last access: 2007-10-04)
4. Alani, H., Kim, S., Millard, D.E., Weal, M.J., Hall, W., Lewis, P.H., Shadbolt, N.R.: Automatic Ontology-Based Knowledge Extraction from Web Documents. IEEE Intelligent Systems 18(1), 14–21 (2003)
5. Knees, P., Pohle, T., Schedl, M., Widmer, G.: A Music Search Engine Built upon Audio-based and Web-based Similarity Measures. In: Proceedings of the 30th Annual International ACM SIGIR Conference on Research and Development in Information Retrieval (SIGIR 2007), Amsterdam, the Netherlands (July 2007)
6. Knees, P., Schedl, M., Pohle, T., Widmer, G.: An Innovative Three-Dimensional User Interface for Exploring Music Collections Enriched with Meta-Information from the Web. In: Proceedings of the 14th ACM Conference on Multimedia 2006, Santa Barbara, CA, USA (October 2006)
7. Pampalk, E., Goto, M.: MusicSun: A New Approach to Artist Recommendation. In: Proceedings of the 8th International Conference on Music Information Retrieval (ISMIR 2007), Vienna, Austria (September 2007)
8. Schedl, M., Knees, P., Pohle, T., Widmer, G.: Towards Automatic Retrieval of Album Covers. In: Lalmas, M., MacFarlane, A., Rüger, S.M., Tombros, A., Tsikrika, T., Yavlinsky, A. (eds.) ECIR 2006. LNCS, vol. 3936, Springer, Heidelberg (2006)

9. Schedl, M., Knees, P., Widmer, G.: A Web-Based Approach to Assessing Artist Similarity using Co-Occurrences. In: Proceedings of the 4th International Workshop on Content-Based Multimedia Indexing (CBMI 2005), Riga, Latvia (June 2005)

10. Schedl, M., Knees, P., Widmer, G.: Discovering and Visualizing Prototypical Artists by Web-based Co-Occurrence Analysis. In: Proceedings of the 6th International Conference on Music Information Retrieval (ISMIR 2005), London, UK (September 2005)

11. Schedl, M., Knees, P., Widmer, G.: Investigating Web-Based Approaches to Revealing Prototypical Music Artists in Genre Taxonomies. In: Proceedings of the 1st IEEE International Conference on Digital Information Management (ICDIM 2006), Bangalore, India (December 2006)

12. Schedl, M., Knees, P., Widmer, G., Seyerlehner, K., Pohle, T.: Browsing the Web Using Stacked Three-Dimensional Sunbursts to Visualize Term Co-Occurrences and Multimedia Content. In: Proceedings of the 18th IEEE Visualization 2007 Conference (Vis 2007), Sacramento, CA, USA (October 2007)

13. Schedl, M., Pohle, T., Knees, P., Widmer, G.: Assigning and Visualizing Music Genres by Web-based Co-Occurrence Analysis. In: Proceedings of the 7th International Conference on Music Information Retrieval (ISMIR 2006), Victoria, Canada (October 2006)

14. Schedl, M., Widmer, G.: Automatically Detecting Members and Instrumentation of Music Bands via Web Content Mining. In: Proceedings of the 5th Workshop on Adaptive Multimedia Retrieval (AMR 2007), Paris, France (July 2007)

15. Zadel, M., Fujinaga, I.: Web Services for Music Information Retrieval. In: Proceedings of the 5th International Symposium on Music Information Retrieval (ISMIR 2004), Barcelona, Spain (October 2004)

16. Zobel, J., Moffat, A.: Inverted Files for Text Search Engines. ACM Computing Surveys 38(2), 6 (2006)

Investigating the Effectiveness of Clickthrough Data for Document Reordering

Milad Shokouhi[1,*], Falk Scholer[2], and Andrew Turpin[2]

[1] Microsoft Research Cambridge, 7 J J Thomson Ave., Cambridge CB3 0FB, UK
[2] School of CS and IT, RMIT University, PO Box 2476v, Melbourne, Australia

Abstract. User clicks—also known as clickthrough data—have been cited as an implicit form of relevance feedback. Previous work suggests that relative preferences between documents can be accurately derived from user clicks. In this paper, we analyze the impact of document reordering—based on clickthrough—on search effectiveness, measured using both TREC and user relevance judgments. We also propose new strategies for document reordering that can outperform current techniques. Preliminary results show that current reordering methods do not lead to consistent improvements of search quality, but may even lead to poorer results if not used with care.

1 Introduction and Related Work

Commercial search engines store various types of information in their logs about how users query for information. Previous queries in the logs, in particular clickthrough data, can be used to improve search for future queries. However, user clicks are subject to many biases and should be interpreted cautiously [9]. For example, users have been found to be significantly biased towards clicking the highly ranked documents (*trust bias*). Therefore, the fact that a document is clicked does not necessarily imply that it is relevant.

Several strategies have been suggested for inferring relevance from user clicks. A significant fraction of previous studies focuses on deriving relative preferences between documents; surprisingly, the actual impact on search results of document reordering using the derived preferences is often ignored. In this paper, we compare the effectiveness of current reordering methods (based on relative preferences) in terms of search quality. We also propose new reordering strategies that can outperform the state of the art.

Clickthrough data has been used in many forms to enhance search quality. Joachims [8] applied an SVM classifier on user clicks to optimize the ranking function of a metasearch engine. Kemp and Ramamohanarao [11] deployed clickthrough for document expansion, while Craswell and Szummer [5] used clickthrough to improve image retrieval. Fox et al. [7] investigated the impact of several implicit measures (including clickthrough) on user satisfaction. Their results suggested an association between more clicks and higher satisfaction rate.

---

* This research was carried out while the author worked at RMIT university.

C. Macdonald et al. (Eds.): ECIR 2008, LNCS 4956, pp. 591–595, 2008.

Similarly, Agichtein et al. [1] used different types of implicit feedback—such as clicks and timestamps—simultaneously to enhance document retrieval.

In a series of papers, Joachims et al. proposed several strategies for accurately interpreting the click data as implicit user judgments. The authors used clicks to obtain relative preferences between documents [9,10,12,13]. We use two of the most successful strategies suggested by them as the baseline of our experiments. These methods are described in the next section.

2 Effective Document Reordering Using Previous Queries

Joachims et al. proposed five strategies for deriving relative preferences between documents by using clicks.[1] Supported by eye-tracking experiments, the authors argued that users are subject to *trust* and *quality* biases. Their proposed strategies take the existing biases into account. Among the methods suggested, the following two showed the highest agreement rates with human judges.

LAST CLICKED > SKIP ABOVE (LCSA): "For a ranking $(l_1, l_2, l_3, ...)$ and a set C containing the ranks of the clicked-on links, let $i \in C$ be the rank of the link that was clicked temporarily last. Extract a preference example $rel(l_i) > rel(l_j)$ for all pairs $1 \le j < i$, with $j \notin C$." [9]

LAST CLICKED > SKIP PREVIOUS (LCSP): "For a ranking $(l_1, l_2, l_3, ...)$ and a set C containing the ranks of the clicked-on links, extract a preference example $rel(l_i) > rel(l_{i-1})$ for all pairs $i > 2$, with $i \in C$ and $i - 1 \notin C$." [9]

Note that instead of deciding whether a document is relevant or not (*absolute judgment*), the authors compared the relative probability of relevance of documents (*relative judgment*). The first strategy (LCSA) is based on the intuition that later clicks are more *informed* than the earlier clicks. Agichtein et al. [1] also argued that later clicks are more important because their likelihood is lower in general. The second strategy (LCSP) is based on eye-tracking experiments that suggest that the snippets of documents around the clicked answers are scanned more carefully by the users, and thus they are more reliable for deriving relative preferences.

The strategies from previous work in this area generally assume that the concept of user relevance is static. That is, the definition of a relevant document remains constant for a user after viewing the clicked answers. However, previous studies [3,4] showed that the notion of relevance can be dynamic. Visiting new documents increases the user knowledge and changes his initial picture of relevance. This leads us to our first new strategy for deriving implicit relevance from clickthrough.

SHIFT-CLICKED: For a ranking $(l_1, l_2, l_3, ...)$ and a set C containing the ranks of the clicked links, let i and j be the ranks of two clicked documents where

---

[1] More recently, the authors have extended their methods by applying them on query-chains [12], and by slightly reranking the results in order to collect clicks for the bottom-ranked documents [13].

$i < j$. Extract a preference example $\text{rel}(l_j) > \text{rel}(l_k)$ for all pairs $i < k < j$, with $\forall k, k \notin C$ (i and j are consecutive clicks).[2]

The definition of relevance for a user varies after visiting each clicked document. Therefore, if i and j are two clicked documents (where i is clicked before j), it is hard to compare the relative relevance of j with that of any document ranked before i. Shift-clicked takes this principle into account, while deriving the relative document preferences.

HYBRID-MODEL: For a given query, some documents may get more clicks than others. For an example query "MSR", if many more users have clicked on "Microsoft Research Home (`research.microsoft.com`)", than "Medical Support in Romania (`www.msr.org.uk`)", the former answer is relatively more *popular* or *preferred* over the other. We consider this form of user behavior as another source of evidence to propose a hybrid model for inferring implicit judgments. In our hybrid model, the returned answers are initially reordered using one of the previously discussed reranking strategies. We then order any two clicked documents in consecutive ranks according to their previous click frequency (for the same query). In the following sections, we investigate the impact of these document reordering strategies on search quality and user satisfaction.

3   Experimental Results

For our experiments, we use the well-known TREC WT10g collection [2]. This collection has 100 associated informational queries (TREC topics 451–550). We compare the performance of document reordering strategies using two sets of relevance judgments: the official TREC judgments, and judgments generated as part of a user study [14].

30 experimental subjects were recruited from RMIT University to participate in a search experiment. To meet the requirements of the user study, the 50 topics with the highest number of relevant documents were selected out of the 100 available items. Users were presented with the description and narrative fields of TREC topics, and asked to find as many relevant documents as possible within five minutes. The search system was modeled closely on popular web search engines: after submitting a query, users were presented with an answer list consisting of summary information for each potential answer document identified by the system, including the document title, URL, and a short query-biased summary. Users could click on the title to view the underlying document; they could then either save the item as being relevant, or continue searching. As in TREC judgments, the degree of bias in our user judgments is unclear. All user interactions with the search system were logged. Full details of the user study are available in the paper by Turpin et al. [14].

TREC judgments. Table 1 shows the results of the different reranking methods based on several well-known retrieval metrics: P@5, P@10, bpref, NDCG, and

---

[2]   Supported by previous research [1,9], we assume that users tend to click from top to bottom.

Table 1. The effectiveness of reranking strategies compared to that of the original ranked lists. TREC topics 451–550 and their judgments are used for evaluation.

	P@5	P@10	bpref	NDCG	NDCG@15
LCSA	0.928‡	0.879	0.399	0.461	0.503
freq + LCSA	0.929‡	0.879	0.398	0.463	0.507
LCSP	0.949	0.883	0.399	0.464	0.506
freq + LCSP	0.950	0.883	0.399	0.465	0.507
Shift-clicked	0.947	0.884	0.400	0.466	0.511
freq + Shift-clicked	0.948	0.884	0.400	0.466	0.512
Original ranking	0.946	0.882	0.399	0.463	0.504

Table 2. The effectiveness of reranking strategies compared to that of the original ranked lists. A document is relevant if more than 50% of judges approve its relevance.

	P@5	P@10	bpref	NDCG	NDCG@15
LCSA	0.362‡	0.313	0.288	0.471†	0.332†
freq + LCSA	0.367‡	0.314†	0.288	0.475‡	0.335‡
LCSP	0.341	0.301	0.288	0.459	0.315
freq + LCSP	0.345	0.302	0.288	0.460	0.316
Shift-clicked	0.373‡	0.316†	0.290	0.477‡	0.339‡
freq + Shift-clicked	0.373‡	0.316†	0.290	0.478‡	0.340‡
Original ranking	0.335	0.299	0.287	0.454	0.307

NDCG@15 (in the table, $\dagger$ and $\ddagger$ indicate statistical significance for a paired t-test at the 0.01 and 0.001 levels, respectively, against the original ranking.) In general, there are no significant changes in retrieval performance for any of the reordering strategies; the exception is LCSA, which performs significantly worse than the original ranking ($p < 0.001$) for P@5.

User judgments. We also evaluate the strategies using user relevance judgments. Since documents may be judged by varying numbers of users, we map these multiple judgments into a binary judgment, deeming a document to be relevant if more than 50% of judges (people in our user study) considered it to be relevant, and irrelevant otherwise. The results are shown in Table 2. Out of the different reranking strategies, LCSP shows the poorest performance, while Shift-clicked and its hybrid-model produce the best results. Overall, taking the click frequency into account slightly (but consistently) improves the quality of reordered lists.

4   Discussion and Conclusions

While a variety of document reordering schemes have been proposed in previous work, the effect of such schemes on actual search results has not been investigated in detail. Our preliminary results show that some existing strategies may actually harm search performance. Further analysis across different collections and topics is needed to investigate whether general conclusions about the relative effectiveness of reordering schemes can be drawn; until then such approaches should be used with care. Our results also show that different relevance judgments can lead to significantly different conclusions about the effect of techniques—the same technique that significantly harms performance for P@5 using the TREC judgments leads to a significant increase in performance based on user judgments. Note that there are far fewer user-based relevance judgements available than there are TREC relevance judgements. Hence, the effectiveness scores are not directly comparable between tables. Instead, the relative performance of systems within a table is important.

The novel shift-clicked model shows consistent higher performance on both TREC and user judgments. Our preliminary results also suggest that (not surprisingly) incorporating click frequency can have positive effects on search performance.

References

1. Agichtein, E., Brill, E., Dumais, S., Ragno, R.: Learning user interaction models for predicting web search result preferences. In: Efthimiadis, et al. (eds.) [6], pp. 3–10
2. Bailey, P., Craswell, N., Hawking, D.: Engineering a multi-purpose test collection for web retrieval experiments. IPM 39(6), 853–871 (2003)
3. Borlund, P.: The concept of relevance in IR. JASIS 54(10), 913–925 (2003)
4. Bruce, H.: A cognitive view of the situational dynamism of user-centered relevance estimation. JASIS 45(3), 142–148 (1994)
5. Craswell, N., Szummer, M.: Random walks on the click graph. In: Proc. ACM SIGIR, Amsterdam, The Netherlands, pp. 239–246 (2007)
6. Efthimiadis, E., Dumais, S., Hawking, D., Järvelin, K. (eds.): Proc. ACM SIGIR, Seattle, WA (2006)
7. Fox, S., Karnawat, K., Mydland, M., Dumais, S., White, T.: Evaluating implicit measures to improve web search. ACM TOIS 23(2), 147–168 (2005)
8. Joachims, T.: Optimizing search engines using clickthrough data. In: Proc. ACM SIGKDD, Edmonton, Alberta, Canada, pp. 133–142. ACM Press, New York (2002)
9. Joachims, T., Granka, L., Pan, B., Hembrooke, H., Radlinski, F., Gay, G.: Evaluating the accuracy of implicit feedback from clicks and query reformulations in web search. ACM TOIS 25(2), 7 (2007)
10. Joachims, T., Radlinski, F.: Search engines that learn from implicit feedback. IEEE Computer 40(8), 34–40 (2007)
11. Kemp, C., Ramamohanarao, K.: Long-term learning for web search engines. In: Proceedings of the 6th European Conference on Principles of Data Mining and Knowledge Discovery, pp. 263–274. Springer, London (2002)
12. Radlinski, F., Joachims, T.: Query chains: learning to rank from implicit feedback. In: Proc. ACM SIGKDD, Chicago, Illinois, pp. 239–248 (2005)
13. Radlinski, F., Joachims, T.: Active exploration for learning rankings from clickthrough data. In: Proc. ACM SIGKDD, San Jose, California, pp. 570–579 (2007)
14. Turpin, A., Scholer, F.: User performance versus precision measures for simple search tasks. In: Efthimiadis, et al. (eds.) [6], pp. 11–18

Analysis of Link Graph Compression Techniques

David Hannah, Craig Macdonald, and Iadh Ounis

Department of Computing Science,
University of Glasgow, G12 8QQ, UK
{hannahd,craigm,ounis}@dcs.gla.ac.uk

Abstract. Links between documents have been shown to be useful in various Information Retrieval (IR) tasks - for example, Google has been telling us for many years now that the PageRank authority measure is at the heart of its relevance calculations. To use such link analysis techniques in a search engine, special tools are required to store the link matrix of the collection of documents, due to the high number of links typically involved. This work is concerned with the application of compression to the link graph. We compare several techniques of compressing link graphs, and conclude on speed and space metrics, using various standard IR test collections.

1 Introduction

The retrieval performance of many Information Retrieval (IR) systems can be benefited by the application of techniques based on the structure of links between documents, such as PageRank [1] (famously applied by Google), and Hypertext Induced Topic Selection (HITS) [2]. To calculate such link analysis techniques, it is necessary to have a matrix of the links between all documents in the collection. However, due to the large nature of such a matrix, it would be impossible to work with it wholly stored in memory. As such, a disk-based structure is required, known as a link database. In particular, a link database stores, for each document, a list of the incoming (or outgoing) links to (from) the documents. In a typical collection of Web documents, there is typically an order of magnitude more links than documents, and the number of incoming links to a document follows a power-law distribution. To quickly and easily compute the various link graph-based query independent features, it is necessary that the Web search engine has timely and efficient access to the link graph.

This paper investigates the use of various state-of-the-art compression techniques, and how they can be applied to the compression of a link graph. We experiment with six recent large Web IR test collections, such as those used in TREC and the very large UK-2006 collection, and conclude on the most efficient representation for the link graph. While these compression techniques have been proposed in literature, they have never been studied extensively in terms of both time and space efficiency, and over as many test collections, of various age, size and domain, and using exactly the same experimental setting.

C. Macdonald et al. (Eds.): ECIR 2008, LNCS 4956, pp. 596–601, 2008.

Number of Links	$\longrightarrow$	$l_1, l_2, l_3, ...$		Number of Links	$\longrightarrow$	$l_1 - 1, l_2 - l_1, l_3 - l_2, ...$

(a) No Encoding   (b) Gamma Encoding

Number of Links	$\longrightarrow$	Number of Intervals, $int_1 : intLeng_1, int_2 : intLen_2,...$	$r_1 - 1, r_2 - r_1,...$

(c) Interval Encoding

Fig. 1. Encoding techniques applied for compressing the link database

2 Link Graph Compression

For the state-of-the-art link graph compression techniques, we base our work on that of Boldi & Vigna [3], who detail various compression techniques for link graphs. Firstly, it is assumed that all documents in the collection have a numerical integer document identifier (document id). We then store for each document, the document ids of each of its incoming links, which we denote inlinks. Moreover, we can without loss, describe the transpose of the inlinks matrix, which we denote outlinks. We experiment with three techniques for compression:

No Encoding: (Fig. 1(a)) A vector that contains the number of links of each document together with pointers to the offset in a second file of the document ids for the links to (from) that document. Links are encoded using 32 bit integers.

Gamma Encoding: (Fig. 1(b)) Again, a vector containing the number of links for each document and a pointer into another file containing the links. However, the links are stored as the differences between the doc id of each link and the previous, written using 'Elias gamma encoding'. To gamma encode a number, it is first written in binary, then a number of zeros (the number of bits required to write the number minus one) is prepended to the number [5]. It follows that each link can take a variable number of bits to encode[1].

Interval Encoding: (Fig. 1(c)) This technique is similar to Gamma Encoding except that it encodes intervals of links. This is based on the intuition that if the documents ids in a collection are stored in lexicographical order of URL, then there will be common 'runs' of links to documents with adjacent document ids. For this compression style, the number of intervals are stored as a gamma encoded integer. Then each intervals of links is stored as the left extreme and the length of the run - again using gamma encoding. Finally, the extra links which are not consecutive are stored using gamma encoding as before. Intervals of less than L_{min} are not encoded.

It is of note that in [3], Boldi & Vigna describe a further compression technique (Reference Encoding), whereby the links for a document are encoded by stating how much it has in common with the links of the previous few documents, iteratively applied. For reasons of brevity, we leave this technique as future work.

---

[1] In contrast to [3], we encode the first link as document id + 1, to ensure uniformity with existing compression used Terrier, the platform on which this work is performed. However this change does not affect any experimental conclusions.

3   Experimental Design

To analyse the effectiveness of the various compression techniques, we use six different Web IR test collections, related to different domains and timescales. In particular, the collections we experiment with are: two older TREC Web test collections WT2G and WT10G, which are small-scale general Web crawls from early 1997; .GOV and .GOV2 are more recent TREC Web test collections, both of which are crawls of the `.gov` domain from 2002 and 2003 respectively - .GOV2 being the largest TREC collection at 25M documents; the TREC CERC collection which is a crawl of the CSIRO website from early 2007; and finally the UK-2006 collection is a large crawl of the `.uk` domain from 2006[2].

For our experiments, we apply Web IR techniques deployed in the Terrier platform [6]. To assess the efficiency of the link database compression methods, we record various metrics for each collection: Firstly, we record the time taken to build the compressed copies of the incoming and outgoing link graphs; Secondly, we record the time taken to compute the PageRank prior using the link database. This is motivated by the fact that the PageRank computation is a realistic application for a link database. However as the PageRank calculation can take many iterations to converge, we normalise the times by the number of iterations to account for the different sizes of collections; Lastly, we record the space required to store the link graph, in terms of the mean number of bits of space required per link. From these metrics, we can conclude in terms of the time to write and read each of the link database compression techniques, as well as their space requirements. Note that we vary the parameter $Lmin$ of interval encoding.

4   Results and Analysis

Table 1 presents the compression level achieved for the inlinks and outlinks tables of the link databases, in terms of mean number of bits per links - the lower the number of bits required per link, the better compression achieved. Observe that, as expected, the No Encoding technique has a stable usage of 32 bits per link.

Applying Gamma Encoding on the same link graphs produce a markedly better level of compression (as low as 4.07 bits per link for the inlinks of the UK-2006 collection, which is similar to that reported by Boldi & Vigna in [4] for their smaller sample of `.uk`). On the unsorted collections, Interval Encoding has comparable but not as good compression as Gamma Encoding. Increasing the $Lmin$ parameter generally improves the compression of the Interval Encoding. Interestingly, similar to that reported by [3], inlinks compresses better than outlinks for most collections, except the domain specific .GOV and .GOV2. On comparing compression techniques across collections, we note that, for inlinks, the collections with the highest number of links per document (i.e. UK-2006 and CERC) exhibit the highest compression, while the older WT2G and WT10G are

---

[2] More information about obtaining the UK-2006 collection is found at `http://www.yr-bcn.es/webspam/datasets/`

Table 1. Comparative compression (bits per link) on Web IR collections. Sorted denotes when the document ids are ordered lexicographically by URL. NB: UK-2006 and CERC collections are initially numbered this way.

Links			Unsorted					Sorted			
	No	Gamma	Interval				Gamma	Interval			
$Lmin =$			2	3	4	5		2	3	4	5
WT2G (247,491 Docs, 1,166,146 Links)											
out	32	**8.95**	11.87	10.69	10.16	9.91	10.32	13.22	11.48	10.96	10.76
in	32	**8.75**	9.70	9.37	9.21	10.38	8.28	10.04	9.18	8.85	8.68
CERC (370,715 Docs, 4,577,312 Links)											
out	32	-	-	-	-	-	**11.67**	14.53	13.30	12.65	12.33
in	32	-	-	-	-	-	5.26	6.08	5.21	6.06	**4.99**
.GOV (1,247,753 Docs, 11,110,989 Links)											
out	32	**18.68**	19.97	19.19	18.97	18.91	7.87	9.76	8.96	8.65	8.45
in	32	**22.68**	22.85	22.89	22.89	22.89	11.78	15.67	13.32	12.61	12.34
WT10G (1,692,096 Docs, 8,063,026 Links)											
out	32	**14.28**	15.51	14.74	14.60	14.55	11.75	15.34	13.33	12.67	12.32
in	32	**13.86**	15.20	14.36	14.22	14.16	10.05	12.09	11.22	10.88	10.58
.GOV2 (25,205,179 Docs, 261,937,150 Links)											
out	32	**20.72**	21.51	20.09	20.83	20.81	21.16	23.51	22.04	21.64	21.47
in	32	**30.87**	31.09	30.90	30.94	30.93	35.14	35.38	35.24	35.22	35.22
UK-2006 (77,741,020 Docs, 2,951,370,103 Links)											
out	32	-	-	-	-	-	**9.36**	12.53	10.61	10.06	9.79
in	32	-	-	-	-	-	4.07	5.13	4.32	4.07	**3.92**

generally worse. It is also of note that in the UK-2006 and CERC collection, the document ids are sorted lexicographically by URL, as recommended in [3], and by increasing $Lmin$ to 5 on these collections, Interval Encoding can achieve higher compression than Gamma encoding. To assess the best compression achievable for the other unsorted Web test collections, we renumber the documents to match the lexicographical order of the URLs, then rebuild the link databases.

On analysing the compression between the unsorted and sorted collections, we note that sorting increases the compression achieved for the WT10G and .GOV collection, as well as for the inlinks of the WT2G collection. For the .GOV2 collection, the compression level decreases, and for the inlinks, both Gamma Encoding and Interval Encoding result is less effective compression than the No Encoding technique. We suggest this is due to a combination of low overall linkage combined with high document ids.

Figure 2(a) plots the build time of the three forms of link database compression across the 6 collections applied. Moreover, Figure 2(b) plots the mean time to perform one iteration of PageRank calculation (Note that due to computational reasons, the PageRank for UK-2006 collection was not computed.). From the figures, we can see that the No Encoding technique takes the longest to write and read, even though this is a simpler technique compared to the Gamma and Interval encoding techniques. We believe that this is due to the markedly

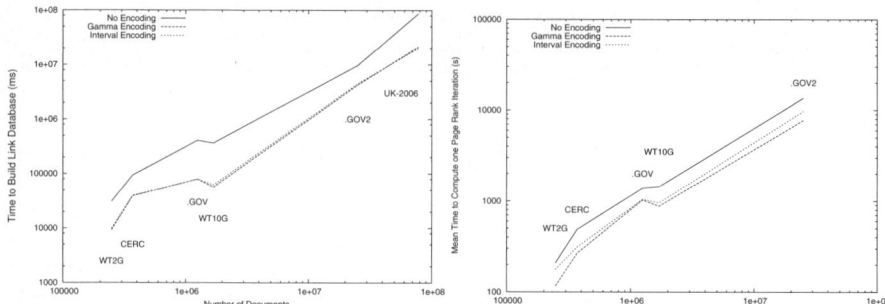

(a) Time to build the link database, for the various tested collections.

(b) Time to compute PageRank, for the various tested collections.

Fig. 2. Timing plots for writing and reading the various link databases (natural collection ordering)

higher number of disk operations required by this technique. Noticeably, the speeds of the Gamma and Interval encoding techniques are equivalent for both read and write operations, reflecting the very similar compression they achieve. Overall, the Gamma and Interval encodings are approximately 3-4 times faster to compress, and 1.5 times faster to decompress.

5   Conclusions

In this work, we thoroughly analysed three techniques for compressing the link graph of six different samples of the Web, of various size. We found that the simple integer (No Encoding) technique suffered from excessive Input/Output overheads. The Gamma Encoding gave best overall compression, and was among the fastest at reading and writing. The Interval Encoding technique does exhibit the high compression promised in [4], but requires an appropriate setting of the *Lmin* parameter. Overall, we conclude that the Gamma Encoding technique, similar to that already used by Terrier for direct and inverted file compression [6], should be deployed for link graph compression.

The large-scale analysis in this work is important as while more effective compression of the link graph may be obtained by the suitable ordering of the document ids in a collection, this ordering may not be compatible with other document id orderings applied by the underlying search engine - for example, some search engine number documents ids by ascending PageRank, or by natural crawl order (which approximates high quality pages first). Another interest of this study is that some conclusions - for instance URL ordering improving compression - do not necessarily generalise to all collections. Moreover, the speed increases shown by applying the compression techniques would benefit a large commercial search engine by allowing less machines to be involved in the computation of PageRank, resulting in data centre power and equipment savings.

References

1. Page, L., Brin, S., Motwani, R., Winograd, T.: The PageRank citation ranking: Bringing order to the Web. Technical report, Stanford Digital Library Technologies Project (1998)
2. Kleinberg, J.: Authoritative sources in a hyperlinked environment. J. ACM 46(5), 604–632 (1999)
3. Boldi, P., Vigna, S.: The WebGraph Framework I: compression techniques. In: Proceedings of WWW 2004, pp. 595–602 (2004)
4. Boldi, P., Vigna, S.: The WebGraph Framework II: Codes for the WWW. Technical Report 294-03, Universit di Milano (2003)
5. Elias, P.: Universal codeword sets and representations of the integers. IEEE Transactions on Information Theory 21(2), 194–203 (1975)
6. Ounis, I., Amati, G., Plachouras, V., He, B., Macdonald, C., Lioma, C.: Terrier: A high performance and scalable IR platform. In: Proceedings of SIGIR OSIR Workshop, pp. 18–25 (2006)

An Evaluation and Analysis of Incorporating Term Dependency for Ad-Hoc Retrieval*

Hao Lang[1], Bin Wang[1], Gareth Jones[2], Jintao Li[1], and Yang Xu[1]

[1] Institute of Computing Technology, Chinese Academy of Science, China
[2] School of Computing, Dublin City University, Ireland

Abstract. Although many retrieval models incorporating term dependency have been developed, it is still unclear whether term dependency information can consistently enhance retrieval performance for *different* queries. We present a novel model that captures the main components of a topic and the relationship between those components and the power of term dependency to improve retrieval performance. Experimental results demonstrate that the power of term dependency strongly depends on the relationship between these components. Without relevance information, the model is still useful by predicting the components based on global statistical information. We show the applicability of the model for adaptively incorporating term dependency for individual queries.

1 Introduction

In most existing retrieval models, documents are scored primarily using occurrences of single query terms in documents, assuming query terms are independent. However, previous studies have shown that incorporating the dependency of query terms in documents into retrieval strategies can improve average retrieval effectiveness on a fixed set of queries [1][2][3]. Moreover, existing retrieval models incorporating term dependency are far from optimal. One problem of current models is that most proposed methods are uniformly applied to all the queries. In fact, we find that not all the queries benefit from taking account of term dependency. Our experimental results in section 3 show that term dependency models fail to improve retrieval performance for around 50% of queries for adhoc title-only topics in the TREC Terabyte Tracks.

Until now, little investigation has been reported on how term dependency information can enhance retrieval performance on an individual query basis. In this paper, we investigate the main features affecting the power of term dependency to enhance retrieval performance. We suggest a novel model which captures the main components of a topic and the relationship between those components and the power of term dependency. We argue, and then show experimentally, that the power of term dependency depends on the relationship between these components. In practice, we do not have the relevance information, thus we cannot compute the components of the model directly. We show that in such cases the proposed model is still useful by

---

* This research is supported by the National Science Foundation of China (grand No. 60603094) and the National Basic Research Program of China (grant No. 2004CB318109).

C. Macdonald et al. (Eds.): ECIR 2008, LNCS 4956, pp. 602–606, 2008.

predicting the components based on global statistical information of the collection. Finally, we show that we can adaptively use the term dependency information on an individual query basis by making use of the proposed model.

2 A Model for the Power of Term Dependency

In this section, we propose a model for predicting the power of term dependency. In this work, the power of term dependency refers to the extent to which retrieval models incorporating term dependency can successfully improve retrieval performance for a given query compared to models based on the standard bag-of-words assumption.

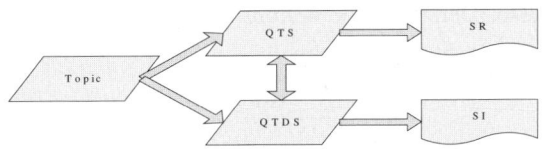

Fig. 1. A general model of a topic based on the *QTS and QTDS*

The effectiveness of an IR model depends on its ability to distinguish relevant documents for a given query from irrelevant ones. In most existing IR models, the main features used to identify relevant documents are various kinds of term statistics such as within-document frequencies, inverse document frequencies, and document lengths. Obviously, if relevant documents have more occurrences of query terms than irrelevant ones, the query tends to achieve better results because of the high quality of the query's term statistics. In this work, the quality of term statistics (*QTS*) of a query refers to the property of the query that determines to what extent relevant documents can be identified from irrelevant ones based on the term statistics of the query.

When IR models are extended to incorporate term dependency, the quality of term dependency statistics (*QTDS*) such as the occurrences of ordered phrases and unordered phrases becomes interesting. In this work, *QTDS* refers to the property of the query that determines to what extent the relevant documents can be identified based on the term dependency statistics of the query.

Therefore, we define the primary object of the model based on the power of term dependency to be a *Topic*. A topic relates to a defined subject. The topic is comprised of two objects: *QTS* and *QTDS*. The topic is also dependent on the set of relevant documents (*SR*) and the set of irrelevant documents (SI), where *QTS* and *QTDS* are computed based on the gap in either term statistics or term dependency statistics from the relevant to irrelevant set. Thus, we denote a topic as:

$$Topic = (QTS, QTDS \mid SR, SI) \tag{1}$$

Figure 1 shows a schema of the model. The two components *QTS* and *QTDS* have high correlation with the retrieval effectiveness of a given query. When *QTS* or *QTDS* is high, term statistics or term dependency statistics tend to be good features to identify relevant documents. *QTDS* is supposed to be positively correlated with the power of term dependency, because IR models tend to benefit from term dependency when *QTDS* is high. Contrarily, *QTS* is supposed to be negatively correlated with the power. When *QTS* is high meaning that features of term statistics are good identifiers for relevant documents. Thus retrieval effectiveness based on term statistics tends to be high, which makes it harder for term dependency to improve the effectiveness. From the other prospective, when *QTS* is high and *QTDS* is low, term statistics tend to be better features than term dependency statistics. Hence in this situation, IR models incorporating term dependency cannot achieve better results than IR models only using good term statistics features.

Now we describe our approach of computing *QTS* and *QTDS*. For *QTS*, we compute the average term frequency of query terms (*TF*) for each document in *SR* and *SI*. *QTS* is the division between the median *TF* of relevant documents and irrelevant documents. For *QTDS,* we compute the average occurrences of ordered phrases (*OF*) in a document instead. Details of the definition of ordered phrases are given [1]. *TF* and *OF* are defined as:

$$TF = \frac{\sum_{w \in Q} tf_{w,D}}{|Q|} \tag{2}$$

$$OF = \frac{\sum_{c \in O} tf_{c,D}}{|O|} \tag{3}$$

where Q is the query, w is the term of the query, $tf_{w,D}$ is the term frequency in a document, O is the set of ordered phrases, c is a kind of ordered phrase, $tf_{c,D}$ is the occurrence frequency of the ordered phrase.

In this work, *SR* refers to the set of documents judged relevant and *SI* refers to the set of documents judged irrelevant. Of course, the judged relevant and judged non-relevant documents are heavily biased because of the pooling procedure used at TREC. However, these statistics still provide valuable information.

In practice when entering a new search query, we do not have relevance information, thus we cannot compute *QTS* and *QTDS* directly. We can though predict the components based on global statistical information of the whole collection. For *QTS*, we compute the average inverse document frequency of the terms in the query (*Avg_IDF*) to predict *QTS*. When a query term has a high IDF value, meaning that the term only appears in a small fraction of documents in the collection, irrelevant documents do not have high chance to have the occurrence of the term. Thus high *Avg_IDF* indicates good quality of term statistics. For *QTDS*, we count occurrences of ordered phrases in the data collection (*OO*) and then we compute the average inverse *OO* (*Avg_IOO*) to predict *QTDS*. Occurrences of ordered phrases are strong evidence of relevance. High occurrences of ordered phrases mean that many relevant documents have ordered phrases. Thus high *OO* indicates high *QTDS*. As a result, *Avg_IOO* is supposed to be negatively correlated to *QTDS*.

3   Validating the Model

In this section, we validate our model by showing the correlation between the components of the model and the power of term dependency. We use the TREC .GOV2

Terabyte test collection, and its associated TREC 2004, 2005 and 2006 adhoc title-only topics and relevance assessment sets. 3 out of the 150 topics were removed since there are no relevant documents in the collection or the topic only has one query term. Thus 147 topics in total were evaluated. For indexing and retrieval we use Indri[1], with Porter's stemming and stop words removal.

Retrieval was performed twice using the full independence (FI) and full dependence (FD) variants of MRF model [1] respectively. FI only uses the term statistics while FD makes use of the term dependency information. The Mean Average Precision (MAP) of FI is 0.2971 while MAP of FD is 0.3298 for the 147 queries. This indicates that incorporating term dependency can improve the average retrieval performance. However, only 87 of 147 queries actually perform better by incorporating term dependency, meaning that the dependence model fails for 41% of them.

The power of term dependency is computed by the division between the Average Precision (AP) of FD and the AP of FI for a given query. We measure the correlation between the components of our proposed model and the power of term dependency by the Spearman rank correlation test, since the power distributions are unknown. The results for correlation are shown in table 1, where bold cases indicate that the results are statistically significant at the 0.05 level.

From these results, we firstly observe that the components of the model QTS and QTDS significantly correlate with the power of term dependency. Queries of high QTDS tend to benefit from term dependency, while for queries with high QTS it becomes harder to achieve better results. Secondly, the combination of the model's two components results in higher correlation, suggesting that the two components measure different properties of a topic. Finally, it can be observed that Avg_IDF and Avg_IOO still work for the model in the absence of relevance information.

Table 1. Spearman correlation coefficients between the components of our model and the power of term dependency. Bold cases indicates that the results are statistically significant at the 0.05 level.

	QTS	QTDS	Combine	Avg_IDF	Avg_IOO
Spearman's	**-0.40**	**0.44**	**-0.528**	-0.18	-0.20

4   Uses of the Model

As shown in the above section, models incorporating term dependency can improve the average retrieval performance on a fixed set of queries. However, the dependency model fails to achieve better results for around 50% of queries, where much more computation resources are required for processing the term dependency information.

Thus, it is not beneficial to use term dependency for every query. Instead, it is advantageous to have a switch that will estimate when term dependency will improve retrieval, and when it would be detrimental to it. In the absence of relevance information, we can use the Avg_IDF and Avg_IOO of our proposed model to predict

---

[1] URL: http://www.lemurproject.org/indri/

Table 2. Improvements in retrieval based on our proposed model

	MAP	GMAP
FI	0.2971	0.2006
FD	0.3298	0.2527
Sel	0.3308	0.2531

whether dependency model can work for a given query. Since both of the two features are statistically negatively correlated to the power of term dependency, queries of high *Avg_IDF* and *Avg_IOO* scores tend not to benefit from term dependency. Thus we try to identify those queries for which the dependency models fail by finding queries of high *Avg_IDF* and *Avg_IOO* scores. For these identified queries, we just use the FI model, while for the other queries we use FD model instead. We name the retrieval results "Sel" in table 2 by adaptively using term dependency on a query basis.

In this work, we label queries of high *Avg_IDF* scores, when the scores are ranked in the top 20% of all the 147 queries. We label queries of high *Avg_IOO* scores in the same way. The overlap of labeled queries of high *Avg_IDF* scores and labeled queries of high *Avg_IOO* scores are identified queries for which dependency model is estimated to fail. The low threshold 20% is chosen, because we want to find those queries when term dependency would be detrimental to retrieval.

In total, 11 out of 147 queries were identified by our proposed model. 10 of the 11 identified queries indeed do not benefit from term dependency, which indicates great prediction power of our model. The retrieval results are shown in table 2. Sel has the best retrieval effectiveness among the three models under the performance measures of MAP and Geometric MAP (GAMP).

5 Summary

This work tries to answer the question of what kind of queries can benefit from term dependency information. We describe a novel model that captures the main components of a topic and the relationship between the components to the power of term dependency. We demonstrate that the power of term dependency strongly depends on those components. Without relevance information, we can predict the components by global statistics information of the index. Finally, we demonstrate the applicability of model to adaptively using the term dependency on a query basis.

References

[1] Metzler, D., Croft, W.B.: A Markov Random Field Model for Term Dependencies. In: Proc. of SIGIR 2005, Brazil (2005)
[2] Peng, J., He, C.M.B., P., V., O., I.: Incorporating Term Dependency in the DRF Framework. In: Proc. of SIGIR 2007, The Netherlands (2007)
[3] Tao, T., Zhai, C.: An Exploration of Proximity Measures in Information Retrieval. In: Proc. of SIGIR 2007, The Netherlands (2007)

An Evaluation Measure
for Distributed Information Retrieval Systems

Hans Friedrich Witschel, Florian Holz, Gregor Heinrich, and Sven Teresniak

University of Leipzig, Germany
{witschel|holz|heinrich|teresniak}@informatik.uni-leipzig.de

Abstract. This paper is concerned with the evaluation of distributed and peer-to-peer information retrieval systems. A new measure is introduced that compares results of a distributed retrieval system to those of a centralised system, fully exploiting the ranking of the latter as an indicator of gradual relevance. Problems with existing evaluation approaches are verified experimentally.

1 Introduction

One of the core requirements when creating an evaluation testbed for either distributed information retrieval (DIR) or peer-to-peer information retrieval (P2PIR) is a realistic distribution of documents onto databases or peers. A common method in DIR is to use TREC ad hoc test collections and distribute them according to source and date (e.g. [3]), with the advantage that human relevance judgments are available. For evaluation of P2PIR, with typically small and semantically more homogeneous collections, this approach is unrealistic.

In P2PIR, distribution of documents is either done in a way that springs naturally from the collection, e.g. via author information [2], built-in categories [1] or domains of web pages [4], or it is established in less natural ways via clustering [6] or even randomly [5]. Since generally these collections lack queries and relevance judgments, queries are either constructed from the documents [4,2,1] or taken from query logs matching the collection [6,8].

Although with both methods a large number of queries can be created, the need remains to assess query results for relevance. This challenge is approached in this article.

2 Related Work

Various approximations of relevance have been studied in P2PIR: assuming documents containing all query keywords to be relevant [2], using "approximate descriptions of relevant material" [1] or comparing results of distributed algorithms to results of a centralised system [6,4,9,8].

The last approach assumes that a distributed system will rarely be more effective than a centralised one. Although some studies (e.g. [7]) show that the

C. Macdonald et al. (Eds.): ECIR 2008, LNCS 4956, pp. 607–611, 2008.

contrary cannot be ruled out, most other studies agree with this (e.g. [4]). In the following, we will therefore concentrate on the last approach.

It is realised by either considering all documents returned by the centralised system (i.e. those with score > 0) relevant [6] – resulting in what is sometimes called *relative recall* (RR) – or just the N most highly ranked documents [4,9,8]. In the latter case, precision at k documents is used as an evaluation measure – we will call it $P_N@k$ in the rest of this work, denoting its dependence on N.

3   Average Ranked Relative Recall

Considering all documents with score > 0 relevant is clearly not what we want to approximate. Assuming the top N documents to be relevant is simple, but also not sufficient since we do not know how to choose N as the number of relevant documents generally depends on the query.

However, the choice of N may influence the evaluation results: consider for example a scenario where the centralised system returns a ranked list $(d_1, d_2, ..., d_{15})$ and two distributed systems A and B, where A returns $(d_1, d_2, d_3, d_4, d_{15})$ and B returns $(d_6, d_7, d_8, d_9, d_{10})$.

This results in a $P_5@5$ of 0 for system B and of 0.8 for system A, i.e. the evaluation predicts system A to perform better than system B. However, $P_{10}@5$ is also 0.8 for A, but 1.0 for B, thus reversing our evaluation result.

In [8], N is chosen equal to k, in [4,9], values of 50 and 100 are used without further justification. Besides the problem of choosing N, this set-based approach also neglects the ranking of the centralised system within the first N documents.

Therefore, we propose *average ranked relative recall* (ARRR), a new evaluation measure that exploits the ranking of the centralised system as an indicator of gradual relevance and does not treat all of its returned documents (or the top N) as equally relevant.

Let $C = (c_1, ..., c_m) \in T^m$ be the ranking of the centralised system, where T is the set of all documents. We assume that the user has specified how many of the top-ranked documents should be retrieved; we call this value k. It plays a similar role as the k in precision at k documents (P@k) commonly used in the IR literature. Further, let $D = (d_1, ..., d_n) \in T^n$ be the ranking returned by the distributed system with $n \leq k$ (we have $n < k$ only if the distributed system retrieves less than k documents in total).

Next, we introduce a function m_D that, for a pair of documents, returns 1 if the first document is ranked ahead of the second within the set D, else 0:

$$m_D : T^2 \rightarrow \{0, 1\}$$

$$m_D(c_j, c_i) = \begin{cases} 1 \text{ if } \exists d_q \in D : d_q = c_j \ \land \ \exists d_p \in D : d_p = c_i \ \land \ q \leq p \\ 0 \text{ else} \end{cases}$$

With this new function, we define

$$\text{ARRR@}k(D, C) = \frac{1}{\min(k, m)} \sum_{i=1}^{m} m_D(c_i, d_n) \frac{\sum_{j=1}^{i} m_D(c_j, c_i)}{i} \tag{1}$$

This measure can be determined by the following algorithm: (1) For each document d_i in ranking D, starting from the top: (a) Mark d_i within ranking C if present, and (b) determine the portion of documents marked so far between the top of C and position j of d_i. (2) Sum up the values obtained for all d_i and divide by k or by m if the centralised system finds less than k documents.

Step 1b corresponds to the recall for the distributed system, considering relevant the first j documents in C. If j is large, i.e. d_i is ranked low in C, the "notion" of relevance becomes looser and it is less likely to achieve good recall. Obviously, ARRR@k becomes 1 iff $D = (c_1, ..., c_k)$, i.e. if the distributed system retrieves exactly the k highest-ranked documents found by the centralised system and ranks them in the same way the centralised system does. On the other hand, ARRR@k becomes small if the distributed system ranks documents highly within its first k documents that have low ranks in C.

As an example, we consider $C = (c_1, c_2, c_3, c_4, c_5, c_6)$. Now let us assume that system A returns $D_A = (c_1, c_3, c_4)$ and system B returns $D_B = (c_3, c_4, c_1)$. This yields ARRR@5 $= \frac{1}{5}(1 + \frac{2}{3} + \frac{3}{4}) = 0.48$ for system A and ARRR@5 $= \frac{1}{5}(\frac{1}{3} + \frac{2}{4} + 1) = 0.37$, penalising B for its "bad" ranking. If $D_B = (c_3, c_4, c_1, c_5)$, we get ARRR@5 $= \frac{1}{5}(\frac{1}{3} + \frac{2}{4} + 1 + \frac{4}{5}) = 0.53$, showing that higher recall can compensate for suboptimal ranking.

4   Experimental Results

Experiments were performed with two IR test collections that provide human relevance judgments:

- Ohsumed: 348,566 medical abstracts, annotated with an average of 10.6 so-called MeSH (Medical Subject Headings) terms each. Each of the 14,596 MeSH terms in the collection was treated as a peer and every abstract was assigned to all peers corresponding to its MeSH terms.
- GIRT: 151,318 German abstracts from the social sciences, annotated with an average of 10.2 controlled terms each which were identified with peers as above, resulting in a total of 7,151 peers.

We will illustrate the flaws of the existing evaluation measures with the following example scenario: for each collection, we consider two peer selection strategies, (1) a variant of the CORI resource selection algorithm [3] and (2) a strategy we call "by-size" that ranks peers by the number of documents that they possess.

Both strategies are applied to ranking peers for the queries. The first 100 peers from the ranking are then visited according to the ranking and after each peer is visited, the quality of the results that have so far been retrieved is assessed using all of the evaluation measures discussed above. All peers use the BM25 retrieval function to rank documents locally; idf values are sampled globally (cf. [10]) so that document scores are comparable across all peers.

Fig. 1 shows the effectiveness of the two strategies and different measures as a function of the number of peers visited using Ohsumed. The curves using GIRT qualitatively resemble Fig. 1 but are not shown to preserve space.

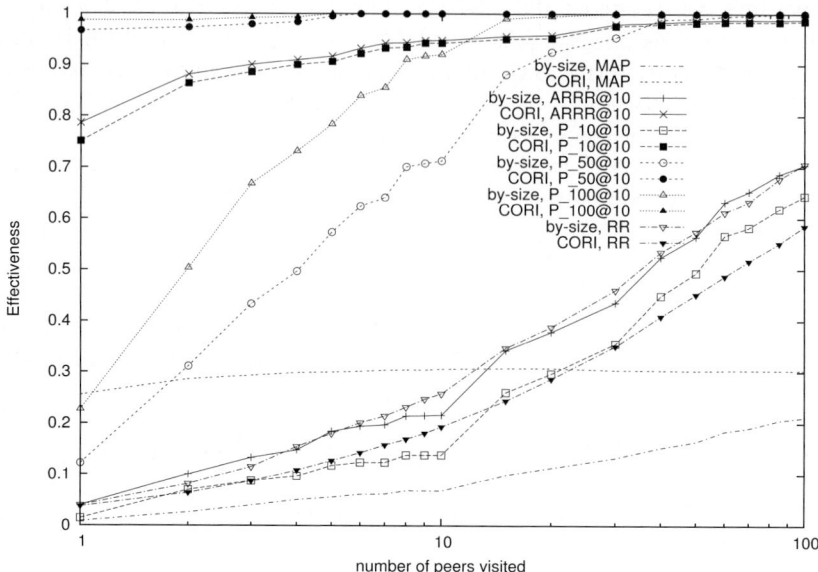

Fig. 1. Effectiveness of "by-size" and CORI as a function number of peers visited in terms of MAP, $P_{10}@10$, $P_{50}@10$, $P_{100}@10$, ARRR@10 and RR for Ohsumed. Note: absolute values of measures may not be compared directly, we need to concentrate on the general shape of the curves.

As one would expect, the CORI strategy is clearly superior to the trivial "by-size" approach when analysed with MAP, ARRR@10 and $P_{10}@10$. However, the values of $P_{50}@10$ and $P_{100}@10$ for the "by-size" strategy catch up with those of CORI rather quickly and RR shows even higher values for by-size than for CORI from the first few peers on.

The results for $P_{50}@10$, $P_{100}@10$ and RR allow the conclusion that "by-size" is competitive with CORI after visiting a relatively small number of peers. However, such a conclusion apparently cannot be drawn from the other measures, especially MAP, which is based on human relevance judgments.

The measure $P_{10}@10$ behaves very similarly to ARRR@10, suggesting that both might be equally trustworthy. It also suggests that the set of the first 10 highest-ranked documents is a better approximation of the set of relevant documents than the first 50 or 100 documents or even the set of all documents with score > 0: since the probability of retrieving some of the N highest-ranked documents increases with N, we will overrate the effectiveness of a strategy as "by-size" (which retrieves *many* documents) at some point. In the case of RR, there is a very high probability of arbitrary documents being considered "relevant"; for Ohsumed, this probability is around 17.5% on average, which, of course, does not mean that the documents are really relevant for the user. Despite the good behaviour of $P_{10}@10$, it is still (at least) theoretically unpleasing that we do not know which choice of N is optimal.

In further experiments not shown here, we also detected that – when ranking retrieval runs using $P_N@k$ – the rank correlation of two run rankings with different values of N is generally high, but often below 1. This indicates that the problem described above does indeed arise in practice: different choices of N may result in different rankings of systems.

5   Conclusions

In this work, we have introduced a new measure – average ranked relative recall (ARRR) – for comparing the retrieval results of a distributed system against a centralised system. As opposed to previous work, this measure fully exploits the ranking of the centralised system as an indicator of gradual relevance. Experimental results confirm problems of existing approaches in ranking systems consistently (something which ARRR avoids by design) and show that – depending on the result set size N – the measures $P_N@k$ and relative recall may lead to wrong conclusions.

References

1. Akavipat, R., Wu, L.-S., Menczer, F., Maguitman, A.G.: Emerging semantic communities in peer web search. In: P2PIR 2006. Proceedings of the international workshop on Information retrieval in peer-to-peer networks, pp. 1–8 (2006)
2. Bawa, M., Manku, G.S., Raghavan, P.: SETS: search enhanced by topic segmentation. In: Proc. of SIGIR 2003, pp. 306–313 (2003)
3. Callan, J.P., Lu, Z., Croft, W.B.: Searching distributed collections with inference networks. In: Proc. of SIGIR 1995, pp. 21–28 (1995)
4. Lu, J., Callan, J.: Content-based retrieval in hybrid peer-to-peer networks. In: CIKM 2003. Proceedings of the twelfth international conference on Information and knowledge management, pp. 199–206 (2003)
5. Michel, S., Bender, M., Ntarmos, N., Triantafillou, P., Weikum, G., Zimmer, C.: Discovering and exploiting keyword and attribute-value co-occurrences to improve P2P routing indices. In: CIKM 2006. Proceedings of the 15th ACM international conference on Information and knowledge management, pp. 172–181 (2006)
6. Neumann, T., Bender, M., Michel, S., Weikum, G.: A reproducible benchmark for P2P retrieval. In: Proc. of First Int. Workshop on Performance and Evaluation of Data Management Systems, ExpDB (2006)
7. Powell, A.L., French, J.C., Callan, J., Connell, M., Viles, C.L.: The impact of database selection on distributed searching. In: Proc. of SIGIR 2000, pp. 232–239 (2000)
8. Puppin, D., Silvestri, F., Laforenza, D.: Query-driven document partitioning and collection selection. In: InfoScale 2006. Proceedings of the 1st international conference on Scalable information systems, pp. 34–41 (2006)
9. Shokouhi, M., Baillie, M., Azzopardi, L.: Updating collection representations for federated search. In: Proc. of SIGIR 2007, pp. 511–518 (2007)
10. Witschel, H.F.: Global term weights in distributed environments. Information Processing and Management (2007), DOI: doi:10.1016/j.ipm.2007.09.003

Optimizing Language Models for Polarity Classification

Michael Wiegand and Dietrich Klakow

Spoken Language Systems, Saarland University, Germany
{Michael.Wiegand|Dietrich.Klakow}@lsv.uni-saarland.de

Abstract. This paper investigates the usage of various types of language models on polarity text classification – a subtask in opinion mining which deals with distinguishing between positive and negative opinions in natural language. We focus on the intrinsic benefit of different types of language models. This means that we try to find the optimal settings of a language model by examining different types of normalization, their interaction with smoothing and the benefit of class-based modeling.

1 Introduction

There has been an increasing interest in opinion mining in recent years, in particular, in polarity text classification. Though Bayesian methods have been widely explored in this context, for example in [1], less attention has been drawn to the impact of language modeling on this classification task. This paper discusses various aspects of language modeling, such as normalization, its interaction with smoothing and class-based modeling.

2 Bayesian Classification and Language Modeling

The Bayesian classifier estimates the optimal class $\hat{c}$ given a sequence of words $w_1 \ldots w_n$ where n is the length of the observation (i.e. a document to be classified) by the *prior* $P(c)$ and the *likelihood* $P(w_1 \cdots w_n | c)$:

$$\hat{c} = \arg\max_i P(w_1 \ldots w_n | c_i) \cdot P(c_i) \tag{1}$$

We model the likelihood as a *Markov Chain*:

$$P(w_1 \ldots w_n | c) = \prod_{i=1}^{n} P(w_i | w_{i-m+1} \ldots w_{i-1}, c) \tag{2}$$

Each word w_i is considered with respect to some short history of preceding content $w_{i-m+1} \ldots w_{i-1}$ where m is the size of the sequence to be modeled. We refer to this as an m-gram. The likelihood is estimated by different language models. Smoothing is essential in Bayesian classification since, otherwise, any unseen event would turn Equation 2 to zero. We experimented with the most common smoothing techniques in IR, as presented in [2], and discovered that *absolute discounting* works best in polarity classification which is why we used it in all subsequent experiments.

C. Macdonald et al. (Eds.): ECIR 2008, LNCS 4956, pp. 612–616, 2008.

3   Methods

3.1   Normalization

In text classification, one usually applies some form of normalization in order to reduce the data sparseness. Typically, one resorts to *stemming* – which is a simple algorithmic approach, where suffixes are removed from words – or to *lemmatization* – in which a base form is looked up in a dictionary. Due to its complexity, lemmatization is less preferred, though it is by far more linguistically accurate. In our experiments, we used Porter stemming and lemmatization as done in WordNet[1].

In order to remove noise, we also examined the effect of omitting *singletons* and *stopwords* being a small set of function words. Moreover, we restricted the vocabulary to the polarity expressions in *General Inquirer (GI)*[2] which is a list of polar adjectives, such as *brilliant* or *horrible*, and verbs, such as *adore* or *hate*, comprising 3440 different words in total. We assume that these polar words are highly discriminative for polar text classification.

In order to investigate the usefulness of negation handling in polar text classification, we examined the impact of two *negation* models which differ in complexity:

1. Each subsequent token of a negation marker, e.g. *not, didn't* or *cannot*, is marked with prefix *NOT_* until the first occurrence of a punctuation mark. This method has been proposed in [3].
2. With the help of regular expressions we disambiguate (potential) negation markers[3]. For this task we have written a small set of regular expressions. Unlike [3] only the negated word is marked with *NOT_*. We identify those words by part-of-speech information. Not only is this linguistically more accurate, but it should also cause less data-sparsity[4].

3.2   Class-Based Language Models

Unlike [1] who attempt to create more generalizing models by manually constructed rules replacing specific words with their respective part-of-speech tags, we try to generalize our training data by applying class-based language models. A mapping $k : V \rightarrow K$ is learned where V is the vocabulary and K is the set of unlabeled classes whose number has to be specified in advance. Unlike [1], this approach is completely *unsupervised* and does not require any form of expensive pre-processing, such as part-of-speech tagging. The objective function of the

---

[1] http://wordnet.princeton.edu/
[2] http://www.wjh.harvard.edu/~inquirer
[3] We observed that, frequently, negation markers do not express negations in certain contexts, e.g. **not** *just ... but ...* or *why* **not** *...*
[4] Consider that each time a prefix *NOT* is added to a word, a new word is created which is different to the unnegated expression.

class-induction is the maximization of the *Likelihood*. The class-based language model is defined by:

$$P(w_i|w_{i-(m+1)}\dots w_{i-1}) = P(w_i|k(w_i)) \cdot P(k(w_i)|k(w_{i-(m+1)})\dots k(w_{i-1})) \quad (3)$$

The first factor is called *emission probability* and the second is called *transition probability*. We use the $O(V \cdot K^2)$ algorithm as presented in [4] to learn the mapping from words to classes.

4   The Data

All our experiments were performed on the *movie review data* set [3]. We chose this dataset since it is commonly regarded as the benchmark dataset for polarity text classification. The dataset comprises 1000 positive and 1000 negative reviews. The classes to be predicted are *positive* and *negative* reviews. We randomly partitioned the dataset into a training set containing 936 documents, a development set for optimizing the language models and a test set both comprising 468 documents[5].

5   Results of the Experiments

We evaluated our experiments on the basis of *accuracy*. Every model has been optimally smoothed on a separate development set. We performed *four-fold cross-validation* meaning that the we generated four different partitions of the dataset as described in Section 4 in order to obtain representative numbers.

5.1   Results of Normalization Experiments

Table 1 displays the performance of the different types of normalizations. On the test set, only lemmatization and porter stemming perform marginally better than the plain unigram model. The remaining models, including both negation models, are worse than the plain unigram model though mostly only marginally. It is also striking that the two worst performing models, the GI model and the singleton model, are exactly those models from which the greatest amount of data has been removed. Apparently, it is fairly impossible to remove noise from the dataset we are using without also omitting meaningful information.

5.2   Interaction Between Smoothing and Normalization

Figure 1 illustrates the interaction between optimizing the smoothing parameter δ and the performance of some normalized models. If a suboptimal smoothing parameter has been chosen, e.g. $\delta = 0.1$, the GI model, which is the second

---

[5] These numbers have been chosen to be consistent with [3]. They use 1400 documents per class, using two thirds for training and one third for testing.

Table 1. Performance of different types of normalizations on unigram models

Model	Test	Devel
Plain	80.4	80.9
Porter Stemming	80.6	80.0
Lemmatization	**81.1**	81.0
Singletons Removed	64.6	64.0
Stopwords Removed	80.2	81.0
GI-Lexicon	78.7	80.6
Negation I	79.6	80.2
Negation II	79.7	81.0

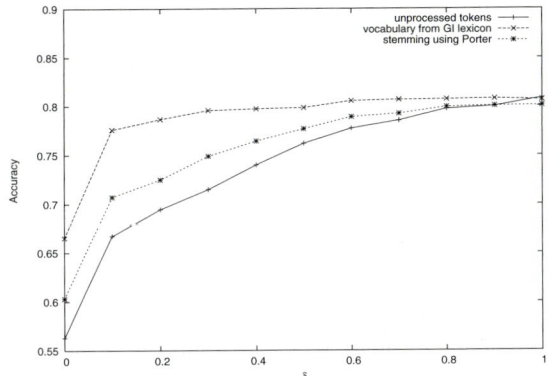

Fig. 1. Iterative optimization of smoothing parameters on unigram models on development set

worst performing model when optimized models are compared (see previous section), significantly outperforms the standard unigram model. Further iterating the smoothing parameter hardly improves the GI model but immensely improves the plain unigram model. The optimally smoothed plain model ($\delta = 0.95$) is, however, not only on a par with the optimal GI model on the development set but also better than the GI model on the test set. We observed a similar but less striking behavior between the model using stemming and the plain model. We conclude from these observations that expensive pre-processing, such as filtering the vocabulary with a manually built task-specific lexicon, does not offer a better performance than a properly smoothed plain unigram model, which is also far cheaper to obtain.

5.3   Results of Class-Based Language Models

In our experiments we found that trigram models work best as transition probabilities. We tested models with 500, 600 and 700 classes. Table 2 displays the results of the different class-based language models we built. For comparison,

Table 2. Performance of class-based language models

Model	Test	Devel
Unigram	80.4	80.9
Bigram	80.9	81.7
Trigram	81.3	81.5
500 Classes	81.7	82.2
600 Classes	**82.4**	81.7
700 Classes	82.3	82.8

we also included the performance of different word-based m-gram models. No normalization was done on any of the models. On the test set, all class-based models performed better than the word-based models, though the improvement is only limited. The class-based model trained on 600 classes with an accuracy of 82.4% is the best model we could generate in our experiments. Considering the error-bars on the results of our experiments at approximately $\pm 1.0\%$ this performance is comparable with SVMs at 82.9 which is the best performance reported in [3].

6   Conclusion

In this paper, we have stated our results on polar text classification using various types of language models. Properly smoothed plain unigram models offer similar performance to normalized models. Pre-processing is only beneficial if one uses insufficiently smoothed models. Removing noise often harms the performance. Class-based language models produce the best Bayesian classifier, though the gap to word-based higher-order m-gram models is small. Since our results are comparable to discriminative methods, such as optimized SVMs, we presume that the inherent noise in the data set does not allow much more room for improvement for Bayesian Classification and presumably any other standard machine learning algorithm.

References

1. Salvetti, F., Reichenbach, C., Lewis, S.: Impact of Lexical Filtering on Overall Opinion Polarity Identification. In: Proc. of the AAAI Symposium on Exploring Attitude and Affect in Text: Theories & Applications, Dordrecht, Netherlands (2004)
2. Zhai, C., Lafferty, J.: A Study of Smoothing Methods for Language Models Applied to Ad Hoc Information Retrieval. In: Proc. of SIGIR, New Orleans, USA (2001)
3. Pang, B., Lee, L., Vaithyanathan, S.: Thumbs up? Sentiment Classification Using Machine Learning Techniques. In: Proc. of EMNLP, Philadelphia, USA (2002)
4. Brown, P., Pietra, V.D., de Souza, P., Lai, J., Mercer, R.: Class-Based n-gram Models of Natural Language. Computational Linguistics 18(4), 467–479 (1992)

Improving Web Image Retrieval Using Image Annotations and Inference Network

Peng Huang, Jiajun Bu*, Chun Chen, and Guang Qiu

College of Computer Science, Zhejiang University, Hangzhou, China
{hangp,bjj,chenc,qiuguang}@zju.edu.cn

Abstract. Currently text-based retrieval approaches, which utilize web textual information to index and access images, are still widely used by many modern prevalent search engines due to the nature of simplicity and effectiveness. However, page documents often include texts irrelevant to image contents, becoming an obstacle for high-quality image retrieval. In this paper we propose a novel model to improve traditional text-based image retrieval by integrating weighted image annotation keywords and web texts seamlessly. Different from traditional text-based image retrieval models, the proposed model retrieves and ranks images depending on not only texts of web document but also image annotations. To verify the proposed model, some term-based queries are performed on three models, and results have shown that our model performs best.

1 Introduction

The boom in the use and exponential growth of the Web have triggered the need for efficient tools to manage, retrieve, and filter information from the WWW, such as web images in this study. Currently prevalent approaches to image retrieval fall into two main categories: content-based and text-based image retrieval. Despite numerous works and research on content-based retrieval, text-based approach is still important and prevailing because of its simplicity and effectiveness. In text-based retrieval, images are indexed and searched in the form of textual descriptions. These descriptions can be produced by human manually or by machines automatically. Due to the huge size of the Web, labeling images manually is a clearly infeasible task. Thus most of modern search engines extract terms from web documents as descriptions of images. However, determining the parts of web documents related to image contents is not a trivial task. Coelho et al. [2] attempted to bypass this issue by combining multiple evidential sources for image retrieval without consideration of image contents.

We propose a novel image retrieval model based on inference network [1] which introduces image annotations produced according to image contents into text-based image retrieval, hoping that can improve solely text-based image retrieval. Different from the models directly utilizing image annotations to index and search images, the proposed model first associates each annotation keyword with certain weight to measure its semantic similarity to image contents, then integrates these weighted annotations as well as other textual evidences into the underlying inference network seamlessly to improve

---

* Corresponding author.

C. Macdonald et al. (Eds.): ECIR 2008, LNCS 4956, pp. 617–621, 2008.

image retrieval. The remainder of the paper is organized as follows. Section 2 details the proposed model. Section 3 describes experimental results and some discussions. Last section concludes the paper with some ideas of future work.

2   The Architecture of Image Retrieval Model

2.1   Evidences Extracted from Web Documents

In general, web documents contain various textual data. Some may be useful for image retrieval, while others may be not. However, it is very difficult to locate specific texts relevant to image because of the nature of unstructured data and diversities of web documents. Coelho et al. [2] attempted to bypass this problem by utilizing belief network to combine multiple evidence sources, including Description tags, Meta tags and text passages, for image retrieval. Description tags include filenames, ALT attribute of IMG tag and anchors; Text passages are extracted from image surrounding passages; Meta tags include the terms located between TITLE tag and META tag. In this study we extract description tags and 40 surrounding words, as recommended by Coelho et al. [2]. Note that we ignore Meta tags because Coelho found that combining description tags and surrounding passages would gain higher precision than including Meta tags in the task of web image retrieval. In a word, in this study description tags and image surrounding passages (40 terms) are used as textual evidences for image retrieval.

2.2   Weighting Image Annotations

We use the Hidden Markov Image Annotation Model (HMIAM) [4] to produce original annotations for images. Since HMIAM is not our focus, we omit the details, and for details please refer to the work of Ghoshal [4]. Without loss of generality, let $C = (c_1, \ldots, c_T)$ be the annotations produced by HMIAM. The annotation set C is usually not perfect and includes some 'noisy' keywords irrelevant to image semantics. To differentiate these 'noisy' keywords from others we use the notions of *coherence* and *relatedness*. The notion of *coherence* assumes that annotation keywords of an image should be semantic similar one another, and it is adopted by Jin et al. [6] to remove noisy keywords from image annotations. Moreover, web documents often include some useful information related to image semantics, thus these texts are used naturally for refining image annotations. We refer to the semantic similarity between image annotation keyword and terms in web document as *relatedness*. In this study we utilize JCN algorithm [5] to measure the semantic similarity between two concepts as:

$$sim_{jcn}(c_1, c_2) = (dist_{jcn}(c_1, c_2))^{-1} = (IC(c_1) + IC(c_2) - 2 \times IC(lcs(c_1, c_2)))^{-1} \quad (1)$$

where $IC(c) = -logP(c)$ and $P(c)$ is the probability of encountering an instance of concept 'c' in WordNet [3]; $lcs(c_1, c_2)$ is the lowest common subsumer (*LCS*) that subsumes both concepts 'c_1' and 'c_2'. Note that all word similarity measures are normalized so that they fall within a 0-1 range, i.e. $0 \leq sim_{jcn}(c_i, c_j) \leq 1$. For an annotation keyword c_i, the measure of *coherence* is defined as:

$$a_i = \frac{1}{\eta_1} \sum_{j=1 \land j \neq i}^{T} sim_{jcn}(c_j, c_i), \quad \eta_1 = \sum_{i=1}^{T} \sum_{j=1 \land j \neq i}^{T} sim_{jcn}(c_j, c_i) \quad (2)$$

where η_1 is normalization factor. Similarly, let $D = (d_1, \ldots, d_n)$ be the terms in web documents, the measure of *relatedness* between c_i and D is defined as:

$$b_i = \frac{1}{\eta_2} \sum_{j=1}^{n} sim_{jcn}(d_j, c_i), \quad \eta_2 = \sum_{i=1}^{T} \sum_{j=1}^{n} sim_{jcn}(d_j, c_i) \tag{3}$$

Now we get two variables, a_i and b_i, as the measure of the importance of concept c_i to the semantics of an image. Note that $\sum a_i$ and $\sum b_i$ are both 1, that is to say, a_i and b_i can be regarded as two independent probability distributions of the quantified importance of concept c_i. We combine these two factors linearly as follows:

$$s(c_i) = \partial a_i + (1 - \partial)b_i, \quad 0 \le \partial \le 1 \tag{4}$$

where $s(c_i)$ is the final score associated with concept c_i. The larger $s(c_i)$ is, the more related to the semantics of corresponding image the concept c_i is. The parameter ∂ is set be 0.5 empirically in this study . Now we can measure the semantic similarity between term t and image I as follows:

$$S(t, I) = S(t, C) = \sum_{c_k \in C} sim_{jcn}(t, c_k) \cdot s(c_k) \tag{5}$$

where C is the set of annotation keywords produced by HMIAM for image I.

2.3   Image Retrieval Using Inference Network

The Bayesian inference network interprets probabilities as degree of belief. It associates random variables with documents, index terms and the user queries. A random variable associated with a document d_j represents the event of observing that document. The observation of d_j asserts a belief upon the random variables associated with its index terms. Without loss of generality, in the remainder of this paper let $K = (k_1, k_2, \ldots, k_t)(k_i \in \{0, 1\})$ be a t-dimensional vector of 2^t possible states representing term space, d_j a document and q the user query. Inference network model would retrieve and rank a document d_j with respect to the user query q as follows [1]:

$$P(q, d_j) = \sum_{\forall K} \left(\prod_{\forall i | k_i = 1} P(k_i | d_j) \times \prod_{\forall i | k_i = 0} P(\overline{k_i} | d_j) \right) \times P(q|K) \times P(d_j) \tag{6}$$

In formula 6 the specifications for $P(d_j)$, $P(k_i|d_j)$ and $P(q|K)$ are as follows. The $P(d_j)$ reflects the probability associated with the event of observing a given document d_j, and is defined as $P(d_j) = 1/|\vec{d_j}|$, where $|\vec{d_j}|$ stands for the norm of the vector $\vec{d_j}$. The $P(k_i|d_j)$ is the measure of the increased belief upon terms k_i caused by the event of observing document d_j, is defined as $\boldsymbol{P(k_i|d_j)}{=}\boldsymbol{W_{i,j}}{=}\boldsymbol{f_{i,j}} \times \boldsymbol{S(k_i, I_j)}$, where $\boldsymbol{f_{i,j}}$, the frequency of term k_i in document d_j, is used to measure the increased belief caused by web textual evidences, while $\boldsymbol{S(k_i, I_j)}$ is used to measure the increased belief caused by image (weighted annotations). By doing this, we can integrate image contents into the classic text-based image retrieval seamlessly. Finally, we define

$P(q|K)$ as follows. Let $\overrightarrow{k_i}$ be a t-dimensional vector as: $\overrightarrow{k_i}|k_i = 1 \wedge \forall_{j \neq i} k_j = 0$. Then we can use $P(q|\overrightarrow{k_i})$ to evaluate $P(q|K)$ in isolation as $P(q|\overrightarrow{k_i}) = idf_i$ if $q_i = 1$, and 0 otherwise. Accordingly, we can rewrite formula 6 as [1]:

$$P(q, d_j) = \left(\prod_{f_{i,j} > 0} (1 - W_{i,j}) \right) \times \frac{1}{|\overrightarrow{d_j}|} \times \left(\sum_{q_i = 1 \wedge f_{i,j} > 0} W_{i,j} \times idf_i \times \frac{1}{1 - W_{i,j}} \right) \quad (7)$$

3   Experiments and Results

We use the Corel Image set, consisting of 5,000 images, as the training set for the HMIAM. All images are associated in advance with up to five keywords drawn from a vocabulary of size 374, denoted by V. In addition, we have downloaded about 137,300 web pages accompanied with images from the WWW as the test set. After the training of the HMIAM, it was used to annotate all web images up to four keywords.

To verify the proposed model, we carried out standard keyword based image retrieval by using the top 25 frequent terms in the training set V listed in table 1. Besides the proposed model, we evaluated other two image retrieval models to make a comparison. One is the solely text-based image retrieval model, shown in Fig. 1 as "Text", which was implemented by setting the parameter $S(k_i, I_j) = 1$ in the specification of $P(k_i|d_j)$, that is, $P(k_i|d_j) = f_{i,j}$. Another model used the weighted annotation keywords to search images, and then attempted to rank results according to the weights associated with keywords, shown in Fig. 1 as "Weighted Annotations". We use *precision* and *recall* to evaluate the three models. Because examining all relevant images to

Table 1. The top-25 frequent terms in training set

water, sky, tree, people, grass, mountains, flowers, snow, clouds, stone, plane, field, bear, sand, birds, leaf, cars, plants, house, bridge, polar, garden, horses, tiger, train

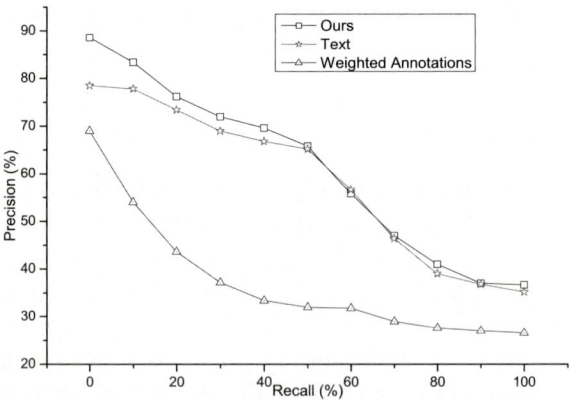

Fig. 1. Average precisions of 25 queries for the three models

a query term is a clearly infeasible task, we adopted the simple and effective strategy in [2]: for a given query, we ran out the image retrieval in the above three models and only evaluated the 40 highest images which were classified by volunteers as relevant or irrelevant with respect to the corresponding query topic. All results were shown in Fig. 1. From the results we can see, our model performed best. By introducing weighted annotations into the underlying inference network, the calculations of increased belief upon relevant terms caused by the observation of documents were more reasonable. Similarly, these terms increased the belief of relevant queries, and finally resulted in the improved retrieval. At the same time we see that retrieving images based on image annotations is inferior to that based on terms extracted from description tags and surroundings, even that annotations had been refined with weights. We think that is mainly caused by the poor annotations produced by the HMIAM.

4   Conclusions and Future Work

In this paper we propose to integrate image contents into text-based image retrieval model in two steps. First, images are interpreted into weighted annotations. Second, these weighted annotations as well as terms extracted from the corresponding web document are used to facilitate image retrieval. The experimental results demonstrate that the proposed model performs best. However, further research is still required. The experimental training data has a limited size of vocabulary, so the annotation results have a low coverage over total keyword space. In addition, the evaluation of image retrieval is conducted on a small set of retrieved results using only top 25 terms, since judging relevancy/irrelevancy to test queries requires substantial human endeavors. A wider evaluation on larger dataset will be carried out in future work. Moreover, we will explore more annotation models besides the HMIAM. We believe that improvement over underlying annotation model can result in further improvement in image retrieval.

References

1. Baeza-Yates, R., Ribeiro-Neto, B.: Modern information retrieval. Addison-Wesley Harlow, England (1999)
2. Coelho, T.A.S., Calado, P.P., Souza, L.V., Ribeiro-Neto, B.: Image retrieval using multiple evidence ranking. Image 16(4), 408–417 (2004)
3. Fellbaum, C.: Wordnet: an electronic lexical database. MIT Press, Cambridge (1998)
4. Ghoshal, A., Ircing, P., Khudanpur, S.: Hidden markov models for automatic annotation and content-based retrieval of images and video. In: Proceedings of the 28th annual international ACM SIGIR conference on Research and development in information retrieval, pp. 544–551 (2005)
5. Jiang, J.J., Conrath, D.W.: Semantic similarity based on corpus statistics and lexical taxonomy. In: Proceedings of International Conference on Research in Computational Linguistics, pp. 19–33 (1997)
6. Jin, Y., Khan, L., Wang, L., Awad, M.: Image annotations by combining multiple evidence & wordnet. In: Proceedings of the 13th annual ACM international conference on Multimedia, pp. 706–715 (2005)

Slide-Film Interface: Overcoming Small Screen Limitations in Mobile Web Search

Roman Y. Shtykh[1,2], Jian Chen[2], and Qun Jin[2]

[1] Media Network Center, Waseda University, Tokyo 169-8050, Japan
[2] Graduate School of Human Sciences, Waseda University,
Tokorozawa, Saitama 359-1192, Japan
{roman@akane.,wecan_chen@fuji.,jin@}waseda.jp

Abstract. It is well known that alongside with search engine performance improvements and functionality enhancements one of the determinant factors of user acceptance of any search service is the interface. This factor is particularly important for mobile Web search mostly due to small screen limitations of handheld devices. In this paper we propose scrolless mobile Web search interface to decrease search efforts that are multiplied due to these limitations, and discuss its potential advantages and drawbacks over conventional one.

Keywords: mobile Web search, mobile Web search interface, slide-film interface, paging, scrolling.

1 Introduction

Information access with wireless devices in any form has been traditionally considered a challenging task due to such limitations as mobile platform heterogeneity, insufficient development of networking infrastructure and limited input-output capabilities. But with the rapid enhancement of handheld device capabilities and wireless networks, cellular phones and other portable devices are becoming more and more suitable for wireless information access. Web search companies have already directed their attention to mobile search market by introducing new search services for mobile operators. Nevertheless, mobile search is still considered a new market with much potential for new entrants to Web search business, taking into account a number of peculiarities and limitations that are not available on personal computers and have to be considered and overcome to introduce successful search in mobile context.

When we consider usability peculiarities of handheld devices, probably the first thing everyone thinks of is the small size of a device screen. Due to this "number one" limitation of information access in mobile context, bringing "normal" (PC-screen-oriented) information to small-screen devices is a very challenging task, especially considering the multiplied user efforts caused by the complexity of Web sites and amount of scroll needed to navigate through multiple information pieces.

A number of different approaches [e.g., 1-2] were proposed to improve the efficacy and efficiency of mobile Web search. In this paper we present our attempt to improve

C. Macdonald et al. (Eds.): ECIR 2008, LNCS 4956, pp. 622–626, 2008.
© Springer-Verlag Berlin Heidelberg 2008

it – not through the enhancements of search engine and its algorithms, but by proposing a specialized interface. The main characteristic of the proposed interface is the abolishment of scroll. We offer a new user experience that, as we hope, will alleviate problems caused by small screen limitations – namely, frustration and eye fatigue one can feel during search result list navigation, increase in time consumption of individual search activities, and decrease of its quality. The proposed interface, called *slide-film interface*, is a kindred of paging found in [e.g., 3-4]. Till now this approach was extensively examined in PC text reading; we bring it into mobile context focusing on Web search. While dividing full text into logical blocks for paging presentation is difficult and such a form of content presentation may be disadvantageous for navigation and reading on PC, we believe this approach is favorable for mobile Web search where all items in the result list are self-contained items linked by one particular information need to satisfy which the search activity is conducted for. In other words, presenting one search result in one page combined with intuitive user interface and its easy operability will not break the information comprehension process; quite the contrary, by removing scrolling it facilitates mobile Web search.

To find out the advantages and potential shortcomings of the proposed approach we conducted an experiment. In this paper we report and discuss its results, in addition to the interface design.

2   Interface Design

One of usability definitions made by Jacob Nielsen says: "usability is also an ideology – the belief in a certain specialized type of human rights" [5]. One of these rights, the right "of people to have their time respected," is especially important for Web search that produces numerous results and takes from seconds to many minutes to find results meeting particular search needs. But when we consider mobile Web search, we believe "the right to simplicity" is even more important because of the extra limitations of handheld devices that are not critical for PC Web search. Therefore, when implementing the proposed interface, we were guided by the belief that the interface design of a mobile Web search system must be as simple and intuitive as possible while preserving all information about search results found in conventional search systems.

As we already mentioned in the previous section, every search result item from the list returned by a Web search engine is a self-contained element associated with a particular information need. Therefore, it is easy to present each item in a separate "slide," or "page," each having the title, URL, summary and the title of the next result to facilitate imaginable comparison of the current and next result item (foreseeing the value of the next result in comparison to the current one). Unlike most mobile Web search services that truncate summary snippets of the search result items to reduce the amount of scroll and in this way facilitate easier navigation through search results that often can lead to difficulties in understanding of the content of a particular result, (owing to the availability of one slide of a screen size for one search result) our approach has an advantage to provide the greater part of one slide screen to place the full summary without any fear to make the search tiresome.

We have implemented a Web-based prototype simulating the behavior of the proposed search interface for handheld devices. The prototype is accessible with a full Web browser from handheld devices and PCs. Page flipping is done with right and left soft keys – pressing right soft key flips search results taking a user to further items and pressing left soft key flips pages backwards (see Fig. 1). In addition, the user can promptly start a new search by pressing a shortcut key (hotkey), and see the title list of navigated search items to be able to return to a specific point in the search and not to be lost in the huge amount of search results.

Fig. 1. Interface in action

3 Experiment

In order to verify the hypothesized advantages and discover the drawbacks of the proposed interface we designed and conducted a preliminary experiment that is the first step of our user study. For this, we had to compare how well a user performs search with the proposed interface and the conventional one.

3.1 Design

In addition to the slide-film interface implementation, we implemented another interface that simulates the behavior of the conventional mobile Web search of Microsoft Pocket Internet Explorer. The experiment was carried out on PCs, but we tried to maximize the similarity of user-system interactions to those where handheld device is used. Therefore we fixed the size of the interface screen to that of Microsoft Windows Mobile Smartphone (176x220 in full screen mode) and forbade the use of the mouse – instead the use of only several keys (arrow keys, *Home* and *Backspace*) of a 10-key device was allowed.

Nine participants of age from 20 to 34 were asked to search and interact naturally with the two interfaces. Two sets of search tasks were prepared with 11 queries in each. Each task contained a question the subject had to find an answer for and a fixed query term[1] to make a query with. The document collection the subject was searching

---

[1] This is done in order to equalize the difficulties of completing tasks for the both interfaces.

on was retrieved from Yahoo! Web search service [6]. The experiment was conducted in two steps:

- Week 1: search with Set A (slide-film interface) and search with Set B (conventional interface)
- Week 2: search with Set A (conventional interface) and search with Set B (slide-film interface) conducted in 5-7 days from the experiments of Week 1.

Conducting experiment in this way, ideally we can obtain search performance results of 22 tasks completed using both interfaces after the experiments are finished. However, the number of successful searches, when the user completes a particular task, differs from user to user. Furthermore, some searches can be recognized invalid. For example, if the answer (result item) to the same question found with the conventional interface (hereafter also referred to as CI for brevity) in Week 1 differs from the answer in Week 2 found with slide-film interface (hereafter also referred to as SFI), the result is recognized to be invalid.

In the experiment we evaluated search with the two interfaces both quantitatively and qualitatively. For qualitative measure we chose five-point Likert interface satisfaction questionnaire scaling from 'very negative' (1) to 'very positive' (5). To understand the efficacy of the proposed interface we measured time spent purely for viewing and navigating search results with the both interfaces when trying to complete a task (i.e. time spent on Web pages opened from the search result list is ignored).

3.2 Results

Comparing the speed of completion of all users over all tasks, on average, the proposed interface fits mobile Web search better – users perform search faster (on average, 48 sec (SFI) vs. 68 sec (CI) for all users to complete all the tasks). Using the Wilcoxon sign rank analysis, there is a significant difference between the search performances of users using the two interfaces (p=0.004).

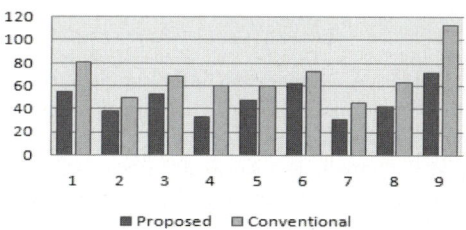

Fig. 2. Mean time spent to complete all tasks for every user

We observed, however, that the number of successfully completed tasks for most subjects grew smaller (by 7 percent) when using the proposed interface. This can be explained by the increased speed and ease of viewing of search results (in comparison with that of the conventional interface) that, in turn, result in omission of relevant items.

Table 1. Mean ratings (with standard deviations) of user satisfaction with the interfaces

Question (*shortened*)	Score (SFI)	*StDev (SFI)*	Score (CI)	*StDev (CI)*
Operations are intuitive	4	*1.26*	3.4	*1.24*
Search results are easily viewable	4.2	*0.98*	3.1	*0.33*
Eye fatigue reduced	3.9	*0.93*	2.3	*1.12*
Less time to complete search	4	*0.5*	3.1	*1.05*

The ratings obtained from the questionnaire are shown in Table 1. As to the usability of the proposed interface, the ratings varied from user to user showing how different user preferences can be. However, in general, the proposed interface was found to be superior by most users and eight of nine preferred it for Web search.

4   Conclusions and Future Work

In this paper we proposed and briefly discussed slide-film interface that, as we found, can be a good alternative to the conventional way to organize and present Web search results in mobile context. Although the scale of the first experiment conducted to verify the advantages of the proposed interface is very small to claim the search we propose with the slide-film interface is more efficient and easier as compared with the conventional one, the results are very promising and encouraging to continue improving the approach.

The research is at the early stage and still in progress. As for a future work, we are planning to continue the user studies with a larger number of participants and bigger variety of experiments. We also consider conducting an eye tracking experiment to get better insight about the peculiarities of the proposed interface.

References

1. De Luca, E.W., Nürnberger, A.: Supporting information retrieval on mobile devices. In: Proceedings of the 7th international conference on Human computer interaction with mobile devices & services, pp. 347–348 (2005)
2. Church, K., Keane, M.T., Smyth, B.: An Evaluation of Gisting in Mobile Search. In: Losada, D.E., Fernández-Luna, J.M. (eds.) ECIR 2005. LNCS, vol. 3408, pp. 546–548. Springer, Heidelberg (2005)
3. Schwarz, E., Beldie, I.P., Pastoor, S.: A comparison of paging and scrolling for changing screen contents by inexperienced users. Hum. Factors 25, 279–282 (1983)
4. Baker, J.R.: The Impact of Paging vs. Scrolling on Reading Online Text Passages. Usability News, 5.1 (2003), http://psychology.wichita.edu/surl/usabilitynews/51/paging_scrolling.htm
5. Nielsen, J.: Usability: Empiricism or Ideology? Jakob Nielsen's Alertbox (2005), http://www.useit.com/alertbox/20050627.html
6. Yahoo! Search Web Services, http://developer.yahoo.com/search/

A Document-Centered Approach to a Natural Language Music Search Engine

Peter Knees, Tim Pohle, Markus Schedl,
Dominik Schnitzer, and Klaus Seyerlehner

Dept. of Computational Perception, Johannes Kepler University Linz, Austria
peter.knees@jku.at

Abstract. We propose a new approach to a music search engine that can
be accessed via natural language queries. As with existing approaches, we
try to gather as much contextual information as possible for individual
pieces in a (possibly large) music collection by means of Web retrieval.
While existing approaches use this textual information to construct representations of music pieces in a vector space model, in this paper, we
propose a document-centered technique to retrieve music pieces relevant
to arbitrary natural language queries. This technique improves the quality of the resulting document rankings substantially. We report on the
current state of the research and discuss current limitations, as well as
possible directions to overcome them.

1 Motivation and Context

While digital music databases contain several millions of audio pieces nowadays,
indexing of these collections is in general still accomplished using a limited set of
traditional meta-data descriptors like artist name, track name, album, or year. In
most cases, also some sort of classification into coarse genres or different styles is
available. Since this may not be sufficient for intuitive retrieval, several innovative
(content-based) approaches to access music collections have been presented in
the past years. However, the majority of these retrieval systems is based on *query-
by-example* methods, i.e. the user must enter a query in a musical representation
which is uncommon to most users and thus lacks acceptance. To address this
issue, recently, different approaches to music search engines that can be accessed
via textual queries have been proposed [4,5,6,9].

In [6], we presented an approach that exploits contextual information related
to the music pieces in a collection. To this end, $tf \times idf$ features are extracted
from Web pages associated with the pieces and their corresponding artist. Furthermore, to represent audio pieces with no (or only little) Web information
associated, also audio similarity is incorporated. This technique enables the user
to issue queries like *"rock with great riffs"* to express the intention to find pieces
that contain energetic guitar phrases instead of just finding tracks that have been
labeled as *rock* by some authority. The general intention of the system presented
in [6] is to allow for virtually any possible query and return the most appropriate pieces according to their "Web context" (comparable to e.g. Google's image
search function).

C. Macdonald et al. (Eds.): ECIR 2008, LNCS 4956, pp. 627–631, 2008.

In this paper, we present an alternative method to obtain a relevance ranking of music pieces wrt. a given query. Instead of constructing vector space representations for each music piece, we apply a traditional document indexing approach to the set of retrieved music-related Web pages and introduce a simple ranking function which improves the overall retrieval performance substantially.

2   Technical Background

Prior to presenting the modified retrieval approach, we briefly review the vector space-based method described in [6]. The first data acquisition step is identical for both approaches.

2.1   Vector Space Model Approach (VSM)

To obtain as much track specific information as possible while preserving a high number of Web pages, for each track in the collection, three queries are issued to Google (at most 100 of the top-ranked Web pages are retrieved per query and joined into a single set):

1. "*artist*" music
2. "*artist*" "*album*" music review
3. "*artist*" "*title*" music review -lyrics

After HTML tag and stop word removal, for each piece, all associated documents are treated as one large document and a weighted term vector representation is calculated using a modification of the $tf \times idf$ function.

In addition to the context-based features, information on the (timbral) content of the music is derived by calculating a Single Gaussian MFCC (*Mel Frequency Cepstral Coefficients*) distribution model for each track. Acoustic similarity between two pieces can be assessed by computing the Kullback-Leibler divergence on their models [8]. Based on the audio similarity information, feature space pruning is performed by applying a modified χ^2 test that simulates a 2-class discrimination task between the most similar sounding and the most dissimilar sounding tracks for each piece. For the evaluation collection used in [6], this step reduces the feature space from about 78,000 dimensions to about 4,700. Beside feature space reduction, the audio similarity measure can also be used to emphasize terms that occur frequently among similar sounding pieces, and – most important – to describe music pieces with no (or few) associated information present on the Web. These two tasks are achieved by performing a Gaussian weighting over the 10 acoustically nearest neighbors' term vectors.

After obtaining a term weight vector for each track in the music collection, natural language queries to the system are processed by adding the constraint *music* to the query and sending it to Google. From the 10 top-ranked Web pages, a query vector is constructed in the feature space. This query vector can then be compared to the music pieces in the collection by calculating cosine distances. Based on the distances, a *relevance ranking* is obtained.

2.2   Rank-Based Relevance Scoring (RRS)

In contrast to the VSM method that relies on the availability of Google to process queries, we propose to directly utilize the Web content that has been retrieved in the data acquisition step. To this end, we create an off-line index of all pages using the open source package *Lucene* [1]. The usage of an off-line index allows to apply an alternative relevance ranking method since all indexed documents are at least relevant to one of the music pieces in the archive. Thus, we can take advantage of this information by exploiting these relations. More precisely, when querying the *Lucene* off-line index, a relevance ranking of the indexed documents according to the query is returned. Since we know for which music pieces these documents have been retrieved (and are thus relevant), we can simply create a set of music pieces relevant to the query by gathering all music pieces that are associated with at least one of the returned documents. Moreover, we can exploit the *ranking* information of the returned documents to introduce a very simple (but effective) relevance scoring function. Hence, for a given query q, we calculate the *rank-based relevance scoring* (RRS) for each music piece m as

$$RRS(m, q) = \sum_{p \in D_m \cap D_q} 1 + |D_q| - rank(p, D_q), \tag{1}$$

where D_m is the set of text documents associated with music piece m, D_q the set of relevant text documents with respect to query q, and $rank(p, D_q)$ a function that returns the rank of document p in the (ordered) set D_q (highest relevance corresponds to rank 1, lowest to rank $|D_q|$). Finally, the relevance ranking is obtained by sorting the music pieces according to their RRS value.

3   Evaluation and Discussion

To examine the impact of RRS on the retrieval quality, we have conducted various experiments on the same test collection as used in [6] for reasons of comparability. This collection consists of 12,601 unique tracks by 1,200 artists labeled with 227 different tags from Audioscrobbler/Last.fm [2,3]. To measure retrieval performance, each of the 227 tags serves as query – music pieces are considered relevant iff they have been labeled with the corresponding tag. Examples for tags (and thus queries) are *hard rock*, *disco*, *soul*, *melancholy*, or *nice elevator music*. More details on the properties of the test collection can be found in [6].

Figure 1 depicts the *precision at 11 standard recall values* curves for RRS, the best scoring VSM approach from [6], and the baseline (giving indication of the "hardness" of the evaluation collection). At the (theoretical) 0.0 recall level, precision is at 0.75, which is 0.13 above the VSM approach. At the 0.1 recall level, the difference is even more evident: while the term vector approach yields around 0.37 in precision, RRS reaches a precision value of 0.66. Similar observations can be made for other IR measures, cf. Table 1. As can be seen, for the VSM approach, on average, five out of the first ten pieces are relevant, using the RRS ranking, seven out of ten pieces are relevant in average.

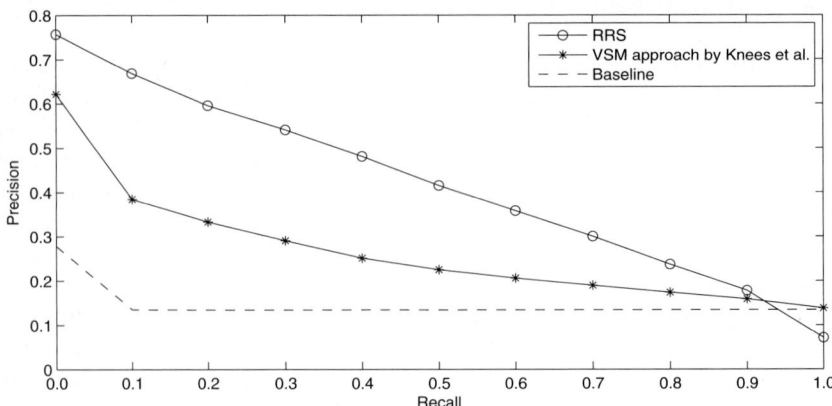

Fig. 1. Comparison of precision at 11 standard recall levels (avg. over all queries)

Table 1. Selected single value summaries averaged over all queries (values in percent)

Ranking method	VSM	RRS
Average precision at seen relevant documents	25.29	**45.70**
R-Precision	26.41	**42.30**
Precision after 10 documents	49.56	**71.59**
Precision	13.47	**15.73**
Recall	**100.00**	87.81

A major difference to the ranking functions based on term vector comparison is that the returned set of music pieces is in most cases only a subset of the whole collection. While term vector approaches always result in an ordering of the complete collection, using RRS returns only pieces that are assumed to be relevant somehow (by ignoring pieces with RRS scores of zero). For the user, it can be valuable information to know the total number of relevant pieces in the collection during examination of results. Furthermore, using the RRS scheme, it is not necessary to construct term vector representations for music pieces, which makes extraction of features as well as feature space pruning obsolete.

4   Conclusions and Future Work

We presented an alternative ranking approach for a natural language music search engine. By creating an off-line Web page index to process arbitrary queries and incorporating a simple ranking method, we were able to improve retrieval quality considerably. A possible explanation for the outcome that a rank-counting method outperforms an ordinary vector space model approach is that important and diverse information contained on the Web pages may be improperly represented by a single term vector created from a merging of documents from different sources. In any case, the reasons for this finding have to be investigated more thoroughly in future work.

A current drawback of the proposed retrieval approach is the fact that it is not applicable to music pieces for which no associated information could be discovered via Web retrieval. As in [6], a possible solution could be to incorporate the audio similarity information, e.g., to associate Web pages related to a music piece also to similar sounding pieces. Furthermore, currently, possible queries to the system are limited by the vocabulary present on the retrieved pages. Future extensions could comprise a method that – again – sends a request to Google for queries containing unknown terms and, for example, finds those pages in the off-line index that are most similar to the top results from Google. From those pages, an RRS-based ranking could be obtained.

As for future work, we will also elaborate methods to incorporate relevance feedback into this new ranking mechanism, like it has already been successfully accomplished for the approach relying on term vector representations [7]. One possibility could be to propagate the feedback information back to the associated documents and perform relevance feedback on the document level. This could also allow for techniques like automatic query expansion. Finally, it is our goal to use a modified focused crawler that is specialized in indexing music related Web pages. Having a search index on our own would improve the applicability of the system and break the dependency on external Web search engines.

Acknowledgments

This research is supported by the Austrian Fonds zur Förderung der Wissenschaftlichen Forschung (FWF) under project number L112-N04.

References

1. Apache Lucene, `http://lucene.apache.org`
2. Audioscrobbler, `http://www.audioscrobbler.net`
3. Last.fm, `http://www.last.fm`
4. Baumann, S., Klüter, A., Norlien, M.: Using natural language input and audio analysis for a human-oriented MIR system. In: Proc. 2nd WEDELMUSIC 2002 (2002)
5. Celma, O., Cano, P., Herrera, P.: Search Sounds: An audio crawler focused on weblogs. In: Proc. 7th ISMIR (2006)
6. Knees, P., Pohle, T., Schedl, M., Widmer, G.: A Music Search Engine Built upon Audio-based and Web-based Similarity Measures. In: Proc. 30th ACM SIGIR (2007)
7. Knees, P., Widmer, G.: Searching for Music Using Natural Language Queries and Relevance Feedback. In: Proc. 5th AMR (2007)
8. Mandel, M., Ellis, D.: Song-Level Features and Support Vector Machines for Music Classification. In: Proc. 6th ISMIR (2005)
9. Turnbull, D., Barrington, L., Torres, D., Lanckriet, G.: Towards Musical Query-by-Semantic-Description using the CAL500 Data Set. In: Proc. 30th ACM SIGIR (2007)

Collaborative Topic Tracking in an Enterprise Environment*

Conny Franke and Omar Alonso

Department of Computer Science
University of California at Davis
One Shields Ave, Davis, CA 95616
franke@cs.ucdavis.edu, oralonso@ucdavis.edu

Abstract. Business users in an enterprise need to keep track of relevant information available on the Web for strategic decisions like mergers and acquisitions. Traditionally this is done by the user performing standing queries or alert mechanisms based on topics. A much richer tracking can be done by providing a way for users to initiate and share topics in a single place. In this paper we present an alternative model and prototype for tracking topics of interest based on a continuous user collaboration.

1   Introduction

In today's enterprises users need to keep track of relevant content in internal and external sources for strategic reasons. These reasons are usually business decisions about acquiring companies, making sense of new technologies, and exploration of partnerships just to name a few. This means that users must take advantage of the wealth of available information on the Web to track content that is business relevant on a daily basis.

Foraging, scanning, and keeping track of vast amounts of information from a variety of sources is a very time consuming task, so there is clearly a need for automation. There are some tools that can help a single user with some tasks, like Google alerts [1], which are email updates of relevant Google content based on queries or topics. The user has to define certain keywords like in standing queries and gets a summary of the results via email. A more collaborative approach is to use a typical mailing list where people who have the same interest can post and respond accordingly. This is powerful as long as there is enough activity on the list. Wikis have also appeared as an alternative platform for collaboration. So far, all these solutions require active participation from users.

We propose an alternative approach where the user seeds the topics of interests and the system, in a very proactive manner, finds relevant content that is shared within the user's enterprise community. Furthermore, this tracking is an ongoing activity until the user decides that the topic is no longer of interest. Detecting new topics of interest among the vast amount of new information from different

---

* Part of this work was performed while the authors were affiliated with SAP Research, 3410 Hillview Avenue, Palo Alto, CA 94304.

C. Macdonald et al. (Eds.): ECIR 2008, LNCS 4956, pp. 632–636, 2008.

sources like news, blogs, etc. that is published daily is a non-trivial task. This is also true for finding interesting themes that are related to topics already popular with an individual or a group of people. Leveraging the collaboration among peers in an enterprise and the increased agility of the group's knowledge that results from the shared interest, is at the center of our approach.

Related Work

The informal network of collaborators and colleagues is one of the most effective channels for dissemination of information and expertise within an organization [4]. In order to extract the most relevant information, concepts of collaborative filtering can be applied to the communication among the members of these networks. Methods like Amazon.com's recommendation engine [5] produce high quality results. However, they only consider content within the same domain as the seeds, i.e., if we applied collaborative filtering on blog posts, the output would only contain related blog posts, not any other kind of articles like news articles in addition to that. The input for traditional topic detection and tracking approaches [2] is a set of keywords. Usually, these keywords are not generated dynamically based on the automated analysis of seed emails, as we do in our method. Recently, the social aspect is having a huge impact in the way people perform information discovery [6] on the Web. Finally, part of this work is based on a sensemaking-based application for technology trends [3].

2   Collaborative Topic Tracking

Our approach aims at getting additional input for accurate topic tracking by utilizing the ongoing discussion about topics of interest within a company. We assume that a group of users has similar interests and goals and thus all members are eventually interested in the same trends. We consider a trend as a general direction expressed in news and blogs that are available on the Web.

In the following sections, we describe how we tap into a discussion to seed the topic tracker, and how we augment the ongoing discussion with results found by the topic tracking system.

2.1   Blok

A pivotal element in our topic tracking system is the "blok", a cross between a blog and a talk/chat client. A blok is similar to a long log session that contains all conversations. The seeds of the blok are individual messages, each consisting of a title or subject, the actual message, a user name to identify the author of the seed, and the time when the seed was created. The idea behind the blok is to enable discussion about enterprise related topics between users with similar interests and augment their discussion with automatic tropic tracking. The blok allows users to see all entries and their associated topics over time. This concept in combination with the variety of filtering options we offer makes it easy to keep track of discussions on specific topics as well as the development of "hot topics" over time.

Users post findings and possibly URLs containing news articles they think are interesting to the blok. Others that read about it can comment on these initial posts and contribute their own facts, links, and findings. To further guide and help the user to explore existing blok posts, we automatically detect the sentiment of each blok post to give the user an indication about whether the contribution is positive, neutral, or negative.

Users can interact with the blok by sending an email. This interface also makes it easy to post to the blok as a byproduct of a discussion on a mailing list or newsgroup. Additionally, we replicate the content that users generate in their enterprise internal blogs in our blok. Users can also seed the system directly by entering a message in the user interface.

2.2   Seeding the Topic Tracker

Given a seed message, we automatically find articles from news and blogs that are related to the message. As new seeds arrive, they undergo a processing pipeline as follows. We use named entity extraction on the blok posts to find names of people and organizations, as these entities represent the seed's content best. Discovered named entities are added to the set of tracked keywords for use in finding related articles. Additionally, we use heuristics on the seed's URLs to determine if they contain worthwhile topic to track, e.g., `http://www.powerset.com`. For the example of `http://www.powerset.com`, "powerset" would be extracted as an additional keyword for seeding the topic tracker.

We augment the content of a blok entry by performing sentiment detection. We do this by first applying a subjectivity analysis to detect neutral blok posts. Then, for all posts that are found to be subjective, we apply a sentiment detector to determine if it is positive or negative. As a last step, we remove all stop words from the seed message and analyze the term frequencies within the message.

2.3   Feedback Loop

After a user posted to the blok, he can explore the additional articles our system found based on his contribution. Ideally, the automatically discovered content contains new and valuable information for the user and he is likely to report his new insights back to the blok or write about them on his private blog. In either case our system picks up his response and includes it in its knowledge base where all other users can pick it up.

3   Prototype Implementation

The main parts of the prototype are the email parser with named entity extractor, URL parser, and sentiment detector, the topic tracker, and the user interface for interaction. Figure 1 shows the processing pipeline. A user's seed message that is input via email or blog post triggers the email parser, which generates a set of topics as input for the automated tracker. Periodically, the topic tracking system is invoked. From a user given set of sources it retrieves articles that

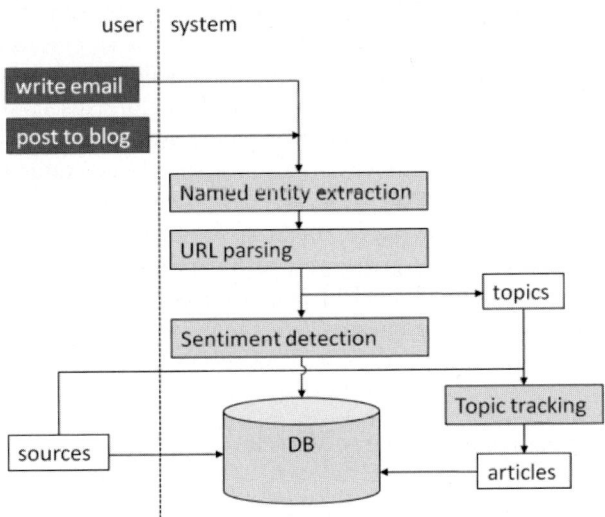

Fig. 1. Text processing pipeline

Fig. 2. Collaborative tracking user interface: blok entries and tracked topics

contain at least one of the tracked topics and stores it into a database for later exploration by the user.

The actual email parser is a Perl script that extracts the content of the email and stores it in a database. The script also recognizes and parses all URLs contained in the email. After storing the relevant content, the Perl script calls a Java program that conducts a more elaborate analysis. First, we apply heuristics to all URLs in the email to extract new topics from the domain name. Then, we use an open source named entity extractor to get seeds for the topic tracker. The last step in processing each incoming email is sentiment detection. We use a combination of subjectivity analysis and sentiment detection to distinguish between positive, neutral, and negative sentiment in the email.

There is a task scheduler that starts the topic tracking to find new articles. In this step, we extract all articles from a set of news sources and blogs and match them with the set of tracked topics. We then store all relevant articles in the database. In addition, we grab all new posts in enterprise internal blogs and include them in our blok. We then apply named entity extraction and sentiment detection on these blok posts just like we do for incoming emails. Once a week, the topic tracking includes an additional step to collect some statistics about the tracked topics. For example, we look up the number of tags listed on the Web site `del.icio.us` as well as the number of blog posts about this topic indexed on `technorati.com`. These numbers give some additional indication about the popularity of a specific topic over time.

Figure 2 shows part of the topic tracking user interface. The current view shows a couple of recent blok posts and an overview over the tracked topics. The automatically generated sparkline next to every topic shows its popularity within the last two weeks, as it summarizes the overall number of news and blog articles found each day.

4   Conclusions and Future Work

We have presented an approach for tracking topics of interest in a collaborative fashion. The proposed technique differs from current standing queries solutions by adding the community factor and by providing a feedback loop to the tracking status. An initial prototype that includes some of the ideas presented was developed in an enterprise environment. At time of writing, the system has been in use with a handful of users, who are tracking business information. The initial feedback on using the blok as input has been very positive. This encourages us to continue working on other aspects that were left out due to time constraints. Future work includes evaluation of the accuracy of the information over significant periods of time.

References

1. http://www.google.com/alerts
2. Allan, J.: Introduction to topic detection and tracking. pp. 1–16 (2002)
3. Alonso, O., James, F., Franke, C., Talbot, J., Xie, S., Klemba, K.: Sensemaking in the enterprise: A framework for retrieving, presenting, and providing feedback on information sources. SIGMOD Industrial Session (submitted, 2008)
4. Kautz, H., Selman, B., Shah, M.: Referral web: Combining social networks and collaborative filtering. Commun. ACM 40(3), 63–65 (1997)
5. Linden, G., Smith, B., York, J.: Amazon.com recommendations: Item-to-item collaborative filtering. Internet Computing, IEEE 7(1), 76–80 (2003)
6. Ramakrishnan, R., Tomkins, A.: Toward a peopleweb. IEEE Computer 40(8), 63–72 (2007)

Graph-Based Profile Similarity Calculation Method and Evaluation

Hassan Naderi and Béatrice Rumpler

INSA of Lyon – LIRIS, Bâtiment Blaise Pascal 7, Av. Jean Capelle,
F69621 Villeurbanne France
{hassan.naderi,beatrice.rumpler}@insa-lyon.fr

Abstract. Collaborative Information Retrieval (CIR) is a new technique for resolving the current problem of information retrieval systems. A CIR system registers the previous user interactions to response to the subsequent user queries more efficiently. But, the goals and the characteristics of two users may be different; so when they send the same query to a CIR system, they may be interested in two different lists of documents. To resolve this problem, we have developed a personalized CIR system, called PERCIRS, which is based on the similarity between two user profiles. In this paper, we propose a new method for User Profile Similarity Calculation UPSC. Finally, we introduce a mechanism for evaluating UPSC methods.

Keywords: User profile, CIR, profile similarity, dynamic community creation.

1 Introduction

CIR is an approach which learns to improve retrieval effectiveness from the interaction of different users with the retrieval system [2]. In other words, collaborative search records the fact that a result d has been selected for query q, and then reuses this information for future similar queries, by promoting results reliably selected during previous sessions. However, the goals and the characteristics of two users may be different, so when they send the same query to a CIR system, they may be interested in two different lists of documents (known as **personalization** problem). Personalization is a common problem encountered by the CIR researchers. For instance, Armin, who has presented three important approaches toward a CIR system in [1], confessed that:

> "*We are aware of the problems of personalization and context, but in our first steps towards techniques we avoid further complexity of CIR by ignoring these challenges*".

Smyth B. et al [6] tried to alleviate the personalization problem by implementing a significant collaborative web search technique as a robust and scalable Meta search engine architecture in the form of I-SPY search engine (http://ispy.ucd.ie). They defined collaborative web search as exploiting repetition and regularity within the query-space of a community of like-minded individuals in order to improve the quality of search results. However they stated that: "*the precise nature of a*

C. Macdonald et al. (Eds.): ECIR 2008, LNCS 4956, pp. 637–641, 2008.
© Springer-Verlag Berlin Heidelberg 2008

community's shared interests may not be so easy to characterize". Because of this difficulty I-SPY can't automatically associate a user to a suitable community. So I-SPY explicitly asks the users to recognize their community among a set of predefined communities at the time of inscription. This approach has several restrictions as:

1. These predefined communities are not exclusive. Thus user often can't find an appropriate community. In I-SPY the user can create his own community, but he will be the first member of his community and he can not profit of CIR advantages.
2. Finding an appropriate community is not an easy task for a user especially when the number of communities rapidly increases (another search problem!).
3. User's interest is dynamic and changes over the time while assigning a user to a predefined community is static (personalization problem again!).
4. The communities are either very general or specific to be helpful in CIR process.

We have developed a PERsonalized Collaborative Information Retrieval System (called PERCIRS) which is able to resolve these problems in a CIR system by dynamically creating user communities [3]. In PERCIRS, creation of user communities is based on the similarity of user profiles. So proposing an efficient method to calculate the similarity between two user profiles is a key factor of PERCIRS's efficiency.

The structure of the used profile in PERCIRS is presented in the section 2. In the section 3 we propose a new graph-based method for User Profile Similarity Calculation (UPSC). We introduce a mechanism for evaluating UPSC methods in section 4. We will finish our paper with a conclusion and future perspectives.

2   User Profile in PERCIRS

In PERCIRS, a user profile P(X) is presented as a set of pairs (q, D_q):

$$P(X) = \{(q_1, D_{q1}), (q_2, D_{q2}), ..., (q_N, D_{qN})\}$$

where q_i is the ith query of profile and D_{qi} is a set of documents the user has marked as relevant to q_i. Thus our ultimate goal of User Profile Similarity Calculation (UPSC) between users X and Y is to calculate the similarity between two following sets:

$$P(X) = \{(q_1^x, D_{q1}^x)..., (q_N^x, D_{qN}^x)\} \text{ and } P(Y) = \{(q_1^y, D_{q1}^y), ..., (q_M^y, D_{qM}^y)\}$$

3   Graph-Based User Profile Similarity Calculation

In [4] we have proposed three different methods to calculate the similarity between two user profiles: equality-based, full-similarity based, and cut-similarity based. In this paper we propose a new graph-based UPSC which has not the weakness of three previous UPSC methods. This approach is based on the **Maximum Weighted Bipartite Matching** [5]. So, before proposing our new UPSC method, we briefly explain the maximum weighted bipartite matching problem.

3.1 Maximum Weighted Bipartite Matching

The bipartite matching problem is a classical subject of study in graph algorithms, and has been investigated for several decades. A matching M in a bipartite graph G=(V=(X,Y),E) is a set of pair-wise non-adjacent edges; that is, no two edges share a common vertex. We say that a vertex is matched if it is adjacent to an edge in the matching. Otherwise the vertex is unmatched. A **maximum bipartite matching** is a matching that contains the largest possible number of edges.

In a weighted bipartite graph, each edge has an associated value. Given a weighted bipartite graph G = (V=(X,Y),E), a **maximum weighted bipartite matching** is defined as a perfect matching where the sum of the values of the edges in the matching has a maximal value. A perfect matching is a matching which covers all vertices of the graph. That is, every vertex of the graph is adjacent to exactly one edge of the matching. If the graph is not a complete bipartite, missing edges are inserted with value zero. For bipartite graphs, we can even assume, if this is needed in some argument, that the two subsets have the same cardinality: we can always add nodes to the smaller subset with only zero-capacity edges to the other subset. The well known and remarkable Hungarian algorithm [5, 7] finds the maximum weighted bipartite matching in $O(|V|^2|E|)$.

3.2 Maximum Weighted Bipartite Matching and UPSC

In our new UPSC method, we use the maximum weighted bipartite matching to find the similarity between two user profiles. This approach tries to find a matching between the elements of one profile and the elements of the other profile. In the first step of this approach, we construct a complete weighted bipartite graph G=(V=(P(X),P(Y)), E) where the elements of profile P(X) create one part of the graph G and the elements of profile P(Y) create the other part of this graph. Each vertex e from P(X) will be connected to each vertex $é$ from P(Y) by the edge $(e,é)$. The weight of edge $(e,é)$ is equal to the similarity between its two adjacent elements e and $é$: $w(e,é)=S_E(e,é)$.

In the second step, we apply the Hungarian algorithm on the created weighted bipartite graph to calculate the maximum weighted matching between two user profiles P(X) and P(Y). We believe that the weight of maximum matching is a good estimation for the similarity between two user profiles P(X) and P(Y) because:

- On one hand, each element of one profile will be matched to exactly one element of the other profile, unlike similarity-based method [4] which matches each element to every elements of the other profile.
- And, on the other side, two non equal but similar elements can be matched, unlike to equality-based method [4] which only matches equal elements.

So this new graph-based UPSC method has the advantages of both equality- and similarity- based methods but not the problems. We believe that this graph-based method can be the most efficient one to calculate the similarity between two user profiles. In the next section, we propose a mechanism to evaluate the efficiency of our proposed UPSC methods in this paper and in [4].

4 Proposed Mechanism to Evaluate UPSC Methods

In this section we propose a user-centric approach to evaluate the efficiency of proposed methods in the previous sections. Because of limitation of the space we can't present our results in this paper.

First we define N different topics. Then we associate each topic to a group of users and we ask each group to consider the associated topic as the needed information. Users of each group execute some queries related to their associated topics, and then they mark the collected relevant documents. The queries and documents which have been marked as relevant to them will be stored in the user profiles. Our approach to evaluate UPSC methods:

1- Let's $f_1, f_2, \ldots, f_R$ be R methods.
2- We define S different topics $t_1, t_2, \ldots, t_S$..
3- We divide the volunteers into S groups: $g_1, g_2, \ldots, g_S$.
4- We associate each topic t_i to a group of volunteers (g_i).
5- We calculate the value of $\overline{F}_k(g_i)$ for each method f_k and for each group g_i. $\overline{F}_k(g_i)$ represents the average similarity between the user profiles in the group g_i calculated by the method f_k:

$$\overline{F}_k(g_i) = \frac{1}{|g_i|^2} \sum_{\forall u_m, u_n \in g_i} f_k(u_m, u_n)$$

where $f_k(u_m, u_n)$ is the calculated similarity between users u_m and u_n by the method f_k, and $|g_i|$ is the number of users in i^{th} group.

6- For each method f_k and each group g_i, we calculate the average similarity between the users of groups g_i and the users of the other groups $g_1, g_2, \ldots g_{i-1}, g_{i+1}, \ldots, g_S$, by the following formula:

$$\overline{F}_k(g_i, U - g_i) = \frac{1}{|g_i| \times |U - g_i|} \sum_{u_m \in g_i, u_n \in U - g_i} f_k(u_m, u_n)$$

where U is the set of all volunteers.

7- Given the formulas 4 and 5, we calculate the following ratio for the method f_k as:

$$\text{ratio}(f_k) = \sum_{i=1}^{S} \frac{f_k(g_i)}{\log[f_k(g_i; U - g_i)]}$$

8- We calculate the average ratio (AR) for the R methods:

$$AR = \overline{\text{ratio}} = \frac{1}{R} \sum_{i=1}^{R} \text{ratio}(f_i)$$

9- The method with the nearest ratio to the average ratio is considered as the optimal method:

$$f_k \text{ is optimal} \Leftrightarrow \forall i : \left| \overline{ration} - \text{ratio}(f_k) \right| \leq \left| \overline{ration} - \text{ratio}(f_i) \right|$$

Because of limited place in this paper we can't explain this mechanism to evaluate the UPSC methods.

5 Conclusion

In this paper we proposed a new graph-based UPSC method to be integrated in our CIR system PERCIRS [3]. Then we have proposed an evaluation methodology to evaluate the user profile similarity calculation methods.

Acknowledgement

This research is partially supported by the French Ministry of Research and New Technologies under the ACI program devoted to Data Masses (ACI-MD), project #MD-33.

References

[1] Armin, H.: Learning Similarities for Collaborative Information Retrieval. In: proceedings of the KI-2004 workshop Machine Learning and Interaction for Text-Based Information Retrieval, TIR-04, Germany (2004)

[2] Fidel, R., Bruce, H., Pejtersen, A.M., Dumais, S., Grudin, J., Poltrock, S.: Collaborative Information Retrieval (CIR). Information Behaviour Research 1, 235–247 (2000)

[3] Naderi, H., Rumpler, B.: PERCIRS: a PERsonalized Collaborative Information Retrieval System. In: INFORSID, Hammamet, Tunisia, June 1-3 2006, vol. 1, pp. 113–127 (2006) ISBN: 2-906855-22-7

[4] Naderi, H., Rumpler, B., Pinon, J.M.: An Efficient Collaborative Information Retrieval System by Incorporating the User Profile. In: Marchand-Maillet, S., Bruno, E., Nürnberger, A., Detyniecki, M. (eds.) AMR 2006. LNCS, vol. 4398, pp. 247–257. Springer, Heidelberg (2007)

[5] Papadimitriou, C., Steiglitz, K.: Combinatorial Optimization: Algorithms and Complexity. Prentice-Hall, Englewood Cliffs (1982)

[6] Smyth, B., Balfe, E., Boydell, O., Bradley, K., Briggs, P., Coyle, M., Freyne, J.: A Live User Evaluation of Collaborative Web Search. In: Proceedings of the 19th International Joint Conference on Artificial Intelligence, Edinburgh, Scotland (2005)

[7] Cormen, T.H., Leiserson, C.E., Rivest, R.L., Stein, C.: Introduction to Algorithms, 2nd edn., pp. 664–669. MIT Press and McGraw-Hill (2001) Section 26.3: Maximum bipartite matching. ISBN 0-262-03293-7

The Good, the Bad, the Difficult, and the Easy: Something Wrong with Information Retrieval Evaluation?

Stefano Mizzaro

Department of Mathematics and Computer Science
University of Udine
Via delle Scienze, 206
Udine, Italy
mizzaro@dimi.uniud.it
http://www.dimi.uniud.it/mizzaro/

Abstract. TREC-like evaluations do not consider topic ease and difficulty. However, it seems reasonable to reward good effectiveness on difficult topics more than good effectiveness on easy topics, and to penalize bad effectiveness on easy topics more than bad effectiveness on difficult topics. This paper shows how this approach leads to evaluation results that could be more reasonable, and that are different to some extent. I provide a general analysis of this issue, propose a novel framework, and experimentally validate a part of it.

Keywords: Evaluation, TREC, topic ease and difficulty.

1 Introduction

As lecturers, when we try to assess a student's performance during an exam, we distinguish between easy and difficult questions. When we ask easy questions to our students we expect correct answers; therefore, we give a rather mild positive evaluation if the answer to an easy question is correct, and we give a rather strong negative evaluation if the answer is wrong. Conversely, when we ask difficult questions, we are quite keen to presume a wrong answer; therefore, we give a rather mild negative evaluation if the answer to a difficult question is wrong, and we give a rather strong positive evaluation if the answer is correct.

The difficulty amount of a question can be determined a priori (on the basis of lecturer's knowledge of what and how has been taught to the students) or a posteriori (e.g., by averaging, in a written exam, the answer evaluations of all the students to the same question). Probably, a mixed approach (both a priori and a posteriori) is the most common choice.

During oral examinations, when we have an idea of student's preparation (e.g., because of a previous written exam, or a term project, or after having asked the first questions), we even do something more: we ask difficult questions to good students, and we ask easy questions to bad students. This sounds quite obvious too: what's the point in asking easy questions to good students? They will almost

C. Macdonald et al. (Eds.): ECIR 2008, LNCS 4956, pp. 642–646, 2008.

certainly answer correctly, as expected, without providing much information about their preparation. And what's the point in asking difficult questions to bad students? They will almost certainly answer wrongly, without providing much information — and incidentally increase examiner's stress level.

Therefore we can state the following principles, as "procedures" to be followed during student's assessment:

Easy and Difficult Principle. Weight more (less) both (i) errors on easy (difficult) questions and (ii) correct answers on difficult (easy) questions.

Good and Bad Principle. On the basis of an estimate of student's preparation, ask (i) difficult questions to good students and (ii) easy questions to bad students.

I am not aware of any lecturer/teacher/examiner which would not agree with the two principles, and which would not behave accordingly, once enlightened by them.

In Information Retrieval (IR) evaluation we are not enlightened, and we do not behave like that, at least so far. In TREC-like evaluation exercises [4], all topics are equal and concur equally to determine IR system effectiveness. If a topic is "easy" (e.g., systems are highly effective on it), and an IR system performs well on that topic, the system gets a boost in its overall effectiveness which is equal to the boost it would get when performing well on a more "difficult" topic. Vice versa, if a topic is "difficult", and an IR system performs poorly on that topic, the system gets a penalty in its overall effectiveness which is equal to the penalty it would get when performing poorly on a more "easy" topic.

The only related approach is to select the difficult topics (a posteriori, on the basis of average systems effectiveness) and to include them in the Robust Track [3]. However, this is of course quite different from the two above stated principles: it would correspond to ask difficult questions only, and anyway all the difficult topics are equally difficult. Also, the effectiveness metric used in the Robust Track (i.e., the GMAP, Geometric Mean Average Precision [2]) gives more weigh to changes in the low end of the effectiveness scale, i.e., to difficult topics, but this is again limited when compared to the two above stated principles.

Furthermore, in IR evaluation we do not take into account ease and difficulty neither at the document level: given a topic, if the relevance estimation of a document by an IR system is "easy" (i.e., it is easy to determine if the document is relevant or nonrelevant — or partially relevant or whatever — to the topic) and an IR system performs well on that document, the system gets a boost in its overall effectiveness which is equal to the boost it would get when performing well on a more "difficult" document. And vice versa. Even worse, when a system is performing well (poorly), it is asked to continue to answer easy (difficult) topics and to rank easy (difficult) documents, which it will likely do with good (bad) performance.

This paper is a first attempt to address these issues. I just concentrate on the first principle at the topic level; the other issues are left for future work.

Table 1. Good, Bad, Difficult, Easy

		Effectiveness (AP)	
		Bad	Good
Difficulty	Difficult	–	++
	Easy	– –	+

2   Ease and Difficulty

A first binary view is represented in Table 1: a good effectiveness on a difficult topic should increase system effectiveness a lot $(++)$; a good effectiveness on an easy topic should increase system effectiveness by a small amount, if any $(+)$; a bad effectiveness on an easy topic should decrease system effectiveness a lot $(--)$; a bad effectiveness on a difficult topic should decrease system effectiveness by a small amount, if any $(-)$.

Effectiveness can be defined, as usual in TREC, by means of AP (Average Precision, the standard effectiveness measure used in TREC): a high AP of a system on a topic means that the system is effective on the topic, although this neglects the ease/difficulty dimension. In a TREC-like setting, difficulty can be defined in a natural way a posteriori, as $1 - \text{AAP}$, where AAP (Average Average Precision [1]) is the average of AP values across all systems for a single topic. Hence, the difficult topics are those with a low AAP, i.e., the topics with a low average effectiveness of the systems participating in TREC. Of course this is just one among all the possible alternatives, since topic difficulty could be defined, e.g., by considering the minimum effectiveness in place of the average, or the maximum effectiveness, or by considering the best systems only, etc.

Therefore, a high AP (Average Precision, the standard effectiveness measure used in TREC) of a system on a topic could mean not only good system (high effectiveness) but also easy topic (low difficulty); conversely, low AP means bad system (low effectiveness) and/or difficulty (high difficulty).

There are several (actually, infinite) ways to turn the binary view into a continuous one. In this paper I stick with a possible choice, i.e., the function shown in Figure 1 and defined as

$$\text{NAP}(e, d) = [(1 - d) \cdot M_E + d \cdot (1 - m_D)] \cdot e^{K^{1-2d}} + d \cdot m_D.$$

This is a function from $[0, 1]^2$ into $[0, 1]$, the two variables being system effectiveness e and topic difficulty d (measured, respectively, as AP and $1 - \text{AAP}$). The result is NAP, a "normalized" version of AP values, that takes into account topic difficulty: $\text{NAP}(e, d)$ has higher values for higher e, and it has higher values, and increases more quickly, for higher d (right hand side of the figure). M_E is the maximum NAP value that can be obtained on an easy ($d = 0$, $\text{AAP} = 1$) topic. m_D is the minimum NAP value that can be obtained on a difficult ($d = 1$) topic. The model could include other 2 parameters m_E and M_D with obvious meanings, but it is natural to set $m_E = 0$ and $M_D = 1$. Also, in the figure and

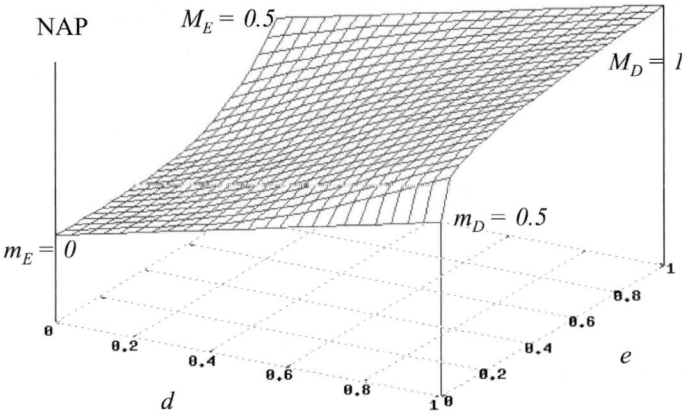

NAP $M_E = 0.5$

$M_D = 1$

$m_D = 0.5$

$m_E = 0$

d

e

Fig. 1. The normalization function

in the following experiments, $M_E = m_D = \frac{1}{2}$. $K \geq 1$ allows different curvatures (in figure, $K = 4$; in the following, $K = 100$; higher K values lead to stronger normalizations, but for lack of space the role of K is not discussed here).

The proposed function is just one among infinite possible choices: Table 1 just sets some constraints on the four corners of Figure 1 (d and $e \in \{0, 1\}$); the chosen parameters values m_D, m_E, M_E, and M_D, satisfy these constraints, but of course their values could be different; and the interpolation of the four corners could be done in infinite ways. The study of variants is left as future work.

3   Experiments and Results

Averaging across topics the NAP values obtained as above described, we obtain a new measure of retrieval effectiveness, that I name NMAP, for Normalized MAP (Mean Average Precision). We can then compare retrieval effectiveness as measured by MAP and NMAP. I use data from TREC 8 (129 systems, 50 topics).

Figure 2 shows the differences in ranking of the 129 systems participating in TREC when their effectiveness is measured by MAP and NMAP. It is clear from the scatterplot that the two rankings are quite different, although related (Kendall's tau correlation is 0.87, linear correlation is 0.92). This means that by using NMAP instead of MAP one would get different rankings of the systems participating in TREC. In other words, what is generally considered an improved version of a system (a version with a higher MAP) would often turn out to be not an improvement at all when using NMAP, which is based on the reasonable assumptions sketched in Section 1. As the figure shows, MAP and NMAP do quite agree on the best systems, those in the first 20 positions or so, with very few exception (see the left hand side of the figure). However, the agreement decreases after the 20th system, with strong disagreement for a dozen of systems (the dots that stand out).

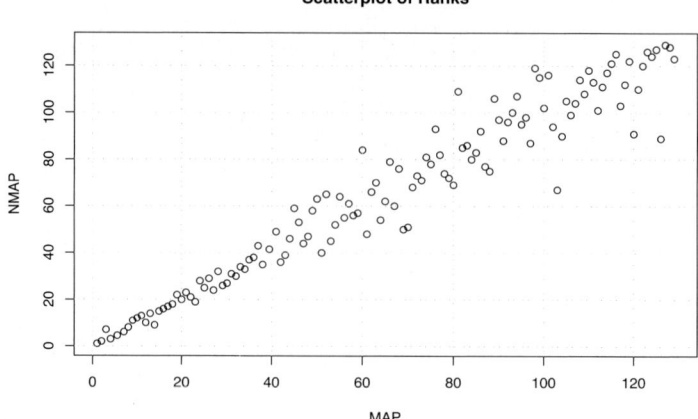

Fig. 2. Differences in systems rankings

4   Conclusions and Future Work

These preliminary experiments do indeed hint that if we followed the first principle stated in Section 1, TREC results could be somewhat different (in terms of both system ranking and absolute effectiveness values): we might be evaluating TREC systems in a wrong way.

This paper can be seen as a research agenda, since further work is needed to confirm these results, on several aspects. The normalization function could be improved (e.g., it could be rewritten with a GMAP [2] flavor, exploiting logarithms). It will be interesting to see what happens when the second principle is considered as well, since this might lead to reduce the number of topics used in TREC-like evaluations, and when the same analysis is extended to the document level. From a different point of view, NMAP is a new metric for retrieval effectiveness; it will be interesting to study its relationships with other metrics (like GMAP, which seems to be a special case of NMAP), its general properties (e.g., stability), and its relationship with user satisfaction (by means of user studies).

References

1. Mizzaro, S., Robertson, S.: HITS Hits TREC - Exploring IR Evaluation Results with Network Analysis. In: 30th SIGIR, pp. 479–486 (2007)
2. Robertson, S.: On GMAP – and other transformations. In: 13th CIKM, pp. 78–83 (2006)
3. Voorhees, E.M.: Overview of the TREC 2005 Robust Retrieval Track. In: TREC 2005 Proceedings (2005)
4. Voorhees, E.M., Harman, D.K.: TREC — Experiment and Evaluation in Information Retrieval. MIT Press, Cambridge (2005)

Hybrid Method for Personalized Search in Digital Libraries

Thanh-Trung Van and Michel Beigbeder

Centre G2I/Département RIM
Ecole Nationale Supérieure des Mines de Saint Etienne
158 Cours Fauriel, 42023 Saint Etienne, France
{van,mbeig}@emse.fr

Abstract. In this paper we present our work about personalized search in digital libraries. The search results could be reranked while taking into account specific information needs of different people. We study many methods for this purpose: citation-based method, content-based method and hybrid method. We conducted experiments to compare performances of these methods. Experimental results show that our approaches are promising and applicable in digital libraries.

1 Introduction and Related Work

Search in digital libraries is usually a boring task. Users have to repeat the tedious process of searching, browsing, and refining queries until they find relevant documents. This is because different users have different information needs, but users queries are often short and, hence, ambiguous. For example, the same query "java" could be issued by a person who is interested in geography information about Java island or by another person who is interested in Java programming language. Even with a longer query like "java programming language", we still do not know which kind of document this user want to find. If she/he is a programmer, perhaps she/he is interested in technical documents about the Java language; however, if she/he is a teacher, perhaps she/he wants to find tutorials about Java programming for her/his course. This problem could be avoided if the system can learn some information about the interests and the preferences of users and use this information to improve their search results. This information is gathered in *user profile*.

The work of Amato et al. [1] presents a user profile model that can be applied to digital libraries. In this model, information about a user is classified in five data categories: i) the *personal data category* ii) the *gathering data category* iii) the *delivering data category* iv) the *actions data category* v) the *security data category*.

In [2], the authors propose some approaches for re-ranking the search results in a Digital Library that contains digitized books. They consider two kinds of search: search for books by querying on the metadata of books (Metadata Search) and search for informations in the pages of book by querying using keywords (Content Search). They use two different profiles corresponding to

C. Macdonald et al. (Eds.): ECIR 2008, LNCS 4956, pp. 647–651, 2008.
© Springer-Verlag Berlin Heidelberg 2008

these two kinds of search. Metadata search results and content search results are re-ranked using these profiles.

The CiteSeer digital library [3] that contains scientific papers uses a heterogenous profile to represent the user interests. If there is a new available paper, CiteSeer will try to decide if this paper would be interesting to the user (i.e. information filtering) using user profile. If so, then the user can be alerted about this paper.

2   Approaches for Personalized Search in Digital Libraries

Our work focus on personalized search in digital libraries of scientific papers. Like in the CiteSeer system [3], the user profile is represented by a set of paper that are interesting to the user. Each time the user issues a query, the first **n** documents[1] will be re-ranked using the original score computed by the search engine and the similarity between the document and the user profile. The similarity between a document and a profile is the sum of the similarity between this document and each document in the user profile:

$$similarity(d, p) = \sum_{d' \in p} similarity(d', d) \qquad (1)$$

The document-profile similarity is computed using two methods: a content-based method and a citation-based method. We use the **zettair**[2] search engine to compute the content-based similarity (under the vector-space model). The citation-based similarity is based on the principle of the co-citation method [4]. In this method, the relatedness between two papers is based on their *co-citation frequency*. The co-citation frequency is the frequency that two papers are *co-cited*. Two papers are said to be *co-cited* if they appear together in the bibliography section of a third paper. However if we want to know this citation information, we have to extract the *citation graph* from the actual library or to get this information from a *citation database*[3]. Both methods are usually limited; i.e. we can only know citing papers of a paper **A** if the citing papers exist within the same digital library or citation database with the paper **A**. Many works [5,6] showed that if the size of a digital library or citation database is not big enough, then the performance of this method will be limited. That is why we propose to use the Web as a citation database to find the similarity between scientific papers. Our method is called *Web co-citation* method.

In our Web co-citation method, we compute the co-citation similarity of two scientific papers by the frequency that they are "co-cited" on the Web. The notion of "co-citation" used here is a "relaxation" in comparison with the traditional definition. If the Web document that mentions two scientific papers is another scientific paper then these two papers are normally co-cited. However,

---

[1] In our experiments n = 300.
[2] http://seg.rmit.edu.au/zettair/
[3] A citation database is a system that can provide bibliographic information of papers.

if this is a table of content of a conference proceeding, we could also say that these two papers are co-cited and have a relation because a conference normally has a common general theme. If these two papers appear in the same conference, they may have the same general theme. Similarly, if two papers are in the reading list for a course, they may focus on the same topic of this course. In summary, if two papers appear in the same Web document, we can assume that they have a (strong or weak) relation. The search engine used in our experiment is the Google search engine. To find the number of time that a paper is "cited" by Google we need only to send the title of this paper (as phrase search using quotation marks) to Google and note the number of hits returned. Similarly, to find the number of times that two papers are "co-cited", we send the titles of these two papers (as phrase search and in the same query) to Google and note the number of hits returned. In our experiments, we use a script to automatically query Google instead of manually using a Web browser. The similarity between two papers is computed by the following formula:

$$cocitation\_similarity(d', d) = ln \left(\frac{cocitation(d', d)^2}{citation(d') + citation(d)} \right) \qquad (2)$$

In Eq. 2, $cocitation(d', d)$ is the number of times that these two papers are co-cited, $citation(d')$ and $citation(d)$ are respectively the citation frequency that papers d' and d received. Note that in the Web co-citation method, the document-profile similarity (cf. formule 1) has a negative value, we convert it into a positive value by this formula:

$$similarity'(d, p) = \frac{1}{|similarit(d, p)|} \qquad (3)$$

The final score that is used for re-ranking is a combination between the following scores: i) the original score computed by the search engine ii) the document-profile similarity computed by the Web co-citation method iii) the document-profile similarity computed by the content-based method. The combination formulas are the two following formulas:

– Linear formula:
$$final\_score = \sum_i \alpha_i \times score_i \qquad (4)$$

– Product formula:
$$final\_score = \prod_i score_i \qquad (5)$$

In the formula 4, α_i are positive coefficients that satisfy the condition $\sum_i \alpha_i = 1$. We tried many different combinations to find the best coefficients. The scores are normalized (divided by the correspondent maximal value) to have the values in the range from 0 to 1. We conducted experiments to evaluate the performance of different combination methods. The experiments are presented in the following section.

3   Experiments and Results

The search engine that we use is the **zettair** search engine, the default model used in **zettair** is the *Dirichlet-smoothing* model. The test collection that we use is the collection used in INEX 2005[4]. This is a collection of scientific papers extracted from journals and transactions of *IEEE Computer Society*. INEX provides also many topics with relevance assessments. Our work simulates the user of user profiles for personalized search. We consider that each topic represents a different information need of one person. The user profile is built from the documents which are judged as relevant. We use a k-fold cross-validation approach [7] for the evaluation. In this approach, the relevant documents of each topic are partitioned into **k** subsets. The documents in a subset are used as test documents and other documents in other **k** − 1 subsets are used as the user profile. The experiment is repeated **k** times, each time a different subset is used as test subset. The evaluation metric is precision at **n** (with **n** = 5 10 15 20 30). Because there are **k** different experiments, with each value of **n** there are **k** different precisions, therefore we have to compute the average value:

$$Average\_of\_precisions\_at\_n = \frac{\sum_{i=1}^{k} precision\_at\_n_i}{k} \tag{6}$$

Table 1. Average of precisions at 5, 10, 15, 20, 30 documents

	Result of zettair	Web Co-citation	Content-Based	Hybrid Approach
5 docs	0.2892	0.3108 (p) (+7,5%) 0.3185 (l) (+10,1%)	0.3185 (p) (+10,1%) 0.3462 (l) (+19,7%)	0.3369 (p) (+16,5%) 0.3631 (l) (+25,6%)
10 docs	0.2123	0.2446 (p) (+15,2%) 0.2477 (l) (+16,7%)	0.2362 (p) (+11,3%) 0.2715 (l) (+27,9%)	0.2661 (p) (+25,3%) 0.2869 (l) (+35,1%)
15 docs	0.1672	0.1944 (p) (+16,3%) 0.1974 (l) (+18,1%)	0.1959 (p) (+17,2%) 0.2174 (l) (+30,0%)	0.2159 (p) (+29,1%) 0.2221 (l) (+32,8%)
20 docs	0.1473	0.1600 (p) (+8,6%) 0.1639 (l) (+11,3%)	0.1677 (p) (+13,8%) 0.1815 (l) (+23,2%)	0.1758 (p) (+19,3%) 0.1781 (l) (+20,9%)
30 docs	0.1154	0.1200 (p) (+4,0%) 0.1215 (l) (+5,3%)	0.1274 (p) (+10,4%) 0.1374 (l) (+19,1%)	0.1297 (p) (+12,4%) 0.1408 (l) (+22,0%)

Results are presented in Table 1. The second column is the original results of **zettair** search engine. The third column is the results of the re-ranking method using two scores: the original score of **zettair** and the citation-based document-profile similarity. The fourth column corresponds to the re-ranking method using the original score of **zettair** and the content-based document-profile similarity. The fifth column corresponds to the re-ranking method using all these three scores. With each method, **p** means product combination (cf. formula 5) and **l** means linear combination (cf. formula 4).

---

[4] http://inex.is.informatik.uni-duisburg.de/2005/

From the results, we can see that all three methods can bring amelioration. The content-based method is better than citation-based method. However, the hybrid approach brings the best performance. Furthermore, the amelioration seems to be more clear with precisions at 5, 10 and 15 documents.

4   Conclusions and Future Work

In this paper, we have present some methods for personalized search in digital libraries. We did experiments on the INEX collection to compare the performance citation-based method, the content-based method and the hybrid method. Experimental results showed that these methods are efficient and the hybrid method is the best method. In the future, knowing that there are similar points between citations and hyperlinks, we intend to do similar experiments on a collection of Web pages to compare the performance of these methods in hyperlinked environment.

Acknowledgement

This work is done in the context of the European Project CODESNET and supported by the Web Intelligence Project of the "Informatique, Signal, Logiciel Embarqué" cluster of the Rhône-Alpes region.

References

1. Amato, G., Straccia, U.: User profile modeling and applications to digital libraries. In: Abiteboul, S., Vercoustre, A.-M. (eds.) ECDL 1999. LNCS, vol. 1696, pp. 184–197. Springer, Heidelberg (1999)
2. Rohini, U., Ambati, V.: A collaborative filtering based re-ranking strategy for search in digital libraries. In: Fox, E.A., Neuhold, E.J., Premsmit, P., Wuwongse, V. (eds.) ICADL 2005. LNCS, vol. 3815, pp. 194–203. Springer, Heidelberg (2005)
3. Bollacker, K., Lawrence, S., Giles, C.L.: A system for automatic personalized tracking of scientific literature on the web. In: Digital Libraries 1999, pp. 105–113. ACM Press, New York (1999)
4. Small, H.G.: Co-citation in the scientific literature: A new measure of the relationship between two documents. Journal of American Society for Information Science 24(4), 265–269 (1973)
5. Huang, S., Xue, G.R., Zhang, B.Y., Chen, Z., Yu, Y., Ma, W.Y.: Tssp: A reinforcement algorithm to find related papers. In: WI 2004, Washington, DC, USA, pp. 117–123. IEEE Computer Society, Los Alamitos (2004)
6. Couto, T., Cristo, M., Goncalves, M.A., Calado, P., Ziviani, N., de Moura, E.S., Ribeiro-Neto, B.A.: A comparative study of citations and links in document classification. In: JCDL 2006 (2006)
7. Kohavi, R.: A study of cross-validation and bootstrap for accuracy estimation and model selection. In: IJCAI, pp. 1137–1145 (1995)

Exploiting Session Context for Information Retrieval - A Comparative Study

Gaurav Pandey and Julia Luxenburger

Max-Planck Institute of Informatics, Saarbrücken, Germany
gpandey_01@yahoo.com, julialux@mpi-inf.mpg.de

Abstract. Hard queries are known to benefit from relevance feedback provided by users. It is, however, also known that users are generally reluctant to provide feedback when searching for information. A natural resort not demanding any active user participation is to exploit implicit feedback from the previous user search behavior, i.e., from the context of the current search session. In this work, we present a comparative study on the performance of the three most prominent retrieval models, the *vector-space*, *probabilistic*, and *language-model based* retrieval frameworks, when additional session context is incorporated.

1 Introduction

With the advent of the *language modeling* [1] paradigm for information retrieval in the late nineties, also the question arose how to integrate relevance feedback in this newly established retrieval framework. Until then, the most common methods for relevance feedback were either an integral part of the model, as for the *probabilistic* retrieval model [5], or in the case of the *vector-space* model [5], a manipulation of the original query by means of query term re-weighting and query expansion according to the *Rocchio* [2] relevance feedback framework. Zhai and Lafferty [8] were the first to propose a feedback approach naturally in consistence with the language modeling approach, and more specifically the Kullback-Leibler divergence framework introduced in [1]. The essence of the KL-divergence framework is to represent both the query and the document by means of language models Θ_Q and Θ_D, and determine the relevance of a document to a given query by measuring the Kullback-Leibler divergence between the two language models

$$D_{KL}(\Theta_Q||\Theta_D) = \sum_{w\in V} P(w|\Theta_Q) \log \frac{P(w|\Theta_Q)}{P(w|\Theta_D)}$$

where V is the vocabulary set. Feedback can be naturally integrated in this framework by updating the query language model.

More recently, *implicit feedback* derived from user interactions with the search interface has attracted more attention [3,6,4,7]. The key idea is to improve retrieval results without imposing any additional burden on the user by exploiting all information extractable from the search process, i.e., the series of query reformulations the user performs until she finds the desired information, as well as,

C. Macdonald et al. (Eds.): ECIR 2008, LNCS 4956, pp. 652–657, 2008.

the documents visisted in that process. [7] incorporates implicit feedback into the probabilistic ranking scheme BM25 [5], [4] utilizes the Rocchio relevance feedback model [2] to update the query representation inside the vector-space model, whereas [3,6] study methods for the integration of implicit feedback into the KL-divergence framework. To the best of our knowledge, there has, however, not been any attempt of comparing the performance of the utilization of implicit feedback across these retrieval models.

In this work, we question the superiority of the Kullback-Leibler divergence framework assumed by [3,6] when studying approaches for exploiting implicit feedback from user search behavior. More precisely, we translate the implicit feedback approach described in [6] to both the probabilistic and the vector-space retrieval model. In contrast to [6], we focus on the *short-term* query context, i.e., the immediate preceding user actions within the current search session, as we believe short-term context to be a better indicator of the current search interest than *long-term* user interests revealed by the farther reaching user search history. Furthermore we study additional variants of the method described in [6]. Our experiments on the TREC Aquaint data set, indeed, strengthen our doubts on the superiority of the language model framework, and favor the classical probabilistic retrieval model BM25. This coincides with our expectation as previous TREC benchmarks prove BM25 to be the best performing retrieval model.

2   Implicit Feedback from the Session Context

In the following we will describe the implicit feedback approach introduced in [6] in the context of the KL-divergence framework, as well as, its translation to the probabilistic and vector-space retrieval model. The key idea in [6] is to update the query language model Θ_Q according to the observed user search behavior. The updated query model Θ_Q^{new} is a linear combination of the old query model and a search history model Θ_H, i.e., $\Theta_Q^{new} = \alpha \cdot \Theta_Q + (1-\alpha) \cdot \Theta_H$ with $\alpha \in [0,1]$. Suppose the current query is the k-th query in the current search session, i.e., the search history H consists of k - 1 queries q_i $(i = 1, \ldots, k-1)$ preceding the current query. Each history query q_i together with its set of clicked (C_i) and non-clicked (NC_i) search results then makes up one *unit history model* Θ_{H_i}, and the overall history model is obtained by a weighted combination of these unit history models as follows.

$$\Theta_H = \frac{\sum_{i=1}^{k-1} \beta_i \cdot \Theta_{H_i}}{\sum_{i=1}^{k-1} \beta_i}$$

Tan et al. estimate the unit history model as follows.

$$\Theta_{H_i} = \lambda \Theta_{Q_i} + (1-\lambda) \cdot \frac{\sigma_C \sum_{d \in C_i} \Theta_d + \sigma_{NC} \sum_{d \in NC_i} \Theta_d}{\sigma_C |C_i| + \sigma_{NC} |NC_i|}$$

That is the unit history model Θ_{H_i} is a linear combination of the query model Θ_{Q_i} and the weighted average of clicked and non-clicked result documents' language models. As Tan et al. report best performance when $\sigma_C = 1, \sigma_{NC} = 0$, i.e., only clickthrough documents as opposed to the whole set of result documents are considered, we choose this parameter setting in our experiments. For the estimation of the basic query and clicked documents' language models used as building blocks in the unit history model we take a MLE (maximum-likelihood estimation) approach, i.e., $\Theta_Q = \frac{c(w,Q)}{|Q|}$ where $c(w,Q)$ is the count of word w in query Q and $|Q|$ is the query length, and $\Theta_D = \frac{c(w,D_{title})}{|D_{title}|}$ where D_{title} denotes the title of document D.

For the decision of how to weight the various unit history models against each other, i.e., how to choose the parameters β_i, we consider four different approaches. (1) **Equal weighting** weights all history queries the same, i.e., $\beta_i = 1 \forall i$. (2) **Cosine similarity** takes the cosine similarity of the current query string with the concatenation of query and search result title strings of each history query into account. In addition to these schemes already proposed in [6], we consider (3) **overlap**, and (4) **time**. Overlap is a simplistic approach that reasons only on the overlap of the current query string and the history query string. In case of overlap β_i is set to 1, otherwise to 0.25, i.e., the corresponding history query is lower-weighted but not completely ignored. We expect overlap to trade a decrease in precision for a decrease in processing time compared to cosine similarity. In the time-based approach we assume that the query formulations are improving with time, i.e., we weight more recent history queries higher. Mathematically this is $\beta_i = \frac{1}{k-i} \forall i$. E.g., if the current query is the 3rd in the session, $\beta_1 = \frac{1}{3-1} = \frac{1}{2}$ and $\beta_2 = 1$.

In the following we describe the simplistic approach we take to compare this reasoning for the incorporation of implicit feedback under the three retrieval models, the probabilistic, vector-space and language-model framework, each represented by its most prominent incarnation. We start by presenting the retrieval formula representing the KL-divergence framework which uses Dirichlet prior smoothing for estimating document language models,

$$\sum_{w \in Q,D} \frac{c(w,q)}{|Q|} \cdot \ln\left(1 + \frac{c(w,D)}{\mu \cdot \frac{c(w|Coll)}{|Coll|}}\right) + \ln\frac{\mu}{|D| + \mu}$$

where $c(w|Coll)$ is the word count of w in the whole collection and μ is typically 2000 (see [9] for more details). When integrating implicit feedback the MLE query model $\frac{c(w,q)}{|Q|}$ in the left summation term is replaced by the updated query model Θ_Q^{new}. In the case of the vector-space model [5] we consider the pivoted normalization scheme

$$\sum_{w \in Q,D} \frac{1 + \ln\left(1 + \ln\left(c(w,d)\right)\right)}{(1-s) + s \cdot \frac{|D|}{avgLen}} \cdot c(w,Q) \cdot \ln\frac{N+1}{df(w)}$$

where the constant s is usually 0.2, N is the collection size, $avgLen$ the average document length, and $df(w)$ denotes the document frequency of word w, i.e., the number of documents in the collection containing word w (see [5] for more details). The ranking of documents stays unaffected if we divide this ranking formula by the constant term $|Q|$, yielding $\frac{c(w,Q)}{|Q|}$ as the only query dependent term in the formula. This term is actually the same as what we called the query language model in the KL-divergence framework. We thus take a very ad-hoc approach of generalizing the implicit feedback approach of Tan et al. to the vector-space model, by substituting this term by Θ_Q^{new}.

We reason along the same line when adopting Tan et al.'s implicit feedback approach to the Okapi BM25 formula, our representative of the classical probabilistic framework.

$$\sum_{w \in Q, D} \ln \frac{N - df(w) + 0.5}{df(w) + 0.5} \cdot \frac{(k_1 + 1) \cdot c(w, D)}{k_1 \cdot ((1 - b) + b\frac{|D|}{avgLen}) + c(w, D)} \cdot \frac{(k_3 + 1) \cdot c(w, Q)}{k_3 + c(w, Q)}$$

where $k_1 = 1.2, b = 0.75, k_3 \in [0, 1000]$ (see [5] for more details). If we consider a large value of k_3 with $c(w, q) << k_3$, the query-dependent part $\frac{(k_3+1) \cdot c(w,Q)}{k_3+c(w,Q)}$ can be approximated by $\frac{(k_3+1) \cdot c(w,Q)}{k_3}$. The constant term $\frac{k_3+1}{k_3}$ does not affect the ranking of documents and can be ignored. With the same argument we can divide the whole formula by the constant $|Q|$ without changing the ranking of documents, and are again left with the query-language-model like expression $\frac{c(w,Q)}{|Q|}$ which we replace with Θ_Q^{new} in the case of implicit feedback.

3   Experiments

Our experiments use the *Aquaint* news data set of TREC (http://trec.nist.gov) which contains 1,033,461 articles . We study 50 topics from the TREC 2004 robust track (topics 651 - 700) which are known to be difficult. The relevance assessments for these topics are available, yet we lack some search session context information. Therefore we ask four volunteers (master students in computer science) to search for the 50 TREC topics. We log the sequence of queries they pose, as well as, the search results they click until they find the desired information. For evaluation, we treat the last query in a logged search session as the current query so that all preceding queries form the search context. The that way obtained search context comprises 5 queries and 4 clicked documents on average with 37.6% of the clicked documents being indeed relevant for the search topic.

Results. We compute the mean average precision (MAP) for all three retrieval models, all combinations of parameter choices for $\alpha = \{0.2, 0.4, 0.6, 0.8\}$ and $\lambda = \{0, 0.2, 0.4, 0.6, 0.8, 1\}$, as well as for all four weighting schemes for choosing the β_i's. We report in the following the MAP of the best performing parameter setting for each combination of weighting scheme and retrieval model.

	VS	BM25	LM
No Feedback	0.133	0.149	0.138
Equal Weighting	0.172	0.18	0.173
Cosine Similarity	0.168	0.186	0.173
Overlap	0.165	0.184	0.17
Time	0.166	0.184	0.17

Clearly, the incorporation of implicit feedback improves the retrieval performance across all ranking schemes. Two-tailed paired t-tests for comparing a particular scheme without as opposed to with implicit feedback according to the cosine similarity method give p-values < 0.02. However, the distinctive choice of only selected parts of the search history does not seem to have a large impact when only the small short-term context of the current session is considered. Comparing the three retrieval models, BM25 outperforms the other schemes in any setting. A two-tailed paired t-test gives a p-value of 0.016 for the comparison between the pure BM25 and vector-space model, and 0.02 when both ranking schemes incorporate feedback with cosine similarity. The comparison between BM25 and the language modeling approach shows a statistically non-significant improvement.

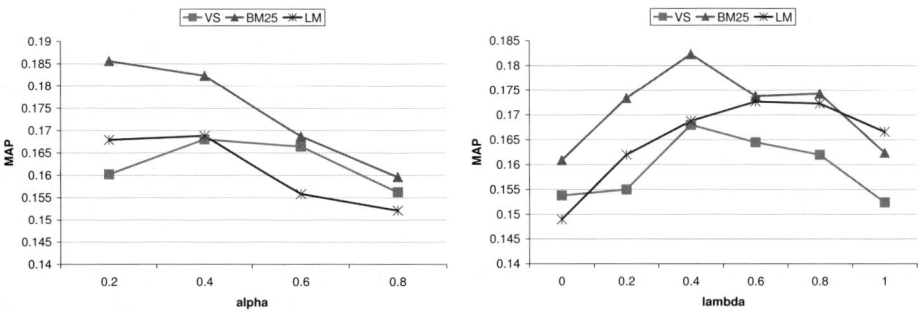

Fig. 1. MAP values when $\lambda = 0.4$ (left) and $\alpha = 0.4$ (right)

When comparing the sensitivity of our results to the choice of the parameters α and λ, all three retrieval model seem to react very similar to different combinations of α and λ, as depicted in Figure 1 which exemplarily shows the MAP values when the β_i's are chosen according to cosine similarity. In each plot we vary either α or λ while the respective other parameter is 0.4. The MAP values decrease when α increases indicating the benefit of incorporting history. Inside unit history models, $\lambda \approx 0.5$ seems to give best performance when query string and clickthrough of the history queries are equally weighted.

References

1. Lafferty, J., Zhai, C.: Document language models, query models, and risk minimization for information retrieval. In: SIGIR 2001 (2001)
2. Rocchio, J.: Relevance feedback in information retrieval. In: The SMART Retrieval System: Experiments in Automatic Document Processing, pp. 313–323 (1971)

3. Shen, X., Tan, B., Zhai, C.: Context-sensitive information retrieval using implicit feedback. In: SIGIR 2005 (2005)
4. Shen, X., Tan, B., Zhai, C.: Implicit user modeling for personalized search. In: CIKM 2005 (2005)
5. Singhal, A.: Modern information retrieval: A brief overview. In: Bulletin of IEEE Computer Society Technical Committee on Data Engineering (2001)
6. Tan, B., Shen, X., Zhai, C.: Mining long-term search history to improve search accuarcy. In: KDD 2006 (2006)
7. Teevan, J., Dumais, S., Horvitz, E.: Personalizing search via automated analysis of interests and activities. In: SIGIR 2005 (2005)
8. Zhai, C., Lafferty, J.: Model-based feedback in the language modeling approach to information retrieval. In: CIKM 2001 (2001)
9. Zhai, C., Lafferty, J.: A study of smoothing methods for language models applied to ad hoc information retrieval. In: SIGIR 2001 (2001)

Structural Re-ranking with Cluster-Based Retrieval

Seung-Hoon Na[1], In-Su Kang[2], and Jong-Hyeok Lee[1]

[1] POSTECH,Pohang,South Korea
{nsh1979,jhlee}@postech.ac.kr
[2] KISTI,Daejeon,South Korea
dbaisk@kisti.re.kr

Abstract. Re-ranking (RR) and Cluster-based Retrieval (CR) have been polar methods for improving retrieval effectiveness by using inter-document similarities. However, RR and CR improve precision and recall respectively, not simultaneously. Thus, the improvement through RR and CR may be different according to whether a query is recall-deficient or not. However, previous researchers missed out this point, and separately investigated individual approaches, causing a limited improvement. To reflect all of positive effects by RR and CR, this paper proposes RCR, the re-ranking with cluster-based retrieval where RR is applied to initially-retrieved results of CR. Experimental results show that RCR significantly improves the baseline, while CR or RR sometimes does not significantly improve the baseline.

1   Introduction

The most popular methods to use inter-document similarities are Cluster-based Retrieval (CR) [1,2] and Re-ranking (RR) [3,4,5]. To recover unseen terms or less-estimated terms from a document, CR refers to other similar documents of the original document, thereby resolving the term-mismatch problem. Thus, CR is helpful to improve the recall. RR re-ranks top n retrieved documents to make similar documents obtain similar scores. Thus, when some relevant documents obtain higher scores in the initial retrieval, RR can highly rank other relevant documents which do not obtain high scores in the initial retrieval. The set of initially-retrieved documents is not changed after RR, thus RR is helpful to the precision, rather than the recall. Due to these different characteristics of CR and RR, an individual use of them cannot effectively deal with all of query types. For example, when a user's query is recall-deficient so that the top retrieved documents seriously suffer from low recall, CR could be effective rather than RR. On the other hand, when the set of initially retrieved documents has high recall but suffer from low precision, RR could be more effective than CR.

None of previous works has focused on the combination of CR and RR, even though the combining deserves to be investigated due to their different roles to the retrieval effectiveness. In this regard, we propose RCR, the naive combining approach where RR is applied to the documents obtained from CR, not from the baseline retrieval. In spite of its simplicity, RCR is expected to produce high-quality retrieval results, since it can simultaneously integrate the recall-improving and precision-improving capabilities from CR and RR.

C. Macdonald et al. (Eds.): ECIR 2008, LNCS 4956, pp. 658–662, 2008.
© Springer-Verlag Berlin Heidelberg 2008

Experimental results on two standard TREC test collections show that CR or RR alone does not always significantly improve the baseline, but RCR significantly improves the baseline for all test collections and produces high retrieval performance over both CR and RR. Considering that most previous approaches have separately investigated either CR or RR, our result is remarkable.

2   Structural Re-ranking with Cluster-Based Retrieval

CR adjusts the similarity score of a document to a query by using the similarity score of its corresponding cluster to the query. Let D, Q and C_D be a document, a query, and a cluster containing document D, respectively. The final score of a document used in CR is calculated by linearly interpolating the score of the document and the score of the cluster as follows.

$$score_{CR}(Q,D) = \alpha score(Q,D) + (1 - \alpha)score(Q,C_D) \tag{1}$$

where α is an interpolation parameter. Here, we assume that a document has a single corresponding cluster. Generally, there can be multiple clusters for each document.

The key-part of CR is to create clusters. A cluster is defined as a set of documents. Different types of clusters can be differently defined according to cluster-based retrieval methods - 1) partitional clustering such as K-means produces non-overlapped clusters, 2) document expansion produces overlapped clusters each of which consists of nearest neighbors of a given document. This paper adopts document expansion instead of partitional clustering due to the following reasons. First, the partitional clustering has many variants of clustering algorithms, and algorithm-dependent parameters, but document expansion takes the number of nearest neighbors as a single parameter. Second, the retrieval effectiveness of document expansion has been verified by more researchers [1,2], than partitional clustering. In document expansion, C_D is defined as

$$C_D = \left\{ D' \mid D' \in NN_k(D) \right\} \tag{2}$$

where $NN_k(D)$ is the set of k-nearest-neighbor documents of D. We use $score(D,D')$ to find the nearest documents of D where document D becomes a query.

Structural re-ranking (RR) (i.e. Re-ranking) is to re-rank the top n documents that an initial retrieval model produces, where the re-ordering utilizes inter-document similarities within the set [4]. Conceptually, RR forces similar documents to have similar scores. Thus, when some relevant documents are assigned to high similarity scores, other relevant documents which do not obtain high scores in the initial retrieval can be highly ranked after RR. The basic approach of RR defines the graph of documents where a node corresponds to a document, and the weight of an edge to the inter-document similarity between documents. Then, the goal of RR can be formulated to determine the scores of document nodes from the structure of a graph. Initially, scores of nodes are set to retrieval scores that the initial retrieval model yields, then RR re-estimates scores of document nodes from the structure of the graph. Zhang et al. applied the random surferring model similar to PageRank on the graph, where the random jumping was utilized to prevent the re-ranked ones from being far from their initial scores [3].

Yang et. al. viewed the re-ranking problem as the problem of semi-supervised learning, and applied the label propagation [5]. The label propagation starts from an incomplete graph where only some nodes' labels are known, and propagates label information from the labeled nodes to the unlabeled nodes on the graph, resulting in label probabilities. To make seed labeled nodes, Yang et. al. assigned relevant labels to the most highly-ranked documents among top-retrieved ones and non-relevant labels to the lowest ranked documents. Then, document nodes are re-ranked according to the finally obtained label probabilities from the label propagation.

We adopted Zhang's random surferring model due to its simplicity. Let $D' \rightarrow_w D$ be the weight of the edge where D is linked from D'. For $D' \rightarrow_w D$, we use $score(D', D)$ where D' is regarded as a query. Using $D' \rightarrow_w D$, the translation probability from a node to node D is calculated as follows:

$$t(D|D') = \frac{D' \rightarrow_w D}{\sum_D D' \rightarrow_w D} \qquad (3)$$

In addition, the prior probability that visits node D is assumed to be proportional to $score(Q, D)$ which the initial retrieval model produces. Then, re-ranking re-calculates the score of document D as $score_{RR}(Q, D)$ as follows:

$$score_{RR}(Q, D) \leftarrow (1 - \beta) \sum_{D'} t(D|D') score_{RR}(Q, D') + \beta score(Q, D) \qquad (4)$$

where β is a random jumping factor. The steady value of $score_{RR}(Q, D)$ is obtained by applying the fixed-point iterations of Eq. (4), until $score_{RR}(Q, D)$ is not further changed. We found that the number of iterations between 10 and 30 is adequate. Note that Eq. (4) can be re-written in terms of probabilities with dividing both sides of Eq. (4) by $\sum_D score_{RR}(Q, D)$, thus $score_{RR}(Q, D)$ is converted to the probability (when $score_{RR}(Q, D)$ is non-negative). In above sections, $score(Q, D)$ is dependent to the retrieval model. We selected the language modeling approaches for $score(Q, D)$. Among popular smoothing methods, we used Dirichlet-prior smoothing due to its relative superiority to others [6]. Let $c(w; D)$ and $c(w; Q)$ be the term frequency of w in D and Q, respectively. And, assume that θ_C is the collection language model. $score(D, Q)$ is defined as follows:

$$score(Q, D) = \sum_{w \in Q} log \left(\frac{c(w; D) + \mu P(w|\theta_C)}{\sum_w c(w; D) + \mu} \right) \qquad (5)$$

where μ is a smoothing parameter. Note that this formula is also utilized for $score(Q, C_D)$, in which C_D is a large document where all elements of documents are concatenated, except for using a different smoothing parameter μ'.

3   Experimentation

For evaluation, we used two TREC test collections - TREC4-AP and WT2G where TREC4-AP is the sub-collection of Associated Press in disk 1 and disk 2 for TREC4. To find high-performing smoothing parameters - μ and μ', we first determined the smoothing parameter for document models then for cluster models, by performing

Table 1. Performances of CR and RR in TREC4-AP and WT2G across different ks

		Baseline	$k = 10$	$k = 20$	$k = 30$	$k = 40$	$k = 50$
TREC4-AP	MAP	0.2560	0.2906‡	0.3043‡	0.3037‡	0.3049‡	0.3021‡
(CR)	Pr@5	0.4531	0.4612	0.4776	0.4816	0.4816†	0.4816†
	Pr@10	0.3612	0.3918	0.4082	0.4061†	0.4163†	0.4041†
TREC4-AP	MAP	0.2560	0.2826‡	0.2855‡	0.2864‡	0.2891‡	0.2898‡
(RR)	Pr@5	0.4531	0.4612	0.4571	0.4449	0.4408	0.4490
	Pr@10	0.3612	0.3878	0.3857	0.3898	0.3918†	0.3816
WT2G	MAP	0.3101	0.3165	0.3168	0.3154	0.3163	0.3161
(CR)	Pr@5	0.5360	0.5360	0.5360	0.5400	0.5360	0.5280
	Pr@10	0.4700	0.4780	0.4660	0.4680	0.4660	0.4660
WT2G	MAP	0.3101	0.3205	0.3246‡	0.3246‡	0.3262‡	0.3270‡
(RR)	Pr@5	0.5360	0.5480	0.5520	0.5400	0.5560	0.5440
	Pr@10	0.4700	0.4920	0.4840	0.4760	0.4680	0.4720

pure document retrieval ($\alpha=1$), and pure cluster-based retrieval ($\alpha=0$), respectively. For comparison, we examined runs of 20 different smoothing parameters between 100 and 30,000, and selected the best one. For RR, we used top 1000 documents. For evaluation measures, MAP, Pr@5, and Pr@10 were used.

Table 1 shows the performances of CR and RR where α and β are selected as the best performed ones among 10 different values. The baseline indicates the result of the pure document retrieval ($\alpha=1$). To check whether or not the proposed method significantly improves the baseline, we performed the Wilcoxon sign ranked at 95% and 99% confidence level, and attached † and ‡ to the performance number of each cell in the table when the test passes at 95% and 99% confidence level, respectively. In TREC4-AP, CR and RR significantly improve the baseline on MAP, but not on Pr@5 and Pr@10. In WT2G, CR does not improve the baseline, as only RR significantly improves the baseline on MAP. In terms of performance values, CR is better than RR in TREC4-AP, and vice versa in WT2G. To further identify why RR is not always effective, we examined the recall in top 1000 documents for each test collection. Table 2 shows the number of retrieved relevant documents and the recall of CR, in TREC4-AP and WT2G.

As shown table 2, the recall of the baseline method indicates that TREC4-AP (58%) has more recall-deficient queries than WT2G (81%). It explains the result of table 1 that CR was highly effective on TREC4-AP but RR was less-effective. In addition, this

Table 2. The number of retrieved documents (*Rel_ret* of trec_eval program's outputs) and the recall on top 1000 retrieved documents of the baseline method and CR in TREC4-AP and WT2G. The recall is given in parenthesises with percent values.

	Baseline	CR ($k = 10$)	CR ($k = 20$)	CR ($k = 30$)	CR ($k = 40$)	CR ($k = 50$)
TREC4-AP	1780	1985	1994	2012	2013	2015
	(58.27%)	(64.98%)	(65.27%)	(65.86%)	(65.89%)	(65.96%)
WT2G	1847	1871	1881	1883	1891	1885
	(81.04%)	(82.10%)	(82.54%)	(82.62%)	(82.97%)	(82.71%)

Table 3. Performances of RCR across different ks

		Baseline	CR ($k=10$)	CR ($k=20$)	CR ($k=30$)	CR ($k=40$)	CR ($k=50$)
TREC4-AP	MAP	0.2560	0.3141‡	0.3153‡	0.3161‡	0.3167‡	0.3155‡
(RCR)	Pr@5	0.4531	0.5020†	0.5020†	0.4980†	0.4980	0.4857
	Pr@10	0.3612	0.4265‡	0.4163†	0.4163†	0.4204‡	0.4184‡
WT2G	MAP	0.3101	0.3281‡	0.3343‡	0.3324‡	0.3313‡	0.3305‡
(RCR)	Pr@5	0.5360	0.5720	0.5640	0.5520	0.5600	0.5440
	Pr@10	0.4700	0.4880	0.5060†	0.4980	0.5040	0.4900

result confirms our discussion in the introduction that CR is basically a recall-improving method, and RR is a precision-improving method.

Table 3 shows the performances of the proposed RCR for different ks. Remark that RCR is highly reliable in TREC4-AP, showing more improved results than CR and RR, and it significantly improves the baseline at even Pr@5 and Pr@10 in many different parameter ks. For WT2G, we can see that the performance numbers are slightly more increased over RR. Thus, we confirm that RCR is more reliable and effective than the separated approach of CR or RR.

4   Conclusion

This paper proposed a simple combining method - RCR, to simultaneously reflect different improving points of CR and RR. Experimental results showed that RCR significantly improves the baseline, with much improvement over CR and RR, while CR or RR sometimes failed to improve the baseline.

Acknowledgement. This work was supported by the Korea Science and Engineering Foundation (KOSEF) through the Advanced Information Technology Research Center (AITrc), also in part by the BK 21 Project and MIC & IITA through IT Leading R&D Support Project in 2007.

References

1. Kurland, O., Lee, L.: Corpus structure, language models, and ad hoc information retrieval. In: SIGIR 2004, pp. 194–201 (2004)
2. Tao, T., Wang, X., Mei, Q., Zhai, C.: Language model information retrieval with document expansion. In: HLT-NAACL 2006, pp. 407–414 (2006)
3. Zhang, B., Li, H., Liu, Y., Ji, L., Xi, W., Fan, W., Chen, Z., Ma, W.Y.: Improving web search results using affinity graph. In: SIGIR 2005, pp. 504–511. ACM Press, New York (2005)
4. Kurland, O., Lee, L.: Respect my authority!: Hits without hyperlinks, utilizing cluster-based language models. In: SIGIR 2006, pp. 83–90 (2006)
5. Yang, L., Ji, D., Zhou, G., Nie, Y., Xiao, G.: Document re-ranking using cluster validation and label propagation. In: CIKM 2006, pp. 690–697 (2006)
6. Zhai, C., Lafferty, J.: A study of smoothing methods for language models applied to ad hoc information retrieval. In: SIGIR 2001, pp. 334–342 (2001)

Automatic Vandalism Detection in Wikipedia

Martin Potthast, Benno Stein, and Robert Gerling

Bauhaus University Weimar, Faculty of Media, 99421 Weimar, Germany
`<first name>.<last name>@medien.uni-weimar.de`

Abstract. We present results of a new approach to detect destructive article revisions, so-called vandalism, in Wikipedia. Vandalism detection is a one-class classification problem, where vandalism edits are the target to be identified among all revisions. Interestingly, vandalism detection has not been addressed in the Information Retrieval literature by now. In this paper we discuss the characteristics of vandalism as humans recognize it and develop features to render vandalism detection as a machine learning task. We compiled a large number of vandalism edits in a corpus, which allows for the comparison of existing and new detection approaches. Using logistic regression we achieve 83% precision at 77% recall with our model. Compared to the rule-based methods that are currently applied in Wikipedia, our approach increases the F-Measure performance by 49% while being faster at the same time.

Introduction. The content of the well-known Web encyclopedia Wikipedia is created collaboratively by volunteers. Every visitor of a Wikipedia Web site can participate immediately in the authoring process: articles are created, edited, or deleted without need for authentication. In practice, an article is developed incrementally since, ideally, authors review and revise the work of others. Till this day about 8 million articles in 253 languages have been authored in this way.

However, all times the Wikipedia and its freedom of editing has been misused by some editors. We distinguish them into three groups: (*i*) lobbyists, who try to push their own agenda, (*ii*) spammers, who solicit products or services, and (*iii*) vandals, who deliberately destroy the work of others. The Wikipedia community has developed policies for a manual recognition and handling of such cases, but enforcing them requires the manpower of many. With the rapid growth of Wikipedia a shift from article contributors to editors working on article maintenance is observed. Hence it is surprising that there is little research to support editors from the latter group or to automatize their tasks. As part of our research Table 1 surveys the existing tools for the prevention of editing misuse.

Related Work. The first attempt to aid lobbying detection was the WikiScanner tool which maps IP numbers recorded from anonymous editors to their domain name. This way editors can be found who are biased with respect to the topic in question. Since there are diverse ways for lobbyists to disguise their identity a manual check of all edits for hints of lobbying is still necessary.

There has been much research concerning spam detection in e-mails, among Web pages, or in blogs. In general, machine learning approaches, possibly combined with

C. Macdonald et al. (Eds.): ECIR 2008, LNCS 4956, pp. 663–668, 2008.

Table 1. Tools for the prevention of editing misuse with respect to the target group, and the type of automation (aid, full). Tools shown gray use the same or a very similar rule set as the tool listed in the line above.

Tool	Target	Type	Status	URL (October 2007)
WikiScanner	lobbyists	aid	active	http://wikiscanner.virgil.gr
AntiVandalBot (AVB)	vandals	full	inactive	http://en.wikipedia.org/wiki/WP:AVB
MartinBot	vandals	full	inactive	http://en.wikipedia.org/wiki/User:MartinBot
T-850 Robotic Assistant	vandals	full	active	http://en.wikipedia.org/wiki/User:T-850_Robotic_Assistant
WerdnaAntiVandalBot	vandals	full	active	http://en.wikipedia.org/wiki/User:WerdnaAntiVandalBot
Xenophon	vandals	full	active	http://en.wikipedia.org/wiki/User:Xenophon_(bot)
ClueBot	vandals	full	active	http://en.wikipedia.org/wiki/User:ClueBot
CounterVandalismBot	vandals	full	active	http://en.wikipedia.org/wiki/User:CounterVandalismBot
PkgBot	vandals	aid	active	http://meta.wikimedia.org/wiki/CVN/Bots
MiszaBot	vandals	aid	active	http://en.wikipedia.org/wiki/User:MiszaBot

manually developed rules, do an excellent spam detection job [1]. The respective technology may also be adequate for a misuse analysis in Wikipedia, but the applicability has not been investigated yet.

Vandalism was recognized as an open problem by researchers studying online collaboration [2,4,5,6,7,8], and, of course, by the Wikipedia community.[1] The former provide statistical or empirical analyses concerning vandalism, but neglect its detection. The latter developed four small sets of detection rules but did not evaluate the performance. Misuses such as trolling and flame wars in discussion boards are related to vandalism, but so far no research exists to detect either of them.

In this paper we develop foundations for an automatic vandalism detection in Wikipedia: (*i*) we define vandalism detection as a classification task, (*ii*) discuss the characteristics by which humans recognize vandalism, and (*iii*) develop tailored features to quantify them. (*iv*) A machine-readable corpus of vandalism edits is provided as a common baseline for future research. (*v*) Finally, we report on experiments related to vandalism detection based on this corpus.

Vandalism Detection Task. Let $E = \{e_1, \ldots, e_n\}$ denote a set of edits, where each edit e comprises two consecutive revisions of the same document d from Wikipedia, say, $e = (d_t, d_{t+1})$. Let $\mathcal{F} = \{f_1, \ldots, f_p\}$ denote a set of vandalism indicating features where each feature f_i is a function that maps edits onto real numbers, $f_i : E \rightarrow \mathbf{R}$. Using $\mathcal{F}$ an edit e is represented as a vector $\mathbf{e} = (f_1(e), \ldots, f_p(e))$; $\mathbf{E}$ is the set of edit representations for the edits in E.

Given a vandalism corpus E which has a realistic ratio of edits classified as vandalism and well-intentioned edits, a classifier c, $c : \mathbf{E} \rightarrow \{0, 1\}$, is trained with examples from E. c serves as an approximation of c^*, the true predictor of the fact whether or not an edit forms a vandalism case. Using $\mathcal{F}$ and c one can classify an edit e as vandalism by computing $c(\mathbf{e})$.

---

[1] http://en.wikipedia.org/wiki/Wikipedia:WikiProject_Vandalism_studies (October 2007)

Table 2. Organization of vandalism edits along the dimensions "Edited content" and "Editing category": the matrix shows for each combination the portion of specific vandalism edits at all vandalism edits. For vandalized structure insertion edits and content insertion edits also a list of their typical characteristics is given. It includes both the characteristics described in the previous research and the Wikipedia policies.

Editing category	Edited content		Link	Media
	Text	Structure		
Insertion	43.9%	14.6%	6.9%	0.7%
	Characteristics: point of view, off topic, nonsense, vulgarism, duplication, gobbledegook	Characteristics: formatting, highlighting		
Replacement	45.8%	15.5%	4.7%	2.0%
Deletion	31.6%	20.3%	22.9%	19.4%

Vandalism Indicating Features. We have manually analyzed 301 cases of vandalism to learn about their characteristics and, based on these insights, to develop a feature set $\mathcal{F}$. Table 2 organizes our findings as a matrix of vandalism edits along the dimensions "Edited content" and "Editing category"; Table 3 summarizes our features.

Table 3. Features which quantify the characteristics of vandalism in Wikipedia

Feature f	Description
char distribution	deviation of the edit's character distribution from the expectation
char sequence	longest consecutive sequence of the same character in an edit
compressibility	compression rate of an edit's text
upper case ratio	ratio of upper case letters to all letters of an edit's text
term frequency	average relative frequency of an edit's words in the new revision
longest word	length of the longest word
pronoun frequency	number of pronouns relative to the number of an edit's words (only first-person and second-person pronouns are considered)
pronoun impact	percentage by which an edit's pronouns increase the number of pronouns in the new revision
vulgarism frequency	number of vulgar words relative to the number of an edit's words
vulgarism impact	percentage by which an edit's vulgar words increase the number of vulgar words in the new revision
size ratio	the size of the new version compared to the size of the old one
replacement similarity	similarity of deleted text to the text inserted in exchange
context relation	similarity of the new version to Wikipedia articles found for keywords extracted from the inserted text
anonymity	whether an edit was submitted anonymously, or not
comment length	the character length of the comment supplied with an edit
edits per user	number of previously submitted edits from the same editor or IP

For two vandalism categories the matrix shows particular characteristics by which an edit is recognized as vandalism: a vandalism edit has the "point of view" characteristic if the vandal expresses personal opinion, which often entails the use of personal pronouns. Many vandalism edits introduce off-topic text with respect to the surrounding text, are nonsense in that they contradict common sense, or do not form a correct sentence from their language. The first three characteristics are very difficult to be quantified, and research in this direction will be necessary to develop reliable analysis methods. Vulgar vandalism can be detected with a dictionary of vulgar words; however, one has to consider the context of a vulgar word since several Wikipedia articles contain vulgar words in a correct sense. Hence we quantify the impact of a vulgar word based on the point of time it has been inserted into an article rather than simply checking its occurrence. If an inserted text duplicates other text within the article or within Wikipedia, one may also speak of vandalism, but this is presumably the least offending case. Very often vandalism consists only of gobbledygook: a string of characters which has no meaning whatsoever, for instance if the keyboard is hit randomly. Another common characteristic of vandalism is that it is often highlighted by capital letters or by the repetition of characters. In cases of deletion vandalism, larger parts of an article are deleted, which explains the high percentages of this vandalism type throughout all content types. Note that a vandalism edit typically shows several of these characteristics at the same time.

Vandalism Corpus. Vandalism is currently not documented in Wikipedia, so that automatic vandalism detection algorithms cannot be compared to each other. The best way to find vandalism manually is by taking a look at the list of the most vandalized pages and then to analyze the history of the listed articles.[2] We have set up the vandalism corpus WEBIS-VC07-11, which was compiled from our own investigations and the results of a study[3] conducted by editors of Wikipedia. The corpus contains 940 human-assessed edits from which 301 edits are classified as vandalism. It is available in a machine-readable form for download at [9].

Evaluation. Within one-class classification tasks one is often confronted with the problem of class imbalance: one of the classes, either the target or the outlier class is underrepresented, which makes training a classifier difficult. In a realistic detection scenario only 5% of all edits in a given time period are from the target class "vandalism" [5]. As a heuristic to alleviate the problem we resort to random over-sampling of the underrepresented class at training time. Nevertheless, an in-depth analysis with respect to domain characteristics of the training samples is still necessary; the authors of [3] have compared alternative methods to address class imbalance.

Using ten-fold cross-validation on the corpus WEBIS-VC07-11 and a classifier based on logistic regression we evaluated the discriminative power of the features described in Table 3 when telling apart vandalism and well-intentioned edits. We also analyzed the effort for computing these features and compared the results to AVB and to ClueBot. Table 4 summarizes the results.

As can be seen, our approach (third row) outperforms the rule-based bots on all accounts. The individual analysis of each feature indicates its contribution to the overall

---

[2] http://en.wikipedia.org/wiki/Wikipedia:Most_vandalized_pages (October 2007)
[3] http://en.wikipedia.org/wiki/Wikipedia:WikiProject_Vandalism_studies/Study1 (Oct. 2007)

Table 4. Vandalism detection performance quantified as category-specific recall and averaged precision values. The first row shows, as the baseline, the currently best performing Wikipedia bot, while the third row (bold) shows the results of our classifier. The right column shows the throughput on a standard PC. The underlying test corpus contains 940 human-assessed edits from which 301 edits are classified as vandalism.

Feature f	Recall			Precision	Throughput
	Insertion	Replacement	Deletion	Average	(edits per second)
Baseline: AVB	0.35	0.53	0.61	0.74	3
ClueBot	0.03	0.29	0.49	1	3
c **with all features**	**0.87**	**0.76**	**0.89**	**0.86**	**5**
char distribution	0.03	0	0.74	0.41	6
char sequence	0.01	0.14	0.2	0.70	43
compressibility	0	0	0.78	0.24	618
upper case ratio	0.13	0.22	0	0.61	656
term frequency	0	0.29	0.01	0.3	4
longest word	0	0.04	0.63	0.54	319
pronoun frequency	0.09	0.1	0	0.53	351
pronoun impact	0	0.04	0.39	0.49	53
vulgarism frequency	0.23	0.35	0	0.65	181
vulgarism impact	0.23	0.41	0.52	0.91	33
size ratio	0.07	0.35	0.54	0.83	8198
replacement similarity	–	0	–	–	9
context relation	0	0	0.13	0.18	3
anonymity	0	0	0	0	8545
comment length	0	0	0	0	14242
edits per user	0.94	0.86	0.96	0.66	813

performance. Note that vandalism detection suggests a two-stage analysis process (machine + human) and hence to prefer high recall over high precision: a manual post-processing of classifier results is indispensable since visitors of a Wikipedia page should never see a vandalized document; as well as that, a manual analysis is feasible because an even imprecisely retrieved target class contains only few elements.

References

1. Blanzieri, E., Bryl, A.: A Survey of Anti-Spam Techniques. Technical Report DIT-06-056, University of Trento (2006)
2. Buriol, L.S., Castillo, C., Donato, D., Leonardi, S., Millozzi, S.: Temporal Analysis of the Wikigraph. In: WI 2006, pp. 45–51. IEEE Computer Society, Los Alamitos (2006)
3. Japkowicz, N., Stephen, S.: The Class Imbalance Problem: A Systematic Study. Intell. Data Anal. 6(5), 429–449 (2002)
4. Kittur, A., Suh, B., Pendleton, B., Chi, E.: He says, she says: Conflict and Coordination in Wikipedia. In: CHI 2007, pp. 453–462. ACM, New York (2007)

5. Priedhorsky, R., Chen, J., Lam, S., Panciera, K., Terveen, L., Riedl, J.: Creating, Destroying, and Restoring Value in Wikipedia. In: Group 2007 (2007)
6. Viégas, F.B.: The Visual Side of Wikipedia. In: HICSS 2007, p. 85. IEEE Computer Society, Los Alamitos (2007)
7. Viégas, F.B., Wattenberg, M., Dave, K.: Studying Cooperation and Conflict between Authors with History Flow Visualizations. In: CHI 2004, pp. 575–582. ACM Press, New York (2004)
8. Viégas, F.B., Wattenberg, M., Kriss, J., van Ham, F.: Talk before you Type: Coordination in Wikipedia. In: HICSS 2007, p. 78. IEEE Computer Society, Los Alamitos (2007)
9. Potthast, M., Gerling, R. (eds): Web Technology & Information Systems Group, Bauhaus University Weimar. Wikipedia Vandalism Corpus WEBIS-VC07-11 (2007),
   http://www.uni-weimar.de/medien/webis/research/corpora

Evaluating Paragraph Retrieval for *why*-QA

Suzan Verberne, Lou Boves, Nelleke Oostdijk, and Peter-Arno Coppen

Department of Linguistics, Radboud University Nijmegen
{s.verberne,l.boves,n.oostdijk}p.a.coppen@let.ru.nl

Abstract. We implemented a baseline approach to *why*-question answering based on paragraph retrieval. Our implementation incorporates the QAP ranking algorithm with addition of a number of surface features (cue words and XML markup). With this baseline system, we obtain an accuracy-at-10 of 57.0% with an MRR of 0.31. Both the baseline and the proposed evaluation method are good starting points for the current research and for other researchers working on the problem of *why*-QA.

We also experimented with the addition of smart question analysis features to our baseline system (answer type and informational value of the subject). This however did not give significant improvement to our baseline. In the near future, we will investigate what other linguistic features can facilitate re-ranking in order to increase accuracy.

1 Introduction

In the current research project, we aim at developing a system for answering *why*-questions (*why*-QA). In earlier experiments, we found that the answers to *why*-questions consist of a type of reasoning that cannot be expressed in a single clause, and that on the other hand 94% of the answers is maximally one paragraph long. Therefore, we decide to consider paragraphs as retrieval units for *why*-QA.

The goal of the present paper is to establish a baseline paragraph retrieval method for *why*-QA, including a proper evaluation method. Moreover, we aim to find out whether a system based on standard keyword based paragraph retrieval can be improved by incorporating our knowledge of the syntax and semantics of *why*-questions in query formulation.

2 Method

2.1 Data

For development and testing, we use a set of 805 *why*-questions that were submitted to the online QA system <u>answers.com</u>, and collected for the Webclopedia project by Hovy et al. [1].

As an answer source, we use the Wikipedia XML corpus [2], which is also used in the context of the Initiative for the Evaluation of XML Retrieval (INEX, [3]). The English part of the corpus consists of 659,388 Wikipedia articles (4.6 GB

C. Macdonald et al. (Eds.): ECIR 2008, LNCS 4956, pp. 669–673, 2008.

of XML data). By manual inspection we found that this corpus contains a valid
answer for about one quarter of the Webclopedia *why*-questions. We randomly
selected 93 questions that have an answer in the corpus and we manually extracted the answer paragraph (reference answer) from the corpus for each of
them. We indexed the complete corpus using the Wumpus search engine [4] in
the standard indexing modus (Wumpus version June 2007).

Thus, we have a development set of 93 questions and corresponding reference
answers, from a corpus of 659,388 Wikipedia articles.

2.2   Baseline Method

Our baseline method consists of four modules:

1. A question analysis module, which applies a list of stop words to the question
   and removes punctuation, returning the set of question content words;
2. A query creation module that transforms the set of question words into one
   or more Wumpus-style queries and sends this query to the Wumpus engine;
3. Ranking of the retrieved answers by the QAP algorithm. QAP is a scoring algorithm for passages that has specifically been developed for question
   answering tasks [5]. It has been implemented in Wumpus;
4. Re-ranking of the results according to three answer features: (a) The presence
   of cue words such as *because*, *due to* and *in order to* in the paragraph; (b)
   the presence of one or more question terms in the title of the document
   in which the retrieved paragraph is embedded; (c) emphasis marking of a
   question term. The corpus contains XML tags for formatting information
   such as emphasis.
   The weights applied in the re-ranking step are variable in the configuration
   of our system.

After re-ranking, the system returns the top 10 results to the user.

2.3   Features for Smart Question Analysis

In this section, we present an extension to our paragraph retrieval system incorporating smart question analysis. In previous experiments, we found that specific
syntactic and semantic features of *why*-questions can play a role in retrieving
relevant answers. We identified two features in particular that seem relevant in
answer selection, viz. answer type and the informational value of the subject.

In factoid QA, the **answer type** is known to be an important parameter
for increasing system precision. The two main answer types for *why*-questions
are 'cause' and 'motivation' [6]. In our question set, we encountered one other
relatively frequent answer type: 'etymology'. Thus we distinguish three answer
types in the current approach: 'cause' (77.4% in our question set), 'motivation'
(10.2%), and 'etymology' (12.4%). We split our set of cue words in four categories: one for each of the answer types (e.g. *in order to* for motivation, *due to*
for cause and *name* for etymology), and a general category of cue words that
occur for all answer types (e.g. *because*). We evaluated answer type prediction

for our question set using earlier defined algorithms and we found a precision of 0.81 (ranging from 0.49 for motivation to 1 for etymology) for this task.

Previous experiments have shown the relevance of a second semantic feature, the **informational value of the subject**. It appears to be a good predictor for deciding which terms from the question are likely to occur in the document title of relevant answer paragraphs. This knowledge can be used for re ranking based on document title (step 4b in the baseline method). We defined three classes of subjects, which are automatically distinguished by our system based on their document frequency. The subjects with lowest informational value are subjects consisting of pronouns only or one of the very general noun phrases *people* and *humans*. In these cases, our re-ranking module only gives extra weight to *predicate* words occurring in the document title. The second class covers those subjects that are not semantically poor, but very common, such as *water* and *the United States*. In these cases, the baseline approach is applied, which does not distinguish between terms from subject and predicate for re-ranking. The third class consists of the subjects that have a low document frequency, and therefore have a large informational value, such as *flamingos* and *subliminal messages*. In these cases our system gives extra weight to paragraphs from documents with one or more words from the *subject* in the title.

We performed a series of experiments in order to find out what the contribution of these features is to the overall performance of our system.

2.4   Evaluation Method

There are no specific evaluation procedures available for *why*-QA, but there is one evaluation forum that includes *why*-questions: the Question Answering Challenge at the Japanese NTCIR Workshop [7]. In NTCIR, all retrieved results are manually evaluated according to a four-level scale of correctness.

We propose a method for the evaluation of *why*-QA that is a combination of the procedure applied at NTCIR and the commonly-used MRR metric. We manually evaluate all retrieved answers according to the four NTCIR correctness scales. Then we count the proportion of questions that has at least one correct answer in the top 10 of the results (accuracy-at-10). For the highest ranked correct answer per question, we determine the reciprocal rank (RR). If there is no correct answer in the top 10 results, RR is 0. Over all questions, we calculate MRR.

3   Results and Discussion

Table 1 shows the results (accuracy-at-10 and MRR) obtained for three configurations: (1) simple paragraph retrieval by QAP, (2) the baseline system and (3) the smart system.

Using the Wilcoxon Signed-Rank Test, we find that there is no significant difference between the baseline results and the results from smart question analysis (Z=-0.66, P=0.5093 for paired reciprocal ranks). The baseline is, however, sightly better than simple retrieval ($Z = 1.67$, $P = 0.0949$).

Table 1. Results per system version

Features	Version	Accuracy	MRR
QAP	Simple retrieval	47.3%	0.25
+Cue words +Title weight +Emph. weight	Baseline	57.0%	0.31
+Answer type +Subject value weight	Smart question analysis	55.9%	0.28

Apparently, the implementation of our question analysis features does not improve the ranking of the results. Since we suspected that some correct answers were missed because they are in the tail of the result list, we experimented with a larger result list (top 20 presented to user). This led to an accuracy-at-20 of 63.4% (MRR unchanged 0.31) for the baseline system and 61.3% (MRR unchanged 0.28) for the smart system.

As regards the answer type feature, we can explain its negligible contribution from the fact that answer type only affects cue word weights. Cue words apparently constitute too small a contribution to the overall performance of the system. As regards the subject value feature, we are surprised by its small influence. Our suspicion is that the ranking algorithm QAP as implemented in the baseline already gives good results with term weighting based on term frequency and inverted document frequency. Another possible explanation to the small influence of the informational value of the subject is that too many errors are still made by our question analysis module in the decision of which question part should be given the position weight.

A further error analysis shows that for 47.5% of unanswered questions, the reference answer is present in the extended result list retrieved by the algorithm (max. 450 results), but not in the top 10 of answers presented to the user. For these questions, re-ranking may be valuable. If we can define criteria that rank the reference answer for this set of questions higher than the irrelevant answers, we can increase accuracy-at-10 (and thereby MRR).

4   Conclusion and Further Work

We developed an approach for *why*-QA that is based on paragraph retrieval. We created a baseline system that combines paragraph ranking using the QAP algorithm with weights based cue words and the position of question terms in the answer document. We evaluated our system based on manual assessments of the answers in four categories according to two measures: accuracy-at-10 and MRR. We get 57.0% accuracy with an MRR of 0.31. We think that both the baseline and the proposed evaluation method are good starting points for the current research and other researchers working on the problem of *why*-QA.

We also implemented and evaluated a system that extends the baseline approach with two features that we obtain from linguistic question analysis: answer type and the informational value of the subject. This smart system does, however, not show significant improvement over the baseline. In section 3, we do some suggestions for explaining these results.

In the near future, we will experiment with adding a number of other linguistic features to the re-ranking module of our system. The features that we consider for re-ranking include the distinction between heads and modifiers from the question, synonym links between question and answer terms, and the presence of noun phrases from the question in the answer. We are currently preparing experiments for selecting the most relevant of these features for optimizing MRR.

In the more distant future, we plan to experiment with smart paragraph analysis. In [8], it is shown that rhetorical relations have relevance for answer selection in *why*-QA; the presence of (some types of) rhetorical relations can be an indication for the presence of a potential answer. Moreover, there is a connection between answer type and type of rhetorical relation; we aim to investigate whether this addition can make answer type more valuable than in the current cue-word based version of the system.

References

1. Hovy, E.H., Hermjakob, U., Ravichandran, D.: A question/answer typology with surface text patterns. In: Proceedings of the Human Language Technology conference (HLT), San Diego, CA (2002)
2. Denoyer, L., Gallinari, P.: The Wikipedia XML corpus. ACM SIGIR Forum 40(1), 64–69 (2006)
3. Clarke, C., Kamps, J., Lalmas, M.: Inex 2006 retrieval task and result submission specification. In: INEX 2006 Workshop Pre-Proceedings, Dagstuhl, Germany, December 2006, pp. 18–20 (2006)
4. Buttcher, S.: The wumpus search engine (2007), http://www.wumpus-search.org/
5. Buttcher, S., Clarke, C., Cormack, G.: Domain-specific synonym expansion and validation for biomedical information retrieval (multitext experiments for trec 2004) (2004)
6. Verberne, S., Boves, L., Oostdijk, N., Coppen, P.: Exploring the use of linguistic analysis for why-question answering. In: Proceedings of the 16th meeting of Computational Linguistics in the Netherlands (CLIN 2005), Amsterdam, pp. 33–48 (2006)
7. Fukumoto, J., Kato, T., Masui, F., Mori, T.: An overview of the 4th question answering challenge (qac-4) at ntcir workshop 6. In: Proceedings of NTCIR-6 Workshop Meeting, Tokyo, Japan, pp. 433–440 (2007)
8. Verberne, S., Boves, L., Oostdijk, N., Coppen, P.: Discourse-based answering of why-questions. Traitement Automatique des Langues. special issue on Computational Approaches to Discourse and Document Processing, 21–41 (2007)

Revisit of Nearest Neighbor Test for Direct Evaluation of Inter-document Similarities

Seung-Hoon Na[1], In-Su Kang[2], and Jong-Hyeok Lee[1]

[1] POSTECH,Pohang,South Korea
{nsh1979,jhlee}@postech.ac.kr
[2] KISTI,Daejeon,South Korea
dbaisk@kisti.re.kr

Abstract. Recently, cluster-based retrieval has been successfully applied to improve retrieval effectiveness. The core part of cluster-based retrieval is inter-document similarities. Although inter-document similarities can be investigated independently of cluster-based retrieval and be further improved in various ways, their direct evaluation has not been seriously considered. Considering that there are many cluster-based retrieval methods, such a direct evaluation method can separate the work of inter-document similarities from the work of cluster-based retrieval. For this purpose, this paper revisits Voorhee's nearest neighbor test as such a direct evaluation, by mainly focusing on whether or not the test is correlated to the retrieval effectiveness. Experimental results consistently verify the use of the nearest neighbor test. As a result, we conclude that the improvement of retrieval effectiveness can be well-predictable from direct evaluation, even without performing runs of cluster-based retrieval.

1   Introduction

Recently, cluster-based retrieval has been successfully applied to improve the retrieval effectiveness, especially in the language modeling approaches [1,2]. For example, Kurland and Lee applied the document expansion where the representation of a given document is refined from its top nearest neighbor documents, showing a significant improvement over the baseline [1]. The core part of cluster-based retrieval is the inter-document similarities. Its importance is confirmed in the following famous van Rijsbergen's cluster hypothesis [3].

CH (cluster hypothesis): "Relevant documents are similar." In fact, there are many different ways to define inter-document similarities, thus CH can be well-satisfied in some similarities but not in other ones. Regarding this, original CH is somewhat ambiguous since it assumes single similarities. To clarify that there are many similarity measures, it would be better to use a predicate statement rather than a proposition as follows.

SH (similarity hypothesis): "There exist inter-document similarities to satisfy CH." SH can be viewed as a more relaxed hypothesis than CH. Along the way to logically connect from CH to the success of cluster-based retrieval, there is an additional hypothesis - RH, which is implicitly agreed among the research community.

C. Macdonald et al. (Eds.): ECIR 2008, LNCS 4956, pp. 674–678, 2008.

RH (hypothesis of cluster-based retrieval): "Using inter-document similarities satisfying CH, there exists a cluster-based retrieval to improve retrieval effectiveness over the baseline."

From SH and RH, we should use inter-document similarities satisfying CH, to improve retrieval effectiveness by cluster-based retrieval. Thus, our goal should be to find inter-document similarities that better satisfy CH, not to prove CH itself. Hence, inter-document similarities can be independently evaluated without cluster-based retrieval, since SH and RH are independent hypotheses. However, recent works of cluster-based retrieval have evaluated the retrieval effectiveness, not focusing on the direct evaluation of inter-document similarities. There are some advantages when using such a direct evaluation.

1. Generally, the evaluation of cluster-based retrievals is inefficient. First, inter-document similarities over an entire collection should be calculated, and some cluster-based retrieval methods may require performing an expensive clustering algorithm over the entire collection. In addition, since there are some parameters in cluster-based retrieval (interpolation parameter or smoothing parameter), several retrieval runs across different parameters should be examined. On the other hand, the direct evaluation requires the partial inter-document similarities on the subset of a collection not on its entire set, and only one time of evaluation.

2. Different inter-document similarities can be compared in a unified measure. As a result, further investigation to inter-document similarities is directly possible, without rejecting good inter-document similarities by misunderstood points from unsuccessful cluster-based retrievals.

Therefore, if a direct evaluation on inter-document similarities is provided, then it becomes more convenient to find better similarities, due to its high efficiency and easier-tunablity. The works on more improved similarities have already been proceeded. Calado et. al. used hyperlinks among web-pages to further improve original term-based similarities for the web document classification task [4]. Also, Bartell et. al. investigated query-centric inter-document similarities [5], and Tombros and van Rijsbergen proposed query-specific similarities [6]. The results of these works could be verified only through a direct evaluation of similarity metric.

One requirement of direct evaluation is to prove its co-relatedness to the retrieval performance. If some evaluation method satisfies the requirement, then using only direct evaluation without the retrieval evaluation would be meaningful. To this end, we noted Voorhee's nearest neighbor test (NNT) to check how many top nearest neighbors of a relevant document are co-relevant of the document to a given query [7]. Intuitively, NNT is related to the retrieval performance, since it is a directly related test to the goal measure of how much a given inter-document similarity satisfies CH. Empirically, we found that NNT is highly co-related to the retrieval performance on standard TREC test collections.

2   Review of Nearest Neighbor Test and Further Extension

For a given document, Voorhee's NNT examines whether or not its top nearest neighbors are co-relevant to the given document, by counting the number of co-relevant documents from the set of neighbors. Originally, Voorhee checked only 5 nearest neighbors, so she called it 5NN.

In fact, Voorhee's NNT defines a well-understandable IR task, since nearest neighbors are generated as a ranked list. NNT defines three necessary elements to define an IR task - a query, a document set, and relevant judgment as follows:

Query	Relevance Judgment	Document set
A relevant document to a given topic	The set of co-relevant documents to a query document	An entire test collection

For each query-document, we can calculate all popular IR evaluation metrics such as AP (average precision), Pr@5 (precision at 5 documents) or Pr@10 (precision at 10 documents). NNT's final value is the average of one evaluation measure over a query set. As a result, NNT's evaluation measure becomes the same as IR evaluation measures. To discriminate NNT's performance measure from the traditional IR measure, we attach "NNT" to the acronym of the measure. Thus, the corresponding NNT's measures of MAP and Pr@5 become NNT-MAP and NNT-Pr@5. Voorhee's 5NN corresponds to NNT-Pr@5.

3   Evaluation of NNT

Our evaluation goal is to examine whether or not NNT is co-related to the retrieval effectiveness of cluster-based retrieval. To this end, we constructed several different inter-document similarities in an automatic manner, calculated NNT's performance and the performance of cluster-based retrieval for each similarity, and evaluated the co-relatedness between two different performances.

For evaluation, we used two TREC test collections - TREC4-AP and WT2G where TREC4-AP is the sub-collection of Associated Press in disk 1 and disk 2 for TREC4. We used language modeling approaches for the retrieval model, since recent successful works of cluster-based retrieval have been derived from them. As a smoothing method, we selected Dirichlet-prior model due to its superiority to other smoothing methods.

For cluster-based retrieval, we used document expansion where the set of nearest documents of a document is used as a cluster. Then, we linearly combined the score of the cluster and a document to produce a final document score.

$$score_{CR}(Q,D) = \alpha score_{CR}(Q,D) + (1-\alpha)score_{CR}(Q,C_D) \tag{1}$$

where $score(Q,D)$ is the similarity score which the retrieval model produces for document D to query Q, α is an interpolation parameter, and C_D is a cluster of D. Our combination method is somewhat different from Kurland's interpolation [1]. Kurland combined them with the mixture model at a probabilistic level, while we combined them at a score level. Thus, our method belongs to the fusion method. We found that the fusion method is less-sensitive to the interpolation parameter than Kurland's probabilistic mixture model, and comparably performs to Kurland's one. Two smoothing parameters are necessary, where one is a cluster-dependent smoothing parameter to calculate $score(Q,C_D)$, and another is a document-dependent parameter for $score(Q,D)$. These parameters are tuned for each test collection.

Inter-document similarities are calculated by $score(D,D')$ where D and D' are documents, and D is a query document. However, we need various inter-document similarities with different qualities. For this, we selectively used contents in the query document D when calculating $score(D,D')$, by using pseudo document D_β as a query document instead of the original document D. Here, β is a parameter to control how many terms of original document D are used in D_β. As β increases, more terms are used in D_β. To construct a pseudo document D_β, assume that θ_D is the document model of D. Initially, D_β is the empty set. We sorted the terms in document D according to $P(w|\theta_D)$. Then, we added the term w to D_β one by one, but stopping the addition if the summation of probabilities of terms in D_β is equal to or larger than β (i.e. $\sum_w P(w|\theta_D) \geq \beta$). Finally, term frequency of w of D_β is proportional to $P(w|\theta_D)$, and the length of document of D_β is adjusted to the length of original document D. In this way, we constructed seven different types of pseudo document D_β, where β is one of 0.1, 0.2, 0.3, 0.4, 0.5, 0.6, 0.8. As a result, we obtained seven inter-document similarities with different qualities. Note that $P(w|\theta_D)$ from MLE makes the probability of common terms not-small. Thus, we used the parsimonious document model for $P(w\theta_D)$ to assign more probabilities to topical terms than common terms [8].

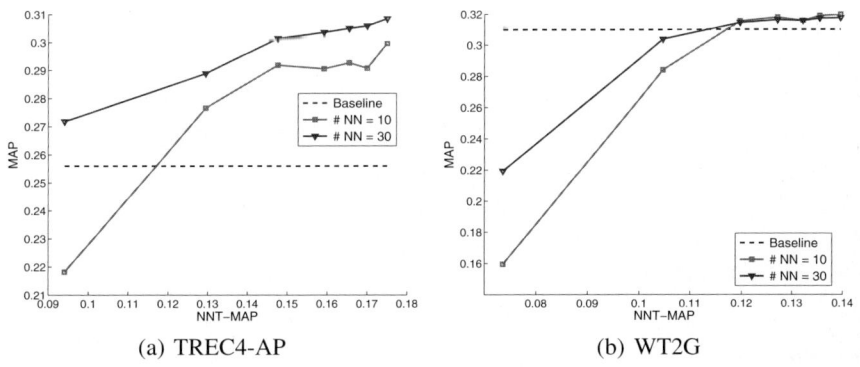

(a) TREC4-AP   (b) WT2G

Fig. 1. NNT-MAP versus MAP in TREC4-AP and WT2G

Figure 1 shows the graphs of plotting NNT-MAP (NNT) and MAP (retrieval performance) for inter-document similarities with different βs, in TREC4-AP and WT2G. #NN indicates the size of the set of nearest neighbors used for cluster-based retrieval, and Baseline the performance of pure document clustering ($\alpha = 1$ in Eq. (1)). As shown in Figure 1, as NNT-MAP increases, MAP becomes larger. From this graph, we can see that NNT's performance is highly related to the retrieval performance.

We performed Spearman's rank correlation test to obtain the correlation coefficients and the statistical confidence value on whether or not the correlation is statistically meaningful. Spearman's rank correlation test is useful when the distribution of data is unknown, and the number of data is small. Spearman's rank correlation coefficient (SRCC) is a value between -1 and 1, meaning that data are highly-correlated if SRCC is close to 1 or -1.

Table 1. SRCC and p-value of NNT's and retrieval performance when NNT-MAP (or NNT-Pr@10) is used for NNT's performance. SRCCs of NNT-MAP and NNT-Pr@10 are the same.

	TREC4-AP		WT2G	
	SRCC	p-value	SRCC	p-value
#NN = 10	0.8214	0.023449	0.9643	0.000454
#NN = 30	1.0000	0.000000	0.9643	0.000454
#NN = 50	1.0000	0.000000	0.8214	0.023449

Table 1 shows SRCC, p-value of the Spearman's test for each test collection, when NNT-MAP and NNT-Pr@10 are selected for NNT's performance. Note that SRCC closes to 1, and its p-value is very low for both of NNT-MAP and NNT-Pr@10. SRCC is the same for two NNT's measures in that their performance values are different but their ranks are the same. It implies the high correlation between NNT's performances and retrieval performance.

4   Conclusion

This paper verified that Voorhee's NNT is an effective method for directly evaluating the inter-document similarity in terms of cluster-based retrieval. As a result, one can directly compare a complicated inter-document similarity with the traditional similarity without performing the inefficient evaluation based on cluster-based retrieval.

Acknowledgement. This work was supported by the Korea Science and Engineering Foundation (KOSEF) through the Advanced Information Technology Research Center (AITrc), also in part by the BK 21 Project and MIC & IITA through IT Leading R&D Support Project in 2007.

References

1. Kurland, O., Lee, L.: Corpus structure, language models, and ad hoc information retrieval. In: SIGIR 2004, pp. 194–201 (2004)
2. Tao, T., Wang, X., Mei, Q., Zhai, C.: Language model information retrieval with document expansion. In: HLT-NAACL 2006, pp. 407–414 (2006)
3. Rijsbergen, C.J.V.: Information Retrieval. Butterworth-Heinemann (1979)
4. Calado, P., Cristo, M., Gonçalves, M.A., de Moura, E.S., Ribeiro-Neto, B., Ziviani, N.: Link-based similarity measures for the classification of web documents. Journal of American Society for Information Science and Technology (JASIST) 57(2), 208–221 (2006)
5. Bartell, B.T., Cottrell, G.W., Belew, R.K.: Representing documents using an explicit model of their similarities, vol. 46, pp. 254–271. John Wiley, New York (1995)
6. Tombros, A., van Rijsbergen, C.J.: Query-sensitive similarity measures for the calculation of interdocument relationships. In: CIKM 2001, pp. 17–24. ACM, New York (2001)
7. Voorhees, E.M.: The cluster hypothesis revisited. In: SIGIR 1985, pp. 188–196 (1985)
8. Hiemstra, D., Robertson, S., Zaragoza, H.: Parsimonious language models for information retrieval. In: SIGIR 2004, pp. 178–185 (2004)

A Comparison of Named Entity Patterns from a User Analysis and a System Analysis

Masnizah Mohd[1], Fabio Crestani[2], and Ian Ruthven[1]

[1] Dept. of Computer and Information Sciences
University of Strathclyde
Glasgow G1 1XH, Scotland, UK
{Masnizah.Mohd,Ian.Ruthven}@cis.strath.ac.uk
[2] Faculty of Informatics
University of Lugano (USI)
via G. Buffi 13, CH-6904 Lugano, CH
fabio.crestani@unisi.ch

Abstract. This paper investigates the detection of named entity (NE) patterns by comparing the results of NE patterns resulting from a user analysis and a system analysis. Findings revealed that there are difference in NE patterns detected by system and user, something that may affect the performance of a TDT system based on NE detection.

Keywords: Named entity, Topic Detection and Tracking (TDT).

1 Introduction

Topic Detection and Tracking (TDT) aims to effectively retrieve and organize broadcast news (speech) and newswire stories (text) into groups of events. Recent approaches to using NE for document representation have been getting more attention in TDT [1, 3, 5, 6]. The detection of NE allows us to characterize and detect events in documents. Recent research in TDT has investigated *NEs* rather than words because TDT investigates the organization of information by *event* rather than by subject [1]. Exploiting NE too has improved the accuracy of the New Event Detection (NED) systems [1, 7].

In this paper, we compare the identification and distribution of NE from a user analysis (a user study) and a system analysis (using ANNIE, A Nearly-New Information Extraction System) [2]. The contributions made in this work are as follows:

- The comparison is important in order to have a correct classification of named entity from user perspective.
- It is also important in guiding ANNIE to extract NE based on user perspective.
- Finally, it gives important information on how to assign the correct weights of NE for document representation in a TDT system.

The rest of the paper is organized as follows. Section 2 presents previous work in TDT using NEs. Section 3 summarises the results of the user study. Section 4

C. Macdonald et al. (Eds.): ECIR 2008, LNCS 4956, pp. 679–683, 2008.

discusses the experimental analysis, the experimental results and the comparison of the user analysis with the system analysis. Finally we wrap up the conclusions in Section 5.

2   Related Work

In recent years, several efforts have been made on exploiting NE for document representation to improve TDT systems. Yang et al. [5] investigated and focused on *location* NE for document representation. The DOREMI research group also looked at *people* and *location* NE to obtain a final confidence score for each story [3].

Kumaran and Allan [1] split document representation into two parts: named entities and non-named entities. It was found that some classes of news such as Elections, Accidents, Violence and War, New Laws, Sports News, Political and Diplomatic Meetings, could achieve better performance using NE representation While some other classes of news could achieve better performance using non-NE representation. Kuo et. al [6] investigated the average correlation between POS and news genre to model New Event Detection (NED) model. They revealed that terms of different types (e.g. Noun, Verb or Person name) have different effects for different genre of stories in determining whether two stories are on the same topic. For example, the names of election candidates (Person name) are very important for stories of election class; the locations (Location name) where accidents happened are important for stories of accidents class.

However no previous research has investigated NE patterns from the user perspective. A better document representation in TDT system comes from having a good understanding of NE patterns from the users. Thus, it is important to identify NE patterns from the user perspective to achieve a better understanding of how system should use NEs. We compare NE patterns from a user analysis and system analysis. This is where we achieve a clear view of what are the NE that people might care most about and at the same time to investigate what are the NE that the system concern most.

3   User Study

We conducted a user study in early February and March 2007. The respondents are the postgraduate students from the Scottish Centre for Journalism Studies (SCJS) of the University of Strathclyde. They were required to give the important keywords that best describe the documents given, classify the keywords according to which type of NE is being mentioned and tick their importance level relative to the content of the document [4].

This study revealed that there is a significant difference in the distribution of NE within domains. *What* and *Who* are the top NE across domains. The sequence of NE distribution was identical across news domains, with *What* more than *Who* NE followed by *Where* NE and the least was *When* NE. This is interesting since the results shows that when humans analyze the documents they pick more *What* NE. We found that there is no significant relationship between the importance level of NE and news domains, indicating that the importance level is domain independent. *What* is the top NE but it is not necessarily *Very Important* across news domains and although *When* has 0% of occurrence in *Government* but it can be *Very Important* type of NE. The level of agreement in the keywords selected among the respondents for the same document varies across genre. The respondents agreed more often for *Entertainment* documents and least often agreed for *Economy* documents. Thus, the keywords given are dependent on the person who is doing the detection based on their interest and familiarity with the genre of documents.

Results from the user study revealed the user perception on *how often NE occurred* and *how important are NE across* domains. This reflects the user interest by showing what are the NE that user care most about and pay less attention to. We believe by comparing results from the user and a system analysis could help to present an ideal NE pattern from the user aspect.

4   System Analysis

We used ANNIE that has been developed using GATE [2]. ANNIE is an information extraction component of GATE which we use for its accurate entity, pronoun and nominal co-references extraction. In this section, we discuss how ANNIE processed the same documents as in our user study to discover NE patterns. The documents were processed with the standard indexing method including stemming and removal of stop words.

4.1   ANNIE (A Nearly-New Information Extraction System)

ANNIE is able to recognize proper nouns, person, organizations, dates and locations and pronominal coreferencing improving the quality of the named entity recognition. We have classified named entities into 4 W's. *Person* and *Organization* is classified as *Who*, *Location* is classified as *Where*, *Date* is classified as *When*, and other token is classified as *What*.

Different words expressing the same entity may also be present where we looked at coreferenced words discovered by ANNIE. For example, for the document shown in Figure 1, ANNIE would create the following coreferenced list: *Egyptian President Hosni Mubarak, Egyptian President, Hosni Mubarak, Mr Mubarak*.

4.2   Comparison of NE Patterns between a User Analysis and a System Analysis

Figure 2 shows the comparison of NE patterns between a user analysis and a system analysis.

...Egyptian President Hosni Mubarak has warned that hanging former Iraqi leader Saddam Hussein will lead to even more bloodshed in Iraq. A Baghdad court condemned Saddam Hussein to death on Sunday for the killing of 148 Shia Muslims after a 1982 assassination attempt against him. *Mr Mubarak* said hanging the former president would only exacerbate ethnic and sectarian divisions between Iraqis.

Fig. 1. Nominal coreference of ANNIE

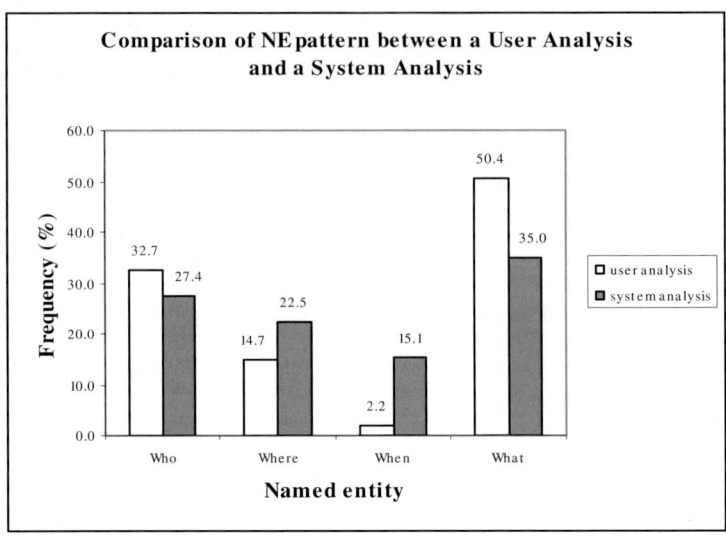

Fig. 2. Comparison of NE pattern

The result is interesting as *What* NE (35.0%) is perceived as being more frequent to be detect in a system analysis. This is followed by *Who* NE (27.4%), *Where* NE (22.5%) and the least was *When* NE (15.1%). The highest percentage of *Who* NE was in *Entertainment* (44.7%) and *When* NE was in *Economy* (32.4%). While *What* NE (37.7%) and *Where* NE (38.9%) has the highest percentage in *Politics*.

Findings revealed that users detect more of *What* NE (50.4%) and *Who* NE (32.7%) to appear in the documents when looking for information or news. Surprisingly *What* NE (35.0%) and *Who* NE (27.4%) are not as popular to occur in the system analysis. This difference is due to the fact that users have a richer notion of *What* rather than the system. They also have a richer notion of *Who* NE but not as much as *What* NE. For example users classify person, location, special occasion, programme as being *What* NE while the system would not. Classification of *What* NE by the users is relative to the content of the documents that makes the notion of *What* even subjective.

However, the user seems to care less about *When* NE (2.2%) compared to the system (15.1%). Users detect figures such as 2006, February and 20 years as *When* NE while system accept a rich notion of date such as *last year, this month, yesterday* to be accepted as *When* NE. This revealed that humans might care about relative dates when searching for recent documents such as newswires but humans usually don't see dates as important descriptors of documents. In addition, reporters and print journalists love to use relative dates for good reasons and it is the nature of journalism.

For *What* and *Who* NE we are in a situation where user would detect more of these NEs but the system detects less of them. However, system detects more *When* and *Where* NE, but users care less about them. One explanation of these differences is related to the importance level of NE given by users whom are more selective in detecting the NE they care about. Figure 3 shows an example of the document in the

> ….. US President George W Bush has made a brief but controversial visit to *Indonesia* for talks with his opposite number, Susilo Bambang Yudhoyono. The two men met in the Javan city of *Bogor* and discussed security, trade and health issues.

Fig. 3. Selection of NE from the importance level

user study and how the detection of NE differs from the system. Users detect *Indonesia* as *Where* NE but not *Bogor*, which they don't see as important. Though, the system would detect *Indonesia* and *Bogor* as *Where* NE.

5   Conclusions and Future Work

This paper compares how identification of named entities differs between human and machine. This evaluation is important to improve the performance of techniques for topic detection as it depends on the correct classification of named entities. It is also important to identify the correct weights to be assign for document representation in a TDT system. The comparison revealed that people care most about *What* and *Who* NE but the system detects less of these. Users have a richer notion of *What* NE compare to the system. The detection of NE also relate to the importance level and the content of the documents. This is how we looked at NE patterns in an interactive way and obtained better news story representation by better understanding of NE patterns.

For future work, we want to refine how we classify things as being *What* and *Who* NE since the user has a rich notion of them. We are not only considering the distribution of NE but also their importance. There should be a procedure to decide whether the term is important by looking at the term frequency and terms that appear on the headlines.

References

[1] Kumaran, G., Allan, J.: Text Classification and Named Entities for New Event Detection. In: Proceedings of the 27PthP Annual International ACM SIGIR Conference, pp. 297–304. ACM Press, New York (2004)

[2] Cunningham, H., Maynard, D., Bontcheva, K., Tablan, V.: GATE: A Framework and Graphical Development Environment for Robust NLP Tools and Applications. In: Proceedings of the 40th Anniversary Meeting of the Association for Computational Linguistics (ACL 2002), Philadelphia (2002)

[3] Juha, M., Helena, A.M., Marko, S.: Simple Semantics in Topic Detection and Tracking. Information Retrieval 7(3–4), 347–368 (2004)

[4] Mohd, M.: Named Entity Patterns across News Domains. In: BCS IRSG Symposium: Future Directions in Information Access 2007, pp. 30–36 (2007)

[5] Yang, Y., Carbonell, J., Brown, R., Pierce, T., Archibald, B.T., Liu, X.: Learning approaches for detecting and tracking news events. IEEE Intelligent Systems, Special Issue on Applications of Intelligent Information Retrieval 14(4), 32–43 (1999)

[6] Kuo, Z., Zi, L.J., Gang, W.: New Event Detection Based on Indexing-tree and Named Entity. In: Proceedings of SIGIR 2007, pp. 215–222. ACM Press, Amsterdam, The Netherlands (2007)

Query-Based Inter-document Similarity Using Probabilistic Co-relevance Model

Seung-Hoon Na[1], In-Su Kang[2], and Jong-Hyeok Lee[1]

[1] POSTECH, Pohang, South Korea
{nsh1979,jhlee}@postech.ac.kr
[2] KISTI, Daejeon, South Korea
dbaisk@kisti.re.kr

Abstract. Inter-document similarity is the critical information which determines whether or not the cluster-based retrieval improves the baseline. However, a theoretical work on inter-document similarity has not been investigated, even though such work can provide a principle to define a more improved similarity in a well-motivated direction. To support this theory, this paper starts from pursuing an ideal inter-document similarity that optimally satisfies the cluster-hypothesis. We propose a probabilistic principle of inter-document similarities; the optimal similarity of two documents should be proportional to the probability that they are co-relevant to an arbitrary query. Based on this principle, the study of the inter-document similarity is formulated to attack the estimation problem of the co-relevance model of documents. Furthermore, we obtain that the optimal inter-document similarity should be defined using queries as its basic unit, not terms, namely a query-based similarity. We strictly derive a novel query-based similarity from the co-relevance model, without any heuristics. Experimental results show that the new query-based inter-document similarity significantly improves the previously-used term-based similarity in the context of Voorhee's evaluation measure.

1 Introduction

The cluster-hypothesis is a widely accepted concept to the community of information retrieval, guiding the study of the cluster-based retrieval [1]. From this, researchers have investigated the study of cluster-based retrievals, implicitly assuming that the inter-document similarity which they use well-satisfies the cluster-hypothesis. Basically, since a retrieval model itself can be directly used to calculate an inter-document similarity, researchers have used the term-based inter-document similarity that the retrieval model defines, without a serious concern, i.e. one document among two documents is regarded as a query [2,3,4,5].

However, inter-document similarity which is adequate to a cluster-based retrieval may not be a term-based similarity. Without any theory of the inter-document similarity, we should not decide whether or not the inter-document similarity is a term-based similarity that a retrieval model defines. This is one of the reasons why we require a theory of an inter-document similarity. Unfortunately, previous works have not investigated it.

C. Macdonald et al. (Eds.): ECIR 2008, LNCS 4956, pp. 684–688, 2008.

This paper theoretically pursues an ideal inter-document similarity which optimally satisfies the cluster hypothesis. For this, we propose a probabilistic principle of inter-document similarities; the optimal similarity of two documents should be proportional to the probability that they are co-relevant to an arbitrary query, i.e. the inter-document similarity is obtained from the estimation of the co-relevance model of documents. Our notable result from this investigation is that an inter-document similarity is not term-based, but should be defined by using a query as its basic unit, namely the query-based inter-document similarity. This result attacks the implicit assumption of previous researches which simply use a retrieval model for inter-document similarity. Beside this theoretical reason, the use of query as a basic unit is intuitively reasonable. Because a term is not a conceptual or semantic unit but a syntactic unit, there are ambiguities of terms when they are represented in a document, where the meaning of a term is dependent to the context.

To formulate a query-based similarity, we develop an approximated estimation of the co-relevance model, and derive a novel query-based similarity without any heuristics. Experimental results show that the newly defined query-based inter-document similarity significantly improves the previously-used term-based similarity in the context of Voorhee's evaluation measure [3]. From an intuitive reason, the theoretical justification and the empirical success, the query-based similarity has a potential to resolve the limitation of previous term-based similarities.

2 Query-Based Inter-Document Similarity from Co-relevance Model

2.1 Probabilistic Principle of Inter-Document Similarity

The probabilistic principle of inter-document similarity is formulated as follows:

Probabilistic principle of inter-document similarity (PPSIM): Optimal inter-document similarity that best satisfies the cluster-hypothesis should be assigned in proportional to the co-relevance probability of two documents for a given query. Formally, let us assume that $sim(D,D')$ is inter-document similarity of two document D and D', and $P(CoRel|D,D')$ is the probability that two documents are co-relevant, namely *probabilistic co-relevance model*. Then, this principle implies that

$$sim(D,D') \propto P(CoRel|D,D') \tag{1}$$

This probabilistic principle of similarity is naturally derived by reversely considering Van's cluster-hypothesis.

2.2 Co-relevance Model: Co-relevant Probability of Two Documents

We will further derive the co-relevance model with some assumptions. To simplify the problem, we assume that a query is the set of terms, thus a query space is a finite set. Because the relevance is dependent to query, let us rewrite $P(CoRel|D,D')$ by introducing the query space.

$$P(CoRel|D,D') = \sum_Q P(CoRel,Q|D,D') = \sum_Q P(CoRel|D,D',Q)P(Q|D,D') \tag{2}$$

Here, we assume that a document and a query have a dependency only via relevance. Thus, $P(Q|D,D')$ is simplified to $P(Q)$ since the relevance is not involved. Thus, Eq. (2) becomes

$$P(CoRel|D,D') = \sum_Q P(CoRel,Q|D,D')P(Q) \qquad (3)$$

where $P(Q)$ indicates the prior probability of a query, which may be dependent on a user and the collection. To estimate $P(CoRel|D,D',Q)$, we consider two different cases.

Relevant documents are known for all queries: When we know the set of relevant documents, the estimation problem is easily resolved via MLE (Maximum Likelihood Estimation) as follows:

$$P(CoRel|D,D',Q) = P(CoRel|D,D',R_Q) = P(R_Q|D,Q)P(R_Q|D',Q) \qquad (4)$$

where R_Q is the set of relevant documents for query Q, and $P(R_Q|D,Q)$ is $\delta(D \in R_Q)$.

Relevant documents are unknown: Let Rel be a random variable to indicate the relevance of a document to a query, and $\overline{Rel}$ be a random variable to indicate the non-relevance of a document. Without relevance set, a reasonable approximation for $P(R_Q|D,Q)$ is $P(Rel|D,Q)$ i.e. the probabilistic relevance model. As a result, the original co-relevance model is decomposed into two separate relevance probabilities of two documents as follows:

$$P(CoRel|D,D',Q) = P(Rel|D,Q)P(Rel|D',Q) \qquad (5)$$

Fortunately, note that the estimation of $P(Rel|D,Q)$ is the basic starting point to derive the ranking formulas of many popular retrieval models. Recently, according to Roelleke's work [6], the probabilistic retrieval model, the language modeling approach, and the Poisson model are the approximation of $P(Rel|D,Q)$ with different assumptions. Here, we will present the approximation of $P(Rel|D,Q)$ for the language modeling approach. From the Bayesian theorem, $P(Rel|D,Q)$ is re-written by

$$P(Rel|D,Q) = \frac{P(D,Q,Rel)}{P(D,Q)} = \frac{P(D,Q,Rel)}{P(D,Q,Rel) + P(D,Q,\overline{Rel})} \qquad (6)$$

Roelleke showed that $P(D,Q,Rel)$ and $P(D,Q,\overline{Rel})$ in the language modeling approaches are decomposed as follows:

$$P(D,Q,Rel) = P(Q|D,Rel)P(D|Rel)P(Rel)$$
$$P(D,Q,\overline{Rel}) = P(Q|D,\overline{Rel})P(D|\overline{Rel})P(\overline{Rel}) \qquad (7)$$

Assume that terms in query Q are independent from each other when Rel or $\overline{Rel}$ is given. Then, $P(Q|D,Rel)$ and $P(Q|D,\overline{Rel})$ are

$$P(Q|D,Rel) = \prod_{w \in Q} P(w|D,Rel) \qquad P(Q|D,\overline{Rel}) = \prod_{w \in Q} P(w|D,\overline{Rel}) \qquad (8)$$

Note that the language modeling approaches discard the relevance concept, i.e. there is no relevance model. Instead, one global collection model which plays dual roles of relevant and non-relevant models is introduced. This dual role of collection model is reasonable, since both set of relevant and non-relevant documents are included in the given collection. However, we can agree that the collection model is too rough to approximate a relevant model. This roughness can be resolved from document language model $P(w|D)$ by regarding $P(Q|D, Rel)$ as the mixture model of $P(w|D)$ and the collection model as follows:

$$P(w|D, Rel) = (1 - \lambda)P(w|D) + \lambda P(w|C) \qquad (9)$$

This mixture model can be understandable if we agree that $P(w|C)$ is introduced for modeling the common terms in relevant documents. Different from relevant model, the collection model $P(w|C)$ is a well-approximation for non-relevant model. Thus,

$$P(w|D, \overline{Rel}) = P(w|C) \qquad (10)$$

where $P(w|C)$ is used for modeling not only topical terms but also common terms in non-relevant documents. In addition, we assume that prior probabilities - $P(D|Rel)$ and $P(D|\overline{Rel})$ are uniformly distributed over all documents, resulting in that $P(D|Rel)$ and $P(D|\overline{Rel})$ are the same as $1/N$. Then, $P(D, Rel)/P(D, \overline{Rel})$ becomes $P(Rel)/P(\overline{Rel})$.

By using all these considerations, $P(Rel|D, Q)$ is obtained by using the following relationship between the posterior probability $P(Rel|D, Q)$ and the likelihood ratio $O(Rel|D, Q)$.

$$P(Rel|D, Q) = \frac{P(D, Q, Rel)}{P(D, Q, Rel) + P(D, Q, \overline{Rel})} = \frac{O(Rel|D, Q)}{O(Rel|D, Q) + 1} \qquad (11)$$

where $O(Rel|D, Q)$ is defined as $P(D, Q, Rel)/P(D, Q, \overline{Rel})$ is formulated as follows:

$$O(Rel|D, Q) = \frac{P(Rel)}{P(\overline{Rel})} \lambda^{|Q|} \prod_{w \in Q} \left(\frac{(1 - \lambda)P(w|D)}{\lambda P(w|C)} + 1.0 \right) \qquad (12)$$

3   Experimentation

We used two small test collections - MED and CISI. Although we require numerous queries, only test query topics are available in test collections, and they are too small. To resolve it, we used randomly generated queries from a given collection. To generate a random query, we select a source document to generate query terms. Then, terms are automatically generated based on a unigram language model and a bigram language model of a document. The unigram language model is used to generate the first term, and then subsequent terms are generated from the bigram model. From this automatic procedure, we constructed 1,000,000 counts of random queries.

For evaluation, we used Voorhee's nearest neighbor test (NNT) to check how many among top nearest neighbors of a relevant document are co-relevant to the document to a given query [3]. Because Voorhee's test is the evaluation of a ranked list, all evaluation measures of information retrieval can be utilized. Thus, we considered MAP (Mean Average Precision), and Pr@X (Precision at X documents). To avoid the confusion with the tradition measure, we call them NNT-MAP and NNT-Pr@X, respectively.

Table 1. NNT-MAP of query-based similarity for the different Ns

Coll.	Similarity Measure	$N = 10$	$N = 30$	$N = 50$	$N = 70$	$N = 100$
MED	*Term-based*			0.4399		
	Query-based (Eq.(4))	0.4585	**0.5159**	0.4938	0.4572	0.4034
	Query-based (Eq.(5))	0.4564	**0.5186**	0.5050	0.4767	0.4323
CISI	*Term-based*			0.1912		
	Query-based (Eq.(4))	0.1752	0.1990	0.2104	0.2180	**0.2243**
	Query-based (Eq.(5))	0.1739	0.1972	0.2089	0.2168	**0.2238**

Table 1 shows the final NNT-MAP after 1,000,000 queries are used to the estimation of co-relevance model. To efficiently calculate the estimation of co-relevance model, we assumed that only top N retrieved documents for each query have non-zero probabilities of $P(Rel|D,Q)$ ($P(Rel|D,Q)$ is zero for other documents). Two query-based similarities are used - 1) a naive query-based similarity using the co-relevance model from MLE (Eq. (4)) is presented by assuming that top N retrieved documents are relevant (all other ones are non-relevant), 2) the proposed query-based similarity using the co-relevance model from the retrieval model (Eq. (5) and Eq. (11)). Both of two query-based similarity measures significantly improve the term-based similarity over most of Ns. Among two query-based similarities, there is no difference between two query-based similarities in CISI. However, in MED, the proposed query-based similarity (Eq. (5)) is more robust than naive one (Eq. (4)) when N is 50, 70 or 100.

4   Conclusion

This chapter proposed the probabilistic principle for inter-document similarity, and derived query-based inter-document similarity metric based on probabilistic co-relevance model. Experimental results show that the query-based metric can significantly improve the traditional term-based metric in terms of Voorhee's NNT.

Acknowledgement. This work was supported by the Korea Science and Engineering Foundation (KOSEF) through the Advanced Information Technology Research Center (AITrc), also in part by the BK 21 Project and MIC & IITA through IT Leading R&D Support Project in 2007.

References

1. Rijsbergen, C.J.V.: Information Retrieval. Butterworth-Heinemann (1979)
2. Croft, W.B.: A model of cluster searching based on classification. Information Systems (5), 189–195 (1980)
3. Voorhees, E.M.: The cluster hypothesis revisited. In: SIGIR 1985, pp. 188–196 (1985)
4. Liu, X.: Cluster-based retrieval using language models. In: SIGIR 2004, pp. 186–193 (2004)
5. Kurland, O., Lee, L.: Corpus structure, language models, and ad hoc information retrieval. In: SIGIR 2004, pp. 194–201 (2004)
6. Roelleke, T., Wang, J.: A parallel derivation of probabilistic information retrieval models. In: SIGIR 2006, pp. 107–114 (2006)

Using Coherence-Based Measures to Predict Query Difficulty

Jiyin He, Martha Larson, and Maarten de Rijke

ISLA, University of Amsterdam
{jiyinhe,larson,mdr}@science.uva.nl

Abstract. We investigate the potential of coherence-based scores to predict query difficulty. The coherence of a document set associated with each query word is used to capture the quality of a query topic aspect. A simple query coherence score, QC-1, is proposed that requires the average coherence contribution of individual query terms to be high. Two further query scores, QC-2 and QC-3, are developed by constraining QC-1 in order to capture the semantic similarity among query topic aspects. All three query coherence scores show the correlation with average precision necessary to make them good predictors of query difficulty. Simple and efficient, the measures require no training data and are competitive with language model-based clarity scores.

1 Introduction

Robustness is an important feature of information retrieval (IR) systems [7]. A robust system achieves solid performance across the board and does not display marked sensitivity to difficult queries. IR systems stand to benefit if, prior to performing retrieval, they can be provided with information about problems associated with particular queries [4]. Work devoted to predicting query difficulty [1, 2, 3, 5, 8] is pursued with the aim of providing systems with the information necessary to adapt retrieval strategies to problematic queries. We investigate the usefulness of coherence-based scores in predicting query difficulty. The *query coherence scores* we propose are inspired by the *gene expression coherence* score used in the genetics literature [6], which functions as a measure of clustering structures. They are designed to reflect the quality of individual aspects of the query, following the suggestion that "the presence or absence of topic aspects in retrieved documents" is the predominant cause of current system failure [4].

We use document sets associated with individual query terms to assess the quality of query topic aspects (i.e., subtopics), noting that a similar assumption proved fruitful in [8]. We consider that a document set associated with a query term reflects a high-quality query topic aspect when it is: (1) topically constrained or specific and (2) characterized by a clustering structure tighter than that of the background document collection. These two characteristics are captured by coherence and for this reason we chose to investigate the potential of coherence-based scores. Like the clarity score [2, 3], our approach attempts to capture the difference between the language usage associated with the query and the language usage in the background collection. Our approach promises

C. Macdonald et al. (Eds.): ECIR 2008, LNCS 4956, pp. 689–694, 2008.

low run-time computational costs. Additionally, our query coherence scores do not require training data as is the case with the method proposed in [8].

We propose three query coherence scores. The first query coherence score, QC-1, is an average of the coherence contribution of each query word and has only the effect of requiring that all query terms be associated with high-quality topic aspects. This score is simple and efficient. However, it does not require any semantic overlap between the contributions of the query words. A query topic composed of high-quality aspects would receive a QC-1 score even if those aspects were never reflected *together* in a collection document. Hence, we develop two further scores, which impose the requirement that, in addition to being associated with high-quality topic aspects, query words must be topically close. The second query coherence score, QC-2, adds a global constraint to QC-1. It requires the union of the set of documents associated with each query word to be coherent. The third score, QC-3, adds a proximity constraint to QC-1. It requires the document sets associated with individual query words to exhibit a certain closeness. QC-2 and QC-3 require more computational effort than QC-1, but fail to demonstrate an improved ability to predict query difficulty.

The next section further explains our coherence-based scores. After that we describe our experiments and results. We conclude with discussion and outlook.

2   Method

Given a document collection C and query $Q = \{q_i\}_{i=1}^{N}$, where q_i is a query term, R_{q_i} is the set of documents associated with that query word, i.e., the set of documents that contain at least one occurrence of the query word. The coherence of R_{q_i} reflects the quality of the aspect of a query topic that is associated with query word q_i. The overall query coherence score of a query is based on a combination of the set coherence contributed by each individual query word. Below, we first discuss set coherence and then present our three query coherence scores.

2.1   The Coherence of a Set of Documents

The coherence of a set of documents is defined as the proportion of "coherent" pairs of documents in the set. A pair of documents is "coherent" if the similarity between them exceeds a given threshold. Formally, given a set of documents $D = \{d_i\}_{i=1}^{M}$ and threshold θ, we have

$$\delta(d_i, d_j) = \begin{cases} 1 & \text{if } similarity(d_i, d_j) \geq \theta, \\ 0 & \text{otherwise.} \end{cases} \quad i \neq j \in \{1, \ldots, M\} \quad (1)$$

where the similarity between documents d_i and d_j can be any similarity metric; here we use the cosine similarity as an example. The *coherence* of the document set D is defined as

$$SetCoherence(D) = \frac{\sum_{i \neq j \in \{1, \ldots, M\}} \delta(d_i, d_j)}{M(M-1)}. \quad (2)$$

Set coherence is a measure for the relative tightness of the clustering of a specific set of data with respect to the background collection. In a random subset drawn

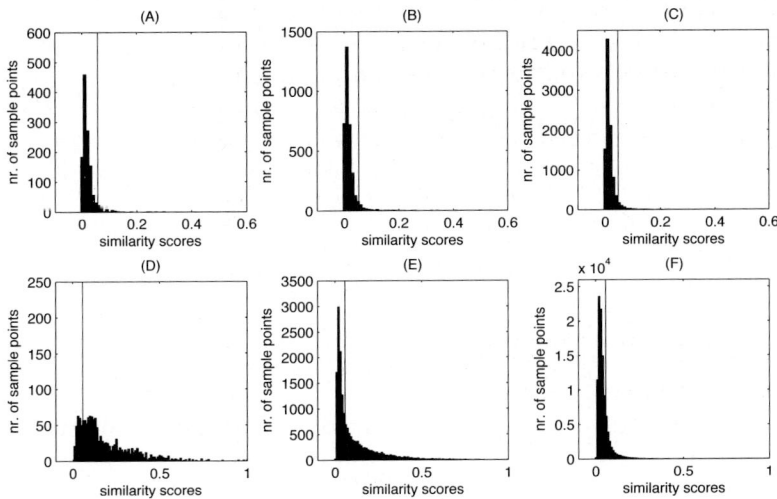

Fig. 1. Distribution of document similarities from subsets of TREC AP89+88. (A)–(C) Randomly sampled 50, 100, and 500 documents, respectively; (D) R_Q determined by query21, $SetCoherence(R_{Q21}) = 0.8483$; AP(Q21)=0.1328; (E) R_Q determined by query57, $SetCoherence(R_{Q57}) = 0.7216$; AP(Q57)=0.0472; (F) R determined by query75, $SetCoherence(R_{Q75}) = 0.2504$; AP(Q75)=0.0027.

from a document collection, few pairs of documents have high similarities. Plots A, B, and C in Figure 1 show that pairs having similarity scores higher than the threshold θ (the vertical line) are proportionally rare cases in a random sample, independently of sample size. Plots D, E and F show the distribution of document similarities for a collection subset associated with a one-word query, which we use to illustrate the properties of the R_{q_i}, the collection subset associated with a single query word q_i. Plots D, E, and F are ordered by decreasing coherence score, which can be seen to correspond to an increasing proportion of dissimilar document pairs. Plot F approaches the distribution of the random samples from the background collection. Initial support for the legitimacy of our approach derives from the fact that across these three queries decreasing set coherence of R_{q_i} corresponds to decreasing average precision.

2.2 Scoring Queries Based on Coherence

For a given query $Q = \{q_i\}_{i=1}^{N}$, we propose three types of query coherence scores. The first requires that each query word have a high contribution to the coherence of the query. This score reflects the overall quality of the aspects of a topic.

QC-1 Average query term coherence

$$QC\text{-}1(Q) = \frac{1}{N} \sum_{i=1}^{N} SetCoherence(R_{q_i}), \qquad (3)$$

where $SetCoherence(R_{q_i})$ is the coherence score of the set R_{q_i} determined by the query word q_i. This score is simple, but leaves open the question of whether query aspects must also be semantically related. Therefore, we investigate whether QC-1 can be improved by adding limitations that would force the R_{q_i}'s to be semantically constrained. The second query coherence score adds a constraint on global coherence, multiplying QC-1 by the coherence of $R_Q = \bigcup_{i=1}^{N} R_{q_i}$.

QC-2 Average query term coherence with global constraint

$$QC\text{-}2(Q) = SetCoherence(R_Q)\frac{1}{N}\sum_{i=1}^{N} SetCoherence(R_{q_i}). \qquad (4)$$

The third query coherence score adds a constraint on the proximity of the R_{q_i}'s, multiplying QC-1 by the average of the closeness of the centers of the R_{q_i}'s.

QC-3 Average query term coherence with proximity constraint

$$QC\text{-}3(Q) = \frac{S}{N}\sum_{i=1}^{N} SetCoherence(R_{q_i}) \qquad (5)$$

$$S = \frac{\sum_{l \neq k}^{N} Similarity(c(q_k), c(q_l))}{N(N-1)} \qquad (6)$$

where S is the mean similarity score of each pair of cluster centers of the R_{q_i}'s. Below, we compare the performance of these three query coherence scores.

3   Evaluation

We run experiments to analyze the correlation between the proposed query coherence scores and the retrieval performance. Following [2], TREC datasets AP88 and AP89 are selected as our document collection. We use TREC topics 1–200 with the "title" field. The threshold θ is determined heuristically: we randomly sample different numbers of documents from the collection, and take the mean of the similarity scores at the top 5% of each sampled document set as the value of θ. For large sets R (e.g., $> 10,000$ documents), we approximate the $SetCoherence$ by using the "collection" score (the threshold θ); a set R with many documents has a $SetCoherence$ similar to the collection.

We use Spearman's ρ to measure the rank correlation between the coherence score and the average precision (AP). The higher this correlation, the more effective the scoring method is in terms of predicting query difficulty. Different retrieval models are applied so as to show stability across models.

Table 1 shows that all three coherence scores have significant correlation with AP. However, QC-2 and QC-3 do not have a substantially stronger predictive ability than QC-1, though they take the semantic relation between query words into account. Since the coherence score is the proportion of the "coherent pairs" among all the pairs of data points, and the similarity score can be pre-calculated without seeing any queries, the run-time operation for QC-1 is a simple counting. The same holds for QC-2, but with more effort for the extra term R_Q. Both are

Table 1. The Spearman's correlation of query coherence scores with average precision. Queries: TREC topics 1–200; document collection: AP89+88.

Model	QC-1		QC-2		QC-3	
	ρ	p-value	ρ	p-value	ρ	p-value
BM25	0.3295	1.8897e-06	0.3389	0.0920e-05	0.3813	2.5509e-08
DLH13	0.2949	2.2462e-05	0.3096	0.8180e-05	0.3531	2.9097e-07
PL2	0.3024	1.3501e-05	0.3135	0.6167e-05	0.3608	1.5317e-07
TFIDF	0.2594	2.0842e-04	0.3301	0.1805e-05	0.3749	4.5006e-08

Table 2. The Spearman's correlation of clarity score (CS) and query coherence score (QC) with AP: the correlation coefficient ρ and its corresponding p-value. The queries are TREC topics 101–200, using title only. AP values obtained by running BM25; scores of column 1 taken from [2].

Score	CS	QC-1	QC-2	QC-3
ρ	0.368	0.3443	0.3625	0.3222
p-value	1.2e-04	4.5171e-04	2.1075e-04	0.0011

much easier to compute than QC-3, which requires the calculation of the centers of the R_{q_i}'s. Therefore, taking into account its computational efficiency, QC-1 is the preferred score. QC-1 is also more efficient at run time than other methods such as the clarity score [2] and has competitive prediction ability; see Table 2.

4   Conclusions

We introduced coherence-based measures for query difficulty prediction. Our initial experiments on short queries, reported here, show that the coherence score has a strong positive correlation with average precision, which reflects the predictive ability of the proposed score. As similarity scores can be computed offline or at indexing time, this method promises run-time efficiency. Moreover, as the only parameter, θ, is obtained from the background collection, the method requires no training data. We plan to evaluate our coherence scores on more and larger data sets, e.g., the collection used in the TREC Robust track, as well as to investigate their behaviors on long queries. We will also use our approach in applications such as resource selection, and selective query expansion.

Acknowledgments

This research was supported by the E.U. IST programme of the 6th FP for RTD under project MultiMATCH contract IST-033104 and by the Netherlands Organization for Scientific Research (NWO) by a grant under project numbers 220-80-001, 640.001.501, 640.002.501.

References

[1] Amati, G., Carpineto, C., Romano, G.: Query difficulty, robustness and selective application of query expansion. In: McDonald, S., Tait, J.I. (eds.) ECIR 2004. LNCS, vol. 2997, pp. 127–137. Springer, Heidelberg (2004)

[2] Cronen-Townsend, S., Croft, W.B.: Quantifying query ambiguity. In: HLT 2002, pp. 94–98 (2002)

[3] Cronen-Townsend, S., Zhou, Y., Croft, W.B.: Predicting query performance. In: SIGIR 2002, pp. 299–306 (2002)

[4] Harman, D., Buckley, C.: The NRRC reliable information access (RIA) workshop. In: SIGIR 2004, pp. 528–529 (2004)

[5] He, B., Ounis, I.: Query performance prediction. Inf. Syst. 31(7), 585–594 (2006)

[6] Pilpel, Y., Sudarsanam, P., Church, G.M.: Identifying regulatory networks by combinatiorial analysis of promoter elements. Nat. Genet. 29, 153–159 (2001)

[7] Voorhees, E.M.: The TREC robust retrieval track. SIGIR Forum 39, 11–20 (2005)

[8] Yom-Tov, E., Fine, S., Carmel, D., Darlow, A.: Learning to estimate query difficulty. In: SIGIR 2005, pp. 512–519 (2005)

Efficient Processing of Category-Restricted Queries for Web Directories

Ismail Sengor Altingovde, Fazli Can, and Özgür Ulusoy

Department of Computer Engineering, Bilkent University, Ankara, Turkey
{ismaila,canf,oulusoy}@cs.bilkent.edu.tr

Abstract. We show that a cluster-skipping inverted index (CS-IIS) is a practical and efficient file structure to support category-restricted queries for searching Web directories. The query processing strategy with CS-IIS improves CPU time efficiency without imposing any limitations on the directory size.

1 Introduction

Web directories typically involve a hierarchy of categories and employ human editors who assign Web pages to corresponding categories. Web surfers make use of such directories either for merely browsing, or issuing a query under a certain category that they have chosen (i.e., a category-restricted search [4, 5]).

In the earlier works [4, 5], a simple way of processing the category-restricted queries is described as follows. The system first determines the categories that are under the user specified category, i.e., the sub-tree (or graph, more generally) rooted at the user's initial category selection (step 1). This set constitutes the *target categories*. Next, the query is processed using an inverted index over the *entire* document collection in the directory and a candidate result set is obtained (step 2). Finally, to obtain the query output, the documents that are not from the target categories found in the first step are eliminated from the candidate result set (step 3). In this paper, this is referred to as the *baseline* method.

The baseline method is not very efficient: The document selection step uses the entire index without making use of the target categories, which is known at that time. This means several accumulators are updated and extracted, just to be eliminated at the very end. Furthermore, the candidate elimination step requires to check the category of (at least) the top-N candidate documents, and this would require N separate disk accesses if the data structure mapping documents to categories is kept on disk. In [1], several alternatives to the above baseline query processing strategy are discussed to allow the use of the target categories as early as possible in the document selection stage. In particular, if it is possible to store the entire mapping of documents to categories in the main memory, then the query processor can avoid computing partial similarities for documents that are not in target categories without making any disk accesses. However, this approach would again suffer if the inverted index is compressed, which is a typical practice. In this case, the query processor would still waste CPU cycles for decoding some postings, just to be discarded when it is realized that they are not from the target categories.

In the literature, cluster-skipping inverted index structure (CS-IIS) is introduced for efficient cluster-based retrieval [2, 3, 6]. In this paper, we propose to adapt the CS-IIS

C. Macdonald et al. (Eds.): ECIR 2008, LNCS 4956, pp. 695–699, 2008.

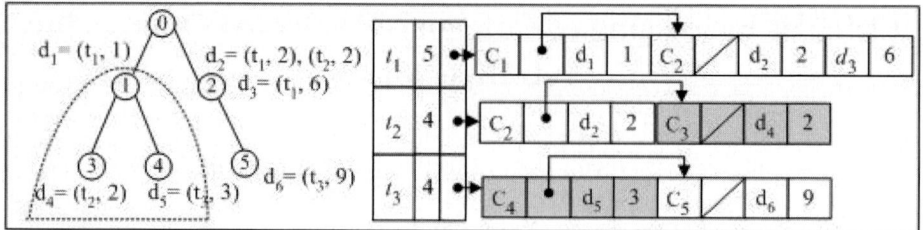

Fig. 1. A hierarchical taxonomy and the corresponding CS-IIS. Given the query = {t_2, t_3} that is restricted to C_1, the query processor first identifies the target categories (C_1, C_3 and C_4, as shown within dotted lines) and then processes posting lists. Note that, only the shaded parts of the posting lists are processed and the rest is skipped.

as a practical and efficient choice to combine the last two steps of the baseline approach. Thus, the major contribution of this work is demonstrating how CS-IIS can be employed in a hierarchical clustering framework, such as a Web directory, and how exactly the gains or costs are affected due to some unique properties of this framework. The experiments are held using the largest available Web directory dataset as provided by Open Directory Project (ODP). This work differs from the earlier works [3, 6] in the following ways: *i)* in the earlier works, an automatic and partitioning clustering structure is assumed, whereas the Web directory domain involves a hierarchical taxonomy, *ii)* the previous works involve moderate number of categories (although they were quite large figures in the automatic text clustering literature) whereas Web directories involve hundreds of thousands of categories, and *iii)* both the data, categorization and queries are *real*, which makes this environment a unique opportunity to show the applicability of the CS-IIS approach.

2   Category-Restricted Query Processing with CS-IIS

In CS-IIS, the *<document, term frequency>* pairs in a posting list are reorganized such that all documents from the same category are grouped together, and at the beginning of each such group an extra element is stored in the form of *<category id, next category address>*. While constructing the CS-IIS for a hierarchy as in the case of a Web directory, documents in a posting list are grouped with the categories under which they immediately appear (see Figure 1). This is different from an earlier proposal where the signature of the full category path is stored for each document [5].

During query processing, the target category set is identified by expanding the initial category given by the user (e.g., see [4]). Next, for each query term, the corresponding posting list is brought to the memory. By comparing the category ids in the posting list and in the target categories in a merge-join fashion, accumulators for only those documents that are from the target categories are updated. That is, if the category id in an inverted list element is not found in the target categories, the succeeding documents in the list (for that category) are skipped and the query processor jumps to the next category pointed by the "next category address." When all query terms are processed, the non-zero accumulators are only from the targets.

3   ODP Dataset Characteristics and Experimental Results

Dataset. For this study, we use the largest publicly available category hierarchy as provided by ODP Web site (www.dmoz.org). After preprocessing and cleaning data files, we end up with a category hierarchy of approximately 719K categories and 4.5 million URLs. For most of the URLs, a one- or two-sentence length description is also provided in the data file. In this paper, we use these descriptions as the actual documents. Note that, this yields significantly shorter documents (with a few words on the average) than usual. Our on-going work involves downloading the actual documents from the Web.

While constructing the hierarchy using the data files, we decided to use *narrow*, *symbolic* and *letterbar* tags in the data file as denoting the children of a category. The resulting hierarchy is more like a graph than a tree in that only 36% of the categories have a single parent. This indicates that, it would be better to keep track of the immediate category of a document as in our adaptation in Section 2 (and also the approach in [4]) with respect to keeping the entire path (e.g., see [5]), as there may be several paths to a particular document.

We find that a great majority of categories are rather small, i.e., 98% of them include less than 50 documents. Furthermore, 93% of the documents (about 4.2 million) belong to only one category, whereas 6% of the documents belong to two parents and only the remaining 1% of the documents appears in three or more categories. These numbers are important for CS-IIS, since a posting list needs to store the same documents as many times as they appear in different categories. The above trends conform the observations in earlier works [4, 5], and show that the waste of storage space due to overlapping documents among categories would not be high.

Indexing. After preprocessing, the document description file takes 2 GB on disk. During inverted index creation, all words (without stemming) are used except numbers and stopwords, yielding 1.1 million terms at the end. The resulting size of the typical inverted file (i.e., to be used by the baseline approach) is 342 MB whereas the size of the CS-IIS file is 609 MB. Note that, the additional space used in CS-IIS is unusually large in comparison to the earlier works (i.e., only 26% more space usage was observed in [6]). We attribute two reasons for this outcome, and state their remedies as follows: *i)* the dataset includes too many categories with respect to the number of documents. In [6], for instance, a collection of approximately 400K documents yields only 1357 clusters, whereas here approximately 4.5M pages are distributed to 719K categories. We believe this situation would change for our benefit in time, as the growth rate of hierarchy may possibly be less than that of the collection. Furthermore, the taxonomy may be populated to reach to a much larger collection size using automatic classification techniques. *ii)* the documents are unusually short, as we use just the summaries in this initial stage of our work.

Queries and query processing efficiency. We use two methods for obtaining category-restricted queries. First, we prepare a Web-based system which allows users (graduate students) to specify queries along with categories and evaluate the results (available at http://139.179.11.31/~kesra/bir2/).

For this paper, we only use 64 category-restricted queries from this system and refer to them as *manual-category* queries. Additionally, we employ the efficiency task

Table 1. In-memory query processing efficiency (all average values)

Query set	Strategy	Query evaluation time (sec)	No. of non-zero accumulators	No. of postings read
Manual-category	Baseline	0.128	17,219	17,339
	CS-IIS	**0.109 (15%)**[*]	11,758	28,913
Automatic-category	Baseline	0.158	19,900	20,367
	CS-IIS	**0.100 (37%)**[*]	250	33,271

[*] Percentage improvement w.r.t. baseline.

topics of TREC 2005 terabyte track. This latter set includes 50K queries, and 46K of them are used in the experiments after those without any matches in the collection are discarded. This set is referred to as *automatic-category* queries.

Notice that, the latter query set lacks any initial target category specification, so we had to match the queries to categories automatically. To achieve this, we use all terms in categories to compute query-category similarities. At this stage, the well-known TF-IDF term weighting with the cosine measure is employed. Next, for each query, we find the top-10 highest scoring category and choose a single one with the shortest distance to the root (i.e., imitating the typical user behavior of selecting a category as shallow as possible [5] while browsing).

For both query sets, this initial target category is then further expanded, i.e., the sub-graph is obtained. In the following experiments, the time cost for obtaining target categories is not considered, as this stage is exactly the same for both of the compared strategies and can be achieved very efficiently by using the method in [4]. The query-document matching stage also uses the TF-IDF based weighting scheme and cosine similarity measure [6]. Top 100 results are returned for each query. The in-memory average query processing (CPU) times are reported in Table 1, as well as the number of non-zero accumulators at the end, and the average length of posting lists read.

Table 1 reveals that for both query sets, using CS-IIS improves the efficiency of work done in main memory. This gain is caused by two factors: first, skipping irrelevant clusters reduces the redundant partial similarity computations. Secondly, but equally importantly, the number of non-zero accumulators at the end of query, which are to be inserted into and extracted from a min-heap, is considerably reduced. We even favor the baseline strategy by assuming that the document-category mapping is in the memory. Note that, the gains would be more emphasized if compression had been used, as skipping would also reduce the burden of decoding operations [2]. A second observation is that, the manual-category queries apparently cover a larger sub-graph and thus process more documents for both strategies. Indeed, in that query set, 55% of the queries are restricted to categories at depth 1. In contrary, the automatic-category queries usually locate the initial target category in a deeper position in the graph. That is why the latter makes much less operations and obtains more gains. Nevertheless, we used the same automatic category computation technique for the manual-category query set, and observed that most of the returned categories are reasonably relevant to queries, but not necessarily the same as the ones as specified by the user. Our current work involves a quantitative analysis of target category selection and using more sophisticated term weighting schemes to represent categories.

For the disk access issues, we assume that posting lists are brought to memory entirely and discarded once they are used (i.e., no caching). It is possible to read only a fraction of the posting lists in the baseline strategy. This is also possible in our approach. Indeed, if the skipping elements are kept at the beginning of each list instead of being intertwined with the document postings, reading only a relevant part from the disk would be straightforward. Furthermore, it is also possible to sort each category's list with respect to, say, frequency, and dynamically prune the search. Lastly, caching (if used) would provide similar improvements for both approaches. In Table 1, the difference between the list lengths fetched from the disk is around 12 K postings (for manual-category query set), adding up to 96 KB (i.e., 8 bytes/posting). Considering a typical disk with the transfer rate of 20 MB/s, the additional sequential read cost is only 5 ms, which is clearly less than the in-memory gains for this case.

4 Discussions and Conclusion

CS-IIS has some other advantages in comparison to the earlier works in the literature. We observe that the real life hierarchies are quite large (in contrast to those in [4, 5]). So, it may be difficult to use the signature-file based system as in [5]. The approach discussed in [4] enforces an upper limit on the number of categories (e.g., 1024). Furthermore, both of these earlier works involve using a part of document id to represent its categories, which would require bitwise operations during query processing and may complicate the use of typical index compression schemes. On the other hand, CS-IIS imposes no limits on neither the size of category nor the number of documents and can be practically used in existing systems, even with compression.

In this paper, the CS-IIS is adapted for hierarchical categories in Web directories to allow efficient processing of category-restricted queries. Our preliminary results show that, despite the use of very short document descriptions and the imbalance between the number of categories and documents, the proposed strategy is quite promising.

Acknowledgments. This work is partially supported by The Scientific and Technical Research Council of Turkey (TÜBİTAK) under the grant numbers 105E024 and 105E065. We thank to İnci Durmaz and Esra Küçükoğuz for processing ODP data.

References

1. Altingovde, I.S., Can, F., Ulusoy, Ö.: Algorithms for within-cluster searches using inverted files. In: Levi, A., Savaş, E., Yenigün, H., Balcısoy, S., Saygın, Y. (eds.) ISCIS 2006. LNCS, vol. 4263, pp. 707–716. Springer, Heidelberg (2006)
2. Altingovde, I.S., Demir, E., Can, F., Ulusoy, Ö.: Incremental cluster-based retrieval using compressed cluster-skipping inverted files. ACM TOIS (to appear)
3. Altingovde, I.S., Ozcan, R., Ocalan, H.C., Can, F., Ulusoy, Ö.: Large-scale cluster-based retrieval experiments on Turkish texts. In: Proc. of SIGIR 2007, pp. 891–892 (2007)
4. Cacheda, F., Baeza-Yates, R.: An optimistic model for searching Web directories. In: McDonald, S., Tait, J.I. (eds.) ECIR 2004. LNCS, vol. 2997, pp. 364–377. Springer, Heidelberg (2004)
5. Cacheda, F., Carneiro, V., Guerrero, C., Viña, Á.: Optimization of restricted searches in Web directories using hybrid data structures. In: Sebastiani, F. (ed.) ECIR 2003. LNCS, vol. 2633, pp. 436–451. Springer, Heidelberg (2003)
6. Can, F., Altingovde, I.S., Demir, E.: Efficiency and effectiveness of query processing in cluster-based retrieval. Information Systems 29(8), 697–717 (2004)

Focused Browsing: Providing Topical Feedback for Link Selection in Hypertext Browsing

Gareth J.F. Jones and Quixiang Li

Centre for Digital Video Processing & School of Computing
Dublin City University, Dublin 9, Ireland
gareth.jones@computing.dcu.ie

Abstract. When making decisions about whether to navigate to a linked page, users of standard browsers of hypertextual documents returned by an information retrieval search engine are entirely reliant on the content of the anchortext associated with links and the surrounding text. This information is often insufficient for them to make reliable decisions about whether to open a linked page, and they can find themselves following many links to pages which are not helpful with subsequent return to the previous page. We describe a prototype *focused browsing* application which provides feedback on the likely usefulness of each page linked from the current one, and a *term cloud* preview of the contents of each linked page. Results from an exploratory experiment suggest that users can find this useful in improving their search efficiency.

1 Introduction

Users approach search engines with a wide range of types of information need; some are very focused, while others are much vaguer. In the latter case the user will often need to browse through multiple documents to address their information need. When browsing between documents returned by a standard search engine users rely on the contents of the item, and in particular the link anchortexts, to determine whether to select a link and progress their search to the linked item. The available information can often be insufficient to enable them to make an informed decision, and they often end up needing to follow a link to determine whether it leads to useful information, and then returning to the previous page when they determine that it does not.

This paper introduces a new enhanced browsing framework of *focused browsing* where the user is provided with feedback on the potential utility of available links from the current page and a summary preview of the content of linked items in the form of a *term cloud*. The principle underlying focused browsing is analogous to the established concept of *focused crawling* [3][1]. The goal of a focused crawler is to selectively seek out web pages relevant to pre-defined topics. Focused browsing aims to focus a user's browsing behaviour towards "on topic" pages that will be of interest to them.

The remainder of this paper describes details of our prototype browsing application and a preliminary pilot user study carried out with this application.

2 Focused Browsing

Conventional information retrieval (IR) systems for hypertext documents do not provide feedback to a user on the potential utility of links appearing in retrieved documents. This

C. Macdonald et al. (Eds.): ECIR 2008, LNCS 4956, pp. 700–704, 2008.

is in contrast to adaptive hypermedia (AH) systems designed for constrained domains where the contents typically comprise manually selected information fragments. These fragments can be composed into personalised presentations based on complex models of the user's goals, preferences and knowledge of the subject at hand. Various navigation support mechanisms are supported by these systems including link ranking, link annotation, and link disabling [4] [2].

A recent emergence in web technology is the facility to enable users to add tags to web content, potentially building up rich descriptions from a community of users. Browsing of these tags is supported using *tag clouds* [5] which provide a means for visualizing groups of tag words. Words to be included in the tag clouds are typically selected based on frequency of user annotation. The significance of individual words can be indicated by varying their font size based on their score. These words are then presented to the user in the form of a simple graphical cloud.

Our focused browsing application provides the user with interactive feedback to support their browsing. It uses IR methods to score document links. While the models used are less sophisticated than those typically used in AH systems, they do not have their domain specific limitations and can easily be applied to any linked documents. A related more complex approach was introduced in [6]. The concept of tag clouds is extended to provide term cloud document summaries.

2.1 Link Scoring

In order to assist the user in determining which link they should follow from the current page, each link is scored with respect to the search request. The intensity of the display of the link is then selected based on its score relative to others leading from the page.

In our current prototype links are scored using a standard $tf \times idf$ approach. After removal of stop words and Porter stemming, terms are weighted as follows. $w(i,j) = tf(i,j) \times \log N/n(i)$, where $tf(i,j)$ is the frequency of term i in the linked page j, N is the no of pages linked from the current page, and $n(i)$ is the number of them which contain i.

A matching score is then computed by summing matching term weights between the user's search query and each linked document. The scores for each linked page are then ranked and the links displayed with their intensity determined by their relative rank score. While the current prototype only uses the query to score the links, many other factors could be used. For example, a user profile based on current or ongoing browsing and search behaviour, or an absolute page significance factor such as PageRank [7] or one based on access behaviour [8].

2.2 Term Clouds

Varying the intensity of the links obviously does not indicate to the user what the linked pages contain. Indeed the score may actually be misleading if the user changes the focus of their information need. To further assist the user in determining whether to follow a link the focused browser makes available a *term cloud* of each linked page. Term clouds are related to the "fluid links" introduced in [10].

Term clouds present a simple summary of the linked page in the form of a cloud of words or phrases taken from the document. Selection of terms is similar to the principles used to form query-biased summaries (QBS) in [9]. Term clouds are more compact than QBSs and are intended to be more efficient in terms of cognitive user interaction and the display space needed, but use of QBSs as an alternative document visualization method will be explored in future work.

The components used to select words for inclusion in term clouds are as follows:

Term Weighting. The first word significance score $ws1(i,j)$ is calculated using the $tf \times idf$ function from Section 2.1. The $ws1(i,j)$ is calculated for each non stop term appearing in each linked document. This is then multiplied by a scaling factor $ws1$.

Title Term Weighting. Words contained in the title of a document are often important to its topic. Words appearing in the title of each document are given a boost of a fixed scalar value $ws2$.

Location Weighting. Words appearing at the beginning of a document are more likely than those later in it to be significant to its topic. Words appearing in the first sentence are given a fixed scalar boost of $ws3$.

Relationship to Current Document. Assuming that the user's browsing is focused on the topic of the current page, we give an additional scalar boost $ws4$ to terms which appear in the current document.

Combining the Term Scores. The final combined score for each term is a simple sum of the four components. The highest scoring ones are then selected for inclusion in the term cloud. The values of $ws1$, $ws2$, $ws3$ and $ws4$ were selected empirically based on experimentation with the small test evaluation site used in our pilot experimental study, with font size varied according to the combined term score. Values used in the current prototype are 0.5, 0.2, 0.1, 0.1 respectively.

3   Prototype Application

Figure 1 shows a sample interface from our prototype focused browsing application. The hyperlinks are highlighted at different levels indicating the estimated utility of each link to the user as described in section 2.1. In the example the user has moused over the link to "document retrieval", in response to this the term cloud shown is produced.

Implementing a practical focusing browsing application requires a tighter integration between the search engine and browser than is generally the case. Computation of link scores and term clouds requires access to term data most easily available from a search engine, thus documents must be displayed from the search engine cache or if they are downloaded from their original sources, they must be automatically rewritten prior to display to include variations in link display and include links to term clouds.

4   Pilot Study

To obtain some initial user feedback on our prototype application a small test web-site was built on the subject of "information retrieval." This was intended for students wishing to explore topics in IR by browsing an online learning resource.

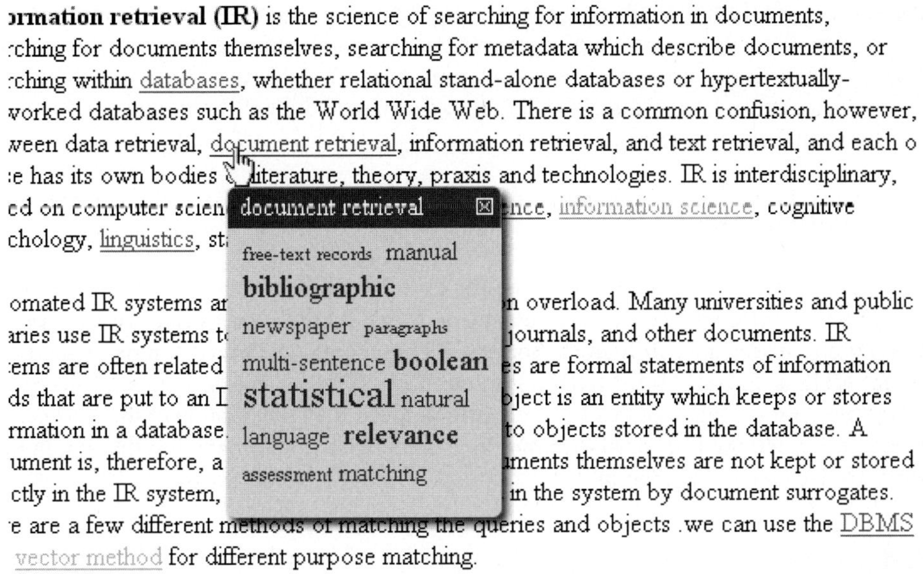

ormation retrieval (IR) is the science of searching for information in documents, :ching for documents themselves, searching for metadata which describe documents, or :ching within databases, whether relational stand-alone databases or hypertextually-vorked databases such as the World Wide Web. There is a common confusion, however, ween data retrieval, document retrieval, information retrieval, and text retrieval, and each o :e has its own bodies literature, theory, praxis and technologies. IR is interdisciplinary, ed on computer scien _____ nce, information science, cognitive chology, linguistics, st:

omated IR systems ar _____ n overload. Many universities and public aries use IR systems t _____ journals, and other documents. IR ems are often related _____ s are formal statements of information ds that are put to an I _____ bject is an entity which keeps or stores rmation in a database. _____ to objects stored in the database. A ument is, therefore, a _____ ments themselves are not kept or stored ctly in the IR system, _____ in the system by document surrogates. e are a few different methods of matching the queries and objects .we can use the DBMS vector method for different purpose matching.

document retrieval ⊠
free-text records manual
bibliographic
newspaper paragraphs
multi-sentence **boolean**
statistical natural
language **relevance**
assessment matching

Fig. 1. Prototype Focused Browsing Application

5 test users each entered 5 search queries on the topic of IR giving a total of 25 queries for this pilot study. 3 queries were judged to be off topic and thus not covered by the test web site; the analysis is thus based on the remaining 22 queries.

Question 1: do the different colour links help you finding target information, how good it is? A Poor, B good C excellent.

Poor: 7 queries, Good: 13 queries, Excellent: 2 queries.

Question 2: Do you find that the summary (keywords) of the documents accurately reflect the content of the documents. A very poor, B poor, C good, D very good, E excellent

Poor: 1 user, Good: 3 users, Very good: 1 user,

These results are reasonably encouraging. However, the test website used for the experiments was very limited in size and scope, meaning that the users had only very limited opportunity to actually browse, and the results of this very simple evaluation are obviously not significant. In order to explore the utility of focused browsing properly, we plan to incorporate focused browsing into a larger domain specific search tool in the near future.

5   Conclusions and Further Development

Focused browsing provides feedback to users engaged in the exploration of hypertext document collections, including the web. The aim being to improve the efficiency with which they can access the information they require. We have developed a prototype application providing feedback using highlighting of available links and term cloud summaries. Potential extension of the application would be to extend it to make use of user tags [5] with the term clouds to form term/tag clouds. In further work we will be incorporating focused browsing into a larger experimental digital library and conducting a more formal experimental evaluation.

Acknowledgement

The authors are grateful to Hyowon Lee for technical assistance with graphics and for the very helpful comments of the reviewers. Work partially supported by European Community under the Information Society Technologies (IST) programme of the 6th FP for RTD — project MultiMATCH contract IST−033104. The authors are solely responsible for the content of this paper. It does not represent the opinion of the European Community, and the European Community is not responsible for any use that might be made of data appearing therein.

References

[1] de Assis, G.T., Laender, A.H.F., Gonçalves, M.A., da Silva, A.S.: Exploiting Genre in Focused Crawling. In: Proceedings of the 14th International Symposium on String Processing and Information Retrieval (SPIRE 2007), Santiago, Chile, pp. 62–73 (2007)

[2] Brusilovsky, P.: Adaptive Hypermedia. User Modeling and User-Adapted Interaction 11(1-2), 87–110 (2001)

[3] Chakrabarti, S., van den Berg, M., Dom, B.: Focused crawling: a new approach to topic-specific Web resource discovery. Journal of Computer Networks 31(11-16), 1623–1640 (1999)

[4] De Bra, P., Brusilovsky, P., Houben, G.-T.: Adaptive Hypermedia: From Systems to Framework. ACM Computing Surveys 31(4) (1999)

[5] Hassan-Montero, Y., Herrero-Solana, V.: Improving Tag-Clouds as Visual Information Retrieval Interfaces. In: Proceedings of the International Conference on Multidisciplinary Information Sciences and Technologies (InSciT2006), Mèrida, Spain (2006)

[6] Olston, C., Chi, E.H.: ScentTrails: Integrating Browsing and Searching on the Web. ACM Transactions on Computer-Human Interaction 10(3), 177–197 (2003)

[7] Page, L., Brin, S., Motwani, R., Winograd, T.: The PageRank Citation Ranking: Bringing Order to the Web. Technical Report, Computer Science, Stanford University (1998)

[8] Soules, C., Ganger, G.: Connections: Using Context to Enhance File Search. In: Proceedings of the 20th ACM Symposium on Operating Systems Principles (SOSP 05), Brighton, U.K, pp. 119–132 (2005)

[9] Tombros, A., Sanderson, M.: Advantages of Query Biased Summaries in Information Retrieval. In: Proceedings of the Twenty First ACM Annual Conference on Research and Development in Information Retrieval (SIGIR 1998), Melbourne, Australia, pp. 2–10 (1998)

[10] Zellweger, P.T., Chang, B.-W., Mackinlay, J.D.: Fluid Links for Informed and Incremental Link Transitions. In: Proceedings of the Ninth ACM Conference on Hypertext and Hypermedia, Pittsburgh, U.S.A (1998)

The Impact of Named Entity Normalization on Information Retrieval for Question Answering

Mahboob Alam Khalid, Valentin Jijkoun, and Maarten de Rijke

ISLA, University of Amsterdam
{mahboob,jijkoun,mdr}@science.uva.nl

Abstract. In the named entity normalization task, a system identifies a canonical unambiguous referent for names like *Bush* or *Alabama*. Resolving synonymy and ambiguity of such names can benefit end-to-end information access tasks. We evaluate two entity normalization methods based on Wikipedia in the context of both passage and document retrieval for question anwering. We find that even a simple normalization method leads to improvements of early precision, both for document and passage retrieval. Moreover, better normalization results in better retrieval performance.

1 Introduction

The task of recognizing named entities in text, i.e., identifying character sequences that refer to items like persons, locations, organizations, dates, etc., has been studied extensively. The Named Entity Recognition (NER) task has been thouroughly evaluated within the Conference on Computational Natural Language Learning (CoNNL) framework in a language-independent setting; techniques applied to NER range from rule-based [9] to machine learning-based [12, 4]. Though significant progress has been achieved, the task remains challenging due to a lack of uniformity in writing styles and domain-dependency. Moreover, NER results are often difficult to use directly, due to high synonymy and ambiguity of names across documents [12]. E.g., the strings *U.S.*, *USA*, *America* can all be used to refer to the concept *United States of America*. Similarly, the string *Washington* can be used to refer to different entities (e.g., *Washington, DC*, or *USA*, or *George Washington*). For information access tasks, such as document retrieval or question answering, these phenomena may harm the performance.

One approach to addressing these problems is Named Entity Normalization (NEN), which goes beyond the NER task: names are not only identified, but also normalized to the concepts they refer to. NEN addresses two phenomena. First, *ambiguity* arises when distinct concepts share the same name; e.g., *Alabama* may refer to the *University of Alabama*, the *Alabama river*, or the *State of Alabama*. This calls for the named entity disambiguation. Second, *synonymy* arises when different names refer to the same entity; e.g., *America* and *U.S.* refering to the *United States of America*.

The multi-referent ambiguity problem was considered at the SemEval Web People Search task [3] and in the Spock Entity Resolution Challenge.[1] Both

---

[1] http://challenge.spock.com

C. Macdonald et al. (Eds.): ECIR 2008, LNCS 4956, pp. 705–710, 2008.

efforts focus on a web search task where the goal is to organize web pages found using a person name as a search engine query, into clusters where pages within a cluster refer to the same person. Cucerzan [8] describes a method for addressing both ambiguity and synonymy; the method uses Wikipedia data and is applied to news texts as well as to Wikipedia itself.

We investigate the impact of NEN on two specific information access tasks: document and passage retrieval for question answering (QA). The tasks consist in finding items in a collection of documents, which contain an answer to a natural language question. E.g., for the question *Who is the queen of Holland?*, an item containing *Beatrix, the Queen of the Kingdom of the Netherlands...* is a relevant response, given that *Holland* is used as a synonym of the *Kingdom of the Netherlands*. Here, NEN may allow a retrieval system to find the answer passage which may have been missed with a standard term-based retrieval.

Specifically, we answer the following research questions: (1) Does NEN improve performance of passage or document retrieval for QA? and (2) To what extent does better entity normalization result in better retrieval for QA? We describe and compare two Wikipedia-based entity normalization methods and evaluate their effectiveness in the setting of passage and document retrieval for QA, using the test collection of the TREC QA track [15].

In Section 2 we review related work. Then, in Section 3, we present two entity normalization methods. Section 4 provides the details of the experimental setup, and shows the results. We conclude in Section 5.

2   Related Work

NEN has been studied both in restricted and in open domains. In the domain of genomics, where gene and protein names can be both synonymous and ambiguous, Cohen [7] normalizes entities using dictionaries automatically extracted from gene databases. Zhou et al. [16] show that appropriate use of domain-specific knowledge base (i.e., synonyms, hypernyms, etc., in a certain domain) yields significant improvement in passage retrieval. For the news domain, Magdy et al. [12] address cross-document Arabic person name normalization using a machine learning approach, a dictionary of person names and frequency information for names in a collection. They apply their method for normalizing Arabic names on the documents related to the situation of Gaza and Lebanon taken from news.google.com. Cucerzan [8] addresses an open domain normalization task, normalizing named entities with information extracted from Wikipedia and machine learning for context-aware disambiguation.

3   Named Entity Normalization

We experimented with two versions of an NEN method based on Wikipedia. Wikipedia is widely used as a rich semantic resource, with natural language processing applications ranging from question answering [2] to text classification [11] to named entity disambiguation [6, 8]. Wikipedia is especially attractive for the task of entity normalization. It covers a huge number of entities (over 2M article titles as of October 2007), most of them named entities. The anchor text of inter-article links allows one to identify different text strings that can be used to refer to the same entity or concept. So-called "redirects" provide information about synonyms or near synonyms (e.g., the article *King of pop* is empty and redirects to the article *Michael Jackson*). Special "disambiguation" pages list possible referents of ambiguous names (such as *George Bush* that lists five persons with that name). Moreover, each Wikipedia entity page has a unique identifier (URL)—a unique and unambiguous way of refering to the entity.

The baseline NEN method in [8] uses this information in the following manner, for each surface form recognized as an NE by an NE recognizer. If there is an entity page or redirect page whose title matches exactly with the surface form, then the corresponding entity is chosen as the normalization result; otherwise the entity most frequently mentioned in Wikipedia using that form as anchor text is selected as the baseline disambiguation. We re-implemented this baseline using the named entity tagger of [10], and refer to it as *MS*.

We also implemented a simple extension of the method by adding a link frequency-based disambiguation algorithm. Whenever a surface form can be resolved to more than one entity using the algorithm above, we select the entity with the highest number of incoming hyperlinks. Our hypothesis of disambiguation is based on the assumption that a more useful and/or popular Wikipedia entity will have many links pointing to it [5]. In other words, we assume that a name found (e.g., "Bush") mostly refers to the most popular compatible Wikipedia entity ("George W. Bush"). We refer to this method as *NN*.

Cucerzan [8] also describes a more sophisticated, context-aware normalization algorithm. We did not use this version of the algorithm in our experiments below because it would have involved classifying each name in the collection—a very computationally expensive step.

We compared the *MS* and *NN* normalization methods, using the evaluation data as described in [8] for intrinsic, stand-alone evaluation of the two methods. The accuracy of *NN* on Wikipedia articles and news articles was 86.5% and 73% respectively, outperforming the accuracy of *MS* (86.1% and 51.7% on Wikipedia and news articles, respectively).

4 Experiments and Results

We performed a number of experiments in a setting similar to [13]. We used a standard set of question/answer pairs from TREC QA tasks of 2001–2003. In addition to using full documents, we split the AQUAINT corpus into 400-character passages (aligned on paragraph boundary). We ran the NER tool of [10] to detect named entities and normalized them using *NN* and *MS*, separately. We used the dump of English Wikipedia from November 2006. Documents and passages were

Table 1. Impact of named entity normalization on document retrieval for QA; * and ** indicate significant improvements over the baseline at p=0.05 and p=0.01

	MRR	s@1	s@5	s@10	p@5	p@10
NONORM	0.532	44.8%	64.6%	72.8%	0.37	0.34
MS	0.511	42.2%	63.4%	71.6%	0.36	0.32
NN	0.523	43.4%	64.4%	72.7%	0.37	0.33
MS+NONORM	0.55	46.4%	67.57%*	74.8%*	0.39**	0.35*
NN+NONORM	**0.56**	**47%***	**68.2%****	**75.3%****	**0.4****	**0.36***

Table 2. Impact of named entity normalization on passage retrieval for QA. *NN+NONORM* outperforms *MS+NONORM* (at p=0.01).

	MRR	s@1	s@5	s@10	p@5	p@10
NONORM	0.411	30.9%	53.6%	**63.3%**	0.26	**0.23**
MS	0.387	29.2%	50%	58.9%	0.23	0.2
NN	0.405	30.7%	51.7%	60.5%	0.24	0.21
MS+NONORM	0.407	30.7%	53%	61.2%	0.24	**0.23**
NN+NONORM	**0.424**	**32.6%**	**54.4%**	62.3%	**0.27**	**0.23**

separately indexed using Lucene [1]. Out of 2,136 question/answer pairs in the TREC QA data, we used only 1,215 whose questions contained a named entity. We normalized named entities in questions in the same way as in the collection. We compared the retrieval performance of the baseline (no normalization, standard vector space retrieval), for both normalization methods and for equally weighted mixture models of the baseline with both methods. Following [13], we measured performance using the Mean Reciprocal Rank (MRR), success at rank n ($s@n$), and average precision at n ($p@n$). For significance testing we applied the McNemar significance test on success evaluations, and Student's t-test on precision evaluations. Tables 1 and 2 show the evaluation results for passage and document retrieval, respectively.

The results show that the combination of NEN with the baseline improves the MRR value, precision and early success of the retrieval system for QA. They also show that *NN* helps more than *MS*, for document and passage retrieval.

An analysis of the effect of NEN on text retrieval shows that for questions where normalization did not improve the retrieval, this was mostly due to NER errors. E.g., for *What river is under New York's George Washington bridge?*, the entity *George Washington* was detected as a person name, while the answer passage contains the entity *George Washington Bridge* correctly detected as LOCATION. Where normalization helped to find relevant passages, this was often due to the correct "gluing" of multiword units: *Buffalo Bill, Crater Lake, Joe Andrew, Andrew Jackson*. Here, without normalization, retrieval failed.

Finally, for the passage retrieval experiments, the difference between *NN* and *MS* is statistically significant (at $p = 0.01$). This indicates that better normalization does indeed lead to better retrieval performance.

5 Conclusion

We described experiments evaluating the impact of name entity normalization on document and passage retrieval for QA. We implemented the normalization method of [8] and a simple refinement. Although our disambiguation methods are not context-aware, we observed improved retrieval performance with entity normalization. Moreover, better normalization has led to better QA performance. The error analysis shows that entity recognition errors are a main source of retrieval errors due to normalization. This indicates an obvious direction for improving the system. Another item for future work is to include surface form context into the disambiguation model in such a way that normalizing a large text collection remains computationally tractable.

Acknowledgements

Mahboob Alam Khalid was supported by the Netherlands Organization for Scientific Research (NWO) under project number 612.066.512, Valentin Jijkoun—by NWO projects 220.80.001, 600.065.120 and 612.000.106. Maarten de Rijke was supported under project numbers 017.001.190, 220-80-001, 264-70-050, 354-20-005, 600.065.120, 612-13-001, 612.000.106, 612.066.302, 612.069.006, 640.001.-501, 640.002.501, and by the E.U. IST programme of the 6th FP for RTD under project MultiMATCH contract IST-033104.

Bibliography

[1] Jakarta Lucene ext search engine (2002), http://lucene.apache.org
[2] Ahn, D., Jijkoun, V., Mishne, G., Müller, K., de Rijke, M., Schlobach, S.: Using Wikipedia at the TREC QA Track. In: TREC 2004 (2005)
[3] Artiles, J., Gonzalo, J., Sekine, S.: The SemEval-2007 WePS Evaluation: Establishing a benchmark for Web People Search Task. In: Semeval 2007 (2007)
[4] Borthwick, A.: A Maximum Entropy Approach to Named Entity Recognition. PhD thesis, New York University (1999)
[5] Brin, S., Page, L.: The anatomy of a large-scale hypertextual Web search engine. Computer Networks and ISDN Systems 30(1-7), 107–117 (1998)
[6] Bunescu, R.C., Pasca, M.: Using encyclopedic knowledge for named entity disambiguation. In: EACL 2006 (2006)
[7] Cohen, A.M.: Unsupervised gene/protein named entity normalization using automatically extracted dictionaries. In: ACL-ISMB Workshop on Linking Biological Literature, Ontologies and Databases, pp. 17–24 (2005)
[8] Cucerzan, S.: Large-scale named entity disambiguation based on wikipedia data. In: EMNLP-CoNLL 2007, pp. 708–716 (2007)
[9] Farmakiotou, D., Karkaletsis, V., Koutsias, J., Sigletos, G., Spyropoulos, C., Stamatopoulos, P.: Rule-based named entity recognition for greek financial texts (2000)
[10] Finkel, J.R., Grenager, T., Manning, C.D.: Incorporating non-local information into information extraction systems by gibbs sampling. In: ACL (2005)
[11] Gabrilovich, E., Markovitch, S.: Overcoming the Brittleness Bottleneck using Wikipedia. In: AAAI 2006 (2006)

[12] Magdy, W., Darwish, K., Emam, O., Hassan, H.: Arabic cros-document person name normalization. In: CASL Workshop 2007, pp. 25–32 (2007)
[13] Monz, C.: Minimal span weighting retrieval for question answering. In: Proceedings of SIGIR 2004 Workshop on Information Retrieval for Question Answering (2004)
[14] Solorio, T.: Improvement of Named Entity Tagging by Machine Learning. PhD thesis (2005)
[15] Voorhees, E.M.: Overview of the trec 2003 question answering track. In: TREC, pp. 54–68 (2003)
[16] Zhou, W., Yu, C., Smalheiser, N., Torvik, V., Hong, J.: Knowledge-intensive conceptual retrieval and passage extraction of biomedical literature. In: SIGIR 2007, pp. 655–662 (2007)

Efficiency Issues in Information Retrieval Workshop

Roi Blanco[1] and Fabrizio Silvestri[2]

[1] University of A Coruña, Spain
rblanco@udc.es
[2] ISTI-CNR, Pisa, Italy
fabrizio.silvestri@isti.cnr.it

Today's technological advancements allow for vast amounts of information to be widely generated, disseminated, and stored. This exponentially increasing amount of information renders the retrieval of relevant information a necessary and cumbersome task. The field of Information Retrieval (IR) addresses this task by developing systems in an effective and efficient way. Specifically, IR effectiveness deals with retrieving the most relevant information to a user need, while IR efficiency deals with providing fast and ordered access to large amounts of information.

The efficiency of IR systems is of utmost importance, because it ensures that systems scale up to the vast amounts of information needing retrieval. This is an important topic of research for both academic and corporate environments. In academia, it is imperative for new ideas and techniques to be evaluated on as near-realistic environments as possible.In corporate environments, it is important that systems response time is kept low, and the amount of data processed high. These efficiency concerns need to be addressed in a principled way, so that they can be adapted to new platforms and environments, such as IR from mobile devices, desktop search, distributed peer to peer, expert search, multimedia retrieval, and so on. Efficiency research over the past years has focused on efficient indexing, storage (compression) and retrieval of data (query processing strategies).

This workshop addresses the efficiency concerns regarding IR applications (both new and traditional):

- Do new applications create novel efficiency problems?
- Can existing efficiency-related technology deal with these new applications?
- About state-of-the-art efficiency: has there been any advance in the last decade, or is it at a stand-still?
- To what extent is efficiency separated from effectiveness? Can this gap be bridged?
- What are the lessons learnt from efficiency research in the last years?. Can any of these be carried across to effectiveness?

Major goals of this workshop are to: a) shed light on efficiency-related problems of modern large-scale IR and new IR environments; b) foster collaboration between different research groups in order to explore new and ground-breaking ideas; c) bearing in mind past research on effciency, sketch future directions for the field.

C. Macdonald et al. (Eds.): ECIR 2008, LNCS 4956, p. 711, 2008.
© Springer-Verlag Berlin Heidelberg 2008

Exploiting Semantic Annotations in Information Retrieval

Omar Alonso and Hugo Zaragoza

A9.com, 130 Lytton Ave, Palo Alto CA 94301
oalonso@a9.com
Yahoo! Research, Ocata 1, Barcelona, Spain
hugoz@yahoo-inc.com

The goal of this workshop is to create a forum for researchers interested in the use of semantic annotations for information retrieval. By semantic annotations we refer to linguistic annotations (such as named entities, semantic classes, etc.) as well as user annotations such as microformats, RDF, tags, etc. The aim of this workshop is not semantic annotation itself, but rather the applications of semantic annotation to information retrieval tasks such as ad-hoc retrieval, classification, browsing, textual mining, summarization, question answering, etc.

In the recent years there has been a lot of discussion about semantic annotation of documents. There are many forms of annotations and many techniques that identify or extract them. As NLP tagging techniques mature, more and more annotations can be automatically extracted from free text. In particular, techniques have been developed to ground named entities in terms of geo-codes, ISO time codes, Gene Ontology ids, etc. Furthermore, the number of collections which explicitly identify entities is growing fast with Web 2.0 and Semantic Web initiatives.

Despite the growing number and complexity of annotations, and despite the potential impact that these may have in information retrieval tasks, annotations have not yet made a significant impact in Information Retrieval research or applications. Further research is needed before we can unleash the potential of annotations!

C. Macdonald et al. (Eds.): ECIR 2008, LNCS 4956, p. 712, 2008.
© Springer-Verlag Berlin Heidelberg 2008

Workshop on Novel Methodologies for Evaluation in Information Retrieval

Mark Sanderson, Martin Braschler, Nicola Ferro, and Julio Gonzalo

Objectives. Information retrieval is an empirical science; the field cannot move forward unless there are means of evaluating the innovations devised by researchers. However the methodologies conceived in the early years of IR and used in the campaigns of today are starting to show their age and new research is emerging to understand how to overcome the twin challenges of scale and diversity.

Scale. The methodologies used to build test collections in the modern evaluation campaigns were originally conceived to work with collections of 10s of thousands of documents. The methodologies were found to scale well, but potential flaws are starting to emerge as test collections grow beyond 10s of millions of documents. Support for continued research in this area is crucial if IR research is to continue to evaluate large scale search.

Diversity. With the rise of the large Web search engines, some believed that all search problems could be solved with a single engine retrieving from a one vast data store. However, it is increasingly clear that evolution of retrieval is not towards a monolithic solution, but instead to a wide range of solutions tailored for different classes of information and different groups of users or organizations. Each tailored system on offer requires a different mixture of component technologies combined in distinct ways and each solution requires evaluation. This workshop will consist of research papers on topics that address evaluation in Information Retrieval, including:

- test collection building for diverse needs
- new metrics and methodologies
- evaluation of multilingual IR and/or multimedia IR systems
- novel evaluation of related areas, such as QA or summarization
- evaluation of commercial systems
- novel forms of user-centered evaluation

Acknowledgements. The workshop is in part supported by the EU 7th Framework Coordinated Action TrebleCLEF, grant agreement No. 215231.

C. Macdonald et al. (Eds.): ECIR 2008, LNCS 4956, p. 713, 2008.

ECIR 2008 Tutorials

Advanced Language Modeling Approaches (Case Study: Expert Search)

Djoerd Hiemstra

University of Twente, The Netherlands

This tutorial gives a clear and detailed overview of advanced language modeling approaches and tools, including the use of document priors, translation models, relevance models, parsimonious models and expectation maximization training. Expert search will be used as a case study to explain the consequences of modeling assumptions. For more details, you can access http://www.cs.utwente.nl/~hiemstra/ecir2008.

Djoerd Hiemstra is assistant professor at the University of Twente. He wrote a Ph.D. thesis on language models for information retrieval and contributed to over 90 research papers in the field of IR. His research interests include formal models of information retrieval, XML retrieval and multimedia retrieval.

Search and Discovery in User-Generated Text Content

Maarten de Rijke and Wouter Weerkamp

ISLA, University of Amsterdam, The Netherlands

We increasingly live our lives online: Blogs, forums, commenting tools, and many other sharing sites offer possibilities to users to make any information available online. For the first time in history, we are able to collect huge amounts of user-generated content (UGC) within "a blink of an eye". The rapidly increasing amount of UGC poses challenges to the IR community, but also offers many

C. Macdonald et al. (Eds.): ECIR 2008, LNCS 4956, pp. 714–715, 2008.

previously unthinkable possibilities. In this tutorial we discuss different aspects of accessing (i.e., searching, tracking, and analyzing) UGC. Our focus will be on textual content, and most of the methods that we will consider for ranking UGC (by relevancy, quality, opinionatedness) are based on language modeling. For more details, you can access http://ecir2008.dcs.gla.ac.uk/tutorial_sd.html.

Maarten de Rijke is professor of information processing and internet at the Intelligent Systems Lab Amsterdam (ISLA) of the University of Amsterdam. His group has been researching search and discovery tools for UGC for a number of years now, with numerous publications and various demonstrators as tangible outcomes. Wouter Weerkamp is a PhD student at ISLA, working on language modeling and intelligent access to UGC.

Researching and Building IR Applications Using Terrier

Craig Macdonald and Ben He

University of Glasgow, UK

This tutorial introduces the main design of an IR system, and uses the Terrier platform as an example of how one should be built. We detail the architecture and data structures of Terrier, as well as the weighting models included, and describe, with examples, how Terrier can be used to perform experiments and extended to facilitate new research and applications. For more details, you can access http://ecir2008.dcs.gla.ac.uk/tutorial_rb.html.

Craig Macdonald is a PhD research student at the University of Glasgow. His research interests includes Information Retrieval in Enterprise, Web and Blog settings, and has over 20 publications with research based on the Terrier platform. He has been a co-ordinator of the Blog track at TREC since 2006, and is a developer of the Terrier platform.

Ben He is a post-doctoral research assistant at the University of Glasgow. His research interests are centered around document weighting models, and particularly concerned about document length normalisation and query expansion. He has been a developer of the Terrier platform since its initial development and has more than 20 publications performed with Terrier.

Author Index

Abdullah, Noorhidawati 246
Alonso, Omar 632, 712
Altingovde, Ismail Sengor 695
Amati, Giambattista 89
Ambrosi, Edgardo 89
Amini, Massih R. 370
Anderka, Maik 522
Artiles, Javier 126
Ayache, Stéphane 187
Azzopardi, Leif 482

Bagchi, Amitabha 422
Balakrishnan, Rahul 422
Balog, Krisztian 296
Barreiro, Alvaro 394
Beigbeder, Michel 647
Belkin, Nicholas J. 1
Bendersky, Michael 162
Bennett, Paul N. 16
Berendt, Bettina 3
Bianchi, Marco 89
Billerbeck, Bodo 28
Blanco, Roi 394, 711
Boves, Lou 669
Braschler, Martin 713
Brunkhorst, Ingo 564
Bruza, Peter 334
Bu, Jiajun 617

Can, Fazli 695
Carmel, David 498
Carterette, Ben 16
Carvalho, Vitor R. 321
Chen, Chun 617
Chen, Hsin-Hsi 77
Chen, Jian 622
Chickering, David Maxwell 16
Chirita, Paul-Alexandru 564
Choudhary, Rohan 422
Clinchant, Stéphane 150
Clough, Paul 126, 570
Cohen, William W. 321
Collier, Rem 358
Coppen, Peter-Arno 669

Crestani, Fabio 679
Croft, W. Bruce 175, 454

David, Gabriel 580
de Rijke, Maarten 296, 689, 705
Delcambre, Lois M.L. 4
Demartini, Gianluca 564
Deschacht, Koen 539
Dumais, Susan T. 16
Dunnion, John 358

Eidoon, Zahra 472
Eliasmith, Chris 514, 575
Erera, Shai 498
Euachongprasit, Waiyawuth 506

Ferro, Nicola 713
Fishbein, Jonathan M. 514, 575
Franke, Conny 632

Gaibisso, Carlo 89
Gambosi, Giorgio 89
Gaussier, Eric 150
Gerling, Robert 663
Gibb, Forbes 246
Gonzalo, Julio 713
Gottron, Thomas 40
Goutte, Cyril 370

Hannah, David 283, 596
He, Jiyin 689
He, Yanxiang 446
Heinrich, Gregor 607
Hiemstra, Djoerd 309
Holst, Anders 531
Holz, Florian 607
Hon, Hsiao-Wuen 346
Hopfgartner, Frank 199
Hotho, Andreas 101
Huang, Peng 617
Huang, Qiang 334, 547

Järvelin, Kalervo 4, 138
Jijkoun, Valentin 705
Jin, Qun 622

Jones, Gareth J.F.   602, 700
Jose, Joemon   199, 490

Kamel, Mohamed S.   222
Kamps, Jaap   270
Kang, In-Su   382, 658, 674, 684
Karlgren, Jussi   531
Kazai, Gabriella   234
Keskustalo, Heikki   138
Khalid, Mahboob Alam   705
Klakow, Dietrich   612
Knees, Peter   585, 627
Koolen, Marijn   270
Krause, Beate   101
Krithara, Anastasia   370
Kurland, Oren   162

Lang, Hao   602
Larson, Martha   689
Lee, Jong-Hyeok   382, 658, 674, 684
Li, Jintao   602
Li, Quixiang   700
Li, Wenjie   446
Li, Xiaoyan   463
Lillis, David   358
Lin, Ming-Shun   77
Liu, Xiaoyong   454
Loponen, Aki   138
Lu, Qin   446
Luxenburger, Julia   652

Ma, Liping   430
Macdonald, Craig   283, 555, 596
MacKinnon, Ian   438
Makrehchi, Masoud   222
Marlow, Jennifer   126
McSherry, Frank   414
Mehta, Sameep   422
Metzler, Donald   175
Mizzaro, Stefano   642
Moens, Marie-Francine   539
Mohd, Masnizah   679
Monz, Christof   406

Na, Seung-Hoon   382, 658, 674, 684
Naderi, Hassan   637
Najork, Marc   414
Nejdl, Wolfgang   564
Nielsen, Marianne Lykke   4
Nunes, Sérgio   580

Ofoghi, Bahadorreza   430
Oostdijk, Nelleke   669
Oroumchian, Farhad   472
Ounis, Iadh   283, 555, 596

Pandey, Gaurav   652
Pehcevski, Jovan   258
Petrakis, Euripides G.M.   65
Pirkola, Ari   138
Pohle, Tim   585, 627
Potthast, Martin   522, 663
Price, Susan L.   4

Qiu, Guang   617
Quénot, Georges   187

Raftopoulou, Paraskevi   65
Ratanamahatana, Chotirat Ann   506
Read, Simon   570
Recuero, Juan Cigarrán   126
Renders, Jean-Michel   370
Ribeiro, Cristina   580
Ritchie, Anna   211
Robertson, Stephen   211
Rüger, Stefan   334, 547
Rumpler, Béatrice   637
Ruthven, Ian   679

Sahlgren, Magnus   531
Sanderson, Mark   713
Schedl, Markus   585, 627
Schnitzer, Dominik   627
Scholer, Falk   28, 52, 591
Serdyukov, Pavel   309
Seyerlehner, Klaus   627
Shokouhi, Milad   28, 591
Shtykh, Roman Y.   622
Silvestri, Fabrizio   711
Singhal, Amit   2
Song, Dawei   334, 547
Song, Ruihua   346
Stathopoulos, Vassilios   490
Stein, Benno   522, 663
Strohman, Trevor   175
Stumme, Gerd   101

Talvensaari, Tuomas   114
Taylor, Michael J.   234, 346
Teresniak, Sven   607

Teufel, Simone   211
Thom, James A.   258
Toolan, Fergus   358
Tsegay, Yohannes   52
Turpin, Andrew   28, 591

Ulusoy, Özgür   695
Urban, Jana   490

Vallet, David   199
Van, Thanh-Trung   647
Vechtomova, Olga   438
Verberne, Suzan   669
Vercoustre, Anne-Marie   258
Vinay, Vishwa   482

Wang, Bin   602
Wei, Furu   446

Wen, Ji-Rong   346
Widmer, Gerhard   585
Wiegand, Michael   612
Witschel, Hans Friedrich   607
Wu, Hengzhi   234

Xu, Yang   602

Yahyaei, Sirvan   406
Yazdani, Nasser   472
Yearwood, John   430
Yu, Yong   346

Zaragoza, Hugo   712
Zhao, Ying   52
Zhu, Zhigang   463

Printing: Mercedes-Druck, Berlin
Binding: Stein+Lehmann, Berlin

Lecture Notes in Computer Science

Sublibrary 3: Information Systems and Application, incl. Internet/Web and HCI

For information about Vols. 1– 4559
please contact your bookseller or Springer

Vol. 4956: C. Macdonald, I. Ounis, V. Plachouras, I. Ruthven, R.W. White (Eds.), Advances in Information Retrieval. XXI, 719 pages. 2008.

Vol. 4947: J.R. Haritsa, R. Kotagiri, V. Pudi (Eds.), Database Systems for Advanced Applications. XXII, 713 pages. 2008.

Vol. 4936: W. Aiello, A. Broder, J. Janssen, E.. Milios (Eds.), Algorithms and Models for the Web-Graph. X, 167 pages. 2008.

Vol. 4932: S. Hartmann, G. Kern-Isberner (Eds.), Foundations of Information and Knowledge Systems. XII, 397 pages. 2008.

Vol. 4928: A. ter Hofstede, B. Benatallah, H.-Y. Paik (Eds.), Business Process Management Workshops. XIII, 518 pages. 2008.

Vol. 4903: S. Satoh, F. Nack, M. Etoh (Eds.), Advances in Multimedia Modeling. XIX, 510 pages. 2008.

Vol. 4900: S. Spaccapietra (Ed.), Journal on Data Semantics X. XIII, 265 pages. 2008.

Vol. 4892: A. Popescu-Belis, S. Renals, H. Bourlard (Eds.), Machine Learning for Multimodal Interaction. XI, 308 pages. 2008.

Vol. 4882: T. Janowski, H. Mohanty (Eds.), Distributed Computing and Internet Technology. XIII, 346 pages. 2007.

Vol. 4881: H. Yin, P. Tino, E. Corchado, W. Byrne, X. Yao (Eds.), Intelligent Data Engineering and Automated Learning - IDEAL 2007. XX, 1174 pages. 2007.

Vol. 4877: C. Thanos, F. Borri, L. Candela (Eds.), Digital Libraries: Research and Development. XII, 350 pages. 2007.

Vol. 4872: D. Mery, L. Rueda (Eds.), Advances in Image and Video Technology. XXI, 961 pages. 2007.

Vol. 4871: M. Cavazza, S. Donikian (Eds.), Virtual Storytelling. XIII, 219 pages. 2007.

Vol. 4858: X. Deng, F.C. Graham (Eds.), Internet and Network Economics. XVI, 598 pages. 2007.

Vol. 4857: J.M. Ware, G.E. Taylor (Eds.), Web and Wireless Geographical Information Systems. XI, 293 pages. 2007.

Vol. 4853: F. Fonseca, M.A. Rodríguez, S. Levashkin (Eds.), GeoSpatial Semantics. X, 289 pages. 2007.

Vol. 4836: H. Ichikawa, W.-D. Cho, I. Satoh, H.Y. Youn (Eds.), Ubiquitous Computing Systems. XIII, 307 pages. 2007.

Vol. 4832: M. Weske, M.-S. Hacid, C. Godart (Eds.), Web Information Systems Engineering – WISE 2007 Workshops. XV, 518 pages. 2007.

Vol. 4831: B. Benatallah, F. Casati, D. Georgakopoulos, C. Bartolini, W. Sadiq, C. Godart (Eds.), Web Information Systems Engineering – WISE 2007. XVI, 675 pages. 2007.

Vol. 4825: K. Aberer, K.-S. Choi, N. Noy, D. Allemang, K.-I. Lee, L. Nixon, J. Golbeck, P. Mika, D. Maynard, R. Mizoguchi, G. Schreiber, P. Cudré-Mauroux (Eds.), The Semantic Web. XXVII, 973 pages. 2007.

Vol. 4822: D.H.-L. Goh, T.H. Cao, I.T. Sølvberg, E. Rasmussen (Eds.), Asian Digital Libraries. XVII, 519 pages. 2007.

Vol. 4820: T.G. Wyeld, S. Kenderdine, M. Docherty (Eds.), Virtual Systems and Multimedia. XII, 215 pages. 2008.

Vol. 4816: B. Falcidieno, M. Spagnuolo, Y. Avrithis, I. Kompatsiaris, P. Buitelaar (Eds.), Semantic Multimedia. XII, 306 pages. 2007.

Vol. 4813: I. Oakley, S.A. Brewster (Eds.), Haptic and Audio Interaction Design. XIV, 145 pages. 2007.

Vol. 4810: H.H.-S. Ip, O.C. Au, H. Leung, M.-T. Sun, W.-Y. Ma, S.-M. Hu (Eds.), Advances in Multimedia Information Processing – PCM 2007. XXI, 834 pages. 2007.

Vol. 4809: M.K. Denko, C.-s. Shih, K.-C. Li, S.-L. Tsao, Q.-A. Zeng, S.H. Park, Y.-B. Ko, S.-H. Hung, J.-H. Park (Eds.), Emerging Directions in Embedded and Ubiquitous Computing. XXXV, 823 pages. 2007.

Vol. 4808: T.-W. Kuo, E. Sha, M. Guo, L.T. Yang, Z. Shao (Eds.), Embedded and Ubiquitous Computing. XXI, 769 pages. 2007.

Vol. 4806: R. Meersman, Z. Tari, P. Herrero (Eds.), On the Move to Meaningful Internet Systems 2007: OTM 2007 Workshops, Part II. XXXIV, 611 pages. 2007.

Vol. 4805: R. Meersman, Z. Tari, P. Herrero (Eds.), On the Move to Meaningful Internet Systems 2007: OTM 2007 Workshops, Part I. XXXIV, 757 pages. 2007.

Vol. 4804: R. Meersman, Z. Tari (Eds.), On the Move to Meaningful Internet Systems 2007: CoopIS, DOA, ODBASE, GADA, and IS, Part II. XXIX, 683 pages. 2007.

Vol. 4803: R. Meersman, Z. Tari (Eds.), On the Move to Meaningful Internet Systems 2007: CoopIS, DOA, ODBASE, GADA, and IS, Part I. XXIX, 1173 pages. 2007.

Vol. 4802: J.-L. Hainaut, E.A. Rundensteiner, M. Kirchberg, M. Bertolotto, M. Brochhausen, Y.-P.P. Chen, S.S.-S. Cherfi, M. Doerr, H. Han, S. Hartmann, J. Parsons, G. Poels, C. Rolland, J. Trujillo, E. Yu, E. Zimányie (Eds.), Advances in Conceptual Modeling – Foundations and Applications. XIX, 420 pages. 2007.

Vol. 4801: C. Parent, K.-D. Schewe, V.C. Storey, B. Thalheim (Eds.), Conceptual Modeling - ER 2007. XVI, 616 pages. 2007.

Vol. 4797: M. Arenas, M.I. Schwartzbach (Eds.), Database Programming Languages. VIII, 261 pages. 2007.

Vol. 4796: M. Lew, N. Sebe, T.S. Huang, E.M. Bakker (Eds.), Human–Computer Interaction. X, 157 pages. 2007.

Vol. 4794: B. Schiele, A.K. Dey, H. Gellersen, B. de Ruyter, M. Tscheligi, R. Wichert, E. Aarts, A. Buchmann (Eds.), Ambient Intelligence. XV, 375 pages. 2007.

Vol. 4777: S. Bhalla (Ed.), Databases in Networked Information Systems. X, 329 pages. 2007.

Vol. 4761: R. Obermaisser, Y. Nah, P. Puschner, F.J. Rammig (Eds.), Software Technologies for Embedded and Ubiquitous Systems. XIV, 563 pages. 2007.

Vol. 4747: S. Džeroski, J. Struyf (Eds.), Knowledge Discovery in Inductive Databases. X, 301 pages. 2007.

Vol. 4744: Y. de Kort, W. IJsselsteijn, C. Midden, B. Eggen, B.J. Fogg (Eds.), Persuasive Technology. XIV, 316 pages. 2007.

Vol. 4740: L. Ma, M. Rauterberg, R. Nakatsu (Eds.), Entertainment Computing – ICEC 2007. XXX, 480 pages. 2007.

Vol. 4730: C. Peters, P. Clough, F.C. Gey, J. Karlgren, B. Magnini, D.W. Oard, M. de Rijke, M. Stempfhuber (Eds.), Evaluation of Multilingual and Multi-modal Information Retrieval. XXIV, 998 pages. 2007.

Vol. 4723: M. R. Berthold, J. Shawe-Taylor, N. Lavrač (Eds.), Advances in Intelligent Data Analysis VII. XIV, 380 pages. 2007.

Vol. 4721: W. Jonker, M. Petković (Eds.), Secure Data Management. X, 213 pages. 2007.

Vol. 4718: J. Hightower, B. Schiele, T. Strang (Eds.), Location- and Context-Awareness. X, 297 pages. 2007.

Vol. 4717: J. Krumm, G.D. Abowd, A. Seneviratne, T. Strang (Eds.), UbiComp 2007: Ubiquitous Computing. XIX, 520 pages. 2007.

Vol. 4715: J.M. Haake, S.F. Ochoa, A. Cechich (Eds.), Groupware: Design, Implementation, and Use. XIII, 355 pages. 2007.

Vol. 4714: G. Alonso, P. Dadam, M. Rosemann (Eds.), Business Process Management. XIII, 418 pages. 2007.

Vol. 4704: D. Barbosa, A. Bonifati, Z. Bellahsène, E. Hunt, R. Unland (Eds.), Database and XML Technologies. X, 141 pages. 2007.

Vol. 4690: Y. Ioannidis, B. Novikov, B. Rachev (Eds.), Advances in Databases and Information Systems. XIII, 377 pages. 2007.

Vol. 4675: L. Kovács, N. Fuhr, C. Meghini (Eds.), Research and Advanced Technology for Digital Libraries. XVII, 585 pages. 2007.

Vol. 4674: Y. Luo (Ed.), Cooperative Design, Visualization, and Engineering. XIII, 431 pages. 2007.

Vol. 4663: C. Baranauskas, P. Palanque, J. Abascal, S.D.J. Barbosa (Eds.), Human-Computer Interaction – INTERACT 2007, Part II. XXXIII, 735 pages. 2007.

Vol. 4662: C. Baranauskas, P. Palanque, J. Abascal, S.D.J. Barbosa (Eds.), Human-Computer Interaction – INTERACT 2007, Part I. XXXIII, 637 pages. 2007.

Vol. 4658: T. Enokido, L. Barolli, M. Takizawa (Eds.), Network-Based Information Systems. XIII, 544 pages. 2007.

Vol. 4656: M.A. Wimmer, J. Scholl, Å. Grönlund (Eds.), Electronic Government. XIV, 450 pages. 2007.

Vol. 4655: G. Psaila, R. Wagner (Eds.), E-Commerce and Web Technologies. VII, 229 pages. 2007.

Vol. 4654: I.-Y. Song, J. Eder, T.M. Nguyen (Eds.), Data Warehousing and Knowledge Discovery. XVI, 482 pages. 2007.

Vol. 4653: R. Wagner, N. Revell, G. Pernul (Eds.), Database and Expert Systems Applications. XXII, 907 pages. 2007.

Vol. 4636: G. Antoniou, U. Aßmann, C. Baroglio, S. Decker, N. Henze, P.-L. Patranjan, R. Tolksdorf (Eds.), Reasoning Web. IX, 345 pages. 2007.

Vol. 4611: J. Indulska, J. Ma, L.T. Yang, T. Ungerer, J. Cao (Eds.), Ubiquitous Intelligence and Computing. XXIII, 1257 pages. 2007.

Vol. 4607: L. Baresi, P. Fraternali, G.-J. Houben (Eds.), Web Engineering. XVI, 576 pages. 2007.

Vol. 4606: A. Pras, M. van Sinderen (Eds.), Dependable and Adaptable Networks and Services. XIV, 149 pages. 2007.

Vol. 4605: D. Papadias, D. Zhang, G. Kollios (Eds.), Advances in Spatial and Temporal Databases. X, 479 pages. 2007.

Vol. 4602: S. Barker, G.-J. Ahn (Eds.), Data and Applications Security XXI. X, 291 pages. 2007.

Vol. 4601: S. Spaccapietra, P. Atzeni, F. Fages, M.-S. Hacid, M. Kifer, J. Mylopoulos, B. Pernici, P. Shvaiko, J. Trujillo, I. Zaihrayeu (Eds.), Journal on Data Semantics IX. XV, 197 pages. 2007.

Vol. 4592: Z. Kedad, N. Lammari, E. Métais, F. Meziane, Y. Rezgui (Eds.), Natural Language Processing and Information Systems. XIV, 442 pages. 2007.

Vol. 4587: R. Cooper, J. Kennedy (Eds.), Data Management. XIII, 259 pages. 2007.

Vol. 4577: N. Sebe, Y. Liu, Y.-t. Zhuang, T.S. Huang (Eds.), Multimedia Content Analysis and Mining. XIII, 513 pages. 2007.

Vol. 4568: T. Ishida, S. R. Fussell, P. T. J. M. Vossen (Eds.), Intercultural Collaboration. XIII, 395 pages. 2007.

Vol. 4566: M.J. Dainoff (Ed.), Ergonomics and Health Aspects of Work with Computers. XVIII, 390 pages. 2007.

Vol. 4564: D. Schuler (Ed.), Online Communities and Social Computing. XVII, 520 pages. 2007.

Vol. 4563: R. Shumaker (Ed.), Virtual Reality. XXII, 762 pages. 2007.

Vol. 4561: V.G. Duffy (Ed.), Digital Human Modeling. XXIII, 1068 pages. 2007.

Vol. 4560: N. Aykin (Ed.), Usability and Internationalization, Part II. XVIII, 576 pages. 2007.